International Conference on Recent Advances in Industrial Engineering and Manufacturing (ICRAIEM 2018)

IOP Conference Series: Materials Science and Engineering
Volume 530

Penang, Malaysia
12 - 13 December 2018

Editor:

Mani Maran Ratnam

ISBN: 978-1-5108-9380-1
ISSN: 1757-8981

Printed from e-media with permission by:

Curran Associates, Inc.
57 Morehouse Lane
Red Hook, NY 12571

Some format issues inherent in the e-media version may also appear in this print version.

Printed by Curran Associates, Inc. (2019)

For permission requests, please contact the Institute of Physics
at the address below.

Institute of Physics
Dirac House, Temple Back
Bristol BS1 6BE UK

Phone: 44 1 17 929 7481
Fax: 44 1 17 920 0979

techtracking@iop.org

Additional copies of this publication are available from:

Curran Associates, Inc.
57 Morehouse Lane
Red Hook, NY 12571 USA
Phone: 845-758-0400
Fax: 845-758-2633
Email: curran@proceedings.com
Web: www.proceedings.com

Table of contents

Volume 530

International Conference on Recent Advances in Industrial Engineering and Manufacturing

12–13 December 2018, Penang, Malaysia

Accepted papers received: 31 May 2019

Preface

Papers

Additive manufacturing/Sustainable/Green manufacturing

Tool condition monitoring/Machining & forming technology

Modeling & simulation

Manufacturing - Others

Manufacturing system

Production improvement methods

Maintenance

Ergonomics

Data mining

IOP Conf. Series: Materials Science and Engineering **530** (2019) 011001 doi:10.1088/1757-899X/530/1/011001

Proceedings of the International Conference on Recent Advances on Industrial Engineering and Manufacturing (ICRAIEM 2018)

Penang, 12-13 December 2019

Preface

The International Conference on Recent Advances on Industrial Engineering and Manufacturing (ICRAIEM 2018) was held from 12-13 December 2018 at Mercure Penang Beach Hotel. Penang, Malaysia. The event, organized by the School of Mechanical Engineering, Universiti Sains Malaysia, was attended by 63 participants. Sixty-one papers were presented in two parallel sessions, namely Manufacturing and Industrial Engineering. The theme of the conference was Towards Industry 4.0 and IoT.

Three renowned keynote speakers were present to deliver the keynote speeches. The first keynote speech entitled 'Current and future trends of advanced manufacturing technologies' was delivered by Prof. Emeritus Andrew Nee from Department of Mechanical Engineering, National University of Singapore. The second keynote speech titled 'Opportunities and Challenges in The Implementation of Industry 4.0' was presented by Mr. Andreas Foerster, general manager of BBS Automation Penang Sdn. Bhd., while the third keynote speech titled 'Industry 4.X Implementation: A Journey of 1000 Miles Begins with a Single Step' was presented by Mr. Abram Ziegelaar, Vice-President (Digital Transformation Office), B. Braun Medical Industries Sdn. Bhd., Penang. The conference was sponsored jointly by Western Digital as a gold sponsor and Vitrox Technologies as a silver sponsor. The co-sponsors were Micro-Modular Systems Sdn. Bhd. and Institution of Engineering and Technology (IET), UK. The collaborators in the conference were University Malaysia Pahang (UMP), Universiti Teknikal Malaysia Malacca (UTeM) and Institution of Malaysia (IEM), Penang Branch.

An exhibition event was also held concurrent with the conference. Two leading companies involved in providing automation and Industry 4.0 solutions, namely BBS Automation Sdn. Bhd. and Advantech Co. Sdn. Bhd. showcased their latest products and services to the conference

IOP Conf. Series: Materials Science and Engineering **530** (2019) 011001 doi:10.1088/1757-899X/530/1/011001

delegates and visitors. A machine vision lighting specialist, TMS Lite Sdn. Bhd., offered LED lighting solutions to industrial automation and inspection applications and problems.

Organizing an international conference for the first time is a challenging task. The organizing committee has been working with great commitment and passion since the first day the idea of the conference was mooted in November 2017. I would like to thank all the members of the committee for their tireless effort in making this conference a success. I would also like to thank the members of the international advisory committee, panel of reviewers, keynote speakers, sponsors and exhibitors for their contribution in this conference. Lastly, I would to express my gratitude to all the authors for submitting their papers and participating in this conference. I will look forward to meeting everyone again in the next ICRAIEM conference.

Prof. Dr. Mani Maran Ratnam
Chief Editor

ICRAIEM 2018

IOP Conf. Series: Materials Science and Engineering **530** (2019) 011001 doi:10.1088/1757-899X/530/1/011001

Organizing committee

Patron
Prof. Datuk Dr. Asma Ismail

Honorary chairman
Prof. Dr. Zainal Alimuddin Zainal Alauddin

General chairman
Prof. Dr. Mani Maran Ratnam

General co-chairman
Assoc. Prof. Datin Dr. Norizah Mohamad

Secretariat
Dr. Hasnida Ab Samat

Finance & sponsorship
Dr. Loh Wei Ping (chairman)
Ir. Dr. Yen Kin Sam

Technical & proceeding publication
Assoc. Prof. Ir. Dr. Chin Jeng Feng (chairman)
Assoc. Prof. Dr. Zahurin Samad
Dr. Md. Sayem Hossain Bhuiyan
Dr. Reza Mahmoodian

Venue & logistics
Ms. Hasnah Hassan
Dr. Hasnida Ab Samat

Conference website, publicity & promotion
Dr. Teoh Yew Heng (chairman)
Dr. Yu Kok Hwa
Ms. Zarirah Karrim Wani (promotion co-chair)

Conference program & schedule
Mr. Mohzani Mokhtar (chairman)
En. Hishamuddin Bin Endan

International advisory committee
Prof. Datuk Dr. Mohd. Razali bin Muhamad (Universiti Teknikal Malaysia Malacca, Malaysia)
Prof. Dr. A. K. Prasada Rao (BML Munjal University, India)
Prof. Dr. Chris Backhouse (University of Loughborough, UK)
Prof. Dr. Kai Cheng (Brunel University, UK)
Prof. Dr. Md. Abdul Wazed (Chittagong University of Engineering and Technology,

IOP Conf. Series: Materials Science and Engineering **530** (2019) 011002 doi:10.1088/1757-899X/530/1/011002

Peer review statement

All papers published in this volume of *IOP Conference Series: Materials Science and Engineering* have been peer reviewed through processes administered by the proceedings Editors. Reviews were conducted by expert referees to the professional and scientific standards expected of a proceedings journal published by IOP Publishing.

IOP Conf. Series: Materials Science and Engineering **530** (2019) 012001 doi:10.1088/1757-899X/530/1/012001

Optimization of The FDM Parameters to Improve The Compressive Strength of The PLA-copper Based Products

N H Huu [1]*, D P Phuoc [1], T N Huu [2] and H T T Thu [1]

[1] Ho Chi Minh City University of Technology, VNU-HCM; 268 Ly Thuong Kiet St., Ward 14, Dist.10, Ho Chi Minh City, 700000, Vietnam

[2] Ho Chi Minh City University of Food Industry (HUFI); 140 Le Trong Tan St., Tay Thanh Ward, Tan Phu Dist., Ho Chi Minh City, 700000, Vietnam

Corresponding author *: hhnghi@hcmut.edu.vn

Abstract. Nowadays, Fused Deposition Modeling (FDM) has accomplished huge proficiency in modelling and wide variety in products shape and size in the industry. Alongside with its development, many problems came up, one of which is to ensure the technical requirements of the parts such as: compressive strength, tensile strength, dimensional accuracy, etc. To solve this problem, new composite materials has been developing and at the same time correlative process parameters has been optimized to achieve the most optimal technical requirements of that material. In this paper, we focus on researching the effect of process parameters consist of: raster angle; build direction and layer thickness on the part's compressive strength made by one of the newest composite material which is the PLA-Copper filament. Central composite design (CCD) methodology is used to lower the experimental runs, the analysis of variance (ANOVA) method is employed to investigate the process parameters in order to achieve optimal compressive strength.

Keyword: *FDM, PLA (Polylactic acid), PLA-Copper, Compressive Strength, ANOVA, RSM, FCCCD.*

1. Introduction

The FDM technology, one of the most commonly used AM methods nowadays due to its simplicity in structure, wide variations in equipment design that suitable for all types of users, from the small design machine for normal using for heavy duty one for fabricating prototype for industrial using. It works on a layer-by-layer principle, the material is liquefied and then laid down in layers on top of each layer until the part is formed (Fig 1). The material used in this case is in a form of a wire filament, there's a wide variety of filament type range from pure plastic like PLA or ABS to plastic with a solid powder composite materials like PLA-copper, ABS-wood, etc.

Along with its development, many problems arise, such as: how to optimize fabricated the technical requirements of FDM products; reduce fabricating times and improve quality of the finished parts, and etc. Many solutions are offered, likes developing a new kind of materials or correcting process parameters. For example, focusing on adjusting the relevant process parameters, such as infill percentage, build orientation or layer thickness, etc to find a set of parameters which will provide the best research output such as: tensile strength; dimensional accuracy; finished surface quality or compressive strength, which is also our main goal in this research.

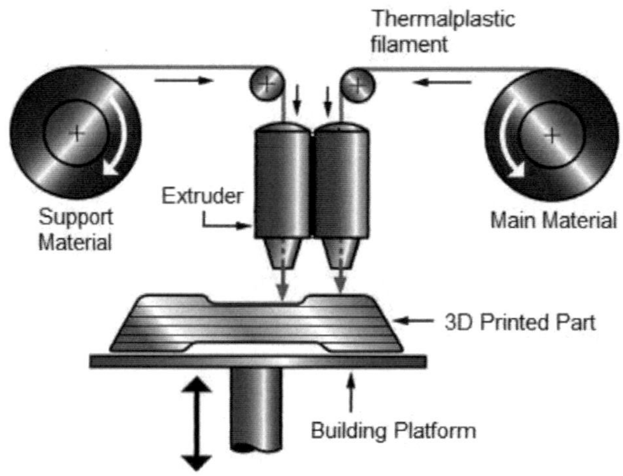

Figure 1. Basic FDM process [11]

Due to the fact that it works on layer-by-layer principle, there is always a separator between the contact surface or two layers, which is made the part weaker against impact force and drastically reduce compressive strength of that part. So in this study, we conducted a compressive strength experiment on a number of samples, each of which was fabricated with a different set of parameters consist of: raster angle; build direction and layer thickness to find out which parameter provide the optimal part's compressive strength made of PLA-copper filament. This experiment was carried out based on Central Composite Design (CCD) methodology to lower the experiment runs and employed the analysis of variance (ANOVA) method to analyze and find out which set of parameters provide the optimal compressive strength.

2. Literature review

Lee et al. [1] have conducted a compressive strength test on 3 sets of sample each of which was fabricated by 3 different methods: FDM, NCDS and binder jetting 3D printing. The measured compressive strength showed that parts made by binder jetting 3D printer had low compressive strength compared to other processes and that FDM parts had high compressive strength. Ahn et al. [2-3] have claimed that among others printing parameters, raster angle and build direction are quite important toward the process and durability of the fabricated parts. Wu et al. [4] have analyzed experiment result on a sample made of ABS and PEEK fabricated with a different set of raster angle and layer thickness, and then compare their tensile, compressive, fractured strength with the other set of samples. The result gave out the optimal which was 0.03mm in layer thickness and 0°/90° raster, this parameter has provided the best sample with optimal mechanical properties. Sood et al. [5] have studied the effect of 5 parameters consist of layer thickness, part build orientation, raster angle, raster width and air gap on compressive strength. The optimal parameters were found through the quantum behaved particle swarm optimization (QPSO) algorithm, the compressive strength of the sample was measured and then compared with the predicted result acquired from using artificial neural network (ANN). Panda et al. [6] have used the particle swarm optimization (PSO) technique to suggest a theoretical combination of parameter setting to achieve optimal parts tensile strength and then compare the result with a different algorithm which is the Differential Evolution algorithm (DEA) in term of prediction accuracy and convergence characteristic. Results show the potential of both argorithm, but the DEA give the better result than PSO. Jinwen et al. [7] has researched 4 specific parameters which are wire-width compensation, extrusion velocity, filling velocity, and layer thickness and the influence on part's dimensional error and warpage deformation, in their research the Taguchi method was used to optimize those 4 parameters, which was then evaluated by the fuzzy evaluation method. Lee et al. [8] have studied the effect of FDM parameters layer thickness, raster angle and air gap on the elastic performance of the ABS sample, the experiment was conducted, measured for the result and then investigated by the signal-to-noise (S/N) ratio measurement and analysis of variance (ANOVA) technique. Rayegani et al. [9] have conducted research to determine the functional relationship between FDM parameters and part's tensile strength, the experiment was carried out on sample with different raster angle and part orientation, the result was submitted to the group method of data handling (GMDH) and then used differential evolution (DE) method to find out the optimal parameters. Mishraa et al. [10] have studied the effect of six controllable process parameters such as layer thickness, part orientation, air gap, raster width, contour number and raster angle on the compressive strength of the FDM builds. It is observed that part orientation, contour number and air gaps between rasters have a significant effect on the compressive strength of the FDM build parts.

3. Material Properties

The main materials used in this case is the PLA-Copper filament provided by Shenzhen 3DSWAY Technology Co. It contains approximately 40% copper by volume mix with PLA and binder substain. With the density around 3.9 ~ 4.0 g/cm^3, melting temperature around 185°C, metal particle size range from 50 μm ~ 150 μm.

Figure 2. PLA-Copper Filament

4. Research Methodology

The compressive strength is the strength of an object when subjected to a single force without breaking. This feature is one of the most important characteristic of the finished part made by FDM process, and its affected by two main elements: the strength of the material itself and the FDM process parameters which are set up to make that part. Which is the thing we focus on optimizing to improve the compressive strength, then find out which set of parameters provide the best output and finally analyze which parameters affect the output the most.

An experiment was conducted based on Face-centered Central Composites Design (FCCCD) a type of Response Surface Method to reduce experiment runs, and then the result will be analyzed by using Analysis Of Variance (ANOVA).

Table 1. Vina FDM 2015 machine specification

Specifications	
Dimension	270x250x250 mm
Layer Height	0,15 mm
Machine Accuracy	±0,45 mm
Working Temperature	225 °C ÷ 280 °C
Extruding Speed	1200 ~ 1 800 mm/min

Figure 3. Vina FDM 2015 machine (Dual Extruder FDM)

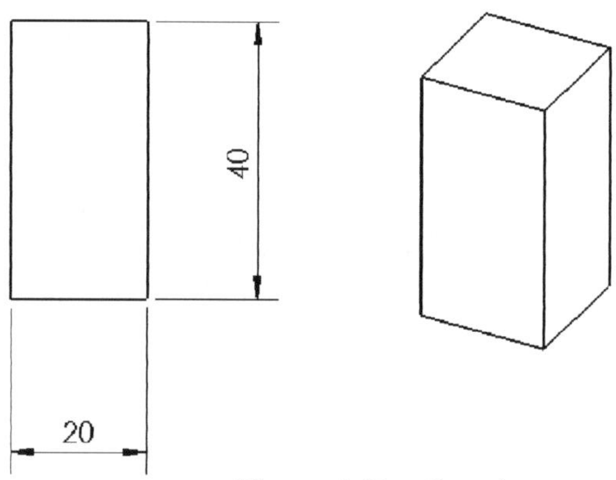

Figure 4. Test Sample

Using Solidworks2016 to design the compressive test sample (Fig. 4) based on ISO 604-2003 "Plastics – Determination of compressive properties" standard. The test sample was fabricated by Vina FDM 2015 machine (Dual Extruder FDM) (Fig. 3), its specs are listed in Table 1. After each sample was brought to the compressive test machine (Fig. 5). An example of a test sample before and after compressive strength test (Fig. 6) All of the experimental procedure is presented in Figure 7.

IOP Conf. Series: Materials Science and Engineering **530** (2019) 012001 doi:10.1088/1757-899X/530/1/012001

Figure 5. JTM tech (Model TM-UTC) compression testing machine

Figure 6. Test sample No. 9 Before and After the compression test

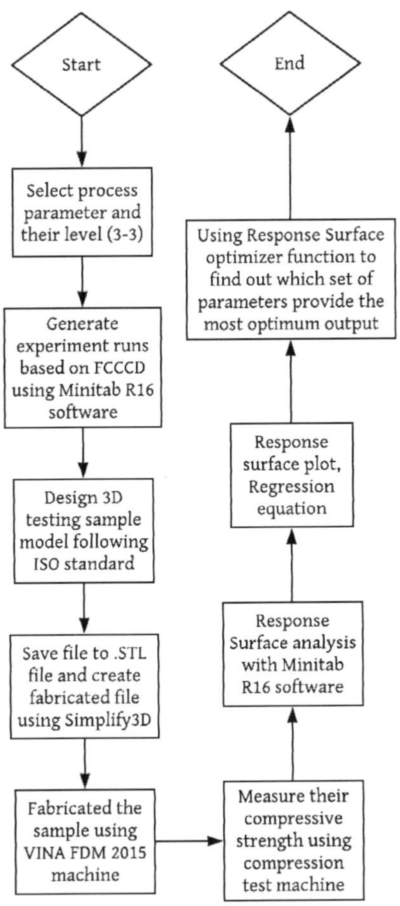

Figure 7. Experimental Procedure

IOP Conf. Series: Materials Science and Engineering **530** (2019) 012001 doi:10.1088/1757-899X/530/1/012001

5. Experimental data

In this research paper, we focus on finding the most optimized set of parameters to improve parts compressive strength made by FDM process. Three main factors were considered in this paper, while other remaining factors are kept unchanged, they are briefly defined as follows:

- Build Orientation: is the part build direction which affect the surface and the structure of the finished products.
- Raster Angle: the angle at which the support material lines will be drawn.
- Layer Thickness: height of each layer laid on top of each other during the printing process

The process factors and its level in coded form are represented in Table 2 below while other factors are unchanged.

Table 2. Levels of process parameters set in coded form

Factors	Symbols	Levels		
		-1	0	1
Build Orientation (degree)	A	0	45	90
Layer Thickness (mm)	B	0.2	0.25	0.3
Raster Angle (degree)	C	0	15	30

Table 3 represented the experiment runs along with respect parameters in coded form, using FCCCD methods we were able to reduce the experiment runs

Table 3. Experimental plan based on FCCCD along with the compressive strength result

Exp Run No.	A	B	C	Compressive strength (MPa)
1	-1	-1	-1	9.33
2	1	-1	-1	10.76
3	-1	1	-1	8.06
4	1	1	-1	9.54
5	-1	-1	1	8.85
6	1	-1	1	11.62
7	-1	1	1	7.97
8	1	1	1	10.43
9	-1	0	0	7.26
10	1	0	0	11.19
11	0	-1	0	6.15
12	0	1	0	5.13
13	0	0	-1	5.62
14	0	0	1	6.83
15	0	0	0	5.17
16	0	0	0	5.60
17	0	0	0	5.94
18	0	0	0	6.41
19	0	0	0	5.36
20	0	0	0	5.13

6. Results and Discussion

Analysis of the experimental data gathered from FCCCD design runs is done by MINITAB R16 analytic software using full quadratic RSM model as given by the equation as follows:

$$y = \beta_0 + \sum_{i=1}^{k} \beta_i x_i + \sum_{i=1}^{k} \beta_{ii} x_i x_i + \sum_{i<j} \sum \beta_{ij} x_i x_j \quad (1)$$

where, y is the response, x_i is i^{th} factor, k is the total number of factors.

T-test was performed to calculate input factors P-value, which is used to identify which input factor is significant to the output result of this experiment, the T-test result (Table 4) are represented below.

Table 4. T-test Result gathered from Minitab

Term	Coef	Se Coef	T	P
Constant	5.32539	0.1550	34.351	0.000
A	1.18150	0.1426	8.285	0.000
B	-0.53400	0.1426	-3.745	0.004
C	0.09175	0.1426	0.643	0.534
A*A	3.88966	0.2719	14.303	0.000
B*B	0.29966	0.2719	1.102	0.296
C*C	0.02591	0.2719	0.095	0.926
A*B	-0.00344	0.1594	-0.022	0.983
A*C	0.32156	0.1594	2.017	0.071
B*C	0.02281	0.1594	0.143	0.889

Table 5. Coefficents of regression

Source	DF	Seq SS	Adj SS	Adj MS	F	P
Regression	9	101.471	101.471	11.2745	55.44	0.000
Linear	3	16.895	16.895	5.6317	27.69	0.000
Square	3	83.744	83.744	27.9148	137.27	0.000
Interaction	3	0.831	0.831	0.2772	1.36	0.310
Residual Error	10	2.034	2.034	0.2034		
Lack-of-Fit	5	1.882	1.882	0.3763	12.37	0.08
Pure Error	5	0.152	0.152	0.304		
Total	19	103.505				

ReferencesS = 0.450958 PRESS = 15.5535
R-Sq = 98.04% R-Sq(pred) = 84.97% R-Sq(adj) = 96.27%

Figure 8. Normal probability plot of residual at 95% confidence level show us that most of the data is distributed normally.

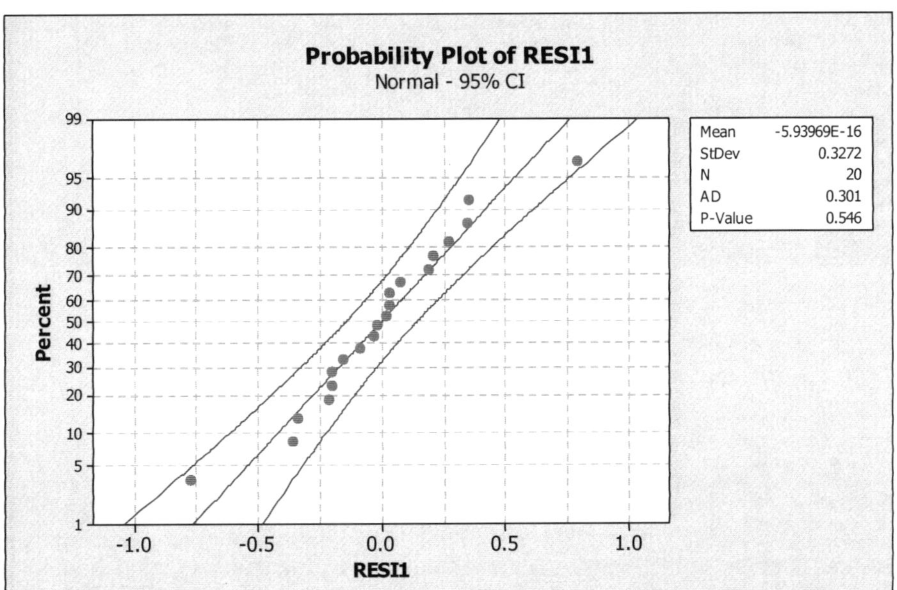

Figure 8. Normal probability plot of residual at 95% confidence level

IOP Conf. Series: Materials Science and Engineering **530** (2019) 012001 doi:10.1088/1757-899X/530/1/012001

As the result represented in Table 4, we can easily identified that factor A, B and A*A are significant to the output result because their P-value is less than 0.05. The lack of fits P-value is larger than 0.05 indicate that the model fits well with this experiment. Based on that we can build the regression plot:

$$\mathbf{Cs = 5.32539 + 1.18150*A - 0.53400*B + 3.88966*A*A}$$

Figure 9 showed the interaction between each factor and their corresponding levels:

- Factor A at 0 level result the lowest output as shown in **AxB** and **AxC** plot, while at 1 level yield the best output as shown in all 3 plots.
- Factor B at 1 level result the lowest output while at -1 level yields the best output as shown in **BxC** plot.

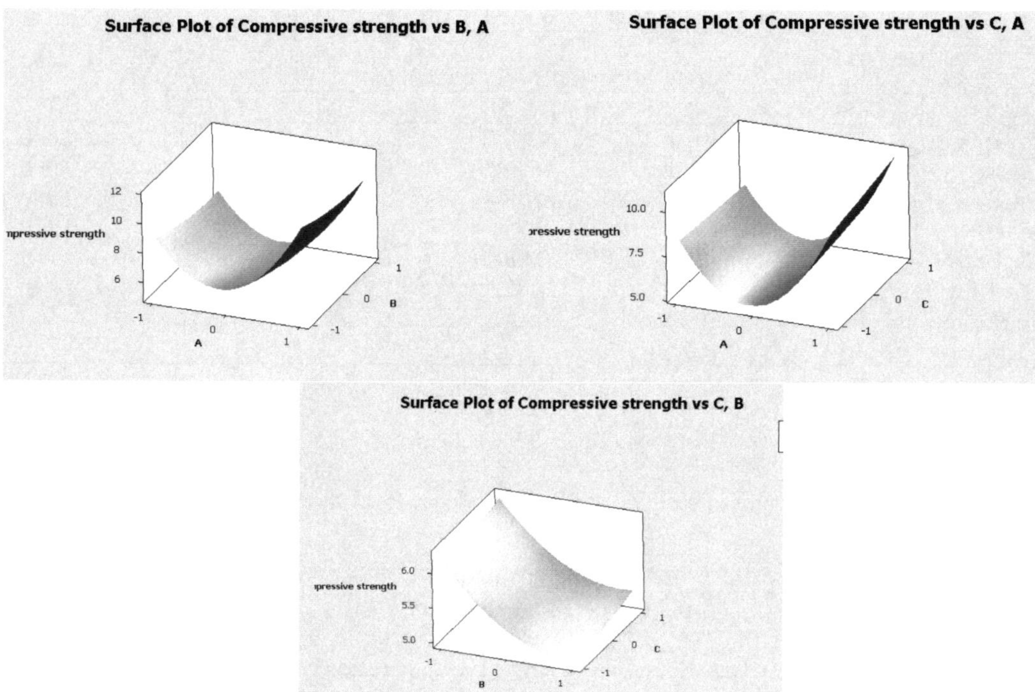

Figure 9. Response surface plot of: AxB; AxC; BxC

Using Response Optimizer function of MINITAB R16, we were able to calculate the optimal parameters, the result is shown in the figure (Fig. 10) as follows:

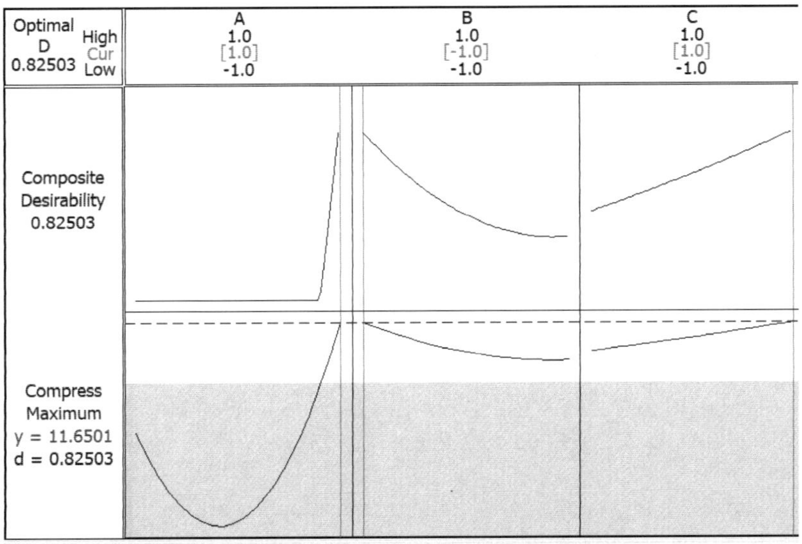

Figure 10. Optimal parameters obtained from MINITAB R16

7. Conclusion

As mention above, factor A; B and A*A are significant to the output. Which means that's part's orientation and layer thickness are significant to the parts compressive strength, while raster angle barely have any effect on it. Based on the data of the implemented experiment, some points of conclusion are carried out as follows:

- 45° part orientation provides the weakest compression strength, due to the fact that all the layer is laid leaning 45 degrees against the compressive force direction, it makes all the part's layer easier to slide on top of each other compare to two of the remaining orientations therefore its easier to break than others.
- 90° part orientation provides the strongest compression strength among 3 parameters, because the layer in this case is laid parallel with the applied force surface of the compressive testing machine, therefore the parts fabricated with this orientation parameter would significantly stronger than others
- As the layer thickness decrease which mean the amount of layer increase, if the number of layers increases, the heat vector increases toward the bottom layers, increasing the diffusion between the fibers, thus increasing the compressive strength.
- The test sample after the compression test experiment (Fig. 6) has broken into pieces, unlike any other pure plastic filament use in FDM, from this we can conclude that the material is very brittle. Due to the presence of metal particle in the composites materials, the material plasticity has been decreased, thus increase its brittleness.

Through FCCCD methodology we have determined the set of parameters which provided the optimal part compressive strength (Fig. 10), the optimal set of parameter settings is: **90°** Build orientation, 0.2 Layer Thickness and 30° Raster Angle.

References

[1] C S Lee, S G Kim, H J Kim and S H Ahn 2007 "Measurement of anisotropic compressive strength of rapid prototyping parts," *Journal of Materials Processing Technology* p 627-630

[2] S H Ahn, M Montero, D Odell, S Roundy and P K Wright 2002 "Anisotropic material properties of fused deposition modeling ABS," *Rapid Prototyping J,* No 8 p 248–257

[3] F Xu, H T Loh and Y S Wong 1999 "Considerations and selection of optimal orientation for different rapid prototyping systems," *Rapid Prototyping J* No 5 p 54–60

[4] W Wu, P Geng, G Li, D Zhao, H Zhang and J Zhao 2015 "Influence of Layer Thickness and Raster Angle on the Mechanical Properties of 3D-Printed PEEK and a Comparative

IOP Conf. Series: Materials Science and Engineering **530** (2019) 012001 doi:10.1088/1757-899X/530/1/012001

Mechanical Study between PEEK and ABS," *Materials 2015 8(9)* p 5834-5846

[5] A K Sood, R K Ohdar and S S Mahapatra 2010 "Experimental investigation and empirical modelling of FDM process for compressive strength improvement," *Journal of Advanced Research* Vol 3 p 81-90

[6] B N Panda, R Bahubalendruni and B B Biswal 2014 "Comparative Evaluation of Optimization Algorithms at Training of Genetic Programming for Tensile Strength Prediction of FDM Processed Part," *Procedia Materials Science* No 5

[7] Z Jinwena and P Anhuab 2014 "Process-Parameter Optimization for Fused Deposition Modeling Based on Taguchi Method," *Advanced Materials Research* Vol 538-541 p 444-447

[8] B H Lee, J Abdullah and Z A Khan 2005 "Optimization of rapid prototyping parameters for production of flexible ABS object," *Journal of Materials Processing Technology* Vol 169 p 54–61

[9] F Rayegani and G C Onwubolu 2014 "Fused deposition modelling (FDM) process parameter prediction and optimization using group method for data handling (GMDH) and differential evolution (DE)," *Int J Adv Manuf Technol* Vol 73 p 509–519

[10] S I Mishraa, K Abhisheka, M P Satapathya and S S Mahapatrab 2017 "Parametric Appraisal of Compressive Strength of FDM Build Parts", *Materials Today: Proceedings 4* p 9456–9460

[11] https://www.researchgate.net/figure/Process-schematic-for-FDM-process_fig2_225072385; accessed on 2018.0729

IOP Conf. Series: Materials Science and Engineering **530** (2019) 012002 doi:10.1088/1757-899X/530/1/012002

Experimental Investigation of Recycled Machining Coolant Mixed with Nanofluids Based Al_2O_3

S K Lim, N Z Mohamad and A R Yusoff *

Faculty of Manufacturing Engineering, Universiti Malaysia Pahang, 26600 Pekan, Pahang, Malaysia.

Corresponding author *: razlan@ump.edu.my

Abstract. Machining coolant is used in manufacturing industry for lubrication, rusting control and cooling which are crucial in operations such as grinding and cutting process. However, the machining coolant lose efficiency in months due to the thermal degradation and contamination. To recycle the used of machining coolant while enhancing the performance, the mixture of nanofluids (N) based Al_2O_3 with better thermal properties to synthesize with the recycled machining coolant (RMC). This study suspended Al_2O_3 nanofluids in four base ratios of recycled machining coolant (i.e. 0:100, 20:80, 40:60 and 60:40) by ultrasonic homogenies. The three main parameters in cooling rate performance and rusting control are thermal conductivity, dynamic viscosity and pH indication, respectively. The thermal conductivity and dynamic viscosity are then measured at temperature range of 30 to 60 °C. The highest enhancement of thermal conductivity was evaluated to be 37.9% higher than machining coolant base fluid at the temperature of 60 °C in 60:40 (N:RMC). However, the enhancement of dynamic viscosity was measured to be 4.5% for 60:40 (N:RMC) at 60 °C. The pH value of 60:40 (N:RMC) at ambient temperature obtained 8.18 value was nearly neutral and decreased the rusting process. Therefore, this study recommends the use of recycled machining coolants based Al_2O_3 as cutting fluid in the ratio of 60:40 for application in machining operations.

1. Introduction

Machining coolant or lubrication is considered as an accessory in a manufacturing process in order to increase the life of the cutting tool and productivity. The surface quality of the mechanical parts and cutting tools are crucial in evaluating the productivity of a manufacturing industry. In 2018, the amount of machining coolants used in manufacturing industry was reported nearly 38 000 tonnes with an increase of 1.2% compared with the year of 2005 [1]. It is reported that European Union alone is consuming about 320 000 tonnes of machining coolants within one year and two-thirds of it needs to be disposed due to the degradation of thermal efficiency of machining coolants [2]. It is expected that cost of machining coolants, including procurement, preparation, operation and the disposal are approximately 16% of the total manufacturing costs of a product. Therefore, recycling of machining coolant can reduce manufacturing cost and solve waste disposal problems. In 2016, Liang et al. [3] conducted an investigation on the tribological behavior of in-situ exfoliated graphene for water-based lubricants and found that graphene-enhanced lubricants improved the surface quality and coolants durability. Syam Sundar et al. [4] experimentally investigated the thermal conductivity of Al_2O_3 in 20:80 EG/water mixtures compared with 60:40 EG/water base fluids. They found that the thermal conductivity was significantly increased compared to the base fluid with higher ratio of ethylene glycol. The augmentation of thermal conductivity increased the heat transfer coefficient value as stated from Kole and Dey [5]. The average enhancement of heat transfer coefficient was up to 22% in their studied.

Few studies are available regarding to the nanofluids as coolant in machining process. The nanofluid based Al_2O_3 is proposed as a surfactant to synthesize with recycled coolant in order to enhance the performance of machining coolant with better thermal properties. The present study focuses on pH indication, thermal conductivity and dynamic viscosity of Al_2O_3 nano-mixtures in four volume ratios of 0:100, 20:80, 40:60 and 60:40 for measured temperature of 30 to 60 °C. The optimum volume ratio of nano-mixtures need to be determined by considering the pH value, dynamic viscosity and thermal conductivity as machining coolant for application in machining process.

2. Experimental Procedures

2.1. Preparation of nanofluid and mixture

Nanofluid was prepared by dispersing the nanoparticles in the mixture of water to ethylene glycol fluid. The aluminum oxide, Al_2O_3 nanoparticles in powder form have an average particle diameter of 13 nm size are suspended in nanofluid by using two-step method proposed by Yu et al. [6]. The nanofluid was mixed in four different ratios of recycled machining coolant as base fluids, which were 0:100, 20:80, 40:60 and 60:40 respectively. The ultrasonic homogenizer was employed to improve the dispersion stability of nanoparticles in the mixture. The mixture of Al_2O_3 machining coolant is synthesized by using a magnetic stirrer and sonicated in ultrasonic bath for two hours, followed the studies by Azmi et al. [6-7]. The stability of the nano-mixture is observed and found to be stable for more than two months.

2.2. pH measurement

The pH values were measured by a water-proof hand-held pH meter, Oakton PCTestr-35 by Eutech Instruments Pte. Ltd. The instrument has a combination of pH, electrical conductivity and temperature probes that are user friendly, as illustrated in Figure 1. It can measure pH values between the ranges of 0.00 to 14.00 at a fluid temperatures of 0 to 60 °C. The instrument was calibrated using the buffer solution provided by the manufacturer. The standard pH buffer solution provided by the manufacturer is at pH values of 1.65, 4.00, 7.00, 10.00 and 12.45. The 5-point pH calibration was done intentionally to ensure accuracy through the pH range [8]. After calibration of the pH meter, the measurement of pH values for Al_2O_3 nano-mixture in different base ratios with 80 ml at ambient temperature were evaluated. The probe as pH sensor was fully immersed into the specimen up to indicator level. The three measurements of each base ratio were collected to generate reliable data, and the average value of the three measurements was taken for analysis. The method of pH measurement was similar to Konakanchi et al. [9] to measure the pH values in their research.

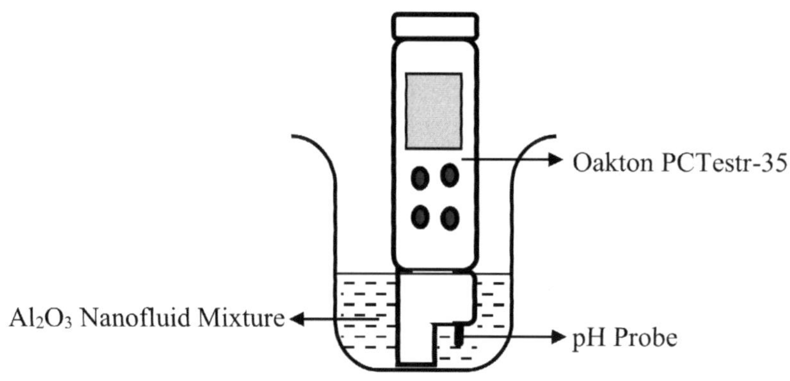

Figure 1. Schematic of pH measurement.

IOP Conf. Series: Materials Science and Engineering **530** (2019) 012002 doi:10.1088/1757-899X/530/1/012002

2.3. Thermal conductivity measurement

KD2 Pro thermal property analyzer (Decagon Devices, USA) measured the thermal conductivity of the Al_2O_3 nanofluid mixture, as depicted in Figure 2. The thermal properties of liquids and solids were determined by the device using the transient line heat source. A single needle sensor (KS-1) in the range of 0.002 to 2.00W/m·K was used, and the sensor was validated by measuring the thermal conductivity of the verification liquid such as glycerine that was provided by the supplier. The measured value of glycerine at 25 °C was 0.286 W/m·K, which agreed with the calibrated data of 0.285 W/m·K and within ±0.4% accuracy [6]. The validation process of the sensor was checked before each measurement of thermal conductivity. A water bath of WNB7L1 model by Memmert was used to maintain a constant temperature of the sample with an accuracy of 0.1 °C [10]. Thermal conductivity of four different ratios of Al_2O_3 nano-mixtures were measured within the temperature range of 30 to 60 °C. The consistency of data measurement can be ensured; a minimum of three values were taken for every base ratio at a specific temperature, and the average of the three values was analyzed.

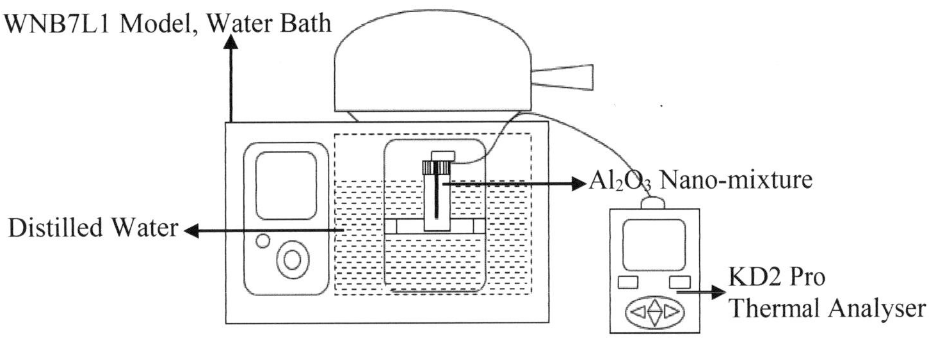

Figure 2. Schematic of thermal conductivity measurement.

2.4. Dynamic viscosity measurement

Dynamic viscosity of Al_2O_3 nanofluid in four different ratios of recycled machining coolant as base fluids were measured by Brookfield (low viscosity digital viscometer) LVDV-III Ultra Programmable Rheometer, as represented in Figure 3. The device is equipped with a RheoCal program in personal computer for data measurement and collection. The viscometer measures the dynamic viscosity of solutions that range from 1 to 6000,000 mPa·s by utilizing an ultra-low adapter. The spindle connected to the viscometer was used to measure nanofluids mixture viscosity [11]. The viscometer drives a spindle that is immersed in nanofluid mixture. Spindle rotation created a viscous drag of the fluid opposite to the spindle, which is measured by the deflection of the calibrated spring. The adapter in this experiment has a provision for temperature circulation of bath fluid. The viscosity of different base ratios of nano-mixtures based Al_2O_3 started from 30 to 60 °C at an interval of 5 °C. Each measurement was conducted three times to generate a reliable data, and the average value of the three values was considered for analysis.

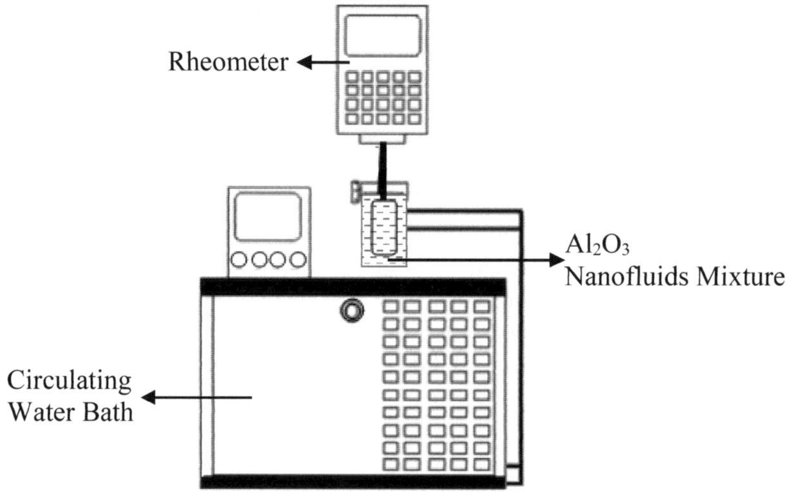

Figure 3. Schematic of dynamic viscosity measurement

3. Results and Discussion

3.1. pH evaluation

According to Scientific, 2015 [12], Al_2O_3 nanoparticles dispersed in mixture of distilled water and ethylene glycol solution have pH measurement in range of 5.0 to 6.5 at ambient temperature condition. However, the nanofluids synthesized with the recycled machining coolants have a pH value of 8.0 to 9.0; and the pH value of machining coolant is between of 8.5 to 9.3 [13]. The percentages of nanofluid based Al_2O_3 applied in this study are 20% to 60% to homogenize with recycled machining coolant as mixture with pH values as shown in Table 1. From the pH indications, it could be seen that the solution chemistry of the nano-mixture is approaching to 8.0, as the percentages of nanofluid increase in the solution. El Baradie, 1996 [14] claimed that pH 7.0 represents neutrality, and the solution with pH value of 8.0 for river water in London readings.

Table 1. pH measurements of several percentages nanofluids mixture

Nanofluids : Recycled Machining Coolant (vol. %)	pH Scale
0 : 100	9.12
20 : 80	8.81
40 : 60	8.57
60 : 40	8.18

3.2. Thermal conductivity of nano-mixtures based Al_2O_3

The thermal conductivity of recycled machining coolant mixed with different ratios of Al_2O_3 nanofluids is shown in Figure 4. As observed from the figure, thermal conductivity is increased with the increasing of temperature and nanofluids percentages, followed the behavior of nanofluid thermal conductivity as stated by Syam Sundar et al. [4] and Lim et al. [6]. However, there is small fluctuation of thermal conductivity occurs at temperature 30 °C where the percentage of nanofluids 20% higher than 40% nanofluids mixture. In overall, all the nanofluids mixture with the percentage of 60% exhibited the maximum thermal conductivity. The observation seem to be related to the Brownian motion and convection heat transfer of nanoparticles where the increases of nanofluid ratio tends to increase the collision of particles with molecules of base liquid at high temperature [15].

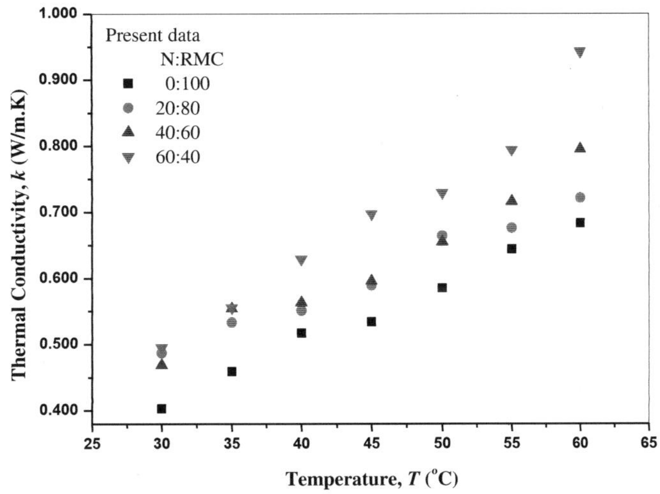

Figure 4. Thermal conductivity of nano-mixtures for different ratios of nanofluid (N) and recycled machining coolant (RMC)

3.3. Dynamic viscosity of nano-mixture based Al_2O_3

The dynamic viscosity measurements of nano-mixtures based Al_2O_3 were conducted in temperature range from 30 to 60 °C as depicted in figure 5. At the base fluids of 60:40 (N:RMC), the dynamic viscosity enhancement percentage was 4.5% at 60 °C compared with that of the 0:100 (N:RMC) base fluids. The dynamic viscosity of the nano-mixture increased as the volume concentration of nanofluid increased. In addition, each volume concentrations have similar increment pattern of viscosity. According to Syam Sundar et al. [16], who found similar pattern of viscosity, it is caused by shear resistance given by the particles onto the fluid. However, the dynamic viscosity of the nanofluid diminished exponentially as the temperature increased. As mentioned by Usri et al. [17] in their research paper, the temperature weakening the inter-particle and inter-molecular adhesion forces.

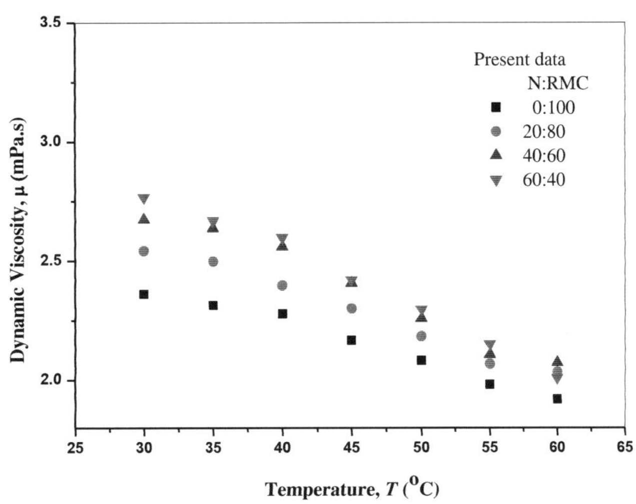

Figure 5. Dynamic viscosity of nano-mixtures for different ratios of nanofluids and recycled machining coolants

4. Conclusion

In this paper, the thermal conductivity, dynamic viscosity and pH indication of Al_2O_3 base nano-recycled machining coolants were investigated. The results showed that pH value of 60:40 (N:RMC) was slightly higher than neutral value of 7.0 and the bacterial growth was prohibited in alkaline condition in order to decrease the rusting process [14]. The experimental analysis found that thermal conductivity increased with the increment of temperature and percentages of nanofluid as base fluids. However, the dynamic viscosity of nano-mixture exponentially decreases with the increase of temperature and the decrease of nanofluid percentages as base fluid.

The thermal conductivity enhancement was determined to be greater than the enhancement of dynamic viscosity. The highest thermal conductivity enhancement increased by 37.9% compared to the machining coolants. However, the enhancement of dynamic viscosity exceed 100% which is not giving benefit to the heat transfer. Therefore, the use of nano-mixtures of 60:40 (N:RMC) based Al_2O_3 is recommended for application as machining coolant to reduce cutting temperatures. Further investigations and experiments on the performance of the machining processes such as milling and turning process by using nano-mixtures as coolants are required to extend the present work.

Acknowledgments

The financial support by Universiti Malaysia Pahang (UMP) under Fundamental Research Grant Scheme of RDU 160142 is gratefully acknowledged.

References

[1] Debnath S, Reddy M M and Yi Q S 2014 Environmental friendly cutting fluids and cooling techniques in machining : a review *J. Clean. Pro.* 83 33-47

[2] Abdalla H S, Baines W, Mclntyre G and Slade C 2007 Development of novel sustainable neat-oil metalworking fluids for stainless steel and titanium alloy machining Part 1 Formulation development *J. Adv. Manuf. Technol.* 34 21-33

[3] Liang S S, Shen Z G, Yi M, Liu L, Zhang X J and Ma S L 2016 In-situ exfoliated graphene for high-performnace water-based lubricants *Carbon* 96 1181-90

[4] Syam Sundar L, Venkata Ramana E, Singh M K and Sousa A C M 2014. Thermal conductivity and viscosity of stabilized ethylene glycol and water mixture Al_2O_3 nanofluids for heat transfer applications: An experimental study *Int. Commun. Heat Mass Transf.* 56 86-95

[5] Kole M and Dey T K 2012 Thermophysical and pool boiling characteristic of ZnO-ethylene glycol nanofluids *Int. J. Therm. Sci.* 62 61-70

[6] Lim S K, Azmi W H and Yusoff A R 2016 Investigation of thermal conductivity and viscosity of Al_2O_3/water–ethylene glycol mixture nanocoolant for cooling channel of hot-press forming die application *Int. Commun. Heat Mass Transf.* 78 182-189

[7] Azmi W H, Sharma K V, Sarma P K, Mamat R, Anuar S and Dharma Rao V 2013 Experimental determination of turbulent forced convection heat transfer and friction factor with SiO_2 nanofluid *Exp. Therm. Fluid Sci.* 51 103-111

[8] Lim S K 2018 Investigations of Nanocoolant Based Al_2O_3 for Improving Cooling Performance in Hot Press Forming (MSc Thesis) *Universiti Malaysia Pahang Malaysia*

[9] Konakanchi H, Vajjha R S, Chukwu G A and Das D K 2015 Measurements of pH of three nanofluids and development of new correlations *Heat Transf. Eng.* 36 (1) 81-90

[10] Zakaria I, Azmi W H, Mohamed W A N W, Mamat R and Najafi G 2015 Experimental investigation of thermal conductivity and electrical conductivity of Al_2O_3 nanofluid in water–ethylene glycol mixture for proton exchange membrane fuel cell application *Int. Commun. Heat Mass Transf.* 61 61-68

[11] Azmi W H, Sharma K V, Sarma P K, Mamat R and Najafi G 2014 Heat transfer and friction factor of water based TiO_2 and SiO_2 nanofluids under turbulent flow in a tube *Int. Commun. Heat Mass Transf.* 59 30-38

[12] Scientific B 2015 Water Stills - Technical Note: T11-001 - pH of distilled water (online)

Retrivedfromhttp://www.stuartequipment.com/adminimages/T11_001_pH_of_Distilled_Water.pdf on 15 June 2015

[13] Gajrani K K, Ram D and Ravi Sankar M 2017 Biodegradation and hard machining performance comparison of eco-friendly cutting fluid and mineral oil using flood cooling and minimum quantity cutting fluid techniques *J. Clean. Pro.* 165 1420-35

[14] El Baradie M A 1996 Cutting fluids: Part II. Recycling and clean machining *J. Mater. Process. Tech.* 56 (1-4) 798-806

[15] Hussein A M, Sharma K V, Bakar R A and Kadirgama K 2014 A review of forced convection heat transfer enhancement and hydrodynamic characteristics of a nanofluids *Renew. Sustain. Energy Rev.* 29 734-743

[16] Syam Sundar L, Venkata Ramana E, Singh M K and Sousa A C M 2012 Viscosity of low volume concentrations of magnetic Fe_3O_4 nanoparticles dispersed in ethylene glycol and water mixture *Chem. Phys. Lett.* 554 236-242

[17] Usri N A, Azmi W H, Mamat R and Abdul Hamid K 2014 Viscosity of aluminium oxide (Al_2O_3) nanoparticle dispersed in ethylene glycol *Appl. Mechanics Mater.* 660 735-739

[18] W H Azmi, K V Sharma, R Mamat, G Najafi and MS Mohamad 2016 The enhancement of effective thermal conductivity and effective dynamic viscosity of nanofluids — a review, *Renew. Sust. Energ. Rev.* 53 1046–1058

Vegetable Oil-Based Lubrication in Machining: Issues and Challenges

M A Mahadi, I A Choudhury*, M Azuddin, N Yusoff, A A Yazid and A Norhafizan

Department of Mechanical Engineering, Faculty of Engineering, University of Malaya, 50603 Kuala Lumpur, Malaysia.

Corresponding author *: imtiaz@um.edu.my

Abstract. The use of vegetable oil as a replacement for conventional mineral-based lubricant has gained many interests in recent years from the machining industries. This is due to the problematic issues with the use of conventional mineral-based lubricant that can cause negative effects on the manufacturing cost, operators' health and the environmental. As vegetable oil is renewable, biodegradable, and non-toxic and has remarkable tribological characteristic from its triglycerides structure, it has been identified as the best replacement for mineral-based cutting lubricant. However, formulating vegetable oil into machining lubricant has its challenges where it must overcome its nature of poor low-temperature properties and low oxidative stability. Therefore, this paper will elaborate the present issues and challenges in formulating green machining lubricant using vegetable oil and its stand under sustainable machining. This includes the method and technique of formulating vegetable-based lubricant and the output performance obtained from vegetable-based lubricant under machining in comparison to the conventional mineral-based lubricant. The result shows that utilizing the vegetable-based lubricant at minimum quantity has comparable or better performance to conventional mineral-based lubricant in terms of surface finish. Therefore, the use of vegetable oil has its challenges and issues as machining lubricant, but it is one of the current best viable alternatives to the mineral-based lubricant in promoting sustainable machining.

Keywords: *Vegetable oil, triglycerides, palm kernel oil, machining lubricant, sustainable machining.*

1. Introduction

The cutting lubricant has three basic actions, which are cooling, reducing friction and minimizing the shear strength [1]. These actions contribute to low cutting force, longer cutting tool life and highly desirable surface finish. The use of lubricant is so important in machining, that it is reported, approximately 320,000 tons of cutting lubricant was being consumed by the European Union annually which mostly were made of mineral oil [2]. Moreover, the quantity of lubricant keeps on increasing over the years. This fact has raised awareness since the use of the mineral-based cutting lubricant can bring negative impact on the environment and the operators' health. Several studies have proven that mineral-based lubricant may contain carcinogenic additives and impurities in their formulation, which causes dermatitis, and skin cancer [3]. Furthermore, mineral oil is very poor in terms of biodegradability, with the range of 15 - 50% biodegradation. The said biodegradation values are below than the acceptable guideline for environmental friendly lubricant, which is at 80% [4]. Thus, to properly dispose of the mineral based cutting lubricant is very difficult and challenging that involves numerous processes at a high cost. Hence, an alternative to mineral oil should be considered and the effort should be concentrated on minimizing the lubricant usage.

The rise in environmental issues and health concerns from the use of mineral oil causes the increased use of vegetable as the base oil of a lubricant. Triglycerides inside the chemical structure of vegetable oils are three long chains of fatty acids that contributed the said oil to have the desirable lubricating properties. Furthermore, vegetable oils have high viscosity index and low volatility making it the best alternative to mineral oil. The main factor of using vegetable oils is that it is safe toward the environment and to the operators handling as the vegetable oils is highly biodegradable and low in toxicity. The said oil has been used as a lubricant since the ancient time, however, the usage stopped due to low oxidation properties and poor low temperature properties.

1.1 Low oxidative stability

Vegetable oils are known to be low in oxidative stability due to the presence of double bond (C=C) in their fatty acid composition. The higher the composition of unsaturated fatty acids (double bond), the more susceptible vegetable oils to the oxidation process. Vegetable oil that undergoes the process of oxidation will increase in their viscosity and produces deposits of varnish and sludge, which degrades the quality of vegetable oils. Table 1 shows the composition of unsaturated and saturated fatty acid inside several types of vegetable oils. Looking at Table 1, most of the vegetable oils have a high composition of unsaturated fatty, which means that most of the vegetable oils are susceptible to the process of oxidation.

Several methods were identified to improve the oxidative stability of vegetable oils. One of the ways is to add additives such as anti-oxidant in the formulation of lubricant. Anti-oxidant works by delaying the viscosity increase of vegetable oils, which consequently helps in decreasing the rate of oxidation. Third-butyl hydroquinone (TBHQ) is one of the additives that can significantly improve the oxidative properties of vegetable oils, by almost 6 times the original performance [5]. Furthermore, the modification can be made on the chemical structure of vegetable oils through ring-opening process to further improve their oxidative stability [6]. Nonetheless, it is important to note that the inclusion of anti-oxidant and modification of the chemical structure will increase the lubricant's formulation cost and time.

Table 1. The composition of fatty acids of vegetable oils

No.	Vegetable oils	Fatty acids (%)	
		Saturated	Unsaturated
1.	Grapeseed	15.2	84.8
2.	Sunflower	15.4	84.6
3.	Soya bean	20.7	79.3
4.	Corn	19.5	80.5
5.	Canola	11.5	88.5
6.	Olive	21.9	78.1
7.	Hazelnut	13.8	86.2

1.2 Poor low temperature properties

The other issue of using vegetable oils as lubricant is that it will have poor low-temperature properties. This can cause the lubricant to exhibit cloudiness, precipitation, poor flowability, and solidification at room condition temperature (24°C). There is contradicting opinion on the source of poor properties of vegetable oils at low temperature. One opinion stated that the composition of saturated fatty acid in the chemical structure of vegetable oils contribute to the poor low-temperature properties [7]. On the other hand, stated that the presence of polyunsaturated fatty acids (multiple double bonds in one chain of fatty acid) is the caused for the said poor performance [8]. However, most researchers are favoring the theory that the composition of saturated fatty acid as the main influence on the poor low temperature performance. This is because, at low-temperature, the carbon atoms of saturated fatty acid chains tend to bundle rapidly than the unsaturated fatty acid, turning into crystalline form. The low temperature properties of vegetable oils are identified by examining the pour point temperature of the vegetable oils.

Efforts have been made in increasing the low temperature properties of vegetable oils. The most popular method is the process of transesterification [9]. This process is conducted by adding alcohol and a catalyst to the vegetable oil. The process helps in reducing the percentage of saturated fatty acid by changing those fatty acids to the monounsaturated fatty acids by increasing the amount of double bond. Thus, the product of this process is vegetable oils that are high in the composition of monounsaturated fatty acids. Tests have been conducted on the product and it is proven that the low temperature properties have increased. However, a new challenge will emerge to this product as it is now high in the composition of unsaturated fatty acid, making the newly produce product prone to the oxidation process.

1.3 Sustainable machining

Sustainable machining refers to improvement in the aspect of environmental protection, profitability and societal benefit. It is an upgrade toward the conventional 3R (reuse, reduce and recycle) concept in promoting green machining (environmentally) to an upgraded 6R (reuse, reduce, recycle, redesign, recover and remanufacturing) for the concept of sustainable machining (innovative). Figure 1 shows the development of sustainable machining over the years. Implementing vegetable oil as the based oil cutting lubricant poses healthy benefits towards the environment and operators' health due to its biodegradability and low toxicity properties [10]. This action achieved the concept of environmental green machining; however, this is not considered as the action under the umbrella of sustainable machining. In terms of the economic view, the use of cutting fluid consumes a significant amount of energy and the production cost. Furthermore, once the cutting fluid is used, the cost is taken to properly dispose of the used cutting fluid. If not properly conducted, it can accumulate bacteria and fungi that can consequently lead to ecological damage and contamination. Therefore, the innovative invention is needed in utilizing the vegetable-based lubricant to meet the expectation of sustainable machining.

Suggestion has been raised to perform machining in dry condition to attain the sustainable machining concept. Dry conditions eliminate the cost of cutting lubricant and its expenses on waste disposal treatment. This would greatly benefit in the economic point of view as it reduces around 15 % of machining cost. However, machining under dry conditions degrades the tool life performance due to the excessive increase in cutting temperature, especially in machining difficult-to-cut materials. Therefore, the best condition in machining hard-to-cut materials is to use cutting lubricant at the least amount, which is called minimum quantity lubrication (MQL) where an average 50 mL/H cutting lubricant delivered to the cutting regions. The amount of lubricant usage under MQL is far less than the conventional flooded cooling at an average of 120,000 mL/H of flowrate. According to Amrita, the utilization of MQL is able to greatly improve the result of a turning process [11]. This is because, under MQL system, the lubricant is forced to penetrate the small gap between the cutting tool and workpiece, which the flooded cooling is unable to achieve. With far less amount of lubricant usage, the system able to perform better than machining under the conventional flooded and dry condition. Although the initial production cost of utilizing vegetable based lubricant under MQL is slightly higher, this is compromised with the superior performance and results that can be obtained, which makes it the best approach in turning under the umbrella of sustainable machining.

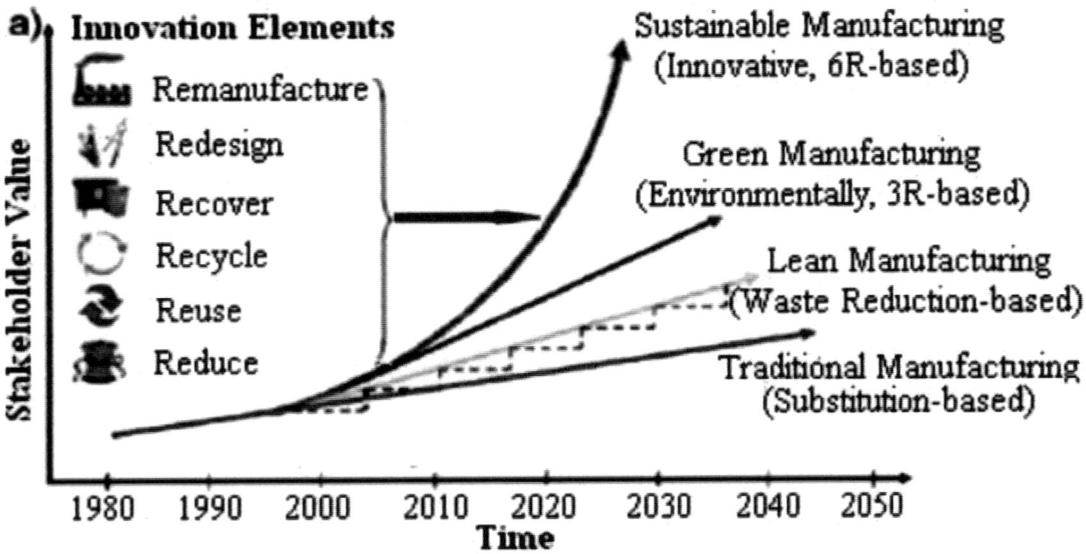

Figure 1. The development of sustainable machining [12]

In this study, the formulation of the vegetable-based cutting lubricant is briefly explained with the brief clarification on the selection of additives. Then, the formulated vegetable oil under MQL mode was tested their performance in comparison to conventional mineral-based cutting lubricant under flooded cooling. The testing was conducted with the help of 2^4 full factorial design of experiment with cutting speed, feed rate, depth of cut and type of lubricant as the variable factors. Then, each experimental run was conducted in turning AISI431 stainless steel and the surface finish of the workpiece was examined.

2. Experimental procedure

2.1 Formulation of lubricant

Vegetable based lubricant for this study was formulated using a mechanical stirrer rotating at 750 – 760 rpm for 1 hour and the materials used were palm kernel oil, several additives, and water. The steps in the formulation process were based on the work of Lawal et al. [13]. The additives used in this formulation were emulsifiers (Polyoxyethylene Sorbitan Monostearate, TWEEN 60), anti-oxidant (2,6-Di-tert-butyl-4-methylphenol), anti-corrosion (ethanolamine) and aluminium oxide nano-powder (20 nm particle size). As the lubricant was a soluble type, emulsifier was required to mix the vegetable oil and water together at the ratio of 1:10. Palm kernel oil was selected as the base oil because of its high saturated fatty acids content. As previously mentioned, a high composition of saturated fatty acids inside the chemical structure of vegetable oils can reduce the rate of oxidation, thus making palm kernel oil as the suitable candidate. Since the temperature and humidity at the current location were high, poor low temperature properties of the vegetable oil should not bring any harm towards the performance of the lubricant. Furthermore, the said oil was selected due to their abundance and accessibility in this country.

Anti-oxidant in this study was used to reduce the rate of oxidation of the palm kernel oil. Even though the oil has a high content of saturated fatty acid, there are still traces of unsaturated fatty acid that can lead to the process of oxidation, thus requiring the use of the anti-oxidant agent. Anti-corrosion was used as a long-term corrosion resistance that provides protection to the ferrous and non-ferrous surface. Furthermore, the additive was used as a stabilizing agent for a stable emulsion and thereby slowing down the rate of separation between palm kernel oil and water. Lastly, aluminium oxide nano-powder was included in the formulation to increase the performance of the vegetable-based lubricant. Researchers have investigated on the utilization of nano-powder in lubricant, and mostly were indicating that the inclusion of nano-powder has significantly increased the lubricant's performance [14]. This is because inclusion of nano-powder increases the thermal conductivity of lubricant, which can extract heat from the cutting region at a much faster rate and thereby increasing the lubricant's cooling capabilities. Moreover, the spherical shape at nano level of the aluminium oxide powder will increase the lubricating properties of the lubricant by providing the rolling effect between the cutting tool and the workpiece.

Before conducting experiments, the formulated vegetable-based lubricant was tested for their wetting capabilities (wetting angle). The wetting angle is the ability of such liquid to cling on to a surface. Low wetting angle (<90°) indicates high wettability, meanwhile high wetting angle (>90°) indicates low wettability. Favourably, the cutting lubricant should have a low wetting angle (high wettability) to provide a better rate of heat transfer from the cutting region to the surroundings [15]. This is because at low wetting angle, a constant volume droplet will exhibit high surface coverage in comparison to high wetting angle. According to Fourier's law, high surface coverage can increase the rate of heat transfer from one position to other. Figure 2 shows the wetting angles of the formulated vegetable-based lubricant and the conventional mineral-based lubricant at constant volume. Observing those figures, the formulated vegetable-based lubricant produces much lower wetting angle in comparison to the conventional mineral-based lubricant. Therefore, the formulated vegetable-based lubricant was expected to provide better cooling than the conventional mineral-based lubricant, which is favorable in the turning process.

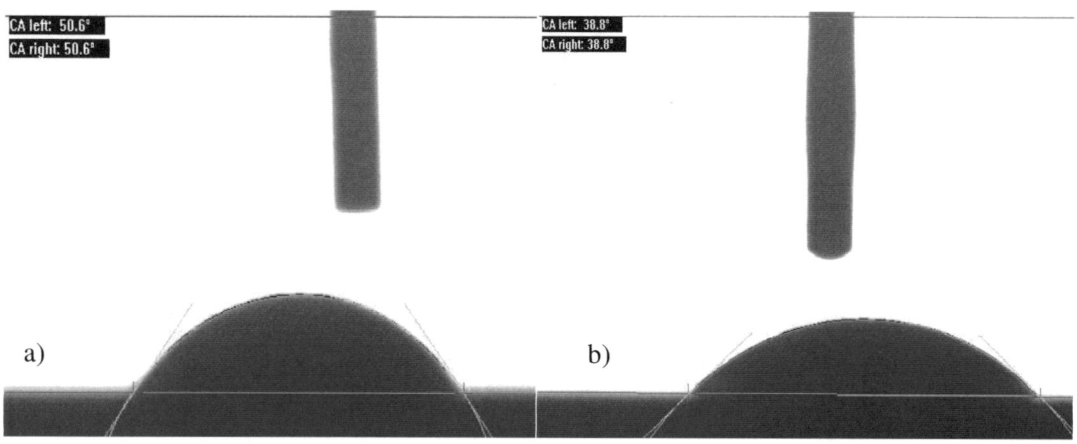

Figure 2. The wetting angle a) conventional mineral based lubricant; b) formulated vegetable-based lubricant.

2.2 Turning process

This experimental process was conducted on a lathe machine (Colchester VS Master 3250) and the AISI 431 stainless steel was used as the work material. The cutting tool used was PVD coated carbide inserts that features multi-layered coating with ultra-thin layers of TiAlN and AlCrN, type: TNMG160408-SU AC520U. The experiment was conducted with the aid of full factorial design of experiment (DOE). The said design will be able to determine the effect of each variable factor at every level of the response, as well as the effects of interactions between factors. In this work, three cutting parameters (cutting speed, depth of cut, and feed rate) and type of lubricants were considered as the variable factors. The lubricants used were the formulated vegetable-based lubricant and the mineral based lubricant. The vegetable-based lubricant was delivered onto the work material by MQL system (KENCO KNPMB 400), while the mineral based lubricant was by flooded cooling. Table 2 shows the parameters and their levels adopted in this experimentation. As this study considered four variable factors at two level experiments, the DOE created a 2^4 matrix that consists of a total of 16 experimental runs. The surface roughness was recorded and analysed for each experimental run using a portable surface roughness tester from Mitutoyo (Surftest SJ-210). Table 3 shows the experimental runs including the variable factors and the recorded surface roughness value.

Table 2. The variable factors and their levels.

No	Variable factors	Unit	Low level (-1)	High level (+1)
1.	Cutting speed, *V*	m/min	150	200
2.	Depth of cut, *d*	mm	1.0	2.0
3.	Feed rate, *f*	mm/rev	0.18	0.24
4.	Type of lubricant, *l*	-	M	V

M = Conventional mineral based lubricant (flooded cooling)
V = Formulated vegetable based lubricant (MQL)

Table 3. The experimental run and the surface roughness results.

No	Variable factors				Surface Roughness, μm
	V	*D*	*f*	*l*	
1.	-1	-1	-1	-1	2.609
2.	+1	-1	-1	+1	2.492
3.	-1	+1	-1	-1	2.451
4.	+1	+1	-1	-1	2.414
5.	-1	-1	+1	-1	4.783
6.	+1	-1	+1	-1	4.707
7.	-1	+1	+1	-1	4.569
8.	+1	+1	+1	-1	4.669
9.	-1	-1	-1	+1	2.470
10.	+1	-1	-1	+1	2.500
11.	-1	+1	-1	+1	2.137
12.	+1	+1	-1	+1	2.362
13.	-1	-1	+1	+1	4.645
14.	+1	-1	+1	+1	4.738
15.	-1	+1	+1	+1	4.372
16.	+1	+1	+1	+1	4.491

3. Experimental results

3.1 Analysis of variance (ANOVA)

The measured values of surface roughness from Table 3 were used to determine the variables that significantly influence the surface roughness. Table 4 shows the ANOVA results for surface roughness. From the ANOVA analysis, factors that significantly influence the surface roughness are the feed rate with 98.6%, followed by the depth of cut (0.7%), type of lubricant (0.3%), and cutting speed (0.04%). The results are similar to other research findings that concluded that the feed rate significantly affects the surface finish of a machined workpiece [16,17]. It appears that the type of lubricants whether flooded or MQL does not affect surface roughness.

Table 4. The ANOVA for surface roughness

Source	DOF	Adj SS	Adj MS	*F*-value	Contribution, %
Model	4	19.429	4.8574	866.76	
Linear	4	19.429	4.8574	866.76	

Cutting speed	1	0.0071	0.0071	1.27	0.036
Depth of cut	1	0.1367	0.1367	24.40	0.701
Feed rate	1	19.226	19.226	3430.6	98.63
Type of lubricant	1	0.0599	0.0599	10.69	0.310
Error	11	0.0056	0.0056		
Total	15	19.491			

3.2 Surface roughness

Figure 3 shows the effect plot of each variable on the surface roughness at every level. Based on the most influential factors, the surface roughness increases as the level of feed rate increase from 0.18 mm/rev to 0.24 mm/rev. At a higher feed rate, a larger cross-sectional area was being cut consequently increasing the friction between the cutting tool and workpiece. Therefore, with the increase of friction from increasing the feed rate, higher surface roughness will be seen on the workpiece. In terms of depth of cut, the surface roughness decreases as their level increases from 1.0 mm to 2.0 mm. This result is contradictory to the other's findings [18]. This is because at a higher depth of cut, the higher amount of material being removed from the workpiece creating more friction as that of higher feed rate. This action consequently will result in higher surface roughness value. For cutting speed, increasing their levels from 150 m/min to 200 m/min slightly increases the value of surface roughness. However, as the influence is so minimal, increasing the cutting speed or decreasing it would not make any difference.

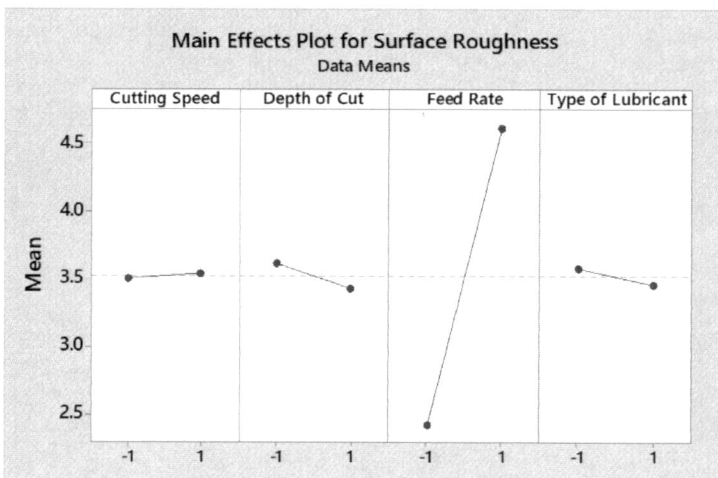

Figure 3. The main effects plot for surface roughness

Focusing on the type of lubricant, better surface finish (low surface roughness) was obtained when utilizing the vegetable-based lubricant under MQL. Even though the percentage of contribution was 0.310 %, it worth noting that the quantity of vegetable-based lubricant used during the study was far less than the mineral based lubricant because of MQL system. With the low amount of lubricant usage, the formulated vegetable-based lubricant was still able to outperform the mineral based lubricant. This result was contributed to the high wettability properties of the vegetable-based lubricant and the presence of nanopowder, which improves the rate of heat transfer from the cutting region to the surrounding due to large surface coverage and high thermal conductivity of nanoparticles [19]. Other than that, the aluminium oxide powder provides additional lubricating properties by exhibiting rolling effects between the cutting tool and machined workpiece, hence reducing the coefficient of friction [20]. Furthermore, the nanopowders usually form a thin film layer on the surface of the workpiece, which reduces the contact pressure and act as a protection layer for the workpiece as the nanopowder will receive the highest contact pressure of the cutting instead of the workpiece. Hence these actions contribute to having low surface roughness value in utilizing the formulated vegetable-based lubricant.

4. Conclusion

The study addresses the current issues and challenges in formulating the vegetable oil into a cutting lubricant for machining applications. An attempt has been made to formulate vegetable base lubricant addressing their wettability issues. A comparison has been made between the formulated vegetable-based oil and conventional mineral-based lubricant in turning process under identical cutting conditions. Based on the results, following conclusions can be drawn.

- The formulated vegetable oil produced low wetting angle (38°) compared to mineral based lubricant (50°). This will ensure better wettability in the cutting zone and improve heat transfer rate from the cutting zone.
- Feed rate is the most significant factor affecting surface roughness. Its contribution is about 98% compared to other factors. The depth of cut and cutting speed have relatively no effects on the surface finish.
- The results from the experimental study showed that the formulated vegetable-based lubricant was able to outperform the mineral based lubricant by 0.31%. Even though, the difference is small, the amount of vegetable-based lubricant (MQL) used was far less than the mineral based lubricant under flooded cooling. Therefore, the study supports that the used of vegetable base oil as cutting lubricant has comparable or even better lubricating performance than the conventional mineral-based lubricant with the least amount of usage.

Acknowledgements

This research was supported by the Ministry of Higher Education Fundamental Research Grant Scheme, Malaysia (FP066-2015A); and the Postgraduate Research Fund (PPP), University of Malaya (PG012-2016A).

References

[1] El Baradie M 1996 Cutting fluids: Part I. Characterisation *J. Mater. Process. Technol.* 56(1-4) 786-97

[2] Shashidhara Y M and Jayaram S R 2010 Vegetable oils as a potential cutting fluid-An evolution *Tribol. Int.* 43(5-6) 1073-81

[3] Dhar N R, Islam S, Kamruzzaman M and Paul S 2006 Wear behavior of uncoated carbide inserts under dry, wet and cryogenic cooling conditions in turning C-60 steel *J Braz. Soc. Mech. Sci. Eng.* 28(2) 146-52

[4] Haus F, German J and Junter G A 2001 Primary biodegradability of mineral base oils in relation to their chemical and physical characteristics *Chemosphere* 45(6-7) 983-90

[5] Ruger C W, Klinker E J and Hammond E G 2002 Abilities of some antioxidants to stabilize soybean oil in industrial use conditions *J. of the American Oil Chemists' Society* 79(7) 733-36

[6] Sharma B K, Adhvaryu A, Liu Z and Erhan S Z 2006 Chemical modification of vegetable oils for lubricant applications *J Am. Oil Chem. Soc.* 83(2) 129-36

[7] Quinchia L A, Delgado M A, Franco J M, Spikes H A and Gallegos C 2012 Low-temperature flow behaviour of vegetable oil-based lubricants *Ind. Crops Prod.* 37(1) 383-88

[8] Rudnick L R 2002 A comparison of synthetic and vegetable oil esters for use in environmentally friendly fluids *Biobased Ind. Fluids and Lubr.* 46-59

[9] Yunus R, Fakhru'l-Razi A, Ooi T L, Omar R and Idris A 2005 Synthesis of palm oil based trimethylolpropane esters with improved pour points *Ind. Eng. Chem. Res.*

[10] Lawal S A, Choudhury I A and Nukman Y 2012 Application of vegetable oil-based metalworking fluids in machining ferrous metals—A review *Int. J. Mac. Tools Manuf.* 52 1-12

[11] Amrita M, Srikant R R and Sita Rama Raju 2015 Performance Evaluation and Economic Analysis of Minimum Quantity Lubrication with Pressurized/Non-Pressurized Air and Nanofluid Mixture *Int. Schol. Sci. Res. & Innovation* 9 1012-17

[12] Gunay M and Yucel E 2013 An evaluation on machining processes for sustainable manufacturing Gazi University *J. Sci.* 26 241-52

[13] Lawal S A, Choudhury I A and Nukman Y 2014 Evaluation of vegetable and mineral oil-in-water emulsion cutting fluids in turning AISI 4340 steel with coated carbide tools *J. Cleaner Prod.* 66 610- 18

[14] Laad M and Jatti V K S 2018 Titanium oxide nanoparticles as additives in engine oil *J. King Saud Univ. – Eng. Sci.* 30(2) 116-22

[15] Nayak V U and Prabhu N K 2016 Wetting Behavior and Heat Transfer of Aqueous Graphene Nanofluids *J. Mater. Eng. Perform.* 25(4) 1474-80

[16] Xavior M A and Adithan M 2009 Determining the influence of cutting fluids on tool wear and surface roughness during turning of AISI 304 austenitic stainless steel *J. Mater. Process. Technol.* 209 900-09

[17] Yang W H and Tarng Y S 1998 Design optimization of cutting parameters for turning operations based on the Taguchi method *J. Mater. Process. Technol.* 84 122-29

[18] Raza S W, Pervaiz S and Deiab I 2014 Tool wear patterns when turning of titanium alloy using sustainable lubrication strategies *Int. J. Precis. Eng. Manuf.* 9 1979-85

[19] Krishna P V, Srikant R R and Nageswara R D 2010 Experimental investigation on the performance of nanoboric acid suspensions in SAE-40 and coconut oil during turning of AISI 1040 steel *Int. J. Mach. Tools Manuf.* 50(10) 911-16

[20] Noordin M Y, Venkatesh V C, Chan C L and Abdullah A 2001 Effect of nanoparticle size on sessile droplet contact angle *J. Mater. Process. Technol.* 116 16-21

Energy Efficiency Assessment in Production Line: An Approach towards Sustainable Manufacturing

C Saleh[1*], R Prasetiyo[1], A P Hendradewa[1] and A B Hassan[2]

[1]Department of Industrial Engineering, Faculty of Industrial Technology, Islamic University of Indonesia, Yogyakarta, Indonesia

[2]Electrical, Electronic and Automation Section, Universiti Kuala Lumpur Malaysian Spanish Institute (UniKL – MSI), Kulim High Technology Park, 09000, Kulim, Kedah

Corresponding author *: chairul@uii.ac.id

Abstract: Global demands of energy which causes a quick depletion of the natural energy reserves become a critical concern. Industrial sector is concerned to be the largest energy consumer compared to other sectors. Thus, an energy efficiency improvement tends to be a proper solution in addressing this problem. This work aims to analyse a prospect energy efficiency improvement within a production line. A discrete-event simulation (DES) was used to investigate the energy saving potential and a case study was conducted in a crumb rubber company. Three parameters were taken into consideration to calculate the energy consumption, i.e. processing time, working shift schedule, and machine states. An initial model which represents an actual condition of the manufacturing production line was developed. From the analysis of the initial model result, a recommendation scenario was built. It was shown that a reduction of 8.22% in the energy consumption is achievable by implementing the suggested scenario.

Keywords: *energy efficiency; discrete-event simulation (DES)*

1. Introduction

The combustion of fossil fuels generates and releases greenhouse gas (GHG) emissions which predominantly composed by CO_2 to the atmosphere as the by-product of the energy use. Industrial sector plays a vital role of the occurred global environmental changes, since it contributes to around a half of the total share of the global energy consumption [1]. Moreover, it is projected that the trends of the energy consumption in this particular sector will exhibit a significant growth to up to 38% between 2017 – 2050 [2].

Responding to such associate environmental drawback which doubtlessly may harm the ecosystem and the life in the future, a conversion from the prevailing production system into more sustainable system is becoming an imperative resolution [3]. To date, discussions associated with the energy measurement, monitoring, modelling, and optimization in machining facilities are significantly high [4]. Cai et al. [5] suggested that in order to boost the energy efficiency in the manufacturing system, a possible solution is by shifting the prevailing energy intensive production facilities into highly efficient machining tools. Jirasuwankul [6] conducted a research on improving energy efficiency use in induction motor drive system using fuzzy logic. Liu et al. [7] proposed an approach for acquiring the real-time energy efficiency (REE) of machining tools. De Carvalho & De Oliveira Gomes [8] investigated the reduction of electrical energy consumption during standby mode. An exploration of possible improvement of the energy consumption through visualization of manufactured parts in the machining systems is provided by [9]. These research are mainly conducted at product and machining level. However, analysis specifically on the products or machining tools do not adequately represent the manufacturing system as an entire. Thus, a more holistic view would be required.

Production line is considered to be the foremost intensive energy consumer in a manufacturing system. Each machine is operated differently and consume energy unequally subject to different products being manufactured, processing time, machine states, etc. Energy consumed in production processes can be divided into two categories: value added energy (VAE) and non-value added energy (NAE) [10]. Non-value added energy can be considered as waste and need to be reduced. Thus, this paper aims to investigate the attainable energy efficiency improvement in the production line by considering the value-added and non-value added energy consumption. The analysis was carried out by employing a discrete-event simulation (DES) approach to simulate the material and energy flows of production line. Three parameters were taken into consideration to calculate the energy consumption, i.e. processing time, working shift schedule, and machine states.

2. Energy Consumption in the Production Line

2.1 *Research Framework*

In order to achieve the goal on measuring the energy consumption and determine the energy efficiency improvement in a production line, a comprehensive research framework is necessary to introduce. The research framework presented here is an adaptation of a proposed steps in simulation study [11].

The initial step of the simulation study is by defining the problem formulation and setting of objectives. Afterwards, the next step is continued to establish a conceptualized simulation model, and the collection of the supported data. The simulation model conceptualization has a strong interplay with the data collection as the complexity of the model changes, it directly changes and affects the required data elements. The fifth step is the translation of the real-world system into a computer-recognizable format or the artificial environment.

This model translation would be executed in a simulation software, namely Tecnomatix Plant Simulation®. The next step is the verification to ensure the simulation model accurately represent the actual system condition. If the input parameters and logical structure of the simulation model is judged acceptable, thus the verification and validation have been completed. In order to investigate the determined goal which in this case is the energy efficiency improvement, the experimental design of alternatives condition should also be developed. The evaluation is used to estimate the improvement of the energy efficiency potential. Finally, the models and the associated results should be documented and reported.

2.2 *Conceptual Simulation Model*

A conceptual simulation model is designed to show the essential features of the problem. Certain bound parameters are considered in order to analyse the energy efficiency throughout the production process. The model takes into consideration three parameters which represent the dynamic characteristic of the production system in terms of energy consumption: processing time, material flow, and machine states (e.g. idle, setting up, stand-by, operational, production, and breakdown).

According to [12] three alternatives are often considered to set up the duration and power for each operational or machine state: a constant value, a predetermined probability distribution function, and a mathematical equation of the existing empirical models of the unit process. A few operational states possess relatively constant trends such as standby, idle, rump up, etc. Hence, the first two options are sufficient to configure the aforementioned conditions in the simulation. The last option is regarded to be the foremost adequate to characterize the energy consumption and process parameters which featuring dynamics behaviour. In order to realize the target and to facilitate a much better understanding upon the proposed methodology, this conceptual model is divided into sections to introduce the interlinked relationships of the mentioned parameters with the quantity of energy consumption.

2.3 *Production energy consumption*

Numerous parameters associated with the machine state would directly specify the performance of machine tools in terms of energy consumption. The production line produces energy profiles dynamically over time that consists of many intense energy consuming components depending on the process parameters and the actual states of the machine [13]. According to [14], the energy consumption at a facility can be formulated with the subsequent equation:

$$E^k = e_r^k T_r^k + e_s^k T_s^k + e_b^k T_b^k + e_i^k T_i^k \tag{1}$$

where:

E^k = Total energy consumption in facility *k*,
e_r^k = Energy consumption per unit time during run state in facility *k*,
e_s^k = Energy consumption per unit time during setup state in facility *k*,
e_b^k = Energy consumption per unit time during breakdown state in facility *k*,
e_i^k = Energy consumption per unit time during idle state in facility *k*,
T_r^k = Run state time in facility *k*,
T_s^k = Setup state time in facility *k*,
T_b^k = Breakdown state time in facility *k*,
T_i^k = Idle state time in facility *k*

2.4 *Machine States*

Every production activity that employs machinery habitually experiences several conditions. Each condition consumes a totally different quantity of energy due to undertaken at different processing times. There are at least five operational states exist during the production period and are defined as follows:

- Setting up: refers to the period of time which correlates with the adjustment for production product changing.
- Production: refers to the state or condition when the machine is performing work to produce desirable products, it is typically associated with the value-adding condition.
- Stand-by: refers to the state or condition when the machine remains activated at the operational readiness for performing tasks.
- Idle: refers to the condition or the state when the machine is not doing any useful work or no material order to be processed, however certain necessary components remain activated (e.g. control panel).
- Breakdown: refers to the condition or the state when the machine fails to produce products and should be deactivated for maintenance due to experiencing certain troubles.

3. Research Methodology

The research framework presented here is an adaptation of a proposed steps in a simulation study by [6]. The initial step of the simulation study is by defining the problem formulation and setting of objectives. The next step is continued to establish a conceptualized simulation model, and the collection of the supported data. The simulation model conceptualization has a strong interplay with the data collection as the complexity of the model changes, it directly changes and affects the required data elements. The fifth step is the translation of the real-world system into a computer-recognizable format or the artificial environment.

In order to investigate the determined goal which in this case is the energy efficiency improvement, the experimental design of alternatives condition should also be developed. The alternatives experimental design is required for the decision making support which the result is compared with the initial system condition. Subsequently, the alternative model is executed. The result of the initial and the recommendation models should be evaluation. The evaluation is used to estimate the improvement of the energy efficiency potential. Finally, the models and the associated results were documented and reported.

3.1 *Initial Model Development*

A discrete-event model simulation is employed to experiment and analyse the energy consumption as well as the energy saving potential of the production line in manufacturing system. The simulation was developed by using Tecnomatix® simulation software. To represent the actual condition of the company's production line, an initial simulation model is developed. Several variables and parameters, such as processing time, working shift schedule, and energy input if machine states have been defined as the input data for the simulation. Figure 1 and figure 2 below depict the layout design of the facilities and the hotspot plotting of the simulation running result indicating which facilities possess larger energy consumption. Working shifts of the wet and dry processes were set up to control the working period of the production line.

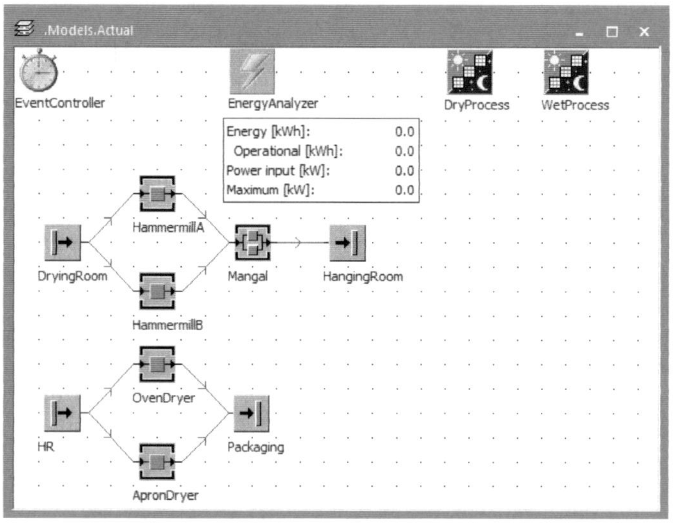

Figure 1. Initial production line simulation model

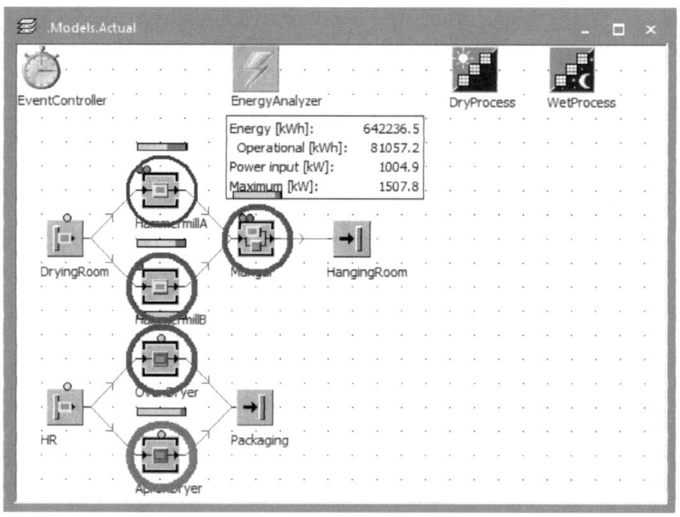

Figure 2. Hotspot plotting

Besides hotspot plotting, the simulation running of the initial model also showed a distribution of the energy consumption in each facility according to the machine states as depicted in figure 3 below.

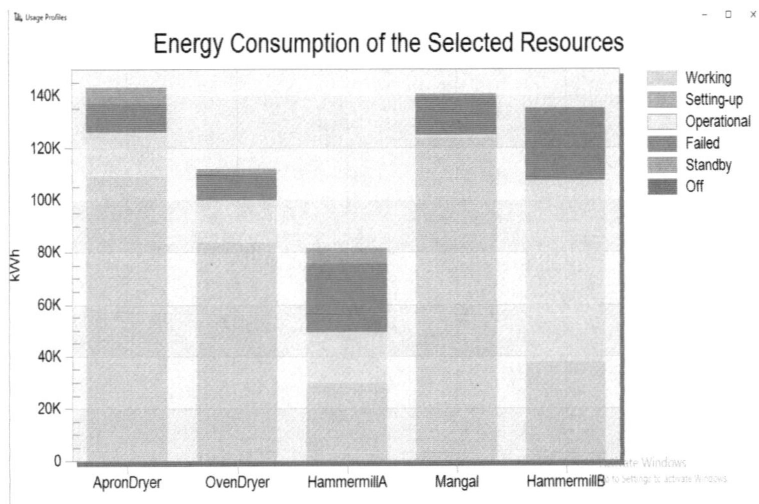

Figure 3. Energy consumption at facilities based on machine states

3.2 *Alternative Scenario Model Development*

The energy efficiency can be attained by reducing the non-added energy expenditures. In the alternative scenario a modification of the parameter setting was done for the hammermill B machine due to it consumed larger energy for non-added activities as indicated in the initial simulation result. The hammermill B consumed the largest portion of the non-added energy value, in the form of operational or idle time. In the actual condition, the hammermill station which consists of two machines, namely hammermill A and hammermill B are utilized interchangeable in random scheduling during the production process. However, both machines are still kept active, even though only one machine is undertaking the production activity. This operational configuration is applied to minimize the pre-production time during a sudden breakdown which occurs quite often of either machine being operated. Besides that, several peripheral elements such as a water pump, which integrated with both machine is activated while one of the machines is turned on. This led to high energy consumption of unfortunately associated with the non-added value . A scenario was suggested to reduce the non-added proposed to improve the energy efficiency in the simulation is by adjusting the associated operational state condition. It is assumed if both machines have separate water pump installation in each machine, so that when one machine is turned off, the operational energy can be lowered.

4. Result and Discussion

This part will discuss the result of the simulation into details. The result consists of the energy consumption estimation and its distribution according to the machine state, which is a parameter that influences the fluctuation of the energy consumption of a facility. The energy efficiency saving potential from the comparison of the initial and the alternative model is also explained.

4.1 *Initial Simulation Result*

A simulation approach is conducted to estimate the energy consumption of the production line in the manufacturing company. Several variables and parameters, such as processing time, working shift schedule, and energy input of machine states have been defined as the input data for the simulation. Figure 4 (a) and (b) below depict the fluctuative power consumption and the hotspot plotting of the simulation running result indicating which facilities possess larger energy consumption. Working shifts of the wet and dry processes were set up to control the working period of the production line. From the simulation result of the case study, it shows that the energy expended in the production line behaves dynamically depending on certain factors as illustrated in figure 4 (a). Energy input, processing times, working schedule, and machine state are among

IOP Conf. Series: Materials Science and Engineering **530** (2019) 012004 doi:10.1088/1757-899X/530/1/012004

the driving factors of the fluctuation of the energy consumption. Figure 4 (b) shows the allocation of the energy consumption distribution based on the machine states of the production facilities.

(a)

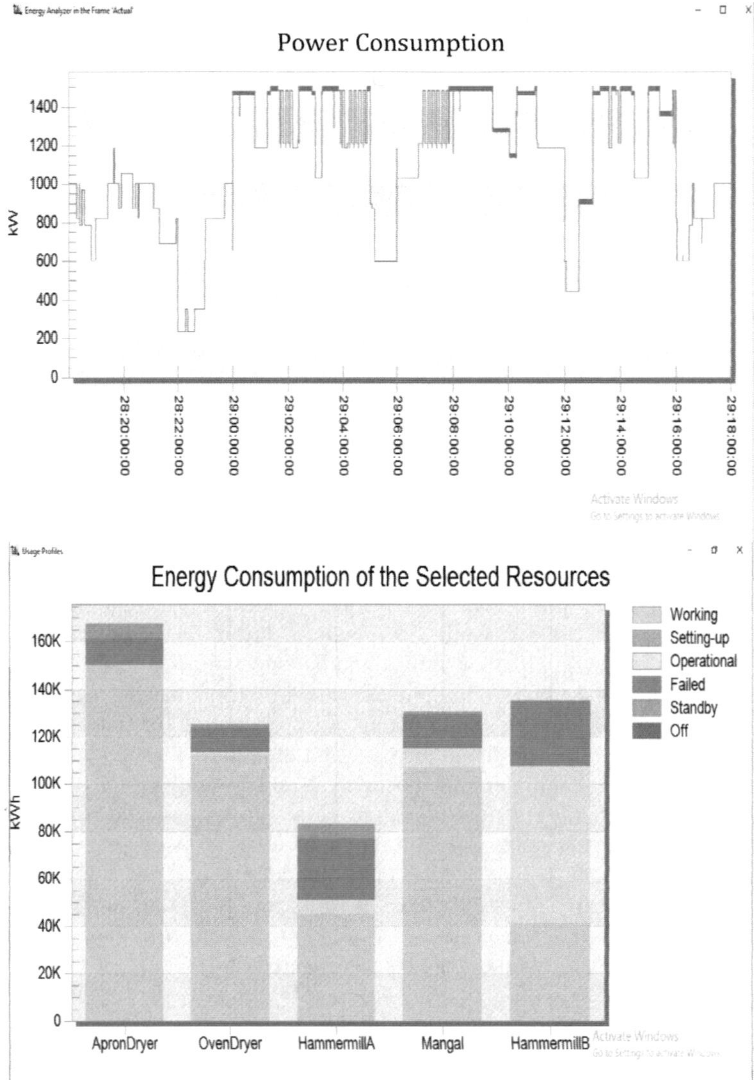

Figure 4. (a) Energy plotter; (b) Energy distribution chart of initial model

From the figures above, it can be seen that the energy was mainly expanded during the working, operational, and failed states. An insignificant amount of energy was expended on the stand-by state. Energy expenditure can be classified into two main categories, namely added-value energy and non-added value energy. The added-value energy is defined as the energy consumed when a machine is running due to the existence of a material being processed. Energy consumed during the working state is considered to be an added-value energy, while on the other hand, the other states are considered to be the non-added value energy since they do not directly correspond to the material processing activities.

Apron dryer, mangal machine, and hammermill B machine were the major consumers of the energy in the production line. However, the mangal machine and the apron dryer consumed energy majorly for the value-added purposes, whilst the energy in the hammermill B machine was much more expended for the non-added value one in terms of operational or idle time.

4.2 *Model Validation*

The model validation was used input-output validation technique. The sample data were collected from the company's historical documentations and the available data from the company were hourly power documentations in February 2018. The The total actual energy consumed is obtained from company's energy bill in February 2018. To validate the model, the total actual energy consumed within the production line is compared with the energy consumption resulted from simulation. The confidence interval is set to 95% for the simulation.

The simulation model has been set to run for a month. The total monthly energy consumed from the simulation result is 642235.50 kWh. This value is generated by summing the total energy consumed during each machine states of each facility, processing time, and working shift schedule. Figure 2 depicts the total simulated energy consumption of the initial model in the production line. In addition, the actual energy consumption as shown on energy bill in February 2018 is 633984.20 kWh. Table 1 shows the comparison of the simulation result and the actual data.

Table 1. Comparison of the simulated result and the actual data

	Actual	Simulation	Difference (Simulation – Actual)	Percentage Error
Energy Consumption (kWh)	633984.20	642235.50	8251.30	1.30%

From the table above it can be seen that the percentage error of the simulation result and the actual data is 1.30%. Thus, the simulation result is considered valid since it is within the targeted percentage error.

4.3 *Alternative Scenario Simulation Result*

The energy efficiency can be attained by reducing the non-added energy expenditures. In the alternative scenario a modification of the parameter setting was done for the hammermill B machine due to it consumed larger energy for non-added activities as indicated in the initial simulation result. The hammermill B consumed the largest portion of the non-added energy value, in the form of operational or idle time. In the actual condition, the hammermill station which consists of two machines, namely hammermill A and hammermill B are utilized interchangeable in random scheduling during the production process. However, both machines are still kept active, even though only one machine is undertaking the production activity. This operational configuration is applied to minimize the pre-production time during a sudden breakdown which occurs quite often of either machine being operated. Besides that, several peripheral elements such as a water pump, which integrated with both machine is activated while one of the machines is turned on. This led to high energy consumption of unfortunately associated with the non-value added energy. A scenario was suggested to reduce the non-value added energy by assuming both hammermill machines have separate water pump installation, so that when one machine is turned off, the operational energy in both machine can be lowered. From the figure 5, it can be seen that by applying the proposed scenario a reduction on the non-added value energy in the hammermill B machine is achievable.

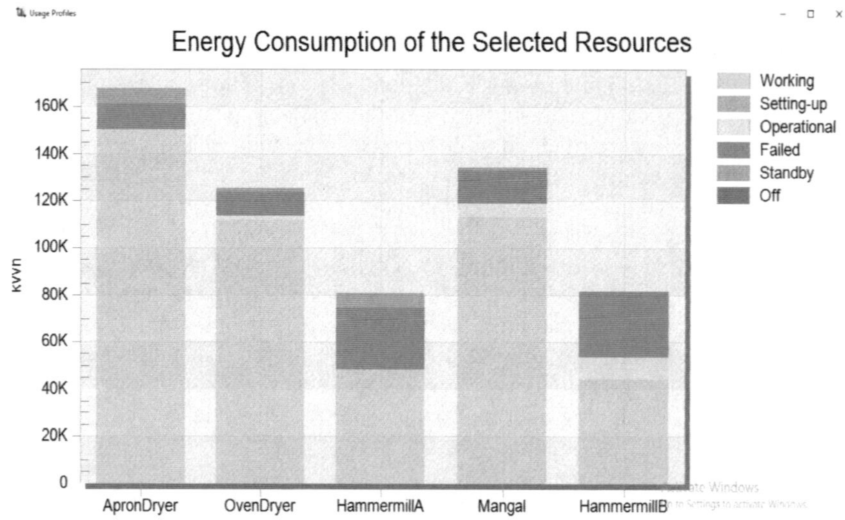

Figure 5. Energy consumption from the alternative scenario model

4.4 *Energy Efficiency Evaluation*

The simulation of both conditions resulted a significance difference. A proposed improvement scenario which can be done to minimize the energy expenditure is by implementing a reduction of non-added value energy. The implementation has been made within the hammermill B machine. After the suggested scenario executed by using the simulation, a promising energy efficiency improvement has been achieved. The initial simulation model generates energy of around 642,236.5 kWh a month, whilst after the alternative scenario has been implemented, the system only consumes 589,418.3 kWh a month. Thus, the energy efficiency can be improved for around 8.22%.

5. Conclusions

An investigation conducted in the system could give a chance to achieve a potential reduction of the consumed energy. In this paper an energy assessment by employing a discrete-event simulation (DES) approach is presented. Initially, a model which represents the actual condition was established, afterwards the simulation was run and analysed. From the analysis of the initial simulation result, an improvement scenario can be developed. Certain parameters, such as processing time, working shift schedule, and machine states was taken into account to see the significant impact of the expended energy quantity. A measure that can be done to minimize the energy expenditure is by reducing the non-added value energy within the production system. In the case study, the hammermill B machine was the major consumer of the non-added value energy, in the form of idle time. Thus, a suggestion proposed to improve the energy efficiency is by separating water pump installation in each machine, so that when one machine is turned off, the operational energy can be lowered. This modification is directly related to an adjustment of the machine states, which is one of the parameters included in the study. By implementing the suggested scenario, 8.22% of the energy consumption reduction can be achieved.

This study was solely conducted to investigate the electrical energy efficiency within a production system, without considering other economic or environmental impacts evaluation. Furthermore, beside energy intensive, crumb rubber manufacturing is also water intensive. An effort to manage the water consumption would be beneficial to pursue the sustainability manufacturing.

References

[1] B P 2018 BP Energy Outlook 2018 Editions *London, United Kingdom*
[2] E I A 2018 Annual Energy Outlook 2018 *Washington, DC, United States*

[3] I Meo, A Papetti, F Gregori and M Germani 2017 Optimization of Energy Efficiency of a Production Site: A Method to Support Data Acquisition for Effective Action Plans *Procedia Manuf.* Vol 11 pp 760–767

[4] W Cai, F Liu, J Xie, P Liu and J Tuo Nov 2017 A tool for assessing the energy demand and efficiency of machining systems: Energy benchmarking *Energy* Vol 138 pp 332–347

[5] P Liu, J Tuo, F Liu, C Li and X Zhang Aug 2018 A novel method for energy efficiency evaluation to support efficient machine tool selection *J. Clean. Prod.* Vol 191 pp 57–66

[6] N Jirasuwankul Feb 2017 Simulation of Energy Efficiency Improvement in Induction Motor Drive by Fuzzy Logic Based Temperature Compensation *Energy Procedia* Vol 107 pp 291–296

[7] P Liu, F Liu and H Qiu Feb 2017 A novel approach for acquiring the real-time energy efficiency of machine tools *Energy* Vol 121 pp 524–532

[8] H M B De Carvalho and J De Oliveira Gomes 2015 Method for increasing energy efficiency in flexible manufacturing systems: A case study *Procedia CIRP* Vol 29 pp 40–44

[9] J Tuo, F Liu, P Liu, H Zhang and W Cai Sept 2018 Energy efficiency evaluation for machining systems through virtual part *Energy* Vol 159 pp 172–183

[10] M Bornschlegl, M Bregulla and J Franke Nov 2016 Methods-Energy Measurement – An approach for sustainable energy planning of manufacturing technologies *J. Clean. Prod.* Vol 135 pp 644–656

[11] J Banks, J Carson II, B Nelson and D Nicol 2014 *Discrete-Event Simulation* 5th ed. Essex: Pearson

[12] K Katchasuwanmanee, R Bateman and K Cheng Jan 2017 An Integrated approach to energy efficiency in automotive manufacturing systems: quantitative analysis and optimisation *Prod. & Manuf. Res.* Vol 5 No 1 pp 90–98

[13] S Thiede, G Posselt and C Herrmann Jan 2013 SME appropriate concept for continuously improving the energy and resource efficiency in manufacturing companies *CIRP J. Manuf. Sci. Technol.* Vol 6 No 3 pp 204–211

[14] T Kobayashi, M Yamaguchi and H Hibino Jul 2017 Lot Size Dependence for Energy Consumption per Unit of Production Throughput Considering Rush Orders in Manufacturing Lines *IFAC-PapersOnLine* Vol 50 pp 5000–5005

IOP Conf. Series: Materials Science and Engineering **530** (2019) 012005 doi:10.1088/1757-899X/530/1/012005

An Application of Cutting-Stock Problem in Green Manufacturing: A Case Study of Wooden Pallet Industry

P Wattanasiriseth* and A Krairit

Engineering and Technology Management Program, Department of Industrial Engineering, Faculty of Engineering at Sriracha, Kasetsart University, Chonburi, Thailand

Corresponding author *: paveepat@hotmail.com

Abstract. The cutting-stock problem is used to minimize the impact of wasting material during the stock cutting process. The linear programming model is developed and implemented on Microsoft Excel to solve for the optimal cutting plan with the criteria of minimizing material waste. The case study of pine wood stock cutting in wooden pallet industry demonstrates the benefit of using the approach in terms of reducing production cost and carbon emission. The results show that waste can be reduced up to 70.86% resulting in a cost reduction of 977,352 Baht/year. The amount of carbon dioxide equivalent emitted from the life cycle of pine wood is reduced by 22,713 kg CO_2-e/year or approximately 5.32% reduction.

1. Introduction

Pine wood is one of the major raw materials in wooden pallet industry of Thailand regardless of its low supply in Thailand. Approximately 60% of wooden pallets produced in Thailand are made of imported pine wood from various sources around the world, mostly America, Finland, Sweden, Brazil, Russia and New Zealand. A cheaper local wood supply such as parawood is the main alternative, however, pine wood pallet is still preferred by customers around the globe forcing Thai wooden pallet industry to manufacture the wooden pallet from the more expensive source of raw material such pine wood.

Beside the incurred cost from importing of pine wood and the risk of currency exchange fluctuation. The greenhouse gas emission related to harvesting, sawmilling, shipping, and manufacturing of pine wood product is another major concern. With the growth of Thailand manufacturing industry, more wooden pallets will be required for carrying products and shipping them to the whole world. One solution to the problem of increasing demand for imported pine wood for wooden pallets industry is to reduce the amount of pine wood wasted during the production process due to poor production management and inadequate manufacturing technology. Minimizing the waste pine wood could result in a saving of production cost and greenhouse gas emissions which is an approach toward achieving green manufacturing.

Based on the observation in Thailand wooden pallet industry, most of pine wood waste are generated during the stock cutting process. The imported pine wood with standard sizes is cut down to the required sizes leaving lots of unusable remaining waste (off-cuts). The pine wood waste is accounted for the cost of pallet and the amount of greenhouse gas emitted to the world. This problem can be easily minimized through improving production management such as optimizing the cutting plan of pine wood based on the production demand.

IOP Conf. Series: Materials Science and Engineering **530** (2019) 012005 doi:10.1088/1757-899X/530/1/012005

Cutting-Stock Problem (CSP) is wildly used to solve this type of problem and it is generally used in lumber and steel industries [1]. This optimization technique can select the optimal cutting pattern to minimize the material waste while meeting the production demand. However, most wooden pallet manufacturers are small and medium enterprises (SMEs) who is lacking of know-how and capital to invest in such optimization system. Thus, they are likely to rely on a simple but inefficient way of stock-cutting resulting in high amount of waste, high production cost and high greenhouse gas emissions. Therefore, in this research, the aim is to develop the Microsoft Excel-based optimization system for solving one dimensional cutting-stock problem with the objective of minimizing material waste. The case study of cutting pine wood in wooden pallet manufacturing is used for an implementation and analysis of both cost and carbon emission.

2. Literature Review

2.1. Green Manufacturing

There has been a focus on the impact of organizations and manufacturing on the overall environment. Stakeholders including customers, shareholders, board members, and employees are asking or requiring organizations to be more environmentally responsible with respect to their products and processes. The reasons for these demands include regulatory requirements, product stewardship, enhanced public image, potential to expand customer base, and potential competitive advantages. [2] Therefore, the term "green manufacturing" was invented to emphasize the focus of manufacturers toward this environmental aspect.

The term "green" manufacturing can be looked at in two ways: the manufacturing of "green" products, particularly those used in renewable energy systems and clean technology equipment of all kinds, and the "greening" of manufacturing including reducing pollution and waste by minimizing natural resource use, recycling and reusing what was considered waste, and reducing greenhouse gas emissions [3]. Growing numbers of manufacturers are finding that reducing resource use, waste, and pollution, along with recycling and reusing what was formerly looked at as waste, yields benefits not only in terms of an improved bottom line, but in terms of employee motivation, morale, and public relations [4].

The concept of cradle-to-cradle product design and production call for products to be designed and produced with an eye towards minimizing, or even eliminating, resource use, waste, and pollution [5, 6]. The drive to reduce fossil fuel use, carbon dioxide emissions, and other greenhouse gas emissions has been at the forefront of this movement. Moreover, people are encouraged to turn to consume products and services that have been produced from environmentally friendly manufacturing processes.

Thailand is also moving toward becoming a green manufacturing base in this region. It sees the need for going green in order for the people to live in a caring and sharing society, in a safe and sound environment. Also, in order to sustainably develop a competitive manufacturing industry, the focus on green manufacturing is now unavoidable.

2.2. Cutting-stock problem

The cutting-stock problem is the problem of filling an order at minimum cost or waste for specified numbers of lengths of material to be cut from given stock lengths of given cost [7]. It is one of the cutting and packing problems which are easy to state but difficult to solve (NP hard). The standard problem is defined as follows: given a set of small and large objects, how the small objects should be obtained from a large one in order to optimized a given criterion.

There are two main approaches to solve this problem: exact algorithm and heuristic methods. Exact algorithm is mainly based on linear programming, dynamic programming and branch-and-bound techniques [8]. Heuristic methods have greater flexibility in taking into account problem specific constraints and offer a trade-off between quality of a solution and its computational effort. Some of the heuristic methods used in solving cutting-stock problem are genetic algorithm [8-9], particle swarm optimization [10-11], simulated annealing [11-12], ant colony [13], etc.

IOP Conf. Series: Materials Science and Engineering **530** (2019) 012005 doi:10.1088/1757-899X/530/1/012005

For one dimensional cutting-stock problem, most of the approaches are based on linear programming models that can be divided into four categories 1) the assignment formulations 2) the pattern-oriented formulation 3) the one-cut formulation and 4) the flow models [8]. The implantations have proven that it can increase yield, decrease production cost, minimize usable waste and reduce carbon emission [1, 8, 13-14].

Various commercially available software was invented for specific applications such as sawmilling, metal and sheetmetal working, plywood cutting, etc. However, most SMEs cannot seek benefits from these available tools due to affordability and complexity of learning and integrating to the existing production planning system.

3. Methodology

In order to generate cutting plan for pine wood stock cutting based on production demand, two steps are required. First, all feasible one dimensional cutting pattern must be generated for all different length of raw pine wood. Second, solving the linear programming model to fine the optimal cutting plan based on the given demands. To facilitate these steps, Microsoft Excel spreadsheet is developed. The add-in OpenSolver, the open source linear, integer and non-linear optimizer for Microsoft Excel, is utilized to solve the linear programming model. The implementation of the optimized cutting plan will lead to reduction of production cost and carbon footprint which then be calculated and compared to the condition before implementation.

3.1. Pattern Generation Algorithm

A pattern is a plan of how to cut a given set of item from a stock material. In order to generate feasible cutting patterns for any length of raw pine wood, the algorithm was developed using logical operation sequence which satisfies the condition stated below:

$$\sum_{i=1}^{n} a_{ijk} l_i \leq L_k \tag{1}$$

where L_k is the length of raw pine wood; $k = 1, 2, \ldots, p$
l_i is the length of item i in descending order; $i = 1, 2, \ldots, n$
a_{ijk} is the number of item i in the cutting pattern j cut from raw pine wood length k
P_{jk} is a cutting pattern j of the raw pine wood length k; $j = 1, 2, \ldots, m$

For each L_k for all $k = 1, 2, \ldots, p$

Step 1: Set $i = 1, j = 1$

Step 2: Cut the maximum number of l_i to generate the first pattern using;
$a_{ijk} = \lfloor L_k / l_i \rfloor$
Calculate the waste material (off-cuts) of pattern j using raw pine wood length k;
$W_{jk} = L_k - a_{ijk} l_i$

Step 3: Compare W_{jk} with other l_i
If $W_{jk} < l_i$
Then $a_{ijk} = 0$,
Else
$a_{ijk} = \lfloor W_{jk} / l_i \rfloor$
$W_{jk} = L_k - \sum_{i=1}^{n} a_{ijk} l_i$
Repeat Step 3 for all $i = i+1, \ldots, n$

Step 4: Generate next pattern $j = j+1$
i_{last} = the last i where $a_{i,j-1,k} \geq 1$ and $i \neq n$
for all $i = 1, 2, \ldots, i_{last}-1$; $a_{ijk} = a_{i,j-1,k}$
for $i = i_{last}$; $a_{ijk} = a_{i,j-1,k} - 1$
for all $i = i_{last}+1, \ldots, n$; $W_{jk} = L_k - \sum_{i=1}^{n} a_{ijk} l_i$ then repeat Step 3
Repeat Step 4 until $a_{ijk} = 0$ for all $i = 1, 2, \ldots, n-1$

The patterns generated are arrange such that;

$$\begin{bmatrix} & l_1 & \cdots & l_n & \\ P_{11} & a_{111} & \cdots & a_{n11} & W_{11} \\ \vdots & \vdots & \ddots & \vdots & \vdots \\ P_{m1} & a_{1mp} & \cdots & a_{nm1} & W_{m1} \end{bmatrix} \cdots \begin{bmatrix} & l_1 & \cdots & l_n & \\ P_{1p} & a_{11p} & \cdots & a_{n1p} & W_{1p} \\ \vdots & \vdots & \ddots & \vdots & \vdots \\ P_{mp} & a_{1mp} & \cdots & a_{nmp} & W_{mp} \end{bmatrix}$$

3.2. Linear Programming Model

The stock cutting process are subjected to two conditions; 1) the available stocks (s_k) have different length and are limited in number 2) the number of item cut from available stocks must meet the production demand (d_i). Since, the objective of the stock cutting is to minimize material waste (off-cuts), the cutting pattern that yield minimum material waste must be selected. Therefore, utilizing mixed-integer linear programing for minimizing waste (off-cuts) in the stock cutting process, the mathematical model can be stated as follow:

$$\text{Minimize} \sum_{k=1}^{p} \sum_{j=1}^{m} W_{jk} X_{jk} \tag{2}$$

Subject to:

$$\sum_{j=1}^{m} \sum_{k=1}^{p} a_{ijk} X_{jk} = d_i \qquad \forall\, i = 1, \ldots.n \tag{3}$$

$$\sum_{j=1}^{m} X_{jk} \leq s_k \qquad \forall\, k = 1, \ldots.p \tag{4}$$

$$a_{ijk} \geq 0 \text{ and integer} \tag{5}$$

$$X_{jk} \geq 0 \text{ and integer} \tag{6}$$

where X_{jk} is a number of times that pattern j is cut in the stock length k

The total waste generated can be defined as:

$$\text{Total waste} = \sum_{k=1}^{p} \sum_{j=1}^{m} W_{jk} X_{jk} \tag{7}$$

The total number of raw pine wood length k used as stocks can be defined as:

$$\text{Total number of raw pine wood length } k = \sum_{j=1}^{m} X_{jk} \tag{8}$$

The total number of item *i* after complete the cutting can be defined as;

$$\text{Total number of item } i = \sum_{j=1}^{m}\sum_{k=1}^{p} a_{ijk}X_{jk} \tag{9}$$

3.3. Spreadsheet Development

The aim of this development is to be used in the small and medium sizes enterprises (SMEs) and also to be easily integrated into the existing production planning and control sequence. Thus, the most commonly used software in production planning of SMEs, Microsoft Excel is selected as a platform for implementation. Due to the limitation of Solver add-in in Microsoft Excel in terms of number of variables, the open-source add-in, OpenSolver is utilized to solve the developed mixed-integer linear programing problem. The following steps are used in order to develop the spreadsheet and solve for the optimal cutting plan;

1) Define the multiple lengths of raw pine wood and item to be cut.
2) Generate all the cutting pattern as a database.
3) Implement the developed linear programming model in one sheet. Calculate the value of objective function and all constraint values.
4) Install the add-in, OpenSolver, into Microsoft Excel and define the model (objective cell, variable cells and constraints).
5) Input the production demands of each item and the number of available stocks, then allow the OpenSolver to solve for the optimal solution.
6) The cutting plan is automatically generated based on the calculated solution from previous step using Microsoft Excel function.

4. Results and Discussions

To demonstrate the applicability of the developed model, the case study related to the cutting problem encountered at an SME wooden pallet production firm in Thailand is presented and discussed.

4.1. The Case Study

Pine wood are imported from New Zealand, shipping as a container. The imported pine wood is cut to the standard rectangular profile of 150 × 100 mm with the length (L_k) of 4.2, 4.8, 5.4 and 6.0 m randomly. The average cost of imported pine wood (include shipping cost) is 265 Baht/foot^3. The pine wood has to be cut down to the item of required sizes depending on the model of pallets to be made. The length of each item (l_i) are shown in Table 1. The numbers of pine wood to be cut to each length are calculated based on the production schedule. The current cutting plan utilizes the simplicity of cutting process by having only single length of cut for each pre-defined stock. This method is good in terms of production rate but generate lots of waste as shown in Figure 1. The analysis of cost and waste generated using current cutting plan for one month is shown in Table 1.

IOP Conf. Series: Materials Science and Engineering **530** (2019) 012005 doi:10.1088/1757-899X/530/1/012005

Figure 1. The raw pine wood (stock) and its waste (off-cuts)

Table 1. The analysis of current cutting plan

No.	Length (mm)	Demand (piece)	Stock length (mm)	Required stock (piece)	Stock volume (foot^3)	Waste volume (foot^3)
1	2316	789	4800	395	1004.36	35.15
2	2200	254	4800	127	322.92	26.91
3	1650	468	5400	156	446.24	37.19
4	1460	554	6000	139	441.79	11.78
5	1140	4686	6000	938	2981.29	149.06
6	1120	1179	4800	295	750.09	50.01
7	980	969	4800	243	617.87	113.28
				Total volume (foot^3)	**6564.56**	**423.38**
				Cost/month (Baht)	**1,739,609**	**112,195**

4.2. Implementing Optimal Cutting Plan

After develop the spreadsheet, using the same production demand and solve for the optimal cutting plan (assuming unlimited number of stocks). The result of the developed model suggests the optimal cutting plan as shown in Table 2.

Table 2. Optimal cutting plan generated by the developed model

Stock length (mm)	Required stock (piece)	No. of item to be cut with the length of (mm)						
		2316	**2200**	**1650**	**1460**	**1140**	**1120**	**980**
4800	394	2						
	1	1				2		
	1					1	3	
5400	107			2			1	1
	304					3		2
6000	254		1	1		1		1
	554				1	3	1	
	371					5		
	103						5	
Number of each item		789	254	468	554	4686	1179	969

IOP Conf. Series: Materials Science and Engineering **530** (2019) 012005 doi:10.1088/1757-899X/530/1/012005

Based on the optimal cutting plan, the required stock volume, cost and waste are calculated and compared as shown in Table 3. It can be seen that, using the new approach can reduce waste up to 70.86% and reduce production cost up to 4.68% or 977,352 Baht/year

Table 3. The comparison of current cutting plan and optimal cutting plan.

List	Current cutting plan	Optimal cutting plan	Change
Volume of stock required ($foot^3$)	6564.56	6257.22	-307.34
Cost of stock (Baht/month)	1,739,609	1,658,163	-81,446 *(4.68%)*
Volume of waste ($foot^3$)	423.38	123.37	-300.01
Cost of waste (Baht/month)	112,195	32,693	-79,502 *(70.86%)*

4.3. Comparison of Carbon Emission

The process-based analysis is used to estimate the carbon emission starting from harvesting of pine wood till finishing the stock cutting process. Based on, the results shown in Table 4, the amount of carbon emission from the pine wood cutting process can be reduced by 22,713 kg CO_2-e/year or approximately 5.32% reduction. The impact of reducing the waste generated during the pine wood cutting process go beyond the pallet manufacturer. The carbon emission from harvesting, sawmilling and transportation are also reduced.

Table 4. The comparison of carbon emission.

Process/Activity	Emission factor (kg CO_2-e/ unit)	Input		Carbon Emission (kg CO_2-e/year)	
		Current cutting plan	Optimal cutting plan	Current cutting plan	Optimal cutting plan
Forest harvest					
Gasoil (l)	2.84	41639	39691	118256	112722
Transport to sawmill					
Gasoil (l)	2.84	11462	10926	32552	31029
Sawmill					
Gasoil (l)	2.84	9135	8708	25944	24730
Electricity (kWh)	0.39	190391	181481	74252	70777
Transport to manufacturer					
Sea freight (Tonne-km)	0.008	13751553	13008226	110012	104066
Truck (Tonne-km)	0.62	100285	94864	62177	58816
Stock cutting					
Electricity (kWh)	0.39	4188	4188	1633	1633
Waste handling					
LPG (l)	1.5	1560	455	2341	682
Total Carbon Dioxide Equivalent Emission				**427168**	**404455**

Note: Emission factors obtained from [15-17]

5. Conclusions

Applying one dimensional cutting-stock problem to the wooden pallet industry can yield benefits in terms of both cost and carbon emission reduction. This is in fact a good practice toward green manufacturing. The developed linear programing model implemented on basic software such as Microsoft Excel and open source add-in OpenSolver has proven to be able to quickly and correctly offer the optimal cutting plan. Raw material waste decrease up to 70.86% resulting in a cost reduction of 977,352 Baht/year. Furthermore, it leads to a reduction of carbon emission in the life cycle of imported pine wood. The amount of carbon dioxide equivalent is reduced by 22,713 kg CO_2-e/year or approximately 5.32% reduction. The method developed in this research is very affordable and can easily be integrated to the existing production planning system.

References

[1] Goulimis C 1990 Optimal solutions for the cutting stock problem *European Journal of Operational Research* 44(2) 197-208

[2] Rusinko C 2007 Green manufacturing: an evaluation of environmentally sustainable manufacturing practices and their impact on competitive outcomes *IEEE Transactions on Engineering Management* 54(3) 445-454

[3] Sangwan K S 2006 Performance value analysis for justification of green manufacturing systems *Journal of Advanced Manufacturing Systems* 5(01) 59-73

[4] Paul I D, Bhole G P and Chaudhari J R 2014 A review on green manufacturing: it's important, methodology and its application *Procedia Materials Science* 6 1644-1649

[5] Nallusamy S, Dinagaraj G B, Balakannan K and Satheesh S 2015 Sustainable green lean manufacturing practices in small scale industries-A case study *International Journal of Applied Engineering Research* 10(62) 143-146

[6] He Y, Liu F, Cao H and Zhang H 2007 Process planning support system for green manufacturing and its application *Frontiers of Mechanical Engineering in China* 2(1) 104-109

[7] Gilmore P C and Gomory R E 1961 A linear programming approach to the cutting-stock problem *Operations research* 9(6) 849-859

[8] Ogunranti G A and Oluleye A E 2016 Minimizing waste (off-cuts) using cutting stock model: The case of one dimensional cutting stock problem in wood working industry *Journal of Industrial Engineering and Management* 9(3) 834-859

[9] Durán-Rosal A M, Guijo-Rubio D, Gutiérrez P A and Hervás-Martínez C 2018 Hybrid Weighted Barebones Exploiting Particle Swarm Optimization Algorithm for Time Series Representation *In International Conference on Bioinspired Methods and Their Applications* pp 126-137 Springer Cham

[10] Cheng C and Bao L 2018 An Improved Artificial Fish Swarm Algorithm to Solve the Cutting Stock Problem *In International Symposium on Neural Networks* pp 165-172 Springer Cham

[11] Sanchez I A L, Vargas J M, Santos C A, Mendoza M G and Moctezuma C J M 2018 Solving binary cutting stock with matheuristics using particle swarm optimization and simulated annealing *Soft Computing* 22(18) 6111-6119

[12] Mundim L R, Andretta M, Carravilla M A and Oliveira J F 2018 A general heuristic for two-dimensional nesting problems with limited-size containers *International Journal of Production Research* 56(1-2) 709-732

[13] Evtimov G and Fidanova S 2018 Ant colony optimization algorithm for 1D cutting stock problem *In Advanced Computing in Industrial Mathematics* pp 25-31 Springer Cham

[14] Brandão J S, Coelho A M and do Carmo F 2018 Study of different setup costs in SingleGA to solve a one-dimensional cutting stock problem *GSTF Journal on Computing* (*JoC*) 2(1)

[15] EPA 2008 Direct emissions from mobile combustion sources *Climate leaders greenhouse gas inventory protocol core module guidance*

[16] McKinnon A and Piecyk M 2010 Measuring and managing CO_2 emissions *Edinburgh: European*

Chemical Industry Council
[17] Alvarez S and Rubio A 2015 Compound method based on financial accounts versus process-based analysis in product carbon footprint: A comparison using wood pallets *Ecological indicators* 49 88-94

Minimization of Cutting Force by Optimizing the Cutting Parameters Using Taguchi Method in Turning

M S H Bhuiyan*

School of Mechanical Engineering, University of Science Malaysia, 14300 Nibong Tebal, Pulau Pinang, Malaysia

Corresponding author *: sayem@usm.my, sayem_um@yahoo.com

Abstract. The changes in magnitude of cutting force demonstrate the different occurrences in turning. The pattern of cutting force signal illustrates the cutting tool conditions and also corresponds to the changes in the cutting parameters. Obtaining an optimal cutting condition would significantly minimize the cutting force in turning, can reduce respective power consumption and consequently, extent the life of cutting tool. Choosing the optimal cutting parameters in turning would safeguard the cutting tool from catastrophe and could ensure a better surface finish of workpiece. From the orthogonal array and S/N ratio, the cutting speed of 120 m/min, feed rate of 0.32 mm/rev and depth of cut of 1 mm is found to be optimum cutting parameters. The magnitude of cutting force is found to be minimal and the surface finish is seen to be comparatively low at this cutting condition. The tool life at this cutting condition is found to be the longest for that particular tool – workpiece combination.

Keywords: *Cutting parameter optimization, Taguchi method, Cutting force, Flank wear, Surface roughness.*

1. Introduction

Turning is one of the mostly used machining techniques in manufacturing industry. In turning, the tool-workpiece combination, and cutting parameters play the most significant role and define the product quality, tool condition and the rate of power consumption in machining. Choosing the optimal cutting speed, feed rate, and depth of cut for machining can reduce tool wear and improve the surface finish of workpiece. Different methods are available to perform this optimization task. Li et al. [1] have used Taguchi method, response surface method (RSM), and multi-objective particle swarm optimization algorithm (MOPSO) to identify the optimal cutting parameters in CNC machining. They have used the Taguchi method to design the experiment, and subsequently employed the signal-to-noise (S/N) ratio to analyze the performance of parameters. In other studies, the cutting speed, feed rate and depth of cut have been optimized with the help of Taguchi design of experiment (DOE), signal to noise (S/N) ratio and analysis of variance (ANOVA) methods [2-3]. Fratila and Caizar [4] have applied Taguchi method to select optimal process parameters for minimum surface roughness and minimum cutting power required in semi-finished cutting. In another study, Camposeco-Negrete [5] has optimized the cutting parameters using Taguchi method focusing on the surface roughness and cutting power consumption in turning. Nalbant et al. [6] have used the Taguchi method to find the optimal cutting parameters for surface roughness in turning. They have employed the orthogonal array, the signal-to-noise ratio, and analysis of variance to study the performance characteristics in turning. Similar method has been applied by and Ravi & Kulkarni [7] and Vijay Kumar et al. [8] in different two investigations to find out the optimal cutting condition in CNC turning. In this study, the Taguchi method has been applied in optimization of the cutting parameters based on the cutting force. Then the results obtained from the Taguchi analysis have been cross-verified with the workpiece surface finish and flank wear data measured during turning.

2. Materials and Methods

A round bar (92 mm diameter and 760 mm long) of ASSAB-705, medium carbon steel (hardness HB270-310) has been used as the workmaterial, which contained carbon (0.35%), chromium (1.40%), iron (95.95%), manganese (0.70%), molybdenum (0.20%), and nickel (1.40%) by weight. A DSBNR 2525M 12 tool holder and TiN coated tungsten carbide, type: SNMG 12 04 08-PM tool insert assembly has been employed to cut the material. The experiment was conducted in dry cutting mode for this investigation. The cutting conditions and tool-workpiece combination have a significant effect on tool wear and on the surface roughness. The tool flank wear has been measured by taking the tool insert off from the tool holder after some 12 minutes of cut. A magnification of 40X has been used to capture the image of flank wear by light source microscope, Model: I CAMSCOPE(G). From the captured image, the average flank wear has been measured using a measuring software, 'Measure IT'. The surface finish of workpiece has been measured with a handy profile-meter type: 'Mahr Perthometer M1/M1 CNOMO 3755350' at the end of 12 minutes of cut.

The turning operation is performed on a COLCHESTER VS MASTER3250 (165 mm × 1270 mm) Gap bed Center Lathe machine. A tri-axial force dynamometer Type: KISTLER 9257A have been used to measure the cutting force during machining. The frequency range of the tri-axial force dynamometer is 10 Hz – 400 kHz. A DEWE- 43 module and a charge amplifier Type: KISTLER 5019 have been employed as signal processing units to process the captured signal. The DEWE-43 module has a low-pass filter of LP: 1000 kHz and the charge amplifier has a high-pass filter of HP: 180 kHz. The signals captured using the force dynamometer have been processed inside these modules before storing for analysis. The schematic diagram of the experiment setup is shown in Figure 1.

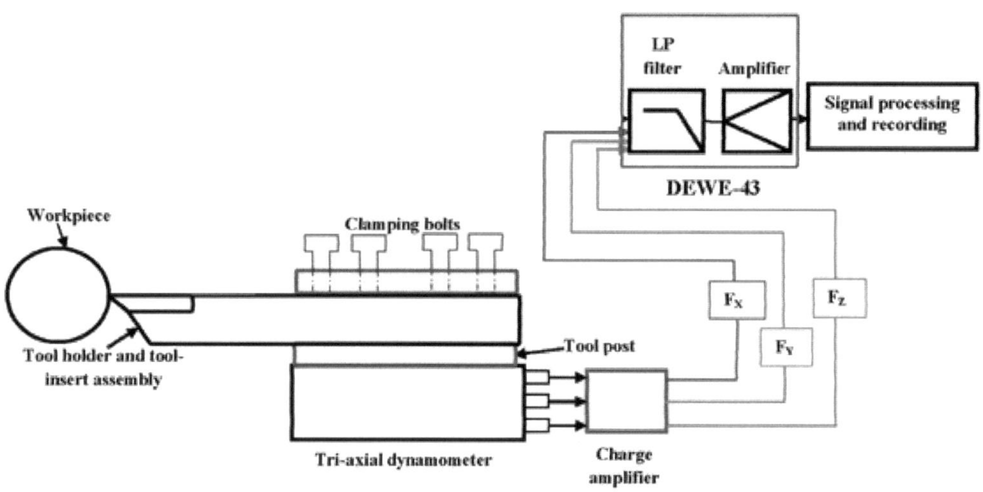

Figure 1. Setup for capturing the cutting force signals in turning.

3. Design of Experiments

The parameter design has been carried out under the design of experiment using the Taguchi method to optimize the machining process. The cutting parameters used in the experiment are presented in Table 1.

IOP Conf. Series: Materials Science and Engineering **530** (2019) 012006 doi:10.1088/1757-899X/530/1/012006

Table 1. Cutting conditions.

Cutting speed (m/min)	120, 150, 170
Feed rate (mm/rev)	0.20, 0.32, 0.50
Depth of cut (mm)	0.5, 1, 2

The design of experiment has been performed with Minitab software. The cutting parameters have been optimized focusing on minimal cutting force requirement in turning. It is expected that the optimal cutting conditions obtained from the parameter design are insensitive to the variation of environmental conditions and other noise factors.

The orthogonal array has been used to study the entire parameter space with a small number of experiments only and thus to identify the optimal cutting condition. A loss function transformed into S/N ratio has been defined to calculate the deviation between the experimental value and the desired value. The 'lower-the-better' category of the performance characteristic for the analysis of the S/N ratio has been used in the design of experiment. The S/N ratio for each level of process parameters is computed based on the S/N analysis.

Finally, a confirmation experiment is conducted to verify the optimal process parameters obtained from the parameter design. The results have been tested by the surface roughness and flank wear measured during machining. In this paper, the cutting parameter design by the Taguchi method is adopted to obtain optimal machining performance in turning.

4. Results and Discussions

The orthogonal arrays of the cutting parameters and the S/N ratios have been obtained from the Taguchi analysis to find the optimal cutting parameters among the cutting conditions. From literature, the larger S/N ratio corresponds to the better performance characteristic regardless of the category of the performance characteristic. Therefore, the optimal level of the process parameters is the level with the highest S/N ratio. It has been also reported that the optimal combination of the process parameters can be predicted with the orthogonal array, and the S/N analyses [6].

A three-level, three factors fractional factorial design, L9 has been used to generate an orthogonal array. Each row of the matrix represents one trial. The orthogonal array obtained from the Taguchi method is shown in Table 2.

Table 2. Orthogonal array.

SN	Cutting speed (m/min)	Feed rate (mm/rev)	Depth of cut (mm)	Flank wear (mm)	Surface roughness (μm)	Cutting force (N)
S1	120	0.20	0.5	0.1165	3.994	45.8
S2	120	0.32	1.0	0.1185	3.912	24.3
S3	120	0.50	2.0	0.1304	4.453	48.5
S4	150	0.20	1.0	0.1145	3.997	28.2
S5	150	0.32	2.0	0.1289	3.896	128.7
S6	150	0.50	0.5	0.1127	4.081	118.5
S7	170	0.20	2.0	0.1601	4.099	151.3
S8	170	0.32	0.5	0.1483	3.998	29.9
S9	170	0.50	1.0	0.1571	4.463	81.1

The tool life, surface roughness and cutting force graphs for these trials are represented in the Figure 2.

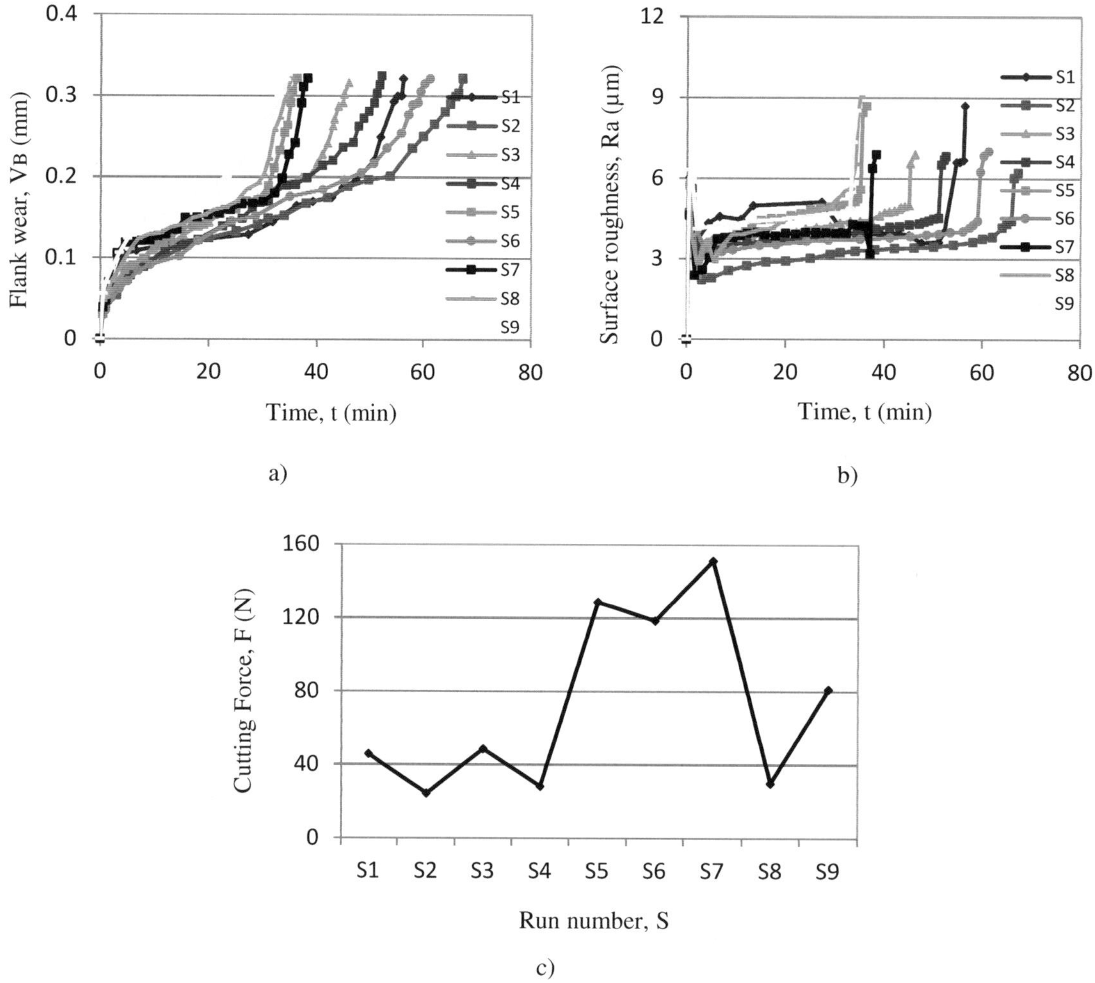

Figure 2. a) Tool life graph, b) Surface rouhness profile, and c) Amplitude of RMS force signal.

Though we can have some idea about the possible optimal cutting conditions for the reduced cutting force from the above graphs, however, it is rather difficult to choose a particular trial condition as the optimal cutting condition. From the S/N ration analysis, the most optimal cutting parameters can be identified.

The main effects and the S/N ratios for different combination of the cutting parameters have been obtained to identify the optimal cutting condition. Figure 3 illustrates the changes of mean data and S/N ratios at different cutting conditions.

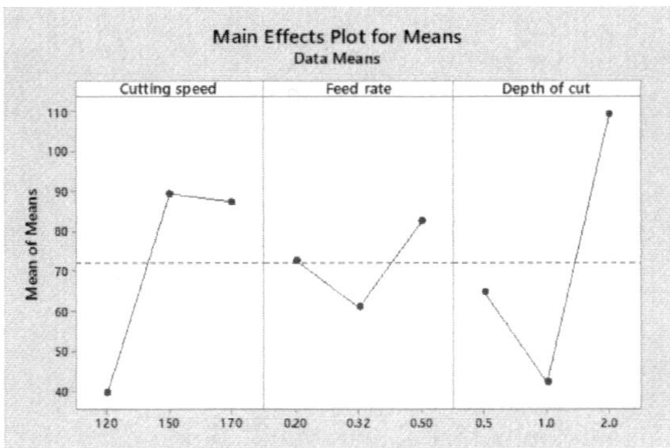

a) Mean data plot for cutting force

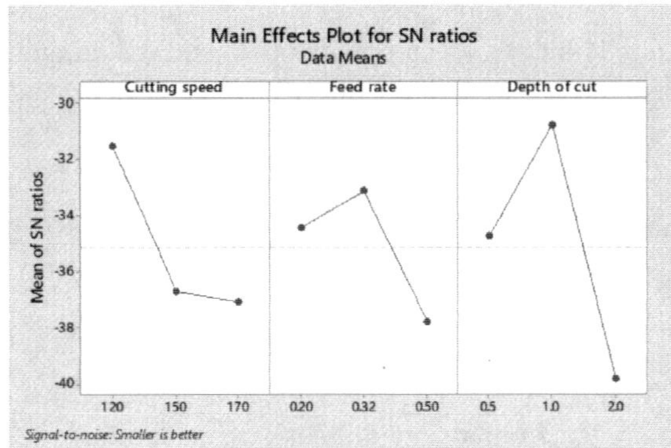

b) S/N ratio plot for cuttingg force

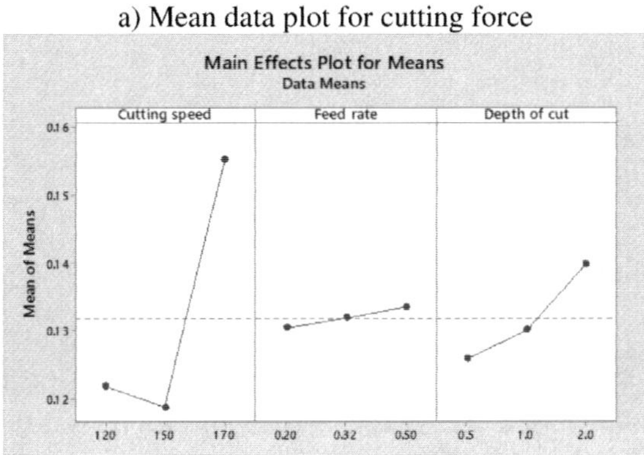

c) Mean data plot for flank wear

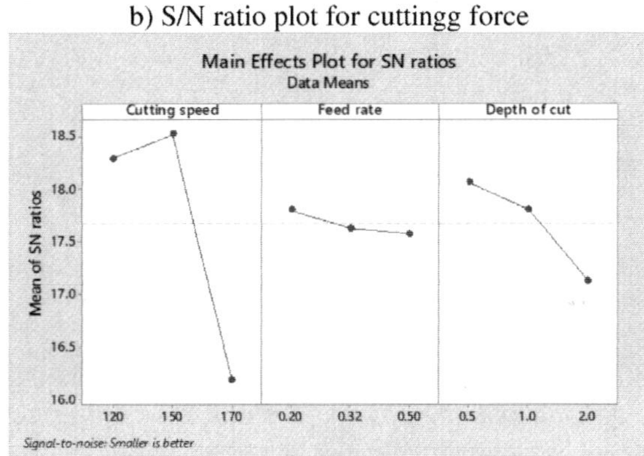

d) S/N ratio plot for flank wear

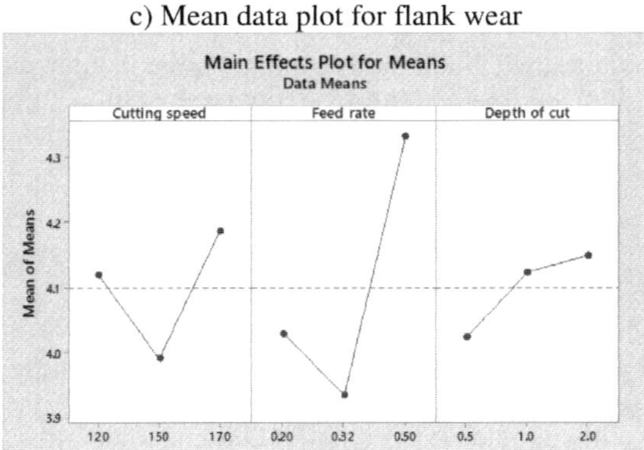

e) Mean data plot for surface roughness

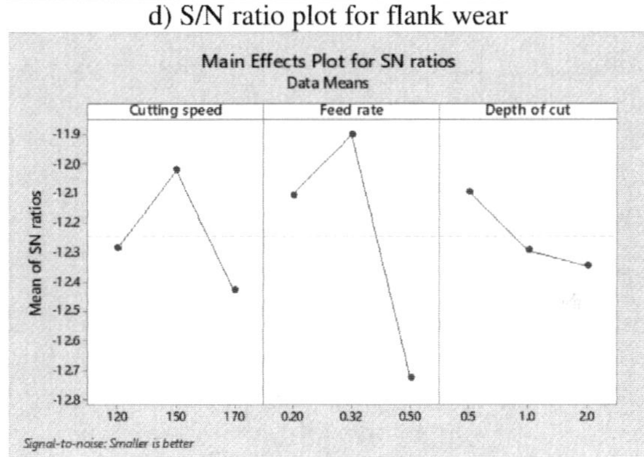

f) S/N ratio plot for surface roughness

Figure 3. The mean data and single-to-noise ratio graphs.

From the Taguchi analysis results, three different optimal cutting conditions have been obtained for cutting force, flank wear, and surface finish optimization. For cutting force optimization, the optimal cutting speed has been obtained 120 m/min, feed rate 0.32 mm/rev and depth of cut of 1 mm; whereas for flank wear optimization, the cutting speed, feed rate and depth of cut have been changed to 150 m/min, 0.20 mm/rev and 0.5 mm respectively. An exceptional drop of tool wear and cutting force has been observed at feed rate of 0.32 mm/rev, which can be attributed to the phenomena of chip breakage taken place during metal-cutting [9]. For surface roughness optimization, the optimal cutting speed and depth of cut have been found to be similar like that of flank wear optimization parameter except the feed rate, which has been changed to 0.32 mm/rev. Therefore, it can be deduced that, when it is to optimize the cutting force and surface roughness, one has to be more concerned about the depth of cut. In case of optimizing the cutting force and flank wear, both the feed rate and the depth of cut are to be given consideration significantly. For the flank wear and surface roughness optimization, the feed rate has to be monitored with great importance.

Figure 4 shows the tool wear and surface roughness graphs obtained from the experiment of these three optimal cutting conditions.

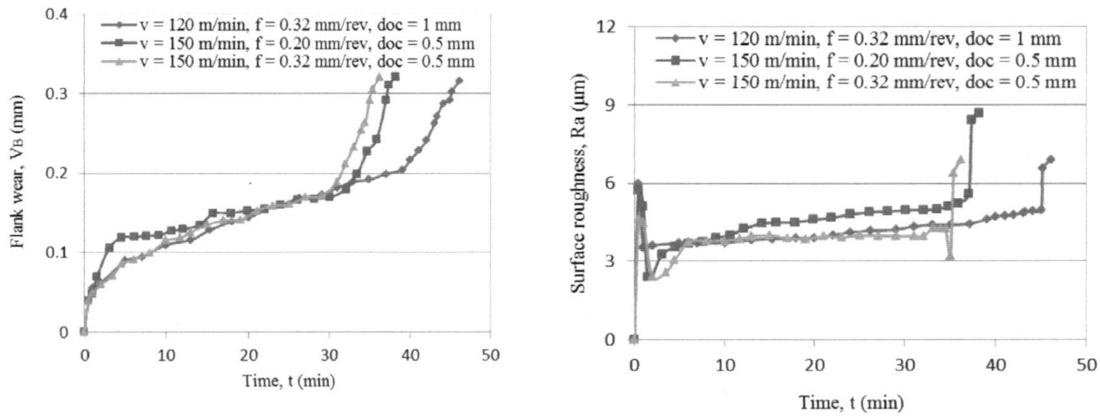

a) flank wear b) surface roughness

Figure 4. a) Tool wear curve, b) Surface roughness graph.

From the tool wear and surface roughness graphs of Figure 4, it has been observed that the rate of flank wear was minimal at the optimal cutting condition found for flank wear optimization, and the rate of surface roughness was the least at the optimal cutting condition for surface roughness optimization. Though, both the tool wear and surface roughness graphs obtained for cutting force optimization have shown some similar trends to those for tool wear and surface roughness optimization, the tool life was the longest for the optimal cutting condition gained for cutting force optimization. Therefore, the optimal cutting condition found for cutting force optimization has been chosen to carry out the experiment to investigate the cutting tool condition using force signal in turning.

5. Conclusions

The optimal cutting parameters obtained from Taguchi analysis have been tested with experiment before recording the force signal in tool condition monitoring during turning. The optimal cutting condition of v = 120 m/min, f = 0.32 mm/rev, and *doc* = 1 mm found for cutting force optimization was the optimum among all three optimal cutting conditions. The flank wear and surface roughness graphs plotted from the experimental results have also supported the optimum cutting condition that obtained for cutting force optimization.

Acknowledgments

The author would like to thank the University of Science Malaysia for providing the facility to carry out this research.

References

[1] Li C, Xiao Q, Tang Y and Li L 2016 A method integrating Taguchi, RSM and MOPSO to CNC machining parameters optimization for energy saving *J. Cleaner Prod.* 135 263-275
[2] Samantaraya D, Lakade S and Keche A, 2018 An Alternate Machining Method for Hardened Automotive Gears *Procedia Manuf.* 20 517-522
[3] Ghani J A, Choudhury I A and Hassan H H 2004 Application of Taguchi method in the optimization of end milling parameters *J. Mater. Process Technol.* 145 84-92
[4] Fratila D and Caizar C 2011 Application of Taguchi method to selection of optimal lubrication and cutting conditions in face milling of AlMg3 *J. Cleaner Prod.* 19 640-645

[5] Camposeco-Negrete C 2013 Optimization of cutting parameters for minimizing energy consumption in turning of AISI 6061 T6 using Taguchi methodology and ANOVA *J. Cleaner Prod.* 53 195-203

[6] Nalbant M, Gökkaya H and Sur G 2007 Application of Taguchi method in the optimization of cutting parameters for surface roughness in turning *Mater. Des.* 28 1379-1385

[7] Ravi Kumar S M and Kulkarni S K 2017 Analysis of Hard Machining of Titanium Alloy by Taguchi Method *Mater. Today: Proc.* 4 10729-10738

[8] Vijay Kumar M, Kiran Kumar B J and Rudresha N 2018 Optimization of Machining Parameters in CNC Turning of Stainless Steel (EN19) By TAGUCHI'S Orthogonal Array Experiments *Mater. Today: Proc.* 5 11395-11407

[9] Bhuiyan M S H, Choudhury I A and Nukman Y 2012 An innovative approach to monitor the chip formation effect on tool state using acoustic emission in turning *Int. J. Mach. Tools Manuf.* 58 19-28

Chipping Detection in Ceramic Insert during Turning by Analysis of Workpiece Surface Profile Using Cross-Correlation

N A Mohamad and M M Ratnam*

School of Mechanical Engineering, Engineering Campus, Universiti Sains Malaysia, 14300 Nibong Tebal, Penang, Malaysia.

Corresponding author *: mmaran@usm.my

Abstract. Chipping in ceramic cutting inserts during machining can cause detrimental effects on the surface finish quality of the workpiece. Current methods of detecting chipping in ceramic inserts do not compare the minute changes in the workpiece profile in subsequent machining passes. In this work, an effective method of detecting chipping in ceramic inserts by analysing the workpiece surface profile using cross-correlation is proposed. In this method, the sub-pixel edge location method was used to extract the surface profile captured using a digital camera. Surface profiles corresponding to subsequent machining passes were cross-correlated at three different sections of the workpiece. An abrupt drop in the correlation coefficient was observed in the middle and end sections of the workpiece after the fifth machining pass, thus providing a reliable measure of tool chipping.

1. Introduction

Ceramic cutting tool inserts are the preferred type of cutting tools for the machining of hard and difficult-to-cut materials such as alloy steels. This is due to their high hot hardness and wear resistance. However, due to their inherent low toughness ceramic inserts have be reported to undergo ladder-like chipping, cracking and fracture during machining [1-3]. Since failure of a cutting tool during machining will cause a detrimental effect on the finished product, the early detection of wear in the cutting tool is an important area of research. Although a significant amount of research has been published on the offline and in-process detection of wear in cutting tools [4-7], most of these researches are limited to carbide tools. The detection of wear and chipping in ceramic cuttings inserts has been investigated only by a few researchers.

Neslušan et al. [8] developed a method to detect the breakage of ceramic cutting tools during hard turning by analyzing acoustic emission (AE) signals. AE signals are known to be sensitive to the plastic deformation processes linked with dislocation slips during machining of ductile materials or brittle cracking. The authors used two AE sensors with different frequency ranges to detect machining changes before and after tool breakage. This method, however, is not sensitive to detect the occurrence of chipping in the ceramic tool. Lee at al. [9] used fast Fourier transform (FFT) to detect the onset of chipping in commercial ceramic cutting inserts. The edge profile of the workpiece was extracted to sub-pixel accuracy and the extracted profile was transformed from the spatial domain to the frequency domain using FFT. During gradual wear the amplitude of the fundamental feed frequency was observed to increase steadily, but after the onset of tool chipping significant fluctuations in the feed frequency was observed. Although this method was shown to be effective in detecting the onset of chipping the edge profiles before and after chipping were not compared quantitatively. In a later publication [10], the authors applied the sub-window FFT and continuous wavelet transform (CWT) methods to detect the onset of chipping in ceramic tools. They reported that compared to the sub-window FFT method the CWT method is more effective in detecting the exact onset of tool chipping. The authors also investigated the effectiveness of the autocorrelation method in detecting chipping in ceramic inserts by analyzing the workpiece profile captured using a digital camera [11]. The authors reported that chipping in the tool caused the peaks of the autocorrelation function to decrease rapidly with increase in lag distance. The correlation was made between profiles recorded from the same section of the workpiece but at different angles. The profiles at various parts of the workpiece were not cross-correlated to detect chipping in the tool.

Besides the work done by Lee et al. [9-11] there is very little published literature on the detection of chipping in ceramic cutting tools. Many researchers have reported the problem of chipping in ceramic cutting inserts but a reliable method of detecting chipping is yet to be developed. The aim of this paper is to propose a vision-based method combined with sub-pixel edge location and cross-correlation to detect the occurrence of chipping in commercial cutting tools during dry turning.

2. Methodology

2.1. Experimental setup and image capture

The dry turning operation was conducted by using a 45 mm diameter AISI 305 stainless steel workpiece of length 200 mm. The machining was done on a *Pinocho S90* conventional lathe machine by using ceramic insert cutting tool from Sandvik Coromant Ltd., Sweden. The workpiece was mounted securely at both ends to ensure there is minimum vibration during turning (Figure 1(a)). The machining was carried out using a cutting speed of 1400 rpm, feed rate of 0.3 mm/rev and a depth-of-cut of 1 mm. A high cutting speed and feed rate were used to accelerate the rate of wear in the ceramic insert. The machining was done for five passes. After each pass the workpiece was removed from the lathe to capture the profile. The cutting tool insert was also removed and observed under that scanning electron microscope. The machined length was divided into three sections, namely start, middle and end, for the surface analysis (Figure 1(b)). A high-resolution DSLR camera (pixel resolution: 5184 × 3456 pixels) fitted with a macro lens was used to capture the edge profile of the machined workpiece with the aid of backlighting (Figure 1(c)). The intensity of the backlighting was controlled appropriately to ensure the image is not too bright or too dark. The surface of the workpiece was cleaned using a jet to compressed air before capturing the images. The workpiece was shifted using a translation stage in order to capture images at various sections. Four images were captured at each section by rotating the workpiece by 90°.

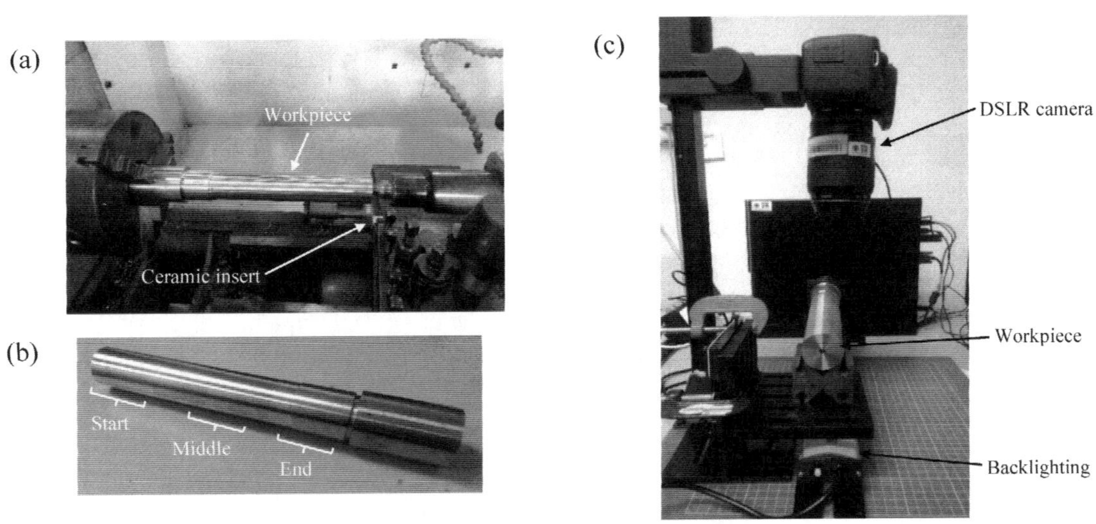

IOP Conf. Series: Materials Science and Engineering **530** (2019) 012007 doi:10.1088/1757-899X/530/1/012007

Figure 1. (a) Machining setup, (b) various section used in analysis, (c) setup for capturing workpiece profile.

2.2. Profile extraction using sub-pixel edge location

Figures 2(a) shows the sample raw image of the workpiece profile captured after the first pass. The profile of the workpiece surface was extracted using the invariant-moment sub-pixel edge detection method originally proposed by Tabatabai and Mitchell [12]. In this method a scan line across a step edge is characterized by a set of numbers x_i, where i= 1, 2, 3, …, n. The edge is defined between a sequence of pixels at one intensity level h_1 followed by a sequence pixels at another intensity level h_2 as illustrated in Figure 2(b). The first three moments m_1, m_2 and m_3 of the input data sequence are given by a threshold independent method based on grey level moment equations, whereby the moments are defined as a sum of the pixel intensity powers. The ith moment of the input grey level data sequence x_j is given by

$$\bar{m}_i = \frac{1}{n}\sum_{j=1}^{n}(x_j)^i \tag{1}$$

where $x_1, x_2, \ldots, x_n$ are pixel intensities and n is the total number of pixels in row j. If k denotes the number of h_1 values in the ideal edge, then the first three samples moments between input and output sequences can be solved in three equations by:

$$\bar{m}_i = \sum_{j=1}^{2} p_j h_j^i \tag{2}$$

where i = 1, 2 and 3, and p_h = number of pixels with grey intensity values h.
With the three unknowns h, k and p_2, the solutions of the edge are calculated by:

$$h_1 = \bar{m}_1 - \sigma\sqrt{\frac{p_2}{p_1}} \tag{3}$$

$$h_2 = \bar{m}_1 + \sigma\sqrt{\frac{p_1}{p_2}} \tag{4}$$

$$p_2 = \frac{1}{2}\left[1 + s\sqrt{\frac{1}{4+s^2}}\right] \tag{5}$$

$$p_1 = 1 - p_2 \tag{6}$$

where $\sigma = \sqrt{\bar{m}_2 - \bar{m}_1^2}$ and s defined as the skewness of the input data sequence given by,

$$s = \frac{2\bar{m}_1 + \bar{m}_3 - 3\bar{m}_1\bar{m}_2}{\sigma^3} \tag{7}$$

Thus, the edge location of the workpiece up to sub-pixel accuracy is determined by:

IOP Conf. Series: Materials Science and Engineering **530** (2019) 012007 doi:10.1088/1757-899X/530/1/012007

$$k = p_1 n \tag{8}$$

The invariant moment method of sub-pixel edge detection allows precise edge location as shown by the red line in Figure 2(c).

2.3. Simulation analysis using cross-correlation

Cross-correlation is essentially a measure of similarity between two signals as a function of displacement of one relative to the other. For two continuous function $f(t)$ and $g(t)$ the cross-correlation is given by the dot product,

$$f * g = \int_{-\infty}^{\infty} f^*(t)g(t+\tau)dt \tag{9}$$

where f^* is the complex conjugate of f and τ is the displacement, also known as the lag distance. For discrete function such as the digitized surface profile, the cross-correlation is defined as,

$$f * g = \sum_{n=-\infty}^{\infty} f^*(k)g(k+l) \tag{10}$$

where k is a measure along the signal length and l is the lag distance.

(a)

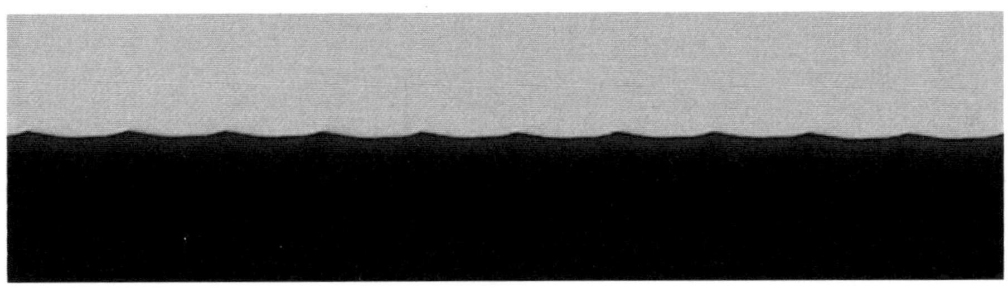

(b)

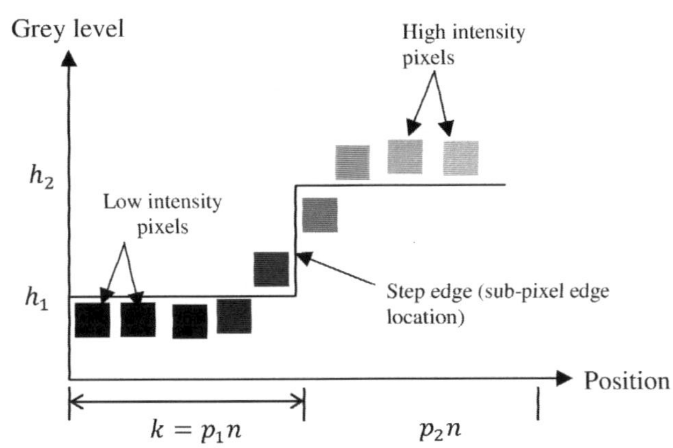

IOP Conf. Series: Materials Science and Engineering **530** (2019) 012007 doi:10.1088/1757-899X/530/1/012007

(c)

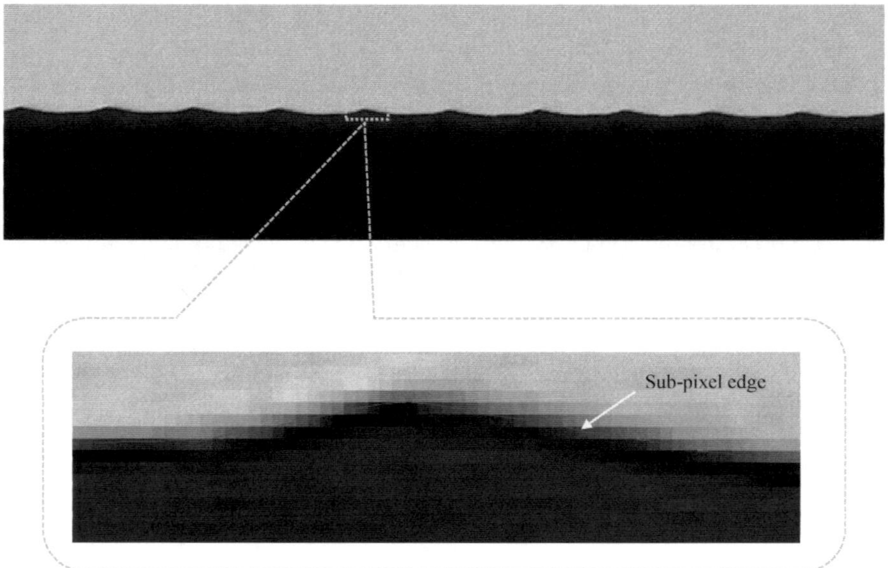

Figure 2. Extraction of workpiece edge profile to sub-pixel level: (a) raw image, (b) grey value variation in the invariant moment method, (c) extracted profile (shown in red).

Figures 3(a)-(c) show three simulated profile, whereby Figure 3(a) is assumed to be shaped by a new cutting tool, Figure 3(b) is formed by a slightly worn cutting insert while Figure 3(c) is formed by a badly worn cutting insert. The corresponding extracted profiles are shown in Figures 4(a)-(c).

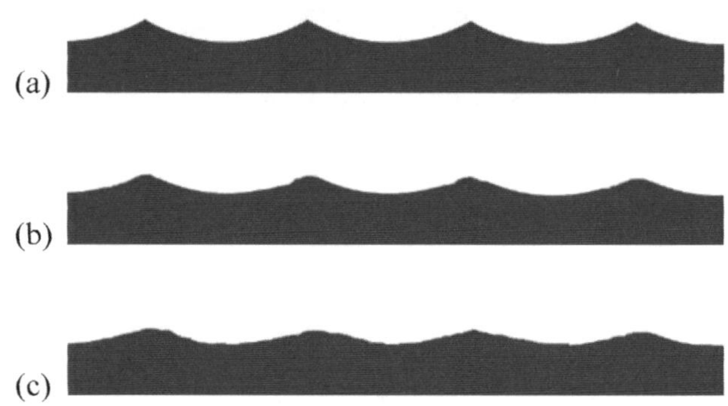

Figure 3. Simulated profiles: (a) formed by new insert, (b) formed by slightly worn inset, (c) formed by badly worn insert.

IOP Conf. Series: Materials Science and Engineering **530** (2019) 012007 doi:10.1088/1757-899X/530/1/012007

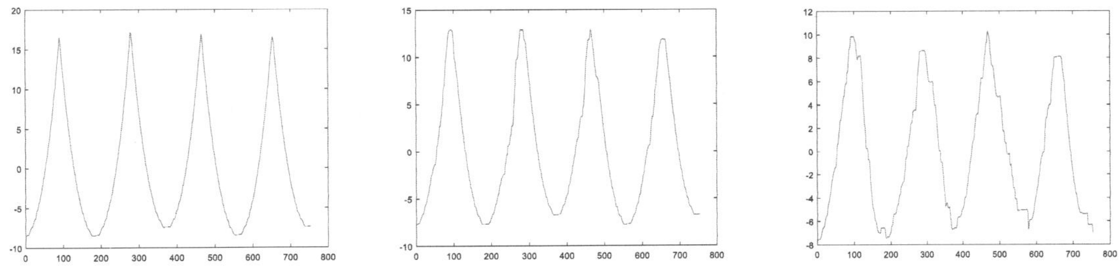

(a) (b) (c)

Figure 4. Extracted profiles corresponding to Figures 3(a)-(c).

The result of cross-correlating the profiles in Figure 4(a) with Figure 4(b) and Figure 4(a) with Figure 4(c) is shown in Figure 5(a) and 5(b). Although the correlation plots look the same the maximum correlation value when the insert is slightly worn is 0.9931 while the corresponding value when the insert is badly worn is 0.9539. The maximum positive correlation coefficient thus gives a measure of the wear condition of the insert.

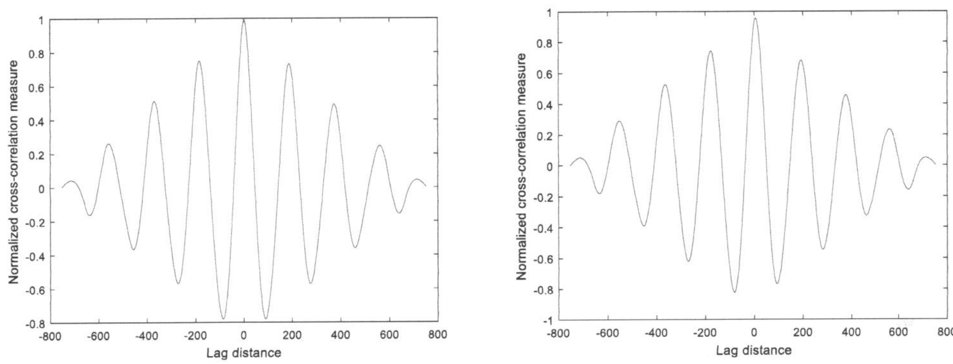

Figure 5. Plot of cross-correlation for (a) slightly worn insert, (b) badly worn insert.

3. Results and Discussion

Figures 6(a)-(e) show the workpiece edge profiles at various stages of machining and the insert image captured at the end of each pass. The gradual change in the profile during the first four passes is clearly noticeable while there is a rapid change in profile from the 4th to the 5th pass. The gradual profile change can be attributed to the gradual loss of tool material during turning while the rapid change in the profile is due to chipping (highlighted in yellow on the insert image).

(a)

(b)

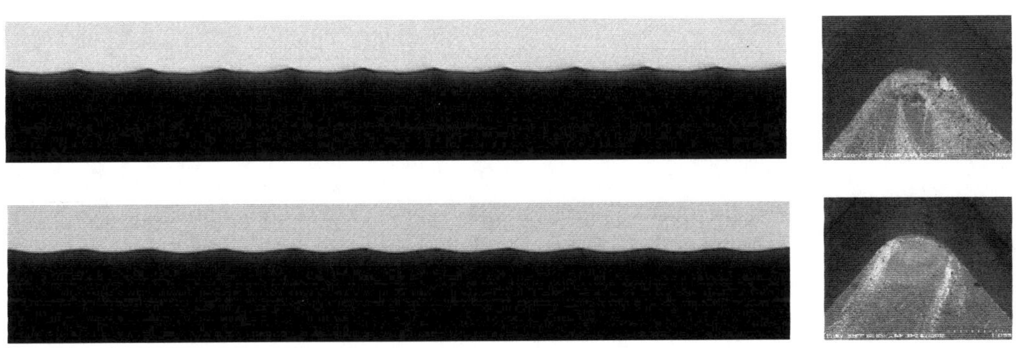

(c) 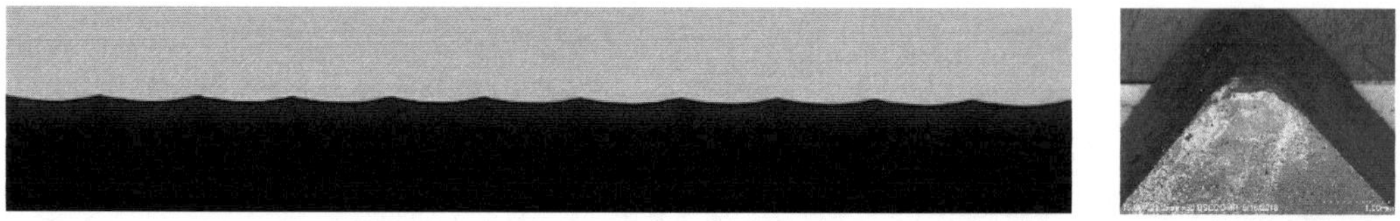

(d) 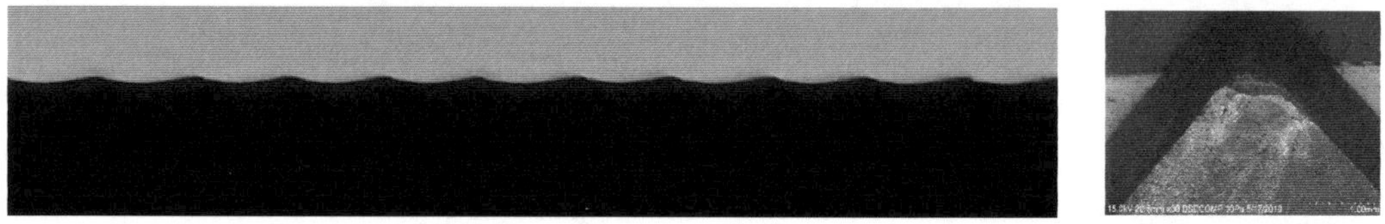

(e) 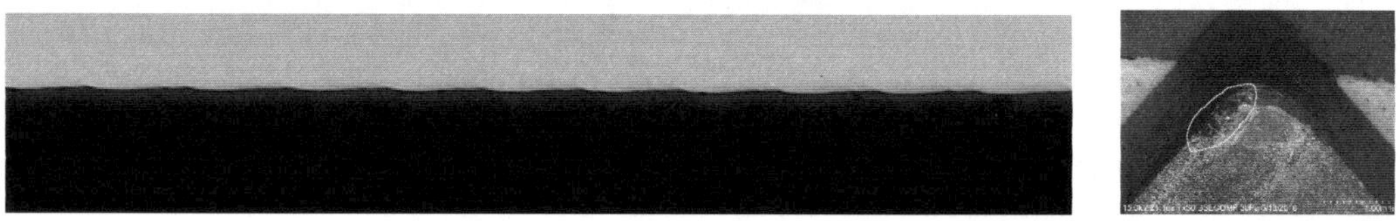

Figure 6. Workpiece edge profile and tool insert after (a) 1st pass, (b) 2nd pass, (c) 3rd pass, (d) 4th pass and (e) 5th pass.

In order to investigate the changes in profile in greater detail the length of profile equivalent to three feed spacing were cropped manually from the original image. Each image was rescaled to accommodate slight variation in the camera-to-workpiece distance after each machining pass. The cropped image was then subjected to sub-pixel edge location. The extracted edge profiles after the 2nd to the 5th passes were compared to the profile after the first pass. The extracted data were realigned so that the peaks in the subsequent passes match those in the first pass. Figures 7(a)-(l) shows the comparison between the 1st pass with subsequent passes at the start (Figures 7(a)-(d)), middle (Figures (7(e)-(h)) and end (Figures 7(i)-(l)). The start, middle sections of the workpiece are shown in Figure 1(b).

Comparison of the profiles shows that a significant change in the profile is visible after the 5th pass in all sections of the workpiece. This can be attributed to the significant loss of cutting tool material due to chipping as seen in insert image in Figure 6(e). Comparison of the changes in the profile between the 1st and subsequent passes at the various sections show that the middle and end sections of the workpiece produce very consistent results compared to the start section. The fluctuation in the profiles in the start section, especially after the 1st and 4th passes show the possible presence of vibration as the tool first comes in contact with the workpiece. As machining proceeds to the middle and end section the vibration effect has diminished and the changes in the workpiece profile due to tool wear and chipping becomes dominant.

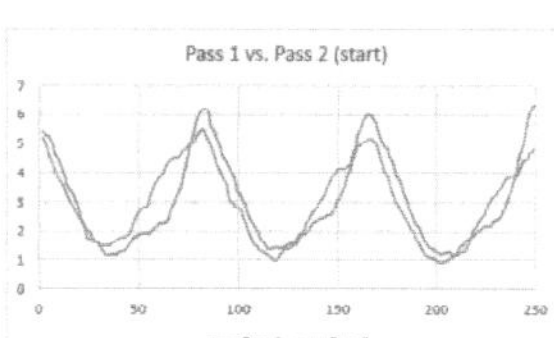

(a)

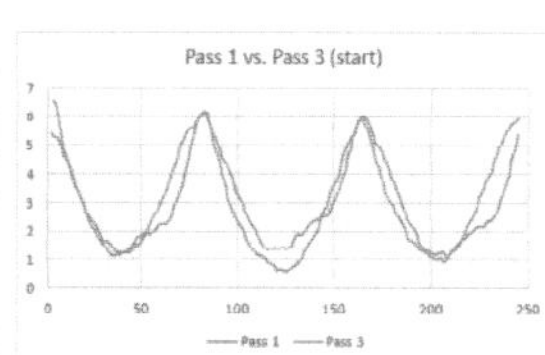

(b)

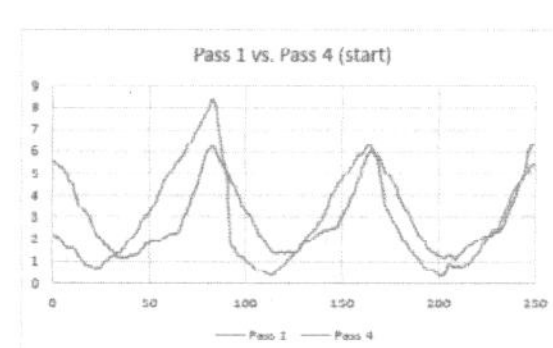

(c)

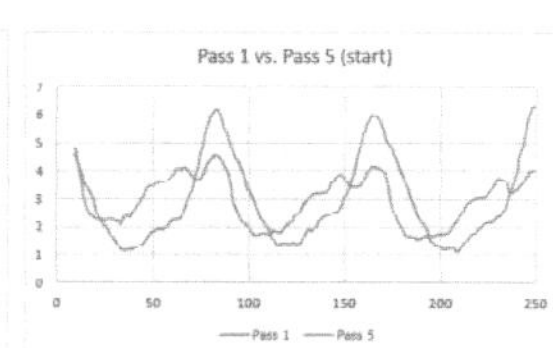

(d)

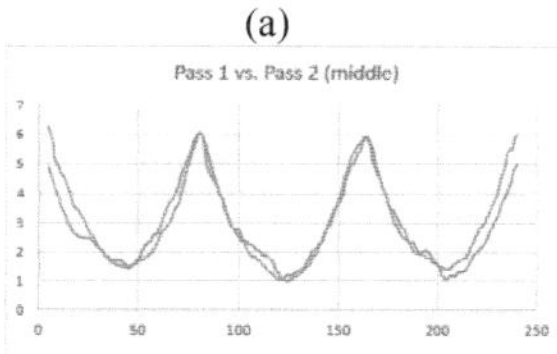

(e)

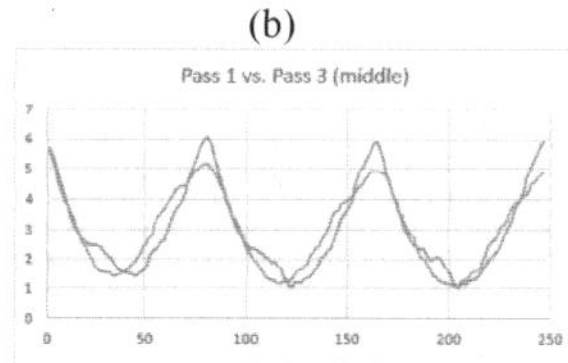

(f)

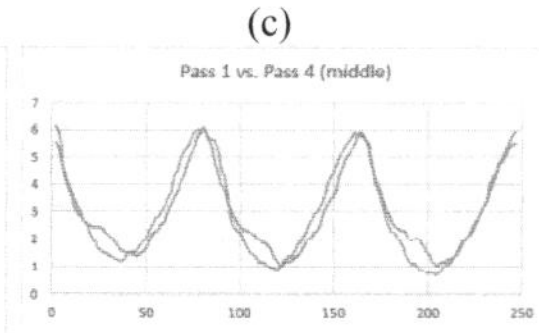

(g)

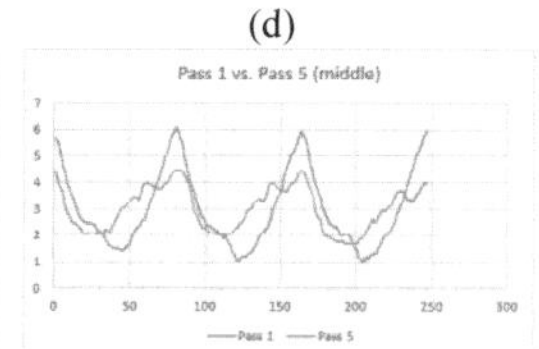

(h)

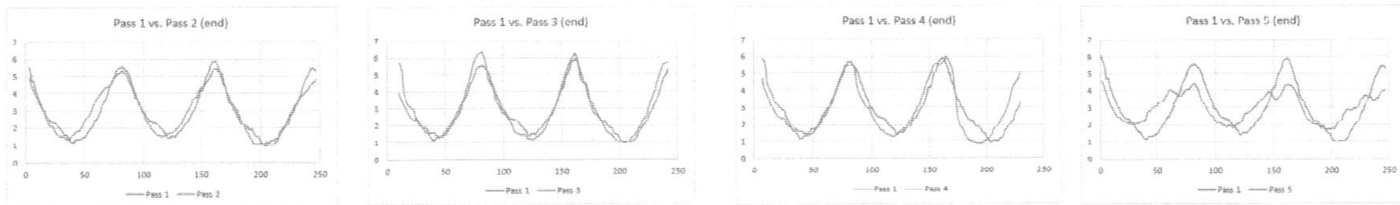

(i) (j) (k) (l)

Figure 7. (a)-(d) Comparison of surface profiles of workpiece between 1st and subsequent passes at: (a)-(d) start,(e)-(h) middle and (i)-(l) end of workpiece.

Figure 7 only shows a visual comparison of the profiles before and after tool chipping. In order to develop an automatic method of detecting chipping in the insert it is necessary to analyse the cross-correlation between these profiles. The maximum cross-correlation coefficients for the profiles are shown in Figure 8. Neglecting the data from the start section it can be seen that the maximum cross-correlation coefficient drops to below 0.8 after tool chipping. Thus, the correlation coefficient criteria for tool chipping can be set to 0.8 for the in-process detection of chipping in ceramic inserts.

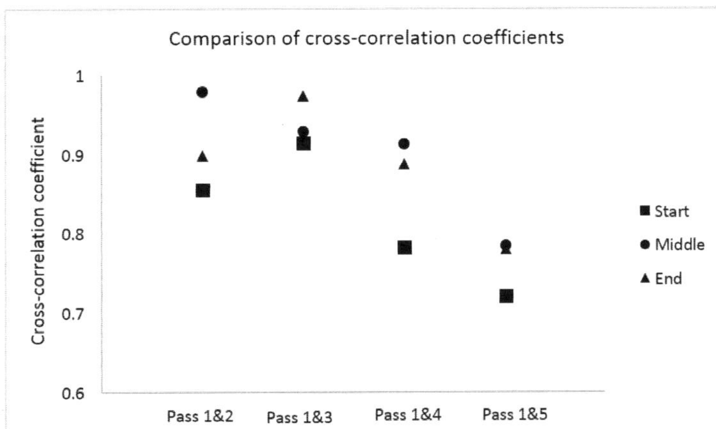

Figure 8. Comparison of cross-correlation coefficients for different machining passes.

4. Conclusion

A simple, yet effective, method of detecting chipping by analysis of the workpiece surface profile is proposed. Comparison of the surface profiles before and after chipping using cross-correlation shows a sharp decrease in the coefficient value after chipping. The study has also shown that the profiles compared should be extracted from the middle or end of each machining pass for better accuracy. The proposed method can be implemented for in-process tool chipping detection by capturing and processing the images in real-time during a turning process.

Acknowledgement

Part of this work was supported by the RU(I) grant (no. 1001/PMEKANIK/8014013). The authors would like to thank Universiti Sains Malaysia for the offer of the grant.

References

[1] Song J, Huang C, Lv M, Zou B, Liu H and Wang J 2014 Cutting performance and failure mechanisms of TiB_2-based ceramic cutting tools in machining hardened Cr12MoV mold steel *International Journal of Advanced Manufacturing Technology* 70 495-500

[2] Lima F F, Sales W F, Costa E S, da Silva F J and Machado A Á R 2017 Wear of ceramic tools when machining Inconel 751 using argon and oxygen as lubri-cooling atmospheres *Ceramics International* 43 (2017) 677–685

[3] Tan D-W, Guo W-M, Wang H-J, Lin H-T and Wang C-Y 2018 Cutting performance and wear mechanism of TiB_2-B_4C ceramic cutting tools in high speed turning of Ti6Al4V alloy *Ceramic International* (online 25 May 2018) In Press

[4] Mikołajczyk T, Nowicki K, Bustillo A and Yu Pimenov D 2018 Predicting tool life in turning operations using neural networks and image processing *Mechanical Systems and Signal Processing* 104 503-513

[5] Henrique L, Maia A, Abrao A M, Vasconcelos W L, Sales W F and Machado A R 2015 A new approach for detection of wear mechanisms and determination of tool life in turning using acoustic emission *Tribology International* 92 519-532

[6] Gutnichenko O, Bushlya V, Zhou J and Ståhl J-E 2017 Tool wear and machining dynamics when turning high chromium white cast iron with pcBN tools *Wear* 390-391 253-269

[7] Rmili W, Ouahabi A, Serra R, Leroy R 2016 An automatic system based on vibratory analysis for cutting tool wear monitoring *Measurement* 77 117-123

[8] Neslušan M, Mičieta B, Mičietová A, Čilliková M and Mrkvica I 2015 Detection of tool breakage during hard turning through acoustic emission at low removal rates *Measurement* 70 1-13

[9] Lee W K, Ratnam M M and Ahmad Z A 2016 Detection of fracture in ceramic cutting tools from workpiece profilesignature using image processing and fast Fourier transform *Precision Engineering* 44 131-142

[10] Lee W K, Ratnam M M and Ahmad Z A 2017 Detection of chipping in ceramic cutting inserts from workpiece profile during turning using fast Fourier transform (FFT) and continuous wavelet transform (CWT) *Precision Engineering* 47 406-423

[11] Lee W K, Ratnam M M and Ahmad Z A 2016 In-process detection of chipping in ceramic cutting tools during turning of difficult-to-cut material using vision-based approach *International Journal of Advanced Manufacturing Technology* 85 1275-1290

[12] Tabatabai A J and Mitchell OR 1984 Edge location to subpixel values in digital imagery *IEEE Transactions on Pattern Analysis and Machine Intelligence* 6(2) 188-201

Effect of Machining Parameters on the Surface Roughness for Different Type of Materials

M H M Ghazali, A Z A Mazlan*, L M Wei, C T Tying, T S Sze and N I M Jamil

School of Mechanical Engineering, Engineering Campus, Universiti Sains Malaysia, 14300, Nibong Tebal, Pulau Pinang, Malaysia

Corresponding author *: zulryan_85@yahoo.com

Abstract. Surface roughness (R_a) is an important parameter in determining the quality of surface finishing of any products obtained from the machining process. Surface roughness can be affected by several factors such as machining parameters, workpiece material properties, cutting tool parameters and machine tool conditions. In this study, the effects of machining parameters such as feed rate and cutting speed on the surface roughness was investigated for aluminium, mild steel and brass materials during the turning process using CNC Lathe machine. The experimental R_a values obtained were compared with the theoretical values with discrepancy ranging from 1.14 % to 113.71 %. From the study, it can be concluded that the surface roughness of materials will increases as the feed rate increases and decrease when higher cutting speed is applied.

1. Introduction

Surface roughness is a measurement of finished surface area where it indicates the state of machined surface. A better surface finishing is one of the desired requirements in industry as it improves fatigue strength and minimizes friction [1]. The parameters that affect the surface roughness can be classified into primary and natural surface roughness effects. Primary surface roughness effect is contributed by cutting tool geometry, feed rate and cutting speed while natural surface roughness effect is contributed by the machine tool and uncontrolled variation in machining process such as tool wear, dynamic unbalance of machining system and chip formation [2-3].

Surface roughness plays an important role in determining how a real object will interact with its environment. There are various factors that will affect the surface roughness of a workpiece such as cutting speed, feed rate and tool nose radius. Figure 1 shows the example of machined components and the affected surface roughness parameters during the machining process.

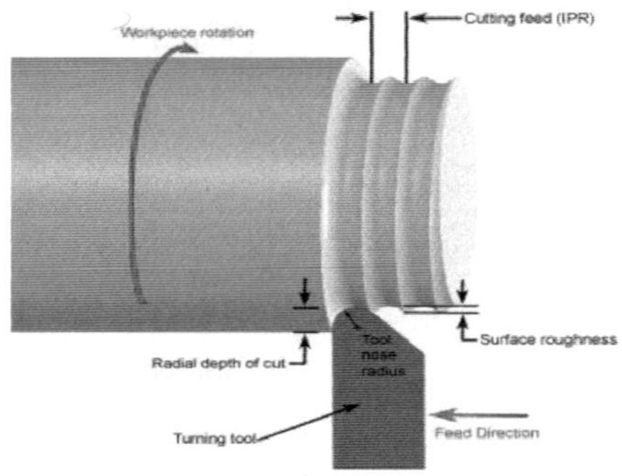

Figure 1. Machined components and the affected parameters during machining process. [4]

The theoretical values of R_a can be calculated using the general exponential modelling Equations (1) and (2) as follows [5]:

$$R_a = 1.22 \times 10^5 M f^{1.004} V_C^{-1.252} \quad (1)$$

$$M = r^{-0.714}(BHN)^{-0.323} \quad (2)$$

Where, R_a = average surface roughness (μm)
V_c = cutting speed (m/min)
f = feed rate (m/min)
r = tool nose radius (m)
BHN = hardness of material

For the turning process, cutting speed (V_c) is related to workpiece diameter D and the revolution rate N [6]. This relation is given by Equation (3) as follows:

$$V_c = \pi D N \quad (3)$$

Previous studies that related to this matter have been carried out over the years. Rajesh et. al., [7] studied the effect of machining parameters such as cutting speed, feed rate and depth of cut (DOC) on Al alloy composite. Experiment was conducted using CNC Turning machine with carbide and polycrystalline diamond (PCD) cutting tools. It can be concluded that the surface roughness of Al alloy is less than Al alloy composite and the optimum machining parameters of cutting speed, feed rate and DOC are 180 to 220 m/min, 0.1 to 0.3 mm/rev and 0.5 to 1.5 mm respectively. A proper combination of cutting speed, feed rate and DOC is very important to achieve good machinability because in conventional machining process, these are the most influential machining parameters [8].

Nexhat Qehaja et. al., [9] conducted a research on the effect of machining parameters and machining time on the surface roughness of cold rolled steel using production lathe machine. Statistical method revealed that the feed rate has the most influence on the surface roughness, followed by nose radius and cutting time. Study on the effect of cutting speed, feed rate, DOC and types of coolant on the surface roughness of stainless steel has been carried out by Sutar and Gujar using CNC Milling machine [10]. Among these parameters, it was found that the most significant parameter is type of coolant, followed by cutting speed, feed rate and DOC.

Dinesh et. al., [11] conducted a study to observe the effect of machining parameters on surface roughness for titanium alloy using 3-axis CNC Milling machine. The result shows that the surface roughness increased with increasing of feed rate and depth of cut but decreased with increasing of cutting speed under wet condition. In addition to the feed rate and DOC, Catherine et. al., [12] have studied the effect of spindle speed, step over and plunge rate on the surface roughness of machined polyurethane (PU) block. Step over is how far between the previous and the next cut is, whereas plunge rate is the speed at which the tool moves in the *z* direction [13]. The study has proved that the step over and silicon content are the most significant parameters that contribute to the surface quality of PU block.

Numerous studies on the effect of machining parameters to the surface roughness has been conducted but the equipment and the material used are different. Also, there are rarely studies that compare the theoretical and experimental roughness values. Thus, there are no discussion behind the percentage difference values. In this study, the effect of machining parameters such as cutting speed and feed rate to the surface roughness of aluminium, brass and mild steel materials will be investigated. Such study on these materials has only been conducted by Suker et. al., [14] but there is no detail discussion on the theoretical and experimental results. In this study, each sample was prepared using Traditional Lathe machine while turning process was done using CNC Lathe machine. Roughness values of each sample is determined through Sufcom-130A surface and roughness tester machine. Both results in terms of theoretical and experimental will be compared and discussed in this study.

2. Methodology

2.1 Workpiece preparation

Each workpiece was set up on the Traditional Lathe machine, as shown in Figure 2. Firstly, the facing process is carried out where the metal was removed from the end of each workpiece with spindle rotation of 540 rpm until flat surface is obtained [15]. After drilling the centre point, the facing process was continue on the other end of each workpiece. Then, the hardness of each workpiece was determined using Rockwell hardness tester based on HRB scale. By referring to the hardness conversion chart, HRB was converted to BHN value and the results acquired were compared with the mechanical properties of hardness (HRB) to obtain the specific classification of materials.

2.2 Turning process

For the turning process, the cutting speed, feed rate, cutting and machine tools are firstly determined. Before setting up the workpiece on the CNC Lathe machine, each workpiece was divided into five sections. Figure 2 shows the CNC Lathe machine used in this study. Turning process was performed with five different values of feed rate which is 0.12 mm/rev, 0.16 mm/rev, 0.2 mm/rev, 0.24 mm/rev and 0.28 mm/rev. The turning process was repeated with different cutting speed value of 75.40 m/min, 113.10 m/min, 150.81 m/min, 188.50 m/min, and 226.19 m/min.

2.3 Determining the roughness value of workpiece

The roughness values of each workpiece for all the five sections were determined by setting up the workpiece on the Surfcom-130A surface and roughness tester with the maximum level of R_a is 400 μm. The experimental values of surface roughness obtained were then compared with the theoretical values form the calculation.

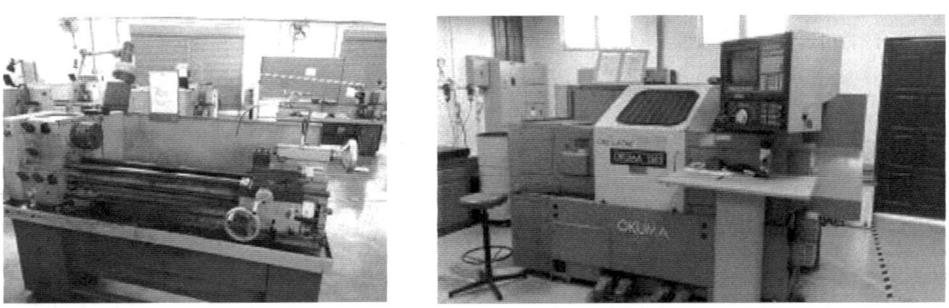

Figure 2. (a) Traditional Lathe machine (left) and (b) CNC Lathe machine (right)

3. Results and discussion

3.1 Feed rate

Theoretical value calculations (this formula applies to all types of workpiece) are shown as follows:

Spindle rotation = 1200 rpm

$V_c = \pi DN = \pi \times (\frac{24}{1000}) \times 1200 = 90.48$ m/min

By combining both equation (1) and (2),

$R_{a\,(Theoretical)} = 1.22 \times 10^5 \times r^{-0.714} \times (BHN)^{-0.323} \times f^{1.004} \times V_c^{-1.252}$ µm

Percentage of error = $\frac{\text{(experimental−theoretical)}}{\text{thereotical}}$ x 100 %

Table 1 shows the effect of feed rate on the surface roughness for aluminium material. The feed rate used in this study are 0.12 mm/rev, 0.16 mm/rev, 0.20 mm/rev, 0.24 mm/rev and 0.28 mm/rev. This feed rate values applies to the mild steel and brass materials as well. The cutting speed is fixed at 90.48 m/min. The highest percentage differences between theoretical and experimental value of surface roughness for the aluminium is 59.11 % at 0.24 mm/rev feed rate.

Workpiece A1 – Aluminium (A16061)

Table 1. Effect of feed rate on surface roughness for Aluminium (A16061).

Diameter of workpiece (mm)				24.00		
Cutting speed, V_c (m/min)				90.48		
Spindle rotation, N (rpm)				1200		
Tool nose radius, r (mm)				0.80		
Workpiece Brinell hardness value, BHN				143		
Depth of cut, a (mm)				0.50		
Feed rate, f (mm/rev)		0.12	0.16	0.20	0.24	0.28
Theoretical roughness value, R_a (µm)		1.642	2.191	2.742	3.292	3.843
Experimental roughness value, R_a (µm)	Reading 1	0.738	0.991	1.242	1.435	2.342
	Reading 2	0.801	0.865	1.258	1.107	1.853
	Reading 3	0.749	0.926	1.134	1.496	2.048
	Average	0.763	0.927	1.211	1.346	2.081
Percentage of error (%)		53.53	57.69	55.84	**59.11**	45.85

Figure 3 shows the graph of surface roughness against feed rate for aluminium material. The differences between theoretical and experimental values can be clearly seen from this figure.

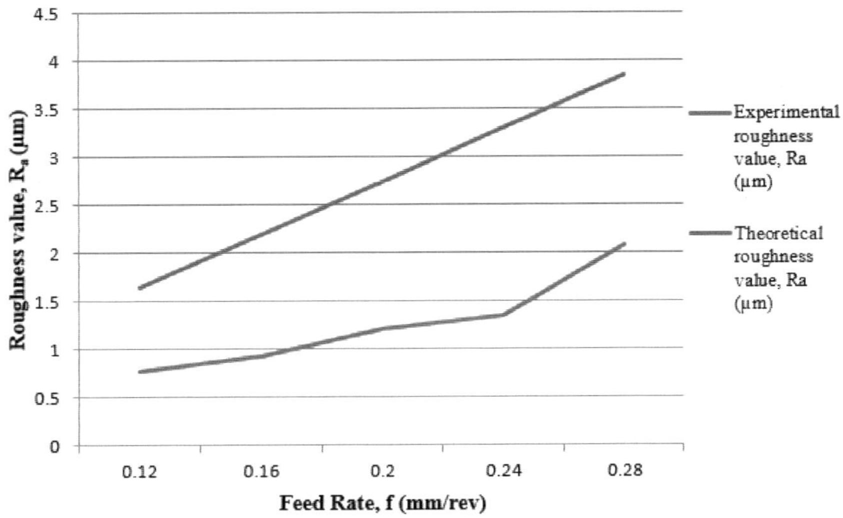

Figure 3. Graph of surface roughness against feed rate for Aluminium (Al6061).

Table 2 shows the effect of feed rate on the surface roughness for mild steel material. The highest percentage differences between theoretical and experimental value of surface roughness is 30.03 % at feed rate 0.28 mm/rev.

Workpiece B1 – Mild steel

Table 2. Effect of feed rate on surface roughness for Mild steel.

Diameter of workpiece (mm)				24		
Cutting speed, V_c (m/min)				90.48		
Spindle rotation, N (rpm)				1200		
Tool nose radius, r (mm)				0.8		
Workpiece Brinell hardness value, BHN				135		
Depth of cut, a (mm)				0.5		
Feed rate, f (mm/rev)		0.12	0.16	0.20	0.24	0.28
Theoretical roughness value, R_a (μm)		1.728	2.232	2.880	3.354	3.916
Experimental roughness value, R_a (μm)	Reading 1	1.706	1.590	2.229	3.168	2.586
	Reading 2	2.156	2.376	2.583	2.886	2.881
	Reading 3	1.859	2.191	3.000	2.751	2.754
	Average	1.907	2.052	2.604	2.935	2.740
Percentage of error (%)		10.36	8.06	9.58	12.50	**30.03**

Figure 4 shows the graph of surface roughness against feed rate for mild steel material.

IOP Conf. Series: Materials Science and Engineering **530** (2019) 012008 doi:10.1088/1757-899X/530/1/012008

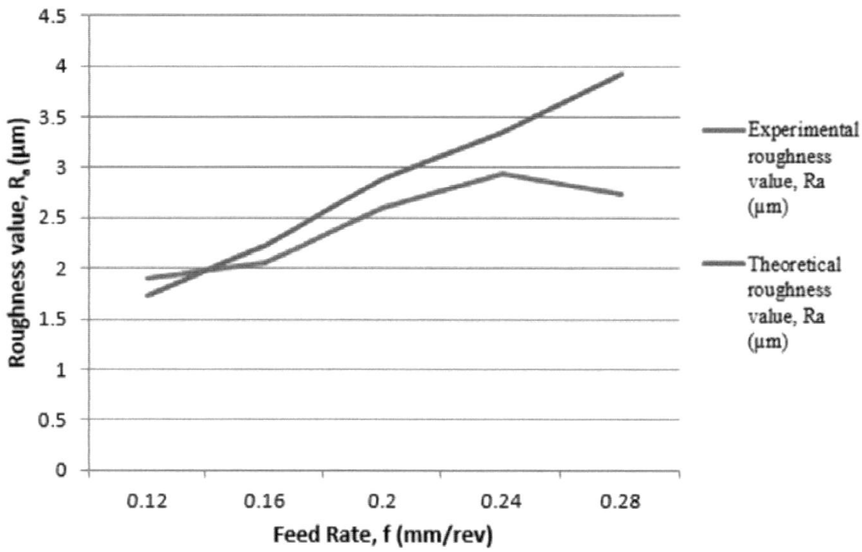

Figure 4. Graph of surface roughness against feed rate for Mild Steel (low carbon steel).

Table 3 shows the effect of feed rate on the surface roughness for brass material. The highest percentage differences between theoretical and experimental value of surface roughness is 51 % at 0.12 mm/rev feed rate.

Workpiece C1 – Brass (C2700)

Table 3. Effect of feed rate on surface roughness for Brass (C2700).

Diameter of workpiece (mm)				24		
Cutting speed, V_c (m/min)				90.48		
Spindle rotation, N (rpm)				1200		
Tool nose radius, r (mm)				0.8		
Workpiece Brinell hardness value, BHN				156		
Depth of cut, a (mm)				0.5		
Feed rate, f (mm/rev)		0.12	0.16	0.20	0.24	0.28
Theoretical roughness value, R_a (μm)		1.596	2.131	2.666	3.201	3.737
Experimental roughness value, R_a (μm)	Reading 1	2.439	2.381	2.113	2.684	3.494
	Reading 2	2.489	2.618	2.143	2.621	3.549
	Reading 3	2.303	2.541	2.266	2.548	3.270
	Average	2.410	2.513	2.174	2.618	3.438
Percentage of error (%)		**51.00**	17.93	18.45	18.21	8.00

Figure 5 shows the graph of surface roughness against feed rate for brass material.

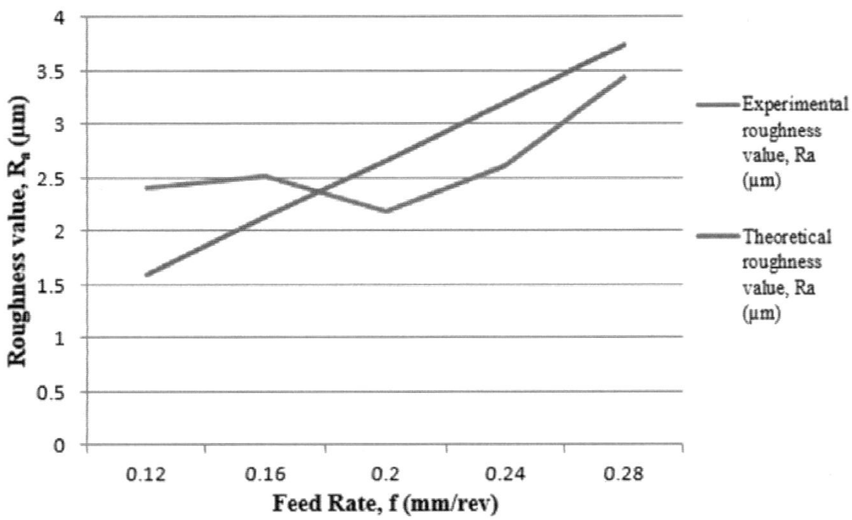

Figure 5. Graph of surface roughness against feed rate for Brass (C2700).

3.2 Cutting speed

Table 4 shows the effect of cutting speed on the surface roughness for aluminium material. For this case, the feed rate is fixed at 0.20 mm/rev. Cutting speed used in this experiment is 75.4 m/min, 113.1 m/min, 150.81 m/min, 188.5 m/min and 226.19 m/min, which applies to all materials. The highest percentage differences between theoretical and experimental value of surface roughness is 82.82 % at 188.5 m/min cutting speed.

Theoretical value calculations (this formula applies to each workpiece) are shown as follows:

Spindle rotation = 1000 rpm

$Vc = \pi DN = \pi \times (\frac{24}{1000}) \times 1000 = 75.40$ m/min

Workpiece A2 – Aluminium (A16061)

Table 4. Effect of cutting speed on surface roughness for Aluminium (Al6061).

Workpiece Brinell hardness value, BHN				143		
Feed rate, f (mm/rev)				0.20		
Spindle rotation, N (rpm)		1000	1500	2000	2500	3000
Cutting Speed, V_c (m/min)		75.40	113.10	150.81	188.50	226.19
Theoretical roughness value, R_a (µm)		3.445	2.073	1.446	1.094	0.871
Experimental roughness value, R_a (µm)	Reading 1	3.572	1.823	1.707	2.184	1.031
	Reading 2	2.765	2.085	1.271	1.476	1.281
	Reading 3	2.809	2.198	2.142	2.342	1.068
	Average	3.049	2.035	1.707	2.000	1.127
Percentage of error, %		11.49	1.83	18.05	**82.82**	29.39

Figure 6 shows the graph of surface roughness against cutting speed for aluminium material.

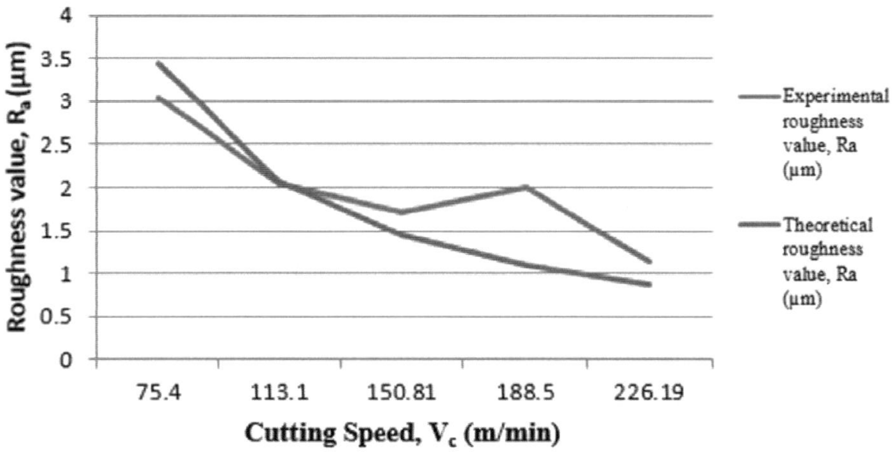

Figure 6. Graph of surface roughness against cutting speed for Aluminium (Al6061).

Table 5 shows the effect of cutting speed on the surface roughness for mild steel material. The highest percentage differences between theoretical and experimental value of surface roughness is 90.08 % at 226.19 m/min cutting speed

Workpiece B2 – Mild steel

Table 5. Effect of cutting speed on surface roughness for Mild steel (low carbon steel).

Workpiece Brinell hardness value, BHN				135		
Feed rate, f (mm/rev)				0.20		
Spindle rotation, N (rpm)		1000	1500	2000	2500	3000
Cutting Speed, V_c (m/min)		75.40	113.10	150.81	188.50	226.19
Theoretical roughness value, R_a (µm)		3.509	2.112	1.473	1.114	0.887
Experimental roughness value, R_a (µm)	Reading 1	4.598	2.576	2.646	1.669	1.583
	Reading 2	4.103	2.283	2.337	2.109	1.767
	Reading 3	4.610	2.673	2.762	1.626	1.708
	Average	4.437	2.511	2.582	1.801	1.686
Percentage of error, %		26.45	18.89	75.29	61.67	**90.08**

Figure 7 shows the graph of surface roughness against cutting speed for mild steel material.

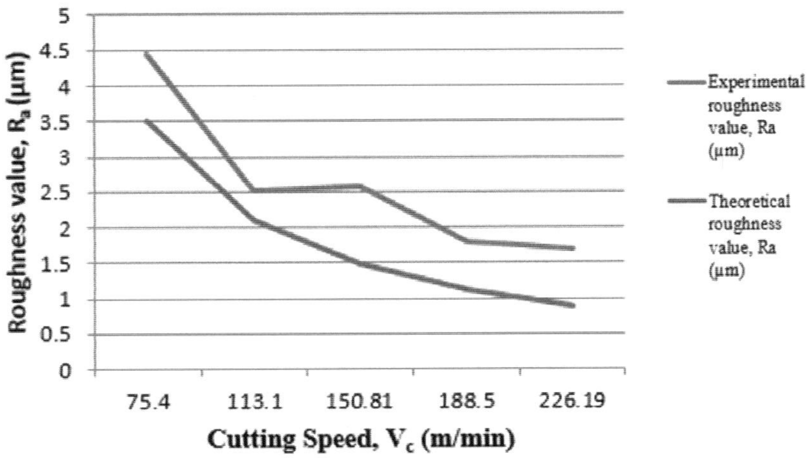

Figure 7. Graph of surface roughness against cutting speed for mild steel

Table 6 shows the effect of cutting speed on the surface roughness for brass material. The highest percentage differences between theoretical and experimental value of surface roughness is 113.71 % at 226.19 m/min cutting speed.

Workpiece C2 - Brass

Table 6. Effect of cutting speed on surface roughness for Brass (C2700).

Workpiece Brinell hardness value, BHN				156		
Feed rate, f (mm/rev)				0.20		
Spindle rotation, N (rpm)		1000	1500	2000	2500	3000
Cutting Speed, V_c (m/min)		75.40	113.10	150.81	188.50	226.19
Theoretical roughness value, R_a (µm)		3.349	2.016	1.406	1.063	0.846
Experimental roughness value, R_a (µm)	Reading 1	2.527	1.742	2.053	1.561	1.731
	Reading 2	2.684	2.057	2.358	1.957	1.877
	Reading 3	2.009	2.318	1.885	1.589	1.815
	Average	2.407	2.039	2.099	1.702	1.808
Percentage of error, %		28.13	1.14	46.29	60.11	113.71

Figure 8 shows the graph of surface roughness against cutting speed for brass material.

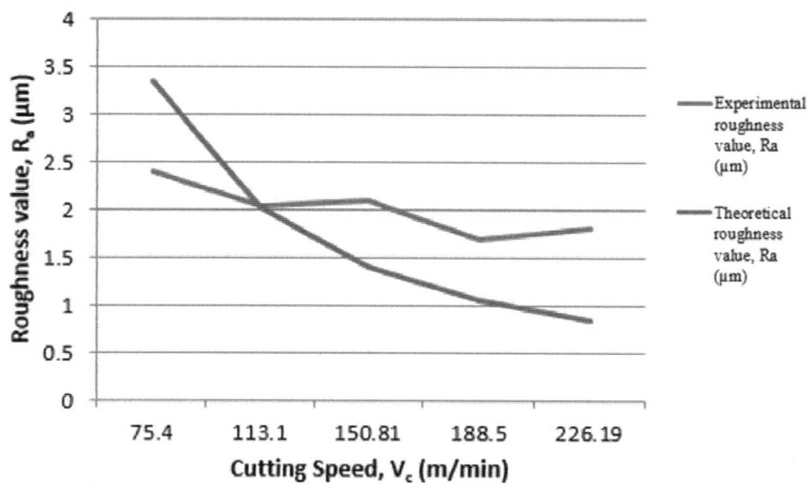

Figure 8. Graph of surface roughness against cutting speed for Brass (C2700).

Based on graphs of surface roughness against feed rate (Figure 3, 4 and 5), the slopes are in positive gradient. This shows that, as the feed rate increases, the surface roughness will also increase which is similar to the trend shows in [16]. This is due to the increased in heat generated that leads to the tool wear. High feed rate also causes the chatter to increase as discussed in [17]. However, the effect of feed rate on chatter is not as critical as spindle rotational speed [18]. In addition, the tool life will decrease, and tool will become dull as the feed rate increases. This can cause the tip of the tool to become larger and rub over the surfaces of workpiece which subsequently inducing the residual stresses, and eventually may cause surface wearing and tearing [19]. In industry, high productivity is always a priority and thus feed rate has to be maximized as much as possible because a maximum material removal rate is desired [20]. This has to be done while maintaining acceptable surface roughness value.

Based on the results in Table 1, 2 and 3, there is a significant difference between theoretical and experimental value for the surface roughness. In Table 1, the percentage of error is between 45.85 % to 59.11 %, in Table 2, the percentage of error is between 8.06 % to 30.03 %, whereas in Table 3, the percentage error is between 8.0 % to 51.0 %. Deviation between theoretical and experimental roughness value may cause by the errors occurred during the roughness measuring process such as operator skills, worn or dirty stylus tip and error in calibration.

By comparing the graphs between the three different material used, for the feed rate of 90.48 m/min, brass (2.410 µm) has highest roughness value, then followed by mild steel (1.907 µm) and aluminium (0.763 µm). In theoretical, brass (156 BHN) which has higher hardness value should have higher roughness value than aluminium (143 BHN) and mild steel (135 BHN). However, the results show that mild steel has higher roughness value than aluminium. This happen due to mild steel has lower elastic recovery rate than aluminium. The presence of elastic recovery will decrease the surface roughness [21]. Thus, the result of roughness value for mild steel is higher than aluminium.

From the experimental roughness values obtained, there are deviations from the theoretical values. The deviations for different workpiece machined at different feed rate ranges from 8.0 % to 59.11 %. This may cause by backlash as the machine uses gear working principle. Backlash may occur during the switching of cutting speed, feed rate and movement of the working table that contributes to the inconsistency of feed rate. The feed rate should be a constant variable, thus the fluctuating feed rate will cause deviation from the theoretical values. Besides, it is no cutting fluid is used while cutting the workpiece. When the workpiece is being cut, the friction between the tool and workpiece increased and increased the temperature of the cutting zone and workpiece. This will increase the thermal distortion of the workpiece and generally increased the surface roughness of the workpiece. On the other hand, it is always variation in z value of the measuring roughness machine. The tester is difficult to be adjusted to zero value due to sensitivity of the measuring machine. It may also due to the presence of dust. Since the measurement of the surface roughness is in micro meter, the presence of dust might affect the measurement of roughness.

Based on the graphs of surface roughness against cutting speed in Figure 6, 7 and 8, the slopes are in negative gradient. This shows that, as the cutting speed increase, the surface roughness will be decreased. This attribute is due to the decrease in the built-up edge size. This built-up edge will generate burr on the machined surface and deteriorating of surface finishing. Therefore, by increasing the cutting speed, the influence of the built-up edge became negligible. Besides, as the cutting speed increase, the turning process become more stable and the vibration during the process become lower and thus a better surface produced [21-22].

Based on the results in Table 4, 5 and 6, there is a significant different between theoretical and experimental value for the surface roughness. In Table 4, the percentage of error is between 1.83 % and 82.82 %, in Table 5, the percentage of error is between 18.89 % and 90.08 %, whereas in Table 6, the percentage error is between 1.14 % and 113.71 %. The deviation between theoretical and experimental roughness values may cause by errors occurred during the roughness measuring process. Roughness tester is a highly sensitivity equipment, even a small residue chip or dust which remain on the surface will cause an error in result.

By comparing the graphs between the three different materials used, for the cutting speed of 75.40 m/min, mild steel (4.437 μm) has the highest roughness value, then followed by aluminium (3.049 μm) and brass (2.407 μm). Based on theory, mild steel (135 BHN) which has lower hardness value should have lower roughness value then aluminium (143 BHN) and brass (156 BHN). However, the results show that mild steel has higher roughness value than aluminium and brass. This happen because the mild steel has a lower elastic recovery rate than aluminium and brass. The presence of elastic recovery will decrease surface roughness. Besides that, the rust on the surface of the mild steel will also affect the roughness value and cause of highest roughness value among these three materials. In general, during turning process of workpiece with high hardness, the temperature of cutting area will increased. According to this temperature, there is thermal softening in cutting area (between tool edge and surface of workpiece) that made the plastic deformation for chips become easier. The hard particles in workpieces will affect the tool wear and this made the surface roughness increases [23].

4. Conclusion

As a conclusion, the feed rate and cutting speed used in metal machining process such as turning process by Lathe machine will give a large effect to the surface finish especially the surface roughness of the products. From this experiment, we can conclude that, the roughness value will increase when the feed rate increases, while the roughness value will decrease as the cutting speed increases. From the graphs, there are some deviations of roughness value between the experimental results and theoretical results. The roughness values that obtained from this experiment is still not acceptable because there are too much of deviations with the theoretical roughness value. This is due to some factors that may affect the accuracy of the roughness value which includes collision between workpieces in storage, rusting and cleanliness of workpiece surface.

Acknowledgement

The authors would like to acknowledge School of Mechanical Engineering, Universiti Sains Malaysia for providing the facilities and financial assistance for this project.

Reference

[1] Yalcinkaya S and Sahin Y 2017 *13th Int. Conf. on Tribology (Galati, Romania)* Vol 174

[2] Vinod M, Neha K, Keshva N, Karanvir S and Ramagopal V S 2015 Experimental investigation on uncontrollable parameters for surface finish during diamond turning *Materials and Manufacturing Processes* Vol 30

[3] Mohammed T H, Montasser S T and Joachim B 2007 A study of the effects of machining parameters on the surface roughness in the end-milling process *Jordan Journal of Mechanical and Industrial Engineering* Vol 1 pp 1-5

[4] Turning Surface Roughness Calculator. Retrieved 15 August, 2018, from http://www.custompartnet.com/calculator/turning-surface-roughness

[5] Lab 6. Retrieved 26 September, 2018, from https://www.scribd.com/document/329537069/lab-6#download

[6] Cutting Data Recommendations. Retrieved 15 August, 2018, from https://www.uddeholm.com/files/Cutting_Data_Corrax_eng.pdf

[7] Rajesh K B, Sudhir K and Das S 2010 Effect of machining parameters on surface roughness and tool wear for 7075 Al alloy SiC composite *Int. Journal of Advanced Manufacturing Technology* (London: Springer-Verlag) vol 50 pp 459-469

[8] Bapi S, Moola M R and Sujan D 2017 *MATEC Web of Conf.* vol 95

[9] Nexhat Q, Kaltrine J, Avdyl B, Mirlind B and Hysni O 2015 Effect of machining parameters and machining time on surface roughness in dry turning process *Procedia Engineering* 100 pp135-140

[10] Pooja A S and Gujar A J 2017 Study the effect of machining parameters on surface roughness in CNC milling of AISI 316L *Int. Journal of Engineering Research and Technology* (In Research Publication House) vol 10

[11] Dinesh K C, Kapil K C and Ganga S S 2015 Effect of machining parameters on surface roughness for titanium alloy *Int. Journal of Engineering and Innovative Technology* vol 4

[12] Catherine L D K, Raja A M, Azrina A and Sangeeth S 2014 Impact of machining parameters On the surface roughness of machined PU block *Int. Journal of Chemical, Molecular Nuclear,Materials and Metallurgical Engineering* (World Academy of Science, Engineering and Technology) vol 8

[13] Speeds and Feeds. Retrieved 26 September, 2018, from https://support.bantamtools.com/hc/en-us/articles/115001658374-Speeds-and-Feeds

[14] Suker D S, Alsoufi M S, Alhusaini M M and Azam S A 2016 Studying the effect of cutting conditions in turning process on surface roughness for different materials *World Journal of Research and Review* 2 pp 16-21

[15] Frank J H Jr 2000 Home facing process. Retrieved 15 August, 2018, from http://www.mini-lathe.com/Mini_lathe/Operation/Facing/facing.htm

[16] Satheesh N K, Ajay S, Ashay S, Ananth K and Harsha S 2012 Effect of spindle speed and feed on surface roughness of carbons steel in CNC turning *Procedia Engineering* 38 pp 691-7

[17] Nurhaniza M, Ariffin M K M A, Mustapha F and Baharudin B T H T 2016 Analyzing the Effect of machining parameters setting to the surface roughness during end milling of CFRP-aluminium composite laminates *Int. Journal of Manufacturing Engineering* 2016

[18] Ten Questions About Chatter. Retrieved 26 September, 2018, from http://www.custompartnet.com/calculator/turning-surface-roughness

[19] Serope K and Steven S 2014 *Manufacturing Engineering & Technology* (London: Pearson) p 996

[20] Kuttolamadom M A, Hamzehlouia S and Mears M L 2010 Effect of machining feed on surface roughness in cutting 6061 aluminium *SAE Int. Journal of Materials and Manufacturing* 3

[21] Abdullah A B, Chia L Y and Samad Z 2008 The effect of feed rate and cutting speed to surface roughness *Asian Journal of Scientific Research* 1 pp 12-21

[22] Khalid A D 2016 Effect of cutting parameters on surface roughness in turning operations *Al-Qadisiyah Journal for Engineering Sciences* 9

[23] Ehsan D D and Ali A A 2009 *Proc. of the World Congress on Engineering* (London) Vol 2

IOP Conf. Series: Materials Science and Engineering **530** (2019) 012009 doi:10.1088/1757-899X/530/1/012009

Study on Effect of Tool Nose Radius Wear on Hybrid Roughness Parameters during Turning Using Vision-Based Approach

B M Kumar and M M Ratnam*

School of Mechanical Engineering, Engineering Campus, Universiti Sains Malaysia, 14300 Nibong Tebal, Penang, Malaysia.

Corresponding author *: mmaran@usm.my

Abstract. In the past, most researchers have investigated the effect of tool flank wear on the average roughness parameter (R_a) during turning. The R_a parameter, however, is sensitive only to the height variations of the machined workiece profile and is insensitive to the spatial changes to the profile caused by gradual increase in tool wear. In this work, a non-contact vision-based method was used to investigate the effect of tool nose radius wear on the hybrid surface roughness parameters during finish turning. The roughness profile of the workpiece surface diametrically opposite the cutting tool was captured during turning of AISI 1035 steel using a digital camera. A high shutter speed (1/4000 s) was used to freeze the motion of the rotating spindle and obtain a blur-free image of the workpiece. The edge of the workpiece profile was extracted to sub-pixel accuracy using the grey level invariant moment method. The hybrid surface roughness parameters were determined from the workpiece profile. The effect on increased nose wear area on these parameters were investigated. Among these roughness parameters the mean wavelength ($R_{\lambda a}$) parameter showed the best correlation with tool nose radius wear with coefficients of determination of 0.9734.

1.0 Introduction

Finish turning is used to produce precision parts whose quality is assessed mainly in terms of the surface roughness. The surface roughness of a finished product is affected by numerous machining factors among which the most critical is the wear of the cutting tool. Three types of wear are dominant during turning, namely flank wear, crater wear and nose wear. Notch wear, which occurs occasionally during the turning of hard materials, is a type of groove wear that appears at the flank where the edge of the cutting tool meets the workpiece. The flank of a cutting tool is divided into the major and minor flanks. The major flank area is involved in bulk material removal while the minor flank plays a greater role in shaping the surface profile of the workpiece. This is due to the direct contact of the minor flank with the freshly cut surface.

In finish turning, where the depth-of-cut is less than the tool nose radius, it is the nose wear that predominantly affects the surface roughness of the finished workpiece. Naiki and Mears [1] reported that tool flank wear does not necessary have a significant effect on the surface roughness of the workpiece. Cantero et al. [2] found that although the flank wear increased significantly during the turning of Inconel 718 the surface roughness R_a of the workpiece only showed a small variation. Kwon et al. [3] concluded that for small depths of cut, such as in finish turning, the flank wear provides little information on the tool wear condition because most of the wear occurs in the nose area of the tool. These findings suggest that the flank wear may not be the right wear parameter to measure when assessing the surface finish quality of the workpiece. Moreover, R_a may not be the right roughness parameter to correlate with tool wear since the amplitude variation in the workpiece surface profile may be too small to be detected. This is evident also from the work of Shahabi and Ratnam [4] where the authors found that at low feed rates (0.2 to 0.3 mm/rev) the value of R_a did not exhibit a significant variation.

Extensive research has been carried out in the past to monitor flank wear from the measured sensor signals but not much attention has been given to the tool nose wear [5-11]. In all these papers, the authors measured the roughness by using a stylus roughness tester, a 3-D optical profiler or a vision system and correlated the roughness parameters with tool flank wear. Coker and Shin [12] used an ultrasonic sensor to detect changes in surface roughness caused by tool wear. A good correlation between the mean sensor signal and average roughness (R_a) measured by a stylus roughness tester was obtained. However, the authors found that there was no straightforward linear correlation between the average roughness R_a and tool wear. Several in-process roughness measurement methods have also been developed in the past. These include light scattering methods [13], ultrasonic sensing [14] and image processing method [15]. However, none of these methods has been applied to monitor the tool nose radius wear from the measured signal.

In a recent work [16] we developed an in-process surface roughness measurement method using machine vision. A digital single-lens-reflex (DSLR) camera was used to capture the image of the workpiece profile diametrically opposite the cutting tool during dry turning. The edge of the workpiece was located using sub-pixel edge location method. From the workpiece surface profile numerous amplitude, spacing, hybrid as well as functional roughness parameters as defined in the ISO 3685 (1993) standard were determined. In addition to the existing parameters three new parameters that are unique to the turning process were introduced. However, the correlation between the tool nose wear and the various roughness parameters was not investigated.

In this work, the previously developed in-process roughness measurement method was applied to investigate the effect of tool nose radius wear on the hybrid roughness parameters during finish turning. The tool nose radius wear occurs in the curved part of the tool and is related to the minor flank wear V_C defined in the ISO 3685 (1993) standard. The nose wear can be determined by digitally subtracting the top (2D) views of the unworn and worn cutting tool and related to V_C by the tool geometry and the cutting angle. In this work, the hybrid roughness parameters were extracted while the cutting was still in progress.

2.0 Methodology

2.1 Hardware setup

The experimental setup is shown schematically in Figure 1 and the actual setup is shown in Figure 2. The lathe used is a conventional lathe machine (*Pinacho S90/200*, Spain). A compressed air hose was provided to blow away chips and cutting fluid during the wet turning. A 10.2 megapixel DSLR camera (*Sony α-230*) mounted diametrically opposite the cutting tool was used to capture the profile of the workpiece during turning. The camera was triggered using a remote controller and the image was transferred to the personal computer via the universal serial bus (USB) cable. A CCD camera (*JAI CV-A1*, Japan) with sensor resolution 1296×1024 pixels was used to capture the images of the cutting tool after each pass. The tool images were used to determine the tool nose wear area to be correlated with the roughness parameters.

Figure 3 shows a sample image captured using the setup during the wet turning and the zoomed-in view of the region-of-interest (ROI) for roughness measurement. The spindle speed used is 420 rpm. Distortions in the images were checked using a multi-frequency distortion target grid as detailed out in Ref. [16]. The scaling factors used for converting image coordinates in pixels to real-world coordinates in millimetres were obtained using a 1.35 mm diameter *Mitutoyo* pin gage. The horizontal and vertical scaling factors for the DSLR camera are 8.44 μm/pixel and 8.43 μm/pixel, respectively, while the corresponding scaling factors for the CCD camera are 1.87 μm/pixel and 2.12 μm/pixel, respectively.

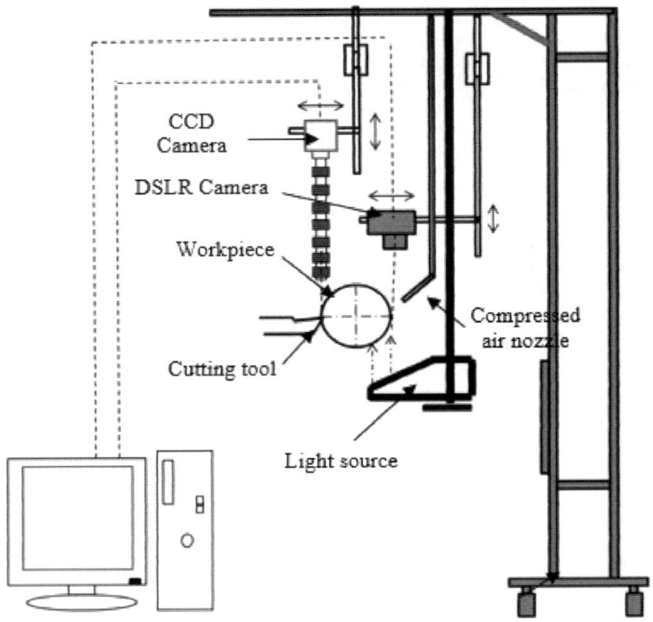

Figure 1. Schematic illustration of in-process tool condition monitoring system.

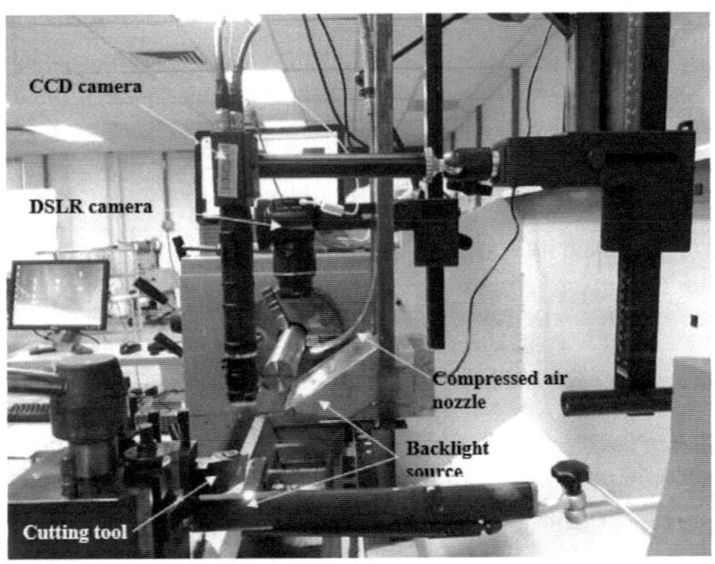

Figure 2. Actual setup of in-process tool wear monitoring system.

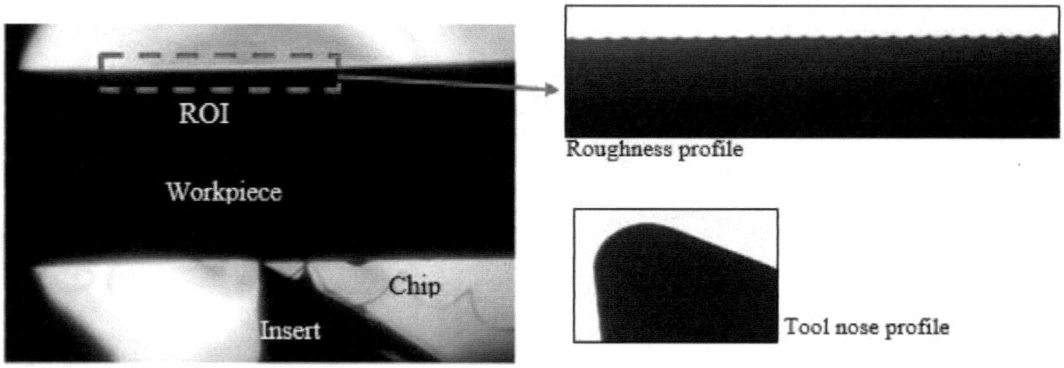

Figure 3. Region of Interest (ROI) captured by DSLR camera (inset image shows the tool nose profile captured by the CCD camera – not to scale).

2.2. Machining condition

Wet turning experiments were carried out on cylindrical bars of AISI 1035 steel of diameter 50 mm and length 250 mm using coated triangular carbide inserts (TNMG 160404R-ST) having nose radius of 0.4 mm. The lead angle is 0°. The feeds and depths-of-cut used were relatively small and are of the same order of magnitude as that of tool edge geometry. Thus, the cutting is confined to a small area on the nose radius of cutting tool and no premature tool insert breakage occurred during the turning tests. The positive cutting edge inclination disposes chips in the cutting direction and thus the chips do not interfere during imaging of the edge of the workpiece. It was found that the coolant initially fouled the Region-of-Interest (ROI) during wet turning. The jet of compressed air supply from nozzle cleared the field-of-view effectively thus producing a good quality image of the workpiece surface profile as seen in Figure 3.

2.3 Acquisition of workpiece surface image using DSLR camera

The DSLR camera was set to manual mode and the focusing ring was adjusted to obtain a sharp focus of the edge of the workpiece opposite the cutting side. The shutter speed of the camera was set to the maximum value of 1/4000 s and the backlight intensity was adjusted to obtain a good contrast image of the workpiece profile. At this high shutter speed blur-free images of the workpiece surface can be captured during the turning. Although the spindle speed used in the turning experiments was only 420 rpm relatively blur-free images can be captured at spindle speeds of up to 4000 rpm using the DSLR camera as demonstrated in our previous work [16]. Vibration of the workpiece did not introduce significant blurring in the image. A sequence of ten images of the rotating workpiece surface was captured remotely during a time interval of 2 s.

2.4. Acquisition of image of cutting tool insert using CCD camera

The contour of the cutting tool nose edge was captured with the aid of backlighting using the CCD camera in-between the cutting cycles without removing the cutting tool from the machine. The images were analysed and the progressive nose wear was calculated using the developed algorithm coded in MATLAB. The algorithm enables automatic correction of tool misalignment.

2.5. Sub-pixel edge location and roughness measurement

The step-edge model applying the concept of invariant moments for sub-pixel edge location proposed by Tabatabai and Mitchell [17] was applied to detect the edge of the workpiece image to sub-pixel accuracy. The details of the sub-pixel edge location method is explained in Ref. [16]. The grey level invariant moment method of sub-pixel edge detection enables accurate determination of the sub-pixel edge of the workpiece profile (Figure 4). The surface contour image was initially read into the MATLAB workspace and the edge image of workpiece was cropped automatically to the required evaluation length for roughness measurement. The threshold independent method of finding the edges with sub-pixel accuracy was then applied. The algorithm starts to scan the surface profile from the first pixel in the first column to find the sub-pixel profile. The starting point of the scan line is taken to be at the upper-left corner point O. The interval ΔD along each pair of neighbouring scan lines $L(i)$ and $L(i+1)$ is one pixel. The variance between the maximum grey level and the minimum grey level data of the scan line $L(i)$ serves as a major characteristic in determining the sub-pixel roughness profile $k(i)$. Since the detected roughness is in pixels the scaling factors were used to convert the roughness to micrometres. Next, the best-fit line of the detected contour data, considered as a mean line, was determined using least-squares fitting. The distance of each pixel on the profile measured from the mean line was used in calculating the roughness parameters defined in Table 1. Detailed definition of the roughness parameters can be found elsewhere [18]. The roughness values obtained using the vision method at different intervals of machining time is shown in Table 2. Figures 5(a)-(d) show the images of workpiece profiles captured during the wet turning for different machining time and the corresponding roughness plots generated using the machine vision method. The deterioration in the surface quality of the workpiece after 30 minutes of machining can be seen from the roughness profiles.

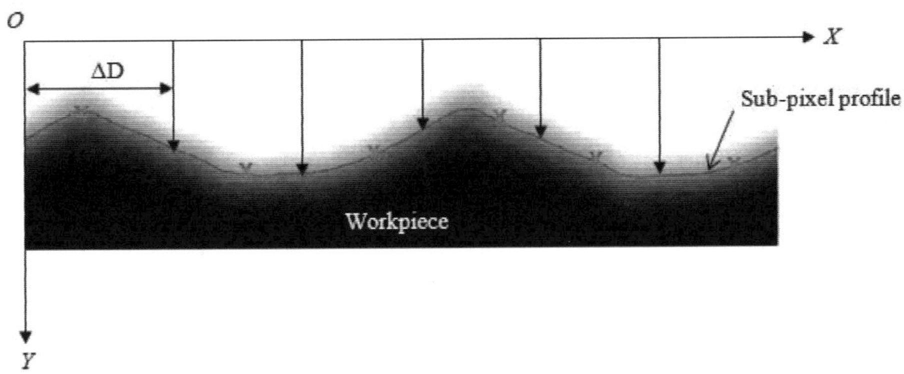

Figure 4. Orthogonal scanning of surface profile with sub-pixel edge location.

Table 1. Specifications of roughness parameters.

Category	Description	Symbol
Hybrid parameters	Mean slope of the profile	$R_{\Delta a}$
	Quadratic mean of all local slopes	$R_{\Delta q}$
	Average wavelength	$R_{\lambda a}$
	Root mean square wave length	$R_{\lambda q}$

Table 2. Roughness data obtained using the vision method at different intervals of machining time.

Roughness Parameters	Machining time (min)														
	1.2	12.2	24.4	36.6	48.8	61.0	73.2	85.4	97.6	109.8	134.2	158.6	183.0	207.4	244.0
$R_{\Delta a}$(°)	23.51	22.36	20.07	19.49	19.49	18.92	18.35	18.35	16.05	14.91	14.91	14.33	14.33	13.76	13.19
$R_{\Delta a}$(°)	32.39	29.24	28.67	28.09	26.95	25.23	24.14	23.57	22.26	21.66	21.02	20.38	19.81	19.75	17.20
λ_a (µm)	198.77	201.86	211.36	232.01	241.51	254.27	268.46	276.74	279.73	280.31	284.42	287.91	284.82	296.57	297.67
λ_a (µm)	179.55	264.23	291.69	275.38	260.15	258.36	275.16	274.21	263.14	263.84	258.65	253.22	247.47	251.10	260.31

IOP Conf. Series: Materials Science and Engineering **530** (2019) 012009 doi:10.1088/1757-899X/530/1/012009

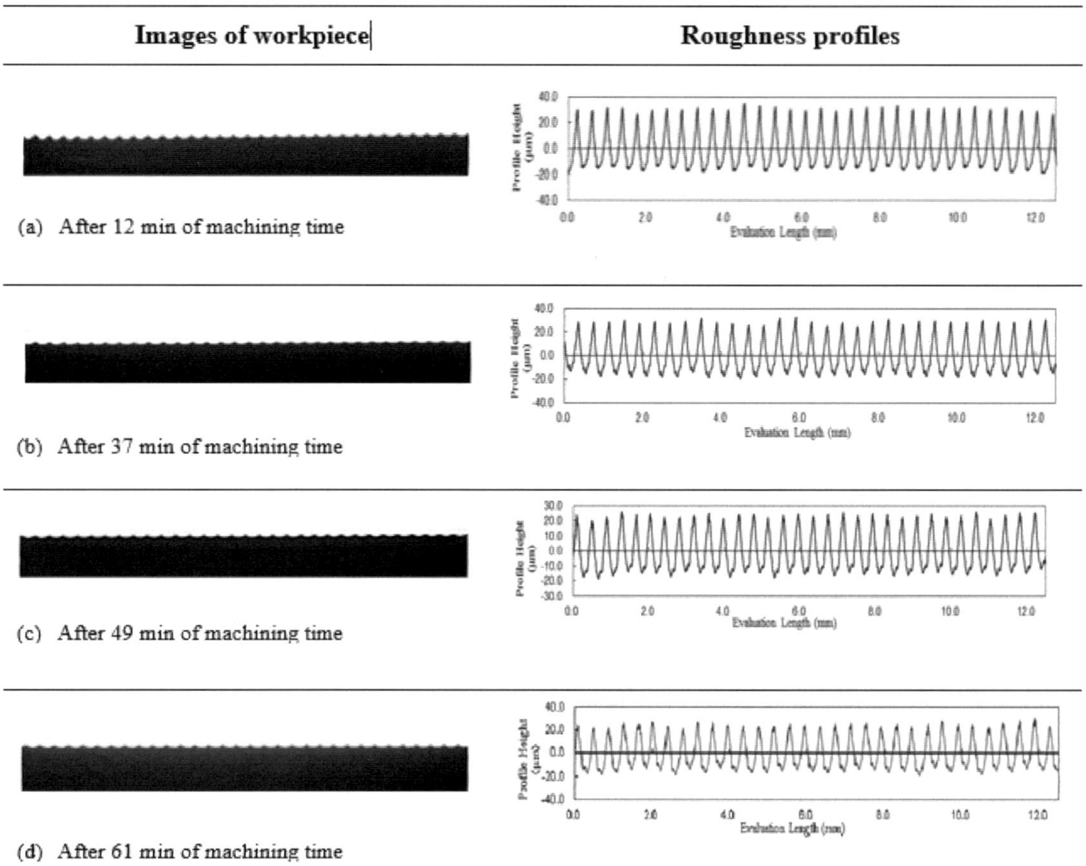

Figure 5. Image of workpiece and roughness profiles at different machining times.

3.0 Results and Discussion

3.1 Verification of roughness parameters determined using the vision method

The hybrid surface roughness data obtained using the wet turning process was verified using a portable stylus surface roughness tester (*Surftest SJ-210*). Table 2 shows the roughness data measured using the vision method. The workpiece surface was also measured using a stylus method ten times at different regions using an evaluation length of 12.5 mm and a cut-off length of 2.5 mm. The averages of these values are tabulated in Table 3. Comparison between the vision and stylus method in measuring the hybrid roughness parameters are shown in Table 4. The maximum deviation for $R_{\Delta a}$, $R_{\Delta q}$, and $R_{\lambda a}$ between vision and stylus method are, respectively 7.69%, 9.68% and 8.74%.

Table 3. Roughness obtained using the stylus method at different intervals of machining time.

Roughness Parameters	Cutting time (min)														
	1.2	12.2	24.4	36.6	48.8	61.0	73.2	85.4	97.6	109.8	134.2	158.6	183.0	207.4	244.0
$R_{\Delta a}$(°)	24.91	23.33	21.62	21.06	20.57	20.19	18.94	19.81	17.39	15.46	15.48	15.30	14.96	14.36	14.28
$R_{\Delta a}$(°)	33.45	30.65	29.58	29.03	28.13	26.71	25.59	24.28	23.74	22.93	23.06	21.58	20.47	21.86	18.31
λ_a (μm)	211.73	215.85	220.70	247.91	251.17	276.88	278.27	299.73	303.64	303.85	308.64	313.51	291.16	316.56	310.49
λ_q (μm)	191.26	270.60	307.73	281.41	281.05	267.79	296.46	287.27	281.95	277.01	262.52	262.45	250.33	259.86	272.43

Table 4. Comparison between hybrid roughness parameters obtained using vision method and stylus method.

Cutting time (min)	$\frac{\lvert R\Delta_{a(s)} - R\Delta_{a(v)}\rvert}{\lvert R\Delta_{a(s)}\rvert}$ (× 100%)	$\frac{\lvert R\Delta_{q(s)} - R\Delta_{q(v)}\rvert}{\lvert R\Delta_{q(s)}\rvert}$ (× 100%)	$\frac{\lvert R\lambda_{a(s)} - R\lambda_{a(v)}\rvert}{\lvert R\lambda_{a(s)}\rvert}$ (× 100%)	$\frac{\lvert R\lambda_{q(s)} - R\lambda_{q(v)}\rvert}{\lvert R\lambda_{q(s)}\rvert}$ (× 100%)
1.2	5.64	3.18	6.12	6.12
12.2	4.15	4.61	6.48	2.36
24.4	7.19	3.10	5.65	5.21
36.6	7.46	3.22	6.41	2.14
48.8	5.26	4.20	3.85	7.44
61.0	6.30	5.55	8.17	3.52
73.2	3.12	5.70	3.53	7.19
85.4	7.41	2.94	7.67	4.54
97.6	7.69	6.25	7.87	6.67
109.8	3.57	5.56	7.75	4.75
134.2	3.70	8.82	7.85	1.48
158.6	6.30	5.55	8.17	3.52
183.0	4.17	3.23	2.18	1.14
207.4	4.17	9.68	6.32	3.37
244.0	7.69	6.06	4.13	4.45
280.6	4.00	3.13	8.74	6.63
Mean diff.	2.09	2.32	2.47	2.24
Std. dev. (σ)	5.49	5.05	6.22	4.41

$R\Delta_{a(s)}$, $R\Delta_{q(s)}$; $R\lambda_{a(s)}$, $R\lambda_{q(s)}$ Roughness measured using stylus method

$R\Delta_{a(v)}$, $R\Delta_{q(v)}$; $R\lambda_{a(v)}$; $R\lambda_{q(v)}$ Roughness measured using vision method

3.2 Assessment of tool nose wear from established roughness parameters

Progressive wear on the tool nose region constitutes a change in the edge geometry at the cutting tool-workpiece interface. Figure 6 shows sample images of cutting tools and the corresponding wear area at different machining times. The cutting tool images were captured in-between the turning passes in order to determine the nose wear area. The nose wear areas were later correlated with the surface roughness parameters determine by the vision method.

Figures 7(a)-(c) show the variation of nose wear area with the hybrid roughness parameters. The decrease in the mean slope $R_{\Delta a}$ and the quadratic mean slope $R_{\Delta q}$ as nose wear increases shows that slopes gradually become less steep as the nose becomes increasing worn. The average wavelength $R_{\lambda a}$, however, shows an increasing trend with increase in nose wear. The average wavelength parameter is a measure of the spacing between local peaks and valleys taking into the account their relative amplitudes and individual spatial frequencies. There appears only a small increase in $R_{\lambda a}$ during the initial stages of machining when the tool is relatively new. Grooves produced by the feed motion of the tool are predominant when the tool is sharp resulting in a small $R_{\lambda a}$ values. As the cutting edge is progressively worn the appearance of the machined surface tends to be smeared and localized peaks and valleys appear. The grooves corresponding to the feed marks become less dominant thus increasing $R_{\lambda a}$. For the mean slope parameter $R_{\Delta a}$ a reasonably good negative exponential fit with R-square value of 0.9588 can be obtain within the range of roughness investigated. On the other hand, a good positive exponential fit with R-squared value of 0.9734 was obtained for the $R_{\lambda a}$ parameter. Although such fits are only approximations they could provide a useful insight into monitoring of the tool nose wear during turning.

IOP Conf. Series: Materials Science and Engineering **530** (2019) 012009 doi:10.1088/1757-899X/530/1/012009

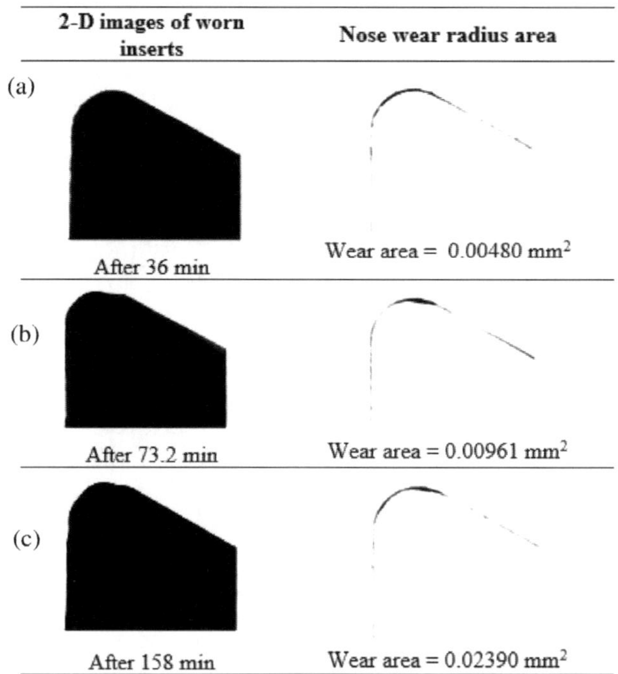

Figure 6. Sample image of cutting tool captured using CCD camera at various machining time and the corresponding nose wear areas: (a) after 36 minutes, (b) after 73.2 minutes and (d) after 158 minutes.

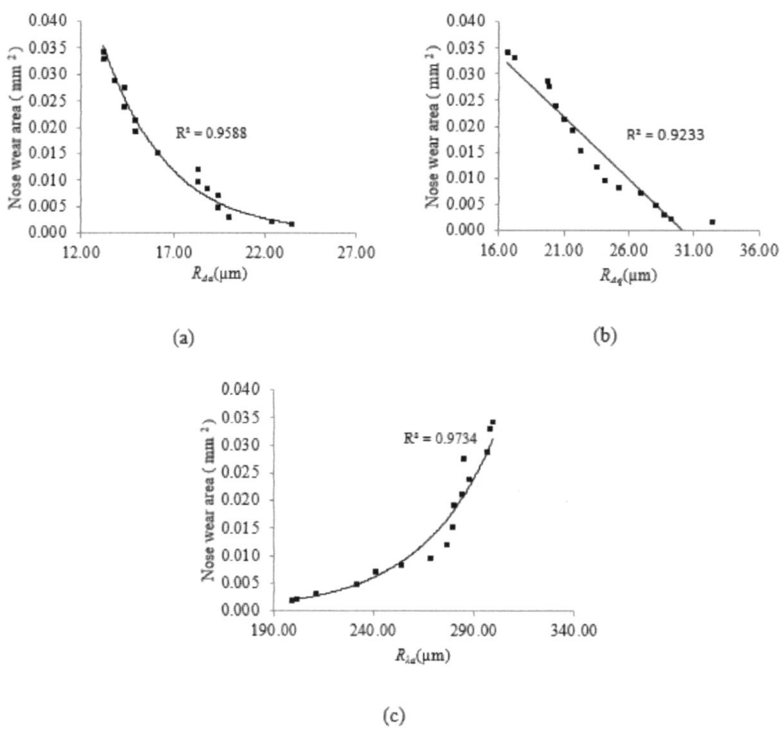

Figure 7. Correlation between tool nose wear with hybrid roughness parameter.

4.0 Conclusion

Among the hybrid parameters investigated the average wavelength ($R_{\lambda a}$) showed the best correlation with the tool nose wear with R-squared values of 0.9734. The average wavelength increased rapidly with increase in nose radius wear due to the smearing effect of the worn tool nose. The mean slope of profile and quadratic mean of all local slopes decreased with increase in nose wear due to the same smearing effect. The fluctuation of the data in all three parameters is suspected to be due to the increased vibration caused by tool wear.

References

[1] Niaki F A and Mears L 2017 A comprehensive study on the effects of tool wear on surface roughness, dimensional integrity and residual stress in turning IN718 hard-to-machine alloy *Journal of Manufacturing Processes* 30 268-280
[2] Cantero J L, Diaz-Alvarez J, Miguelez M H and Marın N C 2013 Analysis of tool wear patterns in finishing turning of Inconel 718 *Wear* 297 885–894
[3] Kwon Y, Ertekin Y and Tseng T-L 2004 Characterization of tool wear measurement with relation to the surface roughness in turning *Machining Science and Technology* 8(1) 39–51
[4] Shahabi H H and Ratnam M M 2009 In-cycle monitoring of tool nose wear and surface roughness of turned parts using machine vision *International Journal of Advanced Manufacturing Technology* 40 1148-1157
[5] Ravindra H V, Srinivasa Y G and Krishnamurthy R 1993 Modelling of tool wear based on cutting forces in turning *Wear* 169(1) 25–32
[6] Mannan M A, Kassim A A and Jing M 2000 Application of image and sound analysis techniques to monitor the condition of cutting tools *Pattern Recognition Letters* 21(11) 969–979
[7] Ozel T and Karpat Y 2005 Predictive modeling of surface roughness and tool wear in hard turning using regression and neural networks *International Journal of Machine Tools and Manufacture* 45(4-5) 467–479
[8] Kassim A A, Mannan M A and Mian Z 2007 Texture analysis methods for tool condition monitoring *Image and Vision Computing* 25(7):1080–1090
[9] Liao Y, Stephenson D A and Ni J 2010 A multifeature approach to tool wear estimation using 3D workpiece surface texture parameters *Journal of Manufacturing Science and Engineering, Transactions of the ASME* 132(6) 0610081–0610087
[10] Venkata Rao K, Murthy BSN and Mohan Rao N 2013 Cutting tool condition monitoring by analyzing surface roughness, work piece vibration and volume of metal removed for AISI 1040 steel in boring *Measurement* 46 4075–4084
[11] Prasad BS, Prasad DS, Sandeep A and Veeraiah G 2013 Condition monitoring of CNC machining using adaptive control *International Journal of Automation and Computing* 10(3) 202–209
[12] Coker SA and Shin YC 1996 In-process control of surface roughness due to tool wear using a new ultrasonic system *International Journal of Machine Tools and Manufacture* 36(3) 411–422
[13] Persson U 1998 In-process measurement of surface roughness using light scattering *Wear* 215(1) 54–58
[14] Shin Y C, Oh S J and Coker S A 1995 Surface roughness measurement by ultrasonic sensing for in-process monitoring *Journal of Manufacturing Science and Engineering* 117(3) 439–447
[15] Wang W, Wong P L, Luo J B and Zhang Z 1998 A new optical technique for roughness measurement of moving surface *Tribology International* 31(5) 281–287
[16] Kumar B M and Ratnam M M 2015 Machine vision method for the non-contact measurement of surface roughness of a rotating work piece *Sensor Review* 35(1) 10–19
[17] Tabatabai A J and Mitchell OR 1984 Edge location to sub-pixel values in digital imagery. *IEEE Transactions on Pattern Analysis and Machine Intelligence* PAMI- 6 (2) 188–201

IOP Conf. Series: Materials Science and Engineering **530** (2019) 012009 doi:10.1088/1757-899X/530/1/012009

[18] Gadelmawla E S, Koura M M, Maksoud T M A, Elewa I M and Soliman HH 2002 Roughness parameters *Journal of Materials Processing Technology* 123(1) 133–145

Effect of Main Cut and Multiple Trim Cuts on Surface Integrity of 2379 Steel

N A Jaafar* and A B Abdullah

Metal Forming Research Group, School of Mechanical Engineering, Universiti Sains Malaysia (Engineeering Campus), Nibong Tebal, Penang, Malaysia

Corresponding author *: noorazam7411@yahoo.com

Abstract. Wire electrical discharge machine (WEDM) is commonly used in producing tool and die parts. However, it rarely achieves the optimum machining condition due to the inappropriate selection of WEDM machining parameters. In this study, 2379 steel which own properties equivalent to SKD 11 was used as die material. Four main wire electrical discharge machine (WEDM) cutting parameters, namely pulse on time, pulse off time, servo voltage, ignition pulse current has been experimentally studied and optimized at both levels, main cut and multiple trim cuts using Taguchi Method. Based on Taguchi Method optimization and analysis of variance (ANOVA), the result found that pulse on time and pulse off time give significant effect on white layer thickness for both main cut and multiple trim cuts. According to Signal to Noise (S/N) ratio response, the optimal combination of A2B3C2D1 and A1B3C3D1 gave the lowest white layer thickness for main cut and multiple trim cuts respectively. It also has been found that the hardness of machined surface with multiple trim cuts is higher compared to main cut.

Keywords: WEDM, 2379 steel, white layer thickness, multiple trim cuts

1. Introduction

Tool steel is a common material that often been used in a tool and die making industries due to the tremendous mechanical properties that it has such as high wear resistance, high compressive strength, high hardness and greater dimensional stability. However, the high carbon and high chrome that consist in this material makes it difficult to be machined. Generally, instead of using conventional machining techniques which are difficult and costly, the non-conventional machining technique likes WEDM is used for producing the parts. No direct contact between electrode and workpiece give the machined area zero stress. WEDM also capable to cut any conductive material regardless of their hardness and ductility with intricate and complex shape. Thus, it makes this machine popular in manufacturing industries [1]. However, since the machine uses the thermal energy as a medium to cut the workpiece, the inappropriate selection of WEDM machining parameters often affects the surface quality of the machined surface [2].

Surface integrity is an indicator of machining quality, which describes the surface condition of a workpiece after it has been modified by a manufacturing process. This performance measure is commonly studied by many researchers in order to evaluate the surface quality of machined surfaces. It reveals the surface condition of a workpiece after being modified by various mechanical and thermal loads during manufacturing process and most likely changes the material's properties. One part of surface integrity of machined surface is about the surface texture which affects the surface roughness. While the other one is surface metallurgy which changed the material properties in the subsurface area

[1]. The thermal effects from wire sparking during machining process produce heat affected zone (HAZ) at the machined surface and it includes the recast layer or white layer which has been studied in this paper. White layer was formed by the re-solidified melted material which adheres to the surface area when repeating rapid heating and cooling cycle occurred during cutting process. Most of the re-solidified debris were stuck back at the edge of the shallow crater that formed due to current spark. Surface defects such as white layer problem will decrease the fatigue strength and wear resistance of the workpiece and it's become main problem for tool life deterioration [3].

Surface integrity of machined surface is influenced by various WEDM machining parameters including pulse on time, pulse off time, servo voltage, ignition pulse current, wire tension, wire speed and dielectric pressure [4-5]. Selection of optimal parameter combination is a great challenge due to the presence of large numbers of variables. Over a decade, various researchers have tried to optimize the machining parameters of various materials to analyze the surface integrity. Rao, et al. [6] used Taguchi Method to optimize the machining parameter of Aluminum 2014T6 alloy for the performance measure of surface properties, surface quality and surface roughness. Kapoor, et al. [7] studied the effect of cryogenic treated brass wire on surface roughness and material removal rate of WEDM by various machining parameters. Dhobe, et al. [1] and Manjaiah, et al. [8] had reported that the white layer thickness was affected by the machining parameter mostly pulse on time current and servo voltage. Singh and Pradhan [9], optimized the process parameter for D2 steel. B.Kiran Kumar [10], Manjaiah, et al. [8] and Patel and Rathod [11] had investigated the effects of four machining control parameters, namely pulse peak current, pulse on time, pulse off time and wire tension on the material removal rate and surface roughness. In other work, Ikram, et al. [12] added another factor i.e. kerf width to be considered and they add more parameters including wire feed velocity, dielectric pressure, open voltage and servo voltage. Lodhi and Agarwal [13] have conducted single objective optimization i.e. surface roughness on D3 steel. Noted that all of these studies have only focused on the main cut of the WEDM process.

White layer elimination is required in order to increase the fatigue strength of tool parts, particularly at machined surface. Many of manufacturers execute secondary process like grinding, polishing, and lapping for improving the surface properties. However, these steps are time consuming and consequently will burden the cost to the customers. WEDM has an operational function, namely multiple trim cuts that capable to cut the previous cut path of the main cut with low discharge energy. In general, the high discharge energy produced during the main cut maximizes the material removal rate but decreased the machined surface quality. The formed of white layer in surface and sub-surface area due to the high discharge energy can be reduced by executing the multiple trim cuts [14-16]. Jangra, et al. [17] investigated the effect of main cut and multiple trim cuts on surface roughness for the four types of material, namely WC-Co composite, HCHCr die steel, Nimonic-90 and Monel-400. The results show that the surface roughness was improved significantly due to lower discharge energy which gives less thermal effect to the machined surface. Jangra [18] also reported that un-machined area named as surface objection which was produced during main cut can be eliminated by multiple trim cuts. An advantage of multiple trim cuts of WEDM also has been used to improve the surface roughness and die corner accuracy of material Monel 400 alloy as reported by Selvakumar, et al. [19].

JIS SKD 11, AISI D2 steel and Assab XW41 are common tool steel for die making [12][20]. In developing countries like Malaysia, local die makers may choose alternative material such as 2379 steel which has similar properties but is more economical. Unfortunately, from the literature survey, only a few studies on multiple trim cuts on 2379 steel are found. It is found that no extensive study has been done on surface integrity of the work material 2379 steel by using multiple trim cuts in order to eliminate or reducing the effect of white layer. Thus, there is a need to investigate the effect and optimization of WEDM machining parameters on surface integrity for main cut and multiple trim cuts by using Taguchi Method. In this study Analysis of variance (ANOVA) and signal to noise (S/N) ratio are used to analyze the significant process parameters for achieving the optimal performance measure. The findings of this study would be useful for the tool and die maker in determining the best combination of WEDM parameters for producing die with high surface quality that would improve production output and extend tool life.

2. Methodology

2.1. 2379 steel preparation

Workpieces size of 25mm x 25mm x 3mm are prepared from 2379 steel bar, which has similar characteristics to SKD 11 steel. The composition of the materials is shown in Table 1. The prepared workpieces are heat treated to obtain a final hardness of 60 ±2 HRC. For the WEDM experiments, 0.2 mm zinc coated brass wire is used as the electrode, capable of sustaining high discharge energies. The WEDM cut is conducted on a 4-axis CNC WEDM (AG400, Sodick Inc).

Table 1. Chemical composition of 2379 steel

C	Si	Mn	P	V	Cr	Mo	S	Fe
1.50-1.60%	0.10-0.40%	0.15-0.45%	0.030%	0.90-1.10%	11.0-12.0%	0.60-0.80%	0.030%	Balance

2.2. Machining parameter

Table 2 shows the level and WEDM machining parameters in this study which have been chosen through extensive review from literatures [1-3, 8, 9]. Two types of cutting methods were investigated, i.e. main cut and multiple trim cuts. In the main cut experiments, the process would consist of only a single cutting pass. While for the multiple trim cuts, the cutting process would consist of a single main cut followed by four trim cuts running along the same cutting path as the main cut, but with additional dimensional offset to obtain the final cutting dimensions. Figure. 1 illustrates the multiple trim cuts for the WEDM process. Table 3 shows combination of the machining parameters for the main cut experiments according to $L_9 3^4$ orthogonal array. While for the multiple trip cuts experiments, the same parameters were used for the 1st main cut, while the subsequent four trim cuts were automatically selected by the machine to use lower discharge energy parameters.

Table 2. Level and WEDM parameters control parameters

No	Parameter	Level 1	Level 2	Level 3
1	Pulse on time (μs)	0.3	0.5	0.7
2	Pulse off time (μs)	5	7	9
3	Servo voltage (V)	15	25	35
4	Ignition pulse current (Amp)	8	12	16

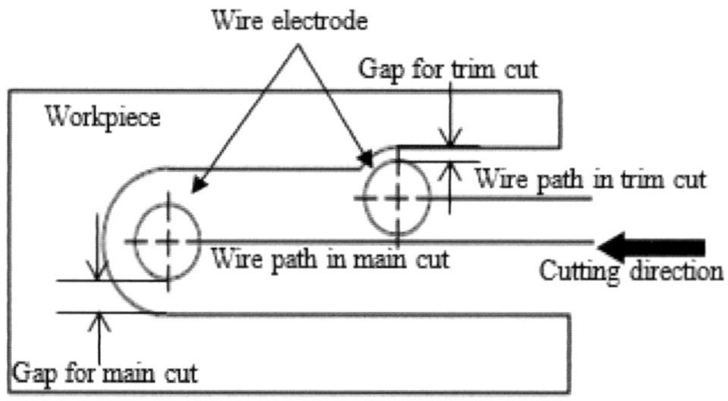

Figure 1. Multiple trim cuts operation

2.3. Design of Experiment

Taguchi method is a powerful statistical technique used to analyze and optimize process parameters. It utilizes orthogonal arrays from design of experiments theory to evaluate the influence of large number of variables on responses with a small number of experiments [10]. In the present work, an $L_9 3^4$ orthogonal array is selected to determine the parameter combinations, as shown in Table 3. The S/N ratio would be calculated to measure the impact of noise factors on the performance characteristics. Three different characteristics of S/N ratio are used in Taguchi Method depending on the target of the experiment, namely, smaller is better, nominal is better and larger is better. A high value of S/N ratio indicates that particular control factor setting has minimum effect on the noise factors [13]. In this study, a smaller is better S/N ratio is considered which signify that the lower value of white layer thickness is desirable. The smaller is better S/N ratio can be expressed as follow:

$$S/N_{\text{ratio smaller is better}} = -10\log\left[\frac{1}{n}\sum_{k=1}^{n} y_{ij}^{2}\right] \quad (1)$$

Where, n is the number of tests, and y_{ij} is the experimental value of the *i*th performance characteristic in the *j*th experiment.

Table 3. Machining parameter (According to $L_9 3^4$ orthogonal array)

Exp. No	A (µs)	B (µs)	C (V)	D (Amp)
1	0.3	5	18	8
2	0.3	7	25	12
3	0.3	9	35	16
4	0.5	5	25	16
5	0.5	7	35	8
6	0.5	9	15	12
7	0.7	5	35	12
8	0.7	7	15	16
9	0.7	9	25	8

Note: A: Pulse on time, B: Pulse off time, C: Servo voltage, D: Ignition pulse current

2.4. Measurement method

For the thickness of the white layer, which existed due to thermal effects during WEDM cutting process, the measurement was carried out with the scanning electron machine (SEM). Procedure such as surface polishing, drying and etching with Nital 5 % (5% HNO_3, 95% Ethanol) was prior conducted in order to measure white layer thickness. The thickness of white layer is measured from the outermost surface to the border, which differentiate the parent material and the altered area. Nine different places were measured before recorded as average thickness of white layer. While for the hardness tests, it was conducted with Mitutoyo ATK-600 for evaluating the hardness of machined surface. Five different places which located on the center line of machined surface were measured and the average values were recorded.

3. Result and discussion

In this section, the influence of various WEDM parameters for both types of cutting method on the surface integrity such as white layer thickness, hardness and surface topography of machined surface will be discussed.

3.1. Effect on white layer thickness

Due to the thermal effect during cutting process with WEDM, machined surface structure usually will be altered. They made up structure consist of three distinctive layer i.e. white layer/recast layer, heat affected zone and unaffected parent metal which is very difficult to be seen unless after etched with the 5% Nital solution and observed by SEM [20]. White layer or recast layer is an area which can't be resolved by any chemical agent and it remain white colour after etched with etchant while heat affected zone area is difficult to be distinguished. In this study, the white layer which exists in the cross section of machined surface has been measured by using SEM. The non-uniform white layer thickness was observed along the cross section of machined surface area (Figure. 2). Nine readings have been taken and the average result of white layer thickness of 2379 steel for main cut and multiple trim cuts and their corresponding S/N ratio values are shown in Table 4 respectively.

Figure 2. White layer structure in cross section of machined surface.

(Multiple trim cuts A= 0.3μs, B=7μs, C=25V, D=12Amp)

Based on Table 4, it shows that when the pulse on time value is 0.7μs the thickness of white layer is the largest compared to 0.3μs and 0.5μs. This is because of workpiece was exposed to the high discharge energy in longer period and its lead to form a thicker white layer. The result was in agreement with Cao, et al. [20] and Manjaiah, et al. [8]. For cutting time shorter than 0.7μs the results show that, even though after multiple trim cuts were being executed, the thickness of white layer for both cutting types is quite similar. This probably due to the effect of higher servo voltage that induced higher discharge energy during second cut before it uses low energy at the subsequent cut at third till to fifth cut. In this study the second to the fifth cut of the multiple trim cuts is generated automatically by default setting provided by the machine tool manufacturer.

Table 4. Experimental design and white layer thickness result

Exp. No	Main cut		Multiple trim cuts	
	White layer thickness (μm)	S/N ratio (dB)	White layer thickness (μm)	S/N ratio (dB)
1	168.7	-44.54	159.50	-44.06
2	185.3	-45.36	165.13	-44.36
3	168.1	-44.51	137.71	-42.78
4	166.4	-44.43	169.78	-44.60
5	162.9	-44.24	177.33	-44.98
6	140.7	-42.97	169.44	-44.58
7	251.9	-48.02	175.67	-44.89
8	304.3	-49.67	194.00	-45.76
9	166.8	-44.45	154.84	-43.80

By using Minitab 16, response table for S/N for main cut and multiple trim cuts were generated (Table 5 and Table 6). These response tables give information about the average S/N ratio for each factor level, delta, and rank. Based on the characteristic of smaller is better, the higher value of S/N ratio represents lower value of white layer thickness, which is good for surface integrity performance. Therefore, of all the three levels, for the two types of machining cuts, the optimum level parameter combination is selected based on higher values of S/N ratio. For the main cut, the optimal set of parameters for optimum white layer thickness is obtained at 0.5μs pulse on time (level 2), 9μs pulse off time (level 3), 25V servo voltage (level 2) and 8Amp ignition pulse current (level 1). Therefore, the optimal combination is A2B3C2D1. Meanwhile, for multiple trim cuts is obtained at 0.3μs pulse on time (level 1), 9μs pulse off time (level 3), 35V servo voltage (level 3) and 8Amp ignition pulse current (level 1). The optimal combination is A1B3C3D1.

Table 5. Response Table for Signal to Noise Ratios (Main cut)

Level	A (μs)	B (μs)	C (V)	D (Amp)
1	-44.80	-45.66	-45.72	-44.41*
2	-43.88*	-46.42	-44.74*	-45.45
3	-47.38	-43.97*	-45.59	-46.20
Delta	3.50	2.45	0.98	1.79
Rank	1	2	4	3

Table 6. Response Table for Signal to Noise Ratios (Multiple trim cuts)

Level	A (μs)	B (μs)	C (V)	D (Amp)
1	-43.73*	-44.52	-44.80	-44.28*
2	-44.72	-45.03	-44.25	-44.61
3	-44.82	-43.72*	-44.22*	-44.38
Delta	1.09	1.31	0.58	0.33
Rank	2	1	3	4

Meanwhile, the effect of control parameter on white layer is shown graphically in Figure. 3(a) and (b). From both figures, it can be observed that white layer thickness increases with increase in pulse on time. Increase in pulse on time will increase the discharge energy that allows higher temperature impose and penetrates into the subsurface. At high-temperature point the material melted and vaporized, then re-solidified and stick back to the machined surface during pulse off time. This makes the white layer thickness thicker. The result of this study is in-line with Li, et al. [14]. Contrarily, for pulse off time, the pause time between discharges allows the debris to solidify and be flushed away by dielectric prior to next discharge which leads to increase the sparking efficiency. The thickness of white layer decreases when the pulse off time was increased. However, there is no such significant effect of servo voltage and ignition pulse current on white layer thickness because there is less change in S/N ratio value.

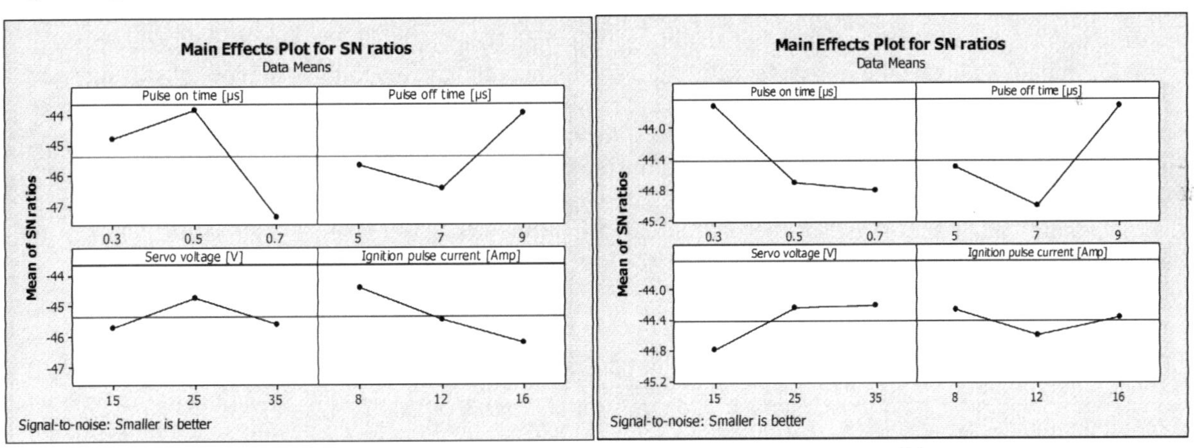

(a) (b)

Figure 3. Mean S/N ratio for white layer thickness at each level for all the parameters. (a). Main cut, (b). Multiple trim cuts

3.1.1 ANOVA. Analysis of variance (ANOVA) has been performed to determine the level of significance of the control parameters on the performance characteristics of white layer thickness. The summary of the result percentage is indicated using R-squared (R-Sq) as shown in Table 7. Considering the value of the percentage that shows the significant factors influenced white layer thickness, it can be determined that the pulse on time with the highest percentage of 60.74% is the most significant factor, followed by pulse off time 27.20%, servo voltage 8.37%, and ignition pulse current 3.69% for the main cut. It can be understood that to get a surface with minimal effect of the white layer thickness, the pulse on time should be set at shorter time which is in agreement with Manjaiah, et al. [8] who mentioned that white layer thickness is related to WEDM parameter mainly pulse on time.

For the multiple trim cuts, the pulse off time is found to be the major factor that affecting the white layer thickness (46.77%). Then followed by pulse on time, servo voltage and ignition pulse current with 38.69%, 11.40%, and 3.15% respectively. The longer pulse off time gave the chance of debris and melted material to be flushed away by the dielectric fluid. It also improves the dielectric fluid circulation for the cutting channel. Additional four trim cuts with a distinctive machining parameter value which available in multiple trim cuts may the reason make each parameter has a various contribution for the surface roughness compared to main cut which has only single rough cut.

Table 7. One-Way ANOVA for white layer thickness

Factors	**Main cut**		**Multiple trim cuts**	
	R-Sq (%)	Rank	R-Sq (%)	Rank
A (µs)	55.29	1	38.69	2
B (µs)	26.33	2	46.77	1
C (V)	4.76	4	11.40	3
D (Amp)	13.61	3	3.15	4

Note: A: Pulse on time, B: Pulse off time, C: Servo voltage, D: Ignition pulse current

3.1.2 Confirmation test. For the Taguchi Method, the confirmation test needs to be executed for validating the result that was determined during the optimization stage. Separate experiments were conducted to investigate the optimal parametric combination and the result of both types of cutting were 86.54µm for main cut and no white layer thickness was seen in multiple trim cuts. The value for white layer thickness for main cut is 38.49% less than the minimum value of white layer thickness obtained (140.70µm) during the parametric test. While for the finish cut, no white layer thickness was seen and it improved significantly. The multiple trim cuts work to eliminate all the recast layer which exists during main cut. It confirms that the optimal combination parameter gave a better result compare to parametric test.

3.2. Effect on surface hardness

The thermal effect of the discharge energy altered the surface mechanical properties. The hardness of machined surface has changed due to the thermal effect [1]. To get a better understanding of the thermal effect on the machined surface, the hardness of it was measured with hardness testing machine Mitutoyo ATK-600. Measurements were conducted at five points along the center line of the machined surface. The average of hardness for main cut and multiple trim cuts is shown in Table 8.

Table 8. Machined surface hardness for main cut and multiple trim cuts

	Cutting condition								
Cutting type	#1	#2	#3	#4	#5	#6	#7	#8	#9
Main cut [HRC]	60.9	60.8	60.6	60.5	60.8	59.9	60.5	60.4	60.6
Multiple trim cuts [HRC]	61.2	60.9	60.9	60.9	61.2	60.7	60.8	61.0	61.0

Delta	0.3	0.1	0.3	0.4	0.5	0.7	0.3	0.6	0.3

The impact of thermal during repetitive heating and rapid cooling, produced a several layers at the machined surface. These formations also depend on the cutting process and workpiece properties. At the outermost area, the changed structure, namely recast layer or white layer altered the hardness of machined surface. As seen from Figure. 4, multiple trim cuts show higher hardness for all cutting conditions compared to the main cut at the machined surface area. This is due to the time for thermal exposure for multiple trim cuts, which is longer about 70% compared to main cut. Increases the time for heating and rapid cooling will increase the carbon content in recast layer. Consequently, increase the surface hardness. At the same time, the oxides element which is formed during machining and adhered to machined surface may explain why the multiple trim cuts have a higher hardness compared to main cut [8].

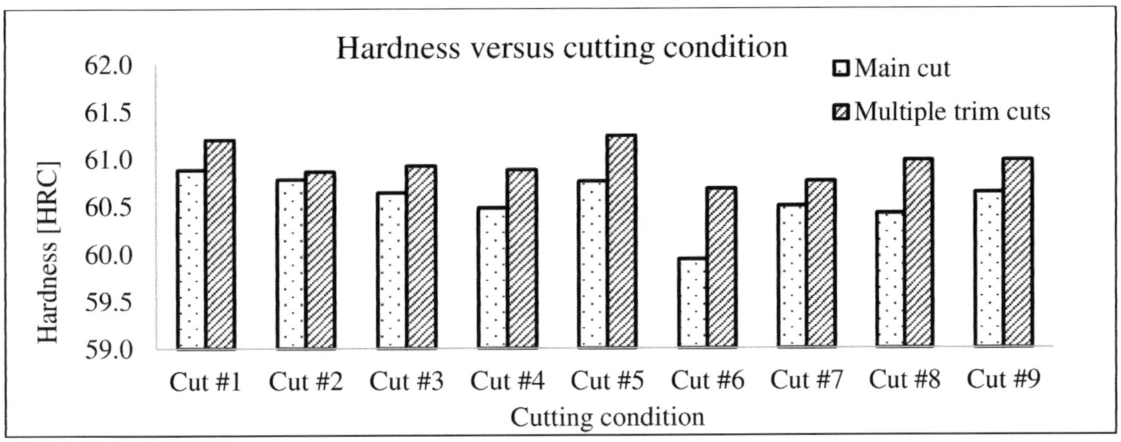

Figure 4. Hardness of machined surface area

3.3. Surface topography

Good machined surface plays a main role in extending the life period of a product. It will determine the fatigue strength of a product. Therefore, understanding the surface characteristics of machined surface is important. In WEDM, the effect of heat from discharge energy causes the deformation of machined surface particularly during main cuts. The surface topography characteristics of the samples of 2379 steel for both types of cutting types are shown in Figure 5.

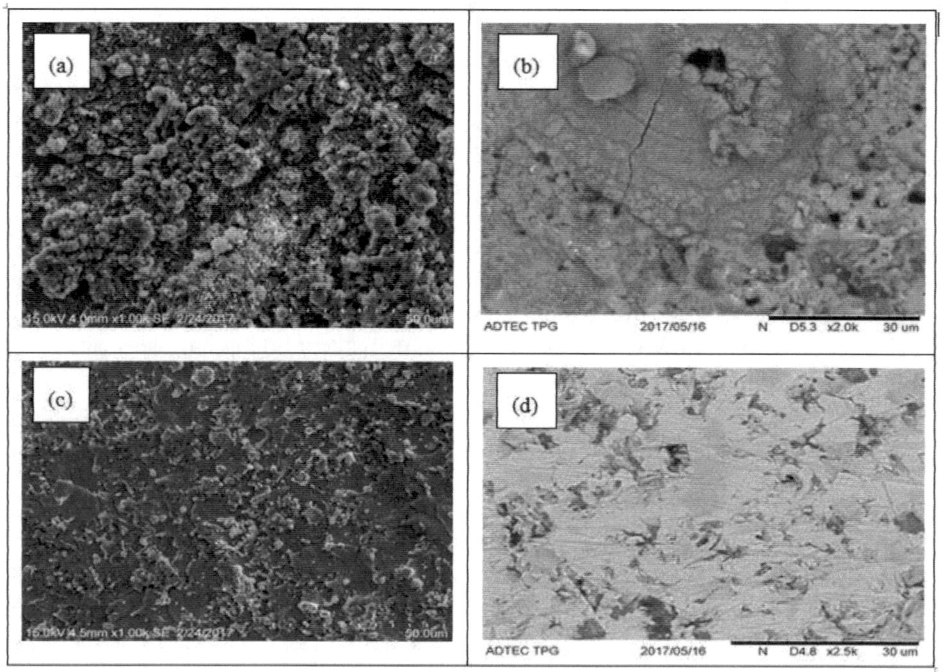

Figure 5. Surface topography characteristics. (a) Main cut (50µm), (b) Main cut (30µm), (c) Multiple trim cuts (50µm), (d) Multiple trim cuts (30µm)

High discharge energy that's been produced during main cut produced the machined surface with porous structure which is consist of shallow craters, voids, micro-cracks and irregular globules (Figure 5 (a) and (b)). High temperature at local region penetrates deeper to subsurface and melt the workpiece material. Molten material later re-solidified due to rapid cooling during pulse off time, stick back and make the surface full with above-mentioned surface characteristics. The coarse surface characteristics of machined surface were decreased after 4th trim cut. Smoother surface was formed with less shallow craters, voids, micro-cracks and irregular globules (Figure 5 (c) and (d)). The low discharge energy used during multiple trim cuts gives less effect to workpiece surface and cut the previous rough surface to the smoother finishing.

4. Conclusion

In this study, effect of four WEDM cutting parameters on the surface integrity of 2379 steel was carried out by using Taguchi method. The following conclusion can be drawn based on the investigation.

1. In the formation of white layer thickness, for the case of main cut, the most significant parameters are pulse on time (60.74%) and pulse off time (27.20%) while the rest was shown no significant effect based on ANOVA analysis. Increase the pulse on time will increase the discharge time, where it will extend the time for current discharged and make more deterioration to the machined surface. Less pulse on time is desirable for lower values of white layer thickness. The optimal machining can be obtained by A2B3C2D1 combination.
2. For multiple trim cuts, the significant factors are as following order, pulse off time (38.69%), pulse on time (46.77%), servo voltage (11.40%) and ignition pulse current (3.15%). For getting the minimum value of white layer thickness the combination is should be set as A1B3C3D1.
3. For the confirmation test, the optimum value of white layer thickness was 86.54µ, which is reduce 38.49% compared to a minimum value during parametric test. While for multiple trim cuts, white layer thickness was clearly eliminated after run with the optimal set of cutting parameter.

4. The hardness of machined surface after multiple trim cuts is harder compared to the main cut machined surface. The increment between 0.1~0.7 HRC was acquired for all cutting conditions. The time exposure to the thermal may the reason of increment of machined surface.
5. Main cut produced surface characteristics consist of shallow craters, voids, micro-cracks and irregular globules and it will reduce the fatigue life of products. While the smooth surface can be achieved by using multiple trim cuts in order to extend the life of products or tools.

References

[1] M Dhobe, I Chopde and C Gogte 2013 Investigations on surface characteristics of heat treated tool steel after wire electro-discharge machining *Materials and Manufacturing Processes* Vol 28 pp 1143-1146

[2] N Azam and M Afendi 2016 Effect of Wire-EDM cutting parameters on material removal rate of titanium alloy (Ti6Al4V) in *AIP Conference Proceedings* p 060001

[3] A Al-Khazraji, S A Amin and S M Ali 2016 The effect of SiC powder mixing electrical discharge machining on white layer thickness, heat flux and fatigue life of AISI D2 die steel *Engineering Science and Technology, an International Journal* Vol 19 pp 1400-1415

[4] M Antar, S Soo, D Aspinwall, D Jones and R Perez 2011 Productivity and workpiece surface integrity when WEDM aerospace alloys using coated wires *Procedia Engineering* Vol 19 pp 3-8

[5] J P Basil Kuriachen and Jose Mathew 2012 Modeling of Wire Electrical Discharge Machining of Alloy Steel (Ti-6Al-4V) *International Journal of Emerging Technology and Advanced Engineering* Vol 13 pp 1989-1995

[6] P S Rao, K Ramji and B Satyanarayana 2011 Effect of WEDM conditions on surface roughness: A parametric optimization using Taguchi method *International journal of advanced engineering sciences and technologies* Vol 6 pp 41-48

[7] J Kapoor, J S Khamba and S Singh 2012 The effect of machining parameters on surface roughness and material removal rate with cryogenic treated wire in WEDM *International Journal of Machining and Machinability of Materials 2* Vol 12 pp 126-141

[8] M Manjaiah, R F Laubscher, A Kumar and S Basavarajappa 2016 Parametric optimization of MRR and surface roughness in wire electro discharge machining (WEDM) of D2 steel using Taguchi-based utility approach *International Journal of Mechanical and Materials Engineering* Vol 11 p 7

[9] V Singh and S Pradhan 2014 Optimization of WEDM parameters using Taguchi technique and response surface methodology in machining of AISI D2 steel *Procedia Engineering* Vol 97 pp 1597-1608

[10] A V S R P B Kiran Kumar and V Sridhar 2013 Parametric Optimization For MRR and Surface Roughness in WIre EDM of D2 Steel Using Taguchi Method *Journal of Engineering and Technology Research* Vol 13 pp 43-52

[11] B Patel and K Rathod 2012 Multi-parameter analysis and modeling of surface roughness in electro discharge machining of AISI D2 steel *International Journal of Scientific & Engineering Research* Vol 3 pp 1-6

[12] A Ikram, N A Mufti, M Q Saleem and A R Khan 2013 Parametric optimization for surface roughness, kerf and MRR in wire electrical discharge machining (WEDM) using Taguchi design of experiment *Journal of Mechanical Science and Technology* Vol 27 pp 2133-2141

[13] B K Lodhi and S Agarwal 2014 Optimization of machining parameters in WEDM of AISI D3 Steel using Taguchi Technique *Procedia CIRP* Vol 14 pp 194-199

[14] L Li, Y Guo, X Wei and W Li 2013 Surface integrity characteristics in wire-EDM of Inconel 718 at different discharge energy *Procedia CirP* Vol 6 pp 220-225

[15] L Li, X Wei and Z Li 2014 Surface integrity evolution and machining efficiency analysis of W-EDM of nickel-based alloy *Applied Surface Science* Vol 313 pp 138-143

[16] J Liu, L Li and Y Guo 2014 Surface integrity evolution from main cut to finish trim cut in W-EDM of shape memory alloy *Procedia CIRP* Vol 13 pp 137-142
[17] K K Jangra, V Kumar and V Kumar 2014 An experimental and comparative study on rough and trim cutting operation in WEDM of hard to machine materials *Procedia Materials Science* Vol 5 pp 1603-1612
[18] K Jangra 2012 Study of unmachined area in intricate machining after rough cut in WEDM *International Journal of industrial engineering Computations* Vol 3 pp 887-892
[19] G Selvakumar, K B Jiju, S Sarkar and S Mitra 2016 Enhancing die corner accuracy through trim cut in WEDM *The International Journal of Advanced Manufacturing Technology* Vol 83 pp 791-803
[20] C Cao, X Zhang, X Zha and C Dong 2014 Surface integrity of tool steels multi-cut by wire electrical discharge machining *Procedia Engineering* Vol 81 pp 1945-1951

IOP Conf. Series: Materials Science and Engineering **530** (2019) 012011 doi:10.1088/1757-899X/530/1/012011

Effect of Rake Angle and Feed Rate on Chip Segmentation in Machining Carbon Steel 1050

M F Zamri* and A R Yusoff

Faculty of Manufacturing Engineering, Universiti Malaysia Pahang, Malaysia

Corresponding author *: fawzizamri@gmail.com

Abstract. A chip breaker plays an essential role for chip breaking as well as to enhance productivity and quality during the turning process. However, chip break as a tool is not able to break the chip and formulate continuous chip, that in turn, causes the tool to wear and the generate excessive heat. These affect the quality of the machined surface. Chip formation is influenced by cutting conditions and tool geometries such as spindle speed, feed rate, depth of cut and rake angle. In this study, experiments were carried out on carbon steel 1050 with chip breaker at a constant cutting speed of 275 m/mm with a depth of cut 0.9 mm. The effect of different feed rate and rake angle towards chip length formation were also investigated in the present investigation. The results obtained from the study indicates that segmented chips could be obtained at a feed rate of 0.4 mm/rev and a rake angle of -9°. Therefore, it could be concluded that the feed rate and rake angle can play a significant role in the formation of the segmented chip through the use of a chip breaker to enhance the productivity and quality in the machining process.

1. Introduction

Machining can be considered as an essential process in manufacturing processes to manufacture high-quality products at low-cost production. One of the most important elements in machining is the tools, and the cost of each tool can vary according to their function as well as its endurance. A new cutting tool performance behaviour test could help businesses gain a competitive edge. Performance of machining process depends on the surface smoothness, and hence it becomes one of the major topics in process planning and machining optimization to increase the productivity of the product and lowering tooling cost. It has been reported in the literature that cutting process parameters such as depth of cut, feed and speed influence product quality [1].

Turning operation is one of the continuous metal cutting processes in which the form and size of chips are among the two factors that indicate the state of a cutting condition. Thus, development of an appropriate chip control method to achieve the desirable chip form concerning cutting force magnitude is necessary. Therefore, chip control plays an import role in metal cutting. Chip control is often governed by chip flow, chip curl, and chip breaking [2]. Moreover, uninterrupted chip flow is required in automated manufacturing systems to reduce human labour, process costs, and eventually to improve productivity in the production systems [3-4]. Rodriguez et al. investigated metal cutting simulation with a particular numerical technique, the Particle Finite Element Method (PFEM) with a new modified time integration algorithm and incorporating a contact algorithm capability. The goal is to reproduce the formation of segmented chip in orthogonal machining [5]. Then, Acunto et al. propose an analytical approach of modelling to determine the chip flow direction in bar turning [6].

Chip breaking is defined as breaking up the chip into a manageable size for a metal removal process where a segmented chip is being formed. In general, this is achieved by causing the chip to be formed into an approximately circular arc [7]. The authors in [8] performed an experimental work which deals with the influence of two design parameters, width and angle of the chip breaker. Mesquita and Barata studied on predicting cutting forces in machining of martensitic stainless steels using coated carbide tools [9]. The technique is based on the measurement of the chip-breaker geometry and the calculation of the effective side-rake angle. The performance of commercial chip breakers was evaluated using artificial neural network (ANN) that was trained through a back-propagation algorithm [10]. Essential independent variables such as depth of cut, land, breadth, and radius that directly influence the chip formation were used as input values of the ANN model. It was demonstrated that the ANN model is able to predict the chip formation well. Gurbuz et al. investigated the influence of different chip breaker geometries on cutting forces and tool stresses developed during turning [11]. Sajgalik et al. deals with the elimination of long chip created when machining of a shaft used for the automotive industry [12]. The impact of chip-breakers' shape on the shape of the chip in various cutting conditions based on production requirements were investigated.

Surface roughness is one of the key indicators of machining quality. Das et al. evaluate the performance of multilayer coated carbide inserts during dry turning of hardened AISI 4340 steel (47 HRC). The effect of machining parameters (depth of cut, feed and cutting speed) on surface roughness (Ra) was investigated by applying ANOVA [13]. Mesquita et al. studied on the effect of cutting speed on chip-breaking when using advanced carbide tooling with sintered chip-breakers [14]. Abbas et al. predicting the surface roughness for different cutting parameters in CNC turning operations by using ANN [15]. In the ultra-precision diamond cutting process, the rake angle of the tool becomes negative because the edge radius of the tool is considerably larger as compared to the sub-micrometre depth of the cut [16]. Puls et al. present an experimental test to analyze friction phenomena within the tool-chip interface in metal cutting. The experimental approach is derived from an orthogonal cutting process, modified to a high-speed forming and friction process by using an extremely negative rake angle [17]. One of the important characteristics of chip morphology is chip segmentation. Devotta et al. focused on the role of tool rake angle (α) on-chip segmentation [18]. A considerable amount of literature suggests that cutting speed and feed rate influences chip segmentation. Therefore, this study aims at investigating the effect of different rake angle and feed rate in the turning process by using a chip breaker towards chip formations.

2. Materials and Methods

The workpiece material used for this investigation is carbon steel 1050 (Fig. 1a). The axial length and the diameter of the cylindrical rod are 90 mm and 125 mm, respectively. The chemical composition of workpiece material is presented in Table 1 [19-22]. The cutting tool used for this study is a carbide insert with chip breaker (CNMG 120412 PM 4325) with Sanvick tool holder (DCLNR 2525M 12) depicted in Fig. 1b and Fig. 1c, respectively. The cutting speed and depth of cut for this project were set at a constant rate of 275 m/mm and 0.9 mm, accordingly. The specimen must be clamped tightly with dead centre during the cutting operation. The experiment was started after, cutting speed and feed rate was set up on the CNC lathe machine. After the cutting operation for the first specimen is completed, the feed rate of was changed to 0.2, 0.3, 0.4 and 0.5 mm/rev. The value was chosen based on the recommendation made by the tool maker. Then, the chips produced from the material were collected and tabulated. The test is repeated for different rake angle values. In this project, -ve rake angle was chosen as recommended by previous investigations [16-18]. Fig. 1d shows the parameters for rake angle which are -3 °, -6 °, -9 °, and -12 °, respectively. The original rake angle from tool holder DCLNR 2525M 12 with flat shim is -6 °. Then, new shim with an angle of 3 ° was fabricated for -3 ° of rake angle. Then, for rake angle of -9 °, and -12 ° the shim with angle -3 ° and -9 ° were fabricated. The turning process was performed under flood cooling cutting conditions. The details of the tool holder and other cutting conditions are presented in Table 2.

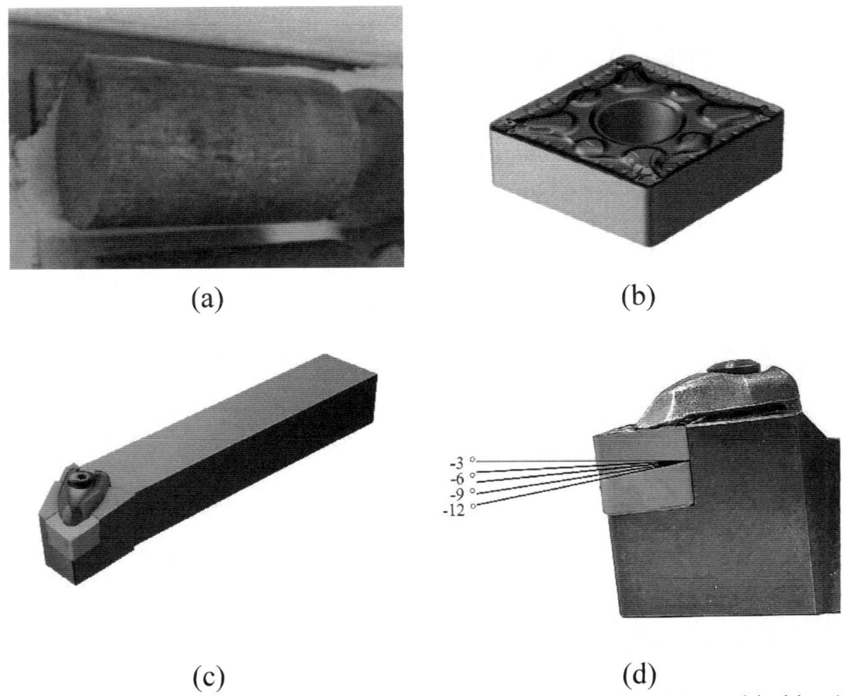

Figure 1. (a) Carbon steel 1050 (b) Carbide insert with chip breaker (c) Tool holder (d) Rake angle

Table 1. The chemical composition of carbon steel 1050

Composition	(Wt. %)	Composition	(Wt. %)
Iron, Fe	98.46-98.92	Sulfur, S	$\leqslant$ 0.050
Manganese, Mn	0.60-0.90	Phosphorous, P	$\leqslant$ 0.040
Carbon, C	0.470-0.55		

Table 2. Cutting conditions

Cutting conditions	Descriptions
Workpieces	Carbon steel 1050
Cutting speed	275 m/mm
Depth of cut	0.9 mm
Feeds	0.2 – 0.5 mm/rev
Cutting conditions	Wet cutting
Cutting tools	CNMG 120412 PM 4325
Tool holder	DCLNR 2525M 12
Rake angle	-3 °, -6 °, -9 °, -12 °,
Responses	Chip length

3. Result and Discussion

The effect of cutting parameters and rake angle on the performance of coated carbide insert is discussed in this section. The chip length obtained from the experiment is present in Fig. 2 and Fig. 3 for varying feed rate and rake angle. The result shows that the chip length decreases with increasing the value of feed rate and rake angle. It could be seen from Fig. 2 that for all rake angles investigated, the chip length decreases along the increasing feed rate, however, a minor increase in the length is observed for the feed rate of 0.5 mm/rev. Nonetheless, it is worth noting that the increase is marginal, with an increase of 1.5

IOP Conf. Series: Materials Science and Engineering **530** (2019) 012011 doi:10.1088/1757-899X/530/1/012011

mm between the feed rate of 0.4 mm/rev and 0.5 mm/rev. In addition, it could also be seen that there is an abrupt increase in the chip length, in particular for the rake angle of -9° and a feed rate of 0.2 mm/rev. Based on the results obtained, it could be concluded that the optimum feed rate for this study is 0.4 mm/rev as it produces the shortest chip length amongst the feed rates regardless of the rake angle. Furthermore, from Table 3, it is evident that the shortest chip length amongst the rake angles evaluated at the optimum feed rate of 0.4 mm/rev is attainable at a rake angle of -9°. This is primarily due to the optimum contact between the chip and the chip breaker. Therefore, it could be concluded that the optimum rake angle and feed rate observed to produce the shortest chip length of carbon steel 1050 in the present investigation are -9° and 0.4 mm/rev, respectively.

Table 3. Cutting conditions

Rake Angle (°)	Feed Rate (mm/rev)	Chip Length (mm)
-3	0.2	500.0
-3	0.3	35.0
-3	0.4	9.0
-3	0.5	10.5
-6	0.2	88.0
-6	0.3	18.0
-6	0.4	6.0
-6	0.5	9.5
-9	0.2	230.0
-9	0.3	8.0
-9	**0.4**	**3.0**
-9	0.5	6.0
-12	0.2	40.0
-12	0.3	4.5
-12	0.4	4.0
-12	0.5	8.0

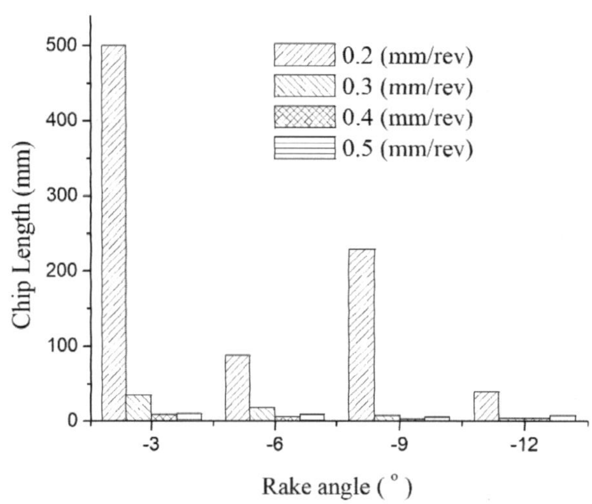

Figure 2. Variation of chip length depending on feed rate

4. Conclusion

This study evaluated the effect of varying the rake angle as well as the feed rate towards the formation of carbon steel 1050 chip. It was demonstrated from the experimental investigation, that the optimum feed rate and rake angle observed for cutting the aforesaid material is 0.4mm/rev and -9°, respectively. It was shown that the selected parameters lead to efficient breaking and effective removal of chips. It is worth noting that the results obtained are based on flood cooling cutting conditions and are limited to this condition. A further investigation is necessary to evaluate the aforesaid cutting parameters at different machining conditions, i.e., dry and minimum quantity lubrication cutting conditions in ascertaining its ability for chip formation. Moreover, the effect of different types of chip breakers towards the formation of carbon steel 1050 chip along with the aforesaid parameters should be investigated.

Acknowledgement

The authors would like to acknowledge Universiti Malaysia Pahang for funding this project via PGRS 180321.

References

[1] Valera H Y and Bhavsar S N 2014 Experimental Investigation of Surface Roughness and Power Consumption in Turning Operation of EN 31 Alloy Steel *Procedia Technology* 14 528–534 https://doi.org/10.1016/j.protcy.2014.08.067

[2] Zhou L 2001 Machining Chip-Breaking Prediction with Grooved Inserts in Steel Turning *Worcester Polytechnic Institute*

[3] Pereira R B D, Braga D U, Nevez F O and Da Silva A S C 2013 Analysis of surface roughness and cutting force when turning AISI 1045 steel with grooved tools through Scott-Knott method *International Journal of Advanced Manufacturing Technology* 69(5–8) 1431–1441 https://doi.org/10.1007/s00170-013-5126-3

[4] Qibiao Y, Zhanqiang L and Bing W 2012 Characterization of chip formation during machining 1045 steel *International Journal of Advanced Manufacturing Technology* 63(9–12) 881–886 https://doi.org/10.1007/s00170-012-3954-1

[5] Rodríguez J M, Carbonell J M, Cante J C and Oliver J 2017 Continuous chip formation in metal cutting processes using the Particle Finite Element Method (PFEM) *International Journal of Solids and Structures* 120 81–102. https://doi.org/10.1016/j.ijsolstr.2017.04.030

[6] D'Acunto A, Le Coz G, Moufki A and Dudzinski D 2017 Effect of Cutting Edge Geometry on Chip Flow Direction-Analytical Modelling and Experimental Validation *Procedia CIRP* 58 353–357. https://doi.org/10.1016/j.procir.2017.03.327

[7] Worthington B 1976 The operation and performance of a groove-type chip forming device *International Journal of Production Research* 14(5) 529–558 https://doi.org/10.1080/00207547608956623

[8] Ali M and Murugan M 2009 Influence of chip breaker location and angle on chip form in turning low carbon steel *International Journal Machinability of Materials 5(January 2009)* 452–475 https://doi.org/10.1504/IJMMM.2009.026903

[9] Mesquita R M D and Marques M J M B 1992 Effect of chip-breaker geometries on cutting forces *Journal of Materials Processing Technology* 31 317–325

[10] Kim H, Sim J and Kweon H 2008 Performance evaluation of chip breaker utilizing neural network *Journal of Materials Processing Technology* 9 647–656 https://doi.org/10.1016/j.jmatprotec.2008.02.064

[11] Gurbuz H, Kurt A, Ciftci I and Seker U 2011 The Influence of Chip Breaker Geometry on Tool Stresses in Turning *Journal of Mechanical Engineering* 57 91–99

[12] Šajgalík M, Czán A, Martinček J, Varga D, Hemžský P and Pitela D 2015 The Impact Of Surface Shape Of Chip-Breaker On Machined Surface *Technological Engineering* XII(2) 4–7 https://doi.org/10.1515/teen-2015-0014

[13] Das S R, Kumar A, Dhupal D and Mohapatra S K 2013 Optimization of Surface Roughness in Hard Turning of AISI 4340 Steel using Coated Carbide Inserts *International Journal of Information and Computation Technology* 3(9) 871–880

[14] Mesquita R M D, Soares F A M and Marques M J M B 1996 An Experimental Study of The Effect of Cutting Speed On Chip Breaking *Journal of Materials Processing Technology* 56 313–320

[15] Abbas A T, Alata M, Ragab A E, Rayes M M El and Danaf E A El 2017 Prediction Model of Cutting Parameters for Turning High Strength Steel Grade-H : Comparative Study of Regression Model versus ANFIS *Advances in Materials Science and Engineering* 12

[16] Son S, Lim H and Ahn J 2006 The effect of vibration cutting on minimum cutting thickness *International Journal of Machine Tools and Manufacture* 46(15) 2066–2072 https://doi.org/10.1016/j.ijmachtools.2005.12.011

[17] Puls H, Klocke F and Lung D 2014 Experimental investigation on friction under metal cutting conditions *Wear* 310(1–2) 63–71 https://doi.org/10.1016/j.wear.2013.12.020

[18] Devotta A, Beno T, Siriki R, Löf R and Eynian M 2017 Finite element modeling and validation of chip segmentation in machining of AISI 1045 steel *Procedia CIRP* 58 499–504 https://doi.org/10.1016/j.procir.2017.03.259

[19] Baday S, Basak H and Gural A 2016 Analysis of spheroidized AISI 1050 steel in terms of cutting forces and surface quality 315–320

[20] Baday S, Basak H and Sonmez F 2017 The Assessment of Cutting Force with Taguchi Design in Medium Carbon Steel – Applied Spheroidization Heat Treatment *Measurement and Control* 50(4) 89–96 https://doi.org/10.1177/0020294017713767

[21] Valera H Y and Bhavsar S N 2014 Experimental Investigation of Surface Roughness and Power Consumption in Turning Operation of EN 31 Alloy Steel *Procedia Technology* 14 528–534 https://doi.org/10.1016/j.protcy.2014.08.067

[22] Gürbüz H, Sözen A and Şeker U 2016 Modelling of Effects of Various Chip Breaker Forms on Surface Roughness in Turning Operations by Utilizing Artificial Neural Networks *Journal of Polytechnic* 19(1) 71–83

Finite Volume Method Study on Contact Line Jump Phenomena and Dynamic Contact Angle of Underfill Flow in Flip-Chip of Various Bump Pitches

F C Ng[1], A Abas[1*] and M Z Abdullah[2]

[1] School of Mechanical Engineering, Universiti Sains Malaysia, Engineering Campus, 14300, Nibong Tebal, Penang, Malaysia.

[2] School of Aerospace Engineering, Universiti Sains Malaysia, Engineering Campus, 14300, Nibong Tebal, Penang, Malaysia.

Corresponding author *: aizatabas@usm.my

Abstract. Three distinct flip-chip underfill cases were considered in this work, with each possess a bump pitch of 0.8 mm, 1.0 mm and 1.2 mm. Contact line jump (CLJ) phenomenon during the underfill flow were successfully visualized with the aided of finite volume method (FVM) based simulation. The current simulated underfill flow fronts were qualitatively validated with the existing experimental data. Generally, the attachment process occurs much faster than the detachment process, by a factor of 14. Within the investigated pitches range, shorter bump pitch would yield a slower underfill flow but a faster-completion underfill process. The voiding mechanism during the underfill process were presented. Furthermore, the dynamic contact angle of underfill meniscus were analytically computed from the numerical data. The contact angle is found to be varies sinusoidally with the filling time, while the impact of bump pitch is minimal. Lastly, it is found that the bump pitch of flip-chip does affect the void formation and propagation.

***Keywords*:** Bump pitch; contact line jump (CLJ); dynamic contact angle; flip-chip encapsulation; underfill process

1. Introduction

To enhance the reliability electronic package and device, underfill encapsulation is performed during the assembly manufacturing process at the chip or substrate levels. During the underfill process, a special type of adhesive fluid is introduced to the gap beneath the chip and substrate, which prior hardening it would form a harden layer. This protective layer would solve the thermal mismatch issue and serves as mechanical resistance to the fragile joints of the chip device. Upon realizing the particular importance of this underfill process, several researches were conducted from various point-views and methodology, with an ultimate aim to optimize and enhance the process and subsequently the package reliability [1 – 7].

Contact line jump (CLJ) is a phenomenon occurred during the underfill meniscus approaching the bump array or detaching the bump. This was firstly discovered by Young [8-9] and later experimentally observed by Lee et al. [10-11]. Later, Yao et al. had incorporate the idea of CLJ in formulating new filling time model [12].

IOP Conf. Series: Materials Science and Engineering **530** (2019) 012012 doi:10.1088/1757-899X/530/1/012012

There are several literatures that studied the effect on solder bump pitch on the underfill flow. They concluded that the decrease in bump pitch of flip-chip will substantially increases the filling time [10-12]. Eventually, Wan et al. had analytically developed the concept of critical clearance of flip-chip for effective underfill process [13]. On other hand, the effect of contact angle on underfill flow is vital [14-15].

This paper is devoted to study the underfill encapsulation process in flip-chip of varying bump pitch using the finite volume method (FVM) numerical scheme. The current numerical model is quantitatively validated to the existing experimental data [10] on the entrance and exit flow fronts. Through simulation, the occurrence of CLJ and the influence of bump pitches can be investigated. Additionally, the contact angle of underfill flow front is determined. These findings may benefit the optimization and enhancement works on underfill process.

2. Numerical simulation

In this paper, the finite volume method (FVM) based commercial software, Ansys is adopted to numerically model the underfill flow during the flip-chip encapsulation process. Navier-Stokes (N-S) and continuity equations were governed and solved using the FVM scheme. On the side of mathematical model of flip-chip, there were three variations with different bump pitches, w of 0.8 mm, 1.0 mm and 1.2 mm (respectively dimensionless pitch, α of 1.6, 2.0 and 2.4). However, all flip-chips have same bump diameter, d of 0.5 mm and gap height, h of 0.45 mm. These flip-chip parameters were defined in Figure 1. Further, the solder bumps were in quadrilateral arrangement and a side length of 0.5 mm. The whole bump array is set at 3 × 2 so the flow meniscus can be securitized. The underfill flow pattern is repetitive along each row of bump, therefore only two rows were considered in current simulation. A Newtonian fluid of density 1600 kg/m^3, viscosity 1.5 Pa-s and surface tension, 0.012 N/m was used as the underfill material. The underfill fluid is continuously dispensed to the flip-chip along one edge side, mimicking the I-type single inlet dispensing method. The computational models were meshed with fine tetrahedral setting with proximity on. Multiphase volume of fluid (VOF) is chosen to track the advancement of flow front of underfill fluid. This underfill simulation is modelled to be transient, laminar and incompressible flow. The transient time step used is 0.01 s with a maximum of 20 iterations per time step. For boundary conditions set are: 0 Pa gauge pressure on both flow inlet and outlet, no-slip on walls and capillary pressure flow front. Lastly, the numerical simulated flow fronts will be compared with the existing experiment data by Lee et al. [10] on the flip-chip case with dimensionless pitch of 2.0. As validation, a qualitative comparison on the flow fronts at different filling stages (i.e. the attachment jump and detachment jump) was performed and presented in Figure 2.

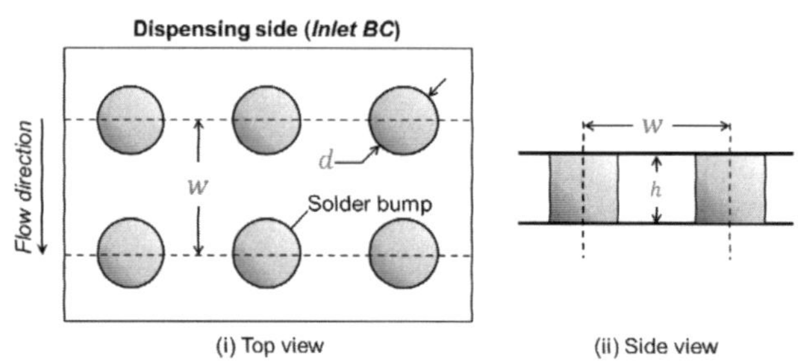

Figure 1. The geometrical representation of a flip-chip.

3. Result and Discussions

Figure 2 depicts the side-by-side comparison of experimental [10] and numerically simulated flow front of underfill fluid in the flip-chip with linear density, $\alpha = 2$.This serves as a validation study to qualitatively verify the accuracy of numerical model and subsequent findings. It is found that the simulated flow fronts are in great consensus with the experimental findings at respective filling stages in terms of meniscus shape and flow front position. Our numerical simulation also able to capture the occurrence of contact line jump in both the entrant and exit (as in Figure 2 (**ii**) and Figure 2 (**iv**) respectively), similarly to that being reported in the experimental works [10].

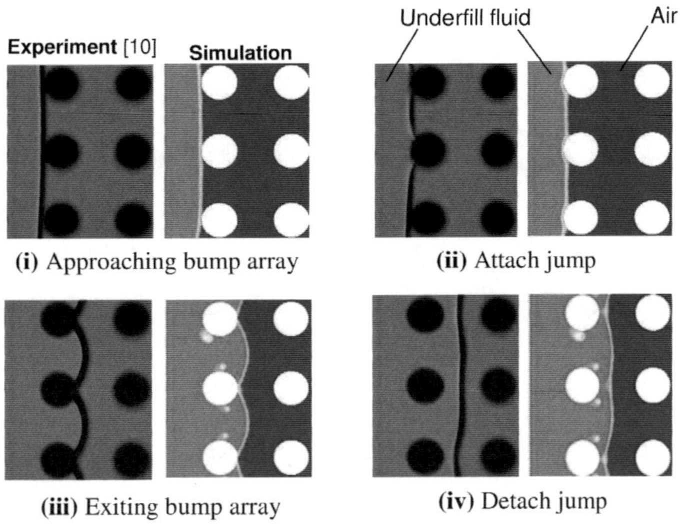

Figure 2. Validation study of simulated flow front (right) of underfill front in flip-chip bump array by comparing to the corresponding experimental flow fronts (left) by Lee et al. [10], for different filling stages of underfill flow.

Figure 3 gives the filling time plots for the underfill flow at various filling stages, for instance attachment, propagation & voiding and detachment. Generally, all filling time plots show monotonically increasing trend. On the trending of bump pitches, smallest bump pitches constitute for the longest filling time as evident in the flip-chip case of 0.8 mm pitch, which is consistent with the conclusions attained by past literatures [10 – 12]. The narrow gap between two adjacent bumps in flip-chip of lower pitch is found to be substantially hindering the detachment process; while the effect of bump pitch on bump attachment process is not significant. Besides, it is found that the attachment jump when the flow front first approach the bump array occurred instantaneously, within 0.05 s (refers Table 1). On contrary, the detachment jump is a slow process, where the flow front took up to 0.70 s to detach away from the bump array (see Table 3). Furthermore, the overall trend of underfill fluid flow across the bump array is decelerating, as being exhibited by the exponential curve in filling time plot in Figure 3.

During the propagation of underfill fluid across the bump array, voids start to form due to the interaction with bump surface by the entrapment of air. Table 2 depicts the voids formed after the underfill flow entered bump array. It is found that the air voids tend to form behind the solder bump, as the flowing underfill fluid is being obstructed by the bump array. Such inelastic collision with the solder bump not only reduces the flow rate, but also cause the formation of undesired defect – air void – which is potent to the package reliability.

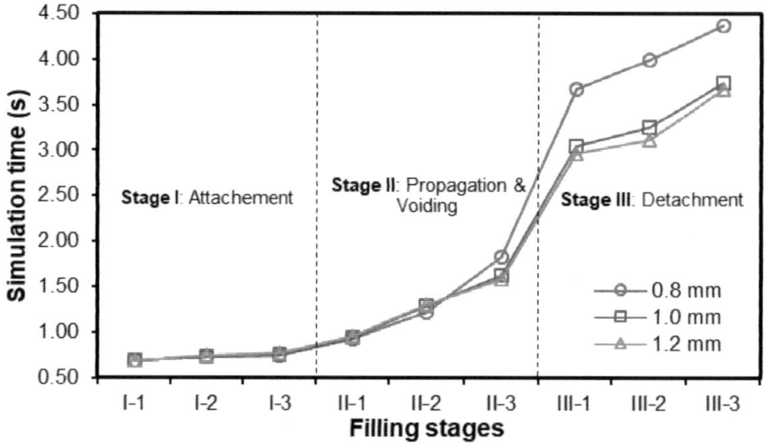

Figure 3. Filling time plots of underfill flow at three filling stages for flip-chips with bump pitches of 0.8 mm, 1.0 mm and 1.2 mm.

Table 1. Stage I – underfill meniscus attachment to the solder bump.

	$w = 0.8$ mm $\alpha = 1.6$	$w = 1.0$ mm $\alpha = 2.0$	$w = 1.2$ mm $\alpha = 2.4$
Stage I-1: Entrance ($x' = 0$)	 0.69 s	 0.69 s	 0.69 s
Stage I-2: Jump (Attach)	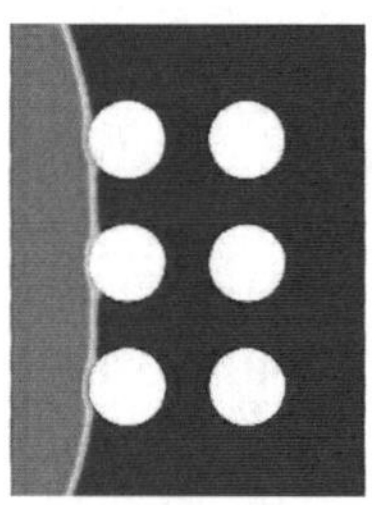 0.73 s	 0.73 s	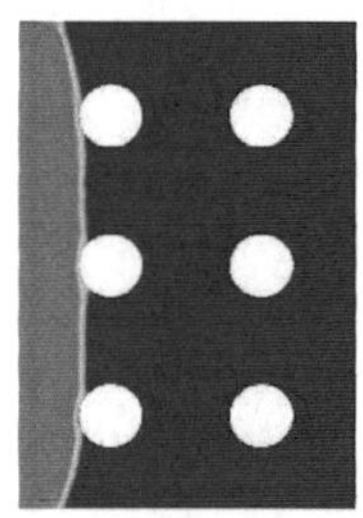 0.74 s
Stage I-3: Meniscus curved	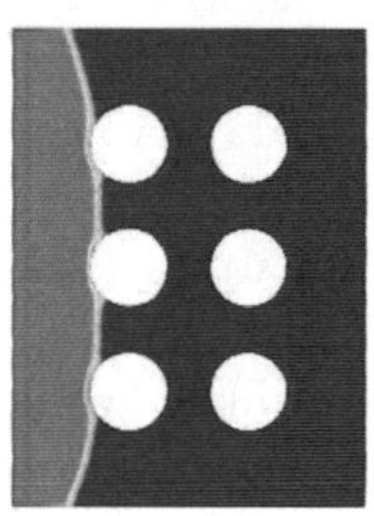 0.75 s	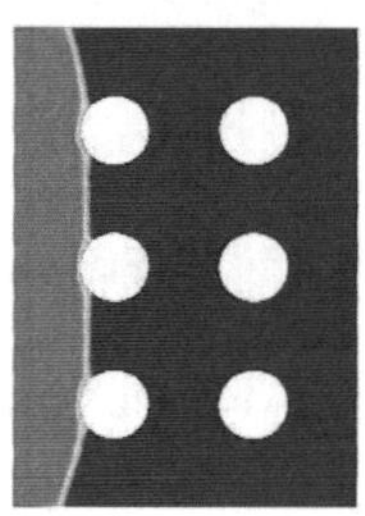 0.76 s	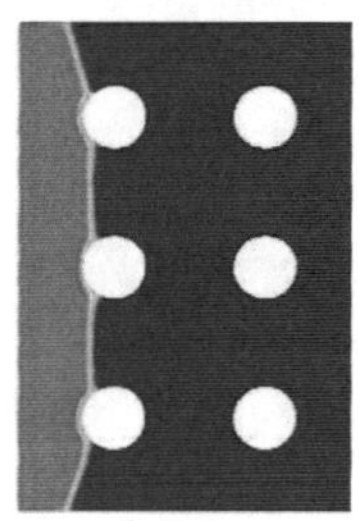 0.77 s

IOP Conf. Series: Materials Science and Engineering **530** (2019) 012012 doi:10.1088/1757-899X/530/1/012012

Table 2. Stage II – propagation of underfill meniscus and void formation.

	$w = 0.8$ mm $\alpha = 1.6$	$w = 1.0$ mm $\alpha = 2.0$	$w = 1.2$ mm $\alpha = 2.4$
Stage II-1: Mid-bump $\left(x' \approx \frac{d}{2}\right)$	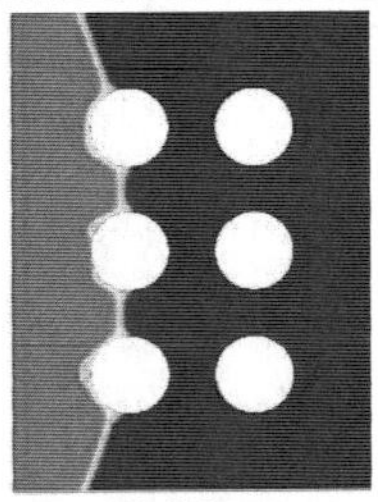 0.93 s	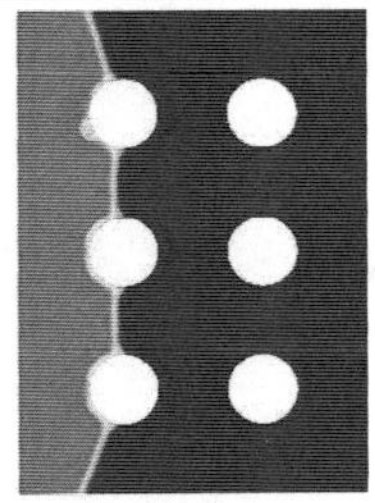 0.94 s	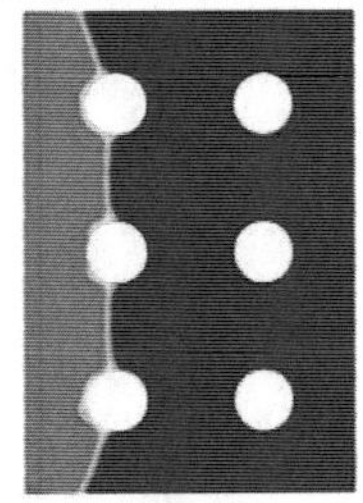 0.96 s
Stage II-2: Voiding	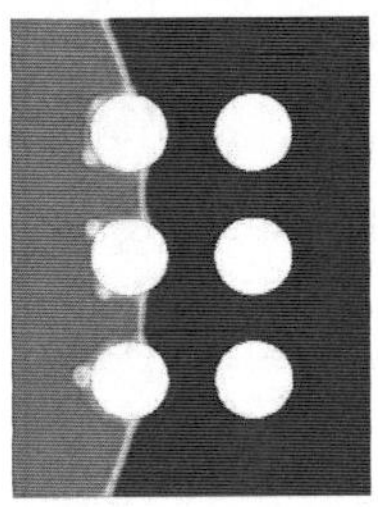 1.22 s	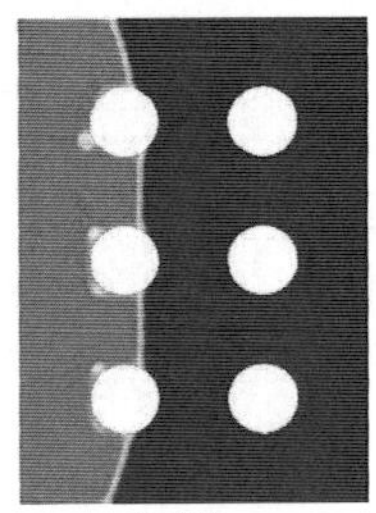 1.29 s	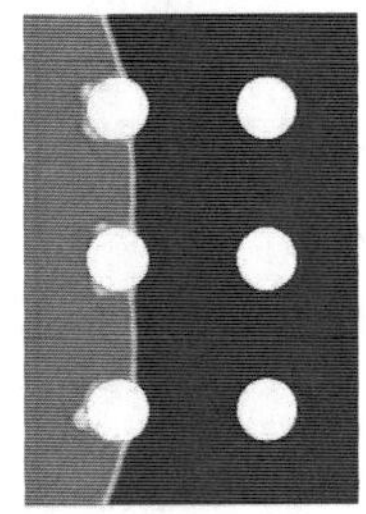 1.31 s
Stage II-3: Exit $(x' \approx d)$	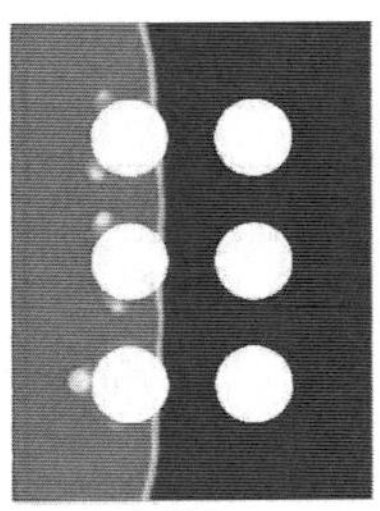 1.84 s	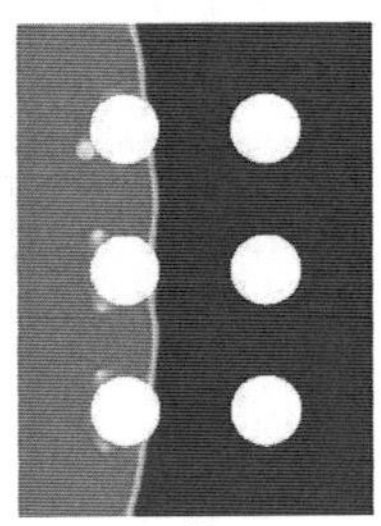 1.62 s	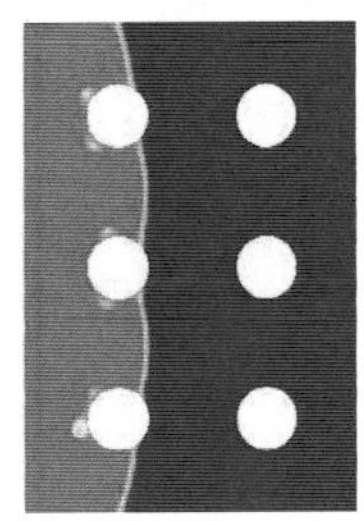 1.58 s

Figure 4 presents the plot of volume fraction filled by the underfill fluid at different filling time for all three flip-chips system. The simulation data were fitted into polynomial equations with a high correlation coefficient of $R > 0.998$. It is observed that in the flip-chip of lowest bump pitch (i.e. 0.8 mm) yields highest volume fraction filled by underfill fluid over the time. This appeared to be contradicted to the finding in Figure 3. Such discrepancy is caused by the fact the total volume of flip-chip with smaller bump pitch is lower. Therefore, upon comparing the volume fraction filled, an opposite trend is being observed instead. It is interesting to observed that the total volume of flip-chip outweighed the influence of bump pitch on the filling time. Therefore, the underfill process of flip-chip with smaller bump pitch can be completed faster, but not account for faster underfill flow.

IOP Conf. Series: Materials Science and Engineering **530** (2019) 012012 doi:10.1088/1757-899X/530/1/012012

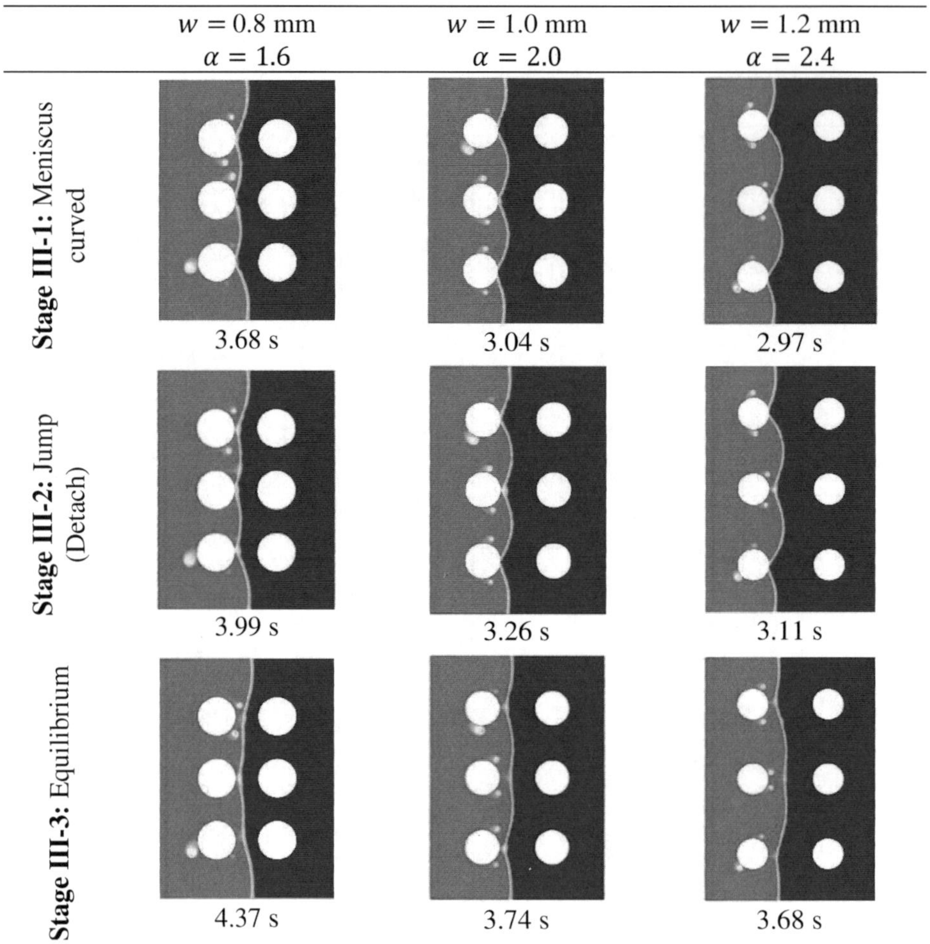

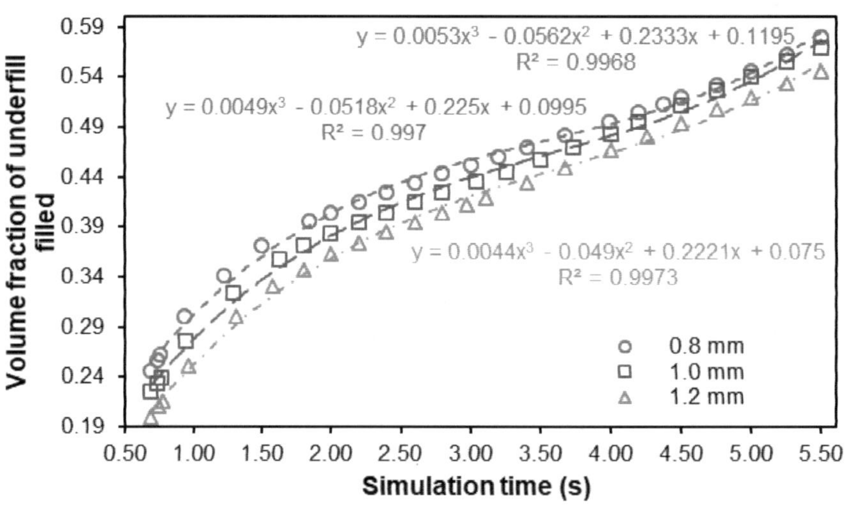

Figure 4. Plot of underfill-filled volume fraction against time for the flip-chip of different bump pitches: 0.8 mm, 1.0 mm and 1.2 mm.

IOP Conf. Series: Materials Science and Engineering **530** (2019) 012012 doi:10.1088/1757-899X/530/1/012012

Figure 5 shows the side-view of the underfill flow in the gap between substrate and flip-chip, together with its geometrical representation to analytically compute the contact angle, θ. By trigonometry, the angle can be mathematically written as follows:

$$\theta = \cos^{-1}\left[\frac{4\beta}{1+4\beta^2}\right], \tag{1}$$

where $\beta = t/h$. The contact angle of the underfill meniscus at different time for all three flip-chips were computed and plotted in Figure 6. A sinusoidal trend is observed for the plot of contact angle against time. This inferred the contact angle of the underfill fluid is not statics but rather dynamic, strongly influenced by the flow time. However, there is no significant effect of bump pitches on the variation of contact angle. The dynamic contact angle is an intrinsic material property of the underfill fluid itself, and not bounded to the geometrical parameter of the flip-chip.

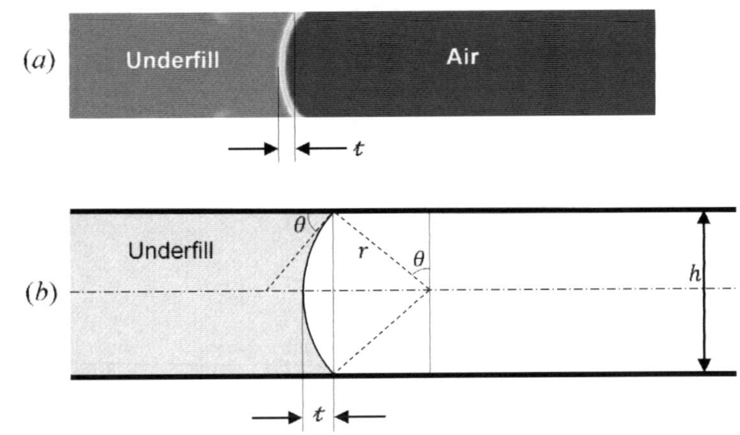

Figure 5. Measurements of meniscus's contact angle, θ from (*a*) the simulated VOF flow front; (*b*) the geometrical interpretation of underfill's meniscus, as viewed from the side.

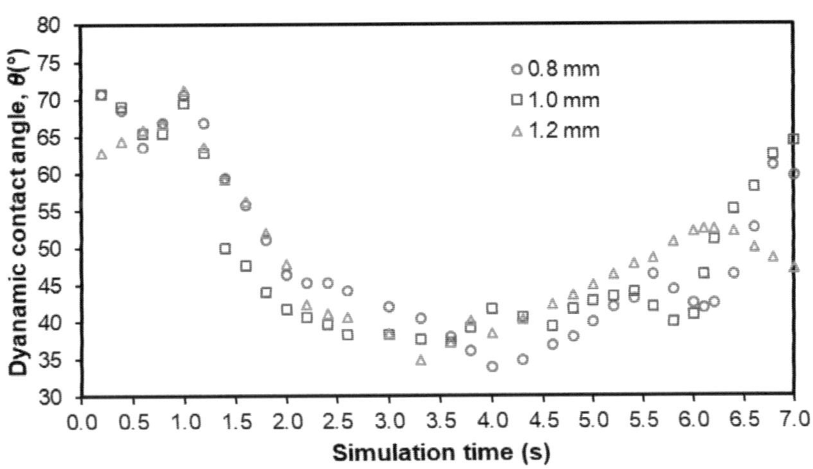

Figure 6. Plot of dynamic contact angle of underfill fluid meniscus at different filling time, for flip-chips with various bump pitches.

Figure 7 depicts the VOF and gauge pressure contours for three flip-chip cases at the time of 5.00 s. Consistent to the boundary conditions applied, the flow front's gauge pressure is negative due to the capillary action while the unfilled air region is at the atmospheric pressure of 0 Pa. The air void is found to possess a positive gauge pressure. Void is a sack of compressed air that being surrounded by the underfill fluid. The dynamic propagation of the air void formed in each flip-chip cases were quantitatively visualized in Figure 8. It is found that the initial position of the void formed significantly affects its propagation, such that the void that formed directly behind the bump ($\varphi < 90°$) would travels slower to escape the vicinity of bump ($\varphi > 180°$). The initial void favours to form at directly behind the bump for the flip-chip configuration with the shortest bump pitch of 0.8 mm (Void 1).

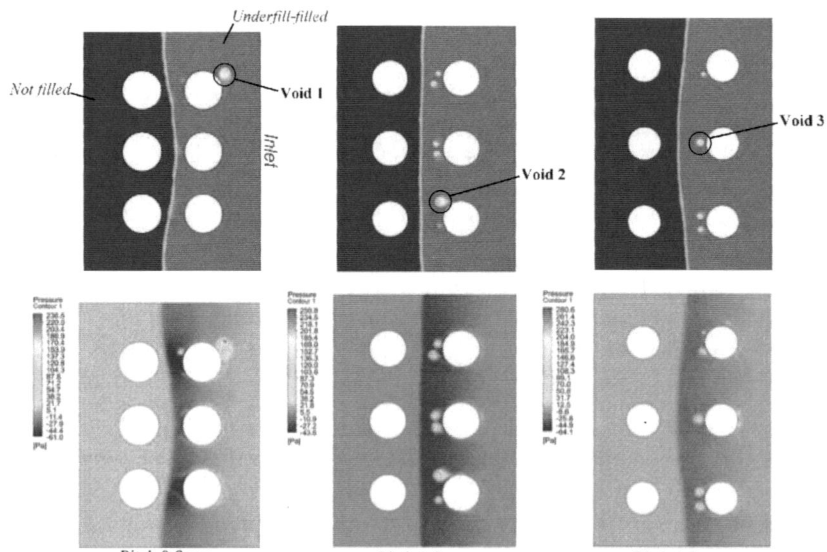

Figure 7. Underfill volume fraction profiles and pressure distributions at the stage IV at the simulation time of 5.00 s. Three air voids, with each from array of different bump pitch, were circled and marked as Void 1, Void 2 and Void 3 respectively.

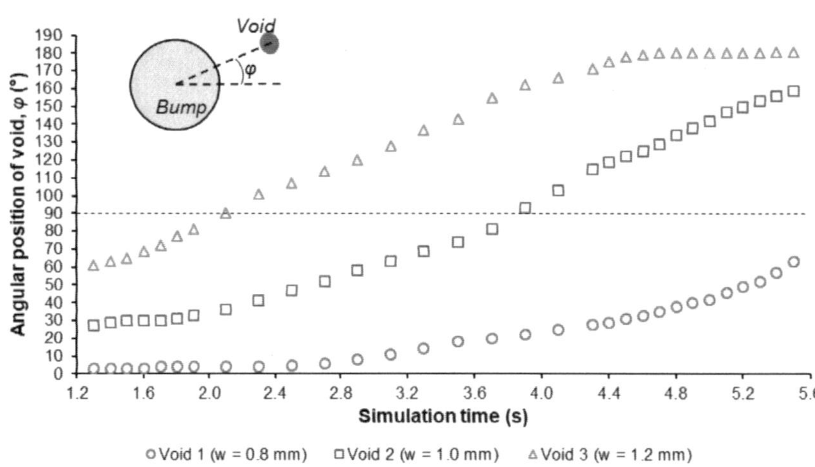

Figure 8. The plot of angular positions of Void 1, Void 2 and Void 3 (respectively from bump array with pitches w = 0.8 mm, w = 1.0 mm and w = 1.2 mm) relative to the center of solder bump.

4. Conclusions

Using finite volume method (FVM) based numerical scheme, the underfill flow across the bump array are successfully simulated alongside with the contact line jump (CLJ) phenomenon during both the attachment and detachment. In term of filling time, detachment process is significantly much slower than the attachment process. The effect of bump pitch on the filling time of underfill flow is investigated by consider the flip-chips of different pitches: 0.8 mm, 1.0 mm and 1.2 mm. It is revealed that lower pitch leads to slower underfill flow but faster completion of the underfill process. The bump pitch is found to affect the detachment process to a significant extend, where the time taken for the meniscus to detach away from the bump is higher in flip-chip with small pitch. Besides, the voiding process during the advancement of underfill along the bump array is observed from the current simulation work. Undesired defect – void – is formed due to the entrapment of air during the underfill fluid first contact with the solder bump. Lastly, the variation of contact angle of underfill fluid along the filling time is analysed and found that the contact angle varies in a sinusoidal manner. These findings justified that the FVM numerical simulation is capable in study the underfill flow from the viewpoints of CLJ, void and dynamic contact angle.

Acknowledgement:

This work was partly supported by the FRGS grant FRGS/1/2015/TK03/USM/03/2 and USM Research University Grant 8014071 from the Division of Research and Innovation, University Sains Malaysia, Engineering Campus, Penang. Additionally, the authors would like to acknowledge the financial support from the USM Fellowship 2017/18 scheme, provided by the Institute of Postgraduate Studies (IPS), Universiti Sains Malaysia (USM).

References:

[1] F C Ng, A Abas, M H H Ishak, M Z Abdullah and M S Abdul Aziz 2016 Effect of Thermocapillary Action in the Underfill Encapsulation of Multi-stack Ball Grid Array *Microelectronics Reliability* Vol 66 pp 143 – 160

[2] F C Ng, A Abas, Z L Gan, M Z Abdullah, F Che Ani and M Yusuf Tura Ali 2017 Discrete phase method study of ball grid array underfill process using nano-silica filler-reinforced composite-encapsulant with varying filler loadings *Microelectronics Reliability* Vol 72C pp 45 – 64

[3] F C Ng, M A Abas, M Z Abdullah, M H H Ishak and G Y Chong 2017 CUF scaling effect on contact angle and threshold pressure *Soldering & Surface Mount Technology* Vol 29 Issue 4 pp 173 – 190

[4] F C Ng, A Abas and M Z Abdullah 2018 Effect of solder bump shapes on underfill flow in flip-chip encapsulation using analytical, numerical and PIV experimental approaches *Microelectronics Reliability* Vol 81 pp 41 – 63

[5] A Abas, F C Ng, Z L Gan, M H H Ishak and G Y Chong 2018 Effect of scale size, orientation type and dispensing method on void formation in the CUF encapsulation of BGA *Sadhana* Vol 43 Issue 4 No 59

[6] J W Wan, W J Zhang and D J Bergstrom 2009 Numerical Modeling for the Underfill Flow in Flip-Chip Packaging *IEEE Transactions on Components and Packaging Technologies* Vol 32 No 2 pp 227 – 234

[7] CY Khor, M Z Abdullah, M Abdul Mujeebu and F Che Ani 2010 FVM based numerical study on the effect of solder bump arrangement on capillary driven flip chip underfill process *International Communications of Heat and Mass Transfer* 37 pp 281 – 286

[8] W B Young 2003 Anisotropic behavior of the capillary action in flip chip underfill *Microelectronics Journal* Vol 34 no 11 pp 1031 – 1036

[9] W B Young 2004 Capillary impregnation into cylinder banks *Journal of Colloid and Interface Science* Vol 273 pp 576 – 580

[10] S H Lee, J Sung and S E Kim 2010 Dynamic flow measurements of capillary underfill through a bump array in flip chip package *Microelectronics Reliability* Vol 50 No 12 pp 2078 – 2083

[11] Y B Kim, J Sung and M H Lee 2011 Micro-PIV measurements of capillary underfill flows and effect of bump pitch on filling process *Journal of Visualization* Vol 14 No 3 pp 237 – 248

[12] X J Yao, Z D Wang and W J Zhang 2014 A new analysis of the capillary driving pressure for underfill flow in flip-chip packaging *IEEE Transactions on Components, Packaging and Manufacturing Technology* Vol 4 No 9 pp 1534 – 1544

[13] J W Wan, W J Zhang and D J Bergstrom 2007 A theoretical analysis of the concept of critical clearance toward a design methodology for the flip-chip package *Journal of Electronic Packaging* Vol 129 pp 473 – 478

[14] W B Young and W L Yang 2006 Underfill of flip-chip: the effect of contact angle and solder bump arrangement *IEEE Transactions on Advanced Packaging* Vol 29 No 3 pp 647 – 653

[15] J Wang 2002 Underfill of flip chip on organic substrate: viscosity, surface tension, and contact angle *Microelectronic Reliability* Vol 42 pp 293 – 299

Research on Determination of Optimum Parameters for The Lower Extremity Exoskeleton

N H Tho[1,*], T T T Vu[2] and H H Nghi[2]

[1] Faculty of Mechanical Engineering and Technology, Ho Chi Minh City University of Food Industry (HUFI), 140 Le Trong Tan St., Tay Thanh Ward, Tan Phu Dist., Ho Chi Minh City, 700000, Vietnam

[2] Ho Chi Minh City University of Technology, VNU-HCM, 268 Ly Thuong Kiet St., Ward 14, Dist. 10, Ho Chi Minh City, 700000, Vietnam

Corresponding author *: tho.nh@hufi.edu.vn or nguyenhuutho99@gmail.com

Abstract. Lower extremity exoskeleton is a device that enables the user to sit down without the need of using a chair. The design of the exoskeleton has to satisfy multiple objectives, many of which are defined by the body of the user. Most CAD (Computer aided design) programs nowadays still have not added multiple-objective evolutionary algorithms (EAs). This paper presents an application of NSGA-II (Non-dominated sorting genetic algorithm II) and AMO (Adaptive-Multiple objectives) in order to solve the problem of multiple-objective optimization in designing Lower extremity exoskeleton. Moreover, the Box-Behnken design (BBD) - based response surface methodology was also used to investigate the effects of the number of initial samples, the number of samples per iteration and the maximum number of iterations on AMO. Then, the Pareto-optimal frontier of feasible points was carried out for the alternative solutions of design problem. In addition, the TOPSIS (Technique for Order Preference by Similarity to an Ideal Solution) method was employed in order to support the decision makers in deciding the engineering optimum parameters. The optimum parameters obtained are $\alpha_{min} = 24.81^\circ$, $\alpha_{max}=115.08^\circ$ and $\alpha_{stand} = 177.2^\circ$. With the same input values, AMO and NSGA-II gave the same output values, however the number of samples which are used to determine convergence of AMO only equals to 0.66 that of NSGA-II. The method in this study is advantageous for multiple-objective optimization in CAD problems.

Keywords: *MOGA, NSGA-II, AMO, exoskeleton, product design.*

1. Introduction

In the 1960s, the USA's Ministry of Defense began to care about expanding human capacities. First, they started developing exoskeletons that allowed soldiers to carry more weight [1]. More than military purposes, researchers also focused on medical services [2]. Nowadays, the rate of older workers or old people in general having bone and joint related problems is rising. According to the population report of General Statistics Office of Viet Nam, Viet Nam has passed population aging period since 2017 [3]. Therefore, the need of developing an exoskeleton for elders has increased over time. At the moment, there is already an exoskeleton developed with this particular purpose in mind [4] (Figure 1), robot HAL [5], a device that helps you sit down, design and manufactured by Noone company [6], Archelis of NITTO [7]. However, the lower extremity of the exoskeleton is not compatible with Vietnamese's body, the price's also too high. Therefore, designing a lower extremity

IOP Conf. Series: Materials Science and Engineering **530** (2019) 012013 doi:10.1088/1757-899X/530/1/012013

that satisfies multiple objectives is necessary. The optimum designs have become more and more important during the last three decades. Multiple objective problems are becoming more popular. The interval method is usually used in multi-objective problems, however, it needs a large amount of calculations and consumption time. In this particular issue, the split method in Solidworks needs to perform 43,281,400 mathematical calculations. Here, the problem has 4 decision variables, 3 objectives and 1 constraint, this number exceeds the program's capacity. In order to solve the problem, the intervals must be large enough and the amount of calculations must be reduced. This process requires careful analysis of the given data, then deciding the intervals to achieve the optimum parameters. The computational capacity of computers and human are not always able to solve any problem. Therefore, algorithms like the Golden-section search method, coordinate search algorithm, method of steepest descent and genetic algorithm (GA) were developed to overcome the above difficulties.

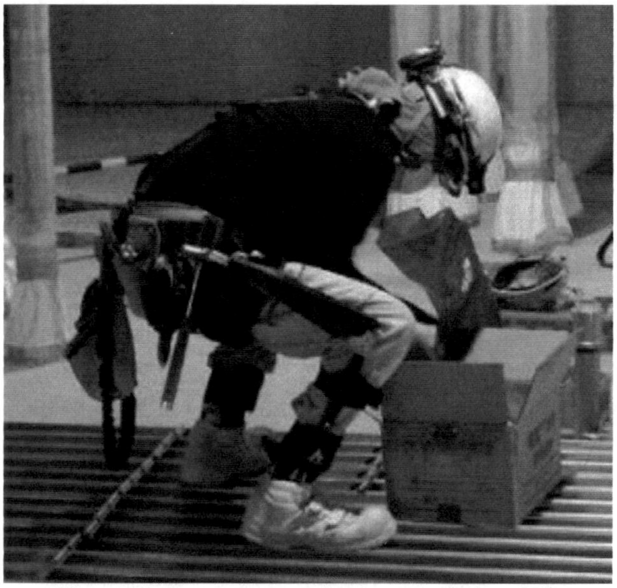

Figure 1. Lower extremity LegX of Ekso [2].

In the mentioned algorithms, the genetic algorithm based on the non-dominated sorting procedure (NSGA-II) has been proven to be most efficient. NSGA-II combined with Kriging is called AMO – adaptive multi-objective optimization, it is available in ANSYS simulation tool of CAE (computer aided engineering). This paper will connect the visual aspect of design in Solidworks and the AMO algorithm of ANSYS, thence solving the problem of optimizing multiple objectives in the design of the lower extremity.

2. Mathematical model

In Viet Nam, Military Technical Academy (MTA) is one of the first institutes to research and develop the exoskeletons. With government investment, the exoskeleton that the MTA developed is focused on military purposes and helping the disabled ones. In order for the exoskeletons to become usable, there must be a large investment and support from all sources to properly research and test in all conditions. However, there are few researches with the purpose of supporting workers and elders in Viet Nam in the particular manner. The exoskeleton can be split into two kinds: The lower extremity and upper extremity. The reason is because of how the two parts of the body operate. The arms – upper extremity are for flexible and exact jobs. The legs – lower extremity are for moving mostly [8].

The main objective of the paper is the lower extremity of the exoskeleton that helps the body to squat or crouch. The exoskeleton must not obstruct the body from moving around; moreover, it must support the body and acts like a moving chair.

IOP Conf. Series: Materials Science and Engineering **530** (2019) 012013 doi:10.1088/1757-899X/530/1/012013

The lower extremity is described in Figure 2. In which, a is the length of the thigh, b is the length of the leg, c is the locking mechanism, and d is the support mechanism. These dimensions are illustrated in Figure 2 and Figure 3.

Table 1. The targets of the lower extremity.

			Target
	c	β	α
Lowest sitting point	215 mm	135°	20°
Highest sitting point	356 mm	135°	120°
Standing point	374.5 mm	45-90°	180°

The exosketon has three targets: the lowest angle of the thigh when squatting is 20° and the highest one is 120°. When the legs are straight, the respective angle is 180°. When you walk, the angle will vary between 180° and 120°, depends on the analysis of stepping sequences [9]. And there is one constraint, when standing the γ angle must be smaller than 180°. Table 1 describes the values and the targets of the mathematical model. The decision factors a, b, d and their limits are shown in Table 2.

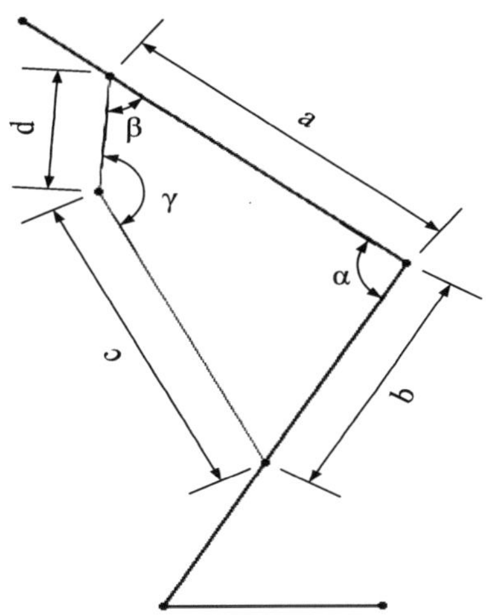

Figure 2. Principle model of the lower extremity.

The objective function α of the three levels: Lowest sitting point, highest sitting point, and standing point are displayed in its most basic form in equation 1.

$$\alpha = \arccos\left(\frac{a^2 + b^2 - f^2}{2 \times a \times b}\right) \tag{1}$$

where, f is the side calculated from equation 2.

$$f = \sqrt{d^2 + c^2 - 2 \times d \times c \times \cos(\gamma)} \tag{2}$$

and γ angle is calculated from equation 3.

IOP Conf. Series: Materials Science and Engineering **530** (2019) 012013 doi:10.1088/1757-899X/530/1/012013

$$\gamma = \arccos\left(\frac{d^2 + e^2 - a^2}{2 \times d \times e}\right) + \arccos\left(\frac{c^2 + e^2 - b^2}{2 \times e \times c}\right) \tag{3}$$

Table 2. Values and their limits.

Parameters	Minimum values (mm)	Maximum values (mm)
a	5	485
b	5	485
d	1	150

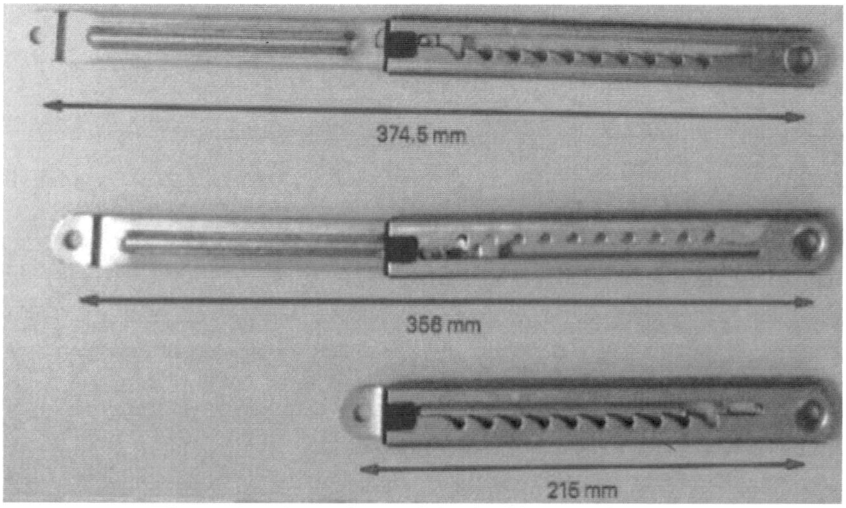

Figure 3. Dimensions of the locking mechanism.

3. Research methodology

3.1 Kriging Algorithm
Kriging [10] is a model algorithm combining with the response surface method (RSM). This method is suitable for changing variables that have a significant effect on the output values. It is an exact two-way interpolation function that combines a polynomial model like response surfaces – giving a global model of the design space and a local deviation, then interpolate the planning points. Kriging has the ability to screen the inputs continuously, including the real values. However, it does not support discrete variables. The efficiency of Kriging depends on the ability to estimate the internal error to improve the quality of the response surfaces by creating the screening points and putting them into the areas that need them the most.

3.2 MOGA (Multi-Objective Genetic Algorithm)
MOGA is a variant of NSGA-II (Non-dominated sorting based genetic algorithm II) that depends on the notation of controlling the dominating individuals [11]. It supports all kinds of input values. The Pareto chart is performed using the rapid classification method. The method of solving the constraints uses the dominated rule and entirely depends on the targets. Therefore, this method does not use penalty functions and Lagrance method. Because of that, the feasible solutions are always ranked based on the priority. The MOGA's flowchart is displayed in Figure 4.

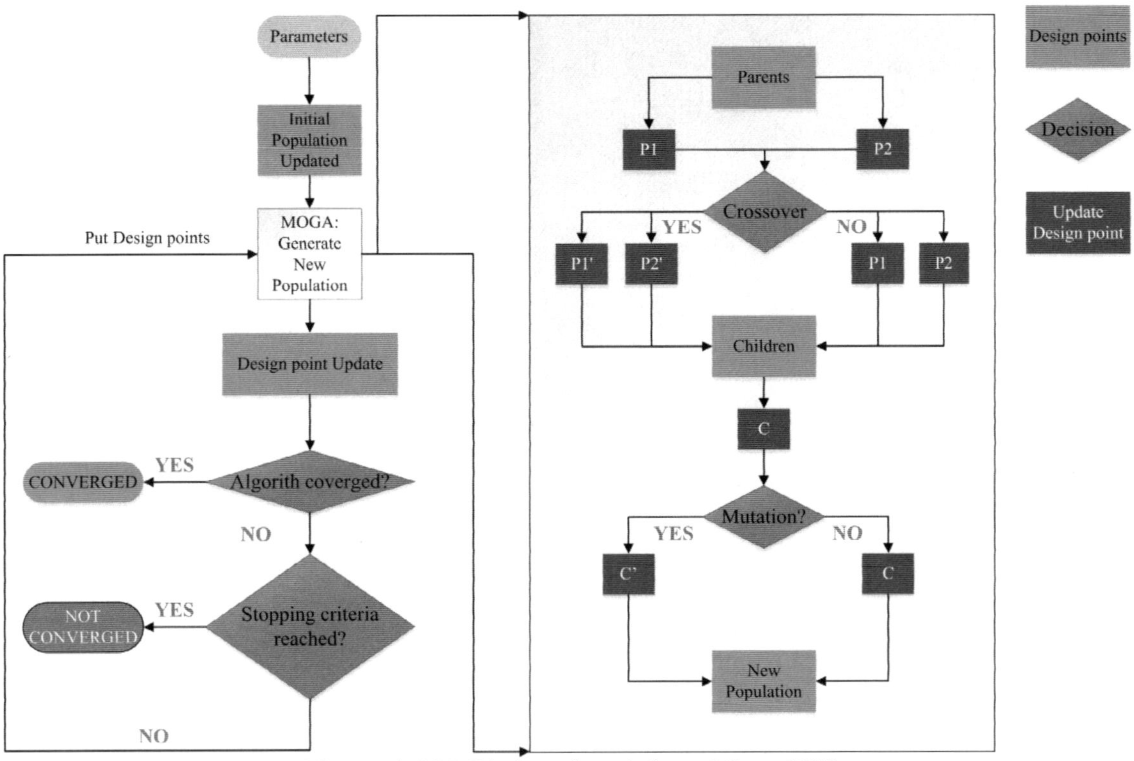

Figure 4. MOGA flowchart (adapted from [12]).

3.3 AMO Algorithm

Adaptive Multiple-Objective (AMO) is a method that incorporates the Kriging and the MOGA. The AMO creates a collection of available samples or uses an available collection in order to approach the problem better than passive searching. This method only rates the designing points when it's necessary. One part of the population can be simulated by rating the Kriging response surfaces. Kriging error prediction, therefore, decreases the amount of necessary evaluations used in finding Pareto peripheral solutions. The AMO flowchart is shown in the second chart in Figure 5.

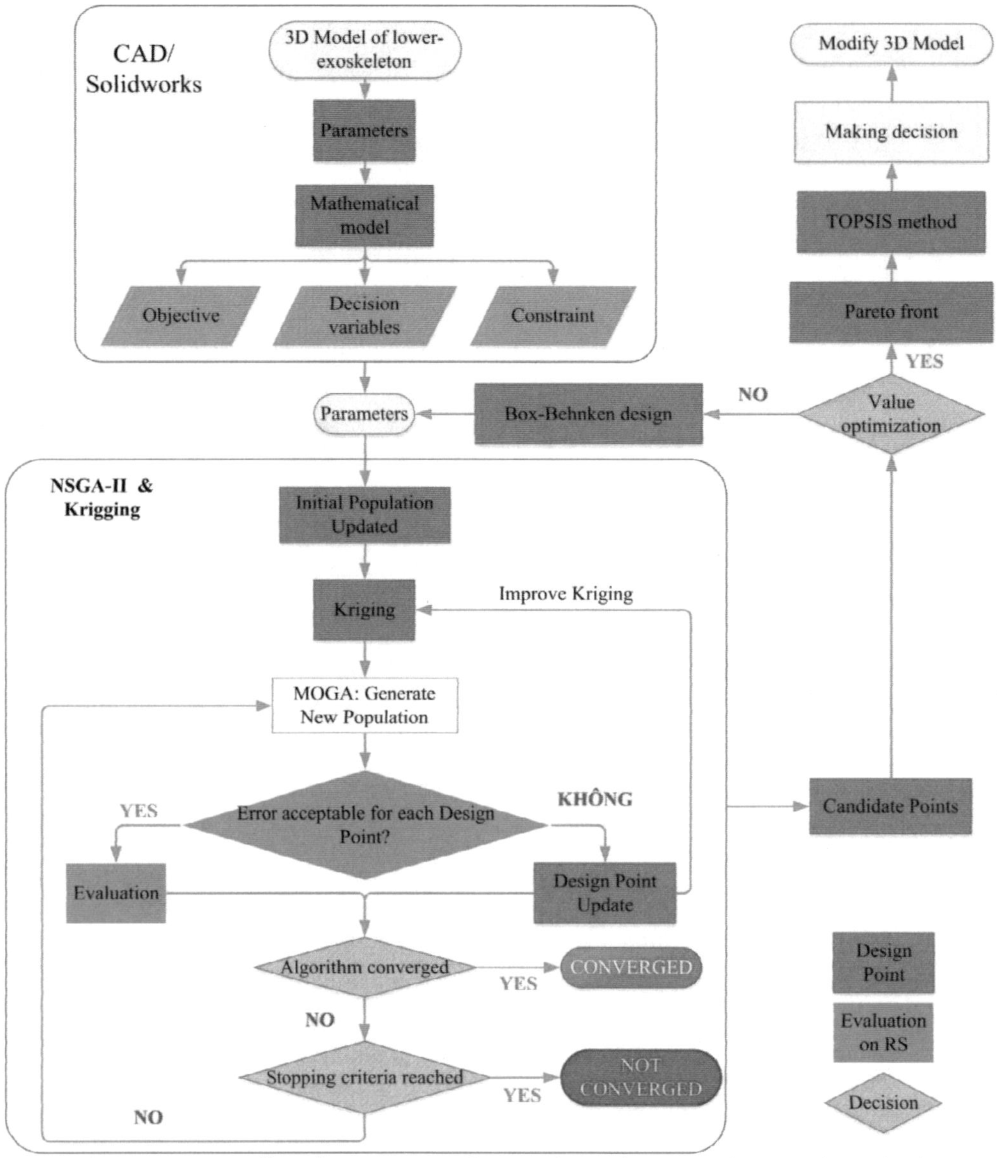

Figure 5. AMO flowchart as general procedure of proposed research method.

3.4 TOPSIS method

TOPSIS (Technique for Order of Preference by Similarity to Ideal Solution) method was developed in 1981 [13]. It is the decision making method used as there are multiple factors affecting the problem. The chosen decision has the shortest distance to the positive ideal solution and furthest distance to the negative ideal solution. TOPSIS is especially suitable in solving multi-criteria decision making problems.

3.5 Analysis procedures

Based on the model principle, we build the CAD (computer aided design) model in Solidworks and create the basic constraints in the 2D design environment. Then, the target variables are assigned, decision variables and constraining variables according to their attributes that can be connected using Solidworks and ANSYS. In ANSYS, we start running the AMO method using Box-Behnken design technique with three factors as shown in Table 3. From the designed experiments, we begin analyzing response surfaces to find out the optimum parameters for the AMO method. Moreover, we replace the AMO's default parameters with the ones that we found to get the optimum values for the lower extremity. The results highlighted that the Pareto chart with optimum values was obtained. Finally, we use TOPSIS decision-making method to select the best value from the feasible ones to use in the CAD model. The research procedure is proposed in Figure 5.

Table 3. Factors and their values.

Factors	Levels		
	Highest (1)	Lowest(-1)	Base (0)
Initial samples	1000	500	750
Maximum samples for each generation	300	100	200
Maximum number of generations	100	50	75

4. Results and discussion

The adaptive multiple objectives (AMO) algorithm implemented the calculation process to find the most feasible solutions. After the number of iterations and the terminated condition obtained, ten most solutions were collected based on the non-dominated sorting procedure from the AMO. These solutions as the non-dominated individuals of evolutionary algorithm were shown in Table 4. Then, the TOPSIS was used to determined the score of ranking of the individual's alternatives based on the multiple criteria of sitting points. Table 5 was the experimental plan for determine the optimal parameters of MOGA and AMO algorithms. The data of Table 5 were created based on the simulation of AMO in the ANSYS environment. The design of experiments (DOE) was used to identify the optimal parameters of the number of iteration, number of samples for the MOGA and AMO as Figure 6. Finally we have run the AMO algorithms with above optimal parameters again, we had got 10 individuals that have optimal values and had been decribed on Pareto frontier in Figure 7. Based on the results obtained in the mentioned 10 individuals, we continued to perform the TOPSIS to choose the best alternative again. In the similar manner with above procedure, we used the MOGA method to identify the optimum results for the best solutions. The results showed that we determined 10 optimum individuals and were presented in the Pareto chart as Figure 8. For the speed of calculation, the AMO showed that the converging results after performing the analysis n_{AMO} = 1179 times, and MOGA converges after performing n_{MOGA}= 1778 times. Thus, we concluded n_{AMO}/n_{MOGA}= 1179/1178 = 0.66, the AMO converged quicker than MOGA algorithm. The connection and interface between the Solidworks (computer aid design system) and ANSYS (computer aided engineering system) to perform the algorithm also played a role to slow down the computational speed. The time needed to switch between the two program needs averagely 15 seconds, so the total time needed to do 15 experiments is quite long and required to be really considered careful at the beginning of the computational process.

Table 4. The 10 optimum individuals after AMO analysis.

Individuals	α Lowest sitting point (α ngoi thap nhat)	α Highest sitting point (α ngoi cao nhat)	α Standing point (α dung)	TOPSIS score	Rank
1	24.82	115.08	177.21	0.913	1
2	26.00	97.89	175.72	0.842	2
3	62.34	114.39	179.88	0.578	7
4	19.44	111.12	156.61	0.563	8
5	31.04	88.64	186.42	0.756	3
6	30.96	88.65	186.75	0.751	4
7	30.96	88.65	186.75	0.751	6
8	72.51	119.96	194.40	0.356	10
9	28.28	87.94	190.05	0.716	6
10	24.11	104.13	149.20	0.466	9

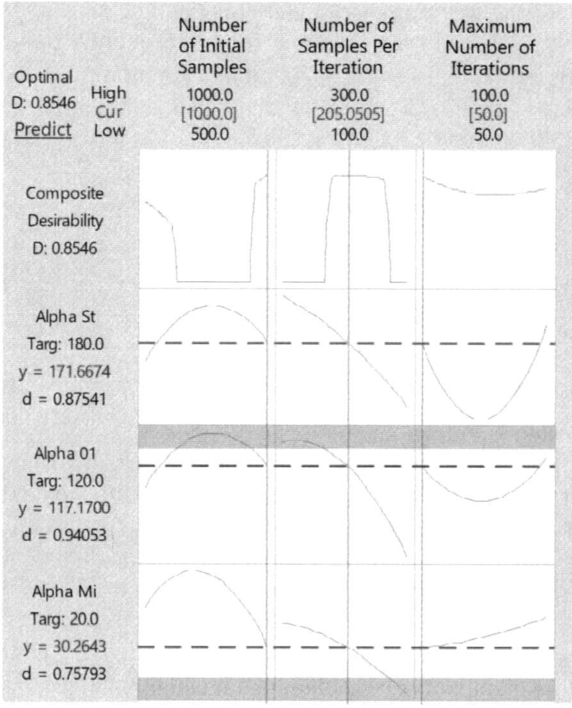

Figure 6. Computation for optimum inputs of AMO method.

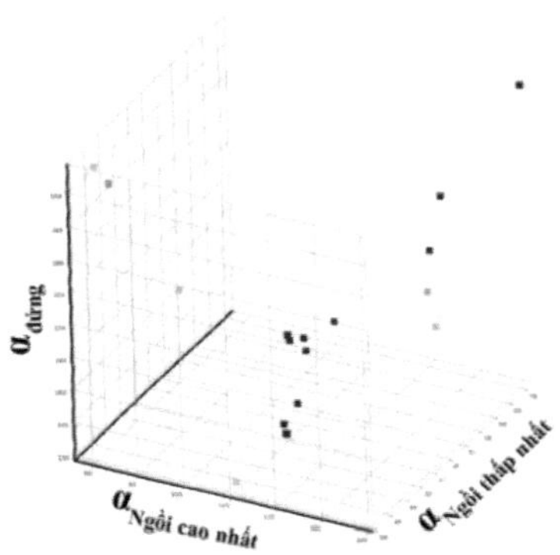

Figure 7. Pareto chart of AMO method.

IOP Conf. Series: Materials Science and Engineering **530** (2019) 012013 doi:10.1088/1757-899X/530/1/012013

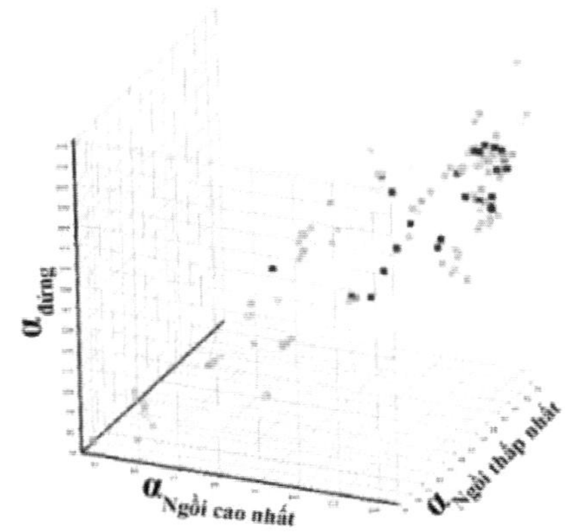

Figure 8. Pareto chart of MOGA method.

Table 5. Results of the experiments.

No	X1 Beginning samples	X2 Maximum samples for each generation	X3 Maximum number of generations	α Lowest sitting point	α Highest sitting point	α Standing point
1	500	100	75	27.086	72.411	113.121
2	1000	100	75	37.289	110.219	172.925
3	750	100	50	62.403	120.866	176.376
4	750	100	100	62.403	120.866	176.376
5	1000	200	100	34.866	115.048	159.849
6	500	200	50	54.064	116.230	183.107
7	1000	200	50	34.866	115.048	159.849
8	500	200	100	54.064	116.230	183.107
9	750	200	75	59.002	117.364	163.735
10	750	200	75	59.002	117.364	163.735
11	750	200	75	59.002	117.364	163.735
12	1000	300	75	26.231	102.432	167.563
13	500	300	75	35.802	115.835	160.059
14	750	300	100	60.061	119.818	184.290
15	750	300	50	37.289	110.219	172.925

5. Conclusion

The exoskeleton can squat at the lowest knee angle of 24.82°, highest at 115.08° and when not affecting the mobility of the user at 177°. These angles have been suitable with collected data and competitive when compared to the parameters in the market. These values are useful for mechanical design of the lower extremity of the exoskeleton that helps the body to squat or crouch. Besides, we can consider another aspect about the applications of interface between CAD and CAE systems. Especially in this study, the AMO method has shown its advantages over MOGA method in regards of both results and converging time, and the main contributions of this work were determined as follows:

- Built a process that helps optimize multiple objective problems using NSGA-II combined with Kriging.
- Compared MOGA and AMO algorithms.
- Employed the design of experiments to investigate the input factors affecting AMO method.
- Based on the input values and the optimization algorithm, the problem of designing the lower extremity of the exoskeleton was solved.

IOP Conf. Series: Materials Science and Engineering **530** (2019) 012013 doi:10.1088/1757-899X/530/1/012013

The advantage of AMO is solving the multiple-objective problem with a faster time of converging. However, because the CAD system has not implemented AMO yet, connecting the two programs prolongs the amount of time needed to calculate and exchange. Changing the parameters only takes 1 or 2 seconds, but switching between two programs takes upwards 13 seconds. Future directions for this research can be implementation of AMO directly integrated into a CAD system such as Solidworks, Autodesk Inventor and Solid Edge. At that time, the computation time can be reduced 10 times for multi-objective design problems compared with the current status.

IOP Conf. Series: Materials Science and Engineering **530** (2019) 012013 doi:10.1088/1757-899X/530/1/012013

References

[1] Mosher R 1967 Handyman to Hardiman SAE Technical Paper 670088
[2] Dollar AM and Herr H 2008 *IEEE Trans Robot* Vol 24
[3] Nguyen AT 2015 VNU Journal of Social Sciences and Humanities Vol 1 (1) (in Vietnamese).
[4] suitX 2017 source: http://www.suitx.com/legx, (accessed on 27/12/2017)
[5] Y. Sankai 2011 "HAL: Hybrid Assistive Limb Based on Cybernics," Robotics Research: The 13th International Symposium ISRR 25-34
[6] Noonee 2017 source: http://www.noonee.com/, (accessed on 27/12/2017)
[7] Archelis 2017 source: https://www.archelis.com/, (accessed on 27/12/2017)
[8] Bruno S and Oussama K 2008 *Springer Handbook of Robotics*, Springer-Verlag Berlin Heidelberg, ISSN: 978-3-540-30301-5, DOI: 10.1007/978-3-540-30301-5 p 773-793
[9] Valiente A 2005 Massachusetts Inst of Tech Cambridge Media Lab
[10] Mukhtar A, Ng K C and Yusoff M Z 2017 *Int J Model Optim* Vol 7 p 92-97
[11] Deb K, Pratap A, Agarwal S and Meyarivan T 2002 *IEEE Transactions on Evolutionary Computation* Vol 6 No 2 p 182-197
[12] ANSYS Help 2017 https://www.ansys.com, accessed on 27/12/2017
[13] Lokare VT and Jadhav PM 2016 " IEEE Proc. *2016 Int. Conf. on Computer Communication and Informatics (ICCCI,* 7-9 Jan. 2016*)* Coimbatore India p 1-6

Numerical Investigation of Continuous Damping of The Semi-Active Suspension System for Passenger Car

M R Ahmed[1*], A R Yusoff [1] and F R M Romlay[2]

[1] Faculty of Manufacturing Engineering, University Malaysia Pahang 26600, Pekan, Pahang, Malaysia

[2] Faculty of Mechanical Engineering, University Malaysia Pahang 26600, Pekan, Pahang, Malaysia

Corresponding author *: razlan@ump.edu.my

Abstract. The suspension of the car is considered an important element in the vehicle. The primary function of the suspension system is to isolate the vehicle structure from shocks and vibration due to irregularities of the road surface. There are two main objectives need to be satisfied which are: ride comfort and road handling. Ride comfort is inversely proportional to the absolute acceleration of the vehicle body, while the road handling is linked to the relative displacement between the vehicle body and the tires. This paper presented an attempted to enhance the performance of the shock absorber by developing a model of continuously variable damping (CVD). To evaluate the effect of the developed semi-active shock absorber on the dynamic behaviour of the vehicle, the model was analyzed and compared with the passive and On/Off sky-hook control strategy in the quarter car using two different types of road (random excitation, bumpy) as input to the quarter car model. Force hysteresis loop with different sets of orifice diameter was generated. The result indicates the CVD shows a reduction in both body acceleration and vertical displacement contrasting with passive and On/Off sky-hook 73.4% and 53.8% respectively and also the selling time by 79% and 59% for a bumpy road. This considered an improvement toward the ride comfort and vehicle stability. The simulated results for the quarter car model are shows similar trends and within range when compared with reference research paper.

Keywords: *Vehicle Suspension System; CVD; controllable; absorbers; orifice.*

1. Introduction

The concentration of researchers and automobile manufactures on suspension system back to decades of time as it improves the car stability, handling characteristics and provide more comfort to both the car and the passenger. The suspension system considered one of the most important parts of the vehicle toward the dynamic behaviour [1-5]. The history drawback that the passenger cars are the main land transportation ever used by mankind. Base on the controllability, the suspension can be divided into three types: passive or conventional, semi-active and active suspension [3]. The passive or conventional suspension comes with a fixed damping ratio by the manufactures where the users cannot change or adjust its dynamics properties, this will draw up a conflict between ride handling and comfort [4-8]. Active suspension is very convenient in term of performance toward the enhancement of vehicle dynamic behaviour and mostly used in luxury vehicles, despite that, it requires high power consumption, control units and high cost due to the complexity, and required hardware and software [7-8]. Lastly, the semi-active suspension system dispenses better performance once compared with the conventional, while with the active in term of cost reduction and simplicity [6-7].

In general, the suspension system is a combination of three parts: springs, shock absorbers and linkages, each one of them have its own contribution to the stability of the car [9]. Shock absorber considered an important element of the suspension system, function as energy dissipation and vertical displacement reduction, it uses mostly by land transportation. The shock absorber evaluated by hysteretic diagram also called force-velocity diagram. The shock absorber can be categorized into two based on their way of dissipating energy: hydraulic and dry friction. Hydraulic shock absorbers are commonly used in current automobiles [10].

Tremendous researchers give their effort toward the development of a numerical model that can observe the damping behavior. Identification methods can be categorized into: parametric and non-parametric also called physical perspective. Parametric modeling regards the output (as for example, the typical force-displacement plot from which the hysteresis phenomenon is evident) more or less like a black box [11]. Despite the fact that this technique is powered by the capacity of machine learning in perceiving conditions among the inputs and outputs, there are practical and theoretical downsides. There is another was proposed called gray box, artificial neural network technique was adopted to overcome the hysteresis in the shock absorber, the hydraulic and friction forces were considered within the parametric model [12]. Parameter identification of non-linear hysteretic model of shock absorber was discussed, the model considered three passes of the fluid flow in monotube charged with gas: flow through the piston orifice, bleed orifice and piston leakage due to gap between the piston and cylinder, beside that the stiffness properties of stock shims and the moving piston that separates the gas the oil in the chamber were included [9] [13]. In a few decades back, Lang made first attempts to formulate the characteristic of shock absorber numerically, the complexity of evaluation of the physical model was occurred due to the tremendous number of parameters reached up to 80 [14]. A few years later, Morman used Lang's model to validate with experimental testing [15]. Later on, a great achievement was occurred in simplifying the physical model by reducing the number of parameters [16]. Chavan carried out a study of the dynamic response of hydraulic damper and its influence by force-displacement hysteresis, and road input to ride comfort and vehicle stability [5].

2. Variable damping and control strategies

Previously, semi-active shock absorbers were manually adjustable by opening or closing a bypass valve. Eventually, electronically adjustable dampers were introduced by researches such as the hydraulic damper [17], the magneto-rheological fluid damper (MR) [2] and the electro-rheological fluid damper (ER) [18]. Mostly the attention is given to the hydraulic damper, due to the simple structure, lower cost, and reliability in a term for performance [9]. The appropriate control technique is delineated as the one that reduces the value of the body vertical displacement, tyre vertical displacement, and the vehicle body acceleration. typically, these quantities are depending on the mass of the sprung and unsprung masses. Different types of control hypothesis have been implemented by many researchers in determining the most appropriate control strategy that can reduce the elevation factors [19].

A few decades back Karnopp et al. as the first to initiate the idea of the semi-active suspension system. An intelligent controller has been utilized to produce an active actuating force and, he introduced sky-hook strategy which is about fixing the adjustable damper between the car body and imagining hook and ground -hook [8] [19]. Swevers et al. have developed a parameterized, model-free control structure for a passenger car equipped with an electro-hydraulic semi-active suspension system [20]. Valášek et al and Ahmadian et al. examined control algorithm called hybrid control that homogenized the advantages of sky-hook and ground-hook. Furthermore, an integrated preference of both the sky-hook (SH) and the acceleration was driven damper control (ADD), in their respective ranges of the best behaviour. As a result, SH-ADD can provide the quasi-optimal performance of ride qualities [21-22]. A continuously variable damper (CVD) was presented by some researchers, based on skyhook and ground-hook control strategy, damper force in the semi-active case must be equal to the damper force in the Skyhook case [23-24]. Used the advantage of mechanical power without hardware or energy consumption to construct a set of four valves that can realize the performance semi-active damper from the passive damper base on the velocity and displacement [25]. Zhao et al. Adapted reversible technique to identifying the roughness of road by using adjustable semi-active suspension, which can lead cost reduction of the production [26].

In addition to the above controlling algorithms and techniques, there are also a large number of another control techniques been originated for the seek of enhancing the ride quality. For example, clipped optimal control [27-28], sliding mode control [29], LQR/LQG [30-31] and so on.

In this paper, an attempt is made to develop a mathematical model to realize semi-continuous variable damping (CVD) characteristics from commercially available monotube shock absorber. This model inspired by the model presented in [4-5] [14]. Therefore, the force-displacement and force-velocity characteristics curves are determined and presented, and furthermore to apprehend the variable damping properties a time response of quarter car simulation is carried out and compared with passive and on/Off sky-hook control to prove the effectiveness.

3. Damper description

The damper consists of several main parts as illustrated in Table1. The tube of the damper houses the standard internal piston. Once assembled, the tube is divided into three chambers: gas, rebound, and compression as the typical monotube damper, shown in figure 1. The compression and rebound chambers are connected to each other through a restricted piston orifice and a bypass valve which is controlled by a stepper motor. When in compression, the fluid flows through two flow paths, through the internal piston and bypass valve. The damping through the internal piston is dictated by the area of the orifices and the stiffness of the shim stacks whereas the flow through the bypass valve depends upon the displacement adjustable pin that is controlled by application of voltage. This takes place the same way but in the opposite direction of the rebound stroke. An electronically controlled semi-active damper mechanism to be implemented. The size of the single semi-active damper was kept as the universal damper size to utilize the advantage of the use of the standard parts that to obtain equivalent results when tested manually and with a controller. Table 1 presents the parts of the damper.

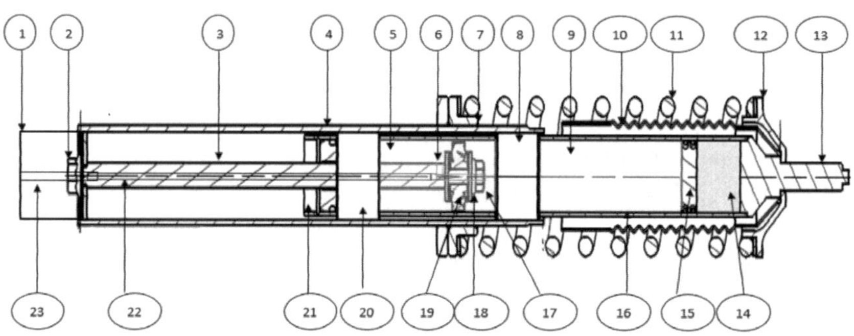

Figure 1. The controllable monotube hydraulic shock absorber

Table 1. Shock absorber parts description.

No	Description	No	Description	No	Description
1	Stepper Motor	9	Compression Chamber	17	Nut
2	Stepper Motor Mounting	10	Dust Cover	18	Shim Stock
3	Piston Rod	11	Spring	19	Piston and Valve
4	Outer Cylinder	12	Cap Bolt	20	Road Guide
5	Rebound Chamber	13	Upper Mount	21	Inner Cylinder Cover
6	Bleed Orifice	14	Gas Chamber	22	Adjustable Rod
7	Adjustable Spring Holder	15	Floating Piston	23	Stepper Motor Shaft
8	Supporting Cover of outer Cylinder	16	Oil Cylinder		

4. Fluid flow modelling

The damping force is generated due to the resistance of hydraulic fluid flow between the different side of the shock absorber through a set of paths. The major assumption for summation of flow is that the damper oil is incompressible and therefore has a constant density. This assumption allows consideration of volumetric rather than mass flow rates [4]. This is expressed in equation 1 and shown in figure 2 is the total volumetric flow rate of the damper, Q in m^2/sec.

$$Q = Q_v + Q_b + Q_{lp} \tag{1}$$

where Q is the overall fluid flow, Q_v flow rate through the valves, Q_b is the fluid flow through the bleed orifice and Q_{lp} fluid flow through the piston boundary. Where boundary can be seen in figure 2. Equivalent flow across boundary due to rod insertion, overall fluid flow, Q, is related to velocity regardless of the compression or rebound. This is shown in equation 2.

$$Q = (A_{\dot{c}} - A_r)\dot{x} \tag{2}$$

where, A_c is the area of the piston on the compression side, while A_r is the area of the piston on the rebound side and $\dot{x}$ is the piston velocity.

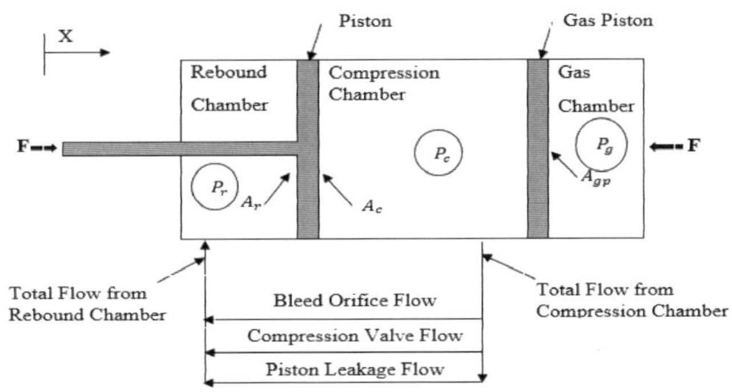

Figure 2. The total volumetric flow rate of the damper diagram.

The pressure difference, $\Delta p=p_c-p_r$, between both side of chambers is considered to establish the partial flow rate for each part. A Bernoulli's equation can be used to model unsteady flow through a passage of area, A. It has the form:

$$Q = AC_D\left(\frac{2\Delta p}{\rho}\right)^{\frac{1}{2}} \tag{3}$$

where the C_D is a coefficient of discharge at steady state and ρ is the density. C_D is a dimensionless parameter including acceleration number. Reynolds number, Cauchy number, and thickness to length ratio have experimentally modified this term by defining a dynamic discharge coefficient, C_D [14].

$$C_D = f\left\{\frac{al}{v^2}, \frac{\mu}{\rho v l}, \beta v^2, \frac{s}{l}\right\} \tag{4}$$

where, the Reynolds number, $Re=\frac{\mu}{\rho v l}$, and Cauchy number, $Ca=\beta v^2$, a is a piston acceleration, l is a piston length, μ is a fluid viscosity, β is a viscous friction coefficient, v is a flow velocity and s is a thickness. The leakage of oil between the piston seal and the cylinder results in a certain amount of fluid flow. Lang modelled this flow using laminar flow through parallel plates [14].

$$Q_{lp} = \left(\frac{(p_c-p_r)b^3}{12\mu l} + \dot{x}\frac{b}{2}\right)\pi D_p \tag{5}$$

where, b is the height of the piston, D_p is the diameter of the piston, μ is a fluid viscosity, l is a piston length, p_c and p_r are compressions and rebound pressure respectively.

5. Variable damper system concept

5.1. Semi-active valve flow (C_b)

The valve is composed of a circular ball detent that allows a change in area from 0 to 5 mm^2. The valve is to be modeled as a bleed hole that allowed a constant amount of fluid at any moment. Equation 6 is formulated to find the flow through the adjustable bleed orifice in the piston.

$$Q_b = A_b(\theta)\, C_b \left(\frac{2(p_c - p_r)}{\rho}\right)^{\frac{1}{2}} \tag{6}$$

Where, A_b, Area of the bleed, was determined through measurement of the damper bleed orifice. Area of the variable bleed orifice as a function of displacement angle can be calculated from equation 7 [27].

$$A_b(\theta) = 2\left[tan\frac{\theta}{2}\left((R^2 - r^2)\left(r^2 sec^2\frac{\theta}{2} - R^2 tan^2\frac{\theta}{2}\right)\right)^{\frac{1}{2}} - r^2 cos^{-1}\left(\frac{(R^2 - r^2)^{\frac{1}{2}}}{r} tan\frac{\theta}{2}\right)\right]\frac{R\,sin\left(sin^{-1}\frac{r}{R} - \frac{\theta}{2}\right)}{r - (R^2 - r^2)^{\frac{1}{2}} tan\frac{\theta}{2}} \tag{7}$$

where, R is the radius of a rotary pin, r is the radius of the bleed orifice and θ is the displacement angle of a rotary pin can be calculated according to equation 8.

$$\theta = N_s \times 1.8^{\circ} \tag{8}$$

where, N_s is the number of steps generated by the stepper motor

6. The mathematical model of quarter car suspension

Quarter car simulation is considered an effective method to demonstrate the influence of damping force on vehicle dynamic behavior due to input waves from irregularity road surface. This simulation can give us chance to evaluate the effect of the controllable bleed orifice in the damper and the vertical acceleration of the vehicle body by the shock absorber. The simulation was done using the SIMULINK toolbox of MATLAB R2015a. The governing equation for a quarter-car model is expressed in equation 9 according to figure 3.

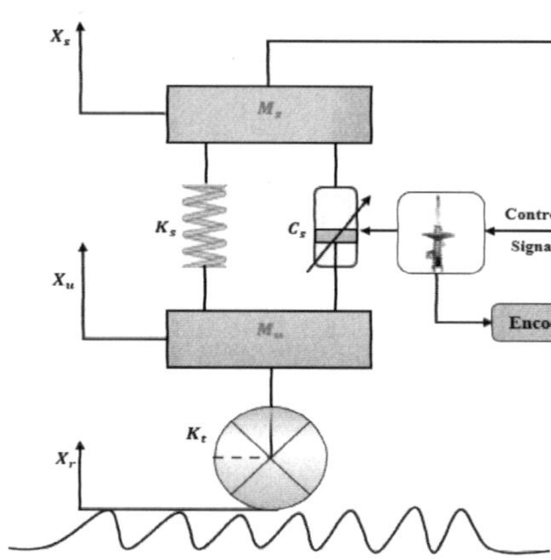

Figure 3. Continuos variable semi active damper.

$$\begin{cases} M_s\ddot{X}_s + C_s(t)(\dot{X}_s - \dot{X}_u) + K_s(X_s - X_u) = 0 \\ M_u\ddot{X}_u + K_t(X_u - X_r) + C_s(t)(\dot{X}_u - \dot{X}_s) + K_s(X_u - X_s) = 0 \end{cases} \quad (9)$$

The sprung mass can be defined as M_s and unsprung mass is represented by M_u. This includes any mass that moves with the wheel and tire. For an independent suspension, a portion of the suspension links and half shafts should be included. K_s and $C_s(t)$ are the spring rate and damper coefficient of the suspension, and K_u is the tire spring properties. The displacement of the sprung and unsprung masses and the road excitation are X_s, X_u, and X_r, respectively. Some assumptions are considered in this model such as neglecting the damping coefficient of the tire and constant spring rate. Vehicle parameters used in this simulation are presented in table 2 according to the exciting literature [33].

Table 2 Properties of quarter car model [33].

Parameter	Symbol	Value	Unit
Sprung mass	m_s	280	kg
Unsprung mass	m_u	35	kg
Shock absorber spring constant	K_s	17600	N/m
Shock absorber damping coefficient	C_s	892-6203	Ns/m
Tire stiffness	K_u	190,000	KN/m

7. Design of Control algorithm

In order to improve the feeling of riding for irregular variation of road surface, an input for a road surface as a disturbance is set and the control algorithm is designed as per the requirements. While designing control algorithm, an ideal skyhook and continuous variable damping (CVD) have been correlated and a following control policy is defined.

$$F_{sky} = C_{sky}\,\dot{X}_s \quad (10)$$

where, $F_{sky}$is skyhook force next is to determine if the semi-active damper is able to provide the same force. If the sprung and unsprung masses are separating, then the semi-active damper is in tension. Thus, the force applied to the sprung mass is:

$$F_{damper} = C_s(t)(\dot{X}_s - \dot{X}_u) \quad (11)$$

where, F_{damper} is the force applied to the sprung mass. Since we are able to generate a force in the proper direction, the only requirement to match the skyhook suspension is,

$$C_s(t) = C_{sky}\frac{\dot{X}_s}{(\dot{X}_s - \dot{X}_u)} \quad (12)$$

The on-off skyhook control strategy switches between high and low damping coefficient in order to achieve body comfort specifications. This control law consists in changing the damping factor of the damper (its fluid flow, fluid viscosity, air resistance, etc) according to the body velocity and the suspension relative velocity by using a logical rule as stated in equation 13[8] [35].

$$C_{sky} = \begin{cases} C_{max} & \dot{X}_s(\dot{X}_s - \dot{X}_u) \geq 0 \\ C_{min} & \dot{X}_s(\dot{X}_s - \dot{X}_u) < 0 \end{cases} \quad (13)$$

It is worth emphasizing that when the product of the two velocities is positive that the semi-active damping force is proportional to the velocity of the sprung mass. Otherwise, the semi-active damping force is at a minimum. Where, C_{max} and C_{min} are the maximum and minimum damping coefficients of a damper, respectively (and usually $C_{max} = C_{sky}$). Equation 13 implies that when the relative velocity across the suspension and the sprung mass absolute velocity have the same sign, a damping force proportional to C_{max} is desired. Otherwise, this control law deactivates the controlled damper when the body velocity and relative velocity have opposite sign.

8. Road profile generation

Two types of road disturbance are assumed as the input for the system. The road profile shown in figure 5 random excitation based on road roughness and a single bump are frequently used by many researchers [1]. The road surfaces are classified into A-H according to the degree of roughness as described in International Organization for Standardization (ISO) 8608 [34] based on power spectral density (PSD) function. Basically, the A-D classes are considered to be paved roads where the vehicle can reach the higher speed and they have a minor degree of roughness. Table 3 presents the road roughness standard deviations for various types of roads.

The random road signal excitation of PSD [33] and the sinusoidal bump can be expressed as in equations 14 and 15 respectively.

$$\dot{X}_r(t) = -\propto VX_r(t) + \eta(t) \tag{14}$$

$$X_r(t) = \begin{cases} \frac{a}{2}\left(\left(1 - cos\left(\frac{2\pi Vt}{\lambda}\right)\right)\right), & 1.0 \le t \le 1.25 \\ 0, & else \end{cases} \tag{15}$$

The solution to the above equation for $Z_o(t)$ is used as road input to quarter-car simulation as presented in figure 5.

where, $\propto$ for the shaping filter and the ISO standard are shown in table 3, V is the car forward velocity on this paper 120 km/h is to be examined and $\eta(t)$ white noise, a is the bump height, λ is the half wavelength of the sinusoidal road undulation.

Table 3. presents $G_\partial(n_0)$values as defined by the ISO 8608 for each pavement category [8] [34].

Road Category	10^3 m	$G_\partial(n_0)(10^6\ \mathrm{m}^3)$, $G_\partial = 1$	$\propto$ (rad/m)
A (very good)	2	1	0.127
B (good)	4	4	0.127
C (average)	8	16	0.127
D (poor)	16	64	0.127
E (very poor)	32	256	0.127

9. Graphic representation using Matlab library blocks

In Figure. 6 is Illustrated the graphical presentation of the semi-active damping based on equations 9-13. Solving was performed using Simulink library blocks computing. In order to analyze the behaviour of the quarter car suspension system, parameters were used as input as shown in table 2.

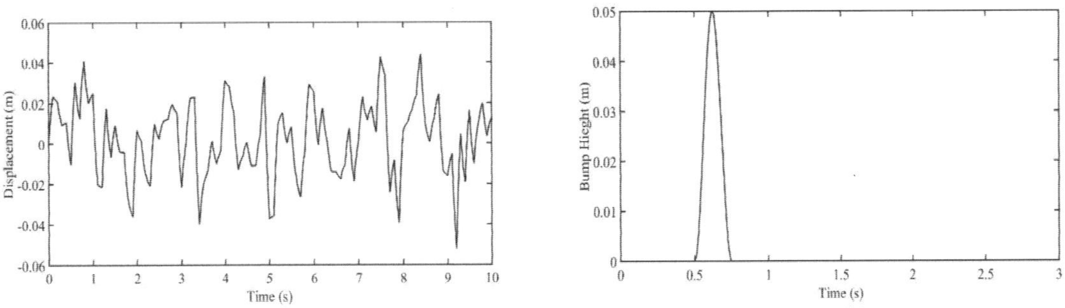

(a) (b)

Figure 4. A segment of the road profile as input for quarter car model (a) input excitation signal, (b) bumpy road input.

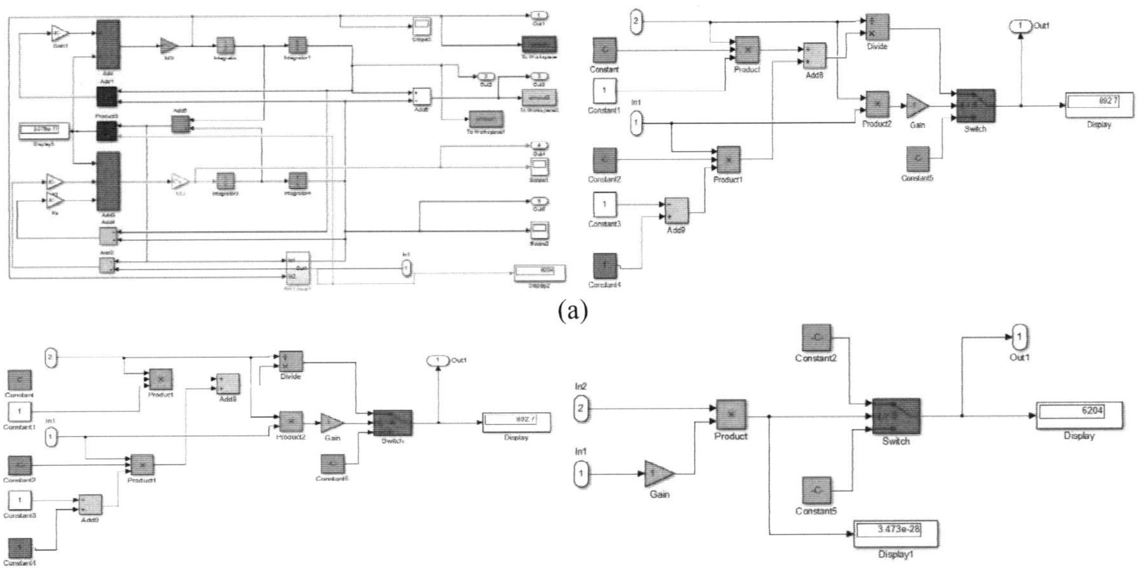

(a)

(b)

Figure 5. Matlab representation of f physical quarter car (a) passive (b) Skyhook and CVD control strategies.

10. Simulation results and discussion

To investigate the effectiveness of the bleed orifice on the generated force, A mathematical modelling for semi-active damper was created and with different setting mode of bleed orifice diameters were examined. The diameter various from 0-2.5 mm, when the setting is zero have the same response as the passive shock absorber. The damping force decreases with increases of orifice damper and verse vice, and this will affect the performance of the shock absorber by varying its softness and stiffness. A force-displacement and force-velocity plots are shown in figure 6.

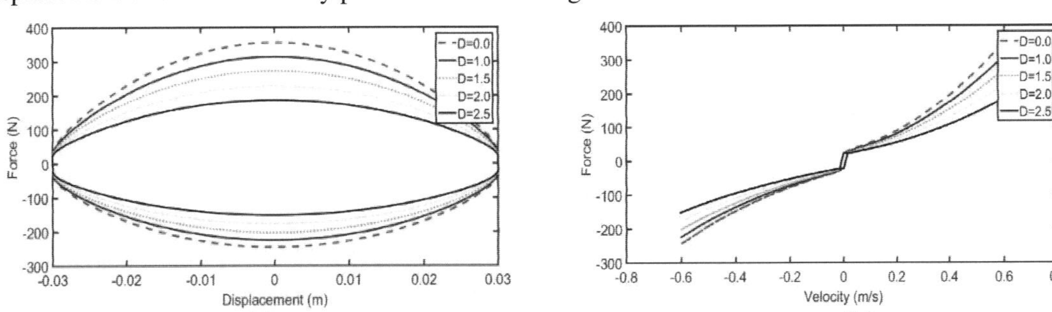

(a) (b)

Figure 6. Details of the predicted damper force from the sine wave excitation with various bleed orifice diameter, (a) Force-displacement curve and (b) force-velocity curve.

The proposed system along with skyhook logic and passive suspension system are simulated to obtain the body acceleration, vertical displacement and suspension deflection for a particular type of road condition. Figure 7(a) shows the sprung mass acceleration response of the passive, On/Off sky-hook and continuous variable damping (CVD) in time domain against the input signal stated in figure 4. CVD has a great decrement in the maximum level of the acceleration and has the best response. The response characteristics of CVD has accomplished a significant reduction in vertical acceleration of the vehicle when compares with both passive and sky-hook control which are measured to be 73.4% and 53.8% respectively. Also, the bumpy road the response of CVD has less overshoot and selling time by 79% and 59% as illustrated in figure 7(b). Therefore, it can be concluded that the ride comfort characteristics of CVD were improved compared with the ones of the passive shock absorber.

Figure 8(a) shows the analysis result of the vertical displacement semi-active damper model in the time domain. Figure 8(b) shows the analysis results of vertical displacement in bumpy road condition. Both these figures provide comparisons between passive, On/Off Sky-Hook and CVD. As shown in figure 8, the response characteristic of the CVD has achieved the lowest amplitude and settling time. CVD shows slightly improved characteristics of ride comfort compares passive shock absorber.

The simulation results also reflect that, the peak to peak suspension deflection of semi active suspension by CVD controller has been decreased tremendously with early dies out of 1.5 sec as illustrated in figure 9. Similarly, the reduction in peak-to- peak velocities and accelerations are also occurred in case of Sky-Hook suspension as compare to passive suspension of sprung mass. However, the benefits achieved in reducing peak-to-peak accelerations are rather low as compared to benefits observed in the peak-to-peak displacements and velocities.

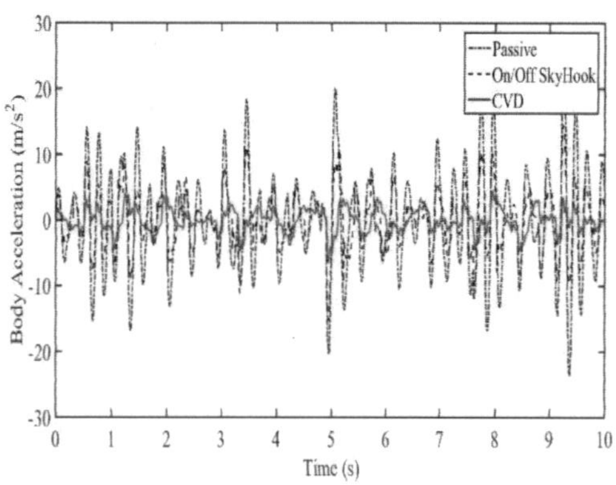

(a)

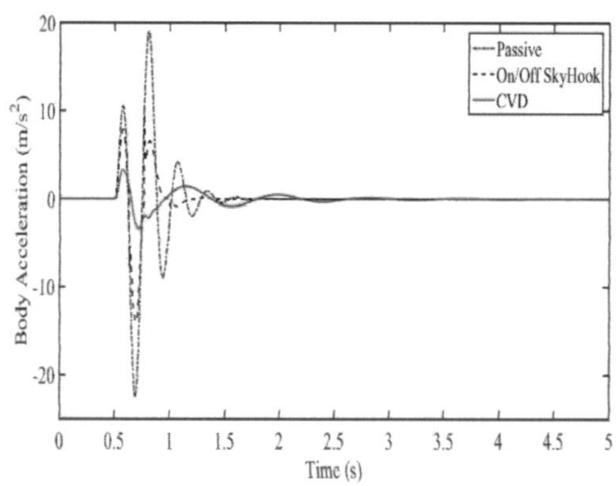

(b)

Figure 7. Sprung mass acceleration response of the semi-active damper model: (a) with input excitation signal and (b) bumpy road as input.

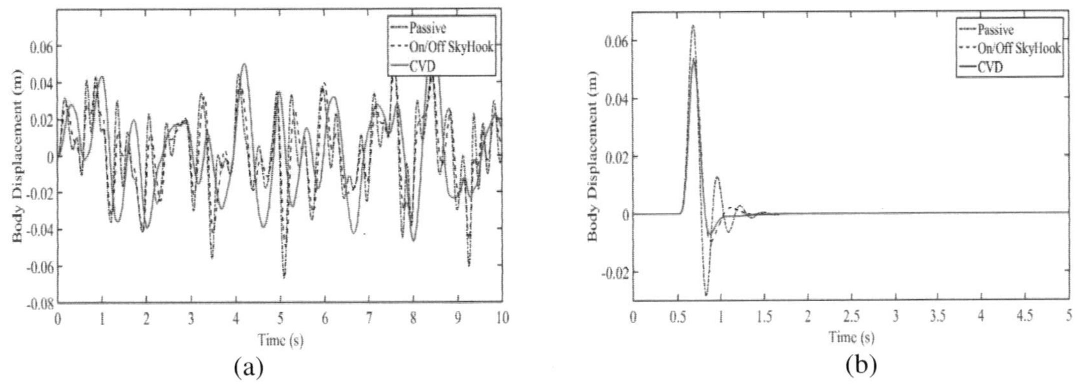

Figure 8. Sprung mass displacement response of the semi-active damper model: (a) with input excitation signal and (b) bumpy road as input.

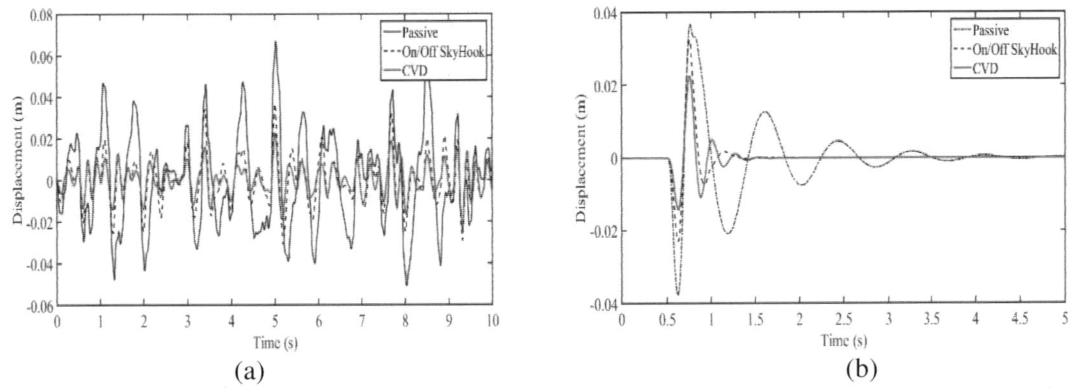

Figure 9. Suspension deflection of the semi-active damper model: (a) with input excitation signal and (b) bumpy road as input.

11. Conclusion

The research presented in this paper demonstrates a mathematical model of an adjustable semi-active damper. The fluid flow rate and the damping force of a semi-active shock absorber of an automotive system were theoretically estimated. The obtained result of the mathematical model was analyzed. It is shown that various damping force can be calculated by adjusting the opening area of the bleed orifice. To ensure the vehicle dynamics performance under the continues variable damping (CVD) semi-active damper, a quarter car model used for analysing. The simulation result of time response was compared with the response of passive shock absorber and the skyhook control strategy on both paved and bumpy road driving conditions. According to the simulation result can be observed the vertical acceleration and displacement are decreased, which inverse proportional with ride comfort. The reported result here is to contribute to an understanding of effect controllable orifice in providing a continuous variable damping (CVD) of the shock absorber.

Acknowledgment

The authors wish to gratefully thank University Malaysia Pahang for financial support received through PGRS180322 grant.

IOP Conf. Series: Materials Science and Engineering **530** (2019) 012014 doi:10.1088/1757-899X/530/1/012014

Reference

[1] Barethiye V, Pohit G and Mitra A 2017 A combined nonlinear and hysteresis model of shock absorber for quarter car simulation on the basis of experimental data *Eng. Sci. Technol. Int J* 20(6) 1610-22

[2] Ata W G and Salem A M 2017 Semi-active control of tracked vehicle suspension incorporating magnetorheological dampers *Veh. Syst. Dyn* 55(5) 626-47

[3] Jugulkar L M and Singh S 2016 Analysis of suspension with variable stiffness and variable damping force for automotive applications *Adv. Mech. Eng* 8(5) 1–19

[4] Talbott M S 2002 An experimentally validated physical model of a high-performance automotive damper *M.S. Thesis* (West Lafayette. Purdue University)

[5] Chacón J L, Boada B L, Boada M J L and Díaz V 2017 Experimental study and analytical model of bleed valve orifice influence of a high-performance shock absorber on vehicle dynamics *Adv. Mech. Eng* 9(9) 1–15

[6] Witters M A and Swevers J 2010 Black-box model identification for a continuously variable, electro-hydraulic semi-active damper *Mech. Syst. Sig. Process* 24(1) 4–18

[7] Phalke T P and Mitra A C 2016 Comparison of passive and semi-active suspension system by MATLAB SIMULINK for different road profiles *IOSR-JMCE* 1 38–43

[8] Hu Y, Chen M Z Q and Sun Y 2017 Comfort-oriented vehicle suspension design with skyhook more inert configuration *J. Sound Vib* 405 34–47

[9] Starkey J and Talbott M S 2002 An experimentally validated physical model of a high-performance mono-tube damper *SAE Technical Paper* 2002-01-3337

[10] Castellani F, Scappaticci L, Bartolini N and Astolfi D 2017 Numerical and experimental investigation of a monotube hydraulic shock absorber *Arch Appl Mech* 87(12)1929–46

[11] Leser C, Renner T E and Salmon D C 2002 Accurate shock absorber load modeling in an all-terrain vehicle using black box neural network techniques *SAE Technical Paper* 2002-01-0581

[12] Beghi A, Liberati M, Mezzalira S and Peron S 2007 Grey-box modeling of a motorcycle shock absorber for virtual prototyping applications *Simul. Pract. Theory* 15(8) 894–907

[13] Duym S, Stiens R and Reybrouck K 1997 Evaluation of shock absorber model *Veh. Syst. Dyn* 27(2)109–27

[14] Lang H H 1997 A study of the characteristics of automotive hydraulic dampers at high stroking frequencies *Ph.D. Dissertation* (Ann Arbor. University of Michigan)

[15] Morman N and Jr K 1984 Modeling and identification procedure for the analysis and simulation of hydraulic shock absorber performance (Ford Motor Company, Michigan)

[16] Reybrouck K 1994 A nonlinear parametric model of an automotive shock absorber SAE Technical Paper 940869

[17] Zhang Y, Chen H, Guo K, Zhang X and Eben S 2017 Electro-hydraulic damper for energy harvesting suspension: Modeling, prototyping and experimental validation *Appl. Energy* 199 1–12

[18] Kishore R, Choudhury S K and Orra K 2018 On-line control of machine tool vibration in turning operation using electro-magneto rheological damper *J. Manuf. Processes* 31 187–98

[19] Karnopp D, Crosby M J and Harwood R A 1974 Vibration control using semi-active force generators *J Eng Ind* 96(2)619-26

[20] Swevers J, Lauwerys C, Vandersmissen B, Maes M, Reybrouck K and Sas P 2007 A model-free control structure for the on-line tuning of the semi-active suspension of a passenger car *Mech. Syst. Sig. Process* 21 (3) 1422–36

[21] Valasek M, Novak M, Šika Z and Vaculin O 1997 Extended ground-hook - new concept of semi-active control of truck's suspension *Veh. Syst. Dyn* 27(5–6) 289-303

[22] Goncalves F D and Ahmadian M 2003 A hybrid control policy for semi-active vehicle suspensions *Shock Vib* 10 59–69

[23] Motta D S, Zampieri D E and Pereira K A 2000 Optimization of a vehicle suspension using a Semi-active damper *SAE Technical Paper* 2000-01-3304
[24] Hwang S, Heo S, Kim H and Lee K 1997 Vehicle dynamic analysis and evaluation of continuously controlled semi-active suspensions using hardware-in-the-loop simulation *Veh. Syst. Dyn* 27(5-6) 423-34
[25] Nie S, Zhuang Y, Wang Y and Guo K 2018 Velocity & displacement-dependent damper: A novel passive shock absorber inspired by the semi-active control *Mech. Syst. Sig. Process* 99 730-46
[26] Zhao L, Yu Y, Zhou C, Wang S and Yang F 2018 A hydraulic semi-active suspension based on road statistical properties and its road identification *Appl. Sci 8*(5) 740
[27] Yokoyama M, Hedrick J K and Toyama S 2001 A model following sliding mode controller for semi-active suspension systems with MR dampers *Proc. 2001 Am. Control Conf. (Cat. No.01CH37148)* 4 2652–57
[28] Tseng H E and Hrovat D 2015 State of the art survey: active and semi-active suspension control *Veh. Syst. Dyn* 53 1034–62
[29] Dixit R and Buckner G 2005 Sliding mode observation and control for semi-active vehicle suspensions *Veh. Syst. Dyn* 43 83–105
[30] Yao J L, Shi W K, Zheng J Q and Zhou H P 2013 Development of a sliding mode controller for semi-active vehicle suspensions *Journal of Vibration and Control* 19(8) 1152-1160
[31] Zhang C-W, Ou J-P and Zhang J-Q 2006 Parameter optimization and analysis of a vehicle suspension system controlled by magnetorheological fluid dampers *Struct. Control. Health Monit* 13 885–96
[32] Yu M, Dong X M, Choi S B and Liao C R 2009 Human simulated intelligent control of vehicle suspension system with MR dampers *J Sound Vib* 319(3-5) 753–67
[33] Barethiye V, Pohit G and Mitra A 2017 A combined nonlinear and hysteresis model of shock absorber for quarter car simulation on the basis of experimental data *Eng. Sci. Technol. Int J* 20(6) 1610–22
[34] ISO 8608:1995-09-01 1995 Mechanical vibration—road surface profiles—reporting of measured data *International Organization for Standardization (ISO)*
[35] Kale P R and Lapalikar S R, 2013 Performance evaluation of quarter car semi active suspension model with skyhook control two state MR damper using MATLAB Simulink *Paripex-Indian Journal of Research -An International Journal* 2(3) 79–81
[36] Yamin A H, Darus I Z, Nor N S and Talib M H 2017 Semi-active suspension system using MR damper with PSO- skyhook and sensitivity analysis of PID controller *J. Adv. Res. Appl. Mech* 33(1) 1–8

IOP Conf. Series: Materials Science and Engineering **530** (2019) 012015 doi:10.1088/1757-899X/530/1/012015

Pre-Forming Inspection System to Detect Deep Drawing Defect Due to Punch-Die Misalignment using Image Processing Technique

A A Ghafar* and A B Abdullah

Universiti Sains Malaysia, Engineering Campus, 14300 Nibong Tebal, Penang, Malaysia

Corresponding author *: alimighafar@gmail.com

Abstract. Deep drawing processes are extensively used in the industry for the fabrication of a wide range of sheet metal products from household appliances to automotive parts. In a deep drawing process, the alignment of the punch and die is extremely important to avoid defects such as variation of thickness and intolerable thinning condition of the drawn part. In this paper, a computer vision system is proposed that is able to identify the degree of misalignment in the punch-die configuration. The apparatus consists of an image capture and analysis system that can be used as a tool for pre-forming inspection setup. The image analysis implements the centroid recognition approach using the Image Processing Toolbox function in MATLAB. The offset value between centre of the punch and the centre of die indicates the severity of misalignment between punch and die. The system was successfully tested for a square profile, with more accurate results obtained for a colored punch as compared to the punch with the original metallic surface. Verification of the system was then conducted to align the centroids of a punch and die, and the results show that a good alignment of the centroids was achieved within two alignment attempts.

1. Introduction

Deep drawing is defined as the forming of a flat stretched sheet metal through a die into various shapes by using a compression force. Four basic components are required to perform a drawing process, namely, the punch, die, blank and blank holder. The drawing process can be conducted in a single step, although for greater depths, multiple drawings are usually needed. The total number of redrawing depends on many factors such as the material used (the blank), the dimensions required, and the final shape of parts. As compared to other production methods, the deep drawing process is able to produce high quality and complex shaped parts with minimal costs [1].

Some of the defects that can occur in deep drawing process are earing, wrinkling (either on the flange or wall of part), tearing and surface scratches [2]. Earing defects occur due to variations in properties because of the anisotropy of the material used as the blank. Anisotropy is a phenomena that is difficult to avoid when working with sheet metals [3]. Wrinkling is another type of defect that usually occur in a deep drawing process, its severity increases by insufficient blank holder force. Wrinkling usually occurs on the flange of the part, although it can also appear on the wall of the part. Tearing defects can occur due to severe thinning of the blank, caused by incorrect geometrical dimension of the die such as die radius, punch nose radius and punch-die clearance. Excessive thinning at the wall of the part is one the major contributing factor in the failure of the draw process. As mentioned by Colgan and Monaghan

IOP Conf. Series: Materials Science and Engineering 530 (2019) 012015 doi:10.1088/1757-899X/530/1/012015

[4], a decrease in the die radius would increase the drawing force. This increase in drawing force creates uneven thickness distribution at the wall of the part and thinning of the part structure. Severe thinning would result in tearing of the part.

Uneven wall thickness would result in thinning of the part. To minimize uneven wall thickness of the drawn part, the alignment between the punch and die should be controlled. Severe misalignment between the punch and die would result in undesired thinning of the part and failure initiation by tearing. This misalignment could be caused by incorrect tolerances used during die fabrication or alignment inaccuracies in the assembly of the die. Such misalignment and dimensional inaccuracies can be reduced by precision workmanship during those stages. Vairavan and Abdullah [5], in their studies on deep drawing of square cups without blank holders, have remarked that that alignment between punch and die is important for obtaining high quality parts. Slavič et al. [6] have found that misalignment between punch and die will increase the drawing force, although their study was limited to circular blanking process only. In contrast to cylindrical cup shape where the pressure is uniform on all area, rectangular shell requires increased force due to additional corners on the shape. This was the conclusion of Tamaoki et al. [7] whom have stated that stress concentration in a drawing process for a rectangular shaped part is higher than a cylindrical cup shape. Younis and Jabber [8] found that thinning of a square cup deep drawing usually appear in the corners of a part due to excessive stretching in those regions. Behrens et al. [9] also concluded that the degree of punch force has a more pronounced influence in a rectangular cup profile as compared to cylindrical cup shape when die radius variation was studied.

In recent years, integration of computer-aided technology in the stamping industry has opened up a new era in the production of high quality stamped components [10]. These technologies are commonly implemented in pre-production as a predictive tool, such as in the use of computer simulation software for deep drawing processes. For examples, Ogawa et al. [11] have utilized LS-DYNA, a finite element analysis (FEA) simulation software, to acquire early knowledge on the formability of square cup deep-drawing of pure titanium, which was then verified by experimentations. Yang et al. [12] were able to propose an optimal blank shape on deep drawing process of trapezoidal and square cups with the use of DEFORM-3D, another commercial FE package. Similarly, Reddy [13] used FEA to determine the formability of warm deep drawing for AA1050-H18 rectangular cup while Zein et al. [14] also used FEA to predict thinning and spring back conditions of sheet metal in a deep drawing process. Results of FEA simulation enable early prediction of the forming process which improves the overall production flow in term of reduced costs, decreased fabrication time and increased part quality. Computer-aided technology can also be used in supervision and surveillance of product quality control. These are usually packaged as a smart automation technology, in which various sensors and computer-aided monitoring equipment are used to detect part quality automatically. In a forming operation, the technology can be implemented before the drawing process (pre-forming), during process (in-situ) or after a process (post-forming). Past researchers have proposed vision and non-visual methods of evaluation which can be integrated in the forming production process. For example, Garcia [15] has used an artificial intelligent method to detect wrinkles and crack formation during a drawing process. In addition, Zoech et al. [16] have used electromagnetic and micromagnetic NDE technique to detect online crack and thinning while Kibe et al. [17] have used a commercial vision system to detect alignment between the punch and die before performing a shearing process. Similarly, Berger and Zussman [18] have used a novel measurement technique based on noncontact ultrasonic-based gauge to detect online wall thickness distribution of circular part during deep drawing processes. Among these methods, the computer-aided vision system is deemed to be the most suitable for integration into a deep drawing process. The vision system can be used as a pre-forming inspection of the alignment between punch and die in deep drawing process to detect the severity of misalignment of the components.

This paper proposes a pre-forming inspection system to detect the misalignment between the punch and the die for a square cup deep drawing process. This inspection procedure would be performed before the actual drawing process takes place. The inspection system consists of a vision system and a computer algorithm developed in MATLAB. The captured images of the punch and die is analyzed using

IOP Conf. Series: Materials Science and Engineering **530** (2019) 012015 doi:10.1088/1757-899X/530/1/012015

MATLAB based algorithm for centroid recognition. The offset values between centre of the punch and the centre of die indicate the severity of misalignment between the punch and die.

2. Model Development

2.1 Equipment Setup

Figure 1(a) shows the schematic diagram of the system developed in this study. The system consists of a digital camera, a back-lighting system, an anti-glare mirror and the die set. A push-through draw dies type with a blank holder was selected for use in this system. This type of die has a blind slot in the die block that enables the blank to be drawn into it via a die insert, as shown in Figure 3. Each component of the die is mounted on a center-post die set type having two guideposts on the right and left of the die shoes. The punch is mounted on the upper die shoe using a punch holder and the die insert is mounted on the lower die shoe by using a die block. The flat surface of the punch was aligned to be parallel with the surface of the die insert using the blank holder. The shape of punch is square with the dimensions of 13.6mm x13.6mm, while the punch nose radius, R_{PN} and the punch corner radius, R_{PC}, are both 2.5mm. For the die insert, the dimensions of square opening size are 15.74mm x 15.74mm, the die radius, R_D, is 5mm and the die corner radius, R_{DC}, is 3.57mm. The clearance value, *C,* between the punch and the die is 1.07mm. Figure 2 shows the overall dimensions of the punch and die insert.

As shown in Figure 1(b), the digital camera is mounted on a tripod facing the front of the die set. The digital camera has a 13 Megapixels sensor with an aperture size of *f*/1.9. In this study, the camera resolution was set to 4128×3096 pixel (4:3, 12.78MP), with focus length of 0.15m, ISO 500 film setting and a shutter speed 1/100 second to ensure clear observation of the die set. The height of camera is adjustable to ensure that the center of focusing lens is always straight and parallel with the center of image reflected off the mirror. The anti-glare mirror, tilted at a 45° angle, provides a clear image of punch and die insert. A mirror holder is mounted below the die block and located at the center of the die block opening. The back lighting increases the ambient illumination, reducing undesired image noise in the image captured. The lighting system consists of two 15 Lumens LED white light is, arranged to illuminate the image capture area. The auto focusing and auto exposure modes were disabled to ensure a consistent image setting was used for all image captures. The captured images are then uploaded into MATLAB software for subsequent image processing.

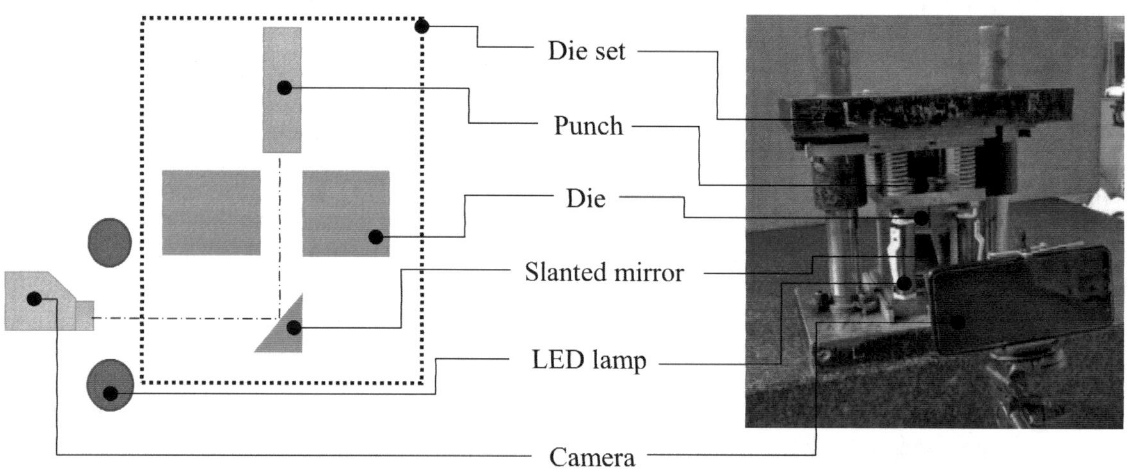

(a) Schematic diagram arrangement of system (b) Actual setup of the system

Figure 1. Arrangement of the system

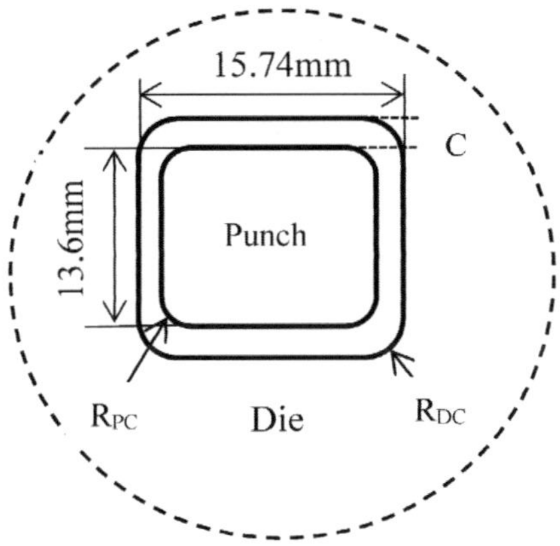

Figure 2. Dimension of punch and die

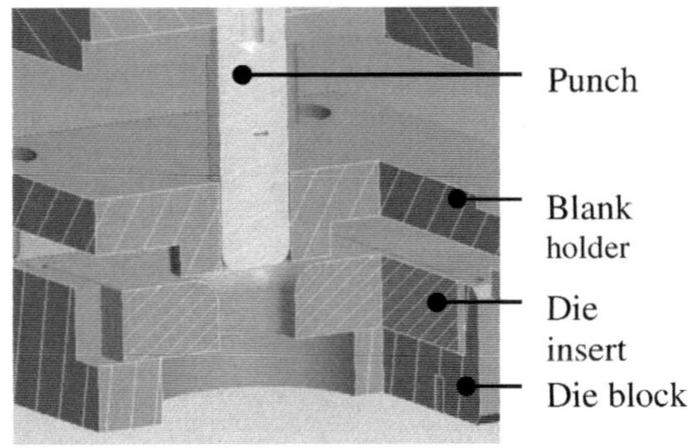

Figure 3. Cross-section view push through draws die

2.2 The Proposed Algorithm

In this study, a centroid recognition method was used to detect misalignments between the punch and die. The Image Processing Toolbox function in MATLAB was used to process the captured image (raw image) to obtain the centroids of the punch and die. Variations of the centroid values between the punch and die indicates the extend of misalignment between the two components. The raw image needs to undergo several image analysis processes prior to the centroid detection procedure. Firstly, the raw image is calibrated and transformed appropriately. Then, image noises are removed through an image restoration process. Color segment detection function is then used to detect the surface of punch while the edge detection function is used to detect the edges of die. The punch is painted red for the color segmentation process. For the die, the canny method of edge detection was used to define the edges of the die. The image is then converted to into a binary image and morphological filtering is performed to remove unwanted pixels and to obtain an improved binary image of the punch and die. The centroid of the punch and die can then be obtained by using image region properties function. Figure 4 shows the process of image processing and analysis to obtain the centroids of the punch and die.

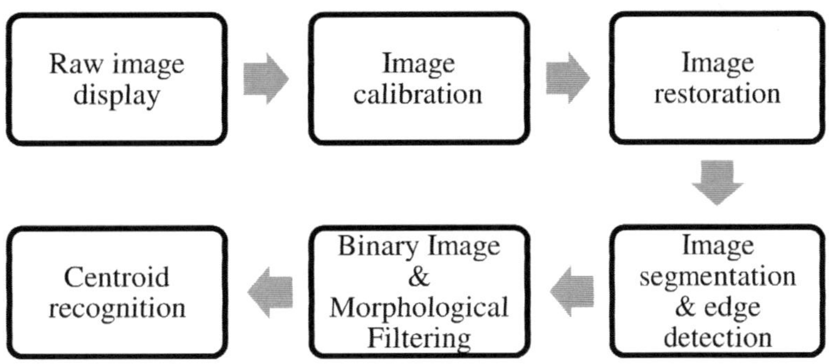

Figure 4. Image processing procedure flow chart to determine centroid

3. Result and Discussion

The captured raw image needs to undergo appropriate cropping and transformation procedures before it can be further processed. Since the position of the camera is not parallel to the image appearing on the 45° tilted mirror, as shown in Figure 5(a), appropriate corrections needs to be conducted to correct the projection of the image. This can be achieved by simple geometrical transformation such as rotating and cropping to obtain an improved image, as shown in Figure 5(b).

(a) (b) (c)

Figure 5. (a) Raw image (b) Image after geometrical transformation (c) Edge detection of die

Several edge detection methods, such as Sobel, Prewitt and Roberts can be used. However in this study, the more efficient Canny edge detection method is selected. Figure 5(c) shows result of Canny edge detection for the die. For comparison, the uncolored metal punch is shown in Figure 6(a). It was seen that an uncolored punch located deep inside the die was very difficult to be observed clearly and the metallic surface was very reflective. This reflectivity affected the visibility of the punch and edge, especially for the edge detection process, with increased image noise resulting in an inaccurate calculation of the centroid. This problem can be overcome by coloring the punch, and then using the color segmentation procedure to identify the punch surface area. Figure 6(b) shows an inaccurate centroid of punch (star mark) due to unclear punch edge identification. In comparison, the centroid of the colored punch is more precise due to distinct edge profile, as shown in Figure 6(c).

The resolution of the raw image also plays a role in the accuracy of centroid detection. Low resolution image also resulted in inaccurate centroid detection. As shown in Figure 7, a low image resolution of 480×640 pixels resulted in a blurry image of the punch edge, and the centroid detected was therefore inaccurate. Using a high-resolution raw image is therefore recommended for accurate centroid detection.

(a) Image of original punch (b) Inaccurate centroid of punch (c) Centroid of colored punch (red)

Figure 6. Comparison accuracy of centroid between original punch and a red colored punch

(a) Raw image of 480×640 pixels (b) Centroid detection of punch

Figure 7 Detection of punch centroid for raw image with 480×640 pixels

IOP Conf. Series: Materials Science and Engineering **530** (2019) 012015 doi:10.1088/1757-899X/530/1/012015

Figure 8 shows the centroid value for an image of a punch and die configuration. The difference between the centroid values of the punch and die were (4.0696, 2.248) pixels. Using a conversion factor, this correspond to a displacement of (0.015,0.05) mm, indicating the punch is slightly to the left and towards the top of the die.

Figure 8. Centroid of punch and die

Table 1 shows the example of applicability of the proposed technique in determining the degree of misalignment of the punch and die for a deep drawing process. Three images were captured and analyzed in an attempt to firstly determine and then to align the centroid of the punch with the centroid of the die using the developed system. The initial offset is calculated from the original image. Once the offset was determined, the first attempt (Attempt 1) was made to align the centroids of the punch and die. The second attempt (Attempt 2) is to fine tune the alignment, based on the offset values calculated from Attempt 1. The results have shown that after the second attempt, the centroids of the punch and die were able to be aligned with minimal offset values.

Table 1. Result of offset values between original condition and after modification

Image Status	**Offset value centroid punch-die, mm**	
	X axis	Y axis
Original Condition 	0.411	-0.326
Attempt 1 	0.204	-0.047

Attempt 2 	-0.008	0.046

4. Conclusion

In this work, a computer vision based pre-forming inspection system to detect misalignments between punch and die for square shape cup has been developed successfully. Preliminary results showed that the proposed system has successfully detected the centroids of the punch and die of a deep drawing die set.

Future works would concentrate on the drawing process of the aligned punch and die using a Universal Tensile Machine (UTM). Result from experimental evaluations such as the punching loads and wall thickness distribution of the part would be compared with the results from FEA simulations.

References

[1] Dwivedi R and Agnihotri G 2017 Study of Deep Drawing Process Parameters *Mater. Today Proc.* 4 820–6
[2] Joshi A R, Kothari K D and Jhala R L 2013 Effects Of Different Parameters On Deep Drawing Process: Review *Int. J. Eng. Res. Technol.* 2 1–5
[3] Hill R 1948 A Theory of the Yielding and Plastic Flow of Anisotropic Metals *Proc. R. Soc. A Math. Phys. Eng. Sci.* 193 281–97
[4] Colgan M and Monaghan J 2003 Deep drawing process: analysis and experiment *J. Mater. Process. Technol.* 132 35–41
[5] Vairavan H and Abdullah A B 2017 Die-punch alignment and its effect on the thinning pattern in the square-shaped deep drawing of aluminium alloy *Int. J. Mater. Prod. Technol.* 54 147–64
[6] Slavič J, Bolka S, Bratuš V and Boltežar M 2014 A novel laboratory blanking apparatus for the experimental identification of blanking parameters *J. Mater. Process. Tech.* 214 507–13
[7] Tamaoki K, Manabe K, Kataoka S and Aizawa T 2013 Continuous Dry Cylindrical and Rectangular Deep Drawing by Electroconductive Ceramic Dies *J. Manuf. Sci. Eng.* 135 031010
[8] Younis K M and Jabber A S 2018 Experimental Evaluation and Finite Element Simulation to Produce Square Cup by Deep Drawing Process *Al-Khwarizmi Eng. J.* 14 39–51
[9] Behrens G, Ruhe M, Tetzel H and Vollertsen F 2015 Effect of tool geometry variations on the punch force in micro deep drawing of rectangular components *Prod. Eng.* 9 195–201
[10] Panicker S S and Kumar Panda S 2017 Improvement in Material Flow During Nonisothermal Warm Deep Drawing of Nonheat Treatable Aluminum Alloy Sheets *J. Manuf. Sci. Eng.* 139 031013
[11] Ogawa T, Ma N, Ueyama M and Harada Y 2016 Analysis of Square Cup Deep-Drawing Test of Pure Titanium *J. Phys. Conf. Ser.* 734 032072
[12] Yang C, Li P and Fan L 2014 Blank shape design for sheet metal forming based on geometrical resemblance *Procedia Eng.* 81 1487–92
[13] Reddy C 2015 Formability of Warm Deep Drawing Process for AA1050-H18 Rectangular Cups *Int. J. Mech. Prod.* 5 85–98
[14] Zein H, Sherbiny M El, Abd-Rabou M and El M 2014 Thinning and spring back prediction of sheet metal in the deep drawing process *J. Mater. Des.* 53 797–808
[15] García C 2005 Artificial intelligence applied to automatic supervision, diagnosis and control in

sheet metal stamping processes *J. Mater. Process. Technol.* 164–165 1351–7
[16] Zoesch A, Wiener T and Kuhl M 2015 Zero Defect Manufacturing: Detection of Cracks and Thinning of Material during Deep Drawing Processes *Procedia CIRP* 33 179–84
[17] Kibe Y, Okada Y and Mitsui K 2007 Machining accuracy for shearing process of thin-sheet metals—Development of initial tool position adjustment system *Int. J. Mach. Tools Manuf.* 47 1728–37
[18] Berger M and Zussman E 2002 On-Line Thinning Measurement in the Deep Drawing Process *J. Manuf. Sci. Eng.* 124 420

IOP Conf. Series: Materials Science and Engineering **530** (2019) 012016 doi:10.1088/1757-899X/530/1/012016

Thermo-Mechanical Simulation of Temperature Distribution and Prediction of Heat-Affected Zone Size in MIG Welding Process on Aluminium Alloy EN AW 6082-T6.

Z Samad*, N M Nor and E R I Fauzi

School of Mechanical Engineering, Engineering Campus, Universiti Sains Malaysia, Seri Ampangan, 14300 Nibong Tebal, Pulau Pinang, Malaysia.

Corresponding Author *: mezahurin@usm.my

Abstract. Welding process is considered as a thermal-mechanical-metallurgical coupled problem. In this study, finite element method (FEM) is adopted for predicting the temperature history in Metal Inert Gas (MIG) welding of 5mm thick aluminium 6082 alloy. The Goldak's double ellipsoidal moving heat source model was used to analyse the influence of peak temperature to the radial distance from the center of the heat source and the thickness of the plate. Temperature-dependent thermal properties of aluminium alloy 6082 in T6 condition and the convective and radiative boundary conditions were included in the model. The finite element code, ANSYS along with APDL command subroutines was employed to obtain the numerical results. The effect of heat input and welding speed on the weld pool shape and temperature distribution were investigated. Finally, the predicted temperature distribution and the size of heat-affected zone were compared with the experimental results. The comparison shows that they are in good agreement.

1. Introduction

Without doubt, the continued development of aluminium can be primarily attributed to this material's many desirable physical characteristics, which are comparatively light weight, high strength, versatility of both extruding and casting, and excellent corrosion resistant. In conjunction with the continually developing environmental issues, the superior recycling capability of aluminium makes it an excellent choice for a wide variety of applications. With increasing demand of aluminium in engineering field, the development of aluminium welding technology is undeniably evolved to improve the structural integrity of aluminum alloys itself.

One of the most severe challenges in aluminium welding is the occurrence of heat-affected-zone (HAZ), which may results in a significantly reduced strength compared to the base metal. Principally, during welding, aluminium properties change in the zone adjacent to the weld seam. This change occurs due to the introduction of heat to the base metal which revokes the treatment of certain alloys. This phenomenon, which also known as softening, is significant factor in the heat-treatable alloys particularly in 6000 and 7000 series, and in 5000 series alloys in a work-hardened temper condition. In case of the 6000 series alloys, the heat dissipation during welding can locally reduce the parent metal strength on the order of 30%-50% and the HAZ normally extends between 10mm-30mm from the centre of the weld [1-3]. However, according to Eurocode 9 [4], the extent of this zone depends on the welding method

IOP Conf. Series: Materials Science and Engineering **530** (2019) 012016 doi:10.1088/1757-899X/530/1/012016

(either Metal Inert Gas (MIG) or Tungsten Inert Gas (TIG)), element thickness, alloying elements and temper designation, while the severity is largely a function of the parent metal used.

Some studies have been focused on prediction of the HAZ of welded joint by using numerical simulation technique. Bjorneklett et al. [5] developed a multi-scale modeling approach where precipitate revolution of Al-Mg-Si alloys can be predicted through finite element simulation. A constitutive model based on microstructure evolution enables realistic prediction of structural performance of aluminium alloy during age-hardening, welding and post weld treatment process. Hori [6] was investigating the effect of HAZ on the joint strength of Al-Mg-Si alloy experimentally. Recently, Dorum et al. [7] presented finite element analyses of plastic failure in the HAZ of 6060 aluminium alloys. The welded connection was modeled using shell elements, solid elements and cohesive-zone elements. The ductility of the welded joints can be estimated with coarser meshes. In this study, a numerical model to predict temperature distribution and the width of HAZ was proposed. The obtained numerical results were compared with experimental data. It can be used in practical simulations of MIG welded aluminium alloy connections to reduce the need of extensive mechanical testing, particularly in mapping the width of HAZ.

2. Experimental procedure

The experimental setup consists of a MIG welding system and a 6-axis high precision robot arm as shown in Figure 1. The welding machine Kemppomat 2500 was used to perform the single-sided welding of T-joint. The robotic arm KUKA KR6 was used to control the welding speed and assist the welding torch along the path with constant distance and angle. The distance of contact tip was maintained at 5mm and the torch angle was 30° inclined from horizontal base specimen. A single pass welding was performed with the presence of pure Argon (99%) as shielding gas to protect the molten pool from atmospheric contamination. The gas flow rate was set at 20 liter/min. A consumable filler of AA 4043 with diameter of 1.2 mm (~ 5% Si) was used in this welding process due to the ability of reducing crack sensitivity. The wire feeding rate was set to 6.5 m/min. During the welding process, welding speed and voltage input were varied; only current input is kept constant. All process parameters were chosen based on preliminary test results and presented in Table 1.

Table 1. The process parameters for MIG welding experiment

Argon flow rate (l/min), *f*	20			
Torch angle (deg), α	30°			
Distance of contact tip (mm), *d*	5			
Wire feeding rate (m/min)	6.5			
Diameter of welding wire (mm)	1.2			
Welding current (A), I	250			
	I	**II**	**III**	**IV**
Welding speed (m/min)	0.35	0.4	0.45	0.5
Voltage (V)	23.0	25.5	28.3	31.4

The material used in this study is aluminium alloy EN AW 6082 in T6 condition. The welding coupons were cut by an abrasive water-jet cutting system with the dimensions of 120 mm x 120 mm x 5 mm and 120 mm x 60 mm x 3mm. The 5 mm-thickness plate was selected to be the base material, whereas the 3 mm-thickness plate was used as the flange material with designed bevel of 45° to ensure full penetration of the weld seam. The flange material was tightly clamped, 90° from the base material and a zero gap was assumed to be achieved. A thermal imaging camera FLUKE Ti32 was used in this analysis to record the temperature history during the welding process.

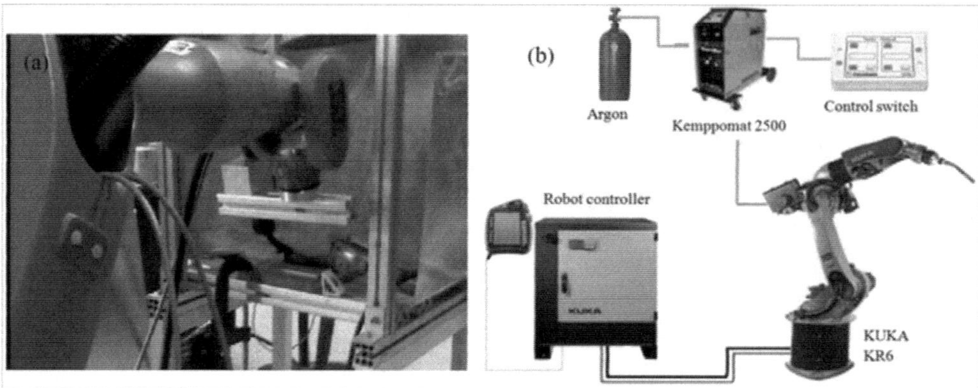

Figure 1. (a) Experimental setup. (b) Schematic diagram of experimental setup

3. Numerical model of thermal analysis

The simulation of thermal analysis during MIG welding process requires a three-dimensional modeling. In this context, the ANSYS non-linear finite element code is adopted due to its flexibility in modeling and its capability in obtaining full field numerical solution. The input files, written in APDL command language of the finite element code, have been inputted for analysis runs and thermal load is applied in time steps, thus defining the analysis as transient. In order to accurately capture the occurring phenomena during thermal cycle, the graded-mesh construction is used for the analysis. This approach is based on a dense mesh division along the weld line and its adjacent zone. The mesh size became coarser as the distance from the weld bead zone increase. It was constructed primarily of solid hexahedral element SOLID70, except in transition areas of thin to coarse mesh, where the tetrahedral SOLID87 element is required.

To reduce the computational time, only half of the actual length and width of specimen was taken into account throughout the analysis. Figure 2(a) and (b) show the considered model of T-joint profile for the analysis and the mesh density transition for the model respectively. A sequence of convergence tests has been conducted to select suitable element size particularly in the thickness direction and at the region near the weld line. Several levels of meshes with varying element numbers from 200,000 to 500,000 in scales of width, length and thickness are chosen to determine the effect of element size on the convergence of computations. The simulated peak temperature becomes independent on the mesh density of 350,000 elements as shown in Figure 3. Therefore, a mesh of over 350,000 elements is employed to the model. Nearby and along the weld bead area, the chosen element size is 0.0005, whereas element size of 0.005 is employed in the coarser region. A free mesh is used in order to connect these two regions.

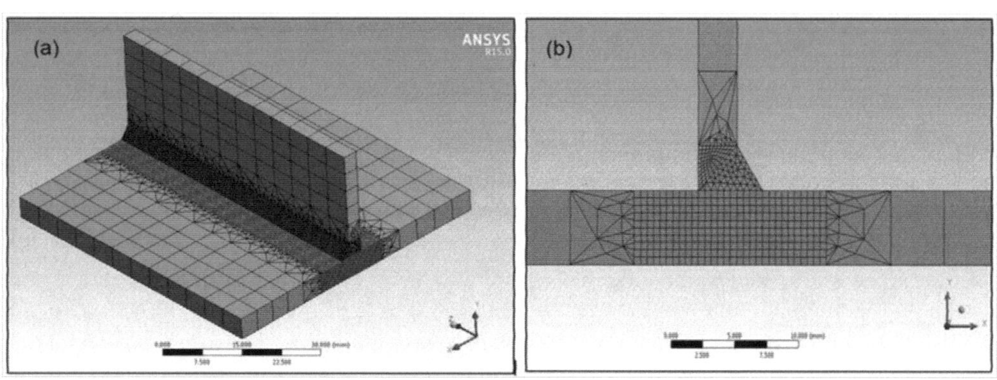

Figure 2. (a) T-joint profile for thermal model analysis. (b) Mesh density transition of thermal model.

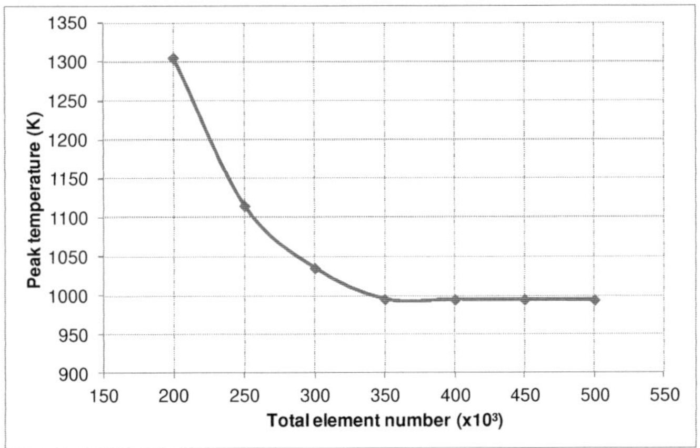

Figure 3. The mesh convergence parametric study

3.1 Model assumptions

In building the model and reasonably simplify the finite element thermal analysis, the following assumptions were introduced:

1) The initial temperature of the model is initially at 25°C.
2) The density of aluminium alloy 6082 is taken to be 2770 kg/m³, while the thermal properties of the material such as conductivity and specific heat capacity are temperature dependent as shown in Figure 4.
3) The fluid flow in the molten pool is neglected.
4) The heat input of welding process obeys the Gaussian distribution of double ellipsoidal heat source model, which was presented by Chen et al. [8].
5) Convection effect is considered.
6) A full penetration depth is assumed and weld seam is incorporated as part of the model.
7) The welding efficiency is taken to be 0.75 [9-12].

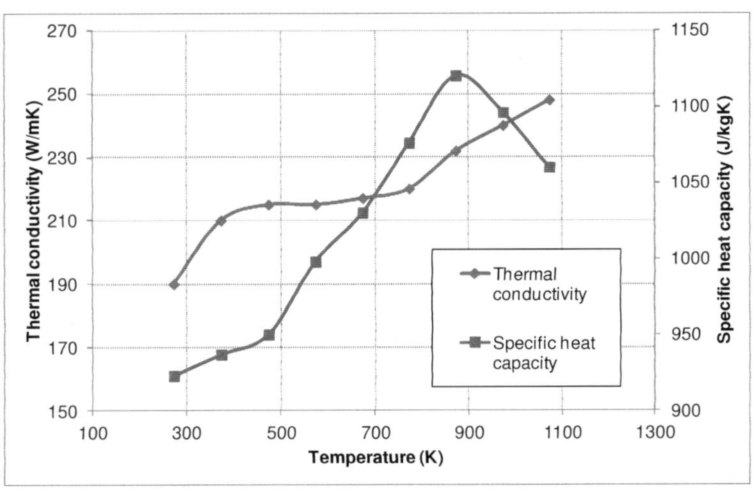

Figure 4. Thermal properties of aluminium alloy 6082 in accordance to ENV 1999-1-2 [13].

3.2 Governing equations and heat source model

The fundamental behavior of heat conduction in the bond area is an energy flux that flows from a hot region to a cooler region, which linearly dependent on the temperature gradient, ΔT

$$Q = -k\Delta T \tag{1}$$

where k is the thermal conductivity in W/mK. It should be noted that the minus sign is necessary due to the direction of the heat transfer to the decreasing temperature. The specific heat, c, can be expressed as the required energy in order to have a temperature gradient in the specimen. Hence, the conservation of energy is expressed in a differential forms in terms of thermal flux, specific heat and a distributed volume heat-source in the Cartesian coordinate system. The governing differential equation for transient thermal analysis can be expressed as [14]:

$$\rho c_p \frac{\partial T}{\partial \tau} = \frac{\partial}{\partial x}\left(k_x \frac{\partial T}{\partial x}\right) + \frac{\partial}{\partial y}\left(k_y \frac{\partial T}{\partial y}\right) + \frac{\partial}{\partial z}\left(k_z \frac{\partial T}{\partial z}\right) + \dot{q} \tag{2}$$

where ρ is the density of conducting medium; c_p is the specific heat of the medium; k_x , k_y , k_z are the thermal conductivities of the medium in the x-, y-, and z-direction, respectively; τ is time; and $\dot{q}$ is the total heat input based on Goldak's double ellipsoidal model [15]. To solve the differential equation, boundary and initial conditions are required. The general form of boundary conditions can be written as:

$$-\left(k_x \frac{\partial T}{\partial x} n_x + k_y \frac{\partial T}{\partial y} n_y + k_z \frac{\partial T}{\partial z} n_z\right) = q_{conv} + q_{rad} - \alpha q_\tau \tag{3}$$

where n_x, n_y, n_z are the direction cosines of the normal to the surface; T_∞ is the ambient temperature; h_{conv} is the convective heat transfer coefficient; α is the absorptivity; and q_τ is the radiant laser heat flux function. The convective heat transfer coefficient was assumed to vary with temperature and defined as:

$$h_{conv} = \frac{kN_u}{L} \tag{4}$$

where k is the thermal conductivity of the material, L is the characteristic length of the surface, and N_u is the Nusselt number:

$$N_u = 5.67 Pr^{1/3} Gr^{1/3} \tag{5}$$

Here, Pr is the Prandtl number and Gr is the Grashof number. Thermal radiation is assumed from the surface to surroundings. Prior to welding, the material is assumed to be at room temperature. The heat losses due to radiation were modeled by Stefan-Boltzmann relation:

$$q_{rad} = \sigma_{st}\varepsilon_\tau (T^4 - T_\infty^4) \tag{6}$$

where σ_{st} is Stephan-Boltzmann constant (equal to 5.67×10^{-8} W/m^{2}K^4) and ε_τ is the emissivity of the surface ($\varepsilon_\tau = 0.05$), in accordance to data published by Mikron Vertrerung of Switzerland [16]. The schematic presentation of applied boundary conditions was illustrated in Figure 5.

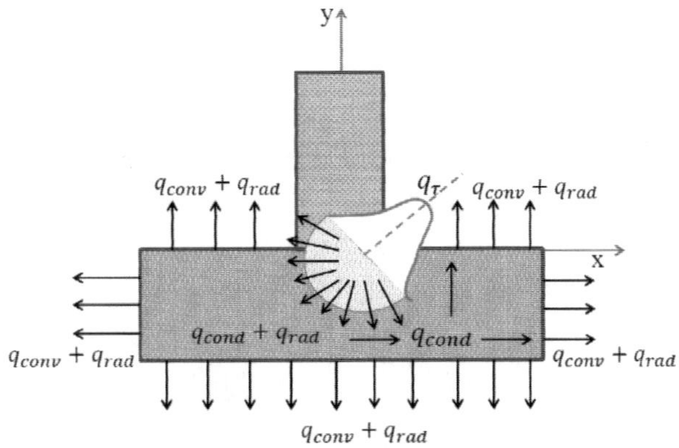

Figure 5. Schematic presentation of applied boundary conditions for MIG thermal model

The power densities of the double ellipsoidal ellipses, that comprise a front and a rear quadrant of the heat source as shown in Figure 6, can be stated as:

$$q_f(x,y,z) = \frac{6\sqrt{3}f_fQ}{bca_f\pi\sqrt{\pi}}.e^{-3\frac{x^2}{a_f^2}}.e^{-3\frac{y^2}{b^2}}.e^{-3\frac{z^2}{c^2}} \tag{7}$$

$$q_r(x,y,z) = \frac{6\sqrt{3}f_rQ}{bca_r\pi\sqrt{\pi}}.e^{-3\frac{x^2}{a_r^2}}.e^{-3\frac{y^2}{b^2}}.e^{-3\frac{z^2}{c^2}} \tag{8}$$

where Q is the energy input rate, f_f and f_r are the fractional factors of the heat deposited in the front and rear quadrant, a_f, a_r, b and c are heat source parameters. The heat source parameters are $a_f = 5$ mm, $a_r = 20$ mm, $b = 5$ mm, $c = 8$ mm and $f_f + f_r = 2$.

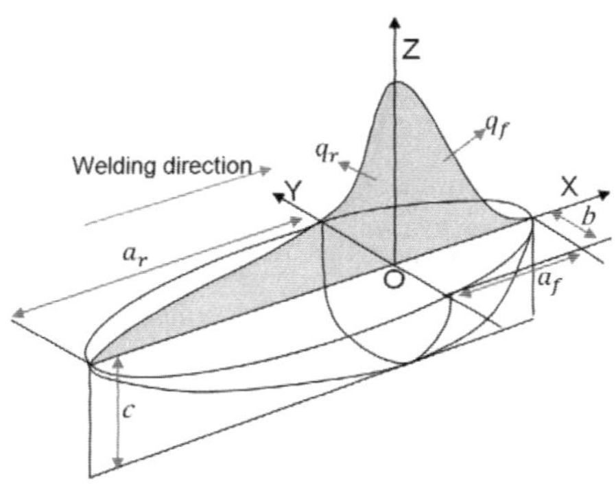

Figure 6. Goldak's double ellipsoidal heat source model (15)

The heat source model was applied based on time-stepped transient analysis by changing their coordinates at every time step. The time steps were defined according to the welding speed and the length of weld pass. An additional time step is required for cooling process afterward. In all simulation case, the welding speed was set to be identical to the heat source movement's speed. Since a relatively small radial flux distance is required to capture the thermal phenomena during simulation accurately, the heat source moved towards the end of the weld seam (120 mm) with a 0.3 mm span-length distance, which was 3 times smaller than the filler wire diameter. The numbers of load steps are resulted from the fraction of the total weld length to the span-length. Hence, 400 load steps were required for a complete welding sequence with additional 100 load step for cooling phase afterward. The time of every load step

IOP Conf. Series: Materials Science and Engineering **530** (2019) 012016 doi:10.1088/1757-899X/530/1/012016

was determined by multiplication of the span length with welding speed. An APDL based subroutine is written to simulate the moving heat source along the weld line. Figure 7 illustrates the heat distribution of the heat source model at the top of the welded plate acquired from the simulation results.

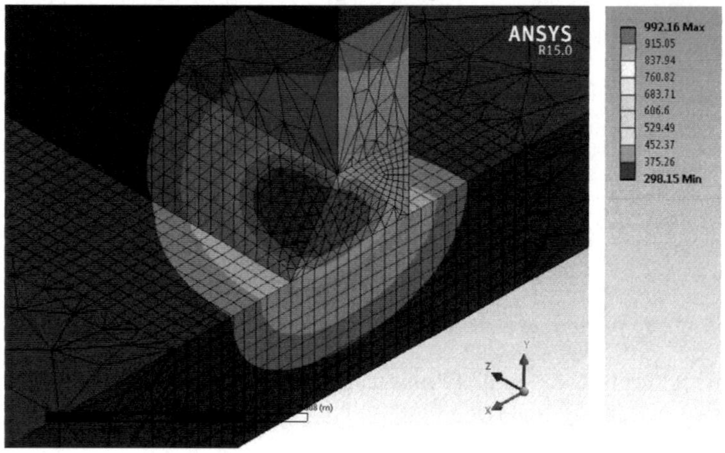

Figure 7. Heat distribution of the heat source model acquired from simulation at t = 0.25s (Heat input = 573.75J/mm).

3.3 Thermal modeling

The thermal analysis is focused on the prediction of the heat transfer in the weld pool, the temperature distribution in the overall weldment, the size of fusion zone (FZ), heat-affected zone and geometry of the molten pool. The heat-affected zone identification relies upon the experimental data and available data from open literature by Matusiak [17] and Myhr and Grong [18], with an assumption that the studied material has an identical chemical composition. From previous studies [19-21], it is evident that the reversion of hardening precipitates occurring during the welding process, where the peak temperature range from 220°C to 500°C as depicted in Figure 8(a). However, a large fraction of alloying elements adjacent to the fusion boundary, undergo natural aging after a period of 5-7 days (Figure 8(b)), resulting in a slight increase of hardness values due to the enhancement of hardening precipitates. According to [18] element which experienced a peak temperature of about 220°C and below will recover to its original hardness (base metal) during the natural aging process. According to this model, the HAZ profile was characterized based upon the quasi-stationary temperature distribution around a moving heat source. Therefore, in this study, the HAZ was divided into 2 regions;

1) Sub-HAZ 1, where the range of peak temperature, $430°\text{C} < T < \mathit{Solidus\ temperature}$ (~545°C).
2) Sub-HAZ 2, where the range of peak temperature, $220°\text{C} < T < 430°\text{C}$.

Aside from those regions (where the range of peak temperature, $T < 220°\text{C}$), the material was assume to have the base metal characteristics. The weld pool was considered as the fusion zone (FZ) during the analysis, by assuming that the material was fully melted at solidus temperature (545°C) and the addition of filler metal will change the Si content in weld zone [22].

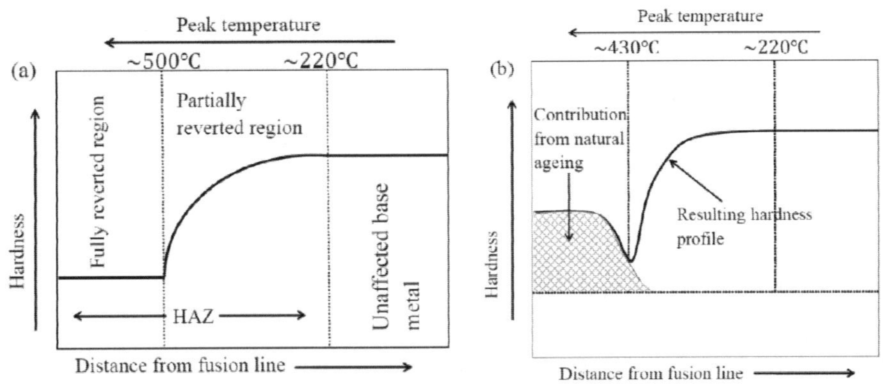

Figure 8. Schematic diagrams presenting the reactions of precipitates in the HAZ of 6082-T6 weldments: (a) hardness distribution during dissolution process (b) hardness distribution following prolonged natural aging in room temperature [18].

The thermal cycle of the process was recorded by the two thermocouples located closer to the weld line. All these data were then compared with the numerical nodal results from nodes of the same coordinates of the thermocouple position. The results were presented in Figure 9. TC1 and TC2 represent the location of thermocouples 10 mm and 20 mm from the center of the weld line. From the figure, the maximum temperatures recorded by both thermocouples have been accurately captured by the simulation model. The measured cooling gradients exhibit almost similar curves with the numerical results of the finite element model. Generally, it can be stated that the developed simulation model was in excellent agreement with the experimental data. The temperature reading for the TC01, which was located closer to the weld line, was recorded the highest value but considered to be moderate in magnitude, since the maximum temperature at 15 mm distance from the weld line was just below 230°C. However, the temperature difference was not very evident even between each thermocouple of TC01 and TC03, at a distance of 15 mm and 25 mm respectively, which have a temperature difference of almost 25°C. This can be attributed to the high thermal conductivity of EN AW 6082, which allow rapid heat dissipation in the bulk of the material. In addition to that, the cooling curve of the thermocouple nearest to the fusion line (TC01) and the thermocouple at 25 mm from the fusion line (TC03) are almost identical. The two curves converge approximately 70 seconds after the arc initiation at 100°C.

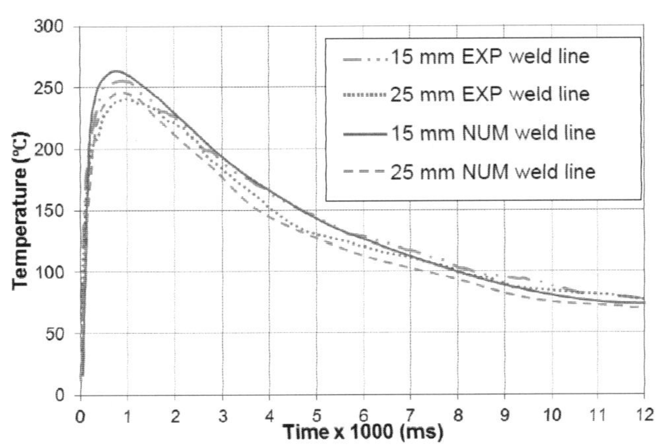

Figure 9. Comparison of thermal cycle during MIG welding of EN AW 6082 experimentally and numerically. (Voltage = 25.5 V, current = 250 A, velocity = 0.35 m/min, heat input = 0.819 kJ/mm)

Figure 10 illustrated the comparison of the molten pool shape after penetration during MIG welding process. The picture on the left depicts the obtained weld pool by experiment, and the right picture illustrates the computed weld pool in terms of the isotherm contour. The region of the molten pool, in which the solidus temperature of the alloy (~545 °C) was exceeded and melting occurred is called the fusion zone. In the left figure, the boundary of the FZ and HAZ was denoted by the dashed white line (fusion line), which also indicates the locus of points whose peak temperature is equal to the solidus temperature. It can be observed clearly here that the FZ exhibit distinctly different microstructural morphology as compared to the adjacent zone. The metallurgical transformation of EN AW 6082-T6 was expected to occur starting from the lower critical temperature of 220°C.

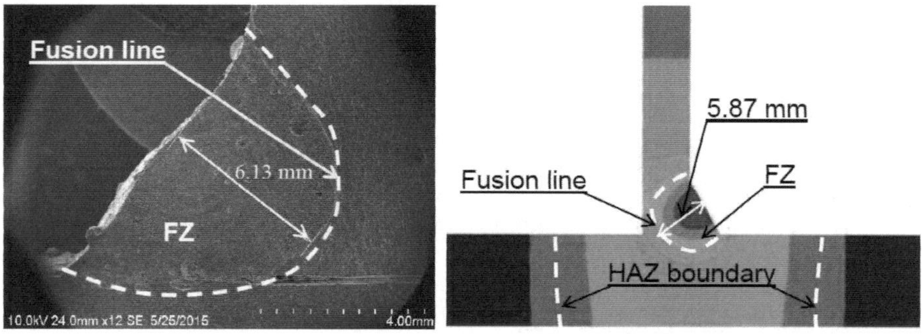

(a) (b)

Figure 10. Weld pool profile obtained by (a) experiment and (b) simulation. MIG welding parameters; Current = 250 A, Voltage = 28.3 V, Welding speed = 0.35 m/min

4. Results and discussions

The temperature distribution profiles are compiled for different welding speed and heat input power. In order to evaluate the effect of heat input on the weld pool dimension of MIG welding, a set of simulation case has been studied by varying the input voltage over a constant value of welding speed at 0.35 m/min. Figure 11 presents a comparison of the simulated weld pool shapes made in the case of M1, M8, and M9. Additionally, simulation cases of M2 and M7 were also analyzed to obtain a clear trend of the corresponding effect. The dimension of the HAZ and penetration depth were measured for each simulation cases and compared with the experimental data obtained, as summarized in Table 2. It can be seen that the simulation cases of M7 and M2 gave almost similar results with the corresponding experiment, with the percentage accuracy of almost 90%. The case of M2 and M7 were done for validation purpose because the experimental data were only available for these two cases. Due to the limitation of time and cost, therefore thermal model was used to simulate the other cases (M1, M8 and M9).

Table 2. List of simulation and experimental results of MIG welding for heat input variation (Constant welding speed = 0.35 m/min)

Case	[a]Heat input (kJ/mm)	*Depth of FZ (mm)*			Width of HAZ (mm)		
		Model	*Exp.*	*% Error*	*Model*	*Exp.*	*% Error*
M1	0.942	5.97	-	-	25.62	-	-
M2	0.849	5.87	6.13	4.24	25.57	26.00	1.65
M7	0.690	5.32	5.92	9.14	25.21	27.00	6.63
M8	0.510	4.77	-	-	23.71	-	-
M9	0.765	5.51	-	-	25.43	-	-

[a]*Based on a constant input current of 250 A and efficiency of 0.70.*

IOP Conf. Series: Materials Science and Engineering **530** (2019) 012016 doi:10.1088/1757-899X/530/1/012016

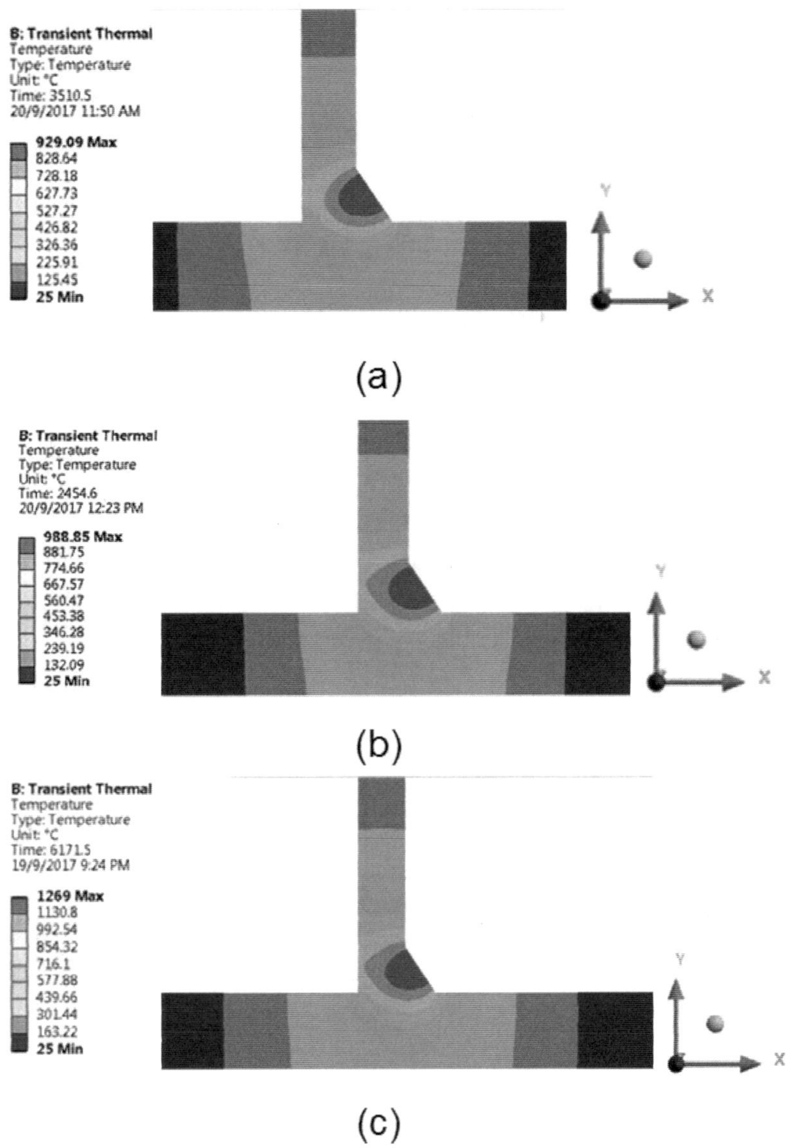

Figure 11. Weld pool profile of MIG welding with varying heat input (a) low heat input (0.51 kJ/mm) (b) intermediate heat input (0.76 kJ/mm) (c) high heat input (0.94 kJ/mm)

Figure 12 illustrates the dependency of penetration depth and width of HAZ on the heat input variation. From the figure, it can be seen that the penetration depth and the size of HAZ have the tendency to increase with the increase of heat input. In general, data fitting resulted in a non-linear relation between the depth of fusion zone and the width of HAZ with the heat input variation.

In the case of penetration depth, the depth of FZ was affected by the power density distribution of the heat source. The weld penetration increases with increasing power density [21]. For all simulation cases, the effective radius of the heat source was kept constant. Therefore, as the heat input increases over the same surface area, the power density increases, and consequently increasing the penetration of weld pool. In all simulation cases, the Marangoni effect and recoil pressure were neglected. From the Figure 11, it can be observed that the dimension of the melt pool was almost identical. However, the HAZ width was widening significantly. The melt pool dimension was affected largely by the heat distribution factor. In contrary, the HAZ width was influenced by the temperature gradient. As the heat input increases, the peak temperature elevates, so does the temperature gradient. Rearranging equation (3) and equation (4), it can be shown that the estimated HAZ size is proportional to the heat input as well as peak temperature (considering the HAZ width in terms of L, the characteristic length of the surface) as follow:

$$\begin{aligned} -Energy &= h_{conv}(T - T_{\infty}) + q_r - \alpha q_\tau \\ -Energy &= \frac{kN_u}{L}\nabla T + q_r - \alpha q_\tau \\ L &= \frac{-1}{Energy}(kN_u\nabla T + q_r - \alpha q_\tau \end{aligned} \tag{9}$$

Another attempt of simulation was made in order to understand the effect of welding speed variation on the size of HAZ and penetration depth of weld pool. In these cases, the input voltage remained unchanged (28.3 V), and the welding speed was increased slightly from 0.35 m/min to 0.75 m/min. The results were presented in Figure 13 and Table 3.

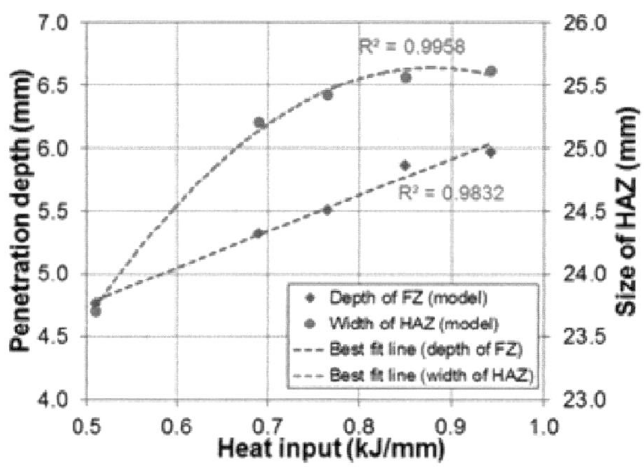

Figure 12. Plots of penetration depth and width of HAZ against heat input variation for MIG thermal model

Table 3. List of simulation and experiment result of MIG thermal model for welding speed variation (Constant voltage = 28.3 V)

Case	Welding speed (m/min)	Depth of FZ (mm)			Width of HAZ (mm)		
		Model	*Exp.*	*% Error*	*Model*	*Exp.*	*% Error*
M2	0.35	5.87	6.13	4.24	25.57	26.00	1.65
M3	0.45	3.51	3.34	5.09	24.78	25.00	0.88
M4	0.55	2.75	-	-	24.53	-	-
M5	0.65	2.58	-	-	23.51	-	-
M6	0.75	2.43	-	-	23.03	-	-

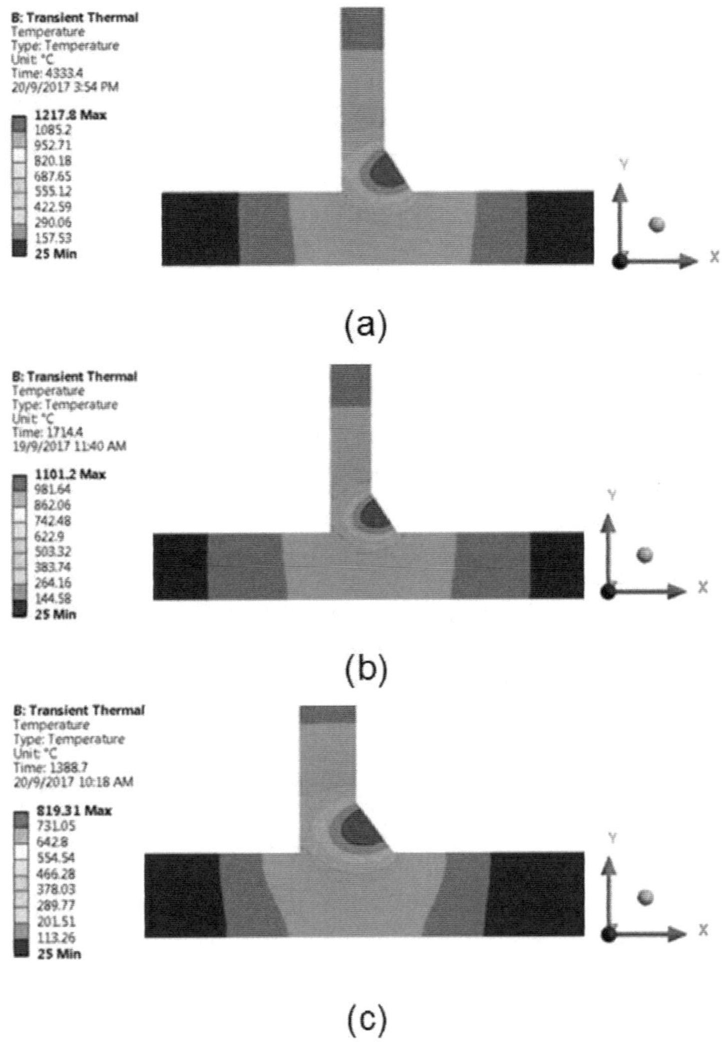

Figure 13. Weld pool profile of MIG welding with varying welding speed (a) low speed (0.35 m/min) (b) intermediate speed (0.55 m/min) (c) high speed (0.75 m/min)

Figure 14 illustrates the effect of the welding speed variation on the penetration depth and width of HAZ of MIG welding model. From the figure, a clear trend can be observed, in which the welding speed has a substantial effect on the penetration depth and the width of HAZ. Higher welding speed will eventually decrease the penetration depth and width of HAZ.

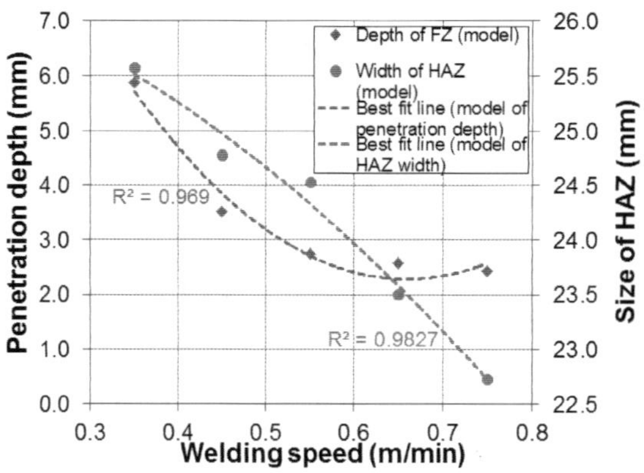

Figure 14. Plots of penetration depth and width of HAZ against welding speed variation for MIG thermal model

5. Conclusion

The experimental and numerical analysis of MIG welded aluminium components were successfully conducted in this study. From the results, it was concluded that heat input and welding speed have a significant effect on the penetration depth and size of HAZ. It was found that in most cases the model could yield an almost accurate result with the prediction error of below 10%. In all cases of simulation, the penetration depth and the width of HAZ have the tendency to increase with the increase of heat input. On the other hand, the welding speed has a substantial effect on the penetration depth and the width of HAZ. Higher welding speed will eventually decrease the penetration depth and width of HAZ.

Acknowledgements

The authors acknowledge the support from Malaysia Ministry of Education under MyBrain15 scheme. This research was funded through a research grant with account number 203.PMEKANIK.6071331 under Fundamental Research Grant Scheme from Malaysia Ministry of Education.

References

[1] Collette M D 2007 The impact of fusion welds on the ultimate strength of aluminium structures *10th International Symposium on Practical Design of Ships and Other Floating Structures* (Houston, Texas, USA)

[2] Duricic D, Aleksic S, Scepanovic B and Lucic D 2016 Experimental, theoretical and numerical analysis of K-joint made of CHS aluminium profiles *Thin-Walled Structures* **119** 58–71

[3] Imam Fauzi E R, Che Jamil M S, Samad Z and Muangjunburee P 2017 Microstructure analysis and mechanical characteristics of tungsten inert gas and metal inert gas welded AA6082-T6 tubular joint: A comparative study *Transactions of Nonferrous Metals Society of China* **27**(1) 17–24

[4] CEN European Committee for Standardization 2007 Eurocode 9: Design of aluminium structures. Part1-1: General structural rules. EN-1999-1-1. Brussels

[5] Bjorneklett B I, Grong O, Myhr O R and Kluken O 1999 A process model for the heat-affected zone microstructure evolution in Al-Zn-Mg weldments *Metallurgical and Materials Transactions A* **30**(10) 2667–77

[6] Hori H 2011 Effect of heat-affected zone on joint strength of welded Al–Mg–Si System *Weld International* **25**(10) 737–41

[7] Dorum C, Lademo O G, Myhr O R, Berstad T and Hopperstad O S 2010 Finite element analysis of plastic failure in heat-affected zone of welded aluminium connections *Computers and Structures* **88**(9–10) 519–28

[8] Chen B Q, Hashemzadeh M and Guedes S C 2014 Numerical and experimental studies on temperature and distortion patterns in butt-welded plates *International Journal Advanced Manufacturing Technology* **72**(5–8) 1121–31
[9] Pepe N, Egerland S, Colegrove P A, Yapp D, Leonhartsberger A and Scotti A 2011 Measuring the process efficiency of controlled gas metal arc welding processes *Science and Technology of Welding and Joining* **16**(5) 412–7
[10] Dupont J N and Marder A R 1995 Thermal efficiency of arc welding processes *Weld Journal* **74** 406
[11] Bosworth M R 1991 Effective heat input in pulsed gas metal arc welding with solid wire electrodes *Weld Journal* **70** 111–7
[12] Giedt W H, Tallerico L N and Fuerschbach W 1989 GTA welding efficiency: calorimetric and temperature measurements *Weld Journal* **68** 28–32
[13] British Standard Institution 1997 ENV 1999-1-2: Design of aluminium structure - structural fire design
[14] Jamil M S C, Fauzi E R I, Juinn C S and Sheikh M A 2015 Laser bending of pre-stressed thin-walled nickel micro-tubes *Optics & Laser Technology* **73**105–17
[15] Goldak J A, Chakravarti A and Bibby M J 1984 A new finite element model for welding heat source. *Metallurgical Transactions B* **15** 299–305
[16] Mikron Instrument Company Inc. 2016 Table of emissivity of various surfaces. *http://www.czlazio.com/tecnica/Tabella%20delle%20Emissivit%C3%A0.pdf* Access date: 18 September 2018
[17] Matusiak M 1999 Strength and ductility of welded structures in aluminium alloys *Ph. D Dissertation* Norwegian University of Science and Technology
[18] Myhr O R and Grong O 1991 Process modelling applied to 6082-T6 aluminium weldments—I. Reaction kinetics *Acta Metall Mater* **39**(11) 2693–702
[19] Enjo T and Kuroda T 1982 Microstructure in weld heat affected zone of Al-Mg-Si alloy *Trans JWRI (Japanese Weld Res Institute)* **11**(61) 61–6
[20] Dumolt S, Laughlin D and Williams J 1984 Formation of a modified β' phase in aluminum alloy 6061 *Scr Metall Mater* **18**(12) 1347–50
[21] Kou S 2003 *Welding Metallurgy* 2nd Edition (Hoboken: John Wiley & Sons)
[22] Ambriz R R and Jaramillo D 2014 Mechanical behavior of precipitation hardened aluminum alloys welds *Light Metal Alloys Applications* 35–59

IOP Conf. Series: Materials Science and Engineering **530** (2019) 012017 doi:10.1088/1757-899X/530/1/012017

Effects of Multi-stack Ball Grid Array on Multi-stack Printed Circuit Board

M A F M Mukhtar[1*], A Abas[2], W M E I W S Bahri[1]

[1]DRB Hicom University of Automotive Malaysia (DHU), Lot 1449, PT 2204 Peramu Jaya Industrial Area 26607 Pekan, Pahang Darul Makmur, Malaysia.

[2]Universiti Sains Malaysia, 11800 Gelugor, Penang, Malaysia

Corresponding author *: fatah@dhu.edu.my

Abstract. In the customary underfill (CUF) epitome process, there are couple of downsides experienced for instance, extended filling time, divided filling and voids improvement. Test and FVM reenactments have been directed to examine CUF dispensing systems for different types of ball grid array (BGA) tendencies out of a solitary layered PCB. In this task, the principle point is to enhance the stream of under filling mold crosswise over multi-stacks BGA. Package on Package (PoP), a strategy for coordinated circuit bundling, is utilized to consolidate BGA bundles vertically to permit higher segment thickness in devices. The L-type liquid stream in multi-stack BGA with edge introduction is contemplated and parameters of CUP exemplification, for example, void development and filling time were examined. The results demonstrated that the BGA sizes with the little weld balls put at the base layer, trailed by the medium size bind balls in the center stack and the biggest patch balls at the top layer has appeared to enhance the stream rate of the encapsulant ideally. Other arrangement in which the medium size solders placed are at the base layer, small solders at the center layer and largest solders on top of it, requires longest total filling time. The result also shows that the racing effect is present at central region and at the side of the layer. Because of the perimeter orientation design, the void formation is minimal and the race effect is not affecting much according experimental and computational result. This investigation additionally uncovers the capability of air gaps in improving velocity and pressure distribution near inlet and outlet to achieve faster total filling time at all layers of the BGA.

1. Introduction

Underfill encapsulation process is a process in which the space between the silicon die and the PCB is filled with the underfill encapsulant that redistributes the induced stress thereby enhancing the solder reliability [1-2]. This underfill process consists of apportioning a managed measure from fabric into an opening among chip and substrate. It is also determined crucial for ensuring and rising unwavering first-class of the digital packaging (EP). It can lessen the global thermal expansion, stresses and pressure within the silicon chip and substrate.Hole among the chip and silicon should totally dispatched with underfilling mildew for the purpose of making certain life of the chip meeting [4]. Affiliation among underfill mould and solder bumps may actuate doable void or air pocket formation and solder bump harmed because of inappropriate dedication of underfill material and processing parameter [2-5]. Thus, superior comprehension at underfilling process is critical and can be done through virtual modeling method [6-7].

In the stream of new technologies nowadays, Electronic industries are seeking for towards smaller scale electronic chip packaging process along with rises in functionality and the demand for multi stack chip is highly desirable[8]. Due to this matters, multi stack BGA are now being under researched. The

IOP Conf. Series: Materials Science and Engineering **530** (2019) 012017 doi:10.1088/1757-899X/530/1/012017

research that have been made by Bart et.al focuses on stress analysis through for the underfilling material and substrate [8]. Meanwhile, the simulation and experiment observation that were conducted by Abdullah et.al showed that the flow was halted as it flows over the chip due to tiny gap height. They also make a calculation to show that the minimum die top clearance is 0.25mm for flip chip to prevent from short shot issues [9]. In other research, Ernest et.al performed a research on flip chip with a series of experiments and simulations to handle the encapsulation flow and chip deformation. These researchers provide some insight on the stress distributions during the encapsulation process [10].

Based on the research that made by Chen et.al, thermal mode analysis is used to optimize thermal management to identify optimal locations and size of multi chip package. The average thermal resistance is analyzed throughout the investigations and found that thermal resistance matrix is independent of power dissipation [11].

This research, focuses on ball grid array (BGA) sizes effect to the fluid flow simulation. There are rarely research and investigation that has been conducted on the size variation of multi stacks BGA. The related Ansys software tutorials are studied to have an understanding on the fluid flow simulation. A modification is done on Ansys Fluent in order to generate the multi-stacks ball grid array (BGA) with perimeter orientation of spherical solder bump array subjected to fluid flow. Various parameters like filling time, flow front pattern, pressure and velocity dissemination are examined based on real world encapsulant liquid. A comparison is then made between simulation and experimental result of the fluid flow.

2. Methodology

2.1. Experiment Detail

In order to enhance the visualization aspect of the micro-sized package a scaled-up experiment is constructed. The BGA models are scaled-up by at least six-time bigger than the real ball grid array (BGA) size for ease of newability. The design parameters of the models are summarized in table 1.

Table 1: Parameter of BGA model

Parameter	**Actual size**	**Scale up size**
Parallel plates (mm)	Not applicable	108 (L) x 108 (W) x 5 (T)
Bump Diameter (mm)	0.5	2.9 and 5.5
Bump Pitch (mm)	1.0	5.8 and 11.0
Bump Height (mm)	0.495	2.4 and 4.9

IOP Conf. Series: Materials Science and Engineering **530** (2019) 012017 doi:10.1088/1757-899X/530/1/012017

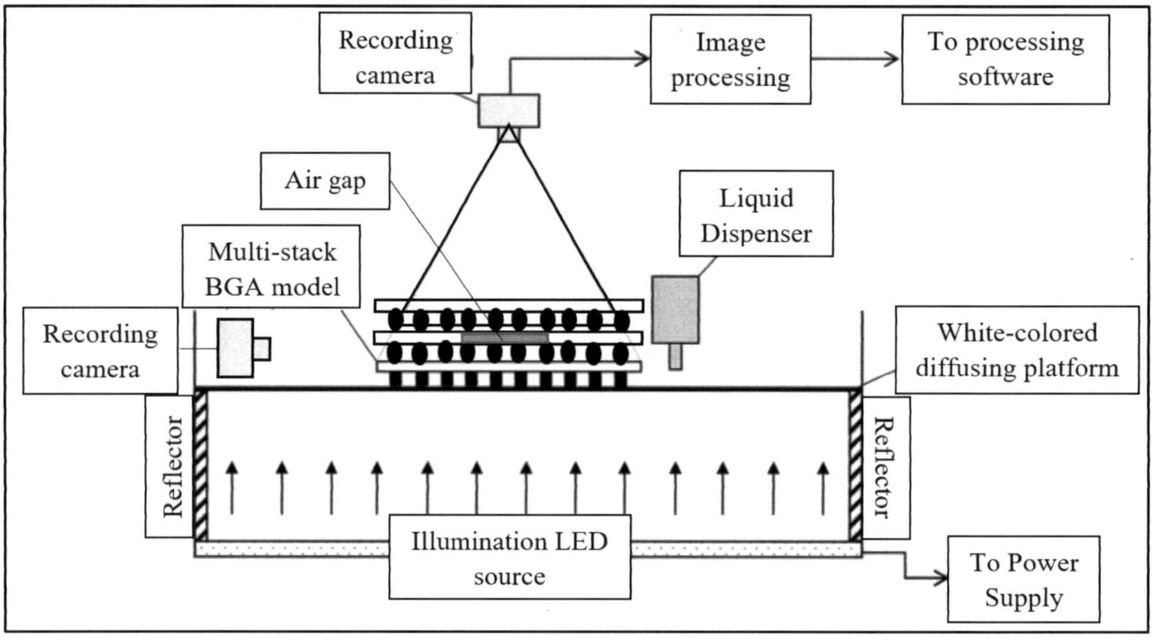

Figure 1: Schematic experimental scaled – up setup for the study of different size of multi stack ball grid array.

The epoxy-molded compound (EMC) used as the encapsulant material with its properties are summarized in Table 2 as stated below.

Table 2: EMC encapsulant material properties

Parameter	Value
Density (kg/m³)	1843
Specific heat capacity (J/kg-K)	1163
Thermal conductivity (W/m-K)	0.8
Molecular weight (kg/kgmol)	44.05358
Dynamic Viscosity (Pa-s)	2.2
Kinematic viscosity (mm2/s)	2111.3
Standard state enthalpy (J/kgmol)	-1.653202e+08
Reference temperature (K)	298.00001

2.2. *Numerical Simulations*

2.2.1. *Equations*

2.2.1.1 *Navier – Stokes Formulation.* The governing equations based on Navier-Stokes equation is used to simulate the encapsulant material and air through the use of conservation of mass and momentum. In the encapsulation system, both fluids are assumed to be incompressible and laminar

Continuity equation:

$$\frac{\partial \rho}{\partial t} + +\nabla.(\rho\vec{u}) = 0 \tag{1}$$

Navier – stokes equation:

$$\frac{\partial}{\partial t}(\rho\vec{u}) + \nabla.(\rho\vec{u}\vec{u}) = -\nabla P + \nabla.\bar{\bar{\tau}} + \rho\vec{g} \tag{2}$$

In the version, EMC and air are differentiated using multiphase formulation. The quantity of fluids (VOF) for each levels is described the use of transport equation at the same time as the distribution of fluid is represented via quantity fraction, f in the range of zero <f <1. Generally, f=0 imply the absence of EMC whilst f=1 imply the cell is completely full of EMC.
Transport equation:

$$\frac{df}{dt} + \nabla.(\bar{u}f) = 0 \tag{3}$$

2.2.1.2 *Capillary Force Formulation.* The underfill fluid flows through the ball grid array (BGA) primarily based on the capillary motion without the assistance of the outside pressure. There are two regions in the fluid capillary waft movement which can be wetting and non-wetting fluid place. The underfill fluid is inside the non-wetting region.The capillary pressure is defined as the pressure difference between wetting and non-wetting fluid. The equivalence to atmospheric pressure can be considered since the non-wetting fluid is air.

$$Capillary\ Pressure,\qquad \Delta P_c = P_{nw} - P_w \tag{4}$$

$$= \frac{2\sigma}{b} \tag{5}$$

Where σ the fluid surface tension and b is is the gap height.

The equation 6 and 7 show that capillary pressure is proportional to surface tension and inversely proportional to the gap height. There are two conditions as shown below:

$$\Delta P_c > \frac{2\sigma}{b} \tag{6}$$

The pressure of non-wetting fluid is greater than the pressure of the wetting fluid. Thus, the fluid in non-wetting region will filled or move to the wetting region (underfilling process phenomenon).

$$\Delta P_c < \frac{2\sigma}{b} \tag{7}$$

The pressure of non-wetting fluid is lower than the pressure of the wetting fluid. Thus, the fluid in the non-wetting region will remain static.

3. Results & Discussion

3.1. Model validation

Two sets of experiments with different solder ball arrangements were carried out for a scaled-up BGA model. The top views of the model are recorded during the filling process of the viscous encapsulant

IOP Conf. Series: Materials Science and Engineering **530** (2019) 012017 doi:10.1088/1757-899X/530/1/012017

into the cavity in the L-type dispensing. The simulation results of the flow front profile obtained from simulation results were compared with the experimental results. Generally, judging from the observation for all two set of experiments of flow front profile, it shows a good relation between the experimental works and computational simulation throughout the CUF encapsulation process at different filling times.

Figure 2 and 3 present the comparison between the flow front profile of experimental works and computational simulation at four distinct filling percentages specifically 20%, 40%, 60% and 80%. This particular case is further emphasized through the data plot of filling time with reference to Figure 2, and 3, the arrangement of beads size among two set of experiments is shown below:

Set		**Size of Beads**	
Layer	**Top**	**Middle**	**Bottom**
A	Small	Medium	Large
B	Large	Small	Medium

Note: Diameter of large, medium and small beads are 5.5mm and 2.9mm respectively.

From the flow front profile shown in Figure 2 and Figure 3 the problem of racing effect is present where the speed of encapsulant varies at different location. At the bumps-free region like central region, it is observed that the flow rate tends to be higher as there is lower resistance to flow whereas flow rate slows down at the bumps region such as the sides of layer, causing the effect of racing effect and possibility of incomplete filling or void formation. Since the orientation of all two sets of experiment are perimeter design, the experimental and computational result are not affected much by the racing effect in terms of filling time and void formation.

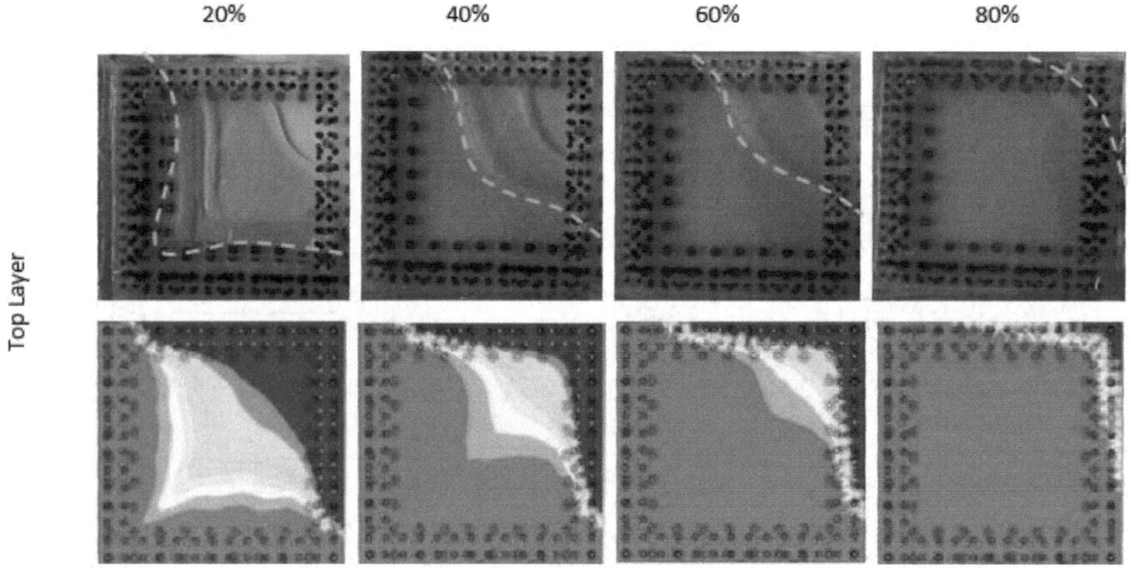

IOP Conf. Series: Materials Science and Engineering **530** (2019) 012017 doi:10.1088/1757-899X/530/1/012017

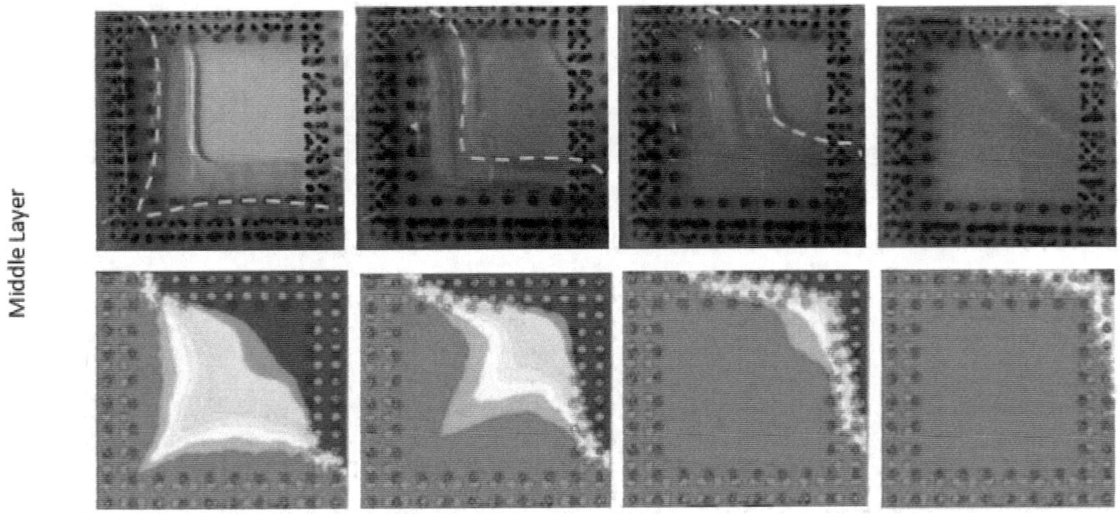

Figure 2: Flow Front Pattern (Experimental vs Simulation) of Set A

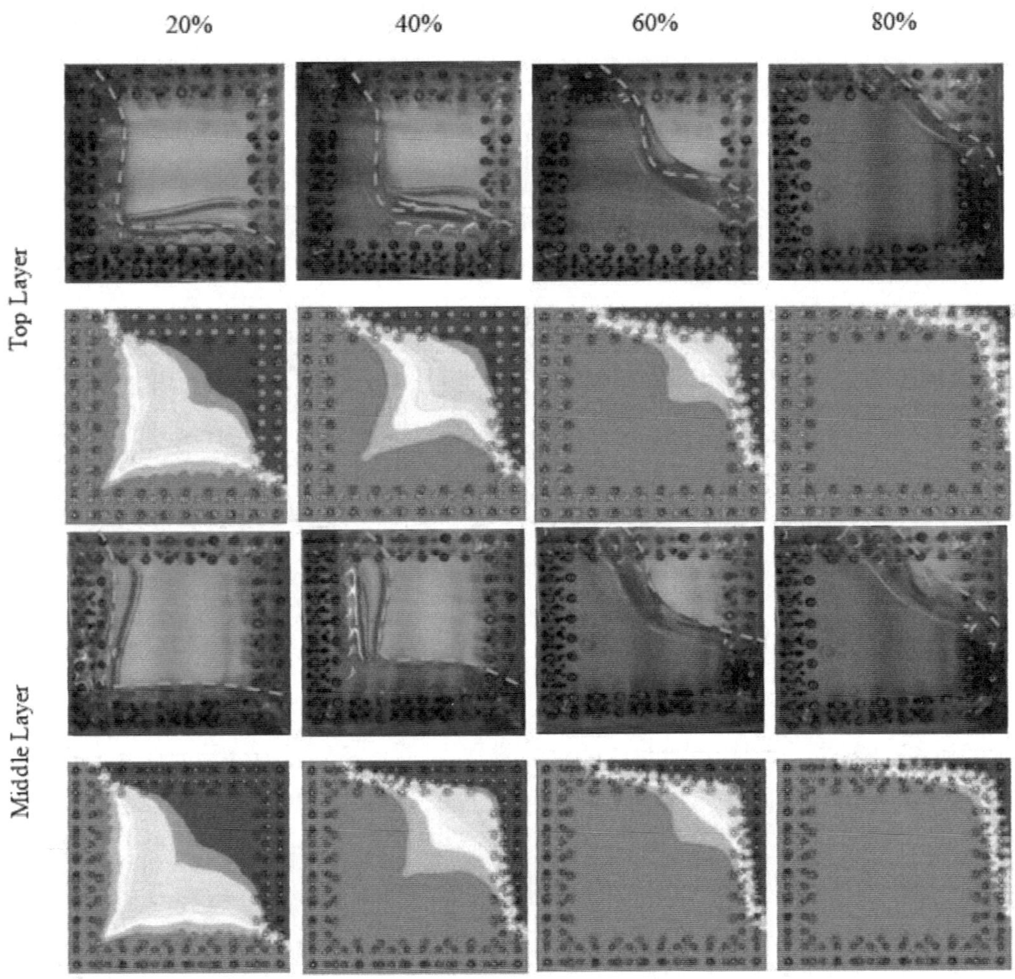

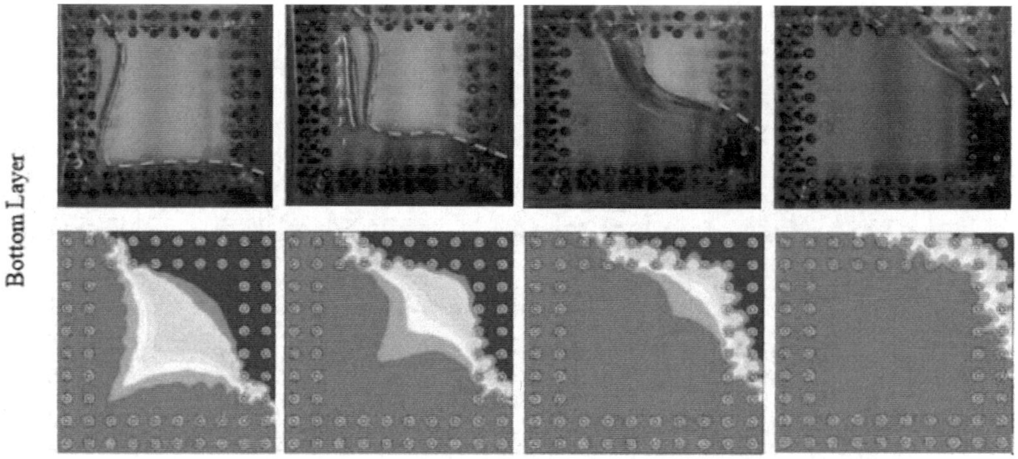

Figure 3: Flow Front Pattern (Experimental vs Simulation) of Set B

3.2. Filling time

According to the experimental and simulation findings, it was found that set A consumes least filling time compared to set B that requires longest time for both layers. Noticeably, longer filling time is expected for larger size of beads as it takes substantial amount of encapsulant to fill up the larger gap height between layers. The filling time for for different percentage of perimeter orientation can be referred to Table 3 and Table 4. Besides, it can be observed that filling rate at set A is very persistant than set B.

Table 3: Experimental filling time

Percentage of filling (%)						
Set		A			B	
Layer	Top	Middle	Bottom	Top	Middle	Bottom
0	0	0	0	0	0	0
20	18	12	6	20	14	17
40	38	25	12	32	22	27
60	57	35	21	42	30	36
80	70	46	31	59	42	46
100	84	59	55	85	60	65

Table 4: Simulation filling time

Percentage of filling (%)						
Set		A			**B**	
Layer	Top	Middle	Bottom	Top	Middle	Bottom
0	0	0	0	0	0	0
20	17	13	8	23	13	19
40	41	28	10	33	24	28
60	58	34	19	42	31	33
80	74	41	29	58	44	40
100	87	57	51	88	59	66

4. Conclusion

This study disclosed the impact of BGA size arrangement filling time, pressure dissemination, and velocity dissemination of CUF encapsulation on BGA packaging. The research was conducted by the approach of experiments and FVM numerical simulations. Two layers of BGA with different sizes (bump diameter of 2.9 and 5.5) in the perimeter orientation are used to determine the best arrangement for constant flow rate and shortest total filling time on L-type dispensing methods. 3-dimensional simulation model with dimensions of 108mm x108mm is generated based on volume of fluid (VOF) method to duplicate the fluid flow phenomenon through BGA which can be used to testify with the data obtained from experiments.

The fluid filling time for BGA size arrangement where the small solders are placed at bottom layer, followed by medium-sized solders and the large solders at the top layer is much faster arrangement since it has smallest gap height to be filled by the fluid at the bottom layer. It also provides the most uniform flow rate at all two layers. Looking at the flow front pattern for all two set of models, it presents a good relation between the experimental works and computational simulation and throughout the CUF encapsulation process at different filling times. Racing effect can be seen as the difference of encapsulant speed at different location. The fluid advances faster in the bump-free region where it has lower resistance to flow, whereas the flow is obstructed by the solder bumps at the bumps region. However, the racing effect is not very obvious and severe for the perimeter orientation.

References

[1] Ardebili H and Pecht M G 2009 Encapsulation Technologies for Electronic Applications. *Encapsulation Technologies for Electronic Applications* Elsevier 129–179 p

[2] Gannamani R and Pecht M 1996 An experimental study of popcorning in plastic encapsulated microcircuits *IEEE Trans Components Packag Manuf Technol Part A* 19(2) 194–201

[3] Ardebili H and Pecht M G 2009 Chapter 3—Encapsulation Process Technology. *In: Encapsulation Technologies for Electronic Applications* Elsevier p 129–79

[4] Khor C Y, Abdullah M Z, Lau C S, Leong W C and Abdul Aziz M S 2014 Influence of solder bump arrangements on molded IC encapsulation *Microelectron Reliab. Elsevier Ltd* 54(4) 796–807

[5] Ong E E, Abdullah M Z, Khor C Y, Leong W C, Loh W K, Ooi C K and Chan R 2013 Numerical Modeling and Analysis of Microbump Pitch Effect in 3D Ic Package With Tsv During Molded

Underfill (Muf) *Eng Appl Comput Fluid Mech* 7(2) 210–22
[6] Wang J 2007 The effects of rheological and wetting properties on underfill filler settling and flow voids in flip chip packages *Microelectron Reliab.* 47(12) 1958–66
[7] Wan J W, Zhang W J and Bergstrom D J 2007 Recent advances in modeling the underfill process in flip-chip packaging *Microelectronics J.* 38(1) 67–75
[8] F C Ng, A Abas, M H H Ishak, M Z Abdullah and M S Abdul Aziz 2016 Effect of thermocapillary action in the underfill encapsulation of multi-stack ball grid array *Microelectronics Reliability 66* 143–160
[9] M Khalil Abdullah, M Z Abdullah, S Kamarudin and Z M Ariff 2007 Study of flow visualization in stacked-Chip Scale Packages (S-CSP) *International Communications in Heat and Mass Transfer 34* 820–828
[10] Ernest E S Ong, M Z Abdullah, C Y Khor, W K Loh, C K Ooi and R Chan 2014 Fluid–structure interaction analysis on the effect of chip stacking in a 3D integrated circuit package with through-silicon vias during plastic encapsulation *Microelectronic Engineering* 113 40–49
[11] Dao-Long C, Tei-Chen C, Ping-Feng Y and Yi-Shao L 2015 Thermal resistance of side by side multi-chip package: Thermal mode analysis *Microelectronics Reliability 55* 822–831
[12] Zienkiewicz O C, Taylor R L and Zhu J Z 2013 The Finite Element Method: its Basis and Fundamentals *Finite Elem Method its Basis Fundam* 493–543
[13] LeVeque R J 2002 Finite Volume Methods for Hyperbolic Problems *publisherName Cambridge University Press*
[14] Morgan M A 1990 1—PRINCIPLES OF FINITE METHODS IN ELECTROMAGNETIC SCATTERING. In: Morgan MABT-FE and FDM in ES, editor *Oxford: Elsevier* p 1–68
[15] Abas A and Abdul-Rahman R 2015 Adaptive FEM with Domain Decomposition Method for Partitioned-Based Fluid–Structure Interaction *Arab J Sci Eng. Springer Berlin Heidelberg* 1–12
[16] Sukop M C and Thorne D T 2006 Lattice boltzmann modeling: An introduction for geoscientists and engineers *Lattice Boltzmann Modeling: An Introduction for Geoscientists and Engineers* p 1–174
[17] Luo L-S 2001 Theory of the Lattice Boltzmann Method: Lattice Boltzmann Models for Non-ideal Gases 62(4) 4982–96
[18] Frisch U and Pazzis D 1986 Lattice as Automata for the Navier-Stokes Equation *Phys Rev Lett.* 1505–8 PMID: 10032689
[19] Mohammed A A 2012 Lattice Boltzmann Method: Fundamentals and Engineering Applications with Computer Codes *AIAA Journal* New York: Springer; 238 p
[20] Bailey C 2005 Numerical modelling for electronic packaging—future requirements *6th Int Conf Electron Packag Technol* 5–6
[21] Chen S and Doolen G D 1998 Lattice Boltzmann method for fluid flows. Annual Review of Fluid Mechanics p 329–64
[22] He X and Doolen G D 2002 Thermodynamic foundations of kinetic theory and Lattice Boltzmann models for multiphase flows *J Stat Phys* 107(1–2) 309–28
[23] Deepak M 2007 VOF based Multiphase Lattice Boltzmann Method using Explicit Kinematic Boundary Conditions at the interface *Georgia Institute of Technology*
[24] Khor C Y, Abdullah M Z, Abdullah M K, Mujeebu M A, Ramdan D, Majid M F M A, Ariff Z M and Rahman M A 2011 Numerical analysis on the effects of different inlet gates and gap heights in TQFP encapsulation process *International Journal of Heat and Mass Transfer* 54(9-10) pp1861-1870
[25] Khor C Y and Abdullah M Z Feb 2013 Analysis of fluid/structure interaction: Influence of silicon chip thickness in moulded packaging *Microelectron Reliab. Elsevier Ltd* 53(2) 334–47
[26] Khor C Y, Abdullah M Z, Che Ani F 2011 Study on the fluid/structure interaction at different inlet pressures in molded packaging *Microelectron Eng. Elsevier B.V* 88(10) 3182–94
[27] Khor C Y, Abdullah M Z, Abdul Mujeebu M and Che Ani F 2010 FVM based numerical study

on the effect of solder bump arrangement on capillary driven flip chip underfill process *Int Commun Heat Mass Transf.* 37(3) 281–6
[28] Orlandini E, Swift M R and Yeomans J 1995 A Lattice Boltzmann Model of Binary-Fluid Mixtures *Europhys Lett* 463
[29] Swift M, Orlandini E, Osborn W and Yeomans J 1996 Lattice Boltzmann simulations of liquid-gas and binary fluid systems *Phys Rev E* 54(5) 5041–52
[30] Zheng H W, Shu C and Chew Y T 2006 A lattice Boltzmann model for multiphase flows with large density ratio *J Comput Phys.* 218(1) 353–71
[31] Yan B and Yan G 2011 A steady-state lattice Boltzmann model for incompressible flows ☆. *Comput Math with Appl. Elsevier Ltd* 61(5) pp 1348–54
[32] Schmieschek S and Harting J 2001 Contact angle determination in multicomponent lattice Boltzmann simulations. Commun Comput Phys. 9:1165
[33] Gunstensen A K, Rothman D H, Zaleski S and Zanetti G 1991 Lattice Boltzmann model of immiscible fluids *Phys Rev A* 43(8) pp 4320–7 PMID: 9905534
[34] Lishchuk S V, Care C M and Halliday I 2003 Lattice Boltzmann algorithm for surface tension with greatly reduced microcurrents *Phys Rev E Stat Nonlin Soft Matter Phys* 67:036701. PMID: 12689196.

IOP Conf. Series: Materials Science and Engineering **530** (2019) 012018 doi:10.1088/1757-899X/530/1/012018

ABAQUS Simulation of Different Critical Porosities Cubical Scaffold Model

Z Wahid[1,2,*], M K A M Ariffin[1], B T H T Baharudin[1], M I S Ismail1 and F Mustapha[3]

[1]Department of Mechanical and Manufacturing Engineering, Faculty of Engineering, Universiti Putra Malaysia, 43400 Serdang, Selangor, Malaysia.

[2]Department of Mechanical and Materials Engineering, Faculty of Engineering, Universiti Kebangsaan Malaysia, 43600 Bangi, Selangor, Malaysia.

[3]Department of Aerospace Engineering, Faculty of Engineering, Universiti Putra Malaysia, 43400 Serdang, Selangor, Malaysia.

Corresponding author *: zaliha@ukm.edu.my

Abstract. Scaffold or porous architectured structure is essential in biomedical application for cell growth or for mass reduction if to be used in automotive and aerospace applications. The structure was once impossible to fabricate but now viable using additive manufacturing techniques. However, in order to reduce weight structure and save on materials using porous structure, strength is one key feature must be maintained to serve the function. The aims of the study are to compare the part strength with different cell repetition and porosity (ranging 56% to 70.45%), and also to determine pattern of stress concentration in different unit cell size. In this study, cubic scaffold-based structures with critical porosity range were simulated using ABAQUS. From the simulation it can be observed generally that strength will reduce with higher porosity, while 58.96% of porosity showed less stress concentration. For each porosity, stress concentration and deflection can be observed and compared. The result of the study is important to help designer to choose suitable design and porosity prior to fabrication.

1. Introduction

The emergence of Additive Manufacturing in 1996 has opened new possibilities for manufacturing world. One of the most prominent advantage is the ability to fabricate complex geometry, practically the complexity that are not feasible before using other conventional manufacturing techniques. There are many techniques for AM but employing similar concept of adding material collectively layer-by layer based on CAD data file. Upon designing, architectured structure also known as scaffold, lattice or porous structure is important mainly to reduce material usage and mass [1-2]. While mostly in medical application, porous structure is purposely employed to enable cell growth in artificial bone structure and also as protective structure for energy absorption [3]. The emerging of additive manufacturing technology has made the objectives to produce complex and lightweight structure possible [4]. Examples of architectured structure are bio inspired bone structure, which initially fabricated as simple porous design, later on the aim was to mimic real bone structure namely trabecular and cortical bones with different strength and currently studies found out that bone is made of hierarchical of multi scale structure [5]. In aerospace part design, topologically optimized brackets which are porous and thus save on material usage was employed for aerospace application [6]. Upon application however, strength of the part is priority over lightweight structure. Strength to weight ratio is nevertheless important and from past research, it is direct translated from the porosity of the structure where the best mechanical properties observed in scaffolds of low porosity [7]. In previous study, ABAQUS simulation was being conducted on cubical designs with wider range of porosity 0-89.6% and as expected, very low porosity made high strength parts with result and stress distribution was similar to the solid cube. At 89.6% porosity, the structure could not even bear one third of the load, and it seems like good mechanical properties lied at some point around 60% porosity. Building orientation is one vital parameter in additive manufacturing process. For ABS parts under compression loading, the sample with a transverse build direction has a lower strength when compared with the sample having the axial build direction, and the compressive strength of the specimen is 80% to 90% of the injection moulded part. In a similar study, experimental approach was found unable to determine compression strength or failure trend of architectured compression test [4]. The objectives of the study are to compare the strength of parts with different cell repetition but same porosity (ranging 56% to 70.45%), and to determine pattern of stress concentration in different unit cell size.

2. Materials and methods

2.1. Design

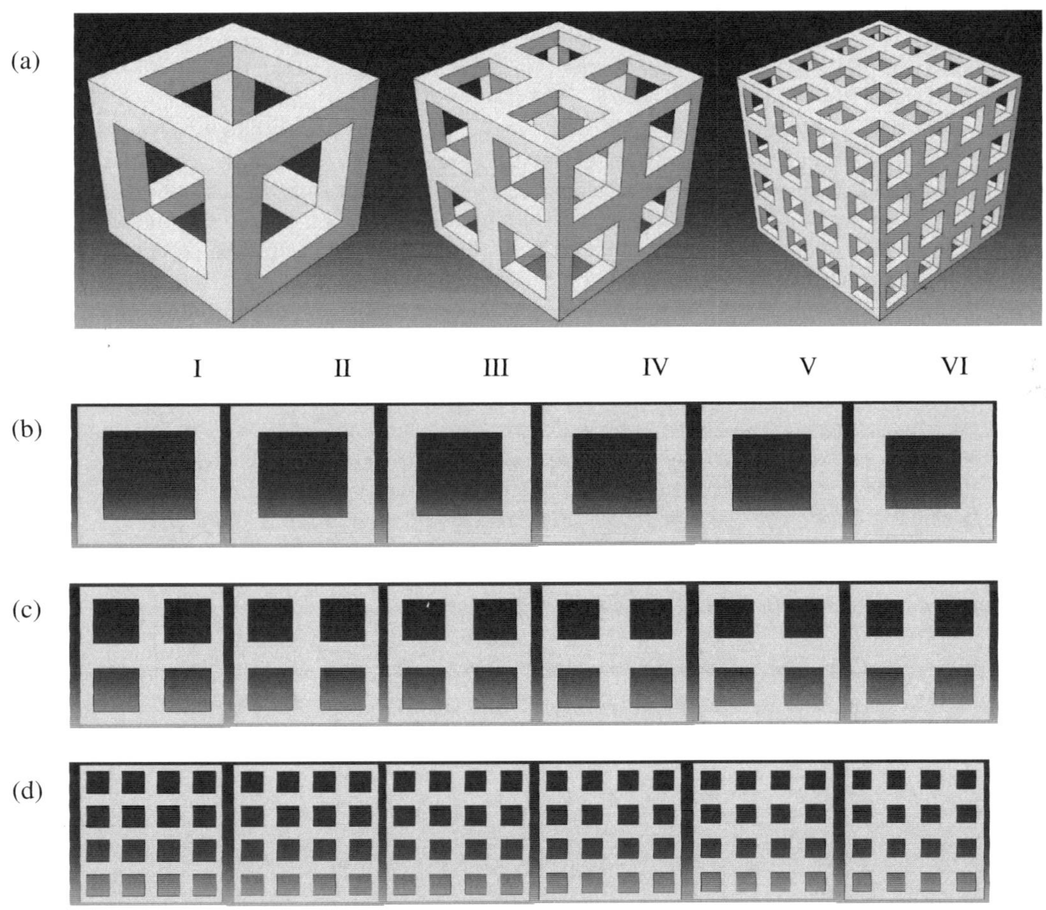

Figure 1. Porosity: (I) 70.45 %, (II) 67.65%, (III) 64.80%, (IV) 61.9%, (V) 58.96% and (VI) 55.99%. (a) Scaffold structure (b) Single Unit cells (c) & (d) Repetitive Unit.

IOP Conf. Series: Materials Science and Engineering **530** (2019) 012018 doi:10.1088/1757-899X/530/1/012018

Architectured structure can be simple and complex polyhedra shapes [8] but in this work, the chosen model was simple lattice cubical cell structure. The part model sized 100mm x 100mm x 100mm in outer dimension whether it represented single or repetitive cell unit. The structure were made with different strut size, thus different porosity range from 70.45% which is very porous to 55.99% porosity. Models designs of open-porous lattice structures with different porosity are as shown in Figure 1, where (I) – (VI) represent range of porosities while (a) - (d) show different representations of design employed in this study which are final model of different unit number, single, double and repetitive units respectively.

All models were designed inside the Abaqus and not outsourced nor imported from other CAD designing software. In order to eliminate the effect size variable, all models are in similar base size of 100mm x 100mm x 100mm whether the cube is a single unit or scaffold.

2.2. Simulation

Analyses were performed using Abaqus Standard 6.14-2 by Dassault Systems Simulia. The analysis was on general static step, and based on previous study using 500N load [4] compressing the specimen. The model was meshed using 10-seed size tetrahedral elements as recommended for lattice design except for solid cubical design which was meshed using 10-seed size hex units. Number of generated nodes and elements varies for all the design.

The works in this study were based on simulation only and material properties were input using collected data from previous experimental works. These data consist of Young Modulus and stress-plastic strain relationship.

The simulation steps are part designing where basically there were five parts, properties input based on material properties such as density and elastic-plastic behaviour information and assembly for combining instances. During the simulation, each part model was loaded with 500N force on top surface and encastre fixed at the base. After all input is done, the model was meshed before simulation job is submitted. From the simple simulation, the result is observed and recorded for each model and will be discussed in the next section.

3. Results and discussions

From the simulation, some important results to expect are to observe the Von Mises plane stress on each model when loaded, to see if similar scaffold design would show pattern in stress distribution and to know if the sharp corner can possibly be stress concentration which may induce failure initiation. For all the result, colour coded legend as shown in each simulation result figure can be referred, where red represents highest value, green is for medium and blue shows lowest value of stress reading. The real value and range for each viewport is actually different and should be referred individually.

3.1. Single unit

From Figure 2, it can be clearly seen that area of stress concentration is on the connecting vertical strut and as the porosity getting lower, higher stress can be observed and green coloured shading has become wider as compared to lower porosity structure. The single cell structure is quite simple and therefore, stress distribution is generally in both vertical and horizontal struts. Only on the base where the structure was encastred, the horizontal struts exerted less stress than the upper did.

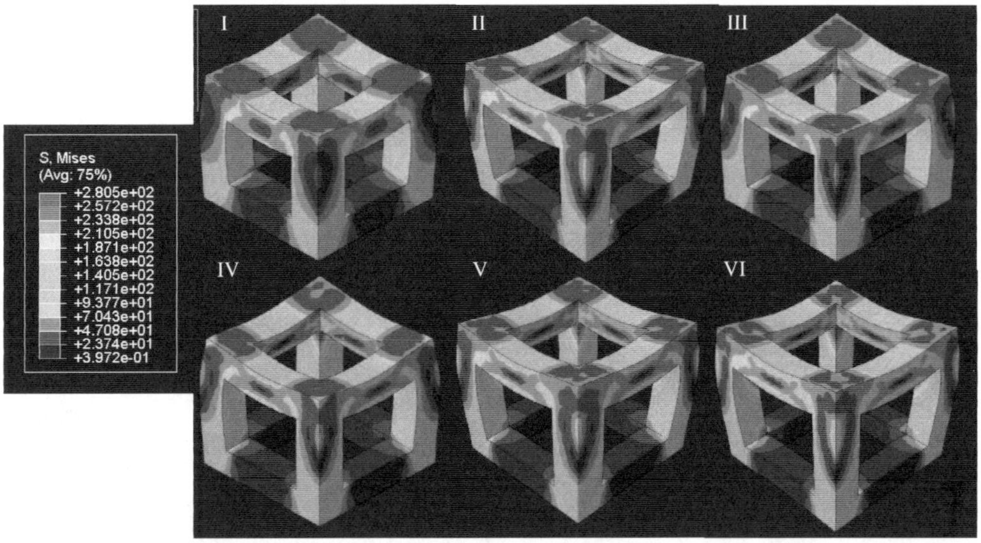

Figure 2. Static loaded for each porosity of single cell: (I) 55.99%, (II) 58.96% (III), 61.9%,(IV) 64.80%, (V) 67.65% and (VI) 70.45%

3.2. Double unit

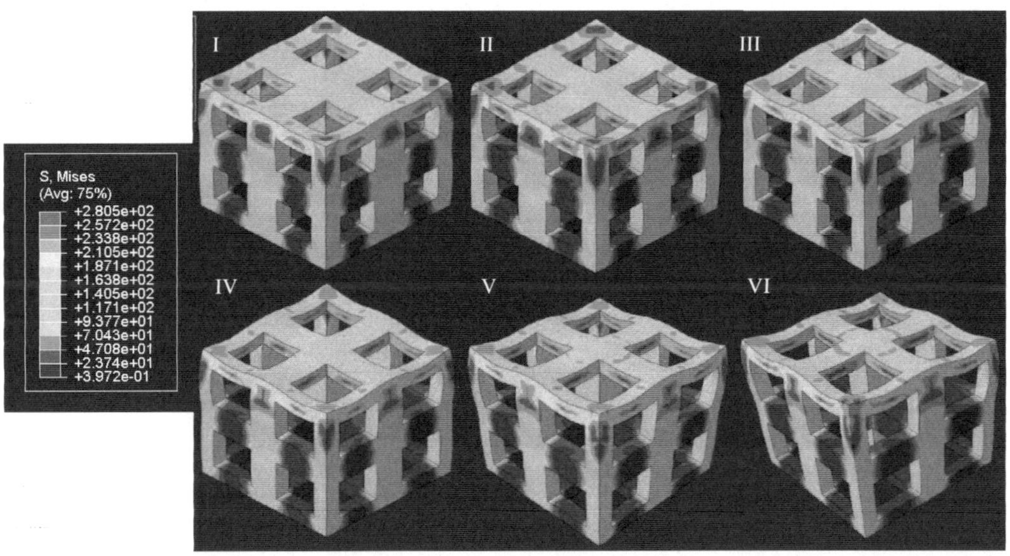

Figure 3. Static loaded for each porosity of double repetitive cell : (I) 55.99%, (II) 58.96% (III), 61.9%, (IV) 64.80%, (V) 67.65% and (VI) 70.45%.

IOP Conf. Series: Materials Science and Engineering **530** (2019) 012018 doi:10.1088/1757-899X/530/1/012018

Interestingly, close observation showed that highest plane stress is on outer struts and less was distributed in centre strut. However, as the strut getting smaller and porosity become higher, the centre strut started to experience stress concentration. As can be seen from Figure 3, pattern of stress distribution is similar as single unit but the stress was not distributed on lower horizontal frame.

3.3. Multiple unit

From figure 4 it showed that all forces are exerted on the top plane and highest stress can remarkably be seen on the four top corners of the cube when the porosity is lower. At highest porosity of 70.45%, the upper vertical strut began to distort and this area is expected to initiate failure upon loaded.

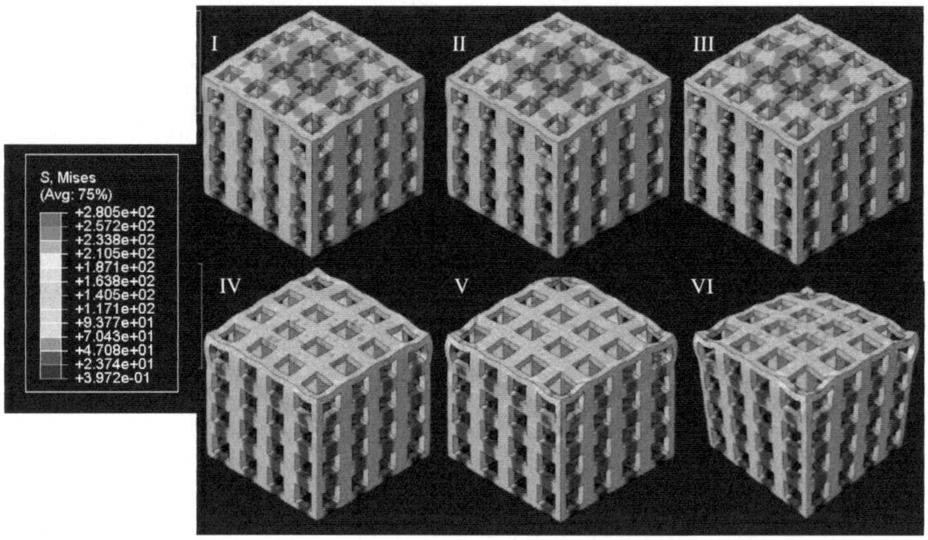

Figure 4. Static loaded for each porosity of multiple cell : (I) 55.99%, (II) 58.96% (III), 61.9%, (IV) 64.80%, (V) 67.65% and (VI) 70.45%

For all porosities and repetitions, it is observable that all structure showed similar stress distribution on the parts. Struts along y-axis exerted stress more than horizontal x-z plane and stress distribution was more on upper than on the base of the structure.

3.4. Deflection and mises stress

Some studies considered porosity of 64% and below is acceptable range in maintaining mechanical and fatigue strength. However, this value depends on several variables such as type of material, processing method and application. Deflection is considered the highest deformed reading taken from the y-axis (or u2 in ABAQUS modelling). From Figure 5(a), deflection pattern for single structure is linearly proportional although the pattern can also be observed in double cell repetitive unit. As the repetitive unit increased, it seems the deflection trend is no more linearly proportional. However, it is suggested in future, study can be further extended by adding more repetitive cell to see the trend.

(a)

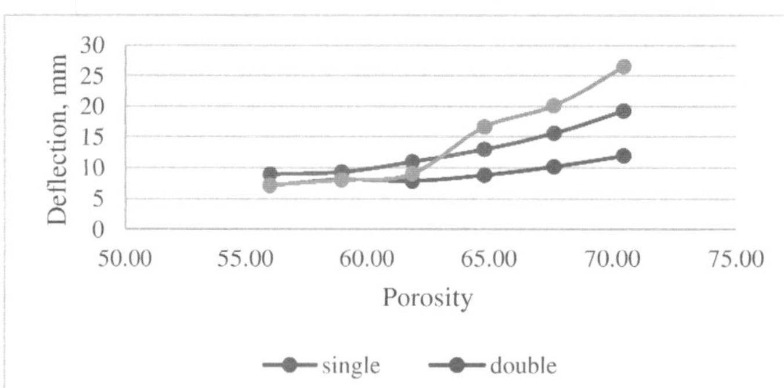

IOP Conf. Series: Materials Science and Engineering **530** (2019) 012018 doi:10.1088/1757-899X/530/1/012018

(b)

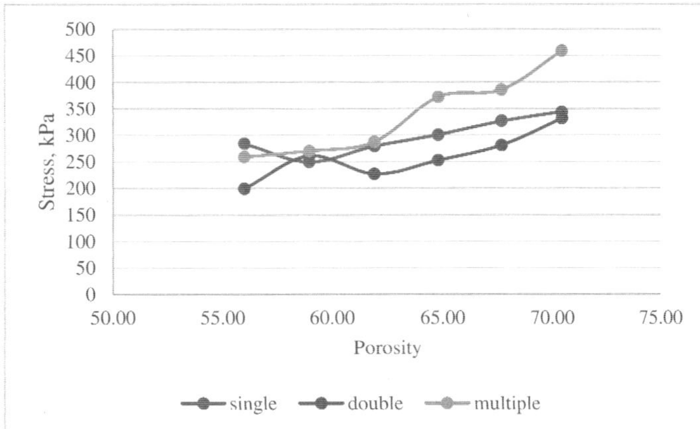

Figure 5 (a) Deflection, (b) Von Mises Stress for different cell type and porosity

Figure 5 (b) shown plot of plane stress (S Mises in ABAQUS) according to cell arrangement (single, double and multiple). From the figure, cell with 58.96% porosity exhibited the best strength for all type of cell arrangement. Von Mises reading for 58.96% porosity is low and congregated disregard of cell arrangement. While for each cell arrangement, the plane stress reading did not according to any specific pattern.

4. Conclusions

Simulation using Abaqus software was conducted successfully and all the results has been analysed and compared. From the study it can be concluded that 58.96% porosity is the best within the range studied in this work (55.99% - 70.45%) in term of Von Mises stress and shows less failure initiation using this porosity.

Acknowledgments

The Authors would like to acknowledge the department of Mechanical and Manufacturing Engineering, University Putra Malaysia and SLAB/SLAI scholarship for funding the research study.

References

[1] M Javaid and A Haleem 2017 Additive manufacturing applications in medical cases: A literature based review *Alexandria J. Med*

[2] L Huynh, J Rotella and M D Sangid 2016 Fatigue behavior of IN718 microtrusses produced via additive manufacturing *Mater. Des.* Vol 105 pp 278–289

[3] T D Ngo, A Kashani, G Imbalzano, K T Q Nguyen and D Hui 2018 Additive manufacturing (3D printing): A review of materials, methods, applications and challenges *Compos. Part B Eng.* Vol 143 pp 172–196

[4] S Limmahakhun, A Oloyede, K Sitthiseripratip, Y Xiao and C Yan 2017 3D-printed cellular structures for bone biomimetic implants *Addit. Manuf.* Vol 15 pp 93–101

[5] J Wang, H Xie, Z Weng, T Senthil and L Wu 2016 A novel approach to improve mechanical properties of parts fabricated by fused deposition modeling *Mater. Des.* Vol 105 pp 152–159

[6] Thompson M K, Moroni G, Vaneker T, Fadel G, Campbell R I, Gibson I, Bernard A, Schulz J, Graf P, Ahuja B and Martina F 2016 Design for Additive Manufacturing: Trends, opportunities, considerations, and constraints *CIRP annals* 65(2) pp 737-760

[7] K C Ang, K F Leong, C K Chua and M Chandrasekaran 2006 Investigation of the mechanical properties and porosity relationships in fused deposition modelling-fabricated porous structures *Rapid Prototyp. J.* Vol 12 pp 100–105

[8] C K Chua, K F Leong, C M Cheah and S W Chua 2003 Development of a tissue engineering scaffold structure library for rapid prototyping. Part 2: Parametric library and assembly program *Int. J. Adv. Manuf. Technol.* Vol 21 No 4 pp 302–312

IOP Conf. Series: Materials Science and Engineering **530** (2019) 012019 doi:10.1088/1757-899X/530/1/012019

Temperature Prediction on Flexible Printed Circuit Board in Reflow Oven Soldering for Motherboard Application

A M Iqbal [1,2], M S A Aziz[1*], M Z Abdullah[1] and M H H Ishak[1]

[1]School of Mechanical Engineering, Universiti Sains Malaysia, Engineering Campus, 14300 Nibong Tebal, Penang, Malaysia

[2]Faculty of Bioengineering and Technology, Universiti Malaysia Kelantan, Jeli Campus, Locked Bag No. 100, 17600 Jeli, Kelantan, Malaysia

Corresponding author *: msharizal@usm.my

Abstract. Flexible printed circuit boards (FPCBs) appoint an interest in current electronics devices with unique capabilities in the form of flexibility, lightweight, and thin thickness. In this paper, the proposed of that utilises FPCB as a substrate for motherboards is investigated. Assembly surface variation temperature in the reflow soldering process is a crucial factor to address the feasibility of substrate for surface mount technology (SMT) applications. For thin FPCB, higher Reynold number (Re) would not give the significant effect of heat transfer on assembly surface temperature. Although the primary focus is in a continuous reflow oven, FPCB findings could enhance the information for other FPCB applications such as designing the soldering thermal profile.

1. Introduction

Flexible printed circuit boards (FPCBs) are referred to as an alternative to a rigid printed circuit board (PCB) for electronic applications since it would have an excellent physical aspect of lightweight, flexibility and twistability. FPCB simplifies a potent of space-efficient design, enhance the inside system presence, reduce weight and installation cost. Over recent years, numerous electronic devices were adapted FPCBs, but in the application as SMT and motherboards are still scarce. Although the FPCB could be very thin in comparison with rigid PCB, the heat transfer performance during reflow soldering process is the primary concern in use as SMT motherboard.

In the microelectronic industry, a reflow oven is widely used in contact to attach surface mount components (SMCs) and devices (SMDs) to PCB. The use of the adjective "reflow'' with the noun "oven'' stems from its hardware being developed and used to fabricate solder interconnects[1]. Prior preparatory steps before the reflow soldering process, a screen printing of solder paste on the copper pad through a stencil or foil. Subsequently, the SMCs insertion on the printed solder paste is through automated component placement machine. Next, the contacting of solder SMCs and copper paste takes place, when PCB passes through the reflow zone (pre-heating, soaking, soldering and cooling). On top selective reflow oven type, continuous type of single or combination of forced convection and infrared (IR) is more preferred for the manufacturing [2]. The air flow in the continuous type of reflow oven required forced air convection for temperature field distribution uniformity within the space.

The understandings of thermal reflow profile are benefits to soldering process control. The defined profiles may influence the solder joint microstructure aspect, such as the thickness of the intermetallic compound (IMC) [3]. Hence, various aspects of reflow oven study had been reported by the scholars. An extensive investigation using experimental and simulation of the reflow oven include oven space heat transfer [4] and gas flow characterisation [5] and the nozzle-matrix convection reflow oven [4-6] were reported in the previous studies.

IOP Conf. Series: Materials Science and Engineering **530** (2019) 012019 doi:10.1088/1757-899X/530/1/012019

Some partial effect of the rigid PCB (RPCB) on continuous reflow oven was examined by [7]. However, the need for further investigation of the effect forced convection velocity towards new FPCB material. Then, the study aims to investigate the temperature prediction on FPCB towards feasibility for electronic motherboards application. Also, thermal impact of reflow soldering process on FPCB assembly, respectively still not available in the literature. The developed method has potential to be used for precise prediction in oven space as well for designing thermal profile to obtain minimal temperature deviation on assemblies.

2. Experimentation

A ready-to-use FPCB was chosen for this study. The FPCB is formed in single-sided consisting of single polyimide and single copper layer with an overall thickness of 53 μm. The complete dimensions and materials of FPCB [8] and epoxy capacitor mould are denoted in Table 1. The gained properties were used in simulation purpose.

Table 1. Materials and thermal properties

Material	Dimensions (mm)	Density, (kg/m^3)	Thermal conductivity,k (W/[mK])	Specific heat, C_p (J/kgK)
FPCB	110 x 80	3839	0.3	421
Mold	4.0 x 4.2	1820	0.6	236

For this study, the soldering process was implemented in the continuous type of BTU Paragon 150 oven. In a particular zone, is equipped with a nozzle-matrix system to form the jet streams as shown in Figure 1. The nozzle-matrix system which uses a centrifugal blower to force heated air toward the assemblies. On top of that, a reflow oven is made up of an entrance, eight heating zones, two cooling zones, and an exit section. The heater is installed on each of top and bottom blower in heating zones. The view of the heater structure is shown in Figure 1a. The soldering agent gas (air) was heated via the heating coil and passed though the small diameter nozzles before contacting the assemblies. The nozzle configuration is in Figure 1b, then the gas pas through the recirculation delivery area to complete the cycle.

According to JEDEC JSTD-020D standard [9], reflow soldering profile typically consists of two regions: the heating region comprises a preheating, soak, and reflow zone, and cooling zone as shown in Table 2. The duration of soaking and reflow-soldering were fall in 60-120 and 60-150 s, respectively. Also, the peak setting for SAC solder material should in 230-240 °C, whereas in the ramp up and ramp down reflow zone temperature was set to be less than 3 °C/s. For this study, duration of 60 s; preheating, 60 s; soaking and 124 s; reflow soldering were considered.

IOP Conf. Series: Materials Science and Engineering **530** (2019) 012019 doi:10.1088/1757-899X/530/1/012019

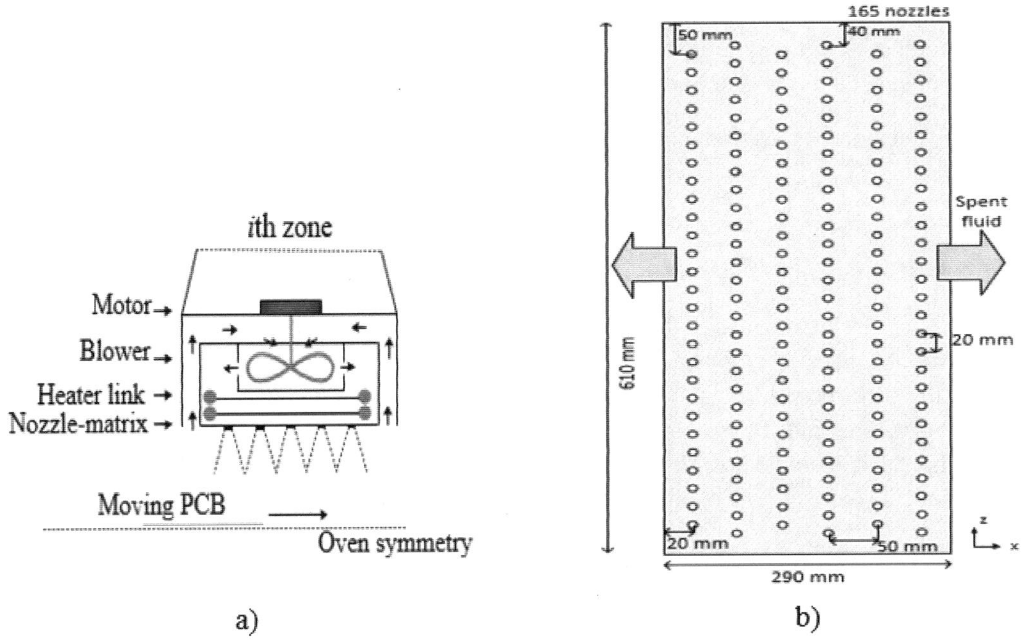

Figure 1. Schematic diagram of a) Upper Blower b) Nozzle configuration

Table 2. Setting temperature and conveyor speed in each setting zone

Oven zone	1	2	3	4	5	6	7	8	9	10
Oven space temperature (°C)	100	120	140	160	180	200	220	240	255	230
Conveyor speed (cm/s)					1.2					

3. Numerical analysis

3.1. Fluid Flow Model

The governing equation is subjected to single-phase flow model, and FLUENT 18 is employed to solve it using velocity and Cartesian spatial coordinate in differential control volume which written in the non-dimensional form [10].

a) Single phase

Eq. (1) denoted the incompressible continuity equation.

$$\frac{\partial \rho u_i}{\partial x_i} = 0 \tag{1}$$

where u_i is the velocity component in x_i direction (for i=1,2,3 reflects to three perpendicular axis)

While, Eq. 2 is the momentum equation with neglecting buoyancy terms as the following;

$$\frac{\partial}{\partial t}(\rho u_i) + \frac{\partial}{\partial x_j}(\rho u_i u_j) = -\frac{\partial P}{\partial x_i} + \frac{\partial \tau_{ij}}{\partial x_j} + \rho g_i + F_i \tag{2}$$

Denoted, ρ is the fluid density, τ_{ij} is the viscous stress tensor, u_j is the velocity component in x_j direction, P is the pressure, g is the gravitational acceleration and F is external body force in the i-direction. Neglecting of buoyancy terms is defined under forced convection from blower to the oven space.

Considering of thermal in the oven space, the energy equation is denoted as Eq. (3),

$$\frac{\partial}{\partial t}(\rho C_p T) + \frac{\partial}{\partial x_i}(\rho u_i C_p T) = -\frac{\partial}{\partial x_i} + \frac{\partial}{\partial x_j}\left(\lambda \frac{\partial T}{\partial x_i}\right) \tag{3}$$

where C_p, T and λ mentioned the specific heat, temperature and thermal conductivity of hot air, respectively.

Flow typically begins at the top and bottom nozzle matrix into oven space, becomes turbulent due to proportional flow is directive on each other correspond to fluid-fluid configuration interaction as jet impingement. As above judgement on flow, Re obtained is the turbulent regime. From the turbulent CFD codes, the standard k-epsilon (SKE) model is selected and performs well in non-large adverse pressure gradient [11] and consist of two transport equation as follows:

$$\frac{\partial}{\partial t}(\rho k) + \frac{\partial}{\partial x_i}(\rho k u_i) = \frac{\partial}{\partial x_j}\left[\left(\mu + \frac{\mu_t}{\sigma_k}\right)\frac{\partial}{\partial x_j}\right] + G_k - \rho\varepsilon \tag{4}$$

$$\frac{\partial}{\partial t}(\rho\varepsilon) + \frac{\partial}{\partial x_i}(\rho\varepsilon u_i) = \frac{\partial}{\partial x_j}\left[\left(\mu + \frac{\mu_t}{\sigma_\varepsilon}\right)\frac{\partial\varepsilon}{\partial x_j}\right] + C_{1_\varepsilon}\frac{\varepsilon}{k}\left(G_k + C_{3_\varepsilon}G_b\right) - C_{2_\varepsilon}\rho\frac{\varepsilon^2}{k} \tag{5}$$

where G_k is the turbulent production term due to mean velocity gradients. The negative term using energy dissipation. The closure constant, $C_{1\varepsilon} = 1.44, C_{2\varepsilon} = 1.92, C_\mu = 0.09; \sigma_k = 1.0; \sigma_\varepsilon = 1.3$ are the used parameter value.

The air properties defined according to the ideal gas equation and for the calculation of the density as follows:

$$\rho = \frac{R}{T} \tag{6}$$

where R is a specific gas constant.

3.2. Modelling strategy

The 3D assembly model of the fluid domain was modelled by SpaceClaim in ANSYS workbench version 18. The half geometry of an actual single zone oven was generated with the dimensions of 290 mm × 305 mm × 90 mm, while the solid assembly model was deducted from the domain as depicted in Figure 2. Figure 2 also illustrated the important boundaries including inlet, outlet, enclosure and capacitor. The model was meshed with 870 000 hexahedral and tetrahedral elements at assembly. For verification accuracy purpose, grid independence test were done as in Table 3.

In FLUENT, 3-dimensional, incompressible and unsteady turbulent type of flow was used. A user-defined function (UDF) script was written in C++ language in order to develop closely to the actual oven atmosphere. The density of air was set to follow the ideal gas properties. Giving in Eq. (7), through which air, ρ is the density, v is the velocity applied as follows:, μ is the dynamic viscosity and D is the nozzle diameter. The boundary setting conditions for the fluid domain were

1) At the enclosure of FPCB walls: $\vec{v}= 0$ (no slip condition).
2) At the inlet: $\vec{v} = \vec{v}_{inlet}$ (inlet flow velocity) with the oven space temperature profile using user define function.
3) At the outlet: Pressure = 0 atm (ambient condition)

$$Re = \frac{\rho v D}{\mu} \tag{7}$$

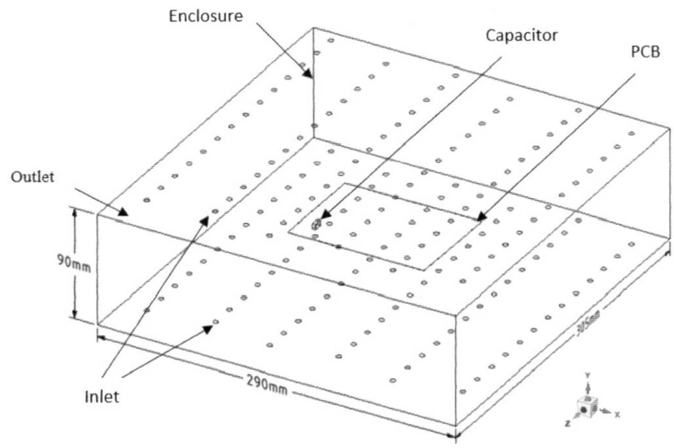

Figure 2. Fluid domain

Table 3. Grid independence test for fluid domain

Case at 10s	A	B	C	D	E
Fluid elements	702836	781853	866928	1034758	1314649
Temperature (K) at 10 s	319.31	326.80	328.25	326.30	328.60
Deviation from case E (%)	2.9	0.6	0.1	0.7	-

The operation mode was run as denoted in Table 4. Different parameters of turbulent flow at Re of (36,611 – 44,088) was carried out with the similar mesh and FLUENT setting for forced convection study.

Table 4. Operating conditions for FPCB reflow soldering

Soldering Agent	Blower pressure (IWC)	*Re* (at 25°C)
Air	0.5	33,260
	1.0	40,221
	1.2	44,088

4. Result and discussions

4.1 Heat transfer performance

The flow induced into reflow oven space could be varied as denoted in Figure 3. The dominant velocity of the jet stream was triggered using blower frequency that in inch water column (IWC). Criteria for selecting the blower IWC would reflect to processing requirement, by means of board assembly soldering quality and throughput. The range of 0.5 to 1.5 IWC were selectived as occupy by oven blower correspond to give uniform velocity on the top and bottom side. Uniformity of forced convection flow induced, is enable to reduce the adverse pressure gradient on oven space. As such explained, a higher IWC lead to induce more heat flux to the oven space, hence alleviate the surface temperature.

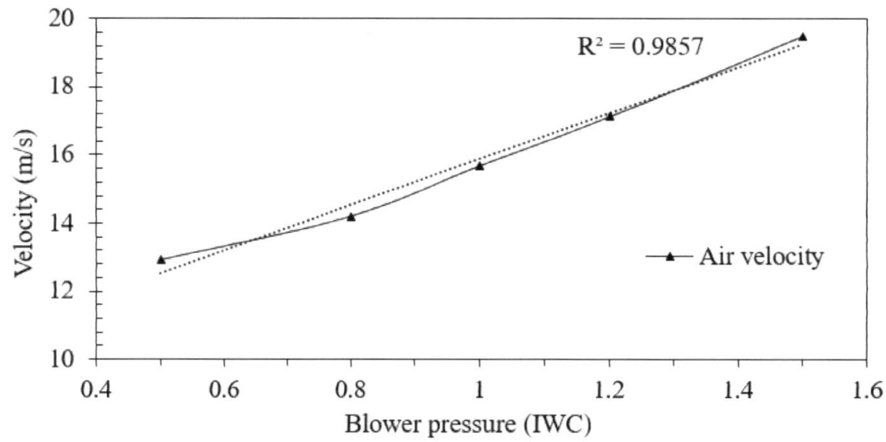

Figure 3. Measured nozzle outlet velocity against blower pressure

Figure 4 shows the board assembly temperature at different zones in a reflow oven. The used initial condition was mimic of situations where the FPCB left the components deposit machine at room temperature. At 10s in preheating, the temperature gradient was up to 15-20 °C at the pushing slope rate of 1.8°C/s. It can be explained that the thermal exchange for FPCB to initially absorb the heat flux from oven space reaching the substrate. Also, heat on the capacitor was high only in moving direction. In the soaking zone, the maximum temperature difference across the assembly board was within 3 C. As predicted, the uniformity or thermal equilibrium was observed at the wetting time, 230 s. For this period, the linear temperature relationship between the assembly surface and molten solder could benefited into a good interconnection. While in 350 s, the gradient of 8 °C can be identified while FPCB was entering the cooling zone process. The implication of proper cooling slope might inhibit the thermal shock towards component and molten solder.

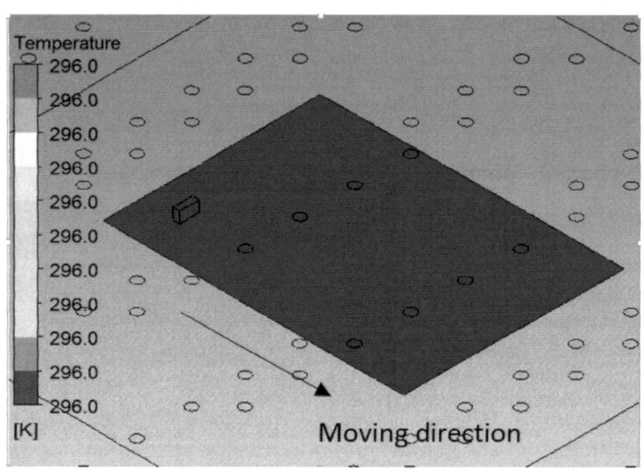

a) Initial

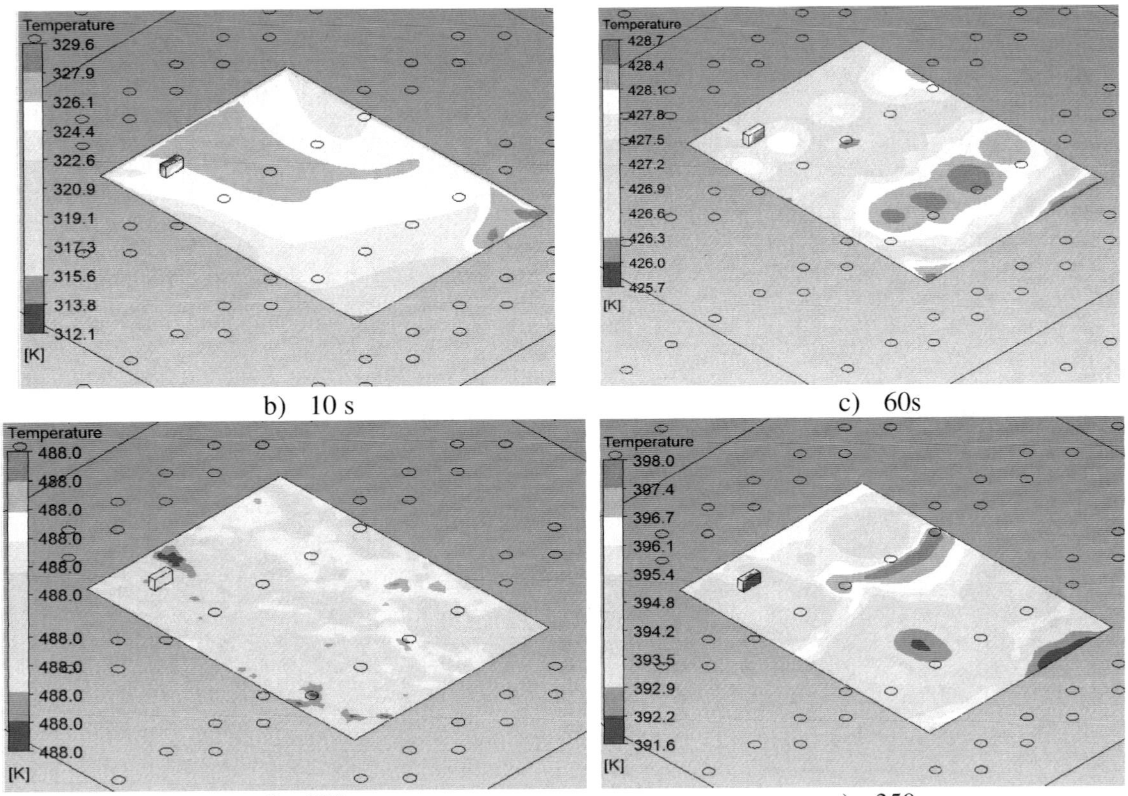

b) 10 s c) 60s

d) 230 s e) 350 s

Figure 4. Zone temperature a) Initial entering (0s) b) pre-heating zone c) soaking zone d) wetting and e) cooling zones in a reflow oven

4.2 Flow characteristics

The complex flow patterns within the array of nozzle jet of the reflow oven in x and z-axis direction are shown in Figure 5. The temperature distribution as mentioned above is governed by the flow field and heat transfer of the oven space. From the flow patterns generated, it has the similarity on those represented by Weigand and Spring [12], which highlights the three types of interactions attribute to nozzle inlet jet stream. To that, the first type of interaction (denoted as 1) is jet-to–jet interactions after approaching the board level surface. A homogenous and gapless heat transfer to the FPCB is observed. While in second type interaction lied (denoted as 2) is the cross-flow interaction when the jet stream is not asymmetrical. The third type of interaction could explain the interaction of neighbouring jet stream in mixing domain upon approach to the board level, which depends on the spacing between each inlet nozzle.

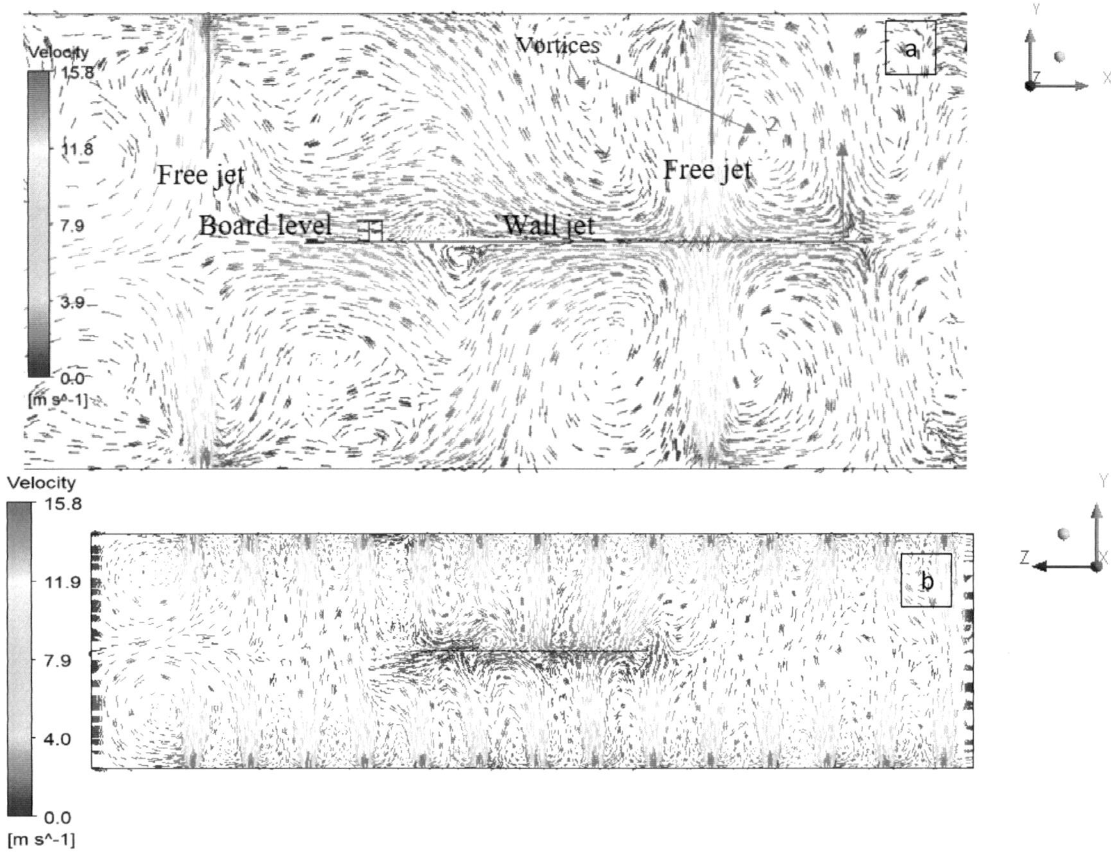

Figure 5. Flow pattern in a) z-axis direction and b) x-axis direction

4.3 Effect of Re

The effect of Re is investigated in the preheating zone. The findings of temperature contours for Re of 33, 260, 40,221 and 44,088 are shown in Figure 6. The comparison was made by the similar speed of FPCB on conveyor speed (1.2 cm/s). Based on the results, the all assembly board surface temperature show the gradient temperature less than 2 °C, with low inconsistency for ensuring uniform heat for a capacitor that attached to it. At this point, successiveness of reflow soldering was incorporated with surface gradient temperature even for RPCB or dealing with FPCB. With higher Re, flow the air from nozzle becomes more vigorous, subsequently produced a higher heat transfer effect on the FPCB surface.

In contrast, the thin thickness of FPCB was not able to capture higher heat flux at the oven space when higher Re was achieved. The discrepancy of heat transfer coefficient, h more or less 5% across the assembly surface where the oven space is dominant with the turbulent flow with wide spreading in all axis. These findings suggest that for thin motherboard as FPCB, Reynold number must be considered in order to predict the surface local temperature. Moreover, the 1.0 IWC or Re at 40,200 was found to suit for low-density components placements.

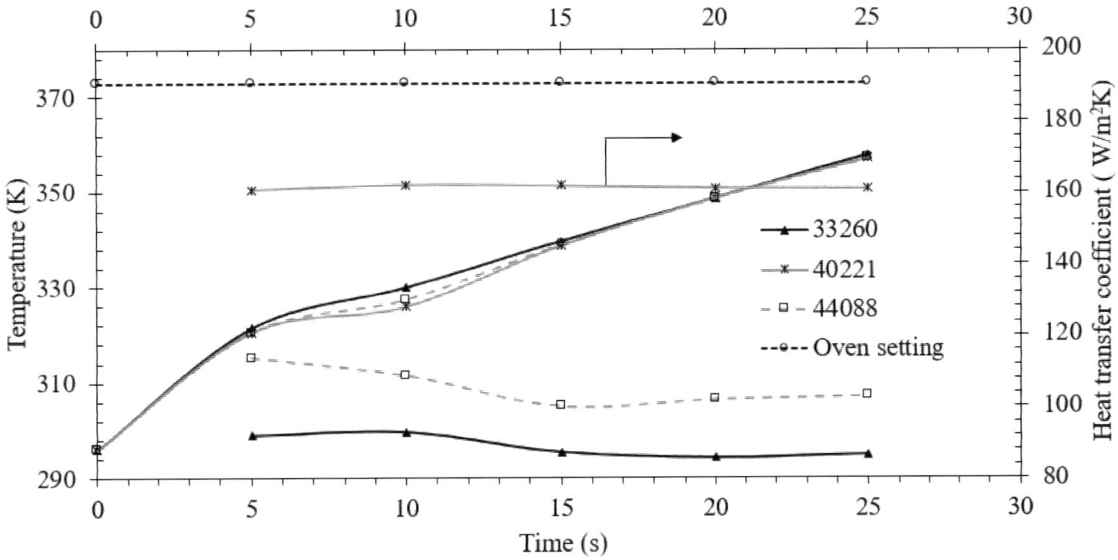

Figure 6. Change of maximum temperature and heat transfer coefficient with Re

5. Conclusion

The current paper has outlined the study on the temperature prediction of FPCB as a substrate for computer motherboard application and performed using FLUENT codes. The predictions model shows the good agreement with oven configuration with minimum deviation. The comparisons were done on different PCB and noticed that FPCB is feasible to have a suitable surface temperature during the reflow soldering process. Even in higher Re, FPCB surface temperature remains almost similar. Although the current findings only focused on board level, the results can be contributed to the advancement of FPCB assemblies.

Acknowledgement

The authors would like to thank Universiti Sains Malaysia, Penang, Malaysia, for the financial support through Research University Grant (RUI) 1001/ PMEKANIK/8014072 and short-term grant 304/PMEKANIK/6315031 and the technical collaboration with Celestica Malaysia Sdn Bhd.

References

[1] A A Wereszczak, B R Chen and B A Oistad 2018 Reflow-oven-processing of pressureless sintered-silver interconnects *J. Mater. Process. Technol.* Vol 255 pp 500–506

[2] A M Najib, M Z Abdullah, C Y Khor and A A Saad 2015 Experimental and numerical investigation of 3D gas flow temperature field in infrared heating reflow oven with circulating fan *Int. J. Heat Mass Transf.* Vol 87 pp 49–58

[3] J Pan, T C Chou, J Bath, D Willie and B J Toleno 2009 Effects of reflow profile and thermal conditioning on intermetallic compound thickness for SnAgCu soldered joints *Solder. Surf. Mt. Technol.*

[4] B Illés and G Harsányi 2009 Investigating direction characteristics of the heat transfer coefficient in forced convection reflow oven *Exp. Therm. Fluid Sci.*

[5] B Illés and G Harsányi 2009 Heating characteristics of convection reflow ovens *Appl. Therm. Eng.*

[6] B Illés and G Kristóf 2011 Gas flow simulations of a fluxless Si solder bonding oven *Microelectronic Engineering*

[7] C S Lau, M Z Abdullah and C Y Khor 2013 Optimization of the reflow soldering process with multiple quality characteristics in ball grid array packaging by using the grey-based Taguchi method *Microelectron. Int.*
[8] L C Hooi, M Z Abdullah and I A Azid 2017 Numerical simulation of fluid-structure interaction on flexible PCB with multiple ball grid array components *AIP Conference Proceedings*
[9] IPC / JEDEC J-STD-020C January 2004 Moisture / Reflow Sensitivity Classification for Non-hermetic Solid State Surface Mount Devices Proposed Standard for Ballot
[10] Inc. ANSYS 2017 ANSYS FLUENT Theory Guide *Release 18.2*
[11] D C Wilcox 1998 *Turbulence Modeling for*
[12] B Weigand and S Spring 2011 Multiple Jet Impingement - A Review *Heat Transf. Res.*

IOP Conf. Series: Materials Science and Engineering **530** (2019) 012020 doi:10.1088/1757-899X/530/1/012020

Optimization of Semiconductor Device Packaging Singulation Process

T K Wei* and M N Mahmud

School of Electrical & Electronic Engineering, Engineering Campus, Universiti Sains Malaysia, 14300 Nibong Tebal, Pulau Pinang, Malaysia.

Correspesping author *: karwei_91@hotmail.com

Abstract. With the advancement of semiconductor technology, the assembly process of semiconductor packaging is driving the devices to become smaller and smaller in size. One of the critical processes in backend assembly is the product's singulation to separate bulk processed products into individual unit. Before the sawing process kick-off, saw blade preparation (blade dressing) has to be performed. Blade dressing is the process of sharpening the saw blade surface to expose the diamond grits to the blade surface that are covered by the blade bonding material (Nickel-based). Inappropriate blade preparation technique induces burn mark defect to semiconductor devices. This research presented the dressing process optimization on critical factors using Silicon carbide (SiC) as part of the effort to enhance the product quality. The 2^k Full Factorial Design of Experiment (DOE) methodology, with consideration of three factors namely Blade Revolution Per Minute (RPM), Feedspeed and Dress Pass. The overall singulation process yields performance improvement by reducing the burn mark to $< 0.1\%$ with optimum saw blade preparation technique. Besides, the optimized process complementarily contributes to less kerf deformation and dressing wear.

1. Introduction

Electronic semiconductor industry has enjoyed consistent growth over the decades. In year 2017, the global electronic semiconductor market grew 21.6% according to the World Semiconductor Trade Statistic (WSTS). This industry growth was driven by two dominant factors which are the electronic products market pull and semiconductor electronic packaging technologies advancement and innovation [1]. This technology evolution has changed the products size to become smaller as Moore's Law explained this scenario well in claiming that, the transistor density doubles every three years.

Numerous packaging technologies have been developed over the years, from individual die packaging, dual in-line package, surface mount technologies, chip-scale packaging, to wafer level packaging and more complicated, 3D package integration, the system-in-package and package-on-package that established in 2000s.[2]. All these advanced technologies process semiconductors products in bulk; hence, singulation process must take place to separate the products into individual part. Singulation precision and quality are the essential criteria. One of the critical processes is the products singulation during the typical opto-semiconductor (LEDs) packaging process as shown in Figure 1, which is the singulation process. It plays an important role in producing a defect-free and highly precise-dimension product.

IOP Conf. Series: Materials Science and Engineering **530** (2019) 012020 doi:10.1088/1757-899X/530/1/012020

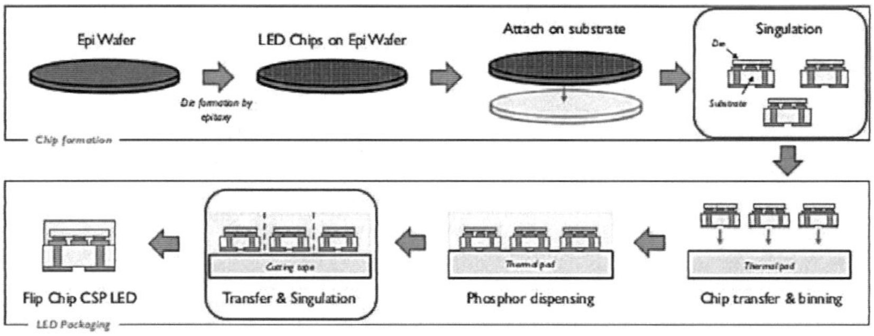

Figure 1. Typical opto-semiconductor (LEDs) process flow that involved singulation

Before a sawing process kick-off to singulate semiconductors assembly parts into individual units, saw blade preparation (blade dressing) have to be performed. Blade dressing is the process of polishing the saw blade surface to expose the diamond grits to the outer layer that is covered by the blade bonding material [3]. Optimized blade dressing recipe is critical to ensure the saw blades give perfect cutting performance and outcome on the semiconductors products.

Inappropriateness in saw blade preparation technique induce burn mark at the contact area between the devices and the saw blades. Burn mark in semiconductor device as shown in Figure 2 considers as rejected unit as it is not only a cosmetic defect (as per most customers requirement) but also affects the functionality of the product, for example the emitter luminance in opto-electronic devices. Approximately 3.88% of burn mark defect rate observed in singulation process involved Film layer, Phosphor-based material [4].

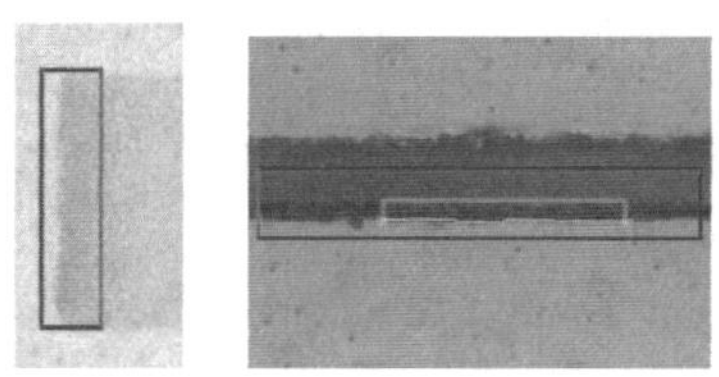

Figure 2. Burn mark defect at the edge of products

2. Advanced Packaging and Dicing Technology

The growth of the semiconductor electronic industries remains to be forcefully impacted by the worldwide major trends. The most significant impact on the semiconductor electronic industry is the introduction and communalized of smartphone globally. Due to huge global interest and volume increased, major innovation of this industry directed by the evolution of smartphone. Besides, environmental awareness, health-care, automotive and connectivity industry are also the megatrend that are impacting the evolution of semiconductor electronic industry.

The package dicing or sawing process is a cutting process that separates assembly parts into individual packaged semiconductor device. The dicing process can be performed by mechanical saw or laser ablation. The area that has been cut away during sawing process is called sawing street [5]. Industry and market requirement of advanced dicing technology have driven the innovation of dicing technology to laser ablation [6], stealth dicing[7], multi-wire sawing [8] and plasma dicing [9] introduced by Disco (Japan) Corporation.

3. Methodology

The method of Design of Experiment (DOE) has been deployed in the optimization of the singulation process in this study. It is a resourceful methodology to review the overall process optimization. DOE is a designing task that aims to describe the variation of information under conditions that are hypothesized to reflect the variation.

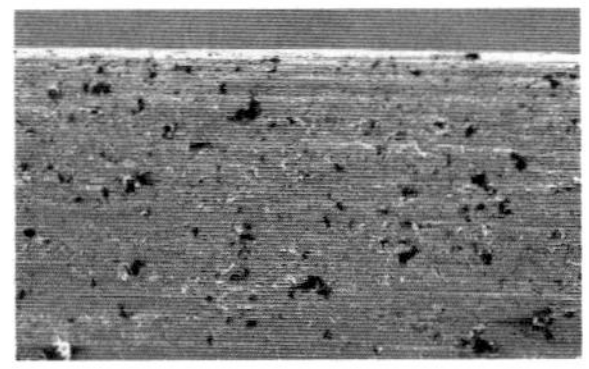

Figure 3. Blade surface topology

A preliminary Scanning Electron Microscope (SEM) imaging has performed on the dressed blade that yet to saw the devices as shown in Figure 3. Pocket or holes noticed from that blade's surface topology. The formation of the holes inferred as the diamond falls out from the blade surface. In short, it is firm that the root cause of burn mark due to the improper blade preparation dressing process.

3.1. Design of Experiment: Factorial Design

Design of Experiment (DOE) [10] is a systematic procedure that is carried out under controlled condition in order to explore the uncertain response to establish a hypothesis or illustration of the effect. When analysing and optimizing a process, experiments are often used to determine which process factors have an important impact on the outcome. Factorial design is a DOE whose design comprises of at least two factors and each carry distinct values or translated into "levels". Those experimentations consider every single feasible combination of the levels sidelong all factors. DOE factorial design permits the investigation into the interaction of each factor (cause) towards the response (effect), also the interception and relationship between factors (multiple causes) on the response (effect) studied.

In this study k=3, 2^3 factorial design is used. Feedspeed, blade RPM, Dress Pass are three factors that have been taken into consideration. The levels of the factors summarized into Table 1 based on the baseline of existing dressing recipe and preliminary technical justification. The Dress Pass study range below 70 lines because the existing recipe with the SEM imaging provides the fundamental that number of Dress Pass is too high. The baseline for Feedspeed and Blade RPM are set at 20mm/sec and 20k/min respectively, hence lower range and upper range are included in this study to understand the factors' behaviour.

Table 1. Factorial level setting

Factors	Factors details	Low (-ve)	High (+ve)
A	Feed-speed (mm/sec)	10	50
B	Blade RPM (k/min)	15	50
C	Dress Pass (lines)	20	70

In 2^3 design, combinations of 2x2x2=8 treatment consist of eight degree of freedom. Three degrees of freedom are correlative with the primary factors of A, B and C. Four degree of freedom are associated with interactions; one each with AB, AC and BC and one with ABC. The main effect of A can be equated by as the difference between opposite faces of the Cube Geometric view as shown in Figure 4, which is the high level of A ($\bar{y}_{A^+}$) deduct by the low level of A effect ($\bar{y}_{A^-}$). Similarly, main effect of B and C apply the same concept to generate the equation of main effect [10].

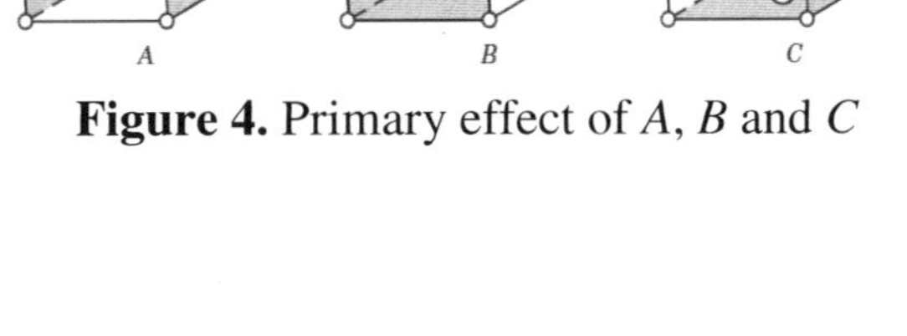

Figure 4. Primary effect of A, B and C

Effect of *A* equation:

$$A = \bar{y}_{A^+} - \bar{y}_{A^-}$$
$$= \frac{a+ab+ac+abc}{4n} - \frac{(1)+b+c+bc}{4n}$$
$$A = \frac{1}{4n}\,[a + ab + ac + abc - (1) - b - c - bc] \quad (1)$$

Effect of *B* equation:

$$B = \bar{y}_{B^+} - \bar{y}_{B^-}$$
$$B = \frac{1}{4n}[b + ab + bc + abc - (1) - a - c - ac] \quad (2)$$

Effect of *C* equation

$$C = \bar{y}_{C^+} - \bar{y}_{C^-}$$
$$= \frac{1}{4n}[c + ac + bc + abc - (1) - a - b - ab] \quad (3)$$

The two-factor interaction effects *AB* defined as half the difference of the average *A* effect at the two level of *B* as shown in Figure 5.

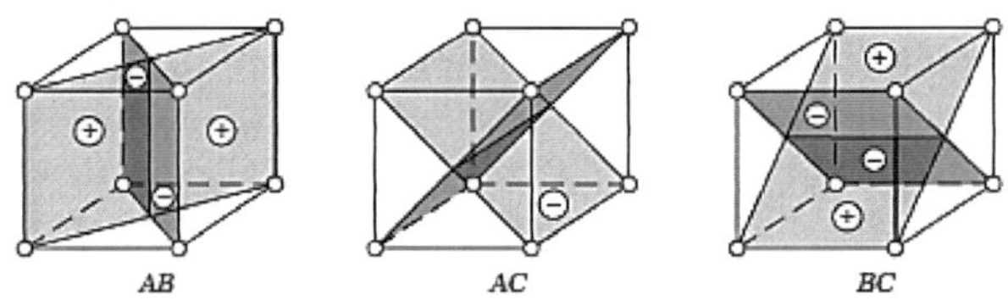

Figure 5. Two-factor interaction

Effect of *AB* interaction is half of this difference, effect of AB equate as:

$$AB = \frac{[abc-bc+ab-b-ac+c-a+(1)]}{4n}$$
$$= \frac{abc+ab+c+(1)}{4n} - \frac{bc+b+ac+a}{4n} \quad (4)$$

Effect of *AC* equation:

$$AC = \frac{1}{4n}[(1) - a + b - ab - c + ac - bc + abc] \quad (5)$$

Effect of *BC* equation:

$$BC = \frac{1}{4n}[(1) + a - b - ab - c - ac + bc + abc] \quad (6)$$

The interaction of *ABC* interaction determined by averaging the contrast between the *AB* interactions at the two different levels of *C*. Effect of *ABC* equation:

$$ABC = \frac{1}{4n}\{[abc - bc] - [ac - c] - [ab - b] + [a - (1)]\}$$
$$= \frac{1}{4n}[abc - bc - ac + c - ab + b + a - (1)] \quad (7)$$

3.2. Experimentation Setup

JMP statistical software is used in this study to generate the pattern of experiment. Randomized pattern of the runs summarized as Table 2. Factors Feedspeed, Blade PRM and Dress Pass are represented as variable *A*, *B* and *C* respectively. All eight combination runs matched with each other to understand the interaction and behaviour of the factors towards output response, burn mark.

Table 2. DOE runs for burn mark response

Run	***A*, Feedspeed (mm/sec)**	***B*, Blade RPM (k/min)**	***C*, Dress Pass (lines)**
1	-	-	-
2	-	-	+
3	-	+	-
4	-	+	+
5	+	-	-
6	+	-	+
7	+	+	-
8	+	+	+

The overview of the experiment flow summarized in Figure 6. The experiments started with mounting a dresser board onto 8-inch wafer ring in order to let the dresser board can be hold by the Disco dicer and process it. Then, a dressing recipe is setup according to the DOE defined parameters before proceed with dressing process using Silicon carbide (SiC) board as shown in Figure 7. After completed dressing process, the same blade perform cutting on substrate, Film layer, example of Phosphor-based material in opto-semiconductor application. Then, followed by performing tape transfer on the substrate for measuring the average width of burn mark (response) by using Nikon high-power smart scope. For confirmation run, the full substrate gone through full inspection via Automatic Optic Inspection (AOI) for determining the overall yield performance.

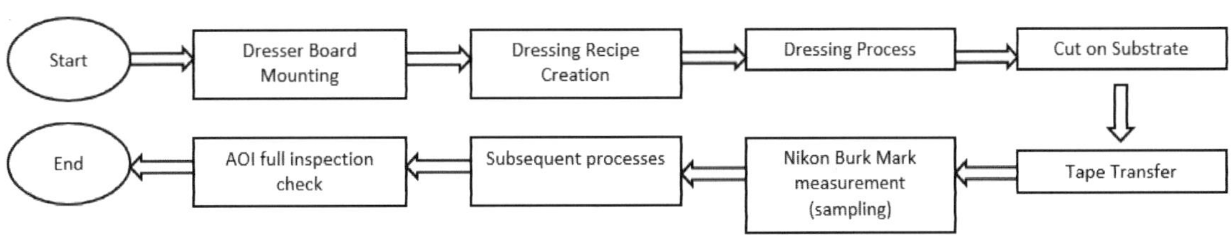

Figure 6. Experimentation flow chart

Dressing process besides exposing the embedded diamond, it also to improve the concentricity of blade edge to spindle rotation and familiarize the blade (metallurgiallyl) before penetrating into the semiconductor devices. The burn mark response induced at the contact area between blade and Film layer. This sawing process is cut-into tape dicing type. The sawing process auto-aligned using pattern recognition based on the die pattern and cut into tape on the centre between die-to-die. The Film layer directly contact with the rotating blade. Improper blade diamond exposure caused low cutting power on the substrate. Hence, higher temperature and blunt blade during dicing induced burn mark to the Film layer. The average width of the burn mark on the Film layer is determined by using the Nikon high-power smart scope.

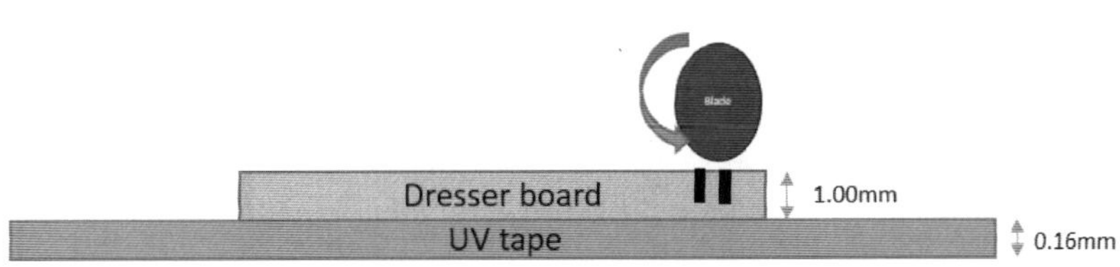

Figure 7. Cutting process for semiconductor device

4. Result and Discussion

The result of burn mark response is analysed in details by using the JMP statistical software. JMP provides in-depth analysis and explanation of the burn mark response with respect to the factors studied. In addition, an optimized set of dressing recipe is proposed by JMP. Confirmation run is conducted and comparison of existing and optimized dressing recipe is made thoroughly.

4.1. Optimization Results and Analysis

Half Normal Plot shown the absolute value of the contrast plotted against the absolute value of quantiles of the half-normal distribution. The blue line passes through the origin with a slope of the Lenth's estimated of sigma. Effects small in values are considered as deviated terms. They are assumed to have a normal distribution with mean zero and standard deviation, hence fall on the blue lines. Significant effects are those that have nonzero means and do not fall on the blue lines. Figure 8 shown that Feedspeed, Dress Pass*Blade RPM and Feedspeed*Blade RPM does not fall on the line.

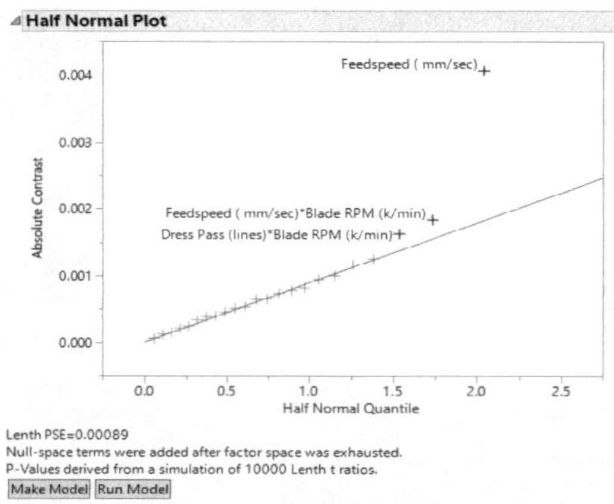

Figure 8. Half Normal Plot of burn mark response

The Fit Least Squares report of this model shown in Figure 9 with p-value<.0001 and Rsq. = 0.78 which shown no disparity in term of model fit and basic assumptions. Since there are no noticeable problem with the model fitness, interpretation of statistical tests is statistically consider valid for this study. The integrated model is significant, as shown in the Analysis of Variance (ANOVA) report in Figure 10.

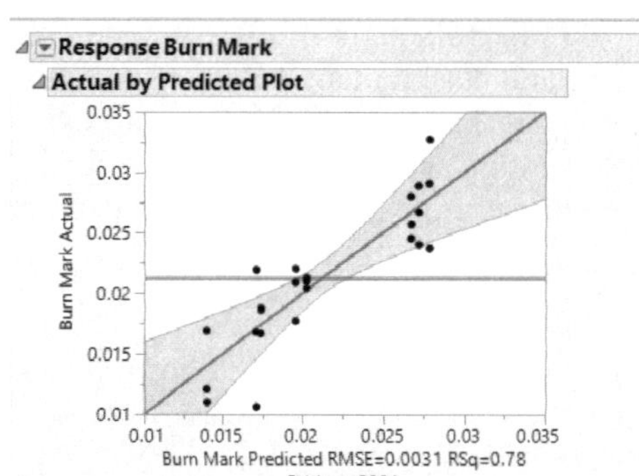

Figure 9. Fit Least Squares Model

Analysis of Variance

Source	DF	Sum of Squares	Mean Square	F Ratio
Model	6	0.00059297	0.000099	10.0823
Error	17	0.00016664	9.802e-6	Prob > F
C. Total	23	0.00075960		<.0001*

Figure 10. Analysis of Variance of burn mark response

Likewise, the Effect Tests in Figure 11 shown that Feedspeed, Feedspeed*Blade RPM and Blade RPM*Dress Pass are significant in explaining burn mark response, but the rest are less significant.

Effect Tests

Source	Nparm	DF	Sum of Squares	F Ratio	Prob > F
Feedspeed (mm/sec)(10,50)	1	1	0.00039854	40.6583	<.0001*
Blade RPM (k/min)(15,50)	1	1	0.00000294	0.2999	0.5910
Dress Pass (lines)(20,70)	1	1	0.00002166	2.2097	0.1555
Feedspeed (mm/sec)*Blade RPM (k/min)	1	1	0.00008140	8.3045	0.0104*
Feedspeed (mm/sec)*Dress Pass (lines)	1	1	0.00002440	2.4894	0.1330
Blade RPM (k/min)*Dress Pass (lines)	1	1	0.00006403	6.5320	0.0205*

Figure 11. Effect Tests of burn mark response

Interaction Profiles consist of matrix cell that shown the relationship between the X-axis and Y-axis effects. Every level of the row effect that indicated by a line plots prediction values of the model. Non-parallel line segments give visual evidence of possible interactions. Figure 12 shown the Interaction Profiles of this burn mark response with respect to the three factors studied.

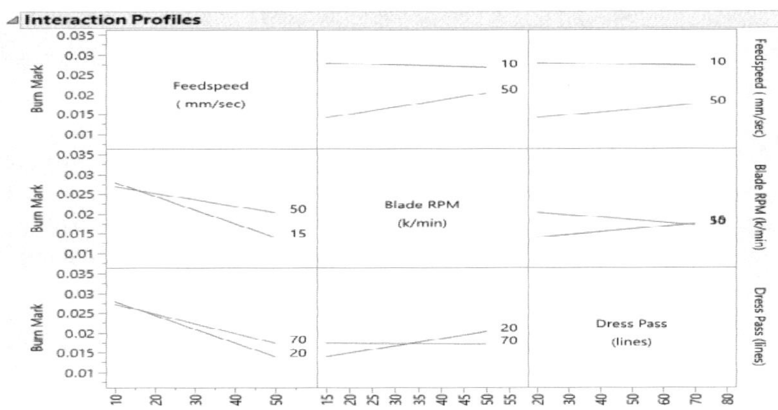

Figure 12. Interaction Profiles

As narrowed down in the initial preliminary analysis, the burn mark response closely related to the blade preparation technique and initial wear of the blade. This indirectly reflects the exposure rate of the embedded diamond within the blade surface. The Interaction Profiles explained at high level Feedspeed of 50mm/sec, the higher the Blade RPM the more obvious the burn mark response, when the blade rotation getting faster, the blade wear lesser which the diamond exposure lower. This is due to the blade does not contact to the new dresser line for dressing purpose while the Blade RPM is rotating at the same position at high speed. At Feedspeed low level of 10mm/sec, it gives relatively higher burn mark response compared to high level Feedspeed, this due to insufficient wear rate for diamond exposure.

The Feedspeed interacts with both Blade RPM and Dress Pass in either in either high or low level as shown in fist column of second and bottom row of Figure 12. This is observed when they both having crossed lines in the graph. In short, Feedspeed factor itself plays significant role to this study, which correlated the Effect Tests p-value<.0001. Both the cells shown one similarity, which is at high

IOP Conf. Series: Materials Science and Engineering **530** (2019) 012020 doi:10.1088/1757-899X/530/1/012020

Feedspeed, they give lesser burn mark as desired. This is understood that sufficient and optimal of Feedspeed give enough diamond expose rate.

Moreover, at Blade RPM high level and low level, both intercept at about Dress Pass of 70 lines, but they both give undesired high burn mark response. At high level of Blade RPM, the higher the Dress Pass, the lower the burn mark. Meanwhile, at low level of Blade RPM, the lower the Dress Pass, the lesser burn mark response. Nonetheless, the least burn mark at low Blade RPM and low Dress Pass.

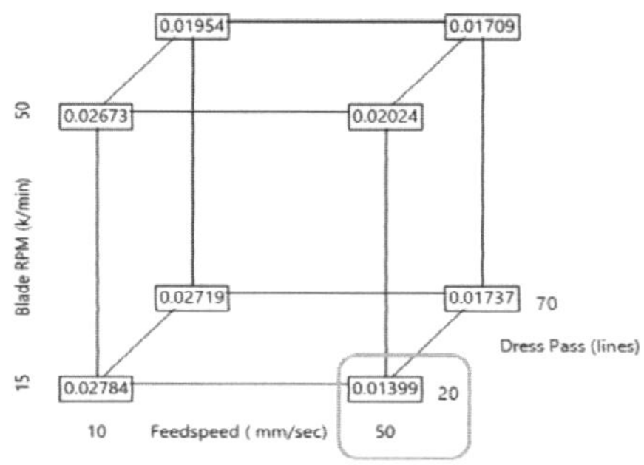

Figure 13. Cube Plot of burn mark response

In overall view, predicted value for the utmost of the factors range are displayed on a Cube Plot. The corners are defined by the smallest and biggest noticed value of factors. Figure 13 shown the study range of burn mark for Blade RPM, Feedspeed and Dress Pass. The response indicated at the eight corners of cube with values of burn mark average width. Cube Plot shown that parameters (20, 50, 15) at vertex of right bottom corner of front plane gives the smallest value of burn mark response among all.

4.2. Optimization Confirmation Run

Based on the above DOE experimentations, combination of three factors have been taken into consideration of establishing a minimum or defect-free of burn mark setting. JMP performed its Full Factorial design calculation and generated an interactive Prediction Profiler as shown in Figure 14. As the value of one factor is changed (drag the red lines), the values of the other two components move in the opposite direction. This is to ensure the ratio and the overall mixture constraint are maintained.

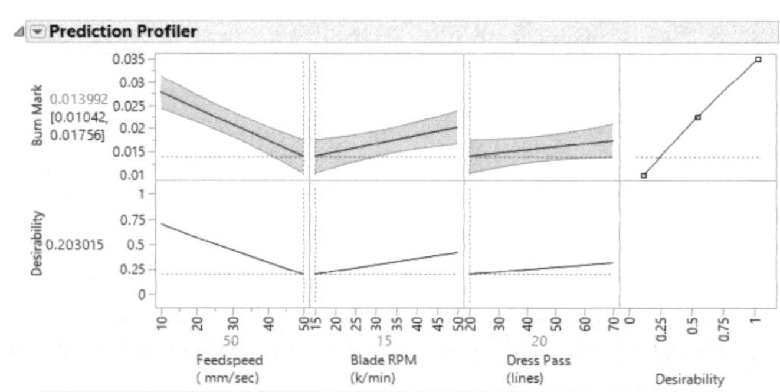

Figure 14. JMP Prediction Profiler for optimized dressing parameters

The red dotted lines in Prediction Profiler are dragged to the least burn mark effect, which is at the bottom of the respective charts. The JMP Prediction Profiler suggested the parameter of Feedspeed 50mm/sec, Blade RPM 15k/min and 20 lines of Dress Pass, which give the least burn mark response as desired. This optimized set of dressing recipe performed confirmation run on the semiconductor devices. These dressing parameters as summarized in Table 3 with the same sawing process recipe have reduced burn mark to 0.03%, almost burn mark defect free.

Table 3. Prediction Profiler Suggested Parameters

Parameters	**Settings**
Feed speed (mm/sec)	50
Blade RPM (k/min)	15
Dress Pass (lines)	20

This is a significant yield improvement to the singulation packaging assembly process as shown in Table 4 for the comparison. Three wafers have been process through with proposed optimized recipe as confirmation run. The burn mark defect rate yield improved from original setting of 3.88% averagely to 0.03% (<0.1% target).

Table 4. Yield performance comparison

Recipe	**Wafer**	**Defect % per Wafer**	**Average %**
Existing	A	3.92%	3.88%
	B	3.83%	
Optimized	C	0.01%	0.03%
	D	0.04%	
	E	0.02%	

Based on the response outcome with respect to the input factors, JMP analyse the responses based on Full Factorial design calculation and give the Prediction Expression of burn mark behaviour. The burn mark response Prediction Expression shown in Figure 15.

Prediction Expression

$$
\begin{aligned}
&0.02125 \\
&+ -0.004075 \cdot \left(\frac{(\text{Feedspeed (mm/sec)} - 30)}{20}\right) \\
&+ -0.00035 \cdot \left(\frac{(\text{Blade RPM (k/min)} - 32.5)}{17.5}\right) \\
&+ -0.00095 \cdot \left(\frac{(\text{Dress Pass (lines)} - 45)}{25}\right) \\
&+ \left(\frac{(\text{Feedspeed (mm/sec)} - 30)}{20}\right) \cdot \left(\left(\frac{(\text{Blade RPM (k/min)} - 32.5)}{17.5}\right) \cdot 0.0018416667\right) \\
&+ \left(\frac{(\text{Feedspeed (mm/sec)} - 30)}{20}\right) \cdot \left(\left(\frac{(\text{Dress Pass (lines)} - 45)}{25}\right) \cdot 0.0010083333\right) \\
&+ \left(\frac{(\text{Blade RPM (k/min)} - 32.5)}{17.5}\right) \cdot \left(\left(\frac{(\text{Dress Pass (lines)} - 45)}{25}\right) \cdot -0.001633333\right)
\end{aligned}
$$

Figure 15. Prediction expression for the optimized dressing recipe

The blade-exposure wear rate is compared at the initial blade dressing. The optimized recipe wear lesser by 10% compared to the existing recipe, yet it successfully expose the embedded diamond out. Hence, this relatively extended the blade life. The blade life extension is one of the complementary contribution to this study as it directly correlated to cost saving, which is important to any industry.

The dresser board first cut and last cut side view profile have been observed. If the measurement on the dresser board for the first dress line and last dress line shown small measurement different, which indicated small blade deformation during the dressing process. This proved that the dressing recipe is good to perform product dicing.

Table 5. Comparison of blade kerf and curvature during first and last cut of dressing

Recipe	Dressing	Kerf (um)	Curvature (um)
Existing	First cut line	59.8	24.9
	Last cut line	64.6	39.4
Optimized	First cut line	64.2	23.1
	Last cut line	65.5	31.4

The kerf measurement referring to the saw street width that made by sawing process and kerf deformation defined as the kerf width difference between the first cut line and last cut line during dressing process. The kerf performance on dresser board correlate to the product cutting saw street performance. From Table 5 noticed that existing dressing recipe having 8.03% of kerf deformed, while the optimized recipe reduced to 2.02% equivalent to 1.3um only. The more consistent the dresser board kerf, the more stable the cutting performance. Hence, with the optimized recipe not only reduced burn mark response but also improved kerf deformation about 6.01%.

Besides, the curvature of the dressed blade are monitored, which are the shape of blades' tip. The length of exposure that having curvature are measured as shown in Table 5, where existing recipe gives 14.5um additional curvature length while the optimized gives 8.3um of induced curvature only. The tip of blades required to be maintained in U-shaped hence they do not produce slant cut during the cutting process.

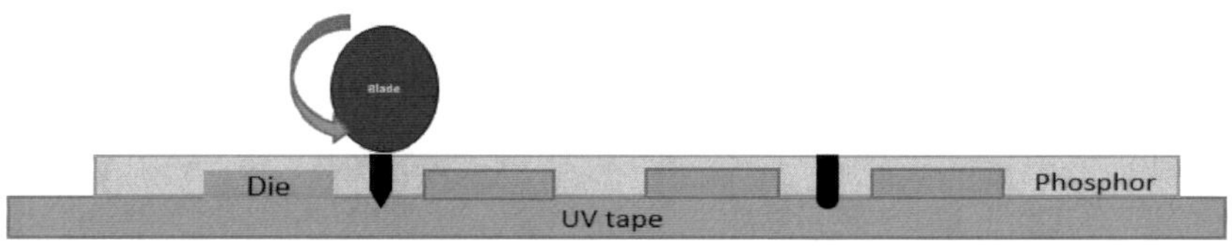

Figure 16. Slant cut (V-shaped) vs. good cut (U-shaped)

The tip of the blade will cause slanted cut (V-shaped) if the blade deformed over a certain angle. This causes the package size to be non-perpendicular at edges as shown in Figure 16 (left side) compared to the right side of Figure 16 illustrated good cutting (U-shaped) without slant cut. Hence, it is important to have a final confirmation of the blade tip profile when qualifying a new set of dressing parameter recipe.

5. Conclusion

In summary, an optimized saw blade dressing have been introduced into the singulation process in High Volume Manufacturing (HVM) opto-semiconductor assembly. In addition, the behaviour of dressing critical parameters have been characterized for the interaction between Feedspeed, Blade RPM and Dress Pass. In short, the overall burn mark defect rate have reduced to <0.1% (~0.03%) where meeting the objective of this study. Besides meeting the main objectives, this study gained several complementary benefits from the optimized recipe. The benefits included smaller dress kerf lost, lesser kerf deformation and reduce in dressing wear compared to the previous recipe. This created a more robust singulation process in semiconductor.

Although several related works have been conducted on blade preparation technique, but that is no interaction factors study been made based on Blade RPM, Feedspeed and Dress Pass. This is the key novelty of this research and contributed significant improvement to the semiconductor singulation field.

References

[1] W T Chen and A Tseng 2001 Overview of recent developments in microelectronic packaging *Advances in Electronic Materials and Packaging 2001 (Cat. No.01EX506)* pp 15–16

IOP Conf. Series: Materials Science and Engineering **530** (2019) 012020 doi:10.1088/1757-899X/530/1/012020

[2] R Beica 2018 Enabling information age through advanced packaging technologies and electronic materials *Pan Pacific Microelectronics Symposium (Pan Pacific)* pp 1–5
[3] K W Shi, K Y Yow and R Khoo 2011 Developments of blade dressing technique using SiC board for C90 low-k wafer sawing *IEEE 13th Electronics Packaging Technology Conference* pp 122–128
[4] Rao H, Ding K, Song J, Xie L, Wang W, Wan X, Zhou L and Liao J 2012 Self-adaptive phosphor coating technology for white LED packaging *Front. Optoelectron.* Vol 5 No 2 pp 147–152
[5] M S Amri, D Liew and F Harun 2010 Chipping free process for combination of narrow saw street (60um) and thick wafer (600um) sawing process *34th IEEE/CPMT International Electronic Manufacturing Technology Symposium (IEMT)* pp 1–5
[6] Kim D, Kim Y, Seong K, Song J, Kim B, Hwang C and Lee C May 2009 Evaluation for UV laser dicing process and its reliability for various designs of stack chip scale package *Electronic Components and Technology Conference 2009* ECTC 2009 59th (pp 1531-1536) IEEE
[7] M Kubota, K Hosaka, M Sugiyama and Y Mita 2013 Evaluation of silicon fracture strength dependence on stealth dicing layers for 'cleave-before-use' MEMS freestanding cantilever probes *Transducers & Eurosensors XXVII: The 17th International Conference on Solid-State Sensors, Actuators and Microsystems (TRANSDUCERS & EUROSENSORS XXVII)* pp 151–154
[8] H J Möller Aug 2004 Basic Mechanisms and Models of Multi-Wire Sawing *Adv. Eng. Mater.* Vol 6 No 7 pp 501–513
[9] F Wei, T Tabuchi, T Lazerand, C Johnston, K Mackenzie and M Notarianni 2017 Plasma Dicing Fully Integrated Process-Flows Suitable for BEOL Advanced Packaging Fabrications *IEEE 67th Electronic Components and Technology Conference (ECTC)* pp 350–357
[10] D C Montgomery 2012 *Design and Analysis of Experiments* USA: John Wiley & Sons, Inc

IOP Conf. Series: Materials Science and Engineering **530** (2019) 012021 doi:10.1088/1757-899X/530/1/012021

Optimisation of Flame Holder Design for Micro Gas Turbine

I I Enagi[1,2], K A Al-attab[2*] and Z A Zainal[2]

[1]Department of Mechanical Engineering, School of Engineering Technology, Federal Polytechnic, P.M.B 55, Bida, Niger State –Nigeria

[2]School of Mechanical Engineering, Universiti Sains Malaysia, Engineering Campus, 14300 Nibong Tebal, Penang, Malaysia

Corresponding author *: khaled@usm.my

Abstract. Micro gas turbine (MGT) is an efficient alternative to costly generation and transmission of electricity especially in combined heat and power (CHP) for remote areas applications. In this study, an optimum geometry of a flame holder of MGT combustion chamber was determined in terms of height and diameter using solid works and ANSYS fluent. To determine and optimise the flame holder geometry, species transport and non-premix with radiation models were used in ANSYS fluent. Optimisation process was performed to select the best geometry of flame holder that produces lowest CO emission. The flame holder was divided in to three zones namely; pre-mix, combustion and dilution zones respectively. The test parameters applied during simulation include: the diameter of holes, number of rows per zone, heat flux through convective heat transfer, percentage excess air and location of dead zone. Optimum diameter of holes for the zones was found out to be 6, 8 and 10 mm with each having four rows of holes. In addition, 60 cm height and 2 inches diameter of flame holder showed a well distributed flame contours. The application of heat flux under radiation model reveals an acceptable CO emission of 102 ppm and average temperature of 1218 °C for the diesel fuel used.

1. Introduction

The most complex technical objects today are modern gas turbine engines and power plants. They are characterized by variety of operating modes under different external conditions, which makes their designing process very complicated [1]. The design of combustion chamber is one of the most difficult process in creating modern gas turbine engines [1]. Combustion chamber plays a significant role in providing proper air-fuel mixing in engines due local fluctuations of the flow field. In addition, it promotes engine performance, combustion and emission formations [2]. In micro gas turbines (MGT), the combustor is a very important part where chemical energy is converted into thermal energy through a combustion process. Many studies on geometry, configuration and dimension of the combustor have been made related to wall cooling, flame stability and emission control [3-4]. Several investigations on MGT by using different design of combustion chamber with different types of fuel were conducted in previous works [5–8]. The application of simulation software has rapidly increased over last decade in improving many complex processes including combustion chambers [9-10]. To explore and design engineering hardware and for the optimisation of combustion chambers, computational fluid dynamics (CFD) [5] has been extensively used as an important design tool [11-12]. Generally, temperature profile is an important characteristics of flame [13]. In addition, the flame location and balance of transport processes affect the performance of micro combustor through its location and shape [3-14]. For the average temperature calculations, only the core air mass flow rate in slide liner at different zones were considered [15-16]. In the present work, a combustion chamber was designed and optimize using ANSYS- FLUENT simulation software. Species transport and non-premixed combustion models were used to determine the optimum flame holder chamber geometries. The designed combustion chamber was investigated through optimisation process, and the effects of different variables and geometries on the combustion performance were studied.

2. Design, development and simulation of combustion chamber

For the combustion chamber design, the initial design parameters were due to constraints concerning compressor exits and turbine inlet temperature. The designed calculations for the dimensions were used to obtain a suitable combustion geometry. The procedure taken for design, development and simulation of the combustion chamber are (1) Design and CAD drawing of the chamber using solid works and (2) CFD simulation using ANSYS Fluent.

2.1. Design and CAD drawing of the chamber using solid works

The chamber is composed of two parts: the jacket and flame holder as shown in Figure 1. The jacket acts as air pre-heater surrounding the flame holder with conical shaped nozzle at the turbine inlet. The flame holder is the inner part where the combustion takes place. Air jacket height was varied in the range of 30-90cm, while flame holder diameter was varied in the range of 2-6 inches. The flame holder is divided in to three zones namely; premix, combustion and dilution zones, each with different diameters of hole. The diameter of holes in each zone are in increasing order starting from the premix zone. For the design, all the three zones in the flame holder geometry was made of 8 number of holes, each having 4 rows.

The solid work drawing of the chamber involves several procedures beginning with 2D transforming in to 3D using boss-extrude. Reference geometry plane was then, created to enable a clear drawing of other salient features of the flame holder. To build and replicate holes in reference for the top edge of the chamber, features like cut extrude and linear pattern were used. Finally, the flame holder body is subtracted from the chamber in order to simplify meshing and simulation process by using combine and subtract feature.

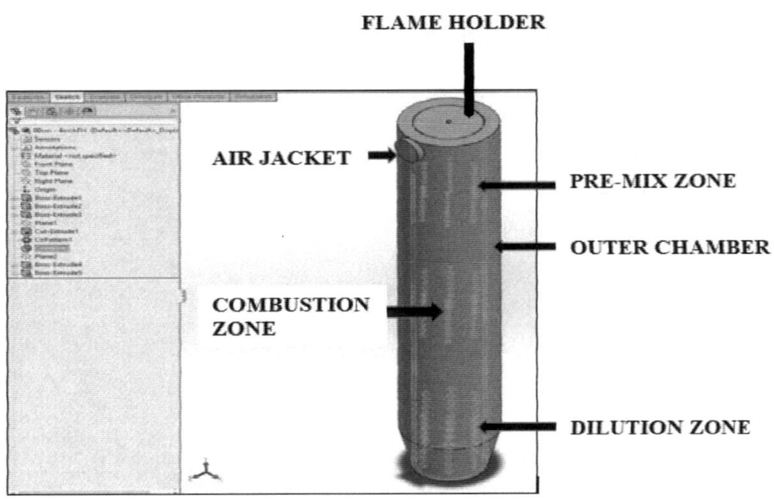

Figure 1. Solid works drawing of the combustion chamber

2.2. Simulation in ANSYS Fluent

Simulation in ANSYS fluent involves the use of different models to achieve optimisation process of selecting the best chamber geometry. Table 1 shows the different models applied and their respective functions.

Table 1. Shows different combustion models applied during simulation

S/N	Models	Functions
1.	Species transport	Used to determine the geometry that will produce combustion
2.	Non-premixed combustion	Used to determine CO released in ppm (parts per million) at the outlet.
3.	Non-premixed with radiation	Used to determine the optimisation in terms of CO emission and outlet temperature.

2.2.1. Species Transport model. Using this model for the first part, each geometry, ranging from 900mm as the maximum to 300mm minimum height of the chamber is simulated in Fluent, while testing each geometry for combustion. After 100 iterations, if any geometry attains temperature of above 2000K, then it is considered to produce combustion, otherwise, combustion is not sustainable.

The process includes simulation of the different geometries of flame holder, starting with minimum diameter of 50.8mm (2 inches) to the maximum of 152.4mm (6 inches). The manipulated parameters that make difference in this process are the height and the diameter of flame holder, however, other variables that were kept constant include number of rows for flame holder's holes, holes diameter, and conic shape of the bottom part of the chamber. The overall stepwise flowchart for simulation performed in ANSYS fluent is shown in Figure 2.

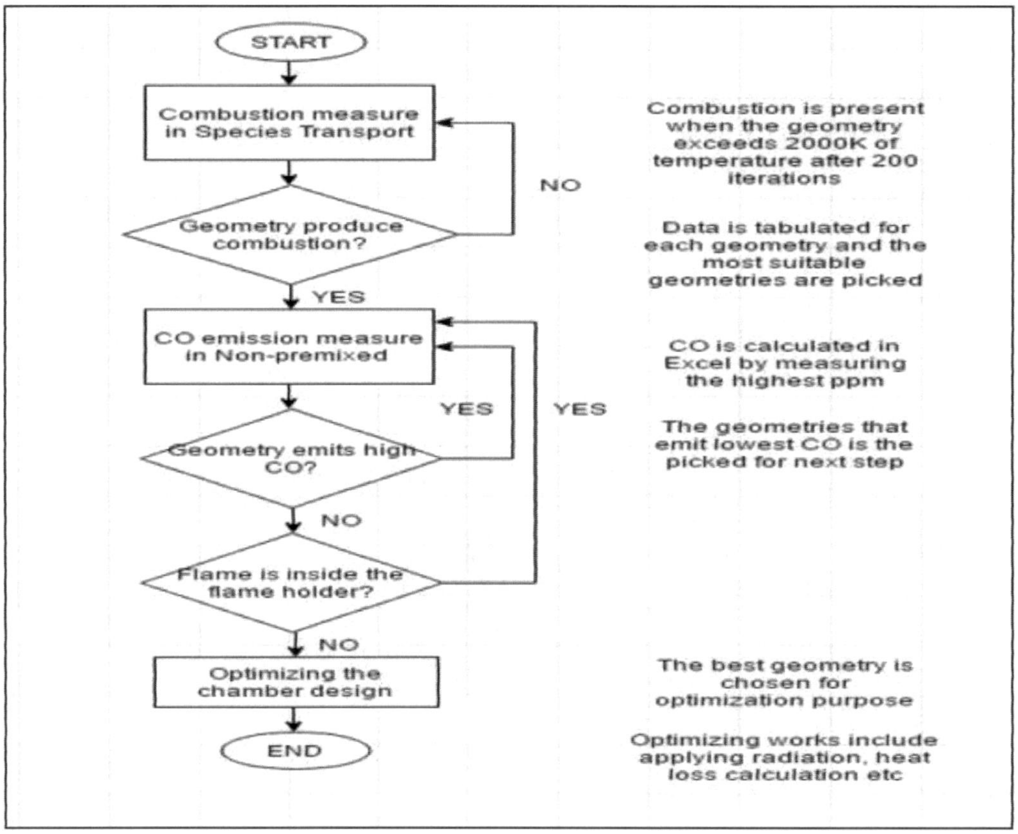

Figure 2. Flowchart of steps taken in ANSYS Fluent

IOP Conf. Series: Materials Science and Engineering **530** (2019) 012021 doi:10.1088/1757-899X/530/1/012021

2.2.2. Non-premixed Combustion model. In order to optimise a geometry that will produce minimum CO emission, species transport model is therefore replaced with non-premixed combustion model in which the PDF files of propane and diesel are generated and calculated.

The same procedure in species transport is also applicable here, but the calculation is running for 300 iterations to ensure the combustion is stable for CO formation at the outlet. The average mole fraction of the CO at the outlet of each geometry is then calculated based on the XY plot and in Microsoft excel.

3. Setting up the ANSYS Fluent

In Setting up the Fluent for the simulation, the parameters are set out for each part named in Mesh feature. For non-adiabatic and flow analysis, energy equation with k-epsilon viscosity model are activated. They are both used for laminar and turbulent flow with species transport and non-premix combustion model. Finite rate and turbulent eddy dissipation equations were also used in volumetric combustion model. The fuel is then set to either propane-air or diesel-air depending on what type of fuel is required for the simulation. The boundary conditions are set out using several parameters as shown in Table 2.

Table 2. Parameters set in Boundary Conditions.

Parameters	Variables
Fuel Inlet	
Temperature	300 K
Pressure	2atm
Mass flow rate (70% excess air)	0.0058kg/s
Air Inlet	
Temperature	530 K
Pressure	1.5atm
Mass flow rate	0.15kg/s
Outlet	
Pressure	1atm
Back flow Temperature	600 K

3.1. Optimization

Optimization of the chamber involves further simulation of the selected geometries that produces best CO emissions. The variables use during optimization include, the number of rows for holes of flame holder and its diameter, percentage of excess air, length of dead zones of the chamber and also height of the chamber.

This is required to reduce the number of CO per mole released at the outlet. Thus, P1 radiation model is introduced which has taken account of heat fluxes around the wall of the chamber. The calculations here therefore, considered the effect of convection and radiation of heat transfer. The value of emissivity applied is 0.5 that is suitable for steel being the material of the existing chamber in the Engine Lab. Parameters set in radiation model are shown in Table 3.

Table 3. Radiation Parameters

Feature	Value
Type of radiation model	P1
Material of the chamber	Steel
Emissivity	0.5
Wall thickness	6mm
Heat fluxes at the outer wall	10800 W

IOP Conf. Series: Materials Science and Engineering **530** (2019) 012021 doi:10.1088/1757-899X/530/1/012021

4. Results and discussions

This section includes results and discussions for simulations in ANSYS Fluent that makes up the optimisation process. Species transport model in ANSYS Fluent shows the rough result of the combustion process that occurred inside the chamber, which is evidently an adiabatic reaction. Since species, transport model only shows the basic and simple model of the chamber, non-premixed combustion model that can provide the formation of CO emission after combustion process was introduced. For optimization process, the use heat flux through convective heat transfer calculation allows for selection of best geometry of the flame holder.

The variables used during the simulation to achieve the optimum geometry are:

- The diameter of flame holder and height of the chamber.
- The number of holes for flame holder's zone.
- The diameter of holes in each zone for flame holder.
- The convective heat flux and radiation model.
- The excess air supplied
- The dead zones of the flame holder.

4.1. Effect of flame holder diameter and chamber height

The results obtained from species transport model are tabulated in a Table 4 for further optimisation of the chamber, which will be determining which geometry to produce the least CO emission. The results shows that chamber heights above 80 cm are not sustainable for combustion of flame holder diameters below 4 inches. However, there was a significant sustainability between 30 to 70 mm chamber heights. For reasons bothering manufacturing cost, and minimum volume and area of flame holder, a preferred choice of 60 cm chamber height and 2 inches flame holder diameter is chosen for further optimisation.

Table 4. Data tabulation of Species Transport model using diesel

Chamber Height (cm)	30	40	50	60	70	80
Flame Holder Diameter (inches)						
2				✓	✓	
3		✓	✓	✓	✓	
4		✓	✓	✓	✓	✓
5	✓	✓	✓	✓	✓	✓
6	✓	✓	✓	✓	✓	✓

4.2. Effect of rows and diameter of holes

For the species transport model simulation, the flame holder diameter of holes was first fixed at 6, 8 and 10 mm and later 8, 10 and 12mm for premix, combustion and dilution zones. In addition, 15 and 4 number of rows per zone were separately varied for the first test considerations. The variables for the chamber height was from 300 to 700 mm and 50.8 to 152.4 mm for flame holder diameter. The temperature contour at 600 mm chamber height for 4 rows of holes in 6, 8, and 10mm configuration shows a well distributed undisturbed flame inside the flame holder, while for 15 rows configurations, the flame is quite disturbed as shown in Figures 3 A and B.

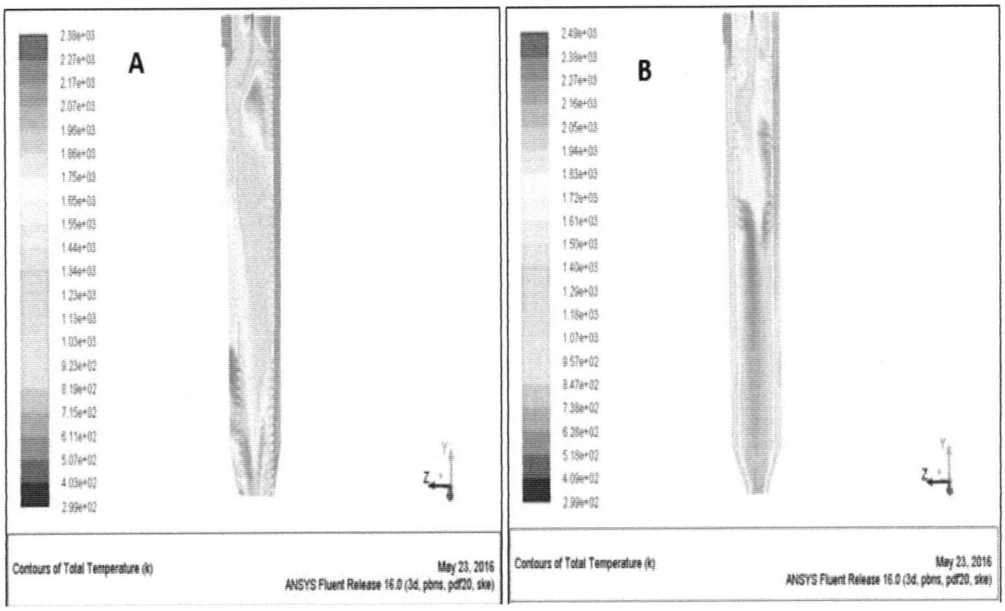

Figure 3. (A) showing 15 and (B) is 4 number of rows configurations

Figures 4 A and B shows the temperature contours for the two set of holes diameters. The configuration with 8, 10 and 12 mm in Figure 4A does not produce good combustion performance as it clearly showed early start of combustion at premix zone rather than combustion zone. On the other hand, configuration with 6, 8 and 10 mm shows a well-distributed combustion starting at the combustion zone as shown in Figure 4 B. While Figures 5A and B show an outlet temperature contours of mole fraction of CO for (6, 8, 10) mm configuration which is better than that of the (8, 10, 12) mm.

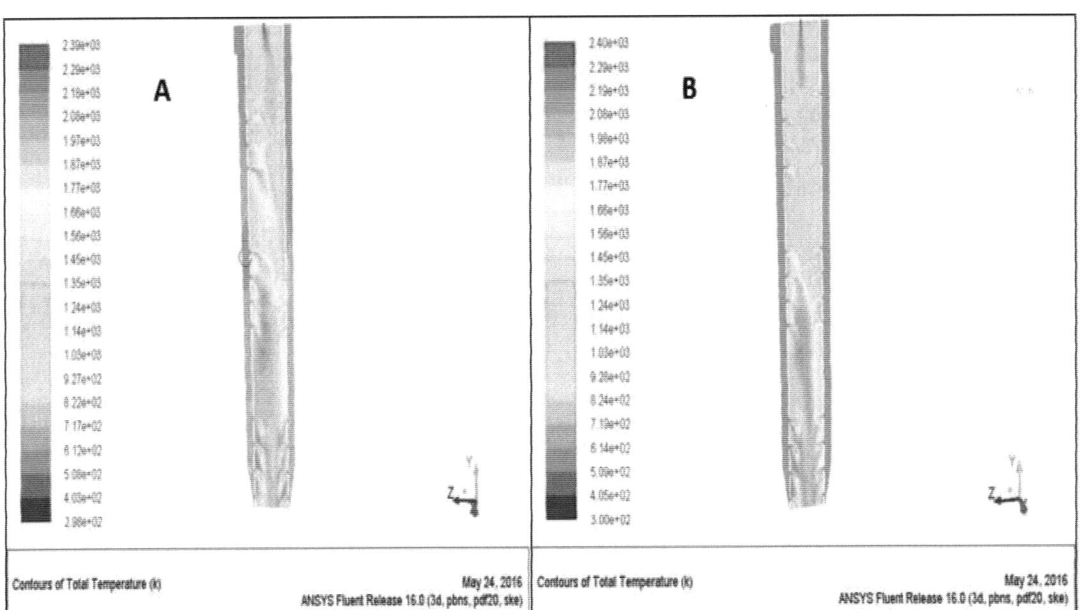

Figure 4. A & B: Temperature contours for (a) 6, 8 and 10 mm and (b) 8, 10 and 12 mm

IOP Conf. Series: Materials Science and Engineering **530** (2019) 012021 doi:10.1088/1757-899X/530/1/012021

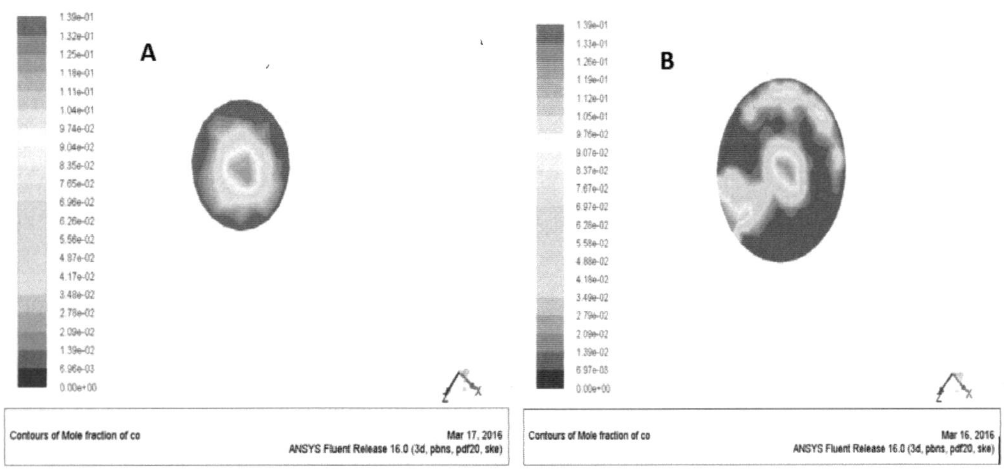

Figure 5. A & B: CO contours for (a) 6, 8 and 10 mm and (b) 8, 10 and 12 mm

4.3. Effect of excess air supplied

The excess air supply to MGT combustion chamber is needed to ensure complete combustion of the fuel used. The excess air supplied were tested from 20% to 70% excess air. The 70% excess air produces the lowest number of CO emissions as compared to others put under investigation.

Figure 6A shows the average mole fraction using diesel fuel with 20% and 70% excess air supplied. The effect of having excess air for combustion is to improve the combustion efficiency. The results reveal that combustion efficiency increases with increased in excess air until the heat loss during the supply of excess air is greater than the heat provided.

4.4. Effect of convective heat flux

Heat transfer surfaces are commonly present in combustion chambers of many types of processes and power equipment. Figure 6B shows the effect of convective heat flux that was applied to further reduce the outlet CO emission. The convective heat flux (heat transfer) of 10800W was applied to the outer wall with the emissivity of 0.5 and 6mm wall thickness of the stainless steel material used.

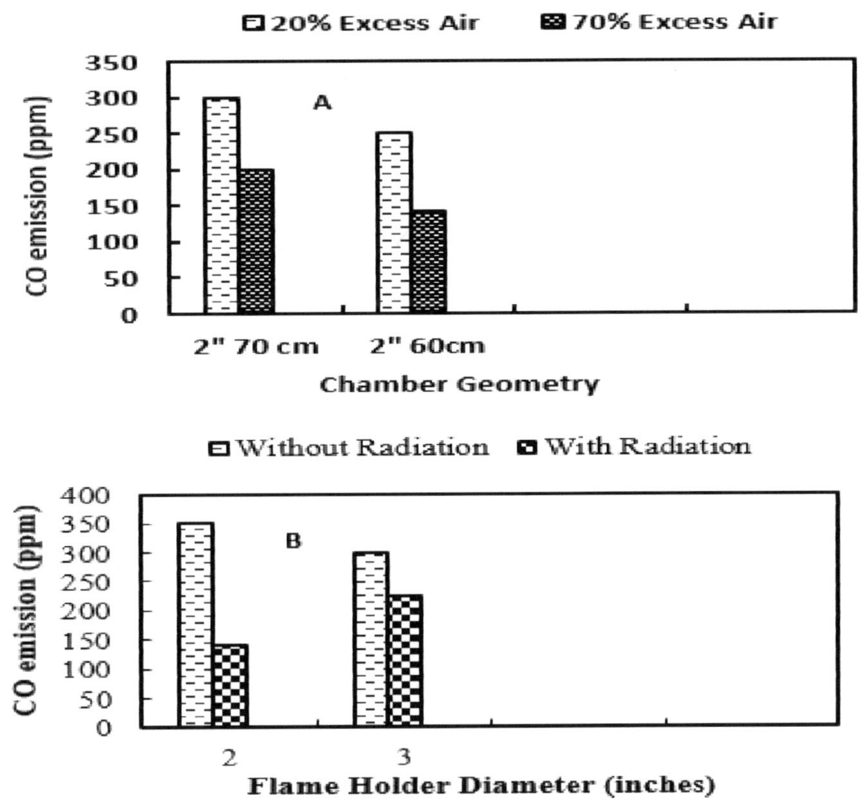

Figure 6. A & B: Average CO emissions for effects of excess air and radiation

4.5. Effect of dead zones

Creation of dead zone between the premix, combustion and dilution zone was also used to improve the combustion. The dead zone between combustion and dilution zones (dead zone B) produced a better combustion in terms of CO emission, than the zone between premix and combustion zones (dead zone A) as shown in Figure 7.

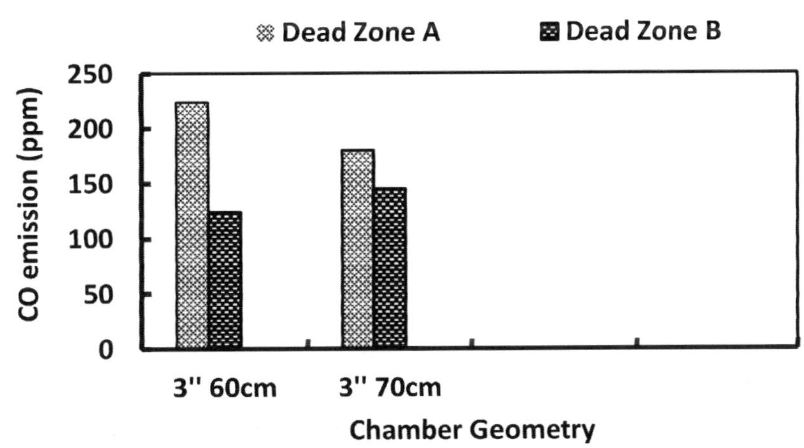

Figure 7. Average CO mole fraction for dead zones A and B

4.6. Optimization

Optimising the combustion chamber geometry promotes the reduction of CO emission and further enhance efficiency of MGT. For the optimisation process, 60, 70 and 80 cm chamber heights and 2, 3 and 4 inches' flame holder diameters were separately simulated. Figures 8A & B show the outlet average CO mole fractions and temperatures.

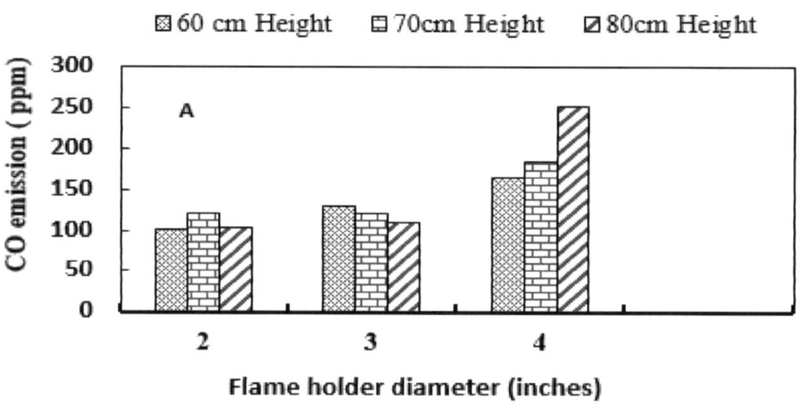

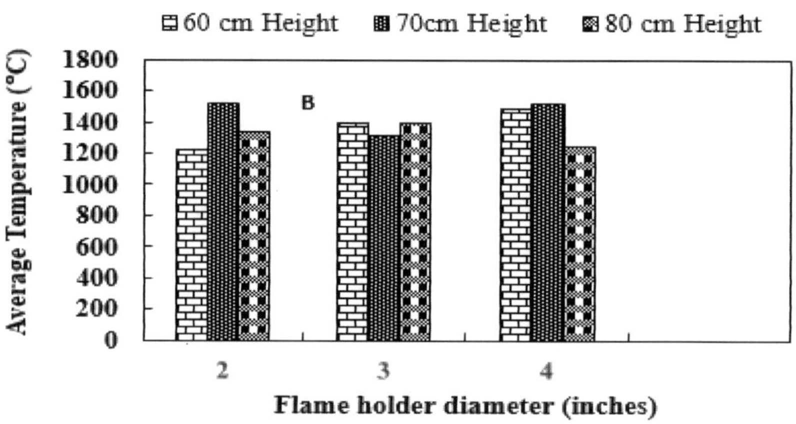

Figure 8. A & B: CO and temperature plots for optimum geometries

4.7. Optimum chamber design

The optimum chamber geometry with dimensions as shown in table 5 and design parameters in Table 6 was therefore adopted after optimisation, in addition, are 4 holes of (6,8,10) mm, and dead zone between the combustion zone and dilution zone. It produces an outlet average CO mole fraction of 102 ppm and temperature of 1218 °C. It also confirms as having the best combustion stability, with the flame well distributed within the flame holder. This shows that the flame does not rush outside the flame holder, confirming that the outer surface temperature of the chamber is cooler than the other parts. Figures 9 (A-D) show the outlet average CO emissions and temperatures of the optimum combustion chamber.

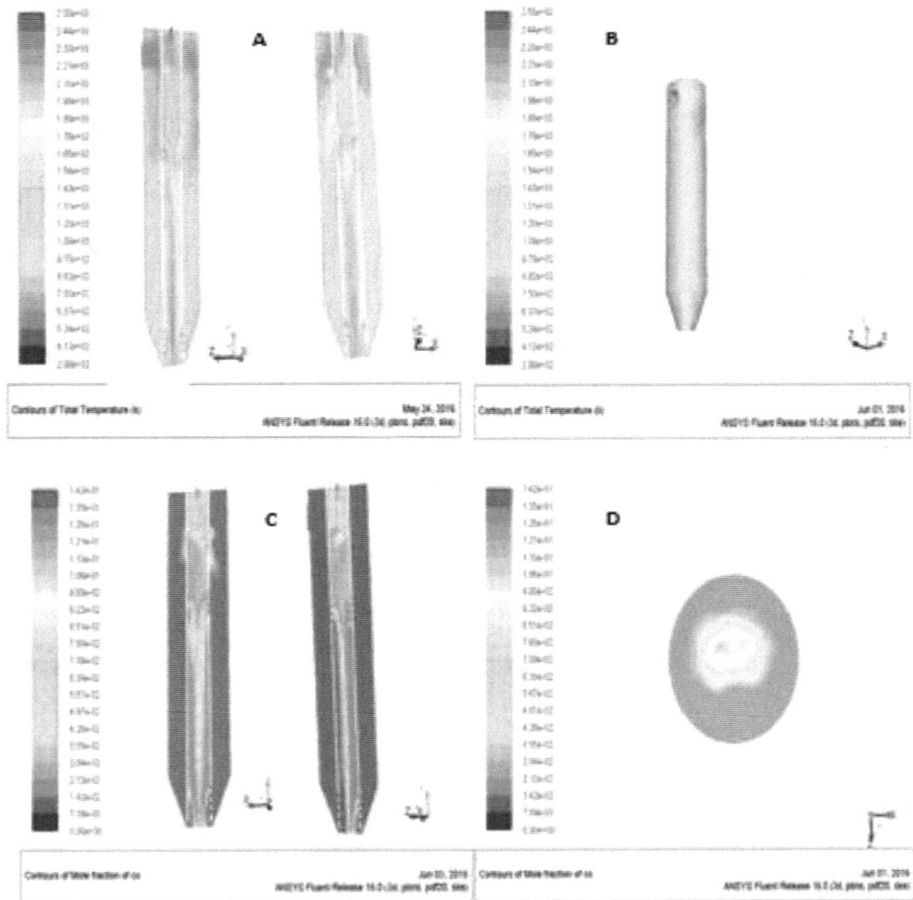

Figure 9. (A-D): Average CO emissions and temperatures of the optimum Combustion chamber

Table 5. Main dimensions of the combustion chamber

Dimensions	Value
Height of the Jacket (mm)	700
Diameter of the Jacket (mm)	152
Height of the flame holder (mm)	690
Diameter of the flame holder (mm)	50
Diameter of air inlet (mm)	50
Diameter of the fuel inlet (mm)	3
Diameter of the Outlet (mm)	50
Number of holes per row	8
Number of rows of holes	4

Table 6. Parameters used in designing the flame holder.

Zones	Diameter (mm)	Hole/ Row	Number of rows of holes	Hole Area (m^2)	Area/Row (m^2)	Total Area (m^2)	Area compared to air inlet (%)
Premix	0.006	8	4	0.000028	0.00023	0.00090	46.08
Combustion	0.008	8	4	0.000050	0.00040	0.00161	81.92
Dilution	0.10	8	4	0.000079	0.00063	0.00251	128

5. Conclusion

The design optimisation of a combustion chamber for MGT was performed, using CAD solid works and ANSYS-FLUENT 16.1 version CFD simulation software. Various design parameters and combustion models were used to produce an optimum design configuration. The simulations were first performed on different 3D geometries of chamber heights/flame holder diameter using species transport combustion model. The non-premixed combustion models, which employs the infinitely fast chemistry assumptions, were further applied to predict the diffusion flames.

The design parameters used with non-premixed transport model include: the effect of holes diameters, excess air supplied, convective heat flux and dead zones. In addition, optimisation was performed to further reduce the number of CO emissions and outlet temperature.

The result revealed that, design of the optimum chamber produced overall combustion stability with least outlet CO emission and well-defined temperature contours. In addition, a 2-inch flame holder diameter, 60cm chamber height, having 4 holes of (6, 8, and 10) mm with dead zone between combustion and dilution zone was adopted as the optimum chamber geometry. The simulations results show that an average CO mole fraction of 102 ppm and temperature of 1218 ^{o}C were produced at the outlet.

Acknowledgments

The authors would like to thank the Universiti Sains Malaysia, RUI grant number: 1001/PMEKANIK/814282, for the financial support of this study.

References

[1] Aleksandrov Y B and Mingazov B G 2017 Optimal design of a combustion chamber of gas turbine engine by a Combustion chamber 1D-2D computer program *IOP Conf. Ser. Mater. Sci. Eng.* 240:012006. doi:10.1088/1757-899X/240/1/012006

[2] Varun, Singh P, Tiwari S K, Singh R and Kumar N 2017 Modification in combustion chamber geometry of CI engines for suitability of biodiesel: A review *Renew Sustain Energy Rev* 79:1016–33. doi:10.1016/j.rser.2017.05.116

[3] Li J, Chou S K, Li Z W and Yang W M 2009 Characterization of wall temperature and radiation power through cylindrical dump micro-combustors *Combust Flame* 156:1587–93 doi:10.1016/j.combustflame.2009.05.003

[4] Wan J, Fan A, Yao H, Liu W. Effect of thermal conductivity of solid wall on combustion efficiency of a micro-combustor with cavities. Energy Convers Manag 2015;96:605–12 doi:10.1016/j.enconman.2015.03.030.

[5] Krieger G C, Paulo S, Filho F L S and Souza R C De 2012 A Swirler Stabilized Combustion Chamber for a Micro-Gas Turbine Fuelled with Natural Gas *J. Braz. Soc. Mech. Sci. Eng.* XXXIV:441–9

[6] Zornek T, Monz T and Aigner M 2015 Performance analysis of the micro gas turbine Turbec T100 with a new FLOX-combustion system for low calorific fuels *Appl Energy* 159:276–84 doi:10.1016/j.apenergy.2015.08.075

[7] Cavarzere A, Morini M, Pinelli M, Spina P R, Vaccari A and Venturini M 2014 Experimental analysis of a micro gas turbine fuelled with vegetable oils from energy crops *Energy*

Procedia 45:91–100 doi:10.1016/j.egypro.2014.01.011
[8] A H Lefebvre 1998 GAS Turbine Combustion, Second Edition -Books.google.com
[9] M M Noor, A P Wandel and T Yusaf 2013 Design and development of mild combustion burner *J. Mech. Eng. Sci.* 5:662–76
[10] K A Al-attab and Z A Zainal 2014 Performance of a biomass fueled two-stage micro gas turbine (MGT) system with hot air production heat recovery unit *Appl Therm Eng* 70:61–70 doi:10.1016/j.applthermaleng.2014.04.030
[11] Computational Fluid Dynamics in Industrial Combustion - Google Books n.d.
[12] Bott R 2014 Applied modelling for bio-and lean gas fired micro gas turbines *Igarss* 2014:1–5 doi:10.1007/s13398-014-0173-7.2
[13] Jiaqiang E, Peng Q, Liu X, Zuo W, Zhao X and Liu H 2016 Numerical investigation on hydrogen / air non-premixed combustion in a three-dimensional micro combustor *Energy Convers Manag* 124:427–38 doi:10.1016/j.enconman.2016.07.048
[14] Li Y, Chen G, Cheng T, Yeh Y and Chao Y 2013 Combustion characteristics of a small-scale combustor with a percolated platinum emitter tube for thermophotovoltaics *Energy* 61:150–7 doi:10.1016/j.energy.2013.09.003
[15] Mark C P and Selwyn A 2016 Design and analysis of annular combustion chamber of a low bypass turbofan engine in a jet trainer aircraft *Propuls Power Res* 2016;5:97–107 doi:10.1016/j.jppr. 2016.04.001
[16] Carlos A and Filho P 2004 Basic design principles for gas turbine combustor. Proc 10o Brazilian Congr Therm Sci Eng -- *ENCIT 2004 Braz Soc Mech Sci Eng -- ABCM, Rio Janeiro, Brazil,* Nov 29 -- Dec 03, 2004 BASIC

IOP Conf. Series: Materials Science and Engineering **530** (2019) 012022 doi:10.1088/1757-899X/530/1/012022

Evaluation of 3D Surface Roughness of Milled Surfaces using Laser Speckle Pattern

S Bharathi[1] and M M Ratnam[2*]

[1]Faculty of Engineering and Computer Technology, AIMST University, Semeling, 08100 Bedong, Kedah, Malaysia

[2]School of Mechanical Engineering, Engineering Campus, Universiti Sains Malaysia, 14300 Nibong Tebal, Penang, Malaysia

Corresponding author *: mmaran@usm.my

Abstract. 3D surface roughness of milled aluminium surface was evaluated using the laser speckle pattern. The milled surface was produced from aluminium alloy Al 6061 material at ten different feed rates using CNC milling machine. Laser beam was directed onto the aluminium surface and the scattered beam from the surface of the sample underwent interference to produce laser speckle pattern. Coefficient of Determination between various characteristic features and 3D surface roughness were determined and it was found that 3D surface roughness parameter S_v have coefficient determination above 0.5 with mean, G_a, roughness, entropy and normalized ACF.

1. Introduction

Surface roughness refers to the finely spaced surface irregularities which are result of machining operation in case of machined surface. The quality of the surface plays a very important role in the performance of milling operation as a good quality milled surface significantly improves fatigue strength, corrosion resistance, or creep life [1]. Surface roughness also affects several functional attributes of parts such as contact causing surface friction, wearing, light reflection, heat transmission, ability of distributing and holding a lubricant, coating, or resisting fatigue. Since surface roughness plays an important role in the quality of the milled surface, therefore surface roughness of milled surface must be verified that it is within the specified roughness.

Measurement of surface roughness can be done either by contact method or non – contact method. The widely used contact method is diamond stylus probe. This conventional approach has some drawbacks, although it is still widely used surface topography measurement technique. There are some critical factors such as stylus tip dimensions and scan parameters (sample length, evaluation length, speed, rate of scanning and force applied on the measured sample).

Optical methods are non-destructive, non – contact and accurate and therefore have advantages over stylus technique. There are various types of optical method and one among them is laser speckle method. When a coherent laser light falls onto a rough surface, the rough surface will act as a secondary source and scatters the laser light. This scattered light will undergo interference and will form randomly arranged granular structures called speckle pattern. There are two basic schemes of speckle pattern formation, one is known as objective speckle and another one is known as subjective speckle as shown in figure 1 [2]. In case of objective speckle, the patterns formed at free space geometry as shown in figure 1(a). In case of subjective speckle, the patterns formed are focused using a lens onto the screen as shown in figure 1(b).

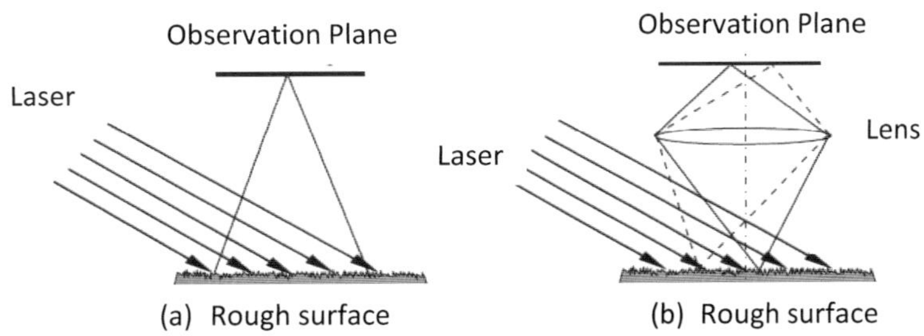

Figure 1. Two basic scheme for speckle formation

For speckle interferometry, the properties of the laser beam has to be monochromatic and coherent and therefore the laser normally used are He-Ne lasers [3]. Laser mode of TEM00 is used as it produces a Gaussian beam.

Several surface roughness measurement method using laser speckle image had been proposed. In speckle contrast method, the relation between speckle contrast and surface roughness was used for determining the surface roughness. Speckle contrast method usually can evaluate surface roughness Ra < 0.3 µm. In case of angular speckle correlation method and polychromatic speckle correlation method, the relation between coefficient of correlation obtained from two speckle image and surface roughness was used for evaluation of surface roughness. Speckle correlation method can work on surface roughness R_a between 1 and 30 µm. Speckle correlation method requires two slightly different speckle image which can be obtained by varying angle of illumination (Angular Speckle Correlation – ASC) or by varying the illumination frequency (Spectral Speckle Correlation – SSC). However other characteristics features had been extracted from the speckle image for correlating with surface roughness. Various techniques which had been used for correlating the characteristics features with the surface roughness are grouped into speckle constrast, speckle correlation, binarisation, phase unwrapping, Hurst exponent, Tsallis Threshold, Neural Network, Gray Level Co-occurrence Matrix (GLCM).

Goh and Ratnam [4] had assessed the 3D roughness parameters of milled surfaces using charged coupled device (CCD) flatbed scanner. From the scanned image, various characteristic features such as average pixel intensity, root mean square pixel intensity, variance, black to white ratio, bright region average area were extracted and correlated with Average height S_a and Root-Mean-Square height S_q. They had obtained good correlation.

Currently available methods for determining 3D roughness parameters are mostly optical methods. Some of them are white light interferometers[5, 6] , focus variation method [7], confocal microscopy [8].As a milled surface is 3D, it is naturally more appropriate to quantify the characteristics of the surface using 3D surface parameters[4].The objective of the study is to propose the evaluation of 3D surface roughness parameters of milled surface from the laser speckle pattern.

2. Experimental Procedure

Milled samples are produced by machining Aluminium alloy Al 6061 material at different feed rate but at constant spindle speed of 6000 rpm and depth of cut of 1 mm using CNC 5 axis milling machine (DMU 40) with 3 flute end mill cutter of carbide type with Ø12 mm, at feed rates of 220, 265, 310, 355, 400, 445, 490, 535, 580 and 625 mm/min. Using Alicona Infinite Focus Microscope measuring

IOP Conf. Series: Materials Science and Engineering **530** (2019) 012022 doi:10.1088/1757-899X/530/1/012022

equipment following 3D surface roughness parameters is measured from the milled surfaces and tabulated in Table 1:-

- Arithmetic Mean Height S_a
- Root-Mean-Square Height S_q
- Maximum Peak Height S_p
- Maximum Valley Depth S_v
- Maximum Height S_z
- Ten Point Height S_{10z}
- Skewness S_{sk}
- Kurtosis S_{ku}
- Root Mean Square Gradient S_{dq}
- Developed Interfacial area ratio S_{dr}

Laser beam from Helium Neon Laser of wavelength 633 nm and max power of less than 35 mW are cleaned using a spatial filter which consist of microscope objective lens (Edmund EFL 8.00) fitted in reverse direction and mounted pinhole (Size 15 μm). The cleaned laser beam falls onto the machined surface at an angle of 26.9° and the scattered beam will undergo interference and forms laser speckle pattern. This laser speckle pattern are captured using CCD camera (JAI CV-M50) fitted with TV zoom lens (1:1.2 / 12.5 – 75) and extension tube of size 20 mm. The CCD camera is placed normal to the machined surfaces. The images are stored in Tag Image File Format (tiff). Figure 2 shows the experimental setup and Figure 3 shows the schematic diagram of the experimental setup.

Figure 2. Experimental Setup

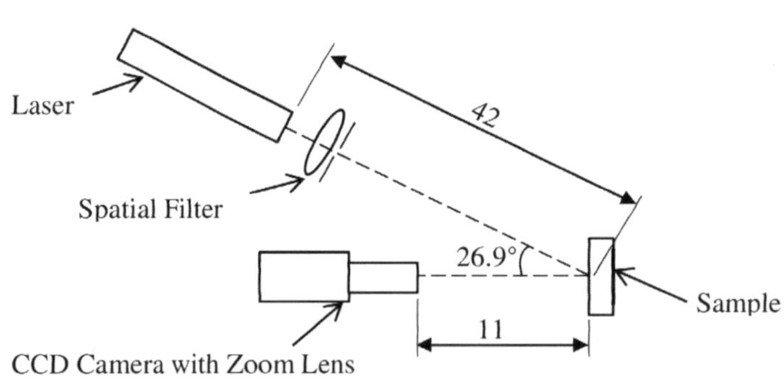

Figure 3. Schematic Diagram of the Experimental Setup

IOP Conf. Series: Materials Science and Engineering **530** (2019) 012022 doi:10.1088/1757-899X/530/1/012022

Using MATLAB software, the image is converted from true color to gray scale image of size 768 pixels x 576 pixels. Figure 4 shows the laser speckle pattern image for surface no 1. Figure 5 shows image of rectangle shape captured using the CCD camera for determining the scaling factor in horizontal direction SF_H and in vertical direction SF_V as given by equation (1) & (2).

Figure 4. Laser Speckle Pattern for Surface No.1

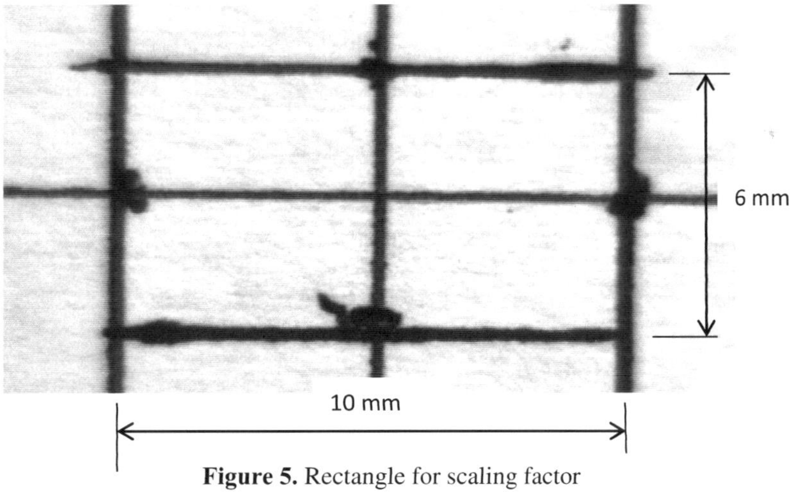

Figure 5. Rectangle for scaling factor

IOP Conf. Series: Materials Science and Engineering **530** (2019) 012022 doi:10.1088/1757-899X/530/1/012022

Table 1. 3D Surface Roughness Parameter for each Milled Surfaces

Surface No.	Spindle Speed (rpm)	Feed rate (mm/min)	Depth of cut (mm)	S_a (µm)	S_q (µm)	S_p (µm)	S_v (µm)	S_z (µm)	S_{10z} (µm)	S_{sk}	S_{ku}	S_{dq}	S_{dr} (%)
1	6000	220	1	0.3926	0.5168	15.449	4.2402	19.689	7.8626	1.0879	25.125	0.1263	0.7841
2	6000	265	1	0.3884	0.5515	20.472	4.1385	24.61	21.99	5.1156	155.38	0.1475	0.9754
3	6000	310	1	0.4803	0.6591	24.538	5.6371	30.176	26.09	3.2782	80.089	0.1836	1.4313
4	6000	355	1	0.5665	0.7051	12.23	33.719	45.95	12.802	0.1977	7.8601	0.1558	1.2117
5	6000	400	1	0.5457	0.6863	16.406	4.2553	20.661	18.418	0.5930	10.744	0.1356	0.8701
6	6000	445	1	0.4387	0.5708	39.646	18.706	58.352	41.507	4.3443	253.81	0.1339	0.7801
7	6000	490	1	0.6156	0.8549	38.124	3.0996	41.224	39.76	6.9088	205.74	0.1992	1.3138
8	6000	535	1	0.8187	0.9762	5.6115	4.3118	9.9233	7.1029	0.1791	2.3002	0.1361	0.9215
9	6000	580	1	0.9114	1.0619	5.351	7.9208	13.272	6.7702	0.0988	2.0208	0.1269	0.8031
10	6000	625	1	0.9262	1.1137	5.4225	6.4135	11.836	8.1326	-0.0964	2.3894	0.1567	1.2109

IOP Conf. Series: Materials Science and Engineering **530** (2019) 012022 doi:10.1088/1757-899X/530/1/012022

$$SF_H = \frac{actual\ length}{length\ in\ terms\ of\ number\ of\ pixels} = \frac{10}{377} = 0.02\ mm/pixel \quad (1)$$

$$SF_V = \frac{actual\ length}{length\ in\ terms\ of\ number\ of\ pixels} = \frac{6}{242} = 0.02\ mm/pixel \quad (2)$$

Scan region of Alicona Infinite Focus Microscope is 544.74 μm by 5.7283 mm and scanning were done at the center of the sample. These scanned region dimension were converted in terms of number of pixels using the scaling factor and found the scanning region to be 27 pixels by 285 pixels which is shown as rectangle box in figure 6. Region for analysis (Cropped region) on the laser speckle pattern was chosen such that its region is smaller and within the Alicona Infinite Focus Microscope scan region.

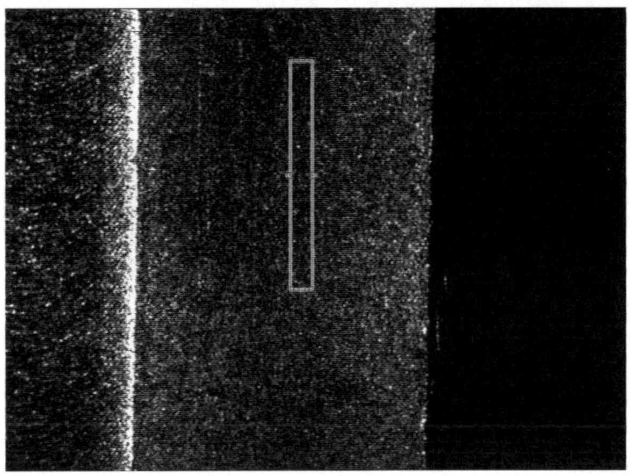

Figure 6. Rectangle box represents the Alicona Infinite Focus scanning area

Using MATLAB software, the 10 surfaces were cropped to region of interest of size 23 pixels x 281 pixels at starting coordinate (354,68) as shown in Figure 7.

Figure 7. Cropped Images for surface No.1

From the cropped image Histogram based (Statistical) features and Texture feature were extracted as follows:-

- Histogram based (Statistical) features

- Mean
  Mean of gray value of the image m obtained from original image $f(x,y)$ of size M x N as in equation (A.1).

- Arithmetic Average of Intensity G_a as in equation (A.2).

- Root Mean Square (RMS)
  Root mean square is square root of mean of each pixel gray value as in equation (A.3).

- Skew
  Skew is a measure of histogram's asymmetry about mean level as in equation (A.4).

- Energy
  The energy descriptor provides another measure on how pixels values are distributed along the gray level range and can be calculated for gray scale image as in equation (A.5).

- Entropy
  Entropy descriptor provides the information about the complexity of the image as in equation (A.6).

- Texture Features
  - Roughness
    Normalised descriptor of roughness R is as in equation (A.7).

- Autocorrelation Coefficient Function (ACF)
  Autocorrelation coefficient is calculated by shifting two copies of the same laser speckle pattern against each other as in equation (A.8).

- Gray Level Co-occurrence matrix (GLCM)
  Histogram based texture descriptors are limited as it does not have any information about the spatial relationship among pixels. These information can be obtained using Gray-Level Co-occurrence Matrix (GLCM). These matrix holds the information of number of times pixels with intensities z_i and z_j occur in image $f(x,y)$ in position specified by displacement vector $\mathbf{d} = (d_x,d_y)$. In this work displacement vector was choosen as $\mathbf{d}=(0,1)$. The matrix is normalized as in equation (A.9).

  - Maximum Probability as in equation (A.10).
  - Correlation as in equation (A.11).
  - Contrast as in equation (A.12).
  - Energy as in equation (A.13).
  - Homogeneity as in equation (A.14).
  - Entropy as in equation (A.15).

- From the binary image following characteristic features are extracted:-
  - Total Black pixels to Total White pixels ratio (B/W)
  - Total White pixels to Total Black pixels ratio (W/B)

3. Result and Discussion

Table 2 is the tabulation of coefficient of determination between characteristic features and 3D roughness parameters.

From the table 2, it is found that S_v has coefficient determination above 0.5 with mean, G_a, Roughness, Entropy and ACF characteristic features. Figure 8 shows the scatter plot between mean, G_a, Roughness, Entropy and ACF characteristic features and S_v.

Table 2. Coefficient of determination for linear regression between characteristic features and 3D roughness parameter

	S_a (µm)	S_q (µm)	S_p (µm)	S_v (µm)	S_z (µm)	S_{10z} (µm)	S_{sk}	S_{ku}	S_{dq}	S_{dr} (%)
mean	0.0119	0.0103	0.0001	**0.5207**	0.1815	0.0005	0.0009	0.0036	0.0499	0.0859
G_a	0.0357	0.0400	0.0032	**0.6743**	0.2881	0.0069	0.0027	0.0000	0.0143	0.0391
Roughness	0.0118	0.0122	0.0020	**0.6106**	0.1885	0.0004	0.0077	0.0115	0.0403	0.0908
Skewness	0.1339	0.1669	0.2684	0.0371	0.2699	0.2132	0.0372	0.2500	0.0721	0.1335
Uniformity	0.0208	0.0189	0.0021	0.4420	0.1890	0.0094	0.0005	0.0004	0.0281	0.0440
Entropy	0.0253	0.0254	0.0011	**0.5428**	0.2184	0.0048	0.0004	0.0001	0.0245	0.0480
RMS	0.0023	0.0092	0.0012	0.1133	0.0304	0.0002	0.0097	0.0018	0.1144	0.1007
B/W	0.1052	0.1052	0.3785	0.0459	0.3699	0.3477	0.1277	0.4110	0.0000	0.0070
W/B	0.0889	0.0825	0.3319	0.0231	0.2921	0.2881	0.1337	0.3580	0.0115	0.0018
ACF	0.0136	0.0143	0.0014	**0.6168**	0.1954	0.0002	0.0073	0.0098	0.0369	0.0849
Max Probability	0.0016	0.0021	0.0452	0.1245	0.0021	0.0246	0.0033	0.0286	0.0019	0.0047
Correlation	0.1135	0.1716	0.1428	0.2625	0.3624	0.1407	0.0018	0.1746	0.2286	0.2750
Contrast	0.1961	0.2030	0.1519	0.2582	0.0000	0.0887	0.0764	0.0830	0.0305	0.1201
Energy	0.0103	0.0063	0.0008	0.3140	0.1284	0.0078	0.0036	0.0000	0.0598	0.0747
Homogeneity	0.1943	0.2167	0.1292	0.1205	0.0052	0.0645	0.0286	0.0632	0.0452	0.0979
Entropy	0.0193	0.0175	0.0000	0.4749	0.1705	0.0012	0.0007	0.0031	0.0414	0.0733

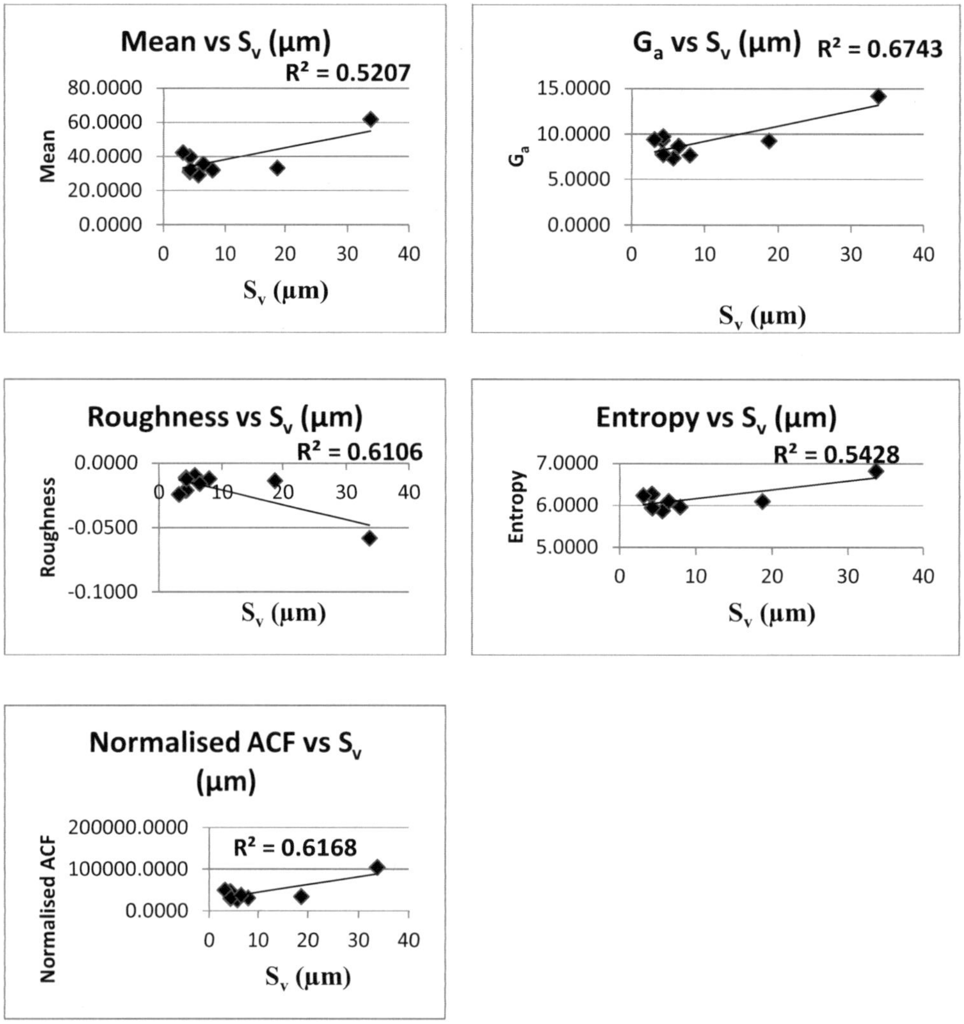

Figure 8. Correlation between Characteristic Features and S_v

From the Table 2 it is clear that there is no correlation between characteristic features and the 3D surface roughness for certain 3D surface roughness. These could be due to: -

- The cropped region is too small
- The size of speckle pattern is too small for analysis. This problem can be resolved by increasing the magnification of the image captured.
- There are some pixels having intensity of value 255 grayscale value. This can be overcome by using Neutral Density filter to cut off the brightness at lower aperture setting.
- There is poor correlation for experimental work using illumination angle of 26.9°. The experiment shall be repeated at different illumination angle and check if there is any improvement in the correlation.

Previous works shows there are good correlation between laser speckle pattern and 2D surface roughness [9-10].

4. Conclusion

From the current experimental work, the result shows that there is only some correlation between S_v and certain characteristic features. Effect of the image crop size on the correlation shall be studied in detail as well the magnification on the correlation. The effect of angle of incident on the correlation shall be studied in detail.

IOP Conf. Series: Materials Science and Engineering **530** (2019) 012022 doi:10.1088/1757-899X/530/1/012022

Appendix A

$$\boldsymbol{m} = \frac{\mathbf{1}}{\boldsymbol{MN}} \sum_{\boldsymbol{x}=\mathbf{1}}^{\boldsymbol{M}} \sum_{\boldsymbol{y}=\mathbf{1}}^{\boldsymbol{N}} \boldsymbol{f}(\boldsymbol{x}, \boldsymbol{y}) \tag{A.1}$$

$$G_a = \frac{1}{MN} \sum_{x=1}^{M} \sum_{y=1}^{N} |f(x, y) - m| \tag{A.2}$$

$$RMS = \sqrt{\frac{f(x, y)^2}{M \times N}} \tag{A.3}$$

$$skew = \frac{mean - mode}{standard\ deviation} \tag{A. 4}$$

$$energy = \sum_{j=0}^{255} p(r_j)^2 \tag{A.5}$$

Where $p(r_j)^2$ is probability of occurrence of jth gray level r_j

$$entropy = -\sum_{j=0}^{255} p(r_j) log_2[p(r_j)] \tag{A.6}$$

$$R = 1 - \frac{1}{1 + \frac{\sigma^2}{255^2}} \tag{A.7}$$

$$ACF = \sum_{s=1}^{M} \sum_{t=1}^{N} \frac{\sum_{s=1}^{M} \sum_{t=1}^{N} A(s,t) A\left(((x-1)+s), ((y-1)+t)\right)}{\sum_{s=1}^{M} \sum_{t=1}^{N} \left(A(s,t)\right)^2} \tag{A.8}$$

$$N_g(i,j) = \frac{g(i,j)}{\sum_i \sum_j g(i,j)} \tag{A.9}$$

The following texture based features are computed using a normalized GLCM, $N_g(i,j)$.

$$Maximum\ probability = max N_g(i,j) \tag{A.10}$$

$$Correlation = \frac{\sum_i \sum_j (i-\mu_i)(j-\mu_j) N_g(i,j)}{\sigma_i \sigma_j} \tag{A.11}$$

$$Contrast = \sum_i \sum_j (i-j)^2 N_g(i,j) \tag{A.12}$$

$$Energy = \sum_i \sum_j N_g^2(i,j) \tag{A.13}$$

$$Homogeneity = \sum_i \sum_j \frac{N_g(i,j)}{1+|i-j|} \tag{A.14}$$

$$Entropy = -\sum_i \sum_j N_g(i,j) log_2 N_g(i,j) \tag{A.15}$$

References

[1] Lou M S, Chen J C and Li C M Nov 1998- Jan 1999 Surface roughness prediction technique for CNC End-Milling *Journal of Industrial Technology* 15(1) 1 - 6

[2] Tian G Y, Lu R S and Gledhill D 2007 Surface measurement using active vision and light scattering *Optics and Lasers in Engineering* 45(1) 131-9

[3] Popov I A, Sidorovsky N V, Veselov I L and Hanson S G 1998 Statistical properties of focal plane speckle *Optics Communications* 156(1-3) 16-21

[4] Goh C S and Ratnam M M 2015 Assessment of Areal (Three-Dimensional) Roughness Parameters of Milled Surface Using Charge-Coupled Device Flatbed Scanner and Image Processing *Experimental Techniques* 1-9

[5] Manojlovic L M, Zivanov M B and Marincic A S 2010 White-Light Interferometric Sensor for Rough Surface Height Distribution Measurement *IEEE Sensors Journal* 10(6) 1125-32

[6] Gao F, Leach R K, Petzing J and Coupland J M 2008 Surface measurement errors using commercial scanning white light interferometers *Measurement Science and Technology* 19(1) 015303

[7] Danzl R, Helmli F and Scherer S 2011 Focus Variation—A Robust Technology for High Resolution Optical 3D Surface Metrology 245-56 p

[8] Al-Shammery H A O, Bubb N L, Youngson C C, Fasbinder D J and Wood D J 2007 The use of confocal microscopy to assess surface roughness of two milled CAD–CAM ceramics following two polishing techniques *Dental Materials* 23(6) 736-41

[9] Jeyapoovan T, Murugan M and Bovas B C 2012 Statistical analysis of surface roughness measurements using laser speckle images *World Congress on Information and Communication Technologies (Trivandrum, India)* (IEEE Computer Society) 378-82

[10] Dhanasekar B, Mohan N K, Bhaduri B and Ramamoorthy B 2008 Evaluation of surface roughness based on monochromatic speckle correlation using image processing *Precision Engineering* 32(3) 196-206

IOP Conf. Series: Materials Science and Engineering **530** (2019) 012023 doi:10.1088/1757-899X/530/1/012023

A 3-Pixel Fuzzy Mask for Edge Detection

N H Seng[1*], Z Samad[2] and N M Nor[2]

[1]Mechanical Engineering Department, Politeknik Seberang Perai, Jalan Permatang Pauh, 13500, Seberang Perai, Pulau Pinang

[2]School of Mechanical Engineering, Universiti Sains Malaysia,14300, Jalan Transkrian, Nibong Tebal, Pulau Pinang

Corresponding author *: nhs167@gmail.com

Abstract. A new group of fuzzy masks is proposed in this study to perform edge detection of images. A group of 3-pixel masks is used to detect edges in four directions. The resulting images retain more edges and show better continuity of the edges compared to the ones produced by the conventional 3 x 3 fuzzy masks. This is accomplished more efficiently using fewer fuzzy rules. Furthermore, thinning functions incorporated yielded edges with good continuity and optimal thickness.

1. Introduction

Edge detection is one of the many important areas in image processing. Edge detection enables interpretation of objects based on their outlines. Literature is abundant concerning the application of edge images in various fields such as medical diagnosis [1], detection of defects [2] and robotics navigation [3]. Edge detection of an image is based on the concept that a line can be detected to separate two regions which show significant changes in grey level values. Pioneering works by Sobel [4] and Canny [5] are well established methodologies today and has inspired many variations of edge detection methods based on them.

It is well known that changes in grey levels between regions are not abrupt but gradual. Furthermore, the changes are ambiguous and imprecise in nature. These characteristics lend themselves well to the application of fuzzy logic which is suitable to address such problems. Russo [6] pioneered the use of a fuzzy mask to classify pixels as edges and non-edges in a software implementation based on the luminance differences between the center pixel and its neighboring pixels. Tao and Thompson [7] used sixteen fuzzy rules corresponding to the various 3 x 3 pixel combinations representing the differences of grey levels of the surrounding pixels with the centre pixel. Begol and Maghooli [8] expanded the fuzzy rule base using the same fuzzy masks to perform similar tasks

Lee, et al. [9] proposed a novel edge detection method which involved automatic fuzzy thresholding of various pixel groups. Their method demonstrated reliability in detecting doorframes along a corridor used by a mobile robot for localization despite changing illumination conditions. In this case no adjustment of parameters was necessary.

The identification of edges based on the concept of competition was introduced by Liang and Rooney [10]. This involved feeding feature vectors, which represented magnitude differences of grey level patterns, into a fuzzy network. The fuzzy network utilized Epanechikov fuzzy sets that classified

IOP Conf. Series: Materials Science and Engineering **530** (2019) 012023 doi:10.1088/1757-899X/530/1/012023

pixels into edge, background and speckle classes. Thinning of edges were also achieved in a shorter processing time compared with Canny's method.

An improvement [11] to fuzzy rule based edge detection methods was made by incorporating a fuzzy rule base to detect continuity in edges. Utilization of maximum entropy principles in the adjustment of the parameters of the fuzzy sets enabled detection of noisy images without the need of filtering.

The concept of generating fuzzy images from gradient magnitudes was introduced by Molina, et al. [12]. Parametric functions for the Triangular, Gaussian, Sigmoidal and Trapezoidal sets were tuned by a histogram based strategy [13] to achieve this task. This novelty was incorporated into the Canny edge detection method. Recently researches have switched to type 2 fuzzy logic [14] to improve the quality of the edges. Interval type 2 fuzzy logic [15] has shown that improvements could still be made in the edge detection of images within the fuzzy logic paradigm.

Type 2 fuzzy sets [16] were also used to determine the magnitude of fuzziness across a gradient histogram to determine the double threshold values of the Canny algorithm for edge detection. A generalized type-2 fuzzy logic [17] was implemented by using the morphological gradient values to determine the parameters of the antecedent and consequent fuzzy sets. It resulted in better detection of images compared to the interval type 2 and type 1 fuzzy logic methodologies. Interval Type 2 Fuzzy Systems coupled with morphological gradient was used by Gonzales, et al. [19] to extract edges from colour images.

Recently, a solution to the optimization of the parameters of interval type 2 fuzzy sets and type 1 fuzzy sets [20] was provided by utilizing the Cuckoo search algorithm and genetic algorithm methodologies. Results showed that the optimized parameters of the interval type 2 fuzzy antecedent sets produced edges which were closer to the ideal compared to those produced by the normally optimized interval type-2, optimized and non-optimized type 1 fuzzy antecedents and Sobel filters.

The conventional fuzzy masks technique uses a multitude of templates which detects edges in various directions but is unable to accommodate all combinations of dark and bright pixel images when convolved together in a fully 3x3 pixel setting. As a result, some edges might not fully be detected or even missed out altogether. To increase the number of templates substantially might be computationally impractical.

In this paper a new fuzzy mask is proposed and its performance in terms of edge detection is compared to one of the conventional fuzzy masks technique [8]. The edge detection is done in four directions viz. north-south, east-west, northwest-southeast and northeast-southwest using a 3-pixel template. In each direction there are six permutations or arrangements of adjacent bright and dark pixels that are used to detect edges. This ensures that no combinations of dark and bright pixels of images will be missed out in each direction when convolved with an image. These permutations are represented as a set of fuzzy rules. A separate set of rule base is added to perform the thinning of edges. Resulting images showed better continuity of the edges and also more details were retained even when thinning was carried out compared to the images yielded by the conventional fuzzy masks.

2. Methodology

Figure 1 shows the 3x3 pixel mask that is used as a fuzzy mask. Pixels i2, i5 and i8 represent the North-South directions. Pixels i4, i5 and i6 represent the East-West directions. Pixels i1, i5 and i9 represent the North-West and South-East directions. Pixels i3, i5 and i7 represent the North-East and South-West directions. Six combinations of pixel arrangements are embedded in each of these directions. This enables detection of edges in their respective directions. Each pixel is categorized into either bright or dark pixels.

Figure 2 shows all the possible combinations of the three-pixel groups for the edge detection task. Since there are three pixels in a certain direction with each pixel categorized as either a bright or dark pixel there will be eight combinations all together. Note that the all consecutive dark and bright pixel are discarded since they do not detect edges.

IOP Conf. Series: Materials Science and Engineering **530** (2019) 012023 doi:10.1088/1757-899X/530/1/012023

Group 1 pixels represent the antecedents associated with the middle intensity pixel value. Group 2 pixels represent the antecedents associated with the bright intensity pixel values. The consequents consist of bright, dark and middle intensity pixels which represent the edges. The fuzzy inferencing incorporating these rules is performed in the aforesaid four directions. The mask on the bottom right represents the fuzzy rule to remove noise or speckles in the image.

The input fuzzy sets representing the bright and dark pixels use Gaussian functions, each with a standard deviation value of 110. This value produces sufficient overlap between the two fuzzy sets. This is shown Figure 3. Figure 4 shows the fuzzy output set representing the dark, edge and bright pixels. Each has a mean grey value of 5, 135 and 250 respectively. The small standard deviation value of 5 provides the separation of pixels into their respective groups. Mamdani min-max inferencing [21] is used in this case. Defuzzification involves using the singleton method instead of the centroid method as the latter is computationally more intensive.

I_1	I_2	I_3
I_4	I_5	I_6
I_7	I_8	I_9

Figure 1. 3x3 Mask

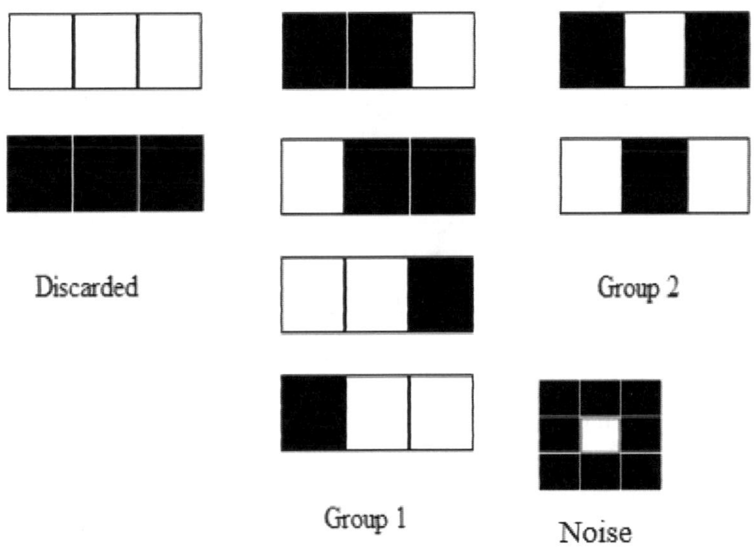

Figure 2. Pixel Combinations for Edge Detection and Noise Removal

IOP Conf. Series: Materials Science and Engineering **530** (2019) 012023 doi:10.1088/1757-899X/530/1/012023

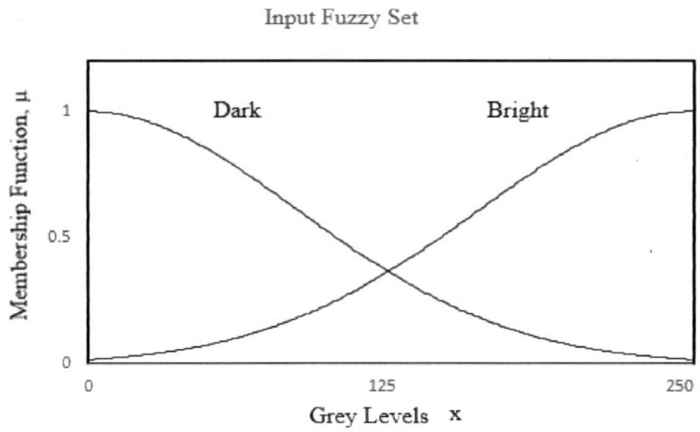

Figure 3. Input Fuzzy Sets

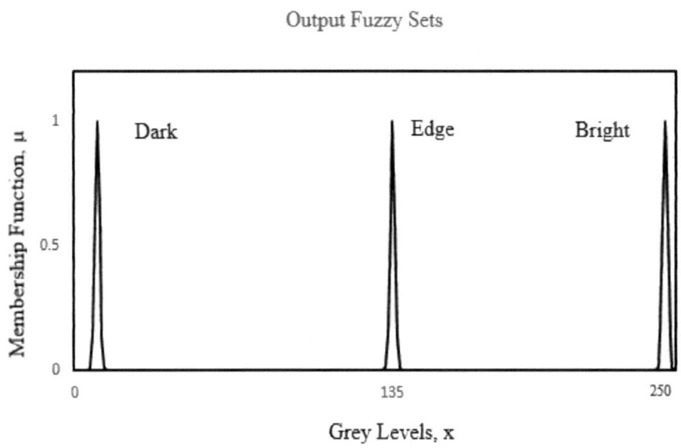

Figure 4. Output Fuzzy Sets

In edge detection, excessively thick lines are undesirable. In such situations thinning is desired while retaining most of the outlines of the objects. To achieve this task, only the maximum membership function value from the six combinations of the rule base is chosen in each respective direction. In addition to this, another 8 rules perform edge linking to minimize any loss of edges. The masks representing those rules are shown in Figure 5.

These masks link any discontinuous lines in all directions. The broken lines may have been falsely discarded in choosing the maximum membership function value in each direction during the edge detection process.

All in all, there are a total of 7 rules used for the edge detection process since the 6 rules for each direction are similar and can be written as reusable subroutines and one rule is used for denoising. As for the thinning process another 8 rules for edge linking are added. The economical use of the fuzzy rules reduces computation time while optimizing the edge detection process.

IOP Conf. Series: Materials Science and Engineering **530** (2019) 012023 doi:10.1088/1757-899X/530/1/012023

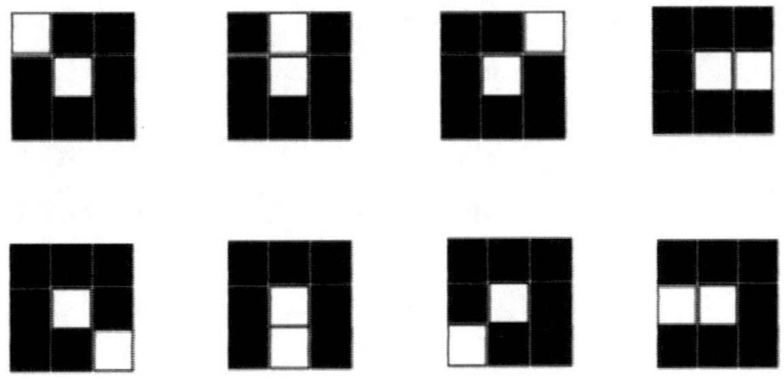

Figure 5. Masks for Edge Linking

Figure 6 shows the histograms for the number of pixels for each grey level retained after defuzzification. The histograms show an overwhelmingly concentrated number of pixels around the grey level value of 166 and 135 for the proposed method and the conventional fuzzy mask technique respectively. The importance of these values is that they are chosen as thresholding values.

Since the performance of the new fuzzy masks are compared to the conventional fuzzy masks [8], it is appropriate to show the conventional fuzzy masks representing all the fuzzy rules. They are reproduced and shown in Figure 7. In the latter case the input fuzzy sets consist of Gaussian functions representing dark and bright pixels with mean values of 0 and 255 respectively. The standard deviation values used for both input fuzzy sets were 60. The output fuzzy sets used were similar to the ones used for the new fuzzy masks. The values used were obtained by exhaustive tuning to obtain the clearest image with the least amount of noise for each case.

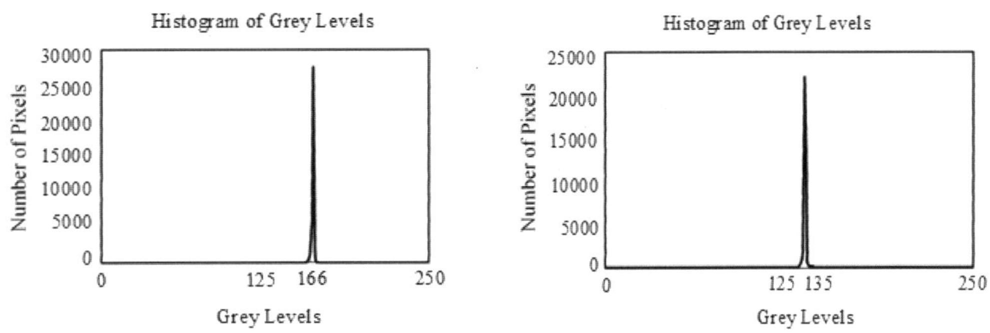

Figure 6. Histogram of Number of Pixels vs Grey Levels

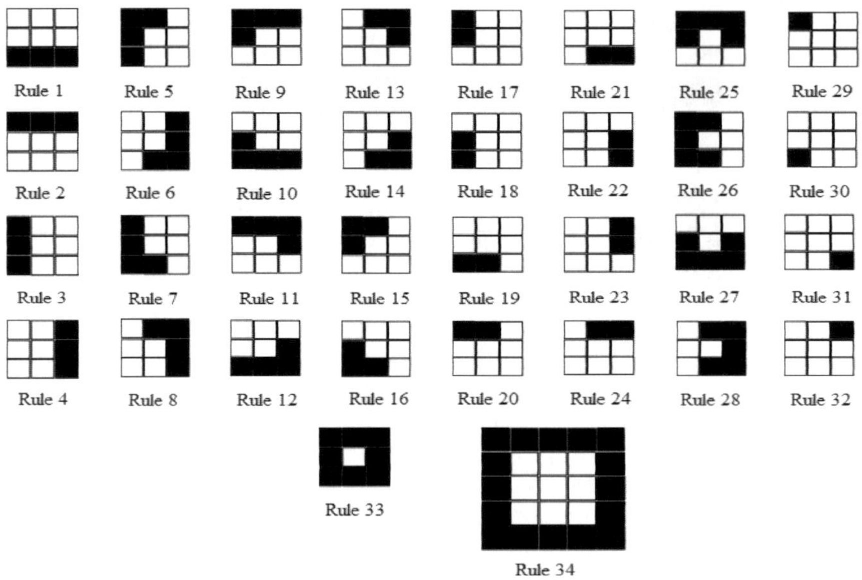

Figure 7. Conventional Fuzzy Masks

Figure 8 shows the representation of the 3-pixel templates in the form of IF-THEN rules with denoising in the North-South direction. Rules 1 to 6 can be repeated as reusable subroutines in the other 3 directions. The denoising rule is used only once.

If I_2 is Dark AND I_5 is Dark AND I_8 is Bright THEN I_5 is an EDGE pixel ____________Rule 1
If I_2 is Bright AND I_5 is Dark AND I_8 is Dark THEN I_5 is an EDGE pixel ____________Rule 2
If I_2 is Bright AND I_5 is Bright AND I_8 is Dark THEN I_5 is an EDGE pixel ____________Rule 3
If I_2 is Dark AND I_5 is Bright AND I_8 is Bright THEN I_5 is an EDGE pixel ____________Rule 4
If I_2 is Dark AND I_5 is Bright AND I_8 is Dark THEN I_5 is a BRIGHT pixel ____________Rule 5
If I_2 is Bright AND I_5 is Dark AND I_8 is Bright THEN I_5 is a BRIGHT pixel ________Rule 6

If I_1 is Dark AND I_2 is Dark AND I_3 is Dark AND
If I_4 is Dark AND I_5 is Bright AND I_6 is Dark AND
If I_7 is Dark AND I_8 is Dark AND I_9 is Dark THEN I_5 is a DARK pixel ____________Rule 7

Figure 8. IF-THEN Rule Implementation

A comparison of results was also made with the images produced by the Sobel and Canny operators. All the algorithms were written using the open source software Processing version 1.5.1 which can be downloaded from https://processing.org/. The threshold grey value for the Sobel operator is 30 and upper and lower threshold values of the Canny operator were 5 and 10 respectively.

3. Results

Figures 9 and 10 show the edge detected images produced from their respective photos. One shows an outdoor image of the Kuala Lumpur tower and its surroundings and the other shows the indoor image of a mobile robot.

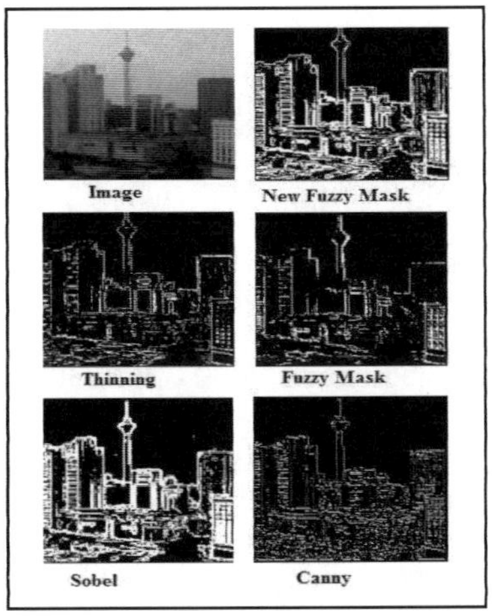

Figure 9. KL Tower and the Resulting Edges

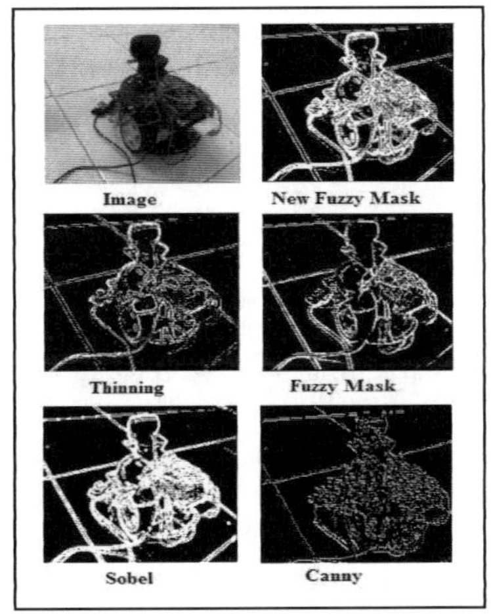

Figure 10. Mobile Robot and the Resulting Edges

In the first case the edges produced by the proposed method is comparable to the one produced by the Sobel operator. Most outline details are retained, bold and continuous. However, the outlines produced by the Sobel operator is slightly thicker. This is clearly borne out by the outlines of the Kuala Lumpur tower shown in Figure 9. The edge detected image resulting from the thinning operation of the proposed method is also shown. The outlines are thinner but still retain most edges uniformly. Although the conventional fuzzy masks [8] uses more rules compared to the new fuzzy mask it is less efficient in capturing the edges. The edges produced from the proposed method is markedly sharper compared to the one obtained through the conventional fuzzy masks. The Canny operator resulted in faint edges. The bottom portion of the image showed that many details were thinned and retained but produced a somewhat scattered edge formation.

The results from the indoor mobile robot photo showed a similar outcome. The new fuzzy mask and the Sobel operator produced detailed outlines but in the latter case the edges were overly thick. The proposed method produced a clearly marked separation between the tiles and clearly shows the edges of the grout. In the case of the Sobel operation the separation of the tiles was not very clear as shown by their overly thick lines.

The edge detected images resulting from the proposed fuzzy mask with the thinning process were slightly fainter than the one produced from the conventional fuzzy masks. However, on closer inspection, the former method retained a more uniform and continuous edge. This is clearly borne out by the continuous edge of the mounted camera. The edges of the camera obtained through the conventional fuzzy mask is broken. The Canny operator in this case produced outlines that were thinned but appearing discontinuous.

To validate the improvement of the proposed method quantitatively a similarity measure analysis is carried out between the convolution of the conventional fuzzy masks and the proposed fuzzy masks with the captured image. Figure 11 shows the sample 3 x 3 image segment with its dark and bright pixel configuration. The labelling of the pixel positions is also shown.

IOP Conf. Series: Materials Science and Engineering **530** (2019) 012023 doi:10.1088/1757-899X/530/1/012023

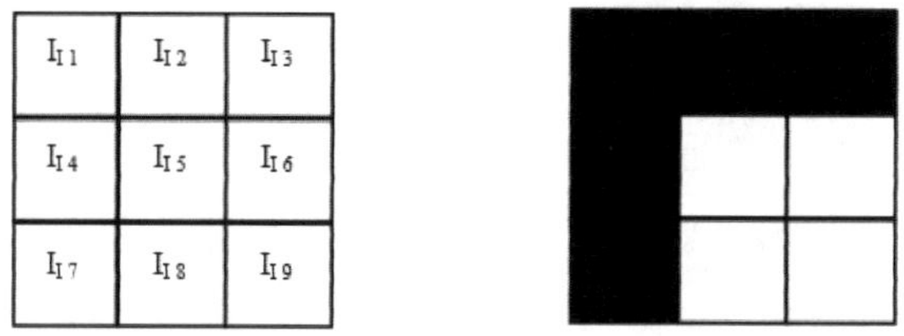

Figure 11. 3 x 3 Image Segment Showing Dark and Bright Pixel Configuration

In the proposed method a 3-pixel template is convolved with a 3 x 3 pixel image segment in the North-South direction as shown in Figure 12. This is repeated for the other three directions. In the conventional method a 3x3 pixel image is convolved with the templates shown in Figure 13. Note that this image does not resemble any of the templates in Figure 13. This is to compare the efficacy of the two methods in detecting edges.

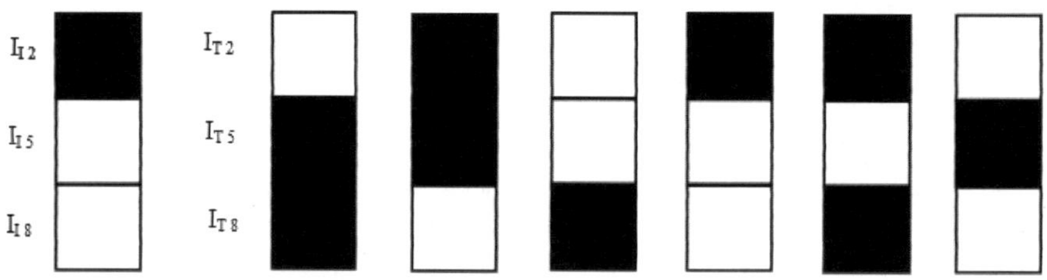

Figure 12. Convolving the Proposed Templates with a Segment of an Image in the North-South Direction

The similarity measure is given by the equations 1 and 2.

$$\text{Similarity measure, } H_j = 1 - \Big(\sum_{x=1}^{x=k} |\text{ pixel } (I_{Tx}) - |\text{ pixel } (I_I)\,|\Big)/k \qquad (1)$$

k = number of pixels
m = total number of templates
I_{Tx} = Template/Mask
I_{Ix} = Image

$$H_S = \max_{j=1}^{j=m} (H_j) \qquad (2)$$

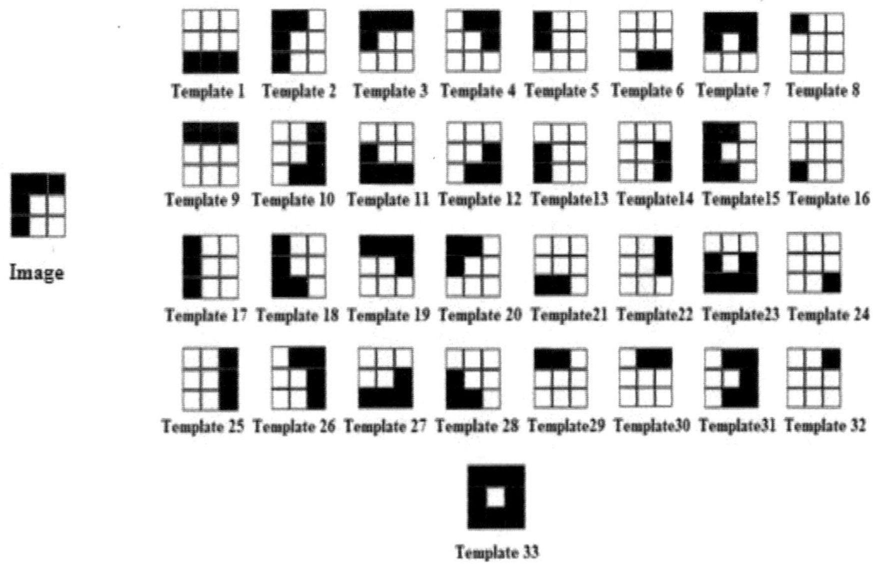

Figure 13. Convolving the Conventional Template with an Image

The similarity measure is easily calculated. If the pixel of the image and its coincident template pixel belong to the same pixel category the difference value is zero. On the other hand, if the image pixel and template pixel belong to different pixel categories a difference value of one is produced. For instance, in the case of the conventional fuzzy mask, the similarity measure between the image and the first template is H_j = 1- ((1+1+1+1+0+0+0+1+1)/9) =3/9.

The same treatment applies to the proposed method. In the convolution between the 3-pixel segment image with the first template in the North-South direction the similarity measure value H_j = 1- ((1+1+1)/3) = 0. The segment images are convolved with all the templates to obtain their respective similarity measure values. Finally, the maximum value is chosen as the value representing the similarity measure value. Tables 1 and 2 show the complete calculated similarity values for the conventional and proposed methods.

In the conventional method the maximum similarity value is 8/9. In the proposed method the maximum similarity value is 1 in all four directions. This shows that the templates in the proposed method can more efficiently detect images in all directions. However, in some templates the similarity values are zero. In the conventional method, none of the templates yielded a zero-similarity measure value but always partially detected an edge although its similarity values could be low. The templates used in the proposed method efficiently detects all permutations of dark and bright pixels in an economical manner without the requirement to use a large number of templates as in the case of the conventional method.

4. Conclusions and Recommendations

The results obtained showed that the proposed fuzzy masks outperformed the conventional fuzzy masks in that they detected edges more sensitively by considering all pixel combinations to capture edges in four directions in a more complete manner. Furthermore, by considering only the maximum membership function value in each direction coupled with a new set of rules for edge linking and noise removal,

IOP Conf. Series: Materials Science and Engineering **530** (2019) 012023 doi:10.1088/1757-899X/530/1/012023

thinning of the edges can be carried out while retaining important edges uniformly. Compared to the conventional fuzzy masks the proposed method uses less rules and the rules for each direction can be reused by encoding them as reusable subroutines.

Table 1. Similarity Measure for the Conventional Method

Similarity Measure Between Template and Image	**Similarity Measure Between Template and Image**
$H_1 = 3/9$	$H_{18} = 6/9$
$H_2 = 8/9$	$H_{19} = 6/9$
$H_3 = 8/9$	$H_{20} = 7/9$
$H_4 = 5/9$	$H_{21} = 4/9$
$H_5 = 6/9$	$H_{22} = 4/9$
$H_6 = 2/9$	$H_{23} = 3/9$
$H_7 = 7/9$	$H_{24} = 3/9$
$H_8 = 5/9$	$H_{25} = 3/9$
$H_9 = 7/9$	$H_{26} = 4/9$
$H_{10} = 2/9$	$H_{27} = 2/9$
$H_{11} = 4/9$	$H_{28} = 5/9$
$H_{12} = 1/9$	$H_{29} = 6/9$
$H_{13} = 6/9$	$H_{30} = 6/9$
$H_{14} = 2/9$	$H_{31} = 3/9$
$H_{15} = 7/9$	$H_{32} = 5/9$
$H_{16} = 5/9$	$H_{33} = 6/9$
$H_{17} = 7/9$	$\mathbf{H_{Max} = 8/9}$

Table 2. Similarity Measure for the Proposed Method

Similarity Measure Between Template and Image- North/South Direction	**Similarity Measure Between Template and Image - West/East Direction**
$H_{N1} = 0$	$H_{W1} = 0$
$H_{N2} = 2/3$	$H_{W2} = 2/3$
$H_{N3} = 1/3$	$H_{W3} = 1/3$
$H_{N4} = 1$	$H_{W4} = 1$
$H_{N5} = 2/3$	$H_{W5} = 2/3$
$H_{N6} = 1/3$	$H_{W6} = 1/3$
$H_{N\,max} = 1$	$H_{W\,max} = 1$
Similarity Measure Between Template and Image- North West /South East Direction	**Similarity Measure Between Template and Image - North East /South West Direction**
$H_{NW1} = 0$	$H_{NE1} = 1/3$
$H_{NW2} = 2/3$	$H_{NE2} = 1/3$
$H_{NW3} = 1/3$	$H_{NE3} = 2/3$
$H_{NW4} = 1$	$H_{NE4} = 2/3$
$H_{NW5} = 2/3$	$H_{NE5} = 1$
$H_{NW6} = 1/3$	$H_{NE6} = 0$
$H_{NW\,max} = 1$	$H_{NE\,max} = 1$
$\mathbf{H_S = max(H_{N\,max}, H_{W\,max}, H_{NW\,max}, H_{NE\,max}) = 1}$	

In the future type 2 fuzzy logic will be applied to improve the edge detection capabilities of the proposed method. Unlike conventional or type 1 fuzzy logic which mostly treats inputs as singleton values, type 2 fuzzy logic also takes into account uncertainties in the input measurements [14].

The yielded results indicate that the proposed method could potentially improve the detection rate of edge and surface defects of manufactured industrial products [22] currently utilizing the standard edge detection methods. For instance, the detection rate of typical glass surface defects [23] such as cracks, scratches, spots, bubbles, dirt and low contrast defective surfaces could be increased. The proposed method could also be extended to the detection of surface crack of metals [24] which is a common occurrence in manufacturing.

References

[1] Pal S K and King R A 1983 On edge detection of x-ray images using fuzzy sets *IEEE Trans. on Pattern Analysis and Machine Intelligence* Vol 5 No1 pp 69-77

[2] Kumar N and Kaur R 2013 Detection of defects in glass using edge detection with adaptive histogram equalization *Int. Journal of Innovative Research in Computer and Communication Engineering* Vol1 pp 1321-1327

[3] Li W, Jiang X and Wang Y 1996 Road recognition and vision navigation of an autonomous vehicle by fuzzy reasoning *Fuzzy Sets and Systems* pp 275-280

[4] Duda R and Hart P 1973 Pattern classification and scene analysis *John Wiley and Sons* pp 271

[5] Canny J 1986 A computational approach to edge detection *IEEE Trans. on Pattern Analysis and Machine Intelligence* Vol 8 No 6 pp 679-698

[6] Russo F 1992 A user friendly research tool for image processing with fuzzy rules *1st IEEE International Conf. on Fuzzy Systems* pp 561-568

[7] Tao C W and Thompson W E 1993 A fuzzy IF-THEN approach to edge detection *Proc. IEEE Conf. on Fuzzy Systems* pp 1356-1360

[8] Begol M and Maghooli K 2011 Improving digital image edge detections by fuzzy systems *World Academy of Science, Engineering and Technology* pp 76-78

[9] W Lee, D Kim and I Kweon 2003 Automatic edge detection method for the mobile robot application *Proc. of the IEEE/RSJ Int. Conf. on Intelligent Robots and Systems* pp 2730-2735

[10] Liang L R and Looney C G 2003 Competitive fuzzy edge detection *Journal of Applied Soft Computing* Vol 3 pp 123-137

[11] Liming H, Cheng H D and Ming Zhang 2007 A high performance edge detector based on fuzzy inference *Information Sciences* pp 4768-4784

[12] Lopez Molina C, Baets B and Bustince H 2011 Generating fuzzy edge images from gradient magnitudes *Computer Vision and Image Understanding* pp 1571-1580

[13] Rosin P L 2001 Unimodal thresholding *Pattern Recognition* pp 2082-2096

[14] Mendel J and John R I 2002 Type 2 Fuzzy sets made simple *IEEE Trans. on Fuzzy Systems* Vol10 pp 117-127

[15] Melin P, Mendoza O and Castillo O 2010 An improved method for edge detection based on interval type 2 fuzzy logic *Expert Systems with Application* Vol 37 Issue 12 pp 8527-8535

[16] Biswas R and Sil J 2011 An improved canny edge detection algorithm based on type 2 fuzzy sets *Procedia Technology* pp 820-824

[17] Gonzalez C, Castro J R, Mellin P and Castillo O 2013 An edge detection method based on generalized type 2 fuzzy logic *IEEE Symp. on Advances in Type 2 Fuzzy Logic Systems* pp 39-44

[18] Gonzalez C I, Castro J R, Mendoza O, Rodriguez Diaz A, Melin P and Castillo O 2015 Edge detection methods based on interval type 2 fuzzy systems for colour images *Fuzzy Information Processing Society (NAFIPS)*, pp 1-6
[19] Gonzalez C I, Mellin P, Castro J R, Castillo O, and Castillo O 2016 Optimization of interval type 2 fuzzy systems for image edge detection *Applied Soft Computing* pp 631-643
[20] Gonzalez C I, Castro J R, Mendoza O, Melin P and Castillo O 2016 Optimization by cuckoo search of interval type 2 fuzzy logic systems for edge detection, *Recent Developments and New Directions in Soft Computing Foundations and Applications* pp 141-154
[21] Mamdani E H and Assilian S 1975 An experiment in linguistic synthesis with a fuzzy logic controller *Int. Journal of Man Machine Studies* Vol 7 pp 1-13
[22] Chavan H L and Shinde S A 201, Defective Product Detection Using Image Processing *Int. Journal of Science and Research* Vol 5 Issue 6 pp 2092-2096
[23] Tiwana S S and Kaur S 2013 Study on Various Glass Defect Using Glass Edge Detection Methods, *Int. Journal of Computer Science and Mobile Computing,* vol.2, no.5, pp 82-85
[24] Sagar S K and Arc X J 2016 Paper Sobel Operated Edge Detection Scheme using Image Processing for Detection of Metal Cracks, *Int. Journal of Comp. Technology and Applications* Vol 9 Issue 37 pp 503-509

IOP Conf. Series: Materials Science and Engineering **530** (2019) 012024 doi:10.1088/1757-899X/530/1/012024

Fuzzy Control of a Low Cost Mobile Robot Based on Vertical Lines in An Edge Detected Image

N H Seng[1*], Z Samad[2] and N M Nor[2]

[1]Mechanical Engineering Department, Politeknik Seberang Perai, Jalan Permatang Pauh, 13500, Seberang Perai, Pulau Pinang

[2]School of Mechanical Engineering, Universiti Sains Malaysia,14300, Jalan Transkrian, Nibong Tebal, Pulau Pinang

Corresponding author *: nhs167@gmail.com

Abstract. Edge detection is a computationally efficient and economical image processing technique as this method retains only the edges of surrounding objects for guiding the navigation and localization of a mobile robot. However, present edge detection based methods are heavily dependent on fusion with odometrical data which gradually accumulates localization errors. Moreover, they lack descriptive capabilities. In this paper, a fuzzy view descriptor which fused the information of the detected adjacent primary colours of the vertical lines in the mobile robot's view with the fuzzified digital compass readings was used to ascertain the actual view. This fusion provided a sense of direction in addition to recognizing the frontal view. The fuzzified distance of the base pixel coordinates of the vertical line of interest determined the attraction of the mobile robot towards it or the repulsion from it. A fuzzy control rule base guided the mobile robot towards the vertical line of interest by constantly reducing its deviation in the edge image. The mobile robot managed to execute its consecutive movements in an intuitive manner without a precise mathematical model. The deviations of the vertical lines were kept small and the detection of the correct views was constantly reliable.

1. Introduction

Movement control is one of the important aspects in the navigational task of a mobile robot. This involves maintaining a steady and reasonably smooth trajectory when approaching a landmark of interest. Various vision systems [1] have been researched and developed to facilitate this task. A number of mapping and localization techniques have been applied in indoor and outdoor navigational tasks in structured and unstructured environments. Recent surveys [2] of vision systems discussed recent advances such as visual simultaneous localization and mapping (VSLAM) based techniques. These systems are largely dependent on the quality of the image processing techniques chosen to accomplish the localization and navigational tasks. Capturing of real-time images are done by several types of cameras such as monocular and stereo cameras [3]. These cameras use perspective projection to represent a three-dimensional coordinates of scene points into two-dimensional image plane. Prior to operation calibration has to be carried out to overcome problems such as radial distortion and determination of the optical centres. Monocular cameras are unable to estimate distances of objects. On the other hand, stereo cameras although able to overcome this problem, often encounter correspondence problems and three-dimensional reconstruction tasks. Epi-polar constraints has to be adopted to ensure that the corresponding points between images lie in a straight line. Computationally intensive algorithms

IOP Conf. Series: Materials Science and Engineering **530** (2019) 012024 doi:10.1088/1757-899X/530/1/012024

such as the eight-point [4] and five-point algorithms [5] require multiple corresponding points which often overload the matrices calculation. They are used to recover the rotation and translation of the cameras attached to the mobile robot.

Several visual processing models have also been introduced to enable a mobile robot to interpret its environment by utilizing point correspondences. For instance, the point detectors introduced by the Harris [6] and SIFT [7] point detectors. The former is not invariant to scale changes while the latter suffers from problems of localization and computational efficiency. Other perceptual models such as the optical flow model [8] encounters erroneous correspondences between moving scene points when encountering obstacles.

Problems such as the need for calibration, radial distortion, computationally intensive algorithms, erroneous identification of corresponding points in successive images make the aforementioned approaches unsuitable for low cost platforms. Moreover, these approaches lack view descriptors that address issues of uncertainty in interpretation of a view.

In this work, an edge detected based technique with descriptive capabilities is used to control the movement of a mobile robot. A fuzzy view descriptor fuses the location information of the vertical lines with their adjacent primary colours with the digital compass readings to ascertain the mobile robot's actual view. It is known that the detection of colours is a straightforward function of a single image and does not need solving problems of correspondences. The fusion of these information offers a less computationally intensive approach while providing reliability pertaining to the view of the mobile robot by accommodating imprecision of the location of the vertical lines of interest and compass directions.

Sugihara [9] pioneered the use of vertical lines in edge detected images for the purpose of localizing a mobile robot. Huttenlocher [10] used the size of objects in the edge images to determine the distance of traverse of the mobile robot. Akihisa et al. [11] used the difference image produced by comparison of the model and captured edge detected images. Matching vertical lines contained in these models were used for navigation, localization and avoidance of static obstacles. Schuster et al. [12] used the vanishing points of straight edges of a ceiling for the translation and rotational movement of a mobile robot. However, this approach was unsuitable for localization tasks. Talluri and Aggarwal [13] imposed a geometric visibility constraint to find a match between a two-dimensional edge and a stored three-dimensional edge in a database to estimate the pose of a mobile robot. A finite state grammar approach [14] was used to combine extracted edge symbols to determine the direction of landmarks. A text extraction method [15] from edge images which was robust to varying lighting in the surroundings and text sizes from various view angles was attempted as a preparation for mobile robot navigation. An efficient matching rate [16] between groups of vertical lines in edge images with a global model was utilized for localization of mobile robots. Zhou and Li [17] utilized vertical lines to estimate the orientation of a camera attached to a mobile robot. Fan [18] applied morphological operation to extract images which yielded edge images that were richer in details compared to standard edge detection methods. Scaramuzza [19] introduced a descriptor in which gradients on both sides of the vertical lines were used to match the vertical lines of consecutive images. This approached demonstrated robustness to changes in illumination and rotation of the mobile robot's view.

The use of edge based images for the purpose of control and navigation of mobile robots also involved the utilization of fuzzy logic. Pan et al. [20] used the extracted vertical lines representing the walls and junctions of a hallway as inputs to a neural network. The outputs were then fed a fuzzy controller to control the direction and distance of traverse of the FUZZY-NAV mobile robot. A method which involved the maximization of the defuzzified values of 3x3 pixel mask [21] for edge detection and the membership function values of triangular output sets were used to detect white line markings on a road. The gradients of edges [22] of a road were used as antecedents of a rule base to control the navigation of a mobile robot along the road. Lee et al. [23] devised an edge detection method with an automatic thresholding capability to detect edges of door frames along a corridor for the purpose of localization. A fuzzy perceptual model was introduced [24] in which a set of fuzzy rules was used to interpret the horizontal and vertical edges of door frames in terms of size, direction and location to localize a mobile robot. Genetic algorithm [25] was used to tune the fuzzy sets and identify the door frames from a large database. Gunes and Baba [26] utilized the 3x3 pixel fuzzy mask to extract the edges of a road. The curvature magnitudes of the edges were used for calculating the speed and speed error to navigate along the road. Kato et al. [27] used the vertical lines in edge images for the purpose of alignment of the position of the mobile robot. An 8x8 pixel mask [28] was used to train a back propagation artificial neural network to detect edges for facilitating the navigational task of a mobile robot. However, sensitivity to vertical edges was unsatisfactory.

The present edge based techniques for the control and navigational tasks lack descriptive capabilities to ascertain a mobile robot's view. For instance, if the mobile robot changes direction after a considerable distance, the viewing of a similar edge structure or features coupled with accumulative odometrical errors may cause erroneous views. Some descriptors [25] have been introduced to describe the edge structures but are dependent on a large database which incurs a computational load. The gradient based [19] descriptor is robust to illumination and has invariance capabilities but the problems lies in the unavailability of the omni-directional camera used.

In this project, a simpler approach is implemented on a low cost mobile robot. A fuzzy view descriptor fuses the location of the vertical lines of interest with the digital compass readings to ascertain its current view. The coordinates of the base pixel of the vertical line of interest in the edge image is used as a reference point to attract and repulse the mobile robot based on the perceived distance.

The integrated fuzzy based system implemented here involves the fuzzy edge detection to detect the vertical lines, the use of a fuzzy rule base to detect the adjacent primary colours of the vertical lines for identification, the fuzzy view descriptor to ascertain the mobile robot' s view and a fuzzy control rule base to steer the mobile robot towards the consecutive vertical lines of interest.

2. Methodology

Figure 1 shows the physical structure of the low cost mobile robot. It is powered up by the DFRduino microcontroller with a built-in direct current motor driver. The 615 LogiTech Web Camera is attached to the front of the mobile robot. Both of these are attached to the USB ports of the M840 Toshiba Sattelite laptop. The Processing software processes the image acquired by the Web Camera. The Arduino software controls the actuation of the pair of wheels of the mobile robot. The HMC 6352 digital compass is attached at the back of the mobile robot to detect the orientation angle of the mobile robot.

Figure1. Mobile Robot

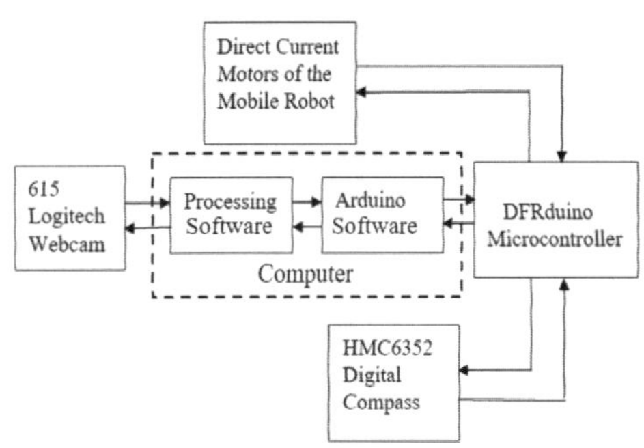

Figure 2. Interfacing of the Complete System

Figure 2 shows the interfacing of the complete system of the mobile robot. The Processing and the Arduino software communicate with each other to enable functionality of the whole system. The real-time image is acquired by the Processing software. The Processing software contains the subroutines and fuzzy rule bases to perform edge detection of the real-time images, to detect the primary colours at the sides of a vertical lines of interest and also to constantly communicate with the Arduino software. It also contains the fuzzy sets for the perception of the distance to the vertical lines of interest. Most importantly it contains the algorithms of the fuzzy view descriptor and the fuzzy rules bases to control the movements of the mobile robot as it approaches its destination.

The Arduino software contains the subroutines to record the orientation readings of the digital compass and the actuation of the direct current motors. The activation signal of the motor subroutines comes from the Arduino software. In the next section, a brief discussion will be given pertaining to the edge detection, primary colour detection process, the fuzzy view descriptor and the control of the motors of the mobile robot.

2.1 Edge Detection

The first step requires the edge detection of the real-time images acquired by the web camera. The detection of the edges of objects, which is a demarcation of different areas of uniform pixel intensities, is a non-linear and ambiguous process. The fuzzy logic based edge detection method proposed here is suitable to address this issue. A group of three-pixel mask is convolved with the real-time image in four directions. These masks consist of arrangements of consecutive bright and dark pixels which can be represented as a fuzzy rule base to detect edges.

Figure 3 shows the group of fuzzy masks on the left side which is convolved with a 3 x 3-pixel segment of the image. The convolution process starts from the upper left and sweeps to the right and subsequent rows and ends at the bottom right segment of the image. The convolution of the fuzzy masks involves the directions associated with the typical pair of compass directions, viz: - the North-South, East-West, North West-South East and the North East-South West directions. Thresholding is performed to retain the edges.

Figure 4 shows the complete edge detection process in which it retains only the edges in the vertical direction. These vertical lines are used as reference edges for the control movements of the mobile robot. A thin line extending from the rightmost vertical line is a reflection of that line on the floor. An algorithm has been incorporated to minimize the appearance of reflected lines. This makes sure that the reflection is minimal and does not affect the descriptive capabilities of the view descriptor.

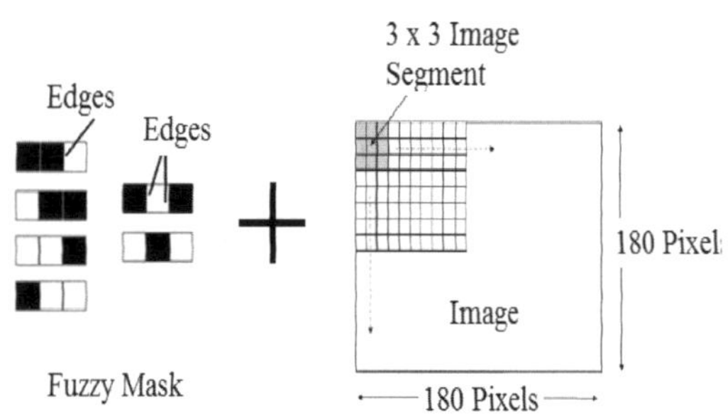

Figure 3. Fuzzy Masks

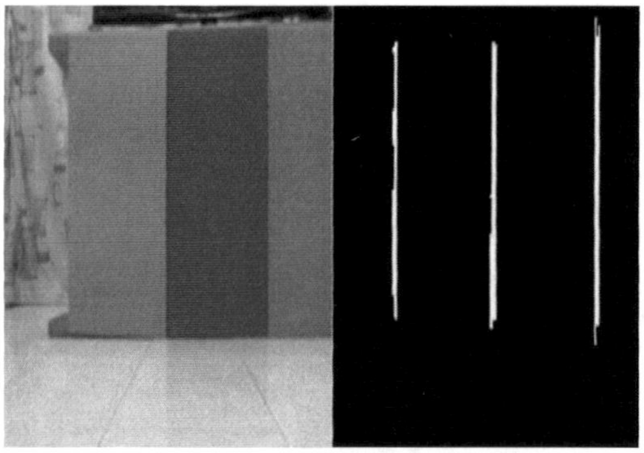

Figure 4. Edge Detection

2.2 Primary Colour Detection

The next stage is the determination of the dominant primary colour of the pixels on the adjacent sides of the vertical lines. The primary colour intensity values are averaged for a width of five pixels on both sides of a vertical line. The left side of Figure 5 shows this procedure clearly. The right side of this figure shows the antecedent sets to fuzzify the primary colour intensity values to obtain their membership functions. Table 1 shows the partial fuzzy rule base to detect the dominant primary colour on both sides of the vertical line of interest. The consequents represent the dominant detected primary colour. There are 32 rules to implement the colour detection task. As can be seen from Table 1, the resulting detected consequent colour is determined by the dominant antecedent colour.

Table 1. Fuzzy If Then Rules for Primary Colour Identification

ANTECEDENTS (IF)				CONSEQUENT
RULE	**RED**	**GREEN**	**BLUE**	**(THEN)**
Rule 1	Low	Low	High	Blue
Rule 2	Low	Low	Very High	Blue
Rule 3	Low	Medium	High	Blue
Rule 4	Low	Medium	Very High	Blue
Rule 5	Low	High	Low	Green
Rule 6	Low	Very High	Low	Green
Rule 7	Low	High	Medium	Green
Rule 8	Low	Very High	Medium	Green
Rule 9	High	Low	Low	Red
Rule 10	Very High	Low	Low	Red
Rule 11	High	Low	Medium	Red
Rule 12	Very High	Low	Medium	Red

Figure 6 shows the fuzzy sets to divide the location of a certain vertical line of interest into the left, right and middle portions of the edge image. The resulting membership function indicates the degree of belongingness of a certain vertical line to a certain location range in the image. These values are used by the fuzzy view descriptor.

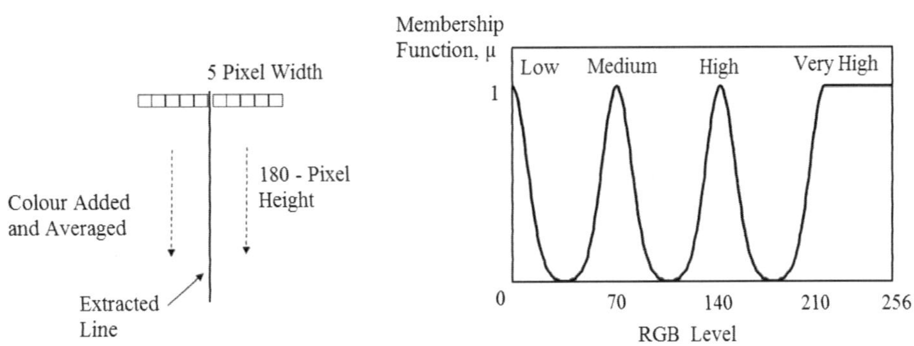

Figure 5. Colour Detection and Antecedents

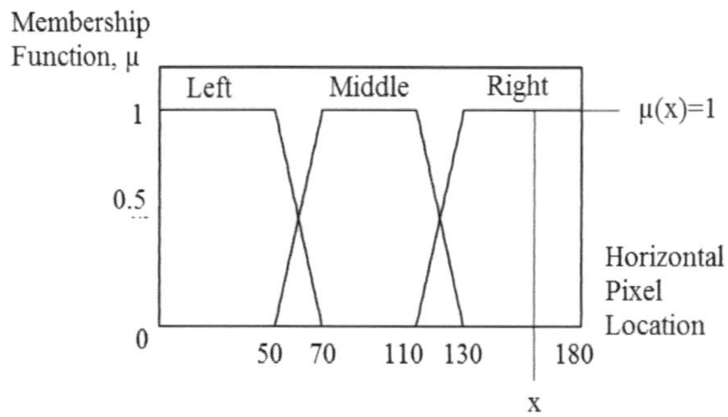

Figure 6. Fuzzification of the Vertical Line Location

2.3 View Descriptor

A fuzzy view descriptor is used to enable the mobile robot to recognize its frontal view. This descriptor fuses the information of the location of the group of vertical lines with similar adjacent primary colours and the fuzzified compass readings. The pseudocodes below show the concept of fusing these two variables to ascertain the actual view. The first group of the pseudocodes detects the presence of the group of vertical lines with the adjacent red colour at the left, middle and right portions of the image.

```
face1a=min(1, midredhigh[1],1);
face1b=min(leftredhigh[1],1,1);
face1c=min(1, 1, rightredhigh[1]);
viewa=max(face1a,face1b,face1c);                (1)
```

In the second group of pseudocodes, the maximum value of one of the left-hand side variables indicate the dominant orientation of the mobile robot. For instance, if the variable 'viewa1' has a unity value and the variables 'viewa2' and 'viewa3' have zero values, this indicates the viewing of a landmark with red colours in the north direction. Provisions could also be made for identifying a certain view from various permissible orientations.

```
viewa1=min(viewa, compassviewnorth);
viewa2=min(viewa, compassviewnorthwest);
viewa3=min(viewa, compassviewwest);             (2)
```

Figure 7 shows the fuzzy sets that fuzzify the readings of the digital compass into their respective membership functions. The trapezoidal fuzzy sets permit a range of values to be associated with a certain direction. A low overlap between the fuzzy sets ensure the unity membership function in a certain direction in most cases.

2.4 Fuzzy Control Algorithms

Once a certain view has been ascertained, a vertical line of interest within the view is chosen as a reference for control. Its horizontal deviation magnitude in pixel counts is sent to the fuzzy control subroutine. The fuzzy control subroutine will steer the mobile robot until the vertical line of interest is brought to the centre of the image. Figure 8 shows the fuzzy sets for the deviation of the vertical line.

Figure 9 shows the deviation of the vertical line of interest from the middle of the image. The vertical line usually deviates to the left or right of the middle axis of the image. The fuzzy control algorithm constantly steers the mobile robot in such a way so as to reduce the deviation of the vertical line of interest from the centre of the image.

IOP Conf. Series: Materials Science and Engineering **530** (2019) 012024 doi:10.1088/1757-899X/530/1/012024

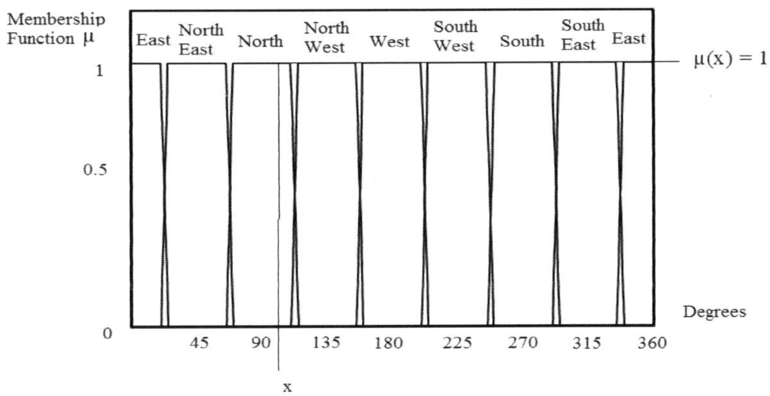

Figure 7. Fuzzification of the Digital Compass Readings

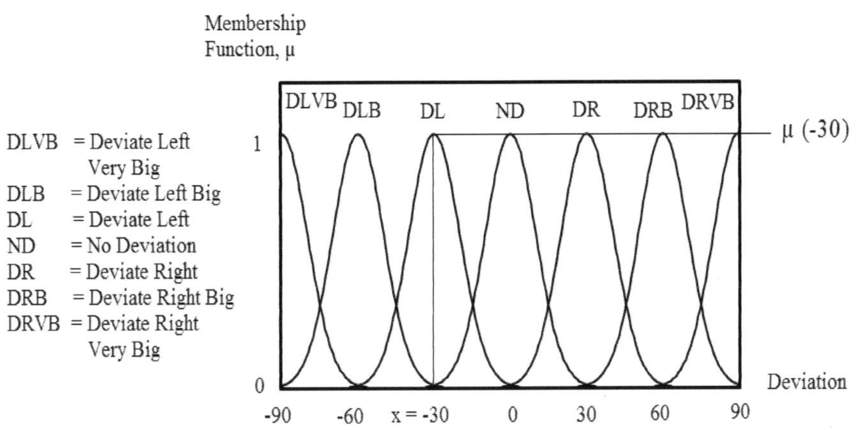

Figure 8. Fuzzification of the Vertical Line Deviation

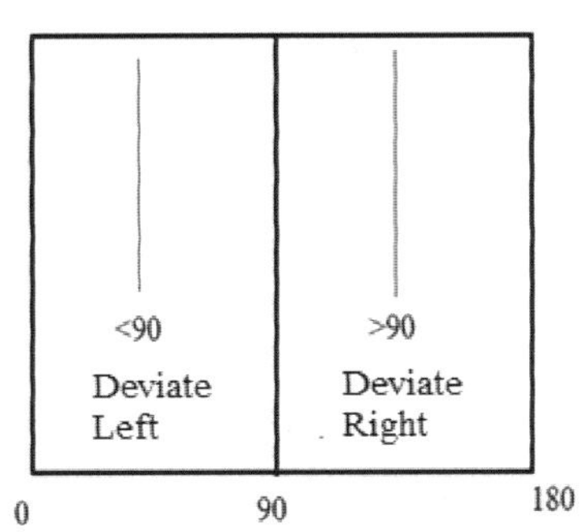

Figure 9. Vertical Line Deviation

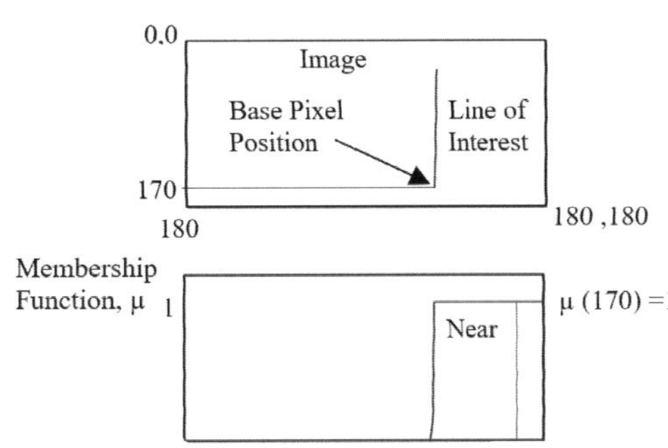

Figure 10. Fuzzification of Base Pixel Position of the Vertical Line

IOP Conf. Series: Materials Science and Engineering **530** (2019) 012024 doi:10.1088/1757-899X/530/1/012024

The next important parameter is the vertical pixel coordinates of the base belonging to the vertical line of interest. The vertical pixel position is fuzzified to enable perception of whether the vertical line is considered as near or far. Figure 10 shows this concept. The upper part of this figure shows the vertical pixel coordinate of the vertical line. When this value is fuzzified by the fuzzy set, the resulting unity membership function indicates that the vertical line is perceived to be near the mobile robot.

Another importance of this fuzzy set is that once a unity membership function has been attained, the mobile robot is deemed to have approached the vertical line of interest and is ready to steer and approach the next vertical line in a new view area. The deviation magnitude of the vertical line of interest and the 'near' membership function value are sent to a fuzzy rule base for inferencing in order to control the steering of the wheels of the mobile robot. Table 2 shows the rule base for steering the mobile robot when approaching a vertical line of interest. When the mobile robot has approached the vertical line, it will steer to the next view area to detect the relevant vertical line. The steering action at this stage is dependent on the orientation of the mobile robot. The activation of the fuzzy control rule base is also dependent on the status of the current view of the mobile robot. Figure 11 shows the consequent fuzzy sets to execute the steering actions.

Table 2. Fuzzy IF THEN Rules for Steering Control

ANTECEDENT (IF)				**CONSEQUENT (THEN)**
Rule Number	**Distance (Pixel Count)**	**Deviation (Pixel Count)**	**Compass Direction**	**Steering Action (Pixel Counts)**
1	**Far**	**None**	**Don't Care**	**No Turning**
2	**Far**	**Right**	**Don't Care**	**Turn Right**
3	**Far**	**Right Big**	**Don't Care**	**Turn Right Big**
4	**Far**	**Right Very Big**	**Don't Care**	**Turn Right Very Big**
5	**Far**	**Left**	**Don't Care**	**Turn Left**
6	**Far**	**Left Big**	**Don't Care**	**Turn Left Big**
7	**Far**	**Left Very Big**	**Don't Care**	**Turn Left Very Big**
8	**Near**	**Don t Care**	**North**	**Turn Left Very Big**
9	**Near**	**Don t Care**	**North East**	**Turn Left Maximum**
10	**Near**	**Don t Care**	**North West**	**No Steer**

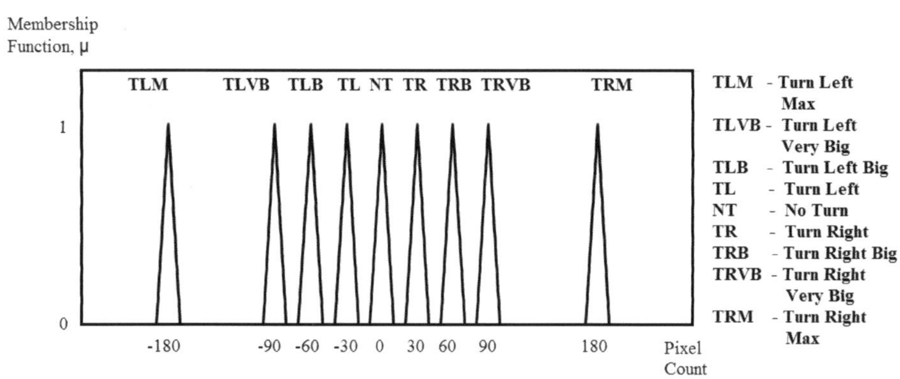

Figure 11. Consequent Sets for Steering Actions

IOP Conf. Series: Materials Science and Engineering **530** (2019) 012024 doi:10.1088/1757-899X/530/1/012024

3. Experiment

Figure 12 depicts the path for the pre-planned movement from the 'start' to 'end' points. The mobile robot moves from its 'start' position and goes towards vertical line 1. When it has approached this vertical line, it then steers to the left and goes to the second vertical line. When it has approached the second vertical line, it then steers to the left again and finally goes towards the location considered as the 'end' area. The mobile robot halts it movement in this area.

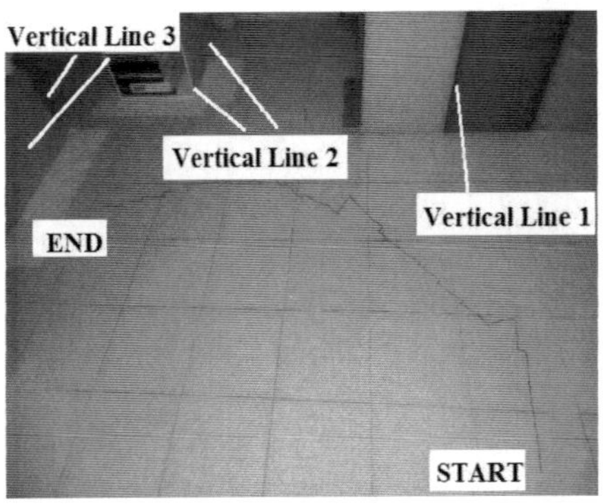

Figure 12. Pre-planned Path

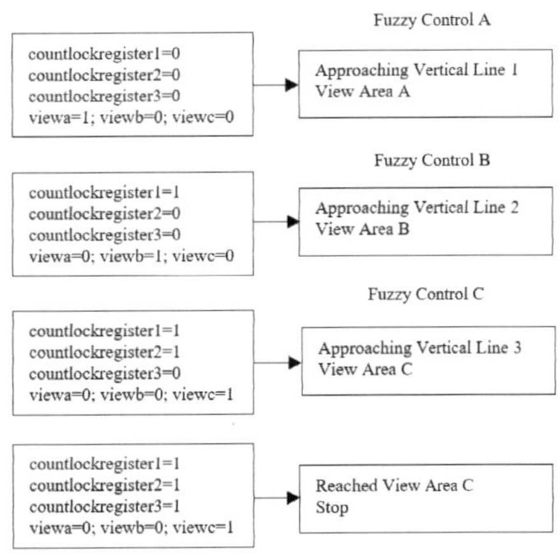

Figure 13. Subroutine Activation

Figure 13 shows the consecutive activation of the fuzzy control subroutines to execute the mobile robot movement from its start to end points. The 'countlockregister' variables play a role in this. In the initial stage all the values of these variables are zero. The fuzzy control 'A' subroutine is activated and the mobile robot moves towards the first vertical line. At this stage the 'viewa' variable value is unity. When it has reached this vertical line, the 'countlockregister1' value changes from zero to one. This indicates that the first vertical line has been reached and is ready to go to the next vertical line. The fuzzy control 'A' subroutine is deactivated and the fuzzy control 'B' subroutine is activated. At this stage, the 'viewb' variable value changes to unity value. When the second vertical line is reached, both the 'countlockregister2'and 'countlockregister 1' values show unity values which indicate that the vertical lines 1 and 2 have been reached. When the second vertical line is reached the mobile robot steers to the left and the fuzzy control 'C' subroutine is activated. The 'viewc' variable becomes unity. Finally, when the vertical line 3 is reached, all the 'countlockregister' variables show unity values which indicate the end of the movement.

4. Results

Figure 14 shows the 10 tracks traced by the mobile robot from the 'start' to the 'end' points. The tracks exhibit a zigzagging path. A closer inspection of track 8, as shown in Figure 15, indicates this clearly. This was caused by the constant corrective steering actions to move the deviated vertical lines of interest back to the centre of the edge image.

Figure 16 shows the consecutive edge images produced as the mobile robot traverses towards the vertical line with the red colour on its right. The bolded data on the right-side display the coordinates of the base pixel belonging to the vertical line of interest. The reduction of the horizontal coordinate of the base pixel shows that the deviation of the vertical line has been reduced. In addition to that, the reduction in the vertical base coordinate values indicates that the mobile robot is approaching the area containing the vertical line of interest. The unity value of the variable 'viewa' indicates that the mobile robot is facing the area which contains the vertical line with the red colour on its right side. The bottom part of the data from photo 1 shows that the mobile robot steers to the right direction by the amount corresponding to a pixel displacement of 5 pixel counts in order to reduce the deviation of the vertical line.

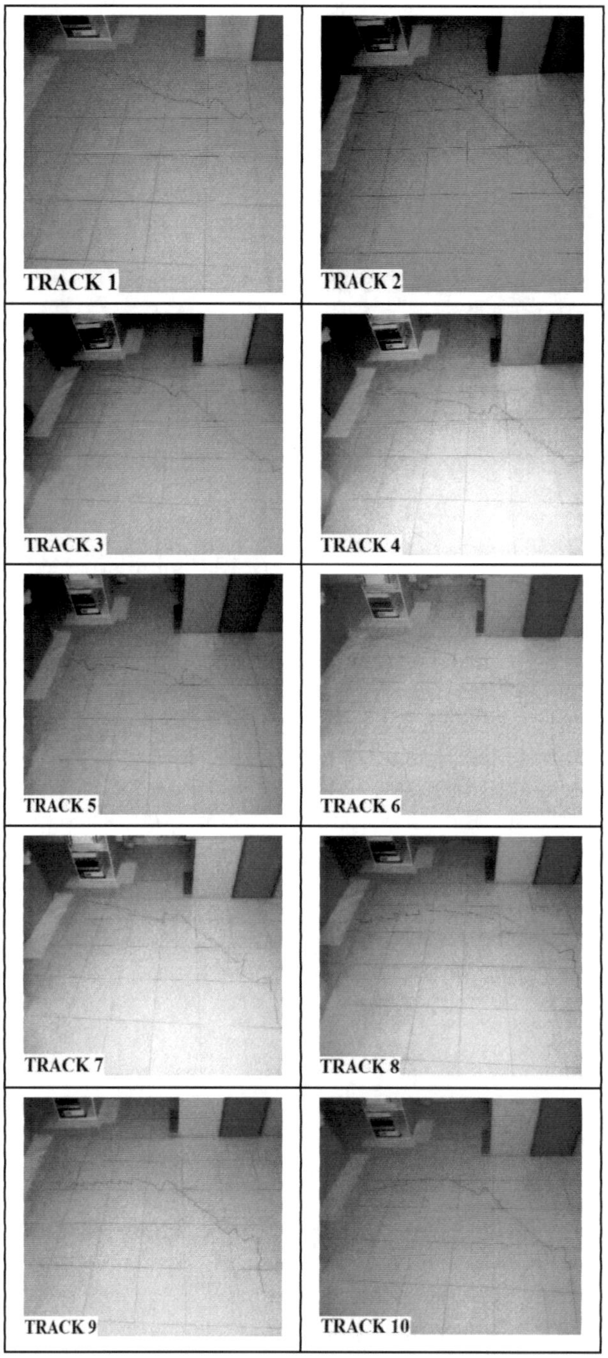

Figure 14. Tracks 1 to 10

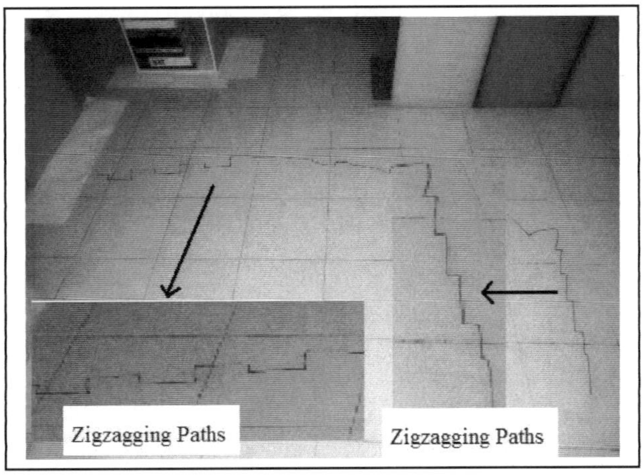

Figure 15. Exploded View of Track 8

Photo No	Edge Detected Photographs	Data
1	(0,0) 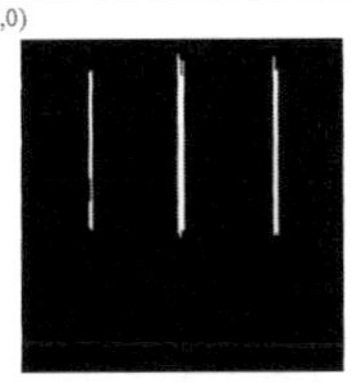 (180,180)	**Setreading** = **1** Compass Angle = 93.0 **xred1[1]=99.5 yred1[1]=118.5** xgreen[1]=8.0 ygreen[1]=71.0 xblue[1]=43.5 yblue[1]=106.75 xblue1[1]=44.0 yblue1[1]=117.333336 **Viewa=1.0** Viewb=0.0 Viewc=0.0 distancea1=118.5 hdistancea1=99.5 Deviation1=9.5 LockRegister1=0.0 LockRegister2=0.0 LockRegister3=0.0 Turn Right =5 Go Straight1 Go Straight1
2	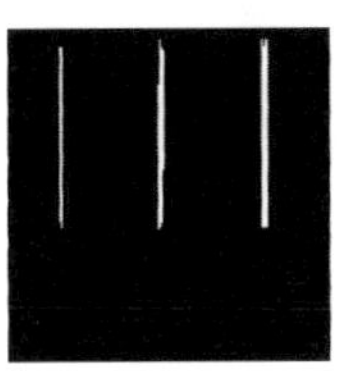	**Setreading** = **2** Compass Angle = 95.0 **xred1[1]=95.5 yred1[1]=112.75** xgreen[1]=159.5 ygreen[1]=122.0 xblue[1]=33.5 yblue[1]=114.25 xblue1[1]=33.5 yblue1[1]=114.25 **Viewa=1.0** Viewb=0.0 Viewc=0.0 distancea1=112.75 hdistancea1=95.5 Deviation1=5.5 LockRegister1=0.0 LockRegister2=0.0 LockRegister3=0.0 Turn Right =2 Go Straight1 Go Straight1

Figure 16. Reduction of Deviation of the Vertical Line of Interest

Photo No	Edge Detected Photographs	Data
4	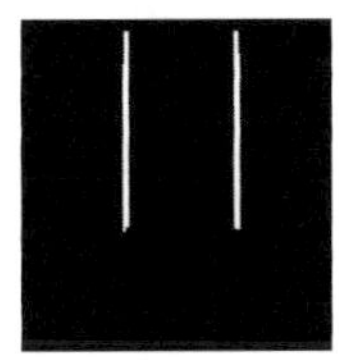	**Setreading** = **4** Compass Angle = 91.0 **xred1[1]=66.5 yred1[1]=128.5** xgreen[1]=138.0 ygreen[1]=128.0 xblue[1]=139.0 yblue[1]=128.0 **Viewa=1.0** Viewb=0.0 Viewc=0.0 distancea1=128.5 hdistancea1=66.5 Deviation1=-23.5 **LockRegister1=1.0 LockRegister2=0.0** **LockRegister3=0.0** Turn Left =2090 Go Straight1 Go Straight1
5	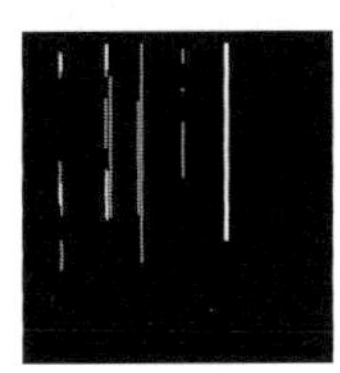	**Setreading** = **5** Compass Angle = 101.0 **xgreen[1]=52.5 ygreen[1]=102.0** **xgreen1[1]=74.0 ygreen1[1]=112.0** xblue[1]=55.0 yblue[1]=117.666664 xblue1[1]=131.5 yblue1[1]=125.75 Viewa=0.0 **Viewb=1.0** Viewc=0.0 distanceb1=112.0 hdistanceb1=74.0 Deviation2=-16.0 LockRegister1=1.0 LockRegister2=0.0 LockRegister3=0.0 Turn Left =2017 Go Straight1 Go Straight1

Figure 17. Transition of Views

Figure 17 shows the consecutive edge images which indicate how the mobile robot steers to the left after approaching the first vertical line. It can be seen that the vertical coordinate of the base pixel belonging to the first vertical line, from photo 4, is 129. This has exceeded the threshold value of 125 which can be considered near to the mobile robot. At this stage, the 'lockregister' value changes to 1 and the mobile robot steers to the left in order to get a new view. From photo 5, it can be seen that the variable 'viewb' becomes1. The pixel coordinates of the vertical lines with the green colour on the left and right sides are displayed. Since the horizontal pixel coordinate of the vertical line of interest, represented by the variable 'xgreen1[1]' has deviated to the left side of the middle axis, the mobile robot steers to the left with a corresponding displacement of 17-pixel counts. (ignore the first 2 digits) to reduce the deviation. It is also noticeable that the coordinates of the base pixels of the vertical lines of

IOP Conf. Series: Materials Science and Engineering **530** (2019) 012024 doi:10.1088/1757-899X/530/1/012024

interest are displayed in decimal points. The decimal points result from the averaging of the clumped lines with similar adjacent colours.

To evaluate the performance of the fuzzy control algorithms, the average line deviations of the10 tracks are calculated. Figure 18 shows the deviation pattern of the vertical line of interest in an edge image for track 8. The zigzagging pattern indicates that the mobile robot always attempts to steer itself in such a way so as to enable the reduction of the vertical line deviations. The deviation values appear to hover around the horizontal pixel coordinate of 90, which is considered the middle position of the edge image.

Table 3 shows the deviations for each stage of movement involving all the 10 tracks. The average deviation values for all the tracks are also shown. Most importantly, the overall average deviation value is a small value of 10 pixels, which indicates that the vertical line positions are constantly adjusted to be kept near the middle of the edge Image. This ensures directedness of the mobile robot towards a destination point.

Table 4 shows how the fuzzy control algorithm responds by producing the steering magnitudes which approximate the deviations of the vertical lines in pixel counts. The steering actions generally alternate between turning right and left on consecutive step movements. It is also noticeable that the magnitude difference between the steering magnitudes and the deviations of the vertical lines are not more than 5-pixel counts.

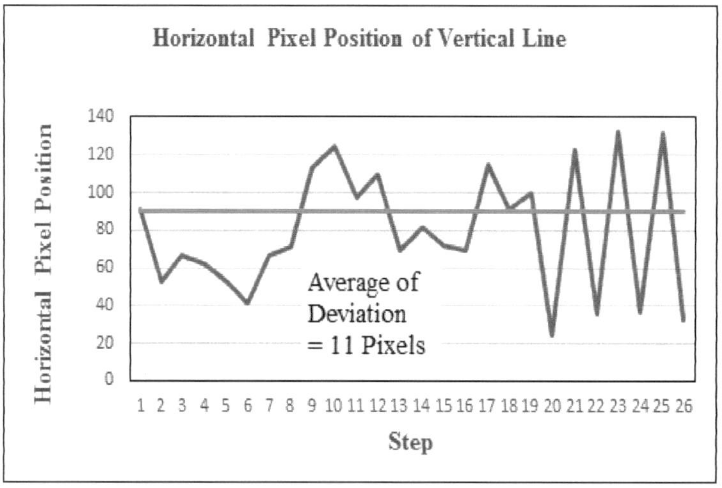

Figure 18. Deviation of Track No.8

IOP Conf. Series: Materials Science and Engineering **530** (2019) 012024 doi:10.1088/1757-899X/530/1/012024

Table 3. Absolute Average of Deviation for Each Stage of Movement

Track Number	Absolute Average of Deviation (Start to Line 1)	Absolute Average of Deviation (Line 1 Line 2)	Absolute Average of Deviation (Line 2 to Line 3)	Absolute Total Average of Deviation
1	7	26	14	10
2	12	7	24	14
3	10	1	5	3
4	16	1	4	5
5	3	24	9	12
6	6	14	0	6
7	21	22	3	13
8	28	3	10	11
9	10	12	27	10
10	9	6	34	15
Absolute Average Deviation	11	7	13	10

Table 4. Deviation and Turn Angle (Track 1)

Step Number	Deviation (In Pixels)	Turn Angle (In Pixels)	Difference (In Pixels)	Absolute Difference (In Pixels)
1	-5.0	Turn Left = -1	4.0	4
2	-62.5	Turn Left = -61	1.5	2
3	28.5	Turn Right=+30	1.5	2
4	1.5	Turn Angle =0	1.5	2
5	36.5	Turn Right=+32	4.5	5
6	35.0	Turn Right=+31	4.0	4
7	15.0	Turn Left = -90	Not Applicable	Not Applicable
8	29.0	Turn Right=+30	1.0	1
9	-52.0	Turn Left = -57	5.0	5
10	-69.5	Turn Left = -65	4.5	5
11	39.0	Turn Right=+34	5.0	5
12	-7.7	Turn Left = -90	Not Applicable	Not Applicable
13	-3.75	Turn Left = -1	2.75	3
14	-27.5	Turn Left = -29	1.5	2
15	8.5	Turn Right=+4	4.5	5
16	-21.5	Turn Left = -26	4.5	5
17	-21.0	Turn Left = -26	5.0	5
18	-46.0	Turn Left = -47	1.0	1
19	9.0	Turn Right=+4	5.0	5
20	-53.75	Turn Left = -58	4.25	4
21	16.75	Turn Right=+19	2.25	2
22	-62.75	Turn Left = -61	1.75	2
23	-13.0	Turn Left = -11	2.0	2

5. Conclusions and Recommendations

An integrated fuzzy system has been implemented in this project. The fuzzy edge detection system processed the captured images and retained the vertical lines of objects. The fuzzy colour detection rule base enabled recognition of primary colours on both sides of the vertical lines in the edge image. The fuzzy view descriptor used the information of the location of the vertical lines of interest with the orientation readings of the digital compass. This enabled the mobile robot to ascertain its view and move towards the selected vertical line of interest within its view.

In the future, the fuzzy view descriptor will be refined to enable recognizing landmarks having multiple vertical lines with various combinations and sequences of the adjacent colours. Invariance capabilities will be incorporated to enable recognition of landmarks from various view angles in more challenging environments.

In addition to these features, localization capabilities may be incorporated to enhance the overall capabilities of the mobile robot. This may include simultaneous localization and mapping operations utilizing vision systems. Type-2 fuzzy logic [29] may also be used to improve the control movements by further reducing the deviations of the tracks.

References

[1] DeSouza G N and Kak A C 2002 Vision for mobile robot navigation: a survey *IEEE Trans. on Pattern Analysis and Machine Intelligence* Vol 24 No 2 pp 237-67

[2] Nawang L, Biswaraj S and Kiran G 2016 Survey: visual navigation for mobile robot *Proc. Int. Conf. on Computing and Communication (Munich)* pp 5-11

[3] Siegwart R, Nourbakhsh I R and Scaramuzza D 2011 *Introduction to autonomous mobile robots* MIT Press Cambridge Massachusetts USA

[4] Higgins H C L 1981 A computer algorithm for reconstructing a scene from two projections *Nature* Vol 293 pp 133-135

[5] Nister D 2004 An efficient solution to the five-point relative pose problem *IEEE Trans. on Pattern Analysis and Machine Intelligence pp 756-770*

[6] Harris C and Stephens M 1988 A combined corner and edge detector *Proc. of the 4th Alvey Vision Conference* pp 147-151

[7] Lowe D G 2004 Distinctive image features from scale invariant key points *Int. Journal of Computer Vision,* Vol 60 Issue 2 pp 91-110

[8] Mai N A and Janschek K 2012 Performance results of qualitative optical flow processing integrated in fuzzy logic based behavioural mobile robot control *10th iFAC Symp. on Robot Control International Federation of Automatic Control* pp 895-901

[9] Sugihara K 1988 Some location problems for robot navigation using a single camera *Computer Vision, Graphics and Image Processing* Vol 42 pp 112-129

[10] Huttenlocher D P, Leventon M E and Rucklidge W J 1990 Visually guided navigation by edge images *Algorithmic Foundation of Robotics* pp 85-96

[11] Akihisa O, Akio K and Avi K 1998 Vision-based navigation of a mobile robot with obstacle avoidance by single camera vision and ultrasonic sensing *IEEE Trans. on Robotics and Automation* Vol14 pp 969-978

[12] Schuster R, Ansari N and Hashemi A B 1993 Steering a robot with vanishing points *IEEE Trans. on Robotics and Automation* Vol 9 pp 491-498

[13] Talluri R and Aggarwal J K 1996 Mobile robot self-location using model-image feature correspondence *IEEE Trans. on Robotics and Automation* Vol12 pp 63-77

[14] Lee J D 1997 Indoor robot navigation landmark tracking *Mathematical Modelling* Vol 26 pp 78-89

[15] Liu X and Samarabandu J 2005 An edge-based text region extraction algorithm for indoor mobile robot navigation *Proc. of the IEEE International Conference on Mechatronics &Automation* pp 701-706

[16] Aider O A, Hoppenot P and Colle E 2005 A model-based method for indoor mobile robot localization using monocular vision and straight-line correspondences, *Robotics and Autonomous Systems* Vol 52 pp 229-246

[17] Zhou J and Li B 2007 Exploiting vertical lines in vision-based navigation for mobile robot platforms *IEEE Int. Conf. on Acoustics, Speech and Signal Processing* Honolulu USA pp 465-468

[18] Fan Y, Cui G and Lei F 2009 Application of edge detection algorithm based on morphology in robot vision system *Int. Conf. on Intelligent Human-Machine Systems and Cybernetics* pp 304-307

[19] Scaramuzza D, Siegwart R and Martinelli A 2009 A robust descriptor for tracking vertical lines in omnidirectional images and its use in mobile robotics *Int. Journal of Robotics Research* pp 149-171

[20] Pan J, Pack D J, Kosaka A and Kak A C 1995 FUZZY-NAV: A vision-based robot navigation architecture using fuzzy inference for uncertainty reasoning *Proc. of World Congress on Neural Networks* Vol 2 pp 602-60

[21] Li Y, Lu G and Wang Y 1997 Recognizing white line markings for vision-guided navigation by fuzzy reasoning *Pattern Recognition* Vol 18 pp 771-780

[22] Castro A P A, Da Silva J D S and Simoni P O 2001 Image based autonomous navigation with fuzzy logic control *Proc. on the Int. Joint Conf. on Neural Networks* Vol 3 pp 2200-2205

[23] Lee W, Kim D and Kweon I 2003 Automatic Edge Detection Method for the Mobile Robot Application *Proc. of the IEEE/RSJ Int. Conf. on Intelligent Robots and Systems* pp 2730-2735

[24] Salinas R M, Aguirre E, Silvente M G and Gonzalez A 2004 Door detection using computer vision and fuzzy logic *Proc. of the 6th WSEAS Int. Conf. on Mathematical Methods and Computational Techniques in Electrical Engineering* pp 1-6

[25] Salinas R M, Aguirre E and Silvente M G 2006 Detection of doors using a genetic visual fuzzy system for mobile robots *Autonomous Robots, Springer Link* Vol 21 pp 123-141

[26] Gunes M and Baba A F 2009 Speed and position control of autonomous mobile robot on variable trajectory depending on its curvature *Journal of Scientific and Industrial Research Robots,* Vol 68 pp 513-521

[27] Kato T, Watanabe K and Maeyama S 2011 Image-based trajectory tracking with fuzzy control for nonnolonomic mobile robots *37th Annual Conf. on IEEE Ind. Electronics Society* pp 3299-3304

[28] Damaryam G K 2016 A back propagation neural network for detection of vertical sub-lines *The Int. Journal of Engineering and Sciences* pp 22-29

[29] Mendel J and John R I 2002 Type 2 Fuzzy sets made simple *IEEE Trans. on Fuzzy Systems* Vol10 pp 117-127

IOP Conf. Series: Materials Science and Engineering **530** (2019) 012025 doi:10.1088/1757-899X/530/1/012025

Study on End Brush Deburring and Sintered Diamond Ball Deburring of Micro-features Milled

T K Sai, K Saptaji* and N Fatchurrohman

Faculty of Manufacturing Engineering, University Malaysia Pahang, 26600 Pekan, Pahang Darul Makmur, Malaysia

Corresponding author *: kushendarsyah@ump.edu.my

Abstract. The presence of top burr in the micro-features produced by milling process can deteriorate and affect the surface quality of the products. Though there are some deburring methods can be used and reported successfully remove the burr in the micro scale features, however simpler and alternative method is still needed. Two of the deburring methods that can be used are end brushing and sintered diamond ball methods. Therefore the aim of this paper is to study the application of the end brush and sintered diamond ball for deburring top burrs exist on micro features produced by milling process. Slot milling experiments were conducted and subsequently deburring process using the two methods was conducted. Micro scale features were also produced using milling process followed by deburring. The deburred results were observed visually using optical microscope and Scanning Electron Microscope (SEM) and surface roughness was measured. Experimental results show that the end brush and sintered diamond ball deburring have successfully remove the top burrs. The end brush deburring method produced better surface quality compared to the sintered diamond ball.

1. Introduction

Burr is defined as the material plastic deformation produced at workpiece edges as a result of machining or shearing process [1]. The presence of burrs affects the quality and the appearance of the machined parts such as the dimension accuracy, surface quality, fit and ease of assembly [2]. The real profile of the burr such as height, width and length; is sometimes difficult to quantify due to the complex shape of the burrs and the limitation of measurement methods such as using stylus method or optical method. In the machining of micro scale features, the size of the burrs is more significant and comparable with the size of the features being produced. Burrs normally occur when mechanical material removal process such as micro-milling used in the production of micro scale features.

The formation of burrs in the micro-milling process can be reduced or even eliminated by two methods; by using the optimum cutting parameters or by deburring process. Changes in process parameters can be adopted to reduce burr formation. Several researchers have reported the effect of cutting parameters and optimization of cutting parameters for various metallic alloys. In micro milling aluminum alloys, the burrs can be minimized by using high spindle speed of about 60,000 rpm and low feed per tooth between 1.0 μm and 2.0 μm [3]. In micro-milling of aluminum alloys Al2124 and stainless steel SS-304, the burr height can be decreased by increasing feed, tool diameter and number of flutes [4]. Lekkala *et al.* [4] argued that the speed has less significant effect on the burr thickness and height whereas tool diameter, depth of cut, number of flutes and the interaction between feed rate and number of flutes have significant effect on the burr height in the micro-milling process.

Deburring is defined as the removal of minute amounts of material from edges after major part features have been produced [5]. Deburring process is difficult to apply on micro-features produced by micro-milling; the process must be carefully conducted to avoid damage to the small features. Incorrect selection of deburring techniques or parameters may also introduce dimensional errors, damage, poor surface finish and residual stresses. The deburring method is normally dependent on the burr locations and has to be based on burr characteristics together with part properties [5].

There are various deburring methods used in the micro-features production especially micro slot using micro-milling reported previously such as using electrochemical polishing [6], diamond milling [6], micro-EDM [7], powder blasting [8], micro-peening [9], supporting material [10] and tapered tool [11]. Alternatively, the end brush [5] and diamond sintered ball can also be used for deburring of micro-features. There are not much literature have reported on the applications of end brush and diamond sintered ball for deburring. Hence, this paper aims to study the effectiveness of two deburring methods namely, stainless steel end brush and diamond sintered ball. These two methods are simple deburring methods that can be applied easily especially on the CNC milling machine.

2. Experimental Setup

In this experiment, two carbide end mill with diameter of 4 mm and 2 mm were used as cutting tool and Aluminum Al6061-T6 was used as workpiece. Experiments were conducted using Makino Ke55 CNC milling machine. At first, the preliminary slotting test was performed in order to have optimum cutting parameters. There are about 16 slot produced using different parameters. Deburring process was performed using 2 different deburring methods namely stainless steel end brush, and diamond sintered ball (Figure 1). The deburring experiments were conducted on the slots without removing the workpiece from the workpiece holder vise. Hence the proposed deburring methods are convenient and efficient to be applied especially for the milling process in producing slots where the deburring can be applied without removing the workpiece. Subsequently, some micro-fluidic device cavities were produced based on the optimal parameters. The micro-fluidic cavity design consists of straight protruded wall cross section and conical protrusion. Visual observation of burr formation and surface roughness measurements were conducted using optical microscope, Scanning Electron Microscope (SEM).

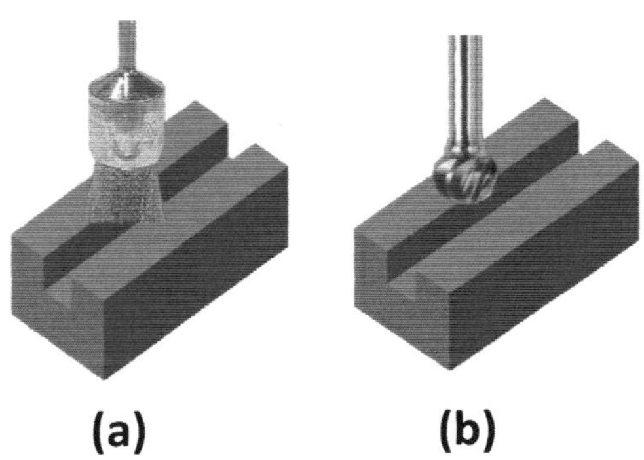

Figure 1. Two deburring methods tested in this experiment, (a) end brush, and (b) diamond sintered ball

2.1. Machining Parameter for slotting

There are four factor used in slotting experiment, tool diameter, depth of cut, spindle speed and feed rate. The slotting parameters with their level were shown in Table 1.

Table 1. Machining parameters and factor levels

Factors	**Unit**	**Factor Levels**	
		1	**2**
A. Tool diameter	mm	4	2

B. Depth of cut	μm	100	200
C. Spindle speed	rpm	3,000	4,000
D. Feed rate	mm/min	10	100

2.2. Deburring Parameters

There are 3 factors that used in deburring which were deburring tool, depth of cut, and spindle speed. The deburring parameters with their level were shown in Table 2.

Table 2. Deburring parameters and factor levels

Factors	Factor levels		
	1	**2**	**3**
A. Deburring Tools	End brush	Diamond sintered ball	Diamond sintered cylinder
B. Depth of Cut (mm)	0.100	0.200	-
C. Spindle Speed (rpm)	2000	4000	-

2.3. Design of experiment

Taguchi orthogonal array (OA) L_{16} was used as the design of experiment for slotting experiment. The experiment layout of L16 was generated by using Minitab software and shown in Table 3. In addition, 1 factor with four level and 2 factor with two level were used in deburring experiment resulted in a total of 12 experiments layout as shown in Table 4.

Table 3. Experiment layour of L_{16} OA for slotting

Slot Number	Tool Diameter (mm)	Depth of Cut (mm)	Spindle Speed (rpm)	Feed rate (mm/min)
1	4	0.1	3000	10
2	4	0.1	3000	100
3	4	0.1	4000	10
4	4	0.1	4000	100
5	4	0.2	3000	10
6	4	0.2	3000	100
7	4	0.2	4000	10
8	4	0.2	4000	100
9	2	0.1	3000	10
10	2	0.1	3000	100
11	2	0.1	4000	10
12	2	0.1	4000	100
13	2	0.2	3000	10
14	2	0.2	3000	100
15	2	0.2	4000	10
16	2	0.2	4000	100

Table 4. Experiment layout for deburring

Deburring Number	Deburring Tools	Spindle Speed (rpm)	Depth of cut (mm)
1	End brush	4000	0.00
2	End brush	4000	0.10
3	End brush	2000	0.00
4	End brush	2000	0.10

5	Diamond sintered ball	4000	0.00
6	Diamond sintered ball	4000	0.10
7	Diamond sintered ball	2000	0.00
8	Diamond sintered ball	2000	0.10
9	Diamond sintered cylinder	4000	0.00
10	Diamond sintered cylinder	4000	0.10
11	Diamond sintered cylinder	2000	0.00
12	Diamond sintered cylinder	2000	0.10

3. Results and Discussion

3.1. Burr formation for slotting

The result of the slots having the most and least burrs are shown in Figure 2. The most burrs was occurred for the slot #5 with the cutting parameters for this slot were tool diameter 4 mm, cutting depth of 200 μm, spindle speed 3,000 rpm and feed rate 10 mm/min. The slot #12 has the least burr formation as shown in Figure 2 (b). The machining parameters for this slot are tool diameter of 2 mm, cutting depth 100 μm, spindle speed 4,000 rpm and feed rate 100 mm/min.

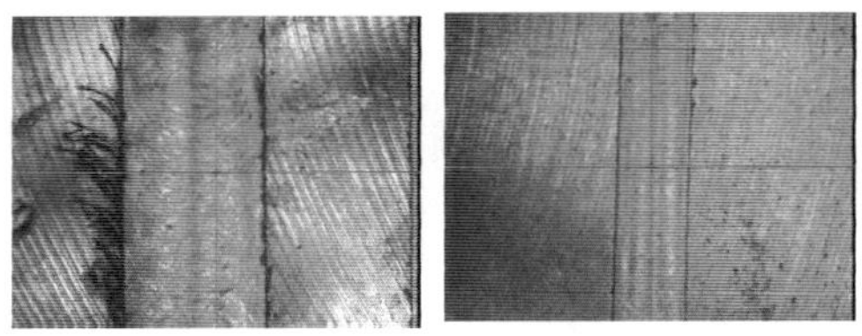

(a) (b)

Figure 2. Slot results having (a) the most burrs and (b) the least burrs

It can be seen that the least burrs formed when the smaller diameter of tool was used and the lower depth of cut was applied. Moreover, the highest the spindle speed with the combination of the highest feed rate will also contribute to the lowest formation of burrs. According to the Wang and Zhang [12], the burr height increased with the depth of cut and feed rate. Besides, the burr height decrease when the spindle speed was increased [13].

3.2. Surface roughness for each slot

The surface roughness measurements result of the slot is shown in Figure 3. It can be seen from Figure 3 that the slot number 12 with the tool diameter 2 mm, cutting depth 100 μm, spindle speed 4,000 rpm and the feed rate 100 mm/min has the lowest surface roughness value which is 0.133 μm. In contrast, the slot number 7 has the highest surface roughness value which is 0.590 μm. The machining parameters for the slot number 7 were tool diameter of 4 mm, depth of cut of 200 μm, spindle speed of 4,000 rpm and feed rate of 10 mm/min. By comparing these two results, it can clearly be observed that with the criteria of bigger tool diameter, higher depth of cut, higher spindle speed and lower feed rate will cause a poor surface finish. This result was in line with the observation of Boswell et. al. [14], stated that the better surface finish can obtain with the combination of high spindle speed and the high feed rate.

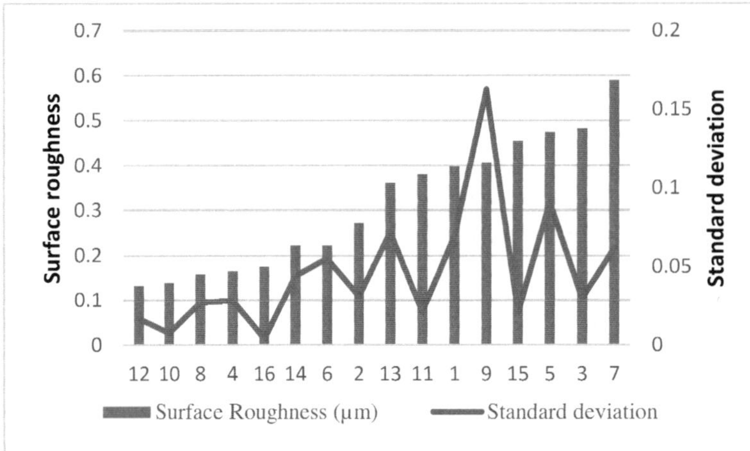

Figure 3. Surface roughness values for all slots

3.3. Deburring slot

Deburring experiments were conducted for the slot produced by 2 mm tool diameter, 4,000 rpm spindle speed, 200 µm depth of cut and 100 mm/min feed rate. 16 slots were produced for the deburring experiments.

a. End Brush

The result of the most and the least burrs remained on the slot after deburring by end brush is shown in Figure 4. The most burrs remain on the wall of slot after deburring using end brush is resulted on deburring number 3 (Table 4). The parameters used for this deburring are spindle speed 2,000 rpm and depth of cut 0.00 mm (the end of brush is levelled with the top surface). In contrast, the least burrs remain on the slot was deburring number 2. The parameters for this slot are the spindle speed of 4,000 rpm and the depth of cut of 100 µm. This can be occurred because when deburring with certain depth of cut, the end brush will cover the top and the wall of the slot this can make sure the top burr and side burr was fully remove.

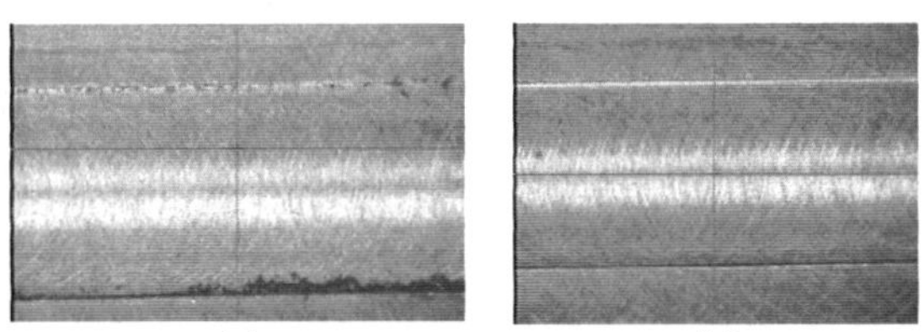

(a) (b)

Figure 4. Comparison of deburring slot using end brush (a) the most burr remain b) the least burr remain

Figure 5 shows the average surface roughness of the slot after deburring by end brush. It can be seen that the deburring slot number 2 has the lowest surface roughness with the parameters of 4,000 rpm for spindle speed and 100 µm depth of cut. This was due to the end brush have contact with the slot with a high speed. In contrast, deburring slot number 1 slot has the highest surface roughness with the parameters of 4,000 rpm spindle speed and 0.00 mm depth of cut. It can be seen that the depth of cut has an important role in the removing of burr and the surface finish.

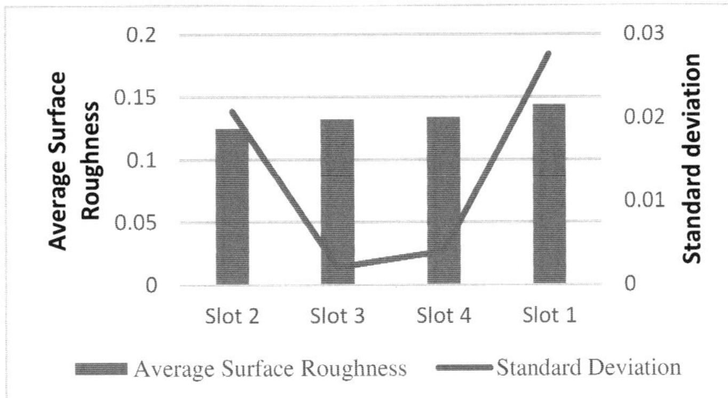

Figure 5. Average surface roughness for slots deburred by end brush.

b. *Diamond Sintered Ball*

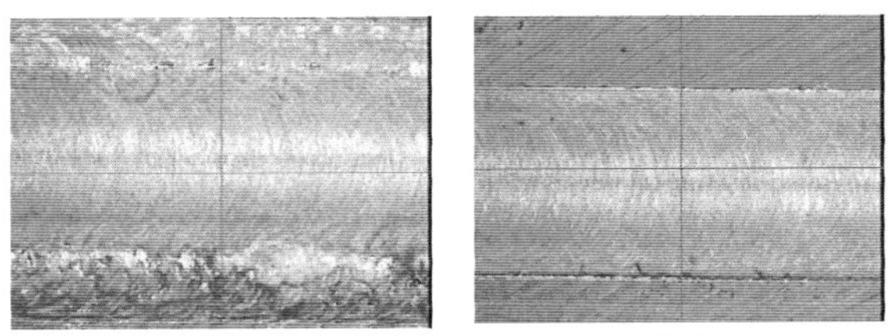

(a) (b)

Figure 6. Comparison of deburring slot using diamond sintered ball (a) the most burrs remain b) the least burrs remain

Figure 6 shows the most and the least burrs remain on the slot after deburring using diamond sintered ball. The most burrs remain on the slot was produced on deburring slot number 6 which produced using 4,000 rpm spindle speed and 100 μm depth of cut. In contrast, the least burrs produced using 4,000 rpm spindle speed and 0.00 mm depth of cut. It can be seen that using minimum depth of cut, the second burrs was formed on the slot and causes more burrs. The result is the same for diamond sintered cylinder used as deburring tools.

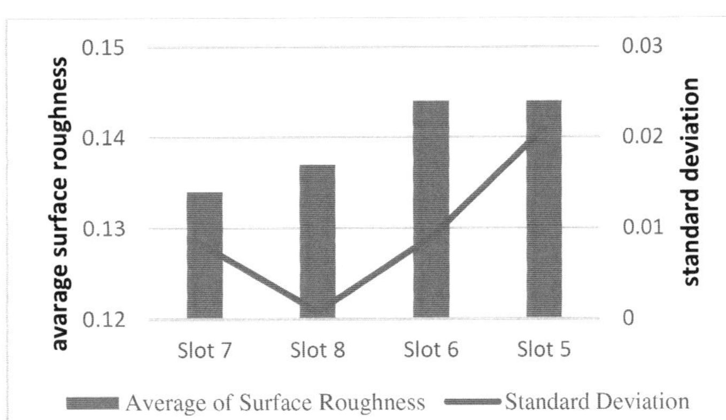

Figure 7. Average surface roughness for slots deburred by diamond sintered ball

Figure 7 shows the average surface roughness of the deburring slots using diamond sintered ball. It can be seen that deburring slot number 7 has the lowest surface roughness. The spindle speed and the depth of cut for this slot are 2,000 rpm and 0.00 mm. On the other hand, the deburring slot number 5 has the highest surface finish with the spindle speed of 4,000 rpm and 0.00 mm for depth of cut. From this result the higher the spindle speed of the higher the surface roughness of the slot.

3.4. Micro-fluidic devices cavity

Based on previous results, the micro-fluidic devices cavities were produced and deburring process by various methods with their optimum deburring parameter was applied. The result of micro-fluidic cavities and the deburred micro-fluidic device cavity are shown in Figure 8. It can be seen that the proposed deburring methods using end brush and diamond sintered ball have successfully removed the burr at the edge of wall. However, the used of diamond sintered ball resulted in the traces of diamond sintered ball mark. Hence, the end brush deburring method shows better results in term of surface quality compared to deburring using diamond sintered ball.

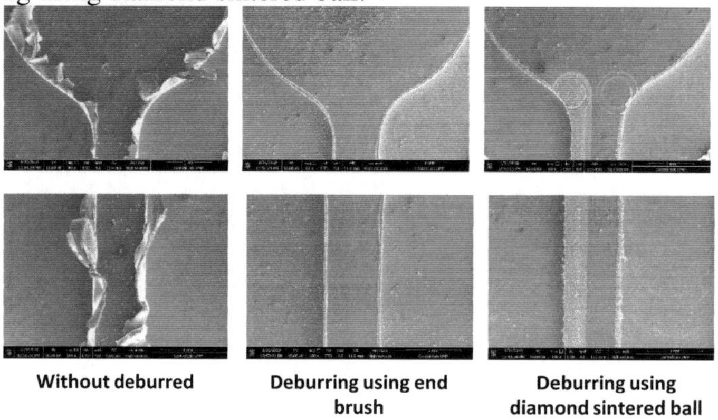

Figure 8. SEM images of the micro-fluidic device cavities produced without deburring and various deburring methods

4. Conclusion

The main conclusions of this paper are:

1. Slotting and deburring experiments have been carried out on aluminium alloys with various parameters. In this paper the optimum slotting and deburring parameters have been figured out.
2. The optimum machining parameters for the least burr formation in this experiment were 2 mm tool diameter, 100 μm depth of cut, 4,000 rpm spindle speed and 100 mm/min feed rate.
3. The optimum deburring parameters by using end brush when using 4,000 rpm spindle speed and 100 μm depth of cut.
4. The optimum deburring parameters by using diamond sintered burr ball are found when using 4,000 rpm spindle speed and 0.00 mm depth of cut.
5. The deburring method using end brush visually shows better result compared to diamond sintered ball method.

Acknowledgements

This research is supported by research grant RDU 180324 from Universiti Malaysia Pahang.

References

[1] L K Gillespie 1999 Deburring and edge finishing handbook *Dearborn, Michigan: Society of Manufacturing Engineers*

[2] C Gwo-Lianq 2006 Experimental observation and analysis of burr formation mechanisms in face milling of aluminum alloys *International Journal of Machine Tools & Manufacture* Vol 46 No 12–13 pp 1517–1525

[3] J Koo, K Park, Y Kim and J Kim 2017 Investigation on the correlation between micro burrs and

AE signal characteristics in micro-scale milling process *International Conference on Applied System Innovation (ICASI)* pp 327–330

[4] R Lekkala, V Bajpai, R K Singh and S S Joshi 2011 Characterization and modeling of burr formation in micro-end milling *Precision Engineering* Vol 35 No 4 pp 625–637

[5] L K Gillespie 1979 Deburring precision miniature parts *Precision Engineering* Vol 1 No 4 pp 189–198

[6] T Schaller, L Bohn, J Mayer and K Schubert 1999 Microstructure grooves with a width of less than 50 μm cut with ground hard metal micro end mills *Precision Engineering* Vol 23 No 4 pp 229–235

[7] Y H Jeong, B HanYoo, H U Lee, B -K Min, D -W Cho and S J Lee 2009 Deburring microfeatures using micro-EDM *Journal of Materials Processing Technology* Vol 209 No 14 pp 5399–5406

[8] D Yun, T Seo and D Park 2008 Fabrication of biochips with micro fluidic channels by micro end-milling and powder blasting *Sensors* Vol 8 No 2 pp 1308–1320

[9] C Horsch, V Schulze and D Lohe 2006 Deburring and surface conditioning of micro milled structures by micro peening and ultrasonic wet peening *Microsystem Technologies* Vol 12 No 7 pp 691–696

[10] Z Kou, Y Wan, Y Cai, X Liang and Z Liu 2015 Burr Controlling in Micro Milling with Supporting Material Method *Procedia Manufacturing* Vol 1 pp 501–511

[11] K Saptaji, S Subbiah and J S Dhupia July 2012 Effect of side edge angle and effective rake angle on top burrs in micro-milling *Precision Engineering* Vol 36 No 3 pp 444–450

[12] G C Wang and C Y Zhang 2003 Study on the Forming Mechanism of the Cutting-Direction Burr in Metal Cutting *Key Engineering Materials* Vol 259–260 pp 860–868

[13] V Bajpai, A K Kushwaha and R K Singh 2013 Burr Formation and Surface Quality in High Speed Micromilling of Titanium Alloy (Ti6Al4V) No 55461 p V002T03A017

[14] B Boswell, M N Islam, I J Davies and A Pramanik May 2015 Effect of machining parameters on the surface finish of a metal matrix composite under dry cutting conditions *Proceedings of the Institution of Mechanical Engineers* Part B: Journal of Engineering Manufacture Vol 231 No 6 pp 913–923

IOP Conf. Series: Materials Science and Engineering **530** (2019) 012026 doi:10.1088/1757-899X/530/1/012026

Production and Verification of RFID MM2 Microchip with Bump and On-Chip Antenna

A B Hassan[1*] and I S A Ali[2]

[1]Electrical, Electronic and Automation Section, Universiti Kuala Lumpur Malaysian Spanish Institute (UniKL – MSI), Kulim High Technology Park, 09000, Kulim, Kedah.

[2]Programme Director, Malaysian Industry-Group For High Technology (MIGHT), MIGHT Partnership Hub, Jalan Impact, 63000, Cyberjaya, Selangor.

Corresponding author*: azmi.hassan@unikl.edu.my

Abstract. This paper describes the special steps taken for the mass production of the Radio Frequency Identification (RFID) MM2 microchip. Since the MM2 is the unique semiconductor product with special on-chip antenna, it requires special production steps that might be quite different from ordinary electronics processes. In order to start the volume production, several preparation steps have to be taken in advance. These tasks are usually not included in the research and development stage. At the initial stage of the production, the production scheduling and quality management are the most important works. Since the availability of the production line varies by the demands, the production sometimes becomes very tight and it will cause the delivery shortage that is critical for customers. In this paper, the steps for the production preparation and production process are described. In order to confirm the proper operation against the process variation by lot, at least three lots (equivalent to 1 million microchips) of the wafers were produced. Very limited number of the microchips were used for the reliability test and the rest were used as the pre-production sample purposes. The uniqueness of the production steps compare to other chip production is the on-chip antenna (OCA) process i.e. build-up very small antenna coil with specified carrier frequency on top of the chips.

1. Introduction

The Radio Frequency Identification (RFID) MM2 microchip has been developed in collaboration with Government of Malaysia with FEC Co. Ltd, Japan. This RFID MM2 microchip is important as it can be embedded in paper [1] for various applications when integrated with the multi-band antenna IC chip [2]. Since the research and development stage of MM2 chip was officially completed in 2008, the mass production of the chip for pilot and full commercial had started. The government incorporated Senstech Sdn. Bhd. to spearhead the applications of the MM2 chip on various RFID systems [3] in the public and private sectors. In this paper, the authors would share the experience carried out by Senstech Sdn Bhd and their partners i.e. Torex Semiconductor Ltd, Japan and Zixsys Inc., Japan in preparing the initial mass production work of the chips.

IOP Conf. Series: Materials Science and Engineering **530** (2019) 012026 doi:10.1088/1757-899X/530/1/012026

2. Production Setup

In order to achieve a stable production, it is very important to maintain the quality of the products. The quality of the products is ensured by the careful design, evaluation of the chip and detailed preparation for the production process. The production preparation activities can be divided into two categories that are: (1) Quality assurance and (2) Setups of the production.

2.1 Quality Assurance

The quality of the products is assured by the various factors such as:

- Design quality (adequate design or margins in parameters)
- Reliability tests for backup data collection
- Tests performance
- Seamless production process control

A reliability tests performed in advance to the production is one of important factors for the quality assurance. In this category, the following works have to be performed:

- Pilot lots production
- Pilot lots evaluations
- Reliability tests

The reliability tests to be performed on the MM2 chip are listed. For the AQL purposes, 22 pieces each from 3 production lots shall be tested in each test category. These lots were produced as the first step of the series of works in this category [4].

- HTST (High Temperature Storage Test): Leave the chip in 125°C environment for 1000 hours.
- HTBT (High Temperature Bias Test): Leave the chip with operating power fed in 125°C environment for 1000 hours.
- HHBT (High Temperature High Humidity Bias Test): Leave the chip with operating power fed in 85°C and 85% humidity environment for 1000 hours.
- HAST (High Temperature High Humidity Acceleration Test): Leave the chip with operating power fed in 125°C, 85% humidity and a pressure of 2 atmospheres environment for 300 hours.
- Temperature Cycles: Cycle the temperature in 30 minutes each between -55°C and 125°C for total 300 hours.
- ESD (Electric Static Discharge): Feed the electric static discharge to the chip of ceramic package.

2.2 Setup of the Production

In order to produce the chip in volume, the production line has to be fully prepared. Although parts of the conventional production line for general semiconductor chips can be used for MM2 production, it still

IOP Conf. Series: Materials Science and Engineering **530** (2019) 012026 doi:10.1088/1757-899X/530/1/012026

require equipment that are specially designed and build for MM2 chip especially at testing routine. Main works in this category are to build the customized test system for the MM2 chip.

Very limited Tester in the world can be used for the RFID chip testing. The Testers for the RFID are very expensive and the performance of those Testers is not sufficient for the MM2 chip testing, since the MM2 employs the original air protocol. Therefore, the testing system shall be a custom design. TOREX and ZIXSYS have designed and build the test board and install it into the production line. Major tasks in this category are listed as follows:

- Design the base Reader/Writer for the testing system
- Design the test board and assembly.
- Build the test board and install between normal Tester and Prober
- Confirm the testing operation
- Install the test board in the production line and make a trial run
- Design of UID numbering application (software on tester)
- Design of Mapping information generator software

3. Production Process

Once the preparation for the production completes, the volume production started. In order to understand the requirements for the production control, it is very important to know the semiconductor production process. Therefore, the explanation of the semiconductor production process is briefly described in this section. The production flow differs depending on form of the chip, either the Bumped die (for external antenna assembly) or the On-Chip Antenna (OCA). The bumped dies are attached on the plastic film and shipped. The back surface of wafer is ground to specified thickness and cut to die size. The production process of the bumped die and On-Chip Antenna is as shown in Figure 1.

IOP Conf. Series: Materials Science and Engineering **530** (2019) 012026 doi:10.1088/1757-899X/530/1/012026

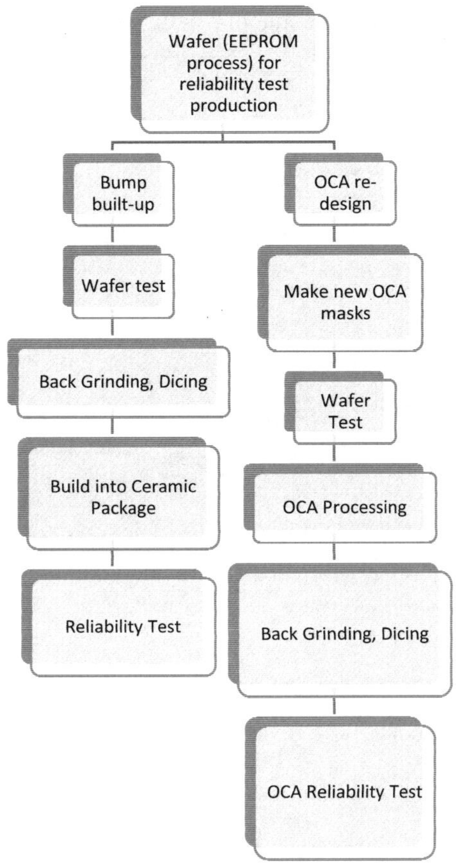

Figure 1. Production flow

The following Table 1 indicates the schedule of the works in the typical condition. The schedule might change depending on the uncontrollable factors such as process time, etc.

IOP Conf. Series: Materials Science and Engineering **530** (2019) 012026 doi:10.1088/1757-899X/530/1/012026

Table 1. Schedule of production works

1st month	2nd	3rd	4th	5th	6th	7th
Wafer production						
		Build Bump				
			Wafer Pre-test			
			BG. Dicing			
			Ceramic Packaging			
	Develop Test Tools for Ceramic Sample			Reliability Test		
					Test Result Evaluation	
		Wafer Pre-test				
OCA Re-design						
	Make Masks for OCA		OCA Processing			
			BG, Dicing			
	Develop Test Tools for Ceramic Sample			Reliability Test		
					Test Result Evaluation	

IOP Conf. Series: Materials Science and Engineering **530** (2019) 012026 doi:10.1088/1757-899X/530/1/012026

4. Work flow of reliability tests

The following steps have to be taken to complete the reliability tests. The descriptions below based on the chip level testing.

4.1 Order of minimum three production lots to the wafer foundry for AQL

In order to confirm the proper operation against the process variation by lot, at least three lots of the wafers shall be produced. Very limited number of the chips will be used for the reliability test and the rest can be used as the pre-production sample purpose.

4.2 Preparation of the R/W for the reliability tests

In order to perform the reliability test, the customized Reader /Writer (R/W) to confirm the chip operation shall be prepared in advance. The normal R/W communicates with the chip via RF. However some tests will be made on the capsulate chip in ceramic package thus it requires the special interface which operates with ceramic package. Some of the tests are executed with the RF power fed into the chip and the chip has to be put into the oven thus the customized structure of the R/W is needed. These R/W's were developed by ZIXSYS [5] and one sample is shown below in Figure 2.

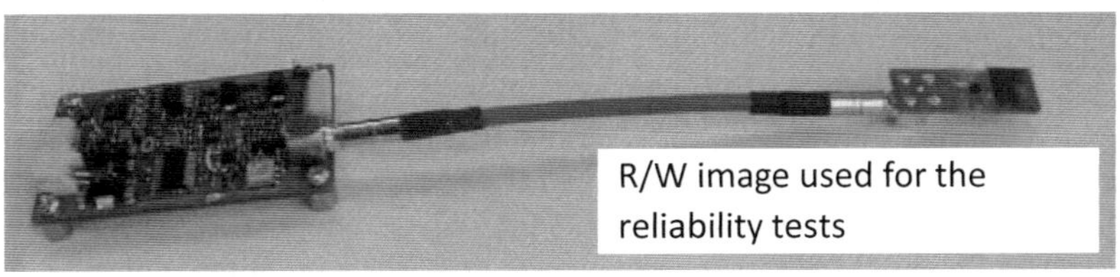

Figure 2. Reader/Writer module for MM2 Chip.

4.3 Building the chip into the ceramic package for test

MM2 chip has an access only through the RF input pads via an antenna. Since there is no method to access to the bare die of MM2 chip without antenna connection, the chip was built into the ceramic package to execute the reliability tests. 22 pieces each from three lots for each test was built into the ceramic packages. Total number of the ceramic samples becomes 396 pieces to cover all six test categories. Samples of the dies capsulated in ceramic package are shown in Figure 3.

Figure 3. MM2 Chip capsulated in ceramic package

IOP Conf. Series: Materials Science and Engineering **530** (2019) 012026 doi:10.1088/1757-899X/530/1/012026

4.4 Execution of the reliability tests
Several equipment are required to execute the reliability tests.

- Heat Chamber to generate high and low temperature
- Heat Chamber equipped with temperature and humidity control
- Heat Shock Chamber with rapid temperature change function
- High voltage electro-static generator for ESD

The longest test will take 1000 hours to complete thus the required duration to complete the tests is estimated around three months. These tests was performed at facilities in TOREX and ZIXSYS.

4.5 Tests result verification
The test results were evaluated and the final reliability specification was defined. The rests of the chips produced for the reliability tests might be supplied in the production stage to the customers.

5. Reliability tests for OCA version

Another reliability test was performed on the OCA, since the OCA was formed on wafer in another process. The reliability tests executed on the OCA were basically the same with those for the chip. The reliability tests on the OCA shown in Figures 4(a) and (b) were conducted separately by Fujikura and ZIXSYS. The equipment used for the OCA reliability tests were the same with the ones used for the chip, therefore most of the resources provided for the chip reliability tests were reused for the OCA reliability tests. Thus, the uniformity of the reliability levels on both chip and OCA will be ensured. Figure 5 shows the brief setup for the MM2 chip OCA testing. The tester have the function to automatically execute the command transmission and response reception by the integrated microchip firmware as shown in Figure 6. The tester has 8 channels of test signals which enable simultaneous tests on 8 chips in order to expedite the total test time. Detailed testing process and specification are not included in this paper. However the example of test program and evaluation behavior of the MM2 chip presented as shown in Figure 7.

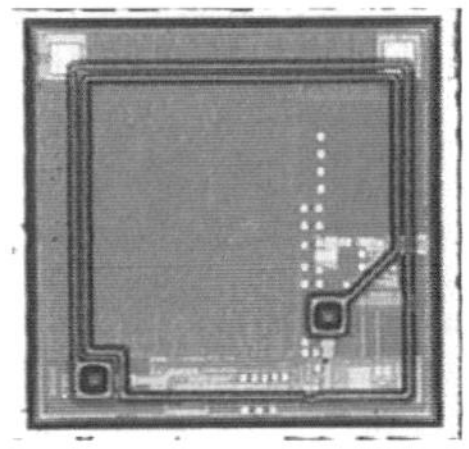

Figure 4(a). MM2 Chip with 2.45GHz

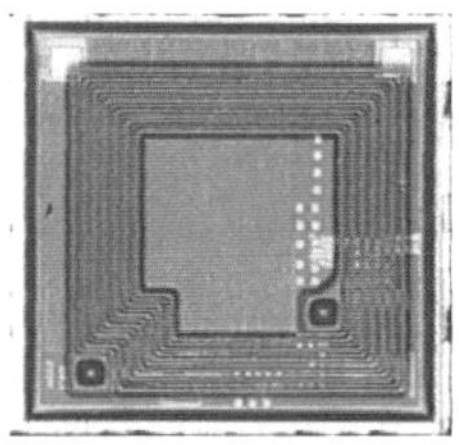

Figure 4(b): MM2 Chip with UHF

Figure 5. Test board connected to the Probe card and 8 channel tester board.

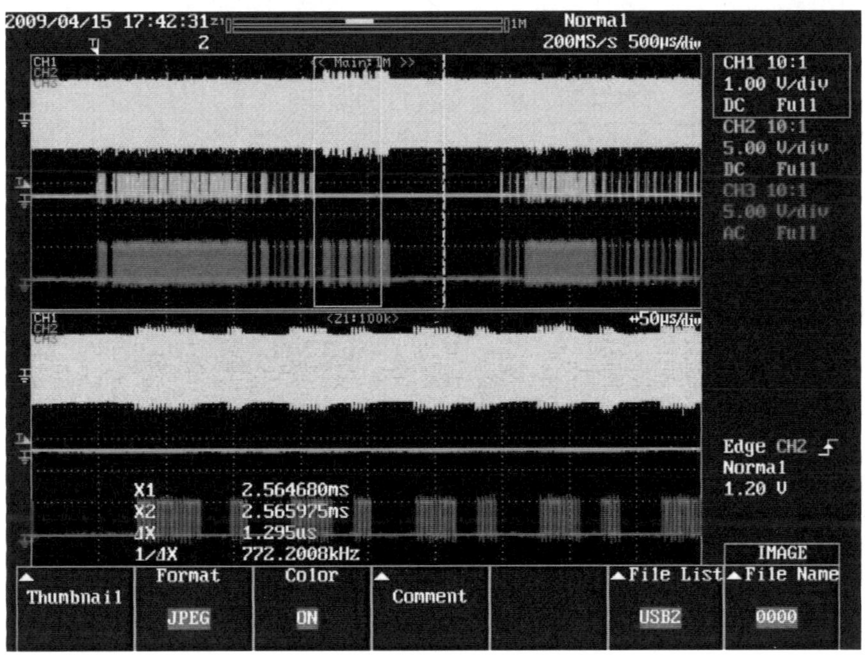

Figure 6. Example of test program

6. Conclusions

The production flow of MM2 chip had been presented in this paper. It were carefully prepared for the seamless operation. The production of MM2 is unique mainly on OCA process and also on the function of data encryption in data connection between tags (chips) and reader/writer module as shown in Figure 6. There are several important factors that must be consider to maintain-stable productions such as required time for the process in each task, factory availability, tightness of production (changing time by time), logistic flow, improvements for further cost down, accurate forecast for the future demands and leveling of the production volume in monthly base. Since the lead time for the semiconductor products are relatively longer, sudden change of the production quantity usually cannot be accepted. Therefore the accurate production scheduling is very important.

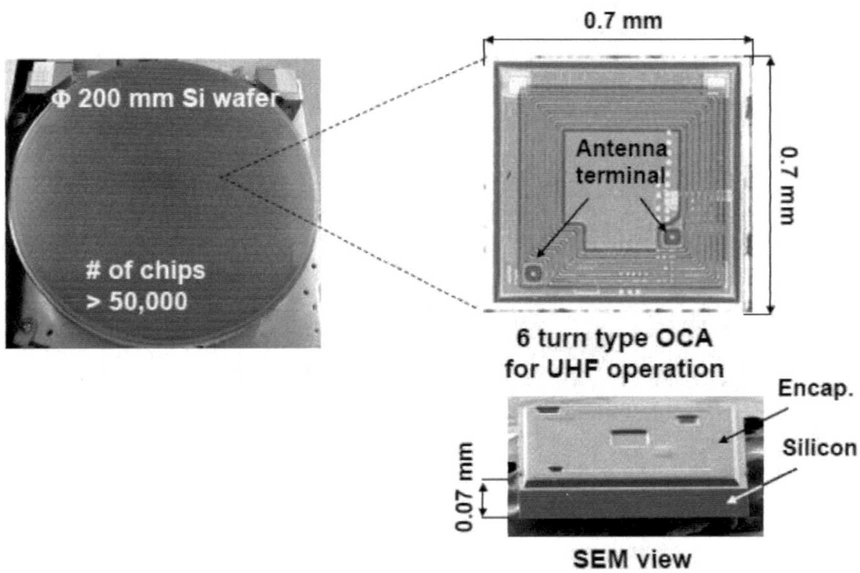

Figure 7. Final wafer of MM2 chip2 with OCA fabrication

Acknowledgments

Authors wish to acknowledge collaboration with FEC Co. Ltd, Torex Ltd., Zixsys Inc., all former technical staff of Senstech Sdn. Bhd., and financial support from Malaysian Industry Government Group for High Technology (MIGHT).

IOP Conf. Series: Materials Science and Engineering **530** (2019) 012026 doi:10.1088/1757-899X/530/1/012026

References

[1] Lehto A, Nummela J, Ukkonen L, Sydanheimo L and Kivikoski M 2009 Passive UHF RFID in Paper Industry: Challenges, Benefits and the Application Environment *IEEE Transactions on Automation Science and Engineering* 6(1) 66

[2] Hong Y, Ng Y S, Chan C F, Guo J, Shi W, Ho M, Ai Y, Mak K L, Leung K N, Choy C S, Pun K P and Leung L K 2009 A passive RFID tag IC development platform *3rd International Conference on Anti-counterfeiting, Security, and Identification in Communication* 286

[3] Choo J and Ryoo J 2014 UHF RFID Tag Applicable to Various Objects *IEEE Transactions on Antennas and Propagation* 62(2)

[4] Utamura K 2010 *Technical Report MM2 Production Preparatory Works April 2009 – 2010* Torex Semiconductor Ltd. Nihonbashi Building *1-13-12 Nihonbashi Kayabacho Chuo-ku Tokyo* 103-0025 Japan

[5] Nakazawa T 2010 *Technical Report and Specifications of ZEAL/ZOOM chip of MM-chip series,* Zixsys Inc. Akari Building *1-312 Kagoharaminami, Kumagaya, Saitama* 360-0847Japan

Numerical Study of Composite Fiberglass Cross Arms under Statics Loading and Improvement with Sleeve Installation

D Mohamad[1], A Syamsir[2], S Beddu[1], A Abas[3*], F C Ng[3], M F Razali[3] and S A H A Seman[3]

[1] Department of Civil Engineering, Universiti Tenaga Nasional, 43000 Kajang, Malaysia.

[2] Institute of Energy Infrastructures (IEI), Universiti Tenaga Nasional, 43000 Kajang, Malaysia.

[3] School of Mechanical Engineering, Universiti Sains Malaysia, Engineering Campus, 14300, Nibong Tebal, Penang, Malaysia.

Corresponding author *: aizatabas@usm.my

Abstract. In this paper, the structural deformation of fiberglass composite cross arms under statics loading were numerically simulated using the ANSYS software. The alternate variant of cross arms with one-meter sleeve span installed on both arms was considered as an improvement study. By comparing both cross arms systems, it is found that the total deformation was reduced from 127.49 mm to 95.367 mm by introducing the sleeve. Besides, the stress level appears to be lowered in the cross arms with sleeve installed. The sleeve is regards as a viable solution to reduce both deformation and stress level of cross arms, thus lengthen its life span and saving maintenance costs.

Keywords: *ACP simulation; cross arms; fiberglass composite; statics structural deformation*

1. Introduction

In electrical transmission tower, cross arm is the structure responsible to hold the transmitting wires in place. The cross arms not only must be rigid enough to sustain the electrical cables, but also able to withstand the dynamic wind load, especially during the windy or stormy days. A failed cross arm will cause disruption to the electrical transmission and incurs additional maintenance costs to the electrical provider. Additionally, broken cross arm will lower the electrical wires near to ground, risking the pedestrians that happened to walk by. Therefore, it is necessary to study the reliability of cross arms of transmission tower to ensure safe and cost-effective electrical transmission [1-2].

This paper is devoted to study the structural rigidity of cross arms system that being used in the typical transmission tower. Numerical simulation is the prime methodology of current research, executed by utilizing the ANSYS Composite PrepPost (ACP) module. In fact, ACP module had been used by previous researchers to simulate the structural behaviour of composite material [3-4]. Furthermore, the sleeve span is introduced as a potential solution to reduce deformation and stress on the cross arms, thus enhancing its reliability. Therefore, two cases of cross arms, with and without sleeve will be simulated and have their results compared side-by-side to justify the effectiveness of sleeve installation.

2. Numerical simulation

Using the commercial available numerical simulation software, ANSYS, the composite structures of the cross arms were modelled and subsequently solved for the applied load. The cross arms' fibreglass composite layers and their orientations were defined in the ANSYS Composite PrepPost (ACP) module; whereas the boundary conditions of statics loading were input in the ANSYS Mechanical module. The materials properties of composite arms, tie members and steel connectors were given in Table 1. Additionally, the composite details (thickness and orientation) were presented in the Table 2.

Table 1. Material properties of the composite and structural steel contained in the cross arms system.

Material properties	Composite	Structural steel
Young modulus in x, E_x (MPa)	16000	200000
Young modulus in y, E_y (MPa)	4800	200000
Young modulus in z, E_z (MPa)	1440	200000
Poisson's ratio, $\nu_{xy} = \nu_{yz} = \nu_{xz}$	0.28	0.3
Shear, modulus, $G_{xy} = G_{xz} = G_{yz}$ (MPa)	4000	76923
Density, ρ (kg/m^3)	1800	7850

Table 2. Thickness and fabric orientation of each composite layer in the cross arms and tie members.

Composite layer	Thickness (mm)	Fabric orientation (°)
First (*outer*)	0.5	45
Second	0.5	–45
Third	0.7	90
Forth	3.6	0
Fifth (*inner*)	0.7	45

The numerical model of cross arms system was meshed with size function adaptive, to generate both hexagonal and quadrilateral meshes as illustrated in Figure 1. Hexagonal meshes were mainly applied on rectangular parts while quadrilateral meshes were adopted in rounded or curved parts. The mesh sizing of the numerical model was varied to attain the optimum grid, as part of mesh independent study (refers Section 3.1).

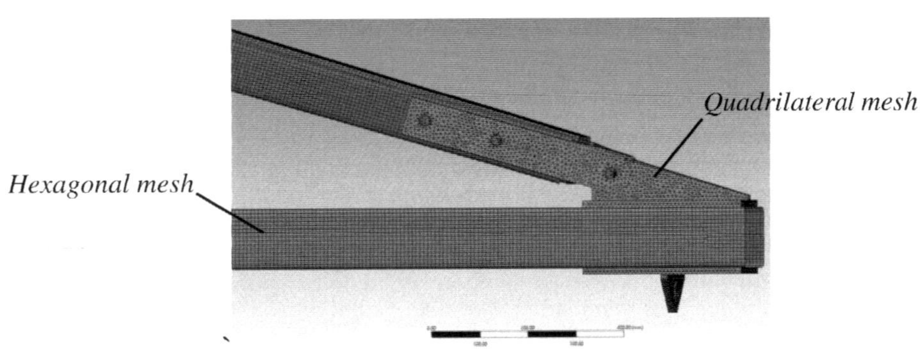

Figure 1. The adaptive hybrid mesh that being implemneted on the cross arms numerical model.

IOP Conf. Series: Materials Science and Engineering **530** (2019) 012027 doi:10.1088/1757-899X/530/1/012027

There are mainly two types of boundary conditions being applied on the cross arms system, for instance fixed support (FS) at the back ends of each cross arms/tie members and force loading at the front end. The force boundary condition applied on the numerical model is equivalent to the loadings of electrical wires on the cross arms. It has the following force components: $F_x = -23208$ N, $F_y = -42496$ N and $F_z = -3262$ N, resultantly acting in the direction indicated by the red arrow as shown in Figure 2.

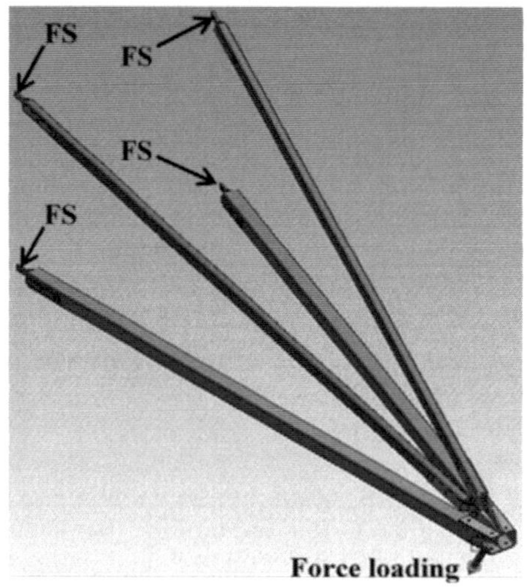

Figure 2. Summary of boundary conditions that being applied on the cross-arms numerical model. The abbreviation "FS" denotes to fixed support boundary condition that applied on the cross arms hinges.

The numerical model of cross arms was later solved to determine the total deformation and von Mises stress at each composite layer. Similar simulation procedures were repeated on cross arms system with one-meter sleeve span installed on both cross arms. Each sleeve has similar thickness and composite layer as that of the cross arm itself. The sleeve will be wrapping around cross arms tightly, such that it appeared the thickness of cross arms were doubled at that particular region.

3. Result and Discussions

3.1 Mesh independency study

A mesh independency study was performed to ensure the accuracy of numerical model is not effected by the mesh details. The results of current independency study were presented in Table 3. Taking the standard cross arms numerical model, its mesh details were manipulated from default to very fine, while observing the changes in their deformation contours.

From Table 3, it is found that both fine and very fine meshes yield similar numerical results with maximal deformation value around 127 mm. Therefore, it is justified that the mesh independent had been achieved in the present simulation. Subsequent deformation and stress data will be based on the numerical model with very fine mesh that being optimized.

IOP Conf. Series: Materials Science and Engineering **530** (2019) 012027 doi:10.1088/1757-899X/530/1/012027

Table 3. The deformation contours obtained using numerical model with different grid resolutions.

Mesh	Grid resolution	Maximal total deformation (mm)	Deformation contour
1	Default	116.69	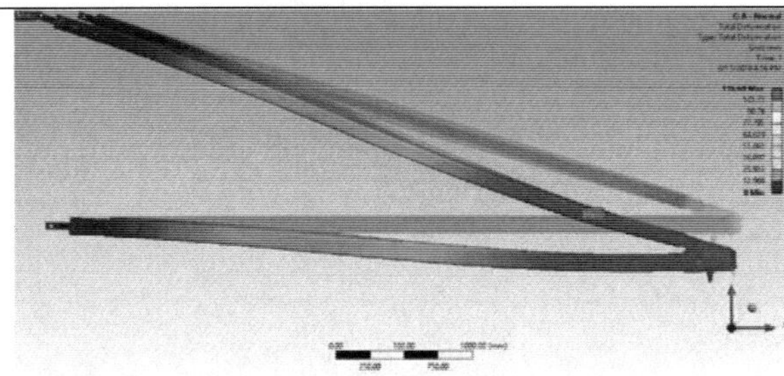
2	Fine	127.61	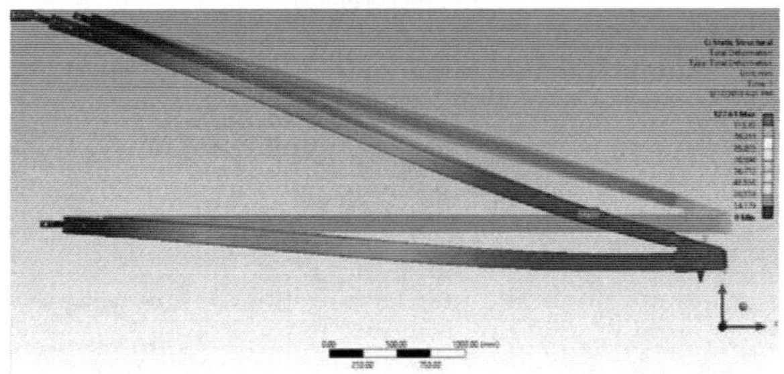
3	Very fine	127.49	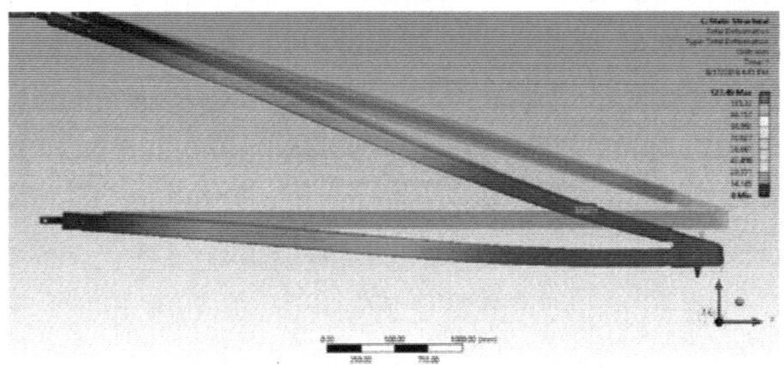

3.2 Deformation

Figure 3 depicts the comparison between the deformation contours of standard cross arm and cross arms with sleeve installed. Table 4 presents the peak deformation and deformation value at the mid-span of the cross arms in both cases. Generally, the deformation gradually increases toward the loading end of the cross arm, where the peak deformation would be observed near the front end where the electrical wires loading is being applied on the cross arm. Besides, it is found that the installation of sleeve on the cross arms effectively reduce the peak deformation by 25.19%. At the mid-span of the cross arms where the sleeve being installed, the deformation reduction is found to be 30.09%.

Table 4. Deformation values of both cross arm systems.

	Peak deformation (mm)	Mid-span deformation (mm)
Standard cross arm	127.49	102.01
Cross arm with sleeve installed	95.37	71.32
Percentage reduction with sleeve installation	**25.19%**	**30.09%**

IOP Conf. Series: Materials Science and Engineering **530** (2019) 012027 doi:10.1088/1757-899X/530/1/012027

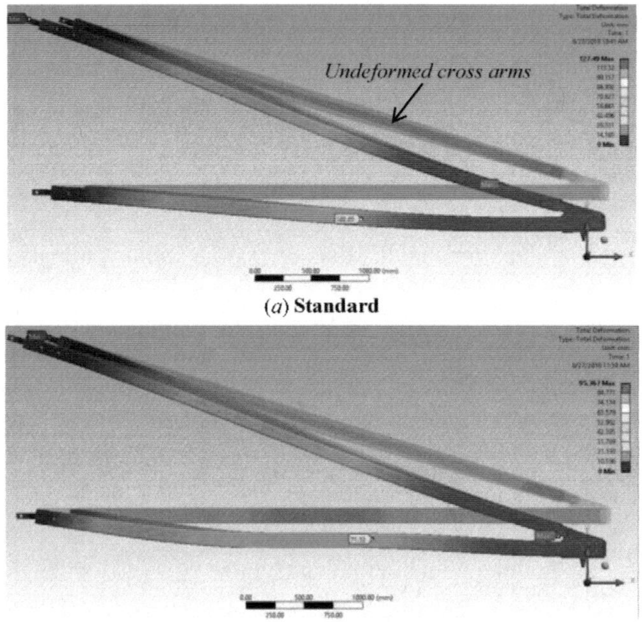

(*a*) **Standard**

(*b*) **With one-meter sleeve**

Figure 3. Comparison of deformation between (*a*) the standard cross arm without sleeve and (*b*) the cross arm with one-meter sleeve installed, upon being subjected to statics loading.

3.3 Stress

Table 5 presents the maximal localized stress at each composite layers. It is found that the third composite fibre layer has the lowest localized stress as compared to other layer, which is due to its 90° fibre orientation. Meanwhile, the forth layer possess the highest localized stress as compared to other. Furthermore, the introduction of sleeve span on the cross arms appeared to not drastically reduce the stress level. Noticeable reduction of stress is found at the third layer, about 36.88%; while for the first and fifth layers, the stress is slightly increased with sleeve introduction.

Table 5. The maximal stress at each composite layers for both cross arms cases.

Composite layer	Maximal localized stress (MPa)	
	Standard	With sleeve
1 (outer)	254.12	255.82
2	210.64	210.31
3	178.30	112.54
4	262.95	269.08
5 (inner)	228.31	228.55

However, the stress value presented in Table 5 is rather localized and occur in small region. Figure 4 depicts the stress contours of whole cross arms. It is found the stress level is averagely in the ranges of 47 – 105 MPa (for standard cross arm) and 37 – 97 MPa (for cross arm with sleeve). Besides, the region underneath the sleeve has significantly lower stress level (about 5 -14 MPa). Overally, the sleeve installation reduced the stress level on the cross arms.

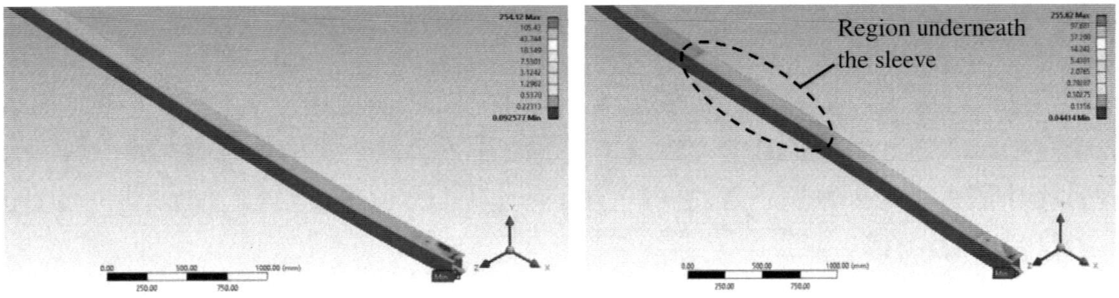

(*a*) Standard cross arm (*b*) Cross arm with sleeve installed

Figure 4. Comparison of stress contours of first (outer) cross arm composite layer between both cases.

4. Conclusions

The deformation of fiberglass composite cross arms subjected to statics loading of electrical wires were successfully simulated using the commercial available software, ANSYS. In particular, two cases of cross arms system were considered: standard cross arms and cross arms with one-meter sleeve span installed on both lower arms. Numerical mesh was optimized to lower potential computational error. It is found that the introduction of sleeve able to reduce both deformation and stress level at the cross arms, making less prone to fatigue failure and thus more reliable for long term usage.

Acknowledgement

TNB Seeding Fund funds this research for Universiti Tenaga Nasional (U-T-S-RD-17-05). Author would like to thank TNB Transmission Line Division and TNB for financial support.

References:

[1] A Nadhirah, S Beddu, D Mohamad, M Zainoodin, S Nabihah, N Mubin, Agusril Z Itam, H Mansor, N Mohd Kamal, M A Alam, Z Che Muda and Z Mohd Yusof 2017 Properties of Fiberglass Crossarm in Transmission Tower - A Review *International Journal of Applied Engineering Research* Vol 12 No 24 pp 15228 – 15233

[2] N Prasada Rao, S J Mohan and N Lakshmanan 2005 A study on failure of crossarms in transmission line towers during prototype testing *International Journal of Structural Stability and Dynamics* 5 03 pp 435 – 455

[3] H Hu, X Bian, X Cao, W Shen and B Li 2017 Impact Properties Study of Fibre Reinforced Resin Matrix Composites *International Journal of Materials Science and Applications* 6(6) pp 309 – 315

[4] Duan C, Huang J, Wu L and Luo X May 2018 Fine modeling of reinforced thermoplastic filament winding container *AIP Conference Proceedings* (Vol 1967 No 1 p 040053) AIP Publishing

Numerical Simulation on the Statics Deformation Study of Composite Cross Arms of Different Materials and Configurations

D Mohamad[1], A Syamsir[2], Z Itam[1], H A Bakar[3], A Abas[3]*, F C Ng[3], M F Razali[3] and S A H A Seman[3]

[1] Department of Civil Engineering, Universiti Tenaga Nasional, 43000 Kajang, Malaysia.

[2] Institute of Energy Infrastructures (IEI), Universiti Tenaga Nasional, 43000 Kajang, Malaysia.

[3] School of Mechanical Engineering, Universiti Sains Malaysia, Engineering Campus, 14300, Nibong Tebal, Penang, Malaysia.

Corresponding author *: aizatabas@usm.my

Abstract. This study focuses on the deformation of cross arms made from composite material under static loading. Two materials properties of fiberglass with are assigned to the cross arms model and it is subjected to a load at the one end of the cross arms. A bracing support is then installed, and the deformation and stress experienced by the cross arms are observed and compared with the case where no bracing support is used. The results obtained show that material B has lower deformation value, but experienced greater stress compared to material A. The bracing managed to reduce the total deformation by about 12% and 20% for material A and B respectively. The reduction of stresses experienced by the cross arms ranges from 4% to 30%.

Keywords*: composite; cross arms; bracing; static structural*

1. Introduction

Fiber-composite is one of the engineering material with high performance in terms of its strength and stiffness [1]. Fiberglass is one of the fiber-composite that is widely used. The applications of fiberglass ranges from food processing to power generation industries. This is due to its characteristic which it is easily moulded and shaped into complex forms. In the power generation industry, fiberglass was introduced as an insulation material for distribution and transmissions system as an additional component to the basic insulator [2]. Grzybowski and Disyadej [2] stated that the main reason is because fiberglass has high dielectric strength. In the case of fiberglass utilized in the distribution line pole, it is considered as a non-uniform composite material with fiber as the main element and resins and filers as the additional components [3]. In this study, the focus will be on the application of fiberglass in the cross arms of the transmission tower. Cross arms, one of many components that build the transmission tower, acts as a support to the transmission conductor [4]. It is constantly subjected to wind loads, its self-weight and other loads such as unexpected broken wire conditions, erections and maintenance [5]. Hence, the common materials used for the cross arms are wood, steel and fiberglass [4]. However, fiberglass is better because of its insulating characteristic and it has a greater mechanical strength to weight ratio, compared to the other two [4].

IOP Conf. Series: Materials Science and Engineering **530** (2019) 012028 doi:10.1088/1757-899X/530/1/012028

Therefore, this study aims at investigating the deformation of the composite cross arms under static loadings using finite element approach. A composite material for the cross arms with different material properties are tested in two different conditions, a cross arms without bracing support and a cross arms with a bracing support. The total deformation and stresses experienced by the cross arms will be evaluated and compared.

2. Numerical simulation

2.1 Finite element analysis (FEA)

Finite element method (FEA) is a numerical technique that benefitted from the computational power to obtain the approximation solutions of boundary value problems or field problems. The field is the domain that is represented by a physical structure. The field variables are the variables that governed by the differential equations and the boundary conditions are the values on the boundaries of the fields. The values of the field variable calculated at the nodes are used to determine the values at non-nodal points by means of interpolation. For a three-node triangle example, the field variable can be represented such as:

$$\varphi(x,y) = N_1(s,y)\varphi_1 + N_2(s,y)\varphi_2 + N_3(s,y)\varphi_3 \quad (1)$$

Where φ_1, φ_2 and φ_3 are the values of the field variable at the nodes, N_1, N_2 and N_3 are the interpolation functions. This Eq.1 is also known as the shape function.

2.2 Shell element

Shell elements are suitable to model thin structures. Thin structure denotes a geometry where one dimension is so small compared to the other two dimensions. Shell elements separate the deformation on the surface and in the normal direction, hence simplify and creating a more efficient simulation. It is usually done by meshing the surface that representing the structure and providing the section data to create the shell thickness. There are three assumptions to be made when implementing the shell element: (1) stresses are assumed to be linear through the thickness; (2) the center surface does not experience any bending and (3) the membrane stresses are uniform over the thickness.

For this simulation, the thickness to be associated with the 3D model is 6 mm. It is set in the ANSYS ACP model setup. The 6 mm thickness later is divided and oriented according to the composite material as stated in Table 2.

2.3 3-D model

Each cross arm was designed and assembled using SOLIDWORKS 2017 software. Figure 1 (a) shows the cross arms design while Figure 1 (b) shows the cross arms with a bracing installed to the four arms. The material properties for the cross arms were later assigned in ANSYS ACP.

IOP Conf. Series: Materials Science and Engineering **530** (2019) 012028 doi:10.1088/1757-899X/530/1/012028

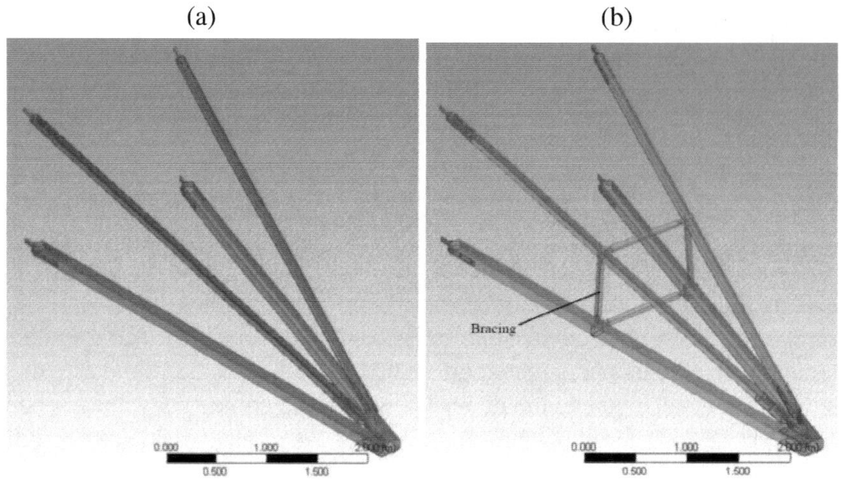

Figure 1. The 3-D model of: (a) the cross arm and; (b) cross arm with bracing

2.4 Material properties

Two composite material properties, A and B are assigned to the cross arms as tabulated in Table 1. Besides that, the material A is made to be only five layers of composite material while material B is made to have nine layers of composite material. Table 2 shows the fiber orientation and thickness for each layer of the material A and B.

Table 1. Material properties for the cross arm

Material	Ex (MPa)	Ey (MPa)	Ez (MPa)	Vxy=Vyz=Vxz	Gxy (MPa)	Gxz (MPa)	Gyz (MPa)	Density (g/cm^3)
A	16000	4800	1440	0.28	4000	4000	4000	1.8
B	36300	10890	3267	0.28	4280	4280	4280	1.9

Table 2. The composite layer and its fiber orientation and thickness for each material

Layer	Material A		Material B	
	Fiber orientation	Thickness (mm)	Fiber orientation	Thickness (mm)
1	45	0.5	0	0.7
2	-45	0.5	45	0.7
3	90	0.7	0	0.7
4	0	3.6	-45	0.7
5	45	0.7	0	0.7
6	-	-	-45	0.7
7	-	-	0	0.7
8	-	-	45	0.7
9	-	-	0	0.7

2.5 Boundary conditions (BCs)

For the simulation, a couple of BCs have been assigned to the cross arms model. Each ends of the cross arms leg is assigned with a fixed support, as shown in Figure 2 (a). The load is applied to the top end of the cross arms as in Figure 2 (b).

IOP Conf. Series: Materials Science and Engineering **530** (2019) 012028 doi:10.1088/1757-899X/530/1/012028

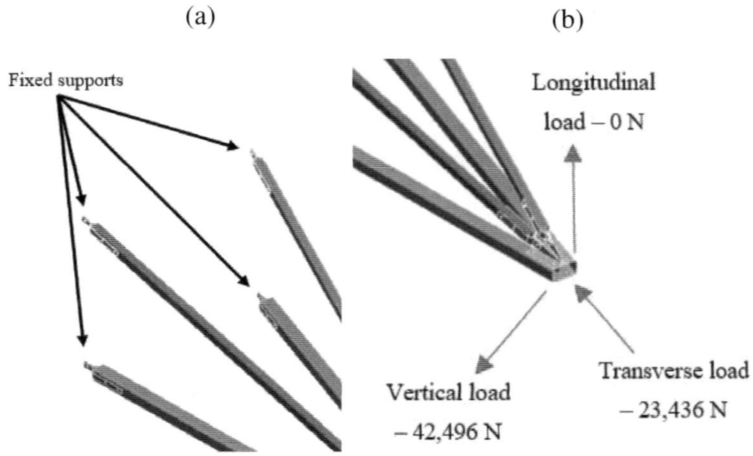

Figure 2. Boundary conditions for: (a) fixed supports; (b) load applied

3.0 Results and Discussion

3.1 Maximum deformation

The total deformations and stresses experienced by the cross arms for material A and B, with and without bracing are tabulated and presented in this section. Table 3 is the total deformation experienced by the cross arms for material A and B, with and without the support of the bracing. The deformation distribution of both cases in each material is similar, where the maximum deformation happens in the region where the load is applied and the minimum deformation occurs in the region where the fixed supports are applied.

For material A, the maximum deformation of the cross arms without the bracing support is around 127 mm, and it is located near the region where the load is applied. However, with the support of the bracing, the maximum deformation managed to be reduced to 111 mm, a reduction of 12%. For material B, the maximum deformation of the cross arms for the standard case is 79 mm, while the case where the bracing is installed is 63 mm. A decrease of 20% in maximum deformation is obtained for the material B. This results shows that the installation of bracing is capable of reducing the maximum deformation experienced by the cross arms.

As for the comparison between material A and B, material B experienced a lower maximum deformation for both cases (standard: 79 vs. 127 mm; bracing: 63 vs. 111 mm). Table 4 shows the total deformation contour for material A and B for the case of with and without bracing.

Table 3. Maximum deformation for material A and B for case of with and without bracing.

Configuration	Maximum deformation (mm)		Percentage reduction (%)
	Standard	Bracing	
Material A	127	111	12.6
Material B	79	63	20.3

IOP Conf. Series: Materials Science and Engineering **530** (2019) 012028 doi:10.1088/1757-899X/530/1/012028

Table 4. Comparison of deformation experienced by different material for cross arm with and without bracing.

Deformation	Standard	Bracing
Material A	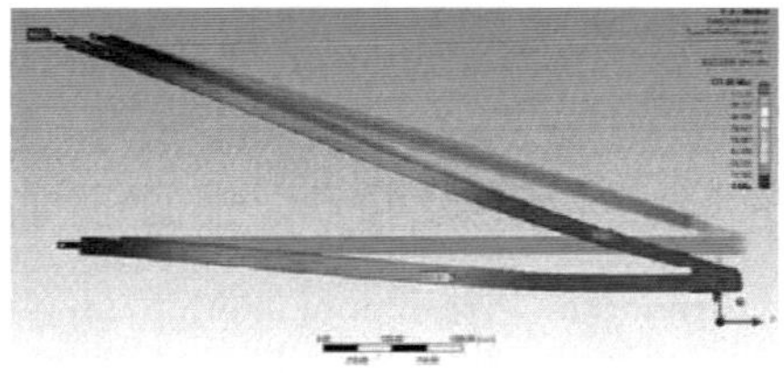	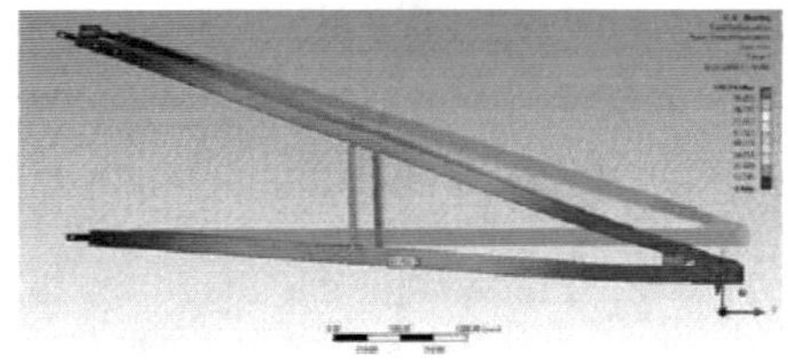
Material B	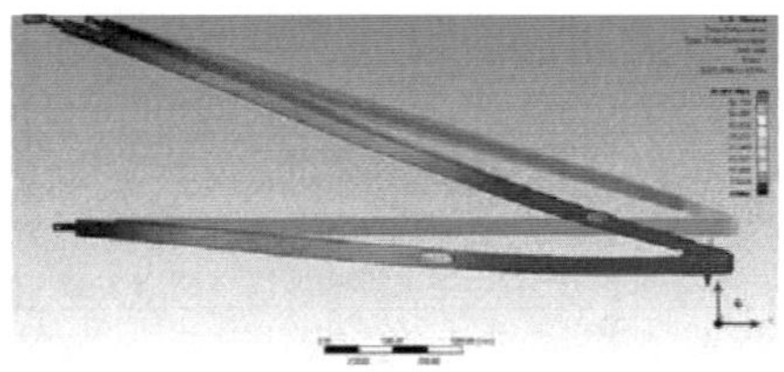	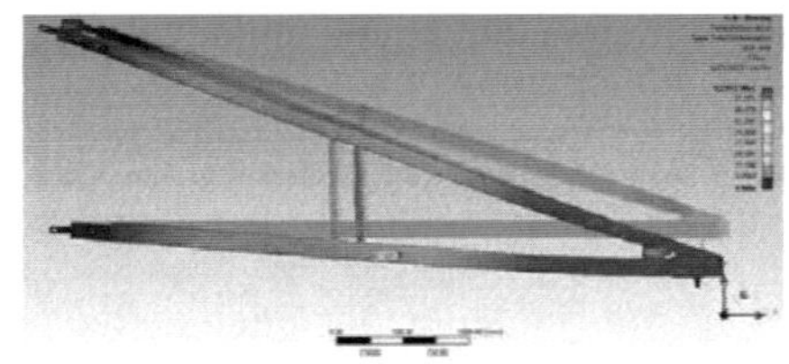

3.2 Maximum stress

The maximum stress for material A and B for cases of with and without bracing support are recorded in Table 5 and 6 respectively. Each layer experienced different stress magnitude and the presence of bracing reduced the stress exerted to the cross arms for both materials. In material A, the used of bracing lower the maximum stress value more than 5% in each layer. The highest reduction achieved is at layer 3, where the maximum stress obtained before and after the placement of bracing is 178.30 MPa and 124.58 MPa respectively, a reduction of approximately 30%.

A similar observation is recorded for material B, where the bracing support reduced the maximum stress subjected to the cross arms with a reduction of more than 5%. The maximum reduction achieved is at layer 5, where the maximum stress obtained before and after the placement of bracing is 153.84 MPa and 143.36 MPa respectively, a reduction of 6.81%. For the comparison between material A and B, material B seems to be experiencing greater maximum stress compared to material B, as tabulated in Table 5 and 6.

Figure 3 shows the plotted maximum stress for material A and B in each layer. The maximum stress experienced at each layer seems to be lower for the layer close to the center and higher at the outer layer of the material. The distribution of stress for both materials for each case seems to be similar where the maximum stress recorded is in the tip region of the cross arms. Figure 4 shows the stress distributions at layer 1 for material A for case of without and with bracing. The distributions of stress in each layer is also similar, in both material, for both cases. Hence, for simplicity, only layer 1 of material A is included in this paper.

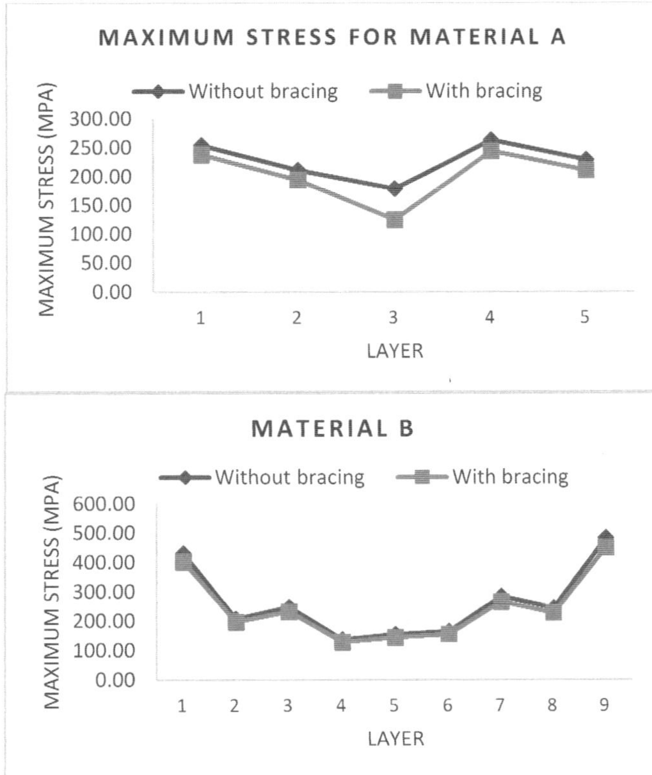

Figure 3. Maximum stress recorded for material A and B at each layer.

Table 5. Maximum stress at each layer, with and without bracing for material A.

Layer	Maximum stress (MPa)		Percentage difference (%)
	Standard	Bracing	
1	254.12	238.17	6.28
2	210.64	193.58	8.10
3	178.30	124.58	30.13
4	262.95	243.96	7.22
5	228.31	210.22	7.92
Bracing	-	2736.80	-

Table 6. Maximum stress at each layer, with and without bracing for material B.

Layer	Maximum stress (MPa)		Percentage difference (%)
	Standard	Bracing	
1	429.99	402.95	6.29
2	207.40	197.09	4.97
3	247.20	231.59	6.31
4	136.50	128.02	6.21
5	153.84	143.36	6.81
6	284.25	265.02	6.77
7	244.41	226.89	7.17
8	482.24	449.34	6.82
9	284.25	265.02	6.77
Bracing	-	1745.20	-

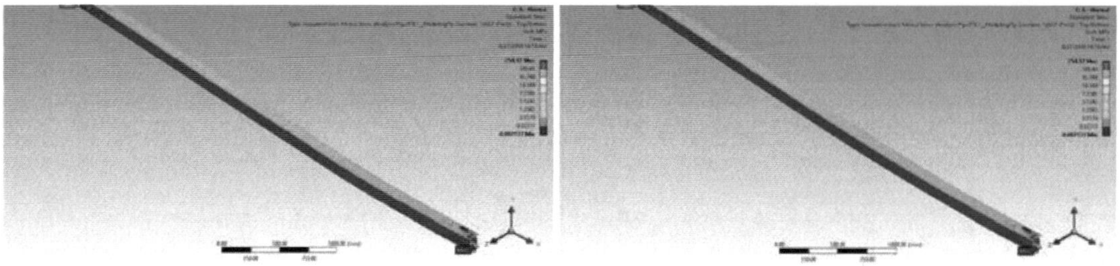

Figure 4. Stress distributions for: (a) without bracing; (b) with bracing.

4. Conclusion

The study aimed at investigating the total deformation and maximum stresses experiencing by the cross arms of different composite material with and without the support of a bracing. The simulations performed concluded that the bracing helped reduced the total deformation and maximum stresses subjected to the cross arms for both materials. The installation of bracing reduced the total deformation around 12% and 20% for material A and B respectively. For the maximum stress parameter, the bracing managed to decrease the stresses underwent at each layer for both materials. The percentage reduction ranges from 4% to 30%. Between the two materials, material B recorded a lower deformation, but experienced greater stresses compared to material A.

Acknowledgement

TNB Seeding Fund funds this research for Universiti Tenaga Nasional (U-T-S-RD-17-05). Author would like to thank TNB Transmission Line Division and TNB for financial support.

References

[1] W Chen, Q Meng, H Hao, J Cui and Y Shi 2017 Quasi-static and dynamic tensile properties of fiberglass/epoxy laminate sheet *Construction and Building Materials* Vol 143 pp 247-258

[2] S Grzybowski and T Disyadej 2008 Electrical performance of fiberglass crossarm in distribution and transmission lines *IEEE/PES Transmission and Distribution Conference and Exposition* pp 1-5

[3] S Grzybowski and X Li 2006 Electrical degradation of fiberglass distribution line pole *IEEE Transactions on Dielectrics and Electrical Insulation* Vol 13 pp 927-934

[4] A Nadhirah, Beddu D Mohamad, M Zainoodin, S Nabihah, N Mubin and et al. 2017 Properties of Fiberglass Crossarm in Transmission Tower-A Review Vol 12

[5] N P Rao, S J Mohan and N Lakshmanan 2005 A Study On Failure Of Cross Arms In Transmission Line Towers During Prototype Testing *International Journal of Structural Stability and Dynamics* Vol 05 pp 435-455

IOP Conf. Series: Materials Science and Engineering **530** (2019) 012029 doi:10.1088/1757-899X/530/1/012029

Effect of Laminate Properties on the Failure of Cross Arm Structure under Multi-Axial Load

D Mohamad[1], A Syamsir[2], S Beddu[1], N L M Kamal[1], M M Zainoodin[1], M F Razali[3]*, A Abas[3], S A H A Seman[3] and F C Ng[3]

[1] Department of Civil Engineering, Universiti Tenaga Nasional, 43000 Kajang, Malaysia.

[2] Institute of Energy Infrastructures (IEI), Universiti Tenaga Nasional, 43000 Kajang, Malaysia.

[3] School of Mechanical Engineering, Universiti Sains Malaysia, Engineering Campus, 14300, Nibong Tebal, Penang, Malaysia.

Corresponding author *: mefauzinizam@usm.my

Abstract. This study investigated the influence of laminate properties toward the collapse of composite cross arm structure under multi-axial quasi-static loading. A three-dimensional finite-element model of a cross arm was developed and integrated with Hashin's failure subroutine to predict the inter-laminar damages of the composite upon its application. The mechanical deformation and failure of the composite structure were evaluated over three laminate properties. This investigation revealed that variation in laminates properties yielded different structural deflection and laminate damages. The cross arm with a greater young modulus and ultimate stresses (laminate B) experienced a single failure mode of fiber buckling in compression at deflection of 0.082 m. For the given multi-axial load, the laminate configuration and properties considered in this study failed to prevent the failure of the cross arm.

1. Introduction

Through years, demand for a stiffer, stronger and lighter laminated composite has grown in many fields such as aerospace, energy and civil construction. For laminated composites made of unidirectional layers, the best performance can be achieved provided that the applied loading direction matches with the reinforcement direction. However, this condition is hard to realize in real application as the loading scenarios are often multiaxial. For such applications, employing laminated composites made of unidirectional layers may provide a poor solution.

Additionally, improper laminated composites are also prone to mechanical damages when they are exposed to various types of loadings such as compressive, tension, and flexural which can lead to material failure [1]. In order to improve the use of laminated composites for high load applications, it is important to understand the mechanical and damage behavior of laminated composite under multidirectional loadings. The best method to refine the damage-resistance characteristics of laminated composites is to use an appropriate combination of fiber orientations, plies thickness, adequate number of laminates and properties that exactly fit the requirement for a particular purpose [2].

Finite-element method has been widely used to study and predict the damage behavior of laminated structures through the implementation of different failure criteria [3-4]. The early generation of failure criteria was based on those criteria developed for isotropic materials, such as von Mises criterion. Unfortunately, most of these early developed criteria are not associated to any type of failure mode. In recent years, Hashin criterion [5] was extensively considered for composite related problems as it associated with different types of failure modes, in specific fiber buckling and matrix cracking.

The purpose of this study is to evaluate the failure modes of different laminate properties cross arms under quasi-static loading condition. The study was conducted using a computational techniques by developing a three-dimensional finite-element model that simulated the standard configuration of cross arm installed on a transmission tower. The failure of cross arm upon loading was evaluated by Hashin's failure criteria, which being employed through user subroutine. The analysis focused on the structural deflection, and the intra-laminar damages exhibited by the cross arm during the application of static load at one end. This finding may assist engineers in selecting the best laminate properties to comply with the severe multi-axial load condition experienced by the composite cross arm.

2. Description of numerical models

2.1 Geometry, mesh and boundary conditions

In this section, the generation of the finite-element models are described following the scope of Abaqus commercial package. A numerical cross arm model was developed by using the finite-element analysis software program Abaqus/CAE. The analysis was carried out with explicit solver, and this solver was selected for quasi-static modeling due to high computational time needed for the whole model analysis. A proper combination of loading rates and mass scaling methods have been applied to reduce the computational time of the explicit time integration method [6].

As shown in Figure 1, the cross arm model consists of four square (0.12 x 0.12 mm^2) composite beam with a length of 4.9 m. The square beam composed 7 stacked plies of glass laminate, each ply having a 1.0 mm thickness. The stacking sequences of the beam are [45/ -45/ 0/ 90/ 0/ 90/ 0]. The cross arm model was modeled by using a total of 73472 linear hexahedral element with reduced integration (C3D8R). The end of the cross arm was loaded with a longitudinal force (F_l) of 6524 N, a vertical downward force (F_v) of 84992 N and a transverse force (F_t) of 46416 N. These force magnitudes reflect the weight of power lines hanged between the cross arms (multiplied by a safety factor of four). The other ends of the cross arm were constrained to be fixed in all directions (encastre).

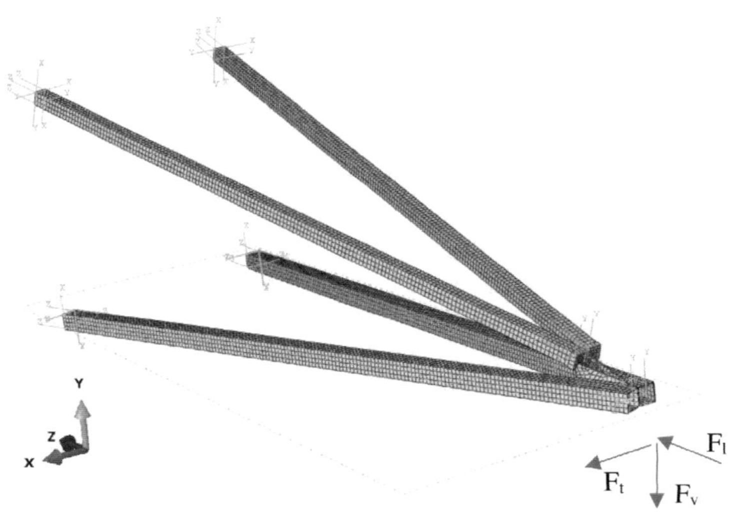

Figure 1. Geometry and mesh of the finite-element model of composite cross arm.

The possible location of the contact between the layer surfaces was defined by surface-to-surface discretization, with a normal contact defined as "hard". For the analysis output, the user-dependent state variable (SDV 1 to SDV 4) was requested from the numerical simulation as these variables represent

IOP Conf. Series: Materials Science and Engineering **530** (2019) 012029 doi:10.1088/1757-899X/530/1/012029

the failure modes of fiber breakage in tension, fiber buckling in compression, matrix cracking in tension and matrix crushing in compression, respectively. Additionally, the deflection (U) of the whole model was also requested for comparison.

2.2 Material properties

The material that constitutes each ply of the cross arm is a glass laminate. The provided mechanical properties of each laminate are presented in Table 1. Three laminate properties, in specific, laminate A, B and C were considered to evaluate the structural deformation of the cross arm. It is worth noting that small difference in laminate properties was recognized between laminate A and C, while greater young modulus and ultimate stresses was recorded for laminate B. This input of material properties was assigned by means of the *USER MATERIAL instruction in Abaqus framework.

Table 1. Mechanical properties of laminate A, B and C.

No.	Parameter	Laminate A	Laminate B	Laminate C
1	Density	1.8 g/cm^3	1.9 g/cm^3	1.8 g/cm^3
2	Young's modulus, E_1	16000 MPa	36300 MPa	17000 MPa
3	Young's modulus, E_2	4800 MPa	10890 MPa	5100 MPa
4	Young's modulus, E_3	1440 MPa	3267 MPa	1530 MPa
5	Poisson's ratio (V_{12}=V_{13}=V_{23})	0.28	0.28	0.28
6	Shear modulus (G_{12}=G_{13}=G_{23})	4000 MPa	4000 MPa	4000 MPa
7	Ultimate tensile stress, X_{1T}	321 MPa	429 MPa	321 MPa
8	Ultimate compressive stress, X_{1C}	150 MPa	320 MPa	150 MPa
9	Ultimate tensile stress, X_{2T}	80 MPa	100 MPa	80 MPa
10	Ultimate compressive stress, X_{2C}	65 MPa	76 MPa	65 MPa
11	Ultimate shear stress, S_{12}	90 MPa	95 MPa	89 MPa
12	Ultimate shear stress, (S_{13} = S_{23})	60 MPa	70 MPa	50 MPa

2.3 Failure criteria

A VUMAT/Hashin subroutine was employed to predict the breaking of the fibers and the progressive damage in the matrix of the cross arm. The Hashin failure criterion is based on the work of Hashin [5]. Unlike the polynomial criteria such as the Tsai-Wu criteria [7], which propose a single equation to predict damage initiation, Hashin damage initiation criterion takes into account four possible failure modes. The four possible failure modes are fiber breakage in tension, fiber buckling in compression, matrix cracking in tension and matrix crushing in compression. Equation 1 to 4 were used to predict the previous failure modes respectively, and failure is expected when one of the following equations is satisfied (equivalent to 1).

$$(\frac{\sigma_{11}}{X_{1T}})^2+(\frac{\sigma_{12}}{S_{12}})^2+(\frac{\sigma_{13}}{F_{13}})^2=1 \tag{1}$$

$$(\frac{\sigma_{11}}{X_{1C}})^2=1 \tag{2}$$

$$\frac{(\sigma_{22}+\sigma_{33})^2}{X_{2t}^2}+\frac{\sigma_{23}-\sigma_{22}\sigma_{33}}{S_{23}^2}+\frac{\sigma_{12}^2+\sigma_{13}^2}{S_{12}^2}=1 \tag{3}$$

IOP Conf. Series: Materials Science and Engineering **530** (2019) 012029 doi:10.1088/1757-899X/530/1/012029

$$\left[\left(\frac{X_{2c}}{2S_{23}}\right)-1\right]\frac{(\sigma_{22}+\sigma_{33})}{X_{2c}}+\frac{(\sigma_{22}+\sigma_{33})^2}{4S_{23}^2}+\frac{(\sigma_{23}^2-\sigma_{22}\sigma_{33})}{S_{23}^2}]+\frac{\sigma_{12}^2+\sigma_{13}^2}{S_{12}^2}=1 \qquad (4)$$

where σ_{11}, σ_{22}, σ_{33}, σ_{12}, σ_{13} and σ_{23} are the applied stresses.

3. Result and discussion

Table 2 summarizes the deflection and failure mode of the numerical model for the three laminate properties. The maximum deflection indicates the highest deflection allows before failure started to occur on the finite-element model. It is seen that laminate C recorded the highest deflection of 0.187 m, followed by laminate A and B. The small deflection recorded with laminate B is related to its higher young modulus, especially for the first direction (E_1). This high young modulus of laminate B cause the cross arm to be a lot stiffer, hence resulting to a smaller deflection for a given bending load.

Additionally, it is worth noting that every laminate properties considered in this study lead to a failure of the cross arm model. The value 1 recorded in Table 2 indicates that the damage initiation criterion was met. It is recognized that both laminate A and C exhibited fiber breakage due to tension (SDV 1) and fiber buckling due to compression (SDV 2). Interestingly, since laminate B result to a smaller deflection upon the loading, only failure due to fiber buckling in compression was recorded.

Table 2. Deflection and failure modes of cross arm of different laminate properties.

No.	Parameter	Laminate A	Laminate B	Laminate C
1	Maximum deflection (m)	0.174	0.082	0.187
2	Fiber breakage in tension (SDV 1)	1	0	1
3	Fiber buckling in compression (SDV 2)	1	1	1
4	Matrix cracking in tension (SDV 3)	0	0	0
5	Matrix crushing in compression (SDV 4)	0	0	0

Figure 2 shows the failure of fiber in compression (SDV 2) of the cross arm for laminate A, B and C. There are two dominant colors can be seen; the blue color indicates the region which free from failure and the red region highlight a region with fiber buckling. It is observed that despite the variation in laminate properties, each case exhibited the failure on similar location, which localized nearly at the middle region of the lower beams. This observation also highlighted that throughout the application of the load, the lower beams experienced greater bending deformation than the upper beams.

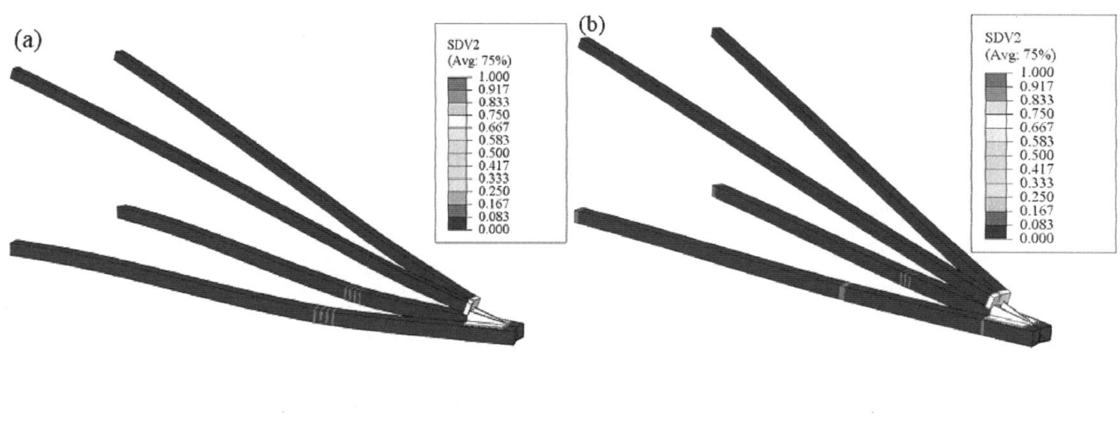

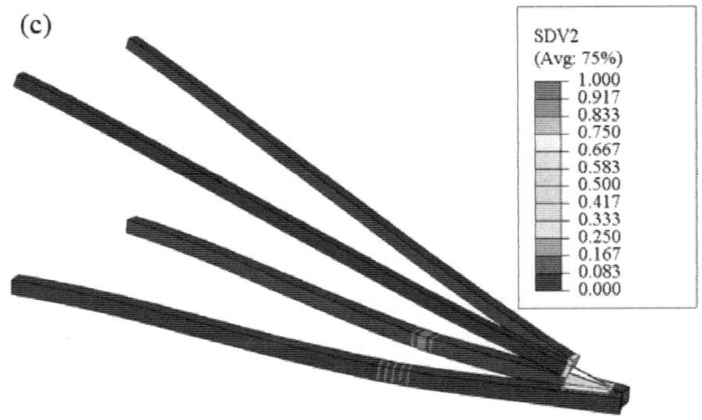

Figure 2. Failure location due to fiber buckling in compression (SDV 2): (a) laminate A, (b) laminate B and (c) laminate C.

4. Conclusion

The implementation of VUMAT/Hashin with finite element enabled the prediction of inter-laminar damages of a composite cross arm. A variation in laminate properties of the cross arm yielded different bending deflection and modes of failure. A cross arm with greater value of young modulus and ultimate stresses (laminate B) exhibited smaller deflection and lesser number of failures upon subjected to multi-axial load condition.

Acknowledgement:

TNB Seeding Fund funds this research for Universiti Tenaga Nasional (U-T-S-RD-17-05). The authors would like to thank TNB Transmission Line Division and TNB for financial support.

References:

[1] N Jauhari, R Mishra and H Thakur 2017 Failure Analysis of Fibre-Reinforced Composite Laminates *Mater. Today Proc.* Vol 4 No 2 Part A pp 2851–2860
[2] F S Almeida and A M Awruch 2009 Design optimization of composite laminated structures using genetic algorithms and finite element analysis *Compos. Struct.* Vol 88 No 3 pp 443–454
[3] Wang Gong-Dong, Melly Stephen Kirwa and Ahmed S K Kafi 2018 Finite element study into the effects of fiber orientations and stacking sequence on drilling induced delamination in CFRP/Al stack *Science and Engineering of Composite Materials* Vol 25 p 555
[4] J Ye, Y Yan, J Li, Y Hong and Z Tian 2018 3D explicit finite element analysis of tensile failure behavior in adhesive-bonded composite single-lap joints *Compos. Struct.* Vol 201 pp 261–275
[5] Z Hashin Jun 1980 Failure Criteria for Unidirectional Fiber Composites *J. Appl. Mech.* Vol 47 No 2 pp 329–334
[6] J Wang, T Chen, D Sleight and A Tessler 2004 Simulating nonlinear deformations of solar sail membranes using explicit time integration in *45th AIAA/ASME/ASCE/AHS/ASC Structures, Structural Dynamics & Materials Conference* p 1580
[7] S W Tsai and E M Wu 1971 A General Theory of Strength for Anisotropic Materials *J. Compos. Mater.* Vol 5 No 1 pp 58–80

IOP Conf. Series: Materials Science and Engineering **530** (2019) 012030 doi:10.1088/1757-899X/530/1/012030

Stacking Sequence Effects on Performance of Composite Laminate Structure Subjected To Multi-Axial Quasi-Static Loading

D Mohamad[1], A Syamsir[2], S N Sa'don[1], N M Zahari[1], S A H A Seman[3]*, M F Razali[3], A Abas[3] and F C Ng[3]

[1] Department of Civil Engineering, Universiti Tenaga Nasional, 43000 Kajang, Malaysia.

[2] Institute of Energy Infrastructures (IEI), Universiti Tenaga Nasional, 43000 Kajang, Malaysia.

[3] School of Mechanical Engineering, Universiti Sains Malaysia, Engineering Campus, 14300, Nibong Tebal, Penang, Malaysia.

Corresponding author *: hertz_zone@yahoo.com

Abstract. In this paper, the effects of the stacking sequence on the mechanical performance and damage formation of composite laminates subjected to multi-axial quasi-static loading is investigated, considering intra-laminar damage onset. The response of the composite laminate configurations characterized by different stacking sequences subjected to multi-axial quasi-static loading have been studied to predict the influence on the static displacement and damage development. A finite element (FE) model has been used to numerically simulate the response of the composite structure. ABAQUS/EXPLICT FE environment has been considered for the analyses and Hashin's failure criteria has been adopted using a VUMAT subroutine to model the intra-laminar damage formation in the analyzed composite structure.

1. Introduction

Through years, the advancement of composite materials has been motivated by the growing need of materials with high specific mechanical properties and low specific weight. Many industries have conducted some exploration in search for ideal structural elements able to bear capacity of multi-axial static loading. Composite laminates consisting of multiple layers or plies which oriented in the desired directions are promising. In general, the industrial requirements and practical manufacturing considerations have restricted the stacking sequence combination of laminates to comply with ply orientations of 0, 90, 45 and -45 degrees. Depending on stiffness requirements, different assembling of plies is expectable. Even though a laminate might possess good stiffness properties, a composite laminate with the same fiber orientations is inefficient and may show a poor structural response when applied with in-plane loads. Therefore, the optimization of composite laminate designs towards a better damage resistance and tolerance can be achieved by varying its stacking sequence.

The influence of stacking sequence of laminated composite structure on the quasi-static response has been studied by several authors [1-4]. In this paper the effect of the stacking sequence on the composite mechanical performance and damage formation has been evaluated. A detailed FE model of the composite laminates have been developed in ABAQUS/EXPLICIT FEM code. The Hashin's [5] failure criteria are implemented using a VUMAT subroutine provided by ABAQUS to predict the intra-laminar damage. Three different stacking sequences have been considered, in order to better understand the influence of stacking sequence on the structural deflection and also intra-laminar damages formation.

2. Damage modelling

For the modelling of intra-laminar damages, such as the breaking of the fibers and the progressive damage in the matrix, the progressive failure approach implemented in the algorithms of the Abaqus Explicit solver has been considered. This approach uses the Hashin's criteria [5] for the detection of damage initiation. The intra-laminar failure mode can be represented by a bi-linear constitutive law as shown in figure 1. In figure 1, the point A shows the damage initiation stress–strain values: before this point the material is undamaged. When the limit values defined was exceeded by the element stress, the damage propagation starts up to point B where the element is totally degraded. The damage initiation criteria are based on the Hashin's theory, which is separated by different intra-laminar failure modes. Four different modes of failure have been considered are fiber rupture in tension and compression and matrix cracking under transverse tension and compression. The damage variable for different failure modes are given by the following expression listed in table 1.

Table 1: Hashin's failure initiation criteria for each failure mode.

Failure criteria	Failure modes
$(\frac{\sigma_{11}}{X_{1T}})^2+(\frac{\sigma_{12}}{S_{12}})^2+(\frac{\sigma_{13}}{S_{13}})^2=1$	Fiber Tensile failure
$(\frac{\sigma_{11}}{X_{1C}})^2=1$	Fiber Compressive failure
$\frac{(\sigma_{22}+\sigma_{33})^2}{X_{2t}^2}+\frac{\sigma_{23}-\sigma_{22}\sigma_{33}}{S_{23}^2}+\frac{\sigma_{12}^2+\sigma_{13}^2}{S_{12}^2}=1$	Matrix Tensile failure
$\left[\left(\frac{X_{2c}}{2S_{23}}\right)-1\right]\frac{(\sigma_{22}+\sigma_{33})}{X_{2c}}+\frac{(\sigma_{22}+\sigma_{33})^2}{4S_{23}^2}+\frac{(\sigma_{23}^2-\sigma_{22}\sigma_{33})}{S_{23}^2}]+\frac{\sigma_{12}^2+\sigma_{13}^2}{S_{12}^2}=1$	Matrix Compressive failure

where X_{1t} , X_{1c}, X_{2t} X_{2c} denotes tensile and compressive failure strength in fibre direction 1 and 2 while S_{12} S_{13} and S_{23} denotes failure shear strength in 1-2, 1-3 and 2-3 planes.

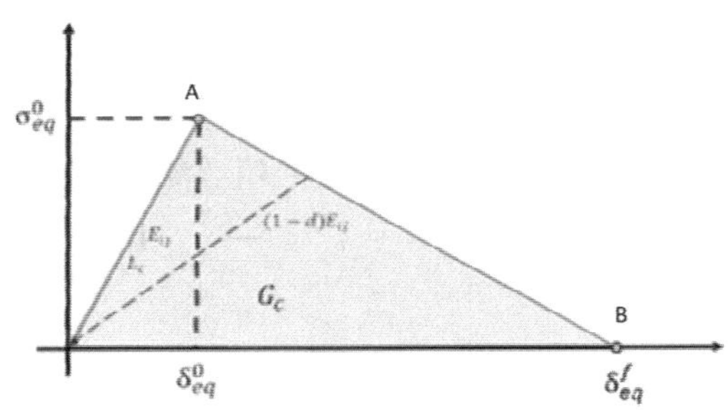

Figure 1. Damage initiation and propagation points.

IOP Conf. Series: Materials Science and Engineering **530** (2019) 012030 doi:10.1088/1757-899X/530/1/012030

3. FE Model

The finite element model has been built in the ABAQUS/EXPLICIT FE code. This alternative approach has been selected for quasi-static modelling due to high computational time needed for whole model analysis. Therefore, a proper combination of loading rates and mass scaling methods have been applied to reduce the computational time of the explicit time integration method [6]. In this study, the composite laminate has been modelled as the electric power grid substation structure. Each arm which made-up of composite laminate was modelled with 7 equal thickness plies. The composite laminate has been modelled by using the solid element formulation available in the Abaqus database. Each element has eight nodes with three degrees of freedom at each node and one integration point. Two solid elements have been modelled across the thickness in each ply in order to improve the prediction of intra-laminar damage. Throughout the analysis, localized stiffness reduction is definite due to the existence of failed elements and subsequent element deletion. For the interface conditions, a perfect bonding was assumed between the plies. The arm structure was pinned on both four end and multi-axial loads were applied on the front end as shown in figure 2. Friction has been introduced between all the contacting surfaces with an average friction coefficient of 0.25. Figure 3 shows the three different stacking sequence schemes used in this paper. For all the numerical models the material properties reported in table 2 have been considered.

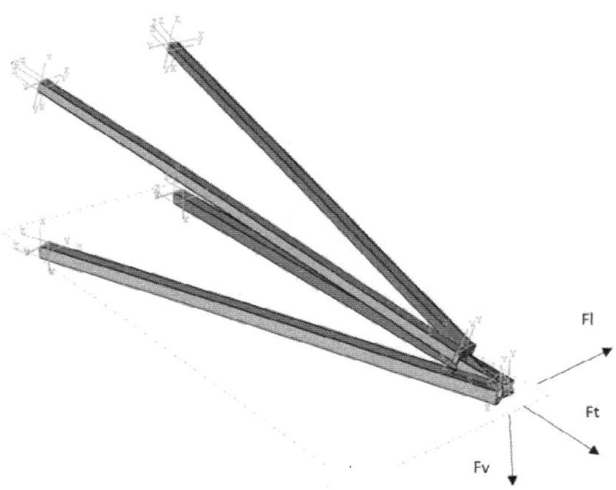

Figure 2. FE model of electric power grid substation structure.

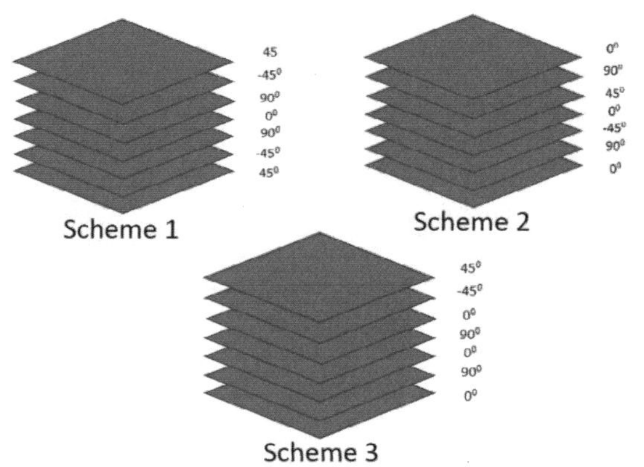

Figure 3. Different stacking sequences.

Table 2. Material properties of each ply.

Intra-laminar model properties	
Density	1900 kg/m^3
Orthotropic properties	$E_1 = 17$Gpa $E_2 = E_3 = 5.1$Gpa
	$G_{12} = G_{13} = G_{23} = 4$Gpa
	$\nu_{12} = \nu_{13} = \nu_{23} = 0.26$
Ultimate stress	$X^T = 321$Mpa $Y^T = 80$Mpa
	$X^C = 151$Mpa $Y^C = 65$Mpa
	$S_{12} = 89$Mpa $S_{13} = S_{23} = 50$Mpa

IOP Conf. Series: Materials Science and Engineering **530** (2019) 012030 doi:10.1088/1757-899X/530/1/012030

4. Result and Discussions

In this section, the simulations performed on the analyzed three different stacking sequences for the considered multi-axial static loading are presented and discussed. The mechanical and damage behavior of the numerical models are represented by the graph of the static displacement contour and damage formation location. In figure 4a and Table 3, the maximum displacement of structure for the different composite schemes are presented. As shown in Figure 4a, the maximum deflection of all three structure with different stacking sequence were predicted at the front end near the applied load area. Based on table 3, the total deflection value of structure was different for each composite laminate schemes. Obviously, the value of maximum deformation is getting bigger from Lay-up Schemes 1 to 3. As proportion of 45-degree plies decreased within the composite laminate structure, amount of deflection has been increased. Besides, results comparison of scheme 2 and 3 show that stacking sequence has only a slight effect on maximum deflection.

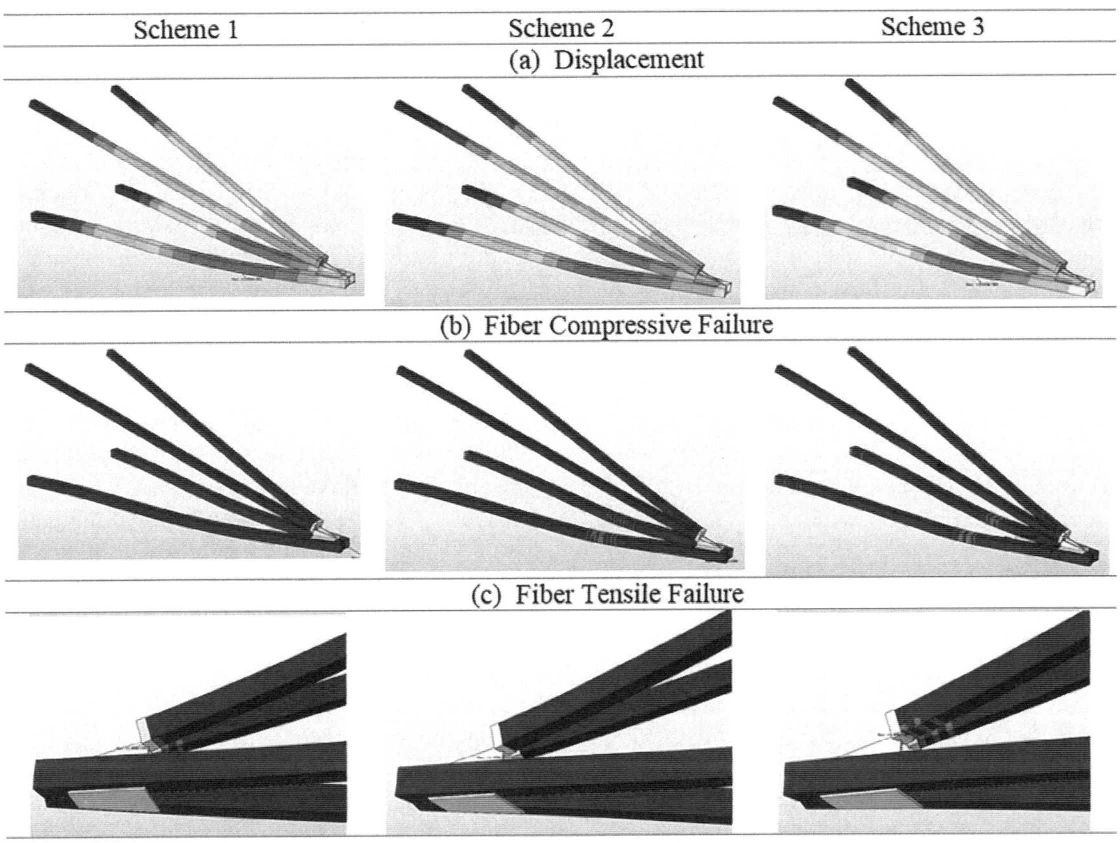

Figure 4. Numerical results (a) Displacement contour (b) Fiber compressive failure location (c) Fiber tensile failure location.

Table 3. Maximum deflection of structure with different stacking sequence

Scheme	Composite laminate lay-up	Maximum deflection of structure (mm)
1	[45/-45/90/0/90/-45/45]	189
2	[0/90/45/0/-45/90/0]	234
3	[45/-45/0/90/0/90/0]	245

The smaller permanent deformation indicates better quasi-static loads characteristics. Scheme 1 with minimal permanent deformation (189 mm) is the optimal lay-up scheme. The results show that stacking sequence has little effect on static displacement, and proportion of layers with different fiber directions has obvious effect on static displacement. The higher the proportion of layers with 45-degree fiber directions is, the lesser the static displacement will be. A good improvement on the impact performance has also been reported by Hu et al. [7] when a higher proportion of layers with ±45° fiber directions was included in the composite laminate.

As can be seen in figure 4b, the localized damage was predicted near the area where the displacement is maximum and concentrated around the area where the arms were pinned. The damage in the structure which is fiber compressive failure was predicted in both lower arms of scheme 2 and 3. However, the predicted results show a wider area of damage on the lower arm of scheme 2. Comparison of results between scheme 2 and 3 show that movement of 45-degree plies from the extreme end has amplified the element normal stresses, σ_{11} which has disrupted the ability of composite laminate to resist fiber compressive failure formation. On the other hand, for scheme 1, due to higher proportion of 45-degree layers located on both ends, element normal stresses, σ_{11} have been reduced which consequently avoided the formation of fiber compressive failure.

In figure 4c, the localized damage was predicted near the area where the arms were attached to a connector. Fiber tensile failure formation was predicted in both upper arms of scheme 1 and 3. However, the predicted results show a broader area of damage on the upper arm of scheme 3. Comparison of results between scheme 2 and 3 show that movement of 45-degree plies from the extreme end has encourage the ability of composite laminate to resist the fiber tensile failure formation. This is because the +45° and -45° plies have resisted the element shear stresses, S_{12} and S_{13} that were localized on the center region of composite laminate.

Referring to figure 4b and 4c, both stacking sequence and proportion of layers with different fiber directions have obvious effect on damage formation. In scheme 2 and 3, the movement of 45-degree plies from the extreme end has increased the formation of fiber compressive failure but able to resist the formation of fiber tensile failure. Additionally, the higher proportion of layers with 45-degree fiber directions was able to reduce the formation of fiber tensile and compressive failure.

Based on the above discussion to determine best stacking sequence for composite laminate, the best stacking sequence is for scheme 1, when two layers of 45 and -45 oriented plies are at both ends and remaining three layers of 0 and 90 are in between them. The position and sequence of arrangement of plies in which the properties come second is for scheme 2 followed by scheme 3.

5. Conclusion

FE analysis have been carried out on the electric power grid substation structure subjected to multi-axial static loading to investigate the effect of stacking sequence on the mechanical performance and damage location of the laminated composites. Based on the above results, stacking sequence has little effect on static displacement but proportion of layers with different fiber directions has obvious effect on static displacement. The higher the proportion of layers with 45-degree fiber directions is, the lesser the static displacement will be. However, both stacking sequence and proportion of layers with different fiber directions have obvious effect on the damage formation. The best stacking sequence (position and orientation) is for scheme 1 which show smaller deformation and lesser area of damage formation.

Acknowledgement
TNB Seeding Fund funds this research for Universiti Tenaga Nasional (U-T-S-RD-17-05). Author would like to thank TNB Transmission Line Division and TNB for financial support.

References

[1] Quinn W J and Matthews F L 1977 *Journal of Composite Materials* 11(2) 139-145
[2] Smith P A and Pascoe K J 1986 *Composite Structures* 6(1) 1-20
[3] Naghipour P, Bartsch M, Chernova L, Hausmann J and Voggenreiter H 2010 *Materials Science and Engineering: A* 527(3) 509-517
[4] Le Riche R and Haftka R T 1993 *AIAA Journal* 31(5) 951-956
[5] Hashin Z 1980 *Journal of Applied Mechanics* 47(2) 329-334
[6] Wang J, Chen T, Sleight D and Tessler A 2004 *45th AIAA/ASME/ASCE/AHS/ASC Structures, Structural Dynamics & Materials Conference* (Palm Springs, California: American Institute of Aeronautics and Astronautics)
[7] Hu Y, Xiao Y, Shang W and Zhang J 2017 *International Journal of Crashworthiness* 22(5) 556-564

IOP Conf. Series: Materials Science and Engineering **530** (2019) 012031 doi:10.1088/1757-899X/530/1/012031

Multi-objective Optimization of Vehicle Speed Control using Gravitational Search Algorithm for Electro-Mechanical Continuously Variable Transmission

K Hudha[1*], N R M Nuri[2,3] and S A Mazlan[3]

[1] Faculty of Engineering, Universiti Pertahanan Nasional Malaysia, Kem Sungai Besi, 57000 Kuala Lumpur, Malaysia

[2] Faculty of Engineering Technology, Universiti Teknikal Malaysia Melaka, Hang Tuah Jaya, 76100 Durian Tunggal, Melaka, Malaysia

[3] Malaysia-Japan International Institute of Technology, Universiti Teknologi Malaysia Kuala Lumpur, Jalan Sultan Yahya Petra, 54100 Kuala Lumpur, Malaysia

Corresponding author *: k.hudha@upnm.edu.my

Abstract. This paper describes the application of the Gravitational Search Algorithm (GSA) for the optimization of the vehicle speed control in the longitudinal vehicle model equipped with Continuously Variable Transmission (CVT) system. In this study, multi-objective procedure which are to minimize the vehicle speed error and fuel usage has been proposed. For comparing and analysing the optimization results, CVT and Automatic Transmission (AT) are placed in the vehicle transmission model separately, and the Proportional, Integral and Derivative (PID) controller is implemented to control the vehicle speed. The simulation work is modelled and performed using MATLAB/Simulink software. Overall, the results show that the vehicle speed transient responses and fuel usage optimized by GSA are better than the values gained from PID controller.

1. Introduction

Latest technologies in the vehicle must satisfy the environmental issues such as excessive pollution emissions and usage of the hydrocarbon fuels. To attain better fuel economy and lower emissions, researchers have taken a lot of effort in improving the transmission system efficiency. Therefore, general type of transmission which is manual transmission (MT), automatic transmission (AT) or continuously variable transmission (CVT) system has been upgraded to fulfill the demand for better vehicle performance. Among these transmission systems, CVT has more capability to achieve good fuel economy and vehicle acceleration. The CVT offers a continuous gear ratio operation within its range without shift shock, thus allow the engine to run in optimum stage for various vehicle load conditions [1].

Most of the vehicle equipped with CVT system use electro-hydraulic-mechanical (EHM) actuation type to change the CVT ratio for smooth and comfortable drive. However, the EHM actuation system requires weighty components, leakage possibility within the hydraulic line system and difficult to control due to the non-linearity of the compressible hydraulic oil [2]. Another option is used electro-mechanical (EM) actuation system, where the system only consists of electronic and mechanical components in getting the desired transmission ratio. In this study, AT and single actuator double acting electro-mechanical (SADAEM) CVT system are used to represent the vehicle transmission model [3]. Meanwhile, for the vehicle performance simulation, lateral forces acting on the vehicle model are neglected, thus the longitudinal vehicle dynamic model is selected to be used in the simulation. For controlling the vehicle performance, several vehicle parameters need to be accessed, such as the vehicle speed and fuel usage.

Previous studies were implemented conventional Proportional, Integral and Derivative (PID) or advanced controllers such as Fuzzy and Fuzzy-PID to control the vehicle speed [4-7]. Additionally, several optimization techniques were introduced such as Particle Swarm Optimization (PSO), Genetic Algorithm (GA), Ant Colony Algorithm (ACA) and Gravitational Search Algorithm (GSA) [8]. By merging one or more of this optimization algorithm, better and optimal of desired results can be obtained [9-10]. All these techniques are used mathematical programming such as MATLAB software for optimizing the desired objectives. Thus, in this study, the GSA that has been proposed by Rashedi et al. [11] is chosen because can performs better results compared to PSO [12]. The objectives of this study are to minimize the vehicle speed error and fuel usage of the developed vehicle longitudinal model.

This paper is structured as follows. Section 1 gives the introduction and several previously and current related works. Next, Section 2 focuses on the development of vehicle longitudinal model which involve several important mathematical equations and systems. Then, Section 3 shows the proposed of the vehicle control strategies using PID and optimization using the GSA. Section 4 discusses the simulation results based on the proposed control strategies and is followed by the conclusion in Section 5.

2. Longitudinal Vehicle Dynamics Model

Consider a hatchback type of vehicle is moving on an inclined road as shown in Figure 1. The longitudinal vehicle dynamic model is presented by a lumped mass, (m) with a forward velocity of (V) along the X-direction. The vehicle wheels rotate at angular velocity presented by (ω_i) with wheel rolling radius of (R_i) and a polar moment of inertia, (J_i). The subscript of $i=f,r$ describes either the front wheel or rear wheel of the vehicle. Then, the coefficient of friction between the road surface and front wheel is donated by (μ_f) and rear wheel is presented as (μ_r), respectively. The dimensions (l_f) and (l_r) represent the distance between the center of vehicle mass and the front axle and rear axle, respectively, with the vehicle wheelbase is denoted by the dimension of (L). The gap between the vehicle center of mass and the ground is presented as (H) and (ϕ) indicates the angle of road inclination.

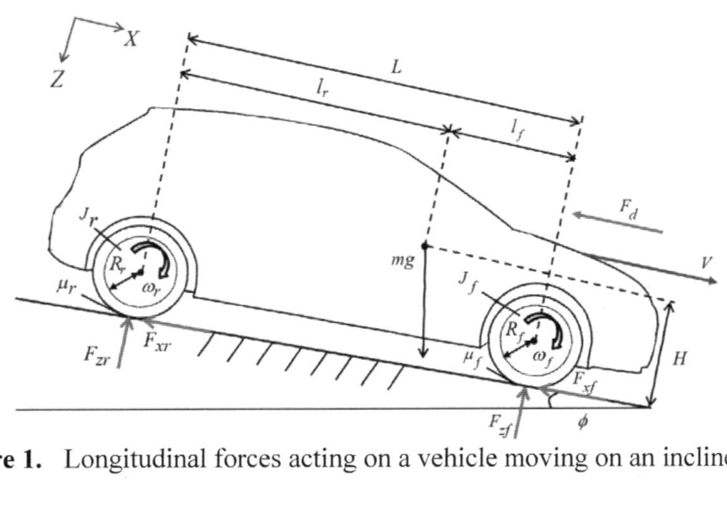

Figure 1. Longitudinal forces acting on a vehicle moving on an inclined road

By applying Newton's second law which is based on the forces balance acting on the vehicle, the vehicle acceleration can be obtained as follow:

$$\ddot{x} = \left[-F_{X_T} - F_d + mg\sin(\phi) \right] / m \tag{1}$$

where total forces that acting at the vehicle body is denoted as (F_{XT}), (F_d) is the aerodynamic resistance force, gravitational acceleration is denoted as (g) and (m) is the mass of the vehicle.

In the longitudinal vehicle model, the main forces that influence the vehicle dynamics are act in the X-direction as well as in the Z-direction. The total forces that act on the vehicle through the vehicle wheels or road surface can be defined as below:

$$F_{X_T} = 2F_{xf} + 2F_{xr} \tag{2}$$

$$F_{xf} = \mu_f F_{zf}; F_{xr} = \mu_r F_{zr} \tag{3}$$

where (F_{zf}) and (F_{zr}) is the normal force that act at the front wheel and rear wheel, respectively. Meanwhile, (F_{xf}) and (F_{xr}) is the reaction force at the front and rear wheel, respectively.

Another important parameters, namely the load distributions, drag forces, wheel dynamics and wheel tractive properties were referred to Aparow et al. [13] and Satar et al. [14]. Then, in term of engine dynamics, 75 kW engine was selected which delivers maximum torque of approximately 160 Nm at engine speed about 3000 rpm.

In this study, two types of transmission were used, namely AT and CVT. A 5-speed AT was selected with the gear shift map data was taken from the CarSim software. In the gear ship map data, there are two corner points at each of the gear shift profile, which at 20% and 80% of throttle opening setting. This allows the engine to operate at the desired torque region. The engine speed will increase when the throttle opening is risen, results the gear numbers to upshift in sequence. In contrast, the gear shift map allows the AT to drop into lower gear if the engine speed is reduced by applying the brake or minimizing the throttle opening. The MATLAB/Simulink block namely *If=else* is used to represents the gear shift map in either for upshifting or downshifting the gears.

On the other hand, another type of transmission is considered, namely SADAEM CVT system which has a pulley gear ratio range from 0.3 to 2.9, that was placed in the transmission model. In simulation, virtually the input shaft of SADAEM CVT system is connected to the ICE via torque converter and the output shaft is coupled with the differential which has a final gear ratio of 4.1 [3].

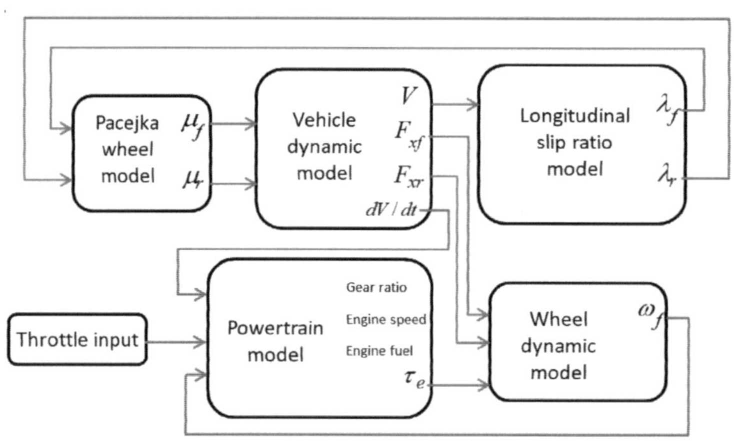

Figure 2. Longitudinal vehicle model in the MATLAB/Simulink software

The developed vehicle dynamics model in longitudinal direction was simulated in the MATLAB/Simulink software where the vehicle was assumed to run on a flat and dry tarmac road surface as shown in Figure 2. A sudden acceleration at 100 % throttle opening was set up to be input to the vehicle model. The simulation was performed using *Bogacki-Shampine* solver for a period of 100 seconds with a fixed step size of 0.01 second. Meanwhile, all the parameters values can be assessed in the CarSim software. All parameters for the vehicle model in the MATLAB/Simulink and CarSim software were kept at the same values and are listed in Table 1. For verification process, both models were used hatchback vehicle type and AT system.

Table 1. Parameters of hatchback type vehicle model

Description	Symbol	Value
Wheelbase	L	2.347 m
Front length from COG	l_f	1.103 m
Rear length from COG	l_r	1.244 m
Height of vehicle from COG	H	0.540 m
Vehicle sprung mass	m	747 kg
Initial vehicle velocity	Vo	1 m/s
Aerodynamic drag coefficient	Cd	0.48
Vehicle frontal area	A	1.6 m^2
Air mass density	ρ	1.206 kg/m^3
Gravitational acceleration	g	9.81 m/s^2
Wheel radius	Rw	0.2786 m
Gear final	η_f	4.10
Gear 1	η_1	3.78
Gear 2	η_2	2.12
Gear 3	η_3	1.36
Gear 4	η_4	1.03
Gear 5	η_5	0.84

3. Vehicle Control Strategies

3.1. PID controller

For the vehicle equipped with AT, outer control loop was considered in evaluating the vehicle speed control of the developed model. Meanwhile, regarding vehicle with CVT, two control loops were designed. An outer loop to control the vehicle speed, while another loop in the transmission model to control the CVT ratio. Both AT and CVT were used same outer loop controller, where a linear, simple and robust controller, namely PID controller was applied as shown in Figure 3. On the other hand, in the transmission model, two types of controller were used which are the PD and PI controllers as shown in Figure 4. To obtain the proportional, integral and derivative gain values, trial and error technique was carried out. For the outer loop, the value of proportional, integral and derivative gain was set to value of 150, 1 and 5, respectively. In contrast, inside the transmission model, the proportional and derivative gain was fixed at 250 and 3, respectively. Meanwhile, proportional and integral gain was set at 53 and 25, respectively.

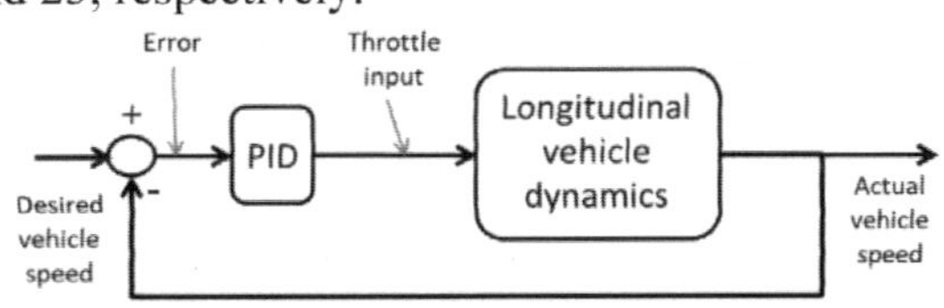

Figure 3. Vehicle speed control structure using PID controller

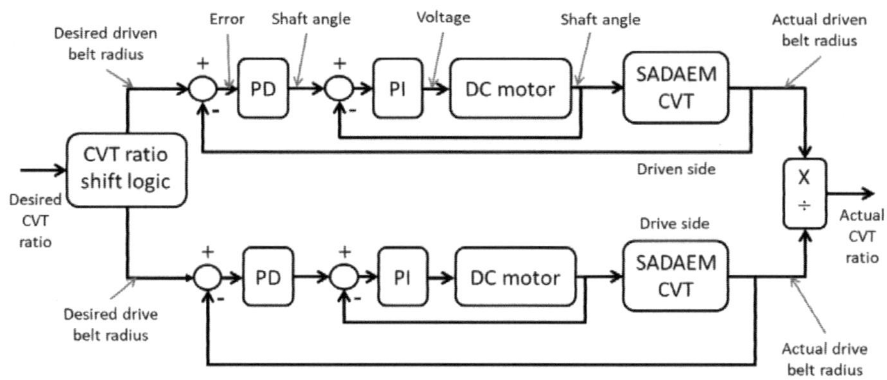

Figure 4. CVT ratio control structure in simulation

3.2. Optimization using the Gravitational Search Algorithm

The GSA utilizes Newton's law of gravity and law of motion to find the optimum values of PID controller parameters. In GSA, there are four parameters, namely position, inertial mass, active and passive gravitational mass for each agent that need to be considered. By using the Newton's law of motion, the acceleration is calculated before the velocity and position of each agent is updated. The updated position will give an update solution for the optimization problem through a fitness function. The updating process is repeated until the criterion is achieved. The GSA flow procedure can be summarized as shown in Figure 5. All the parameters and equations related to the GSA have been studied as stated by Sabri et al. [8], Rashedi et al. [11] and Amer et al. [15].

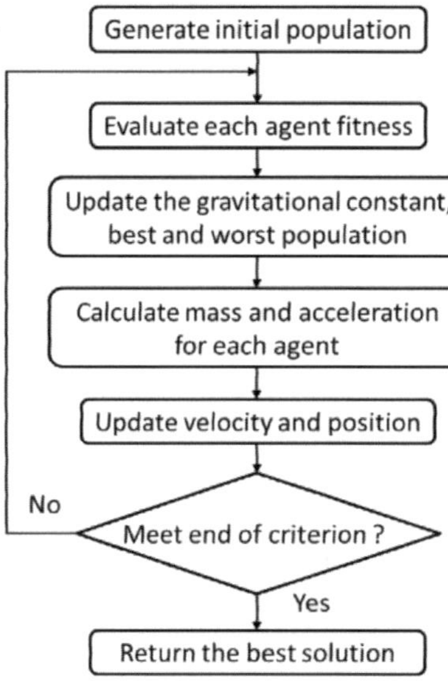

Figure 5. An algorithm of the GSA [8]

In this study, since the vehicle speed error and fuel consumption were intended to be minimized, the fitness function (j) related to these aims can be expressed by:

$$j = \sqrt{\left(\frac{e_v}{e_{v\max}}\right)^2 + \left(\frac{f_c}{f_{c\max}}\right)^2} \tag{4}$$

where the vehicle speed error is expressed as (e_v), the maximum error of vehicle speed is denoted as (e_{vmax}), the fuel consumption is stated as (f_c) and the maximum fuel consumption is represented as (f_{cmax}). The optimization process only concentrates on outer loop only. Moreover, initial value of the parameters using the GSA are listed in Table 2. On the other hand, both PID and optimization using GSA control strategies are using *Bogacki-Shampine* solver with a step size of 0.01 second for a period of 100 seconds.

Table 2. Initial parameters for GSA simulation

Parameter	Value
Agents number, N	50
Iteration number, T	50
Bheta, β	0.1
Epsilon, ε	0.01
Dimension number, D	3
Upper bound limit [Kp Ki Kd]	[200 10 20]
Lower bound limit [Kp Ki Kd]	[0 0 0]

4. Results and Discussion

4.1. Verification of vehicle longitudinal with CarSim model

Verification results are presented through comparison between vehicle model in MATLAB/Simulink and CarSim vehicle model in term of vehicle speed and fuel usage as shown in Figure 6. These vehicle responses are analysed by evaluating the percentage difference of the root mean square (RMS) for each model. As seen in the graphs, response trends between CarSim and simulation models are relatively in good agreement with minor deviation. For CarSim model, the vehicle speed and fuel consumption have the RMS value of 187.964 km/h and 0.372 L, respectively. Meanwhile for vehicle model, vehicle speed records RMS value of 187.999 km/h and fuel consumption shows RMS value of 0.371 L. Therefore, RMS percentage difference is up to 0.035 % for vehicle speed and 0.269 % for fuel consumption. The small RMS percentage difference and minor deviation of responses pattern are caused by various simplified model assumptions in designing the vehicle longitudinal model. However, the developed vehicle model is adequate to be used for determining another desired vehicle responses.

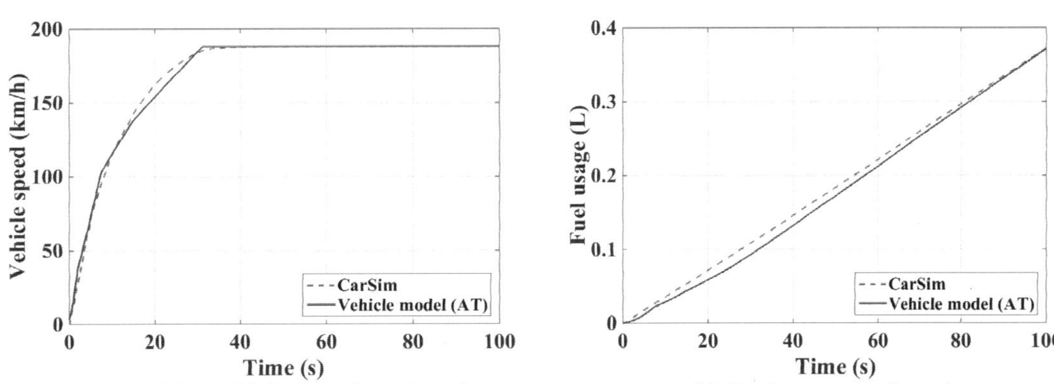

(a) Vehicle speed against time (b) Fuel usage against time

Figure 6. Verification results between developed vehicle and CarSim model

4.2. Optimization results

Figure 7 (a) and (b) show the vehicle speed and fuel usage responses using PID controller in AT and CVT system, while optimization using GSA has been performed for CVT system. From the graph, with same controller which is PID, the vehicle model consists of CVT is perform better than AT in term of rise time, settling time and fuel usage. Despite higher value of overshoot (OS) percentage and steady state error compared with AT system, the CVT system inside the vehicle model still can performed with acceptable level. In addition, optimization using GSA will improve the fuel usage up to 0.04 % when compared with CVT (PID) model. Moreover, the results proven that the vehicle model with GSA managed to obtain better transient responses and fuel usage results. The analysis of vehicle speed transient responses for each transmission and controller are tabulated in Table 3.

Table 3. Transient analysis for vehicle speed response

Transmission / Controller	Rise time (s)	Percentage of OS (%)	Settling time (s)	Steady state error (%)
AT / PID	10.90	2.54	16.00	1.36
CVT / PID	9.27	3.91	15.80	1.55
CVT / GSA	9.25	2.73	15.70	0.91

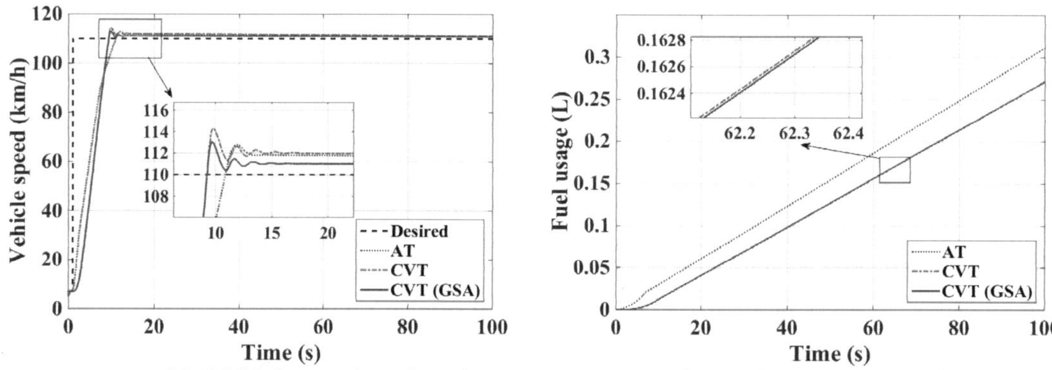

(a) Vehicle speed against time (b) Fuel usage against time

Figure 7. Vehicle responses with different transmissions and controllers

5. Conclusion

The developed vehicle longitudinal model was verified against the vehicle performance obtained from CarSim software in term of vehicle speed and fuel usage. The verification results show the pattern of the responses were almost similar but varies in term of RMS percentage difference values, approximately below 0.3 %. Moreover, the vehicle with CVT system shows better performance in term of acceleration and transient responses compared with AT system. Furthermore, in optimizing the vehicle speed and fuel usage in the CVT, the vehicle speed control using GSA has better characteristics and good dynamic behaviour than PID, which less response time, smaller overshoot, shorter settling time and less steady state error. Therefore, the proposed GSA method can be used in solving complex and highly non-linear of vehicle system.

Acknowledgement

This project is supported by Ministry of Education Malaysia through the research grant LRGS (LRGS/B-U/2013/UPNM/DEFENCE & SECURITY-PI).

References

[1] R Pfiffner, L Guzzella, C H Onder and C H 2003Fuel-optimal Control of CVT Powertrains *Control engineering practice* 11(3) pp 329-336
[2] M L H A Rahman, K Hudha, Z A Kadir, N H Amer and V R Aparow 2018 Modelling and Validation of a Novel Continuously Variable Transmission System using Slider Crank Mechanism *International Journal of Engineering Systems Modelling and Simulation* 10(1) pp 49-61
[3] N R M Nuri, K Hudha and S A Mazlan Design and Simulation of a New Single Actuator Double Acting Electro-Mechanical Continuously Variable Transmission *International Journal of Mechanical Engineering and Robotics Research* Accepted, waiting to publish
[4] O Munyaneza, B B Munyazikwiye and H R Karimi 2015 Speed Control Design for a Vehicle System using Fuzzy Logic and PID controller *IEEE International Conference on Fuzzy Theory and Its Applications (iFUZZY)* pp 56-61
[5] K Mahmud and L Tao L 2013 Vehicle Speed Control through Fuzzy Logic *IEEE In Global High Tech Congress on Electronics (GHTCE)* pp 30-35
[6] H M Li, X B Wang, S B Song and H Li 2016 Vehicle Control Strategies Analysis based on PID and Fuzzy Logic Control *Procedia engineering* 137 pp 234-243
[7] S Ariyono, K B Tawi, H Jamaluddin, M Hussein and B Supriyo 2007 Adaptive Neural Network Optimisation Control of ICE for Vehicle With Continuously Variable Transmission *IEEE International Conference In Intelligent and Advanced Systems (ICIAS)* pp 257-262
[8] N Mohd Sabri, M Puteh and M R Mahmood 2013 A Review of Gravitational Search Algorithm *Int. J. Advance. Soft Comput. Appl* 5(3) pp 1-39
[9] X Wu, B Cao, J Wen and Y Bian 2008 Particle Swarm Optimization for Plug-in Hybrid Electric Vehicle Control Strategy Parameter *IEEE Vehicle Power and Propulsion Conference (VPPC)* pp 1-5
[10] S Mirjalili, S Z M Hashim and S Z M 2010 A New Hybrid PSOGSA Algorithm for Function Optimization *IEEE International Conference on Computer and Information Application (ICCIA)* pp 374-377
[11] E Rashedi, H Nezamabadi-Pour and S Saryazdi 2009 GSA: A gravitational search algorithm *Information sciences* 179(13) pp 2232-2248
[12] L C Cagnina, S C Esquivel and C A C Coello 2008 Solving Engineering Optimization Problems with the Simple Constrained Particle Swarm Optimizer *Informatica* 32(3) pp 319-326
[13] V R Aparow, K Hudha, M M Hamdan and H Jamaluddin 2016 Development and Verification of a 9-Dof Armored Vehicle Model in the Lateral and Longitudinal Directions *Jurnal Teknologi* 78(6) pp 117-137
[14] M I Satar, K Hudha, W A Mat, R N Othman, M Murrad, V R Aparow and M S Salleh 2015 Modelling and Verification of 5 Degree of Freedom Vehicle Longitudinal Model *IEEE Control Conference (ASCC)* pp 1-6
[15] N H Amer, N Ahmad and A F Zainal Abidin 2015 Weight Minimization of Helical Compression Spring Using Gravitational Search Algorithm (GSA) *Applied Mechanics and Materials* 773 pp 277-281

Comparison of Control Strategy on Magneto-Rheological Fluid Damper Performance for Impact Reduction

M S Rahmat[1], K Hudha[1*], Z A Kadir[1], N R M Nuri[2], N H Amer[1] and S Abdullah[3]

[1]Department of Mechanical Engineering, Faculty of Engineering, National Defence University of Malaysia, Kem Sg. Besi, 57000 Kuala Lumpur, Malaysia.

[2]Department of Mechanical Engineering Technology, Faculty of Engineering Technology, Universiti Teknikal Malaysia Melaka, Hang Tuah Jaya,76100 Melaka, Malaysia.

[3]Research and Innovation Management Centre (RIMC), SEGi University, No. 9 Jalan Teknologi, Taman Sains Selangor, Kota Damansara, PJU 5, 47850 Petaling Jaya, Selangor.

Corresponding author *: k.hudha@upnm.edu.my

Abstract. This paper investigates the effects of control strategy selection on the effectiveness of impact reduction control for a magneto-rheological fluid (MRF) damper-based system. The control performance of a passive damper, an MRF damper controlled by skyhook and an MRF damper controlled by a fast practical control are compared. A mechanical system model of a single degree-of-freedom (SDOF) impact isolation system using MRF dampers with different controllers is constructed, and the governing equation for the SDOF impact isolation system is derived. The FPC control algorithm is used to verify the MRF damper performance of the systems. The control performance of the systems under impact loading from various horizontal impact energies is evaluated. The results indicate that the MRF damper with the FPC controller provides much better performance for impact reduction control.

Keywords. *Magneto-rheological fluid damper; impact reduction; control strategy; hybrid control; fast practical control.*

1. Introduction

The magneto-rheological fluid (MRF) damper is a semi-active damper that provides a controllable damping isolation system and has attracted much attention during the past two decades [1-2]. The MR fluid-based actuators and their semi-active control systems are typically employed to absorb or dissipate unwanted energy, such as vibration suppression [3–5] and shock mitigation [6–9]. The MRF damper has been applied to a variety of isolation systems, such as stability and comfort improvement in gun recoil systems [10–12], suspension and engine mounting systems in automobiles [13–16], suspension in railway vehicles [17–19] and seismic isolation systems in civil structures [20]. A semi-active control system based on the MRF damper has also proved useful in mitigating shock and vibration excitations to the system, such as in crashworthiness, gun recoil and seat suspension systems.

However, in high speed impact loading applications, the control strategy of the MRF damper is a significant factor in the performance of controllable damping to mitigate unwanted force. The available literature shows that most research on MRF damper and control performance has focused on noise and vibration attenuation, vibration absorption for large-scale buildings and related issues. Therefore, while discussion of control strategy of the MRF damper abounds, there are few studies on the control strategy of the MRF damper for an impact loading application. A few researchers, such as Hu et al.[20], have studied the control strategy of the MRF damper for an impact loading application. They invented a large-scale, single-ended MRF damper and developed a fuzzy logic controller capable of reducing recoil displacement by 40%. Bai et al. [14] developed a semi-active shock and vibration isolation system using an MRF damper and also invented a bidirectional-controllable mechanism using a skyhook control algorithm to improve shock and vibration performance. Bai & Wereley [13] studied the impact seat suspension for ground vehicles in a simulation analysis. They concluded that the proposed MRF damper can provide a larger damping force and wider dynamic range by using Optimal Bingham control.

Liu and Chen [21] designed a control strategy for the MRF damper for high-impact loads. In their research, a fuzzy proportional integral derivative (PID) control algorithm was developed, and the results significantly improved the system's dynamic performance under high impact damping conditions. Li and Wang [8] designed a full-scale gun recoil buffering system using an MRF damper and formulated an optimal control strategy with no feedback system (open-loop control). The study discovered that optimal control was better than passive control because it produces a smaller variation in recoil force. Fu et al. [22] invented a high viscosity MRF damper with a radial flow mode and design using a high viscosity linear polysiloxane magnetorheological fluid (HVLP MRF) as a control medium to produce excellent impact stability. Based on the literature, most of the research has focused on control strategy for MRF dampers used in impact applications. In summary, the control strategies for MRF dampers under impact loading that have been developed use a control algorithm with high computational cost. Such high computational cost can delay response of a control strategy, making it less viable in real-time applications. The present study seeks to formulate a simple control algorithm for the MRF damper for impact reduction control.

In the present study, MRF damper control characteristic performance is presented. First, MRF damper characteristics due to impact loadings are investigated using an impact pendulum test rig. The characteristics of the MRF damper are determined by applying various impact forces of the impact pendulum on the test rig. These characteristics data are used to generate a non-parametric model to represent MRF damper behaviour in MATLAB Simulink software using the adaptive neuro-fuzzy inference system (ANFIS) black box. The model is then used to design a simple controller, namely a fast, practical control (FPC) for the MRF damper. This controller is formulated for the MRF damper during impact by reducing the mathematical computational cost. In evaluating control strategy performance, a dynamic model of a single-degree of freedom (SDOF) impact isolation system is developed. The dynamic model is utilised in the controller development with the aim of reducing unwanted impacts such as jerk, acceleration and transmitted force.

This paper is organised as follows: The first section contains a brief introduction and summary of previous developments in control strategy for the MRF damper. Section two presents the modelling of an MRF damper. This is followed by a description of the proposed control strategy for the MRF damper due to impact loading application in Section three. Section four elaborates the experimental setup to measure the behaviour of the MRF damper. Section five discusses the simulation results to evaluate the effectiveness of the control strategy. Finally, Section six presents the conclusion and findings of this study.

2. Modelling of Magneto-Rheological Fluid Damper

The magneto-rheological fluid (MRF) was modelled by simulating its MRF damper characteristics using an interpolated multiple adaptive neuro-fuzzy inference system (ANFIS) approach. Detailed explanation to train ANFIS has been stated in Rahmat et al. [23]. To develop the mathematical model, the MRF damper was characterized using the impact pendulum test rig, considering various impact energies produced by the pendulum mass and numerous values of constant input current. The physical figure for experimental instrumented characterization test rig is shown in Figure 1. Furthermore, the data obtained from experimental testing was then used to train ANFIS in predicting MRF damper behaviour. Three characterization data were used as model inputs for ANFIS training namely displacement, input current and impact energy. However, damping force was set as the model output in training data for the ANFIS toolbox in MATLAB Simulink software. Then, the ANFIS model was used in control strategy development to improve the performance of MRF damper. The training procedure has been stated in Rahmat et al. [23] and similar approach as proposed in [24-25].

3. Modelling of Impact Isolation System

Since the MRF damper is used to rapidly dissipate impact energy, a fast, practical control (FPC) controller is proposed in the present study. Fast, practical control is a practically and non-complex algorithm controller which is based on skyhook control combined with active force control (AFC) to improve stability and sustainability mechanical system performance of skyhook control. To evaluate the performance of the proposed controller, a single degree of freedom (SDOF) impact dynamics model of the MRF damper was formulated in the MATLAB Simulink simulation software. Figure 1 shows the SDOF model of the impact dynamics system for the MRF damper considered in this study.

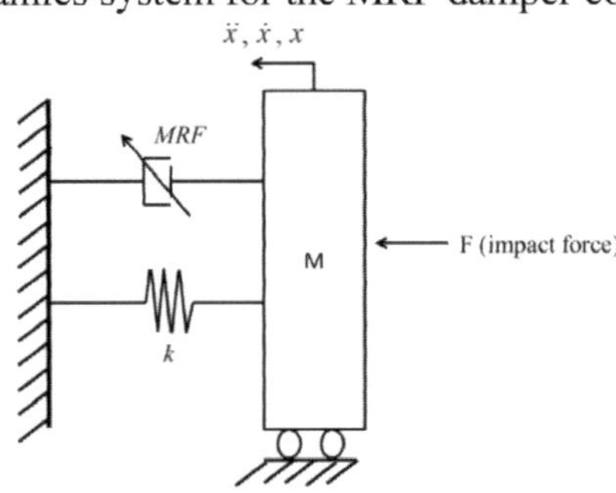

Figure 1. SDOF of impact dynamics system

According to the Newton's second law, the dynamic model can be derived as:

$$M\ddot{x} + MRF\dot{x} + kx = F_{impact} \quad (1)$$

where M is mass, k is the stiffness effect on the MRF damper, and MRF is the controllable damping force. The variables $\ddot{x}$, $\dot{x}$ and x are acceleration, velocity and displacement of the mass, respectively. It should be noted that the damping force of the MRF damper is related to the magnitude of the current applied to the MRF damper and exhibits a time response for real time applications.

4. Control Strategy

In this study, the FPC is proposed to verify the impact reduction control performance for the SDOF impact isolation system as shown in Figure 2. The FPC controller is invented by combining skyhook and AFC controllers. These controller approaches were integrated to improve the force tracking performance of the MRF damper and to reduce the impact force to the body-impact isolation system. Therefore, each controller has an individual task to mitigate impact force. According to the mechanism of the skyhook control task, the damping force of the MRF damper should be adapted to pull down the mass since the relative velocity is positive. However, if the relative velocity is negative, the damping force pushes up the mass. AFC control assists skyhook control by providing better force-tracking control and using the output controller as the input for the current generator. This is accomplished through force estimation. Force estimation can be obtained from the force actuator by comparison to the force from the impact isolation system. The governing equation for FPC based on skyhook control can be formulated as:

$$F = C_{sky} \times \dot{x} \tag{2}$$

Where F, C_{sky} and $\dot{x}$ denote force sky, proportional gain and velocity of the system, respectively. Therefore, based on the general form of the skyhook control algorithm, the FPC control algorithm can be written as:

$$FPC = (C_{sky} \times \dot{x}) + F_e \tag{3}$$

where F_e represents force estimation. The proportional gain can be obtained through trial and error tuning methods, while the force estimation can be obtained by comparing the force actuator and the force of the system. Therefore, to estimate the force, the equation is derived as:

$$F_e = \frac{1}{k}(F_a - (Em \times \ddot{x})) \tag{4}$$

where k, F_a m and $\ddot{x}$ denote a weighting function, force actuator, mass and acceleration of the impact isolation system, respectively. In the simulation analysis, the weighting function k and the estimated mass Em were obtained by trial and error method [26–28].

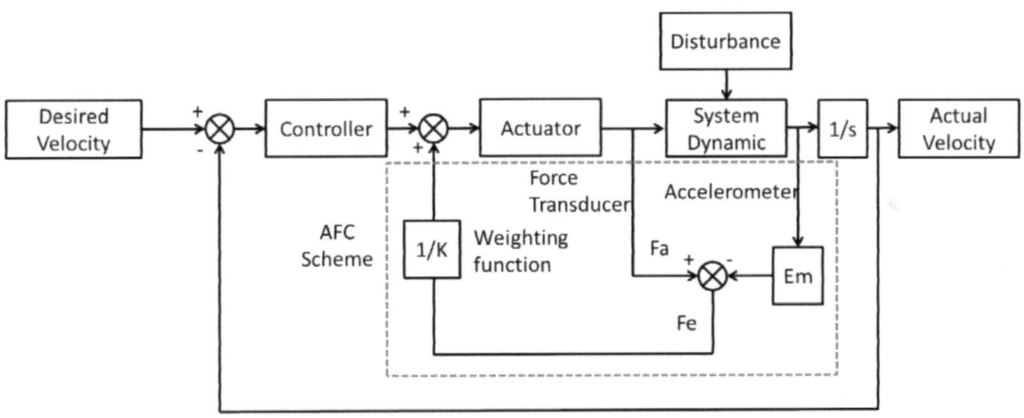

Figure 2. Schematic diagram of FPC

Table 1. Controller Parameter

Parameter	Value
C_{sky}	906
k	0.01
Em	0.2

5. Simulation Results

To analyse the performance of the MRF damper with FPC, the semi-active control response of the MRF damper was evaluated by considering with three different impact energies. The FPC control performance was then compared with the passive damper and the MRF damper controlled by skyhook control. Figure 3 presents the simulation results in term of jerk, acceleration and force in the time domain under low-impact, energy.

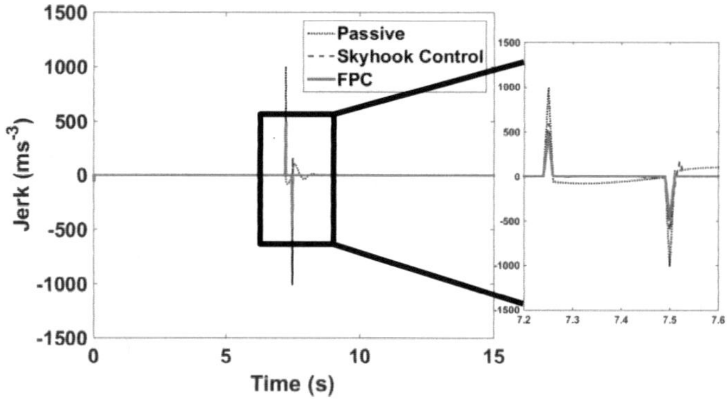

(a)

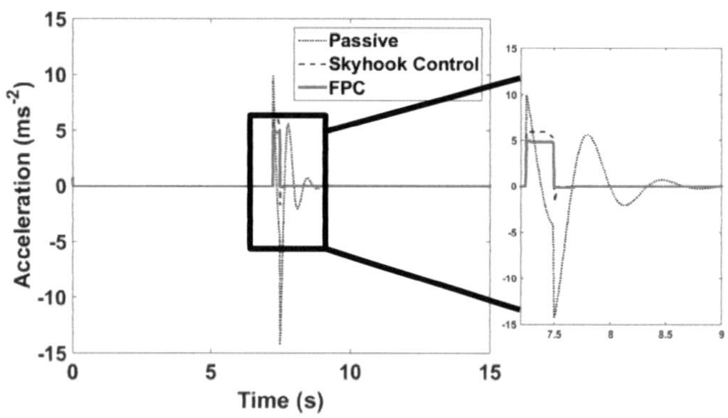

(b)

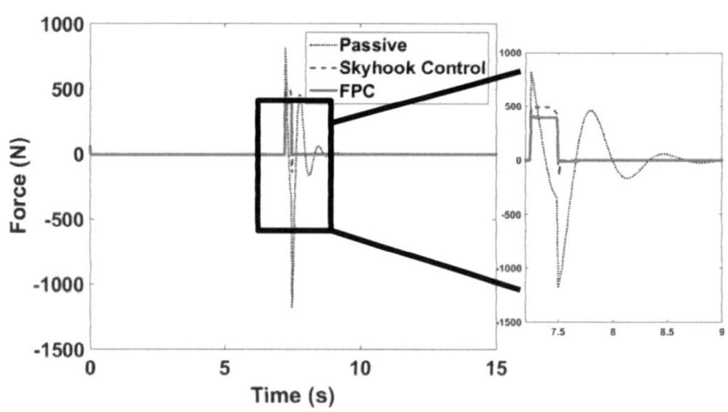

(c)

Figure 3. Simulation results under low impact energy (a) jerk response, (b) acceleration response and (c) force response

The simulation results shown in Figure 3 for the control performance of the passive damper, MRF damper controlled by skyhook control and FPC were evaluated and compared. The jerk response, acceleration response and force response of SDOF impact isolation are better in reducing impact energy using the FPC controller for MRF damper. The FPC controller reduced impact energy by 41% more than the passive damper. The MRF damper with FPC controller produced an acceleration response of 4.81 ms^{-2}, which is 18.7% reduction than skyhook control.

IOP Conf. Series: Materials Science and Engineering **530** (2019) 012032 doi:10.1088/1757-899X/530/1/012032

Next, the transmitted force of SDOF impact isolation by the MRF damper was demonstrated in Figure 3(c). The FPC controller showed by 18.7% greater effectiveness in acceleration reduction than the skyhook control. Since the controller performance was analysed with various impact energies, medium impact energy is considered in this study. The medium impact energy was used as simulation input for evaluation of the control strategy on the MRF damper performance. The analysis results under medium impact energy are illustrated in Figure 4.

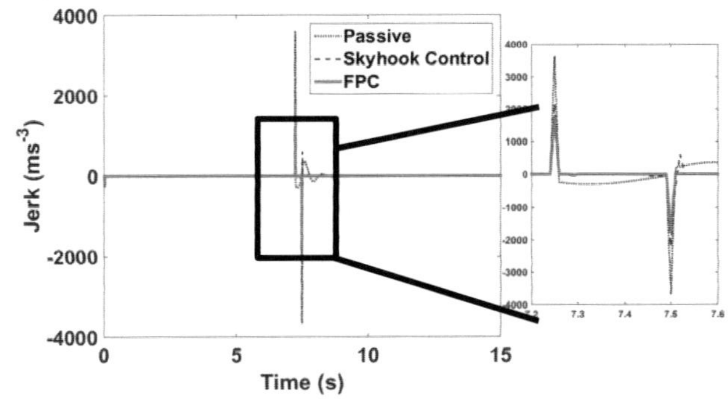

(a)

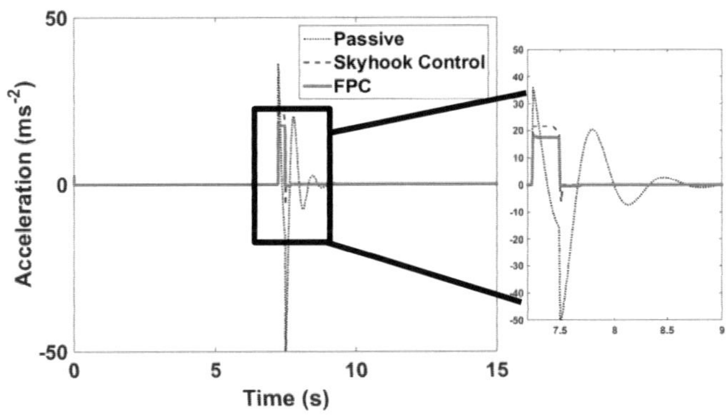

(b)

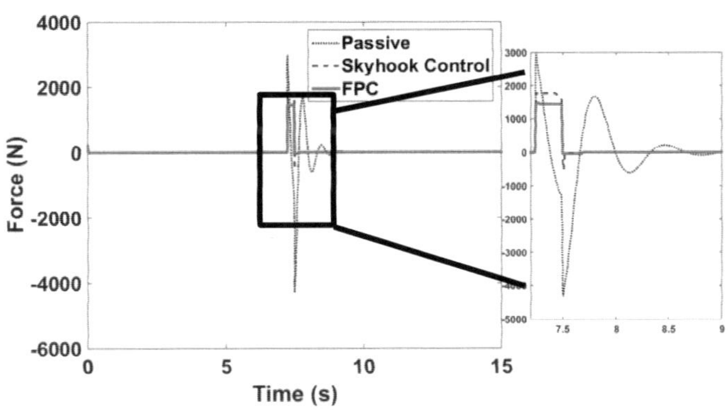

(c)

Figure 4. Simulation results under medium impact energy (a) jerk response, (b) acceleration and (c) force response

Based on these results, the performance criteria in term of jerk, acceleration and damping force of the SDOF impact isolation system were evaluated. Figure 4(a) shows the jerk response of the SDOF impact isolation system in the time domain by comparing the performance of passive damper, MRF damper with skyhook control and FPC control. Under medium impact energy, the jerk response 2165ms^{-3} produced by MRF damper with skyhook control and 1796 ms^{-3} produced by MRF damper with FPC controller. Based on that, the FPC controller has reduced 17.1% more than controlled by using skyhook control. The FPC controller exhibited good performance by providing input current to the MRF damper to produce damping force based on controller feedback. The acceleration response of the SDOF impact isolation system was substantially improved by the MRF damper control using skyhook control and the FPC controller.

The FPC controller provided the most stable of SDOF impact isolation system; it reduced acceleration by 18% as compared to the MRF damper controlled by the skyhook controller. The transmitted force was also analysed to measure the impact force that was dissipated by the MRF damper under medium impact energy. The FPC exhibited a better performance that the other approaches in force tracking. Next, the simulation results under high impact energy are indicated in Figure 5.

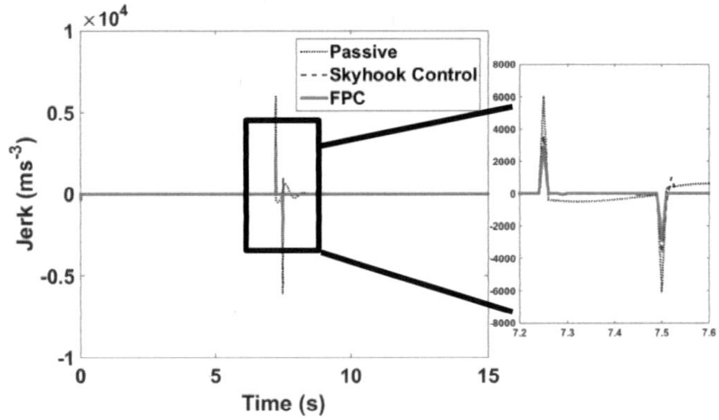

(a)

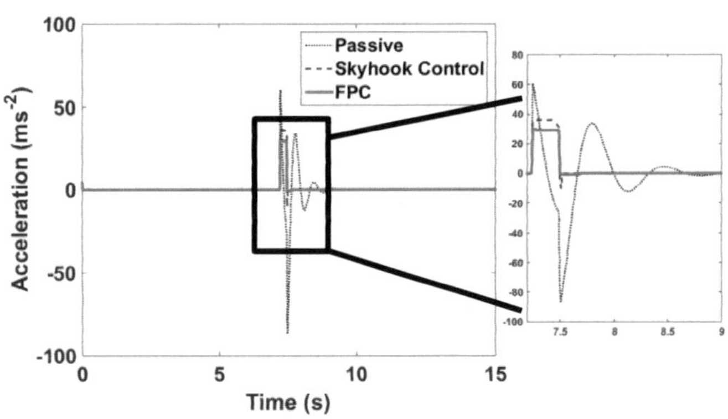

(b)

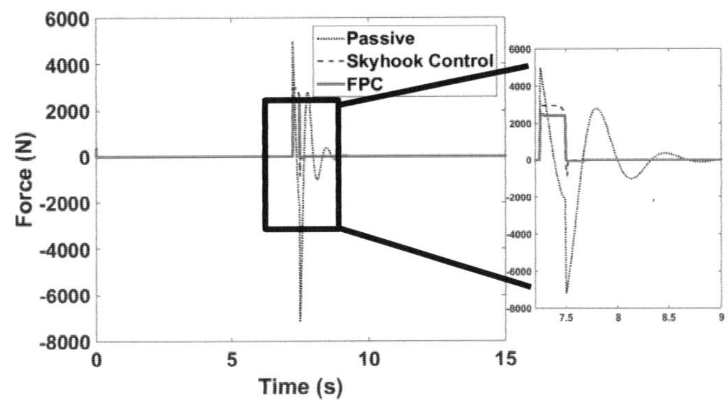

(c)

Figure 5. Simulation results under high impact energy (a) jerk response, (b) acceleration response and (c) force response

The results demonstrate the performance of the controller after being evaluated under high-impact energy. The responses of jerk, acceleration and transmission force SDOF impact isolation were used as the performance criteria to evaluate the effectiveness of the controller. As the results show, the passive damper was able to reduce the impact force; however, it needed a longer settling time to stabilize than the other control strategies. The MRF damper controlled by skyhook displayed in 0.8 s compared to other. The FPC tested in this analysis indicated better performance than that of the skyhook control. The FPC demonstrated a 0.35 s stabilization time while simultaneously reducing jerk, acceleration and transmitted force to the SDOF impact isolation system and achieving by 18.6% reduction in acceleration as compared with MRF controlled by skyhook control. The performance of improvement of control strategy based root mean square (RMS) peak to peak for impact reduction is summarized in Table 2, Table 3 and

Table 4.

Table 2. RMS analysis in term of jerk responses

Impact Energy	RMS (peak to peak)		Percentage improvement (%)
	Skyhook Control	FPC	
Low impact energy	$595 ms^{-3}$	$495 ms^{-3}$	16.7%
Medium impact energy	$2165 ms^{-3}$	$1796 ms^{-3}$	17.1%
High impact energy	$3598 ms^{-3}$	$2994 ms^{-3}$	16.8%

Table 3. RMS analysis in term of acceleration response

Impact Energy	RMS (peak to peak)		Percentage improvement (%)
	Skyhook Control	FPC	
Low impact energy	$5.92 ms^{-2}$	$4.81 ms^{-2}$	18.7%
Medium impact energy	$21.51 ms^{-2}$	$17.50 ms^{-2}$	18.6%
High impact energy	$35.90\ ms^{-2}$	$29.20\ ms^{-2}$	18.7%

Table 4. RMS analysis of force response

Impact Energy	RMS (peak to peak)		Percentage improvement (%)
	Skyhook Control	FPC	
Low impact energy	488.5 N	397.5 N	18.6%
Medium impact energy	1776.9 N	1445.7 N	18.6%
High impact energy	2961.5 N	2409.5 N	18.6%

6. Conclusion

The effectiveness of control strategies for impact reduction control in an MRF damper-based system was investigated in this study. Control performance of a passive damper, an MRF damper controlled by skyhook and an FPC were compared. Furthermore, a mechanical system model of SDOF impact isolation systems using an MRF damper with different controllers was constructed. The governing equation for SDOF impact isolation system was derived. An FPC control algorithm with less computational intensity was formulated and used to verify the impact reduction control performance of the systems through simulation. The control performance of the impact isolation systems under impact loading due to various horizontal impact energies was evaluated. Various impact energies were considered in this analysis for evaluating the capability of the controller to adapt to various inputs. Based on the results of the simulation analysis, the FPC demonstrated better control performance for the MRF damper, achieving 18.7% more reduction in acceleration than the skyhook control for SDOF impact isolation system.

Acknowledgement

This work is supported by the National Defence University of Malaysia and the Ministry of Higher Education of Malaysia through a My PhD scholarship and financial support under the Long-term Research Grant Scheme (L.R.G.S.) project number LRGS/B-U/2013/UPNM/DEFENSE & SECURITY-P1, led by Associate Professor Dr. Khisbullah Hudha. This financial support is gratefully acknowledged.

References

[1] Sherman S G, Becnel A C and Wereley N M 2015 Relating Mason number to Bingham number in magnetorheological fluids *J. Magn. Magn. Mater.* 380 98–104

[2] Xia Y, Dai L, Fu M, Li C and Wang C 2014 Application of active disturbance rejection control in tank gun control system *J. Franklin Inst.* 351 2299–314

[3] Choi Y-T and Wereley N M 2005 Biodynamic Response Mitigation to Shock Loads Using Magnetorheological Helicopter Crew Seat Suspensions *J. Aircr.* 42 1288–95

[4] Mao M, Hu W, Choi Y T, Wereley N M, Browne A L, Ulicny J and Johnson N 2013 Nonlinear modeling of magnetorheological energy absorbers under impact conditions *Smart Mater. Struct.* 22 1–12

[5] Sariman M Z, Harun H, Ahmad F, Yunos R and Mat Yamin A K 2015 Vibration control of a passenger car engine compartment model using passive mounts systems *ARPN J. Eng. Appl. Sci.* 10 7472–6

[6] Fu B, Liao C, Li Z, Xie L, Zhang P and Jian X 2017 Impact behavior of a high viscosity magnetorheological fluid-based energy absorber with a radial flow mode *Smart Mater. Struct.* 26 1–17

[7] Goncalves F D, Ahmadian M and Carlson J D 2006 Investigating the magnetorheological effect at high flow velocities *Smart Mater. Struct.* 15 75–85
[8] Li Z C and Wang J 2012 A gun recoil system employing a magnetorheological fluid damper *Smart Mater. Struct.* 21 1–10
[9] Woo D, Choi S B, Young Tae Choi and Wereley N M 2007 Frontal crash mitigation using MR impact damper for controllable bumper *J. Intell. Mater. Syst. Struct.* 18 1211–5
[10] Li Z, Gong Y and Wang J 2018 Optimal control with fuzzy compensation for a magnetorheological fluid damper employed in a gun recoil system *J. Intell. Mater. Syst. Struct.* 1–12
[11] Li Z and Wang J 2010 Experiment Investigation of Magnetorheological Damper for High Impulsive Load *Proceedings of the ASME 2010 Conference on Smart Materials, Adaptive Structures and Intelligent Systems* pp 1–5
[12] Singh H J and Wereley N M 2014 Optimal control of gun recoil in direct fire using magnetorheological absorbers *Smart Mater. Struct.* 23 1–12
[13] Bai X X and Wereley N M 2014 Magnetorheological Impact Seat Suspensions for Ground Vehicle Crash Mitigation *Active and Passive Smart Structures and Integrated Systems* vol 9057 pp 1–12
[14] Bai X, Wereley N M, Wang D and Hu W 2012 A Bidirectional-Controllable Magnetorheological Energy Absorber For Shock And Vibration Isolation Systems *Proceedings of the ASME 2012 Conference on Smart Materials, Adaptive Structures and Intelligent Systems* pp 1–11
[15] Harun M H, Azhari M A and Yunos R 2018 Characterization of a Magnetorheological Fluid Damper Applied to Semi-Active Engine Mounting System *J. Mech. Eng.* 5 248–59
[16] Hudha K, Jamaluddin H, Samin P M and Rahman R a. 2005 Effects of control techniques and damper constraint on the performance of a semi-active magnetorheological damper *Int. J. Veh. Auton. Syst.* 3 230
[17] Harun M H, Zahmani Q F, Hudha K, Sariman M Z, Harun M H, Ahmad F and Azhari M A 2017 Stability Enhancement Of Railway Vehicle Dynamics Bogie-Based Skyhook Control *ARPN J. Eng. Appl. Sci.* 12 4219–24
[18] Harun M H, Jamaluddin H, Hudha K, Rahman R A and Zailimi W M 2014 Application of Stability Augmentation System for Semi-active Railway Vehicle Secondary Vertical Suspension *Int. J. Math. Model. METHODS Appl. Sci.* 8 294–301
[19] Ubaidillah N a., Hudha K and Kadir F A A 2011 Modelling, characterisation and force tracking control of a magnetorheological damper under harmonic excitation *Int. J. Model. Identif. Control* 13 9
[20] Hu H, Wang J, Qian S, Li Y and Jiang X 2009 Investigation on controllability of a magnetorheological gun recoil damper *IEEE Int. Conf. Inf. Autom. ICIA 2009* 7493 1044–9
[21] Liu B and Chen J 2011 Control System Design of Magneto-rheoloical Damper under High-Impact Load *Mod. Appl. Sci.* 5 253–8
[22] Fu J, Li P, Wang Y, Liao G and Yu M 2016 Model-free fuzzy control of a magnetorheological elastomer vibration isolation system: Analysis and experimental evaluation *Smart Mater. Struct.* 25 0
[23] Rahmat M S, Hudha K, Kadir Z A, Nuri N R M and Abdullah N H A S 2018 Characterization and Modelling of a Magneto-Rheological Elastomer Isolator Device under Impact Loadings using Interpolated Multiple Adaptive Neuro Fuzzy Inference System Structure *Int. J. Mater. Struct. Integr.* In press
[24] Al-Hmouz A, Jun Shen, Al-Hmouz R and Jun Yan 2012 Modeling and Simulation of an Adaptive Neuro-Fuzzy Inference System (ANFIS) for Mobile Learning *IEEE Trans. Learn. Technol.* 5 226–37
[25] Flower Queen P, Sasi Kumar M and Babu Aurtherson P 2012 ANFIS technique applied to the control of a robot manipulator with disturbances *Procedia Eng.* 38 1984–93
[26] Rahmat M S, Ahmad F, Yamin A K M, Aparow V R and Tamaldin N 2013 Modelling and torque

tracking control Of permanent magnet synchronous motor for hybrid electric vehicles *Int. J. Automot. Mech. Eng.* 7
[27] Ahmad F, Hitam A L, Hudha K and Jamaluddin H 2011 Position tracking of slider crank mechanism using PID controller optimized by Ziegler Nichol's method *J. Mech. Eng. Technol.* 3 27–41
[28] Ahmad F, Hudha K and Jamaluddin H 2009 Gain scheduling PID control with pitch moment rejection for reducing vehicle dive and squat *Int. J. Veh. Saf.* 4 45–83

End-of-life Electrical and Electronic Equipment Remanufacturing Prospects in Malaysia

A Shamee[1*] and A Shamsuddin[2]

[1]Department of Industrial & Production Engineering, Shahjalal University of Science and Technology, Sylhet, Bangladesh.

[2]Department of Mechanical and Chemical Engineering, Islamic University of Technology (IUT), Dhaka, Bangladesh.

Corresponding author *: shameem.ipe@gmail.com

Abstract
Organized remanufacturing has to be an important alternative option instead of throwing-away of used electrical and electronic equipment's (EEE) in order to save our mother-nature and recover useful resources. The increase in awareness of environmentally conscious manufacturing making the remanufactured products most popular. The significant benefits of remanufacturing are: greater availability of products and low costs to customers, new employment and industrial skills training to workers, and conservation of material and energy resources to society. Remanufacturing not only focus on economic profits but also keeping contribution to minimize environmental impacts by utilizing end-of-life (EOL) products. This study attempts to find out remanufacturing opportunities, and main challenges, along with the dominant influencing factors in a fast-developing country like Malaysia based on authentic literature review and expert opinions. The paper presented a short description of e-waste status, potential remanufacturing areas, its current practices, possible policy measures, and the present barriers in remanufacturing. The findings and observations provide a foundation and clear idea about the prospective of remanufacturing in Malaysia and probable further research directions.

1. Introduction

At the present time, the whole world is facing resources shortage and environmental pollution threats, while lots of recyclable and remanufacturable end-of-life products are abandoned, or are just recycled in low level, such as smashing or melting them into new raw materials. High demand of electronic products and their short life cycles are generating enormous e-wastes every year all over the world, which are harmful for the environment and human health. United Nations Environment Programme (UNEP) estimates that between 20 and 50 million tonnes of e-waste are generated annually around the world [1]. This implies that the growth in global electronic production of 4.4 percent in 2002 and 6.8 percent in 2003 will result in similar growth in e-waste generation [2]. A recent annual estimation for Wastes from electric and electronic equipment (WEEE) was almost 6.5 million tonnes, and it has been predicted that by 2015 the figure could be as high as 12 million tonnes [3]. The increase of waste generation results is the permanent loss of valuable material, energy and housing or agricultural lands. E-waste also increases air, water and land pollution all over the world [4]. Similarly, the high purchasing ability of Malaysian population and rapid growth of electrical and electronics industries in Malaysia tend to produce serious e-waste and landfills' saturation. A research has shown that e-waste generation has been estimated to be 706,000 tonnes in 2010 and finally to about 1.2 million tonnes by 2020 in Malaysia [5]. Another estimation by Department of Environment (DOE) shown seven categories wastes where the personal computer waste projection will be total volume of 608.191 million metric by 2020 [6]. A short description of e-waste status of Malaysia is discussed in section 2. It is an alarming issue for Malaysian government need to be seriously think how to minimize those wastes from different kind of end-of-life EEE. Simply if we consider only an EOL personal computer, then we have seen that it has a serious environmental impact by producing hazardous wastes such as (a) Cathode ray tubes are generating barium, lead and other heavy metals leaching into the ground water and release of toxic phosphor; (b) Printed circuit board (PCB) is responsible for air emissions, water pollution and landfilling. PCBs are generating beryllium, cadmium, lead, tin, mercury, brominated dioxin and dust; (c) Chips and other gold plated components are producing hydrocarbons, heavy metals, brominated substances discharged directly into rivers acidifying fish and flora. Tin and lead contamination of surface and groundwater. Air emissions of brominated dioxins, heavy metals and hydrocarbons; (d) Plastics from printers, keyboards, monitors, etc. are producing emissions of brominated dioxins, heavy metals and hydrocarbons; and (e) Computer wires are generating hydrocarbon ashes released into air, water and soil [7]. Therefore, when a product reached its end-of-life there are several recovery options are available like reuse, remanufacturing, recycling, and energy recovery. EOL product recovery can minimize the environmental effects of disposal, raw material extraction, and environmentally damaging substances. Thus, e-waste recycling, reusing or remanufacturing is an important issue, not only from the aspect of waste management, but also aspect of valuable materials recovery and reuse again and again. The remanufacturing prospects and challenges in Malaysia are discussed in this research because remanufacturing is environmentally desirable and industrially profitable by reducing energy and raw material consumption. Remanufacturing is dissembling, cleaning, refurbishing, replacing parts (as necessary) and reassembling a product to meet or exceed the original equipment manufacturer's quality standards and performance specifications. According to a statistics, remanufacturing products can save 60 percent of the energy, 70 percent of the materials, and 50 percent of the cost compared to the new products [8]. Indeed, remanufactured products are sold at 60–70 percent of the new products' price and production costs only 35–60 percent of the original casts [9]. Based on extensive literature review and a few field visits, this work used discussions and interviews with Malaysian remanufacturing, recycling and original equipment manufacturers (OEMs) companies. The purpose of this paper is to provide an overview of the existing remanufacturing practices to understand the current state or future remanufacturing business scopes, opportunities and its challenges in Malaysia to increase the use of EOL EEE and reducing environmental impacts.

2. E-waste status in Malaysia

In Malaysia, waste generation is significantly increasing where solid waste generation is estimated to increase from about 10.9 million tonnes in 2010 to about 15.6 million tonnes in 2020 [5]. Malaysian scheduled waste (hazardous waste) generation based on the Department of Environment (DOE) was 1.1 million tonnes in 2006, increased to 1.7 million tonnes in 2011 [6]. Another estimation by DOE, based on the projection of selected seven categories EEE products project that the amount of e-waste will be increased by an average of 14 percent annually and a total of 1.17 billion units (21.38 million tons) will be generated by the year 2020 [10]. According to past years (1981 -2007) data, the total amount by weight of WEEE generation is calculated based on WEEE discarded by the household, business entities and institutions in metric tonne unit (Figure 1). It shows that the total amount of the seven categories of WEEE (Television=TV, Personal Computer=PC, Mobile Phone=MP, Refrigerator=REF, Air Condition=AC, Washing Machine=WM, and Mobile Phone Rechargeable Batteries=MPRB) will be 11.199 million metric tonnes by the year 2020. As per estimation from the year 1981-2020, the highest e-waste generated by personal computer with a cumulative total volume of 608.191 million metric tonnes where television sets are 244.226 million metric tonnes only. It seems that day by day, personal computer waste is increasing continuously all over the country. The air condition, refrigerator and washing machine waste are also gradually increasing with each year.

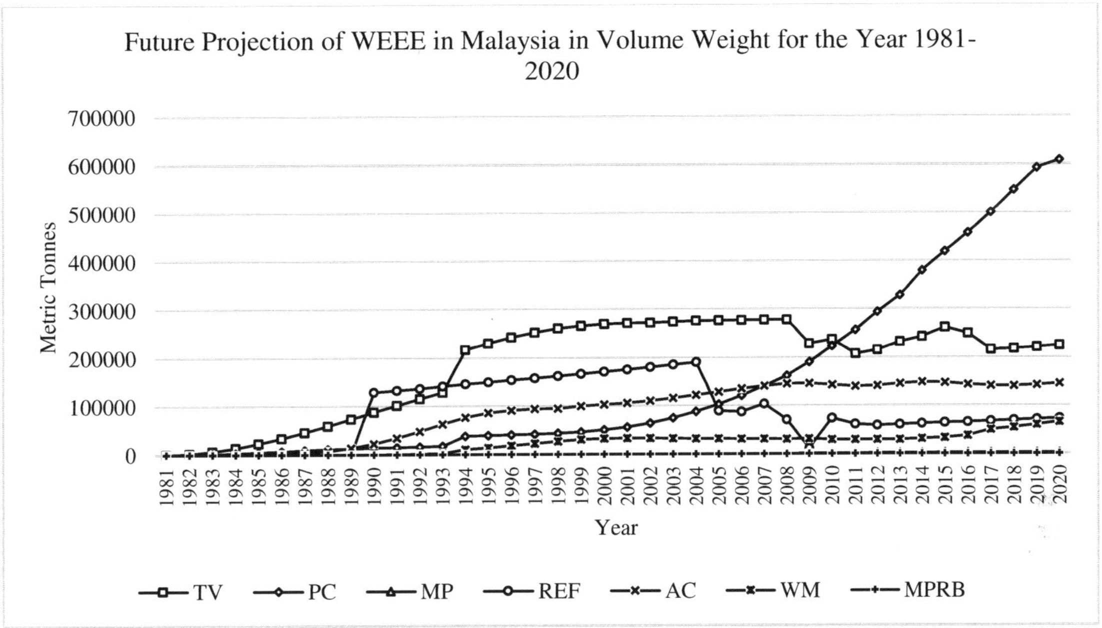

Figure 1. Projection of WEEE in Malaysia in Volume Weight for the Year 1981-2010 [10]

3. Remanufacturing scenario and possible potential areas in Malaysia

Malaysian economic growth is consistently and constantly increasing due to its Government's transformation policy and lots of local and foreign investments. Obviously, economic growth is an outcome of increasing production (and better services), which increases consumption and thus piles wastes. After discussion and interviewing a few manufacturing experts, academia, and reading journal articles or newspapers related to remanufacturing in Malaysia we have found that remanufacturing practices in Malaysia is still in its commencement stage. There are some automotive companies and third parties (tonner, printer, and computer etc.) practicing remanufacturing business in Malaysia to a small range. But the growth of remanufacturing activities in Malaysia is still modest due to absence of e-wastes business analysis, complete ecosystem and awareness. According to the Department of Environment in Malaysian remanufacturing possible potential sectors are (a) Computer and telecommunication equipment and office photocopiers (laser toner cartridges); (b) Aerospace and robots; (c) Automatic transmissions and rolling stock (railway vehicles); (d) Bakery equipment, defense equipment, industrial food processing equipment, and excavation equipment; (e) Gaming machines and machine tools; (f) Light and heavy duty automobile engines; (g) Carpet tiles, office furniture, and musical instruments; and (h) Compressors, pumps, starter motors, electrical motors and apparatus, vending machines, power bearings, and air-conditioning units [11]. Its represent that there have lots of opportunities for remanufacturing business in Malaysia. But according to the below statistics (Figure 2) it is found that remanufacturing product by sectors is not increasing in Malaysia. Number of companies along the *X*-axis and remanufacturing products based on the companies' claim along the *Y*-axis are shown that a few companies or third-party industries are doing remanufacturing activities here. So Malaysians are not only losing valuable product recovery, energy and profit, but also generating lots of wastes all over the country every year which will be hazardous and environmentally harmful.

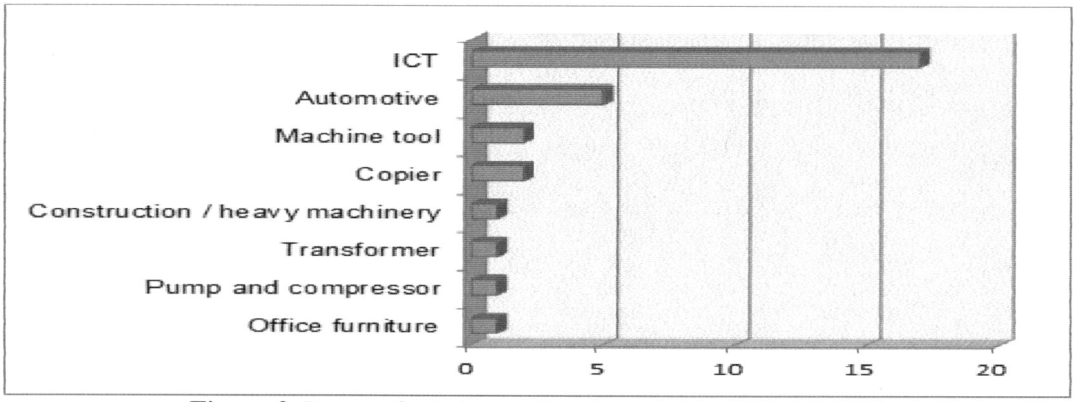

Figure 2. Remanufacturing product by sector in Malaysia [12]

Another conceptual graphical presentation (Figure 3) on remanufacturing and fast-growth industries in Malaysia is done by the Ministry of International Trade and Industry (MITI). This portrays that compares the growth of remanufacturing industries is very slow compared with the fast-growth industries in the country.

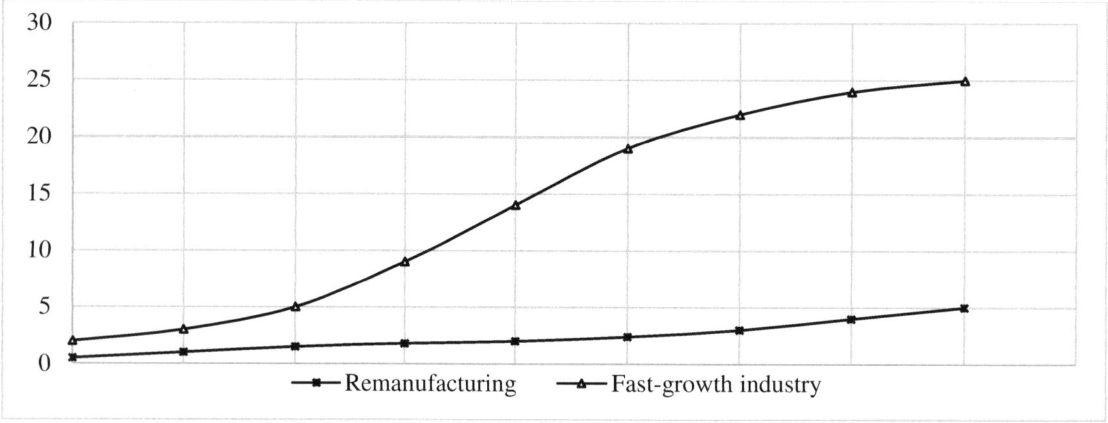

Figure 3. Comparison of growth remanufacturing industries [12]

Remanufacturing activities of several companies and other relevant features are summarized in Table 1. The Table 1 shown that remanufacturing products have market demand and the companies cannot fulfill the customer requirements due to Malaysian remanufacturing industries are practicing only initial techniques of remanufacturing like inspection, disassembly, cleaning, and testing. Merely a few manufactures are using heavy machineries for giving the second life of the product. Even though remanufacturing has lots of benefits, only lack of remanufacturing business analysis and market information are the primary barriers for remanufacturing growth in Malaysia. After analyzing our expert opinions and remanufacturing related papers published in Malaysia, we come to the point that most of the original equipment manufacturers (OEM) are not practicing remanufacturing due to lack of business information and remanufacturing business analysis for remanufacturing decision making with consideration of organizational factors. Remanufacturing business has challenging activities due to its products have to conform the same or similar quality specifications of the respective new products. Hence, lack of analysis about remanufacturing economic profits and performance of manufacturing industry is responsible for slow growth of remanufacturing industries in Malaysia. We tried to find out remanufacturing barriers in Malaysia and major problems being faced by manufacturing industries to make remanufacturing decision as follows: (1) product verity and product selection problem for remanufacturing; (2) technological and tools problem for remanufacturing; 3) lack of specific procedures and policies for remanufacturing; (4) lack of awareness for remanufacturing and raw material (EOL) collection; (5) lack of design for recovery; (6) insufficient skill and expert personnel; (7) lack of market research of remanufactured product; (8) lack of study for remanufacturing challenges consideration; and (10) lack of remanufacturing industrial master plan.

Table 1. Some remanufacturing companies' activities in Malaysia

Company Name	Remanufacturing Product	Process	Market	Remarks
Malaysia Transformer Manufacturing Sdn Bhd	Power and Distribution Transformer	Inspection, disassembly, cleaning, drying, repair/ replace of cores, repainting, testing, and upgrading (if needed)	Domestic and International	OEM products are 80 percent and remanufacturing products are 20 percent. Fulfilling remanufacturing product criteria
Hitachi Construction Machinery (M) Sdn Bhd	Hydraulic pumps and motors	Inspection, disassembly, cleaning, replacement of cores, reassembly, and testing	Domestic and International (Europe)	Fulfilling remanufacturing product criteria. Stiff competition with other OEM brand such as Komatsu (Japan), Caterpillar (US), Soosan (Korea).
Motor Teknologi Industri Sdn Bhd	Highway and off highway diesel engines. Army and Police trucks, cylinder head, crank shaft, cam shaft and crank case	Inspection, disassembly, cleaning, repair/ replace of cores, repainting, and testing	Domestic and International	Fulfilling remanufacturing product criteria
Prota Remanufacturing Sdn Bhd	Laser toner of HP, Canon, Samsung, Brother and Lexmark.	Inspection, disassembly, cleaning, testing, and refill services	Domestic Market	Fulfilling remanufacturing product criteria
Esyink Remanufacturing Sdn Bhd	Cartridge remanufacturing (Inkjet and Toner) of HP, Canon, Samsung, Brother , Lexmark, and Xerox	Inspection, disassembly, cleaning, testing, and refill services	Domestic Market	Fulfilling remanufacturing product criteria
Shin Etsu Denso Co., Ltd.	Remanufacturing of Automotive Rotating Electrics Automotive Parts. Alternators, Starters, A/C Compressors, Power steering pump, and Transmission	Core sorting, disassembling, cleaning, painting, sub-assembling, assembling, testing, and packing	Domestic and International (North America, Europe, Middle East, South East Asia, South America, and Oceania)	Fulfilling remanufacturing product criteria

4. Remanufacturing - challenges in Malaysia

We tried to find out the present remanufacturing challenges in Malaysia from the outcomes of interview of manufacturing experts and literature. The identified potential factors are (a) raw materials collection, (b) technology & method, (c) product design, (d) skill manpower & experts, (e) marketing & competition, and (f) environments & government which are causing an overall effect.

(a) Raw materials or cores collection: Raw materials or cores are the heart of remanufacturing business. It is the most essential and important issue for remanufacturing business decision. In order to collection of cores require some polices or system to get back EOL product from the end user. After discussing with some recycling and remanufacturing industries (at Penang) in Malaysia we have found that they don't have any standard system for collecting cores. At present, they are collecting cores by directly buying from end customers, but they don't have any standard collection system. If customers are willing to inform them, then they contact if it is nearest distance. Some multinational companies like Dell, Nokia and Motorola etc. are trying to get back their EOL products by offering a promotion. Price of the EOL product is the major issue for returning back to remanufacture or OEMs. So OEMs can offer promotions to customers to return back EOL product by proposing discount on a new product. The cost or price of EOL product, quality of EOL product, uncertainty cores supply and demand, cores inventory management and control, cores collection system, transportation, public awareness, and core collection cost are making cores collection so challenging for Malaysian remanufacturing and recycling business.

(b) Technology and method: Technology and method are certainly challenging for remanufacturing business. The initial stage of remanufacturing is fault identification and inspection of product for remanufacturing decision. That's why standard fault identification technologies are required to make the remanufacturing business profitable and reliable. A high level of remanufacturing technologies and methods are helping remanufacturing industries to make remanufacturing decision properly. The technical test (like visual testing, mechanical testing, electrical testing, audio testing, display testing, and logical testing etc.) at initial stage and final stage is essential due to remanufacturing product warranty and quality have to be as like new manufacturing product. Thus the degree of organizational remanufacturing technology, remanufacturing testing cost, technology cost, and the method of remanufacturing are making remanufacturing business more challenging day by day.

(c) Product design: The efficiency and effectiveness of the remanufacturing process significantly depend on product design features like joining and fastening methods, product architecture, and material choice can have an effect upon ease of testing, ease of disassembly, ease of reassembly, ease of reprocessing, and so on. Recently, environmentally friendly design and design for remanufacturing are the greatest challenges for engineers to provide potential benefits to the society by increasing the use of recordable materials. Lack of designing factors consideration makes the remanufacturing process more complex, very labor intensive, high cost, low quality, time consuming, and finally EOL product produce hazardous waste. So Malaysian government also can make some plans for OEMs to inspire or force them to produce as per consideration of designing factors.

(d) Skill Manpower and experts: Skill manpower and experts are another important challenges for remanufacturing business in Malaysia. Remanufacturing processes are essentially labor intensive. Remanufacturing business is profitable if the organization can analyze their business efficient way from the remanufacturing decision making stage to deliver remanufactured product to the end customer. The inspection and identification of faults, disassembly of product, cleaning and storage of parts, replace parts or remanufacturing operations, reassembly of product, testing to verify functioning as new product are the main process of remanufacturing. Thus, every phase of the remanufacturing process requires technically skilled operators and expert's decision to satisfy or exceed the original equipment manufacturers (OEMs) product qualifications, warranty and performance specification.

(e) Marketing and competition: Original equipment manufacturers already have almost all required technologies to remanufacture a product at low cost to make competition with lower end product manufacturers. If OEMs are not starting remanufacturing, then other competitors will inspire to start to control market share. Most of the time OEMs is giving their product remanufacturing to third party industries, but the poor quality of remanufacturing product may damage their brand reputation. Sometimes customer is not willing to pay for remanufactured product due to lack of advertising and marketing, even though the product function and quality are same as new. Market research may increase the profitability of the company by forecasting market demand, return flow of EOL product and change of technology. Remanufactured product acceptance to the customer depends on product price, quality, warranty and early launching. So marketing is an important challenge to create a remanufacturing product demands inside to the market by delivering essential information to the customer.

(f) Environmental and government: Remanufacturing business is economically profitable and environmentally desirable to all as well as the environmental issues are most important for every country to minimize the waste of end-of-life electrical and electronic equipment. So during remanufacturing or manufacturing a product need to consider its environmental impacts because most of the EOL products contain hazard substance and its disassembly can pollute the environment. Green manufacturing or remanufacturing, green supply chain and environmental friendly design can help to e-waste reduction. Some governmental policies may inspire and increase remanufacturing business activities by providing some facilities like ordinary take-back laws, public information campaigns, investment opportunities, tax free or low taxes for raw materials or technology import, financial supports for new business, and incentive and subside for EOL product collection.

5. Discussion and conclusion

Since decades, e-waste and waste material has been treated as a cost addition in reproducing and waste management for this reason original equipment manufacturers (OEMs) was not interested to recycle or remanufacture EOL products. But today, sustainable practices and environmental issues are legally forced by governments for OEMs to recycle and remanufacture EOL products. Malaysian government has been implementing some policy and regulations by establishing different departments for sustainable waste management system like the Action Plan for a Beautiful and Clean Malaysia (ABC) in 1988 followed by the Environmental Quality Act (EQA) 1974 in 1996, the National Strategic Plan for Solid Waste Management (NSP) in 2005, the National Solid Waste Management Policy (NSWMP) in 2006, and the Master Plan on National Waste Minimization (MWM) in 2006 [5]. Those organizational activities show that the Malaysian policy- makers only focused on waste management of both solid and scheduled wastes on the basis of EOL product collection, recovery, and disposal [13]. But policy makers should think about parallel alternative solutions to reduce the e-waste generation and valuable resource recovery. Recovering resources through remanufacturing will improve the overall organizational efficiency, cash flow as well as remanufacturing can minimize environmental impacts. The relevant authority can make some policies to inspire responsible organizations or manufacturer for product repairing, reusing and remanufacturing. So Malaysian authority may consider some policies for increasing the growth of remanufacturing business, such as: (1) specific policies and procedures to ensure that the industry is developed strategically and in accordance to the domestic legislations; (2) raise awareness - promoting remanufactured products as new products; (3) involve OEMs for evaluating impacts - comparing life-cycle impacts of remanufactured and manufacturing new goods for key product lines; (4) determine appropriate factors and criteria for standardizing EOL product management legislation; (5) force original equipment manufacturers for ensuring design for recovery during new product manufactured; (6) force the OEMs

IOP Conf. Series: Materials Science and Engineering **530** (2019) 012033 doi:10.1088/1757-899X/530/1/012033

to increase the use of recyclable materials and to reduce the use of hazardous substance; (7) force OEMs to take-back EOL products (such as returning with new, promotion, and offer); (8) government facilities for industrial collaboration; (8) financial support providing for recovery business exploration; (9) regulatory framework strongly monitor; and (13) EOL product recovery industrial master plan. This paper also addresses remanufacturing practices, barriers and challenges to take advantage of regional-scale benefits of remanufacturing to reduce the e-waste generation. In summary, remanufacturing business is still in its early stage in Malaysia. Indeed, the remanufacturing barriers which are identified by the interview and discussion with remanufacturing and recycling companies are also confirming that those issues are influencing remanufacturing business growth in Malaysia. This paper also highlights the present challenges of remanufacturing in Malaysia. Expectation of quality, raw material supply limitation, competition, product design and change of pace of technology are the key factors requiring consideration before starting remanufacturing businesses. Remanufacturing can be grown industries in Malaysia due to its potential scale of local and foreign investments. Common trade and tax policies for remanufacturing industries need to be formulated to enhance and facilitate its growth.

References

[1] M Eugster, D Huabo, L Jinhui, O Perera, J Potts and W Yang 2008 Sustainable electronics and electrical equipment for China and the world International Institute for Sustainable Development Switzerland

[2] J A Kalana 2010 Electrical and Electronic Waste Management Practice by households in Shah Alam Selangor Malaysia International Journal of Environmental Sciences 1 132-144

[3] M P Luda Recycling of Printed Circuit Boards

[4] A M King, W Ijomah and C McMahon September 2004 Reducing end-of-life waste: Repair, recondition, remanufacture or recycle, in: Proceedings of the ASME Conference .

[5] P Agamuthu and D Victor 2011 Policy trends of extended producer responsibility in Malaysia Waste Management & Research 29 945-953

[6] DOE 2011 Environmental Quality Report 2006-2009 Department of Environment, Malaysia

[7] Wikipedia 2013 Electronic waste See http://en.wikipedia.org/wiki/Electronic_waste . Accessed on 8 October 2013

[8] W Xiaoyan 2012 Research on design management based on green remanufacturing engineering Systems Engineering Procedia 4 448-454

[9] R Giutini and K Gaudette 2003 Remanufacturing: The next great opportunity for boosting US productivity Business Horizons 46 41-48

[10] DOE 2009 E-waste Inventory Project in Malaysia Department of Environment Malaysia & EX Corporation Japan (Executed by Perunding Good Earth Sdn Bhd)

[11] DOE 2010 Guidelines for the Classification of Used Electrical and Electronic Equipment in Malaysia Department of Environment

[12] R Afroz, M M Masud, R Akhtar and J B Duasa 2013 Survey and analysis of public knowledge, awareness and willingness to pay in Kuala Lumpur, Malaysia – a case study on household WEEE management Journal of Cleaner Production 52 185-193

IOP Conf. Series: Materials Science and Engineering **530** (2019) 012034 doi:10.1088/1757-899X/530/1/012034

State of The Art Review of Quality Control Method in Automotive Manufacturing Industry

M Hafizi[1*], S N S Jamaludin[2] and A H Shamil[3]

[1] Master Student DRB-Hicom University of Automotive Malaysia

[2] Research and Innovation Head DRB-Hicom University of Automotive Malaysia

[3] Advance and Distance Learning Dean DRB-Hicom University of Automotive Malaysia

Corresponding author *: mejmohdhafizi@gmail.com

Abstract. Quality control plays a vital role in automotive industry to ensure the standard of the end product are accepted by customer and stay competitive in this industry. Company must remain updated to enhance and improve method in defining their product quality and maintain the continuous quality improvement. The purpose of this paper is to provide an overview of implementation quality control method used in automotive industry in reducing defects. Methodology used for this paper is extensive review of existing paper and literature. This paper provides an overview of quality control method, advantages and disadvantages of the method. The literature in this subject is fairly limited. There are very limited discussions of quality control method in the automotive industry. Therefore, great effort to carry out the study was done on the manufacturing environment particularly in the automotive part.

1. Introduction

Quality control in the automotive industry played a significant role in an ever growing need to implement continuous improvement methodologies in automotive companies. Every companies setting up higher quality standard to attract customer into purchase their product and maintain the customer expectation. The selection and application of quality control and continuous improvement method depends on companies structuring of the quality system, classification of faultless operation methods and techniques. Each company need to establish and follow quality system methodologies to help ensure their product consistently meet applicable requirement and specifications.

In order to remain competence and survive in the business, each company focus on improving product quality and reduce rework to save time and cost subsequently fulfils customer needs. From product development to final product delivery to the customer, the application of quality tools remains vital in manufacturing as a standard monitoring [1-2] and increasing productivity and profitability. How industries evolve in managing customer quality perfection demand using quality tools? In fulfilling customer demand on quality, it is important step in each manufacturing process to make sure the quality of product are above expectations. Manufacturing industries are required to be able to give quality guarantee of the manufactured products and services. This quality guarantee to the highest is able to retain competitiveness in the consumers markets [3]. Poor handling of quality issue and

IOP Conf. Series: Materials Science and Engineering **530** (2019) 012034 doi:10.1088/1757-899X/530/1/012034

procedure and below par performance of quality personnel neglecting quality effort is the main culprit for most of the defect in the manufacturing [2].

2. Review on previous research

In Product Reject Analysis using Weibull Distribution Model - A Case Study in Automotive Parts Manufacturing Industry [4], author proposed product analysis method using application of Weibull Distribution in analyzing characteristics pattern of product reject data. This method aim is to pinpoint the root cause of the defect product and providing statistical value and distribution plotting graph from analysis to determine level of production performance. This method divides the most common root cause into four groups. This group are design reject, manufacturing reject, lack of warning and instruction reject. Material, design and processing method is the most common factor contributing to product defect in every manufacturing. The result from this method could identify the inconsistency and abnormality practice in the production line. This inconsistency of working mostly caused by human error [4]. However, the Weibull Distribution models are considered as too complex and esoteric. The accuracy of Weibull analysis depends on the quality of the data. Weibull distribution generally provides a good fit to data when the data quality is understood. The data must include item-specific failure (times to failure) for the population being analyzed, data for all item that did not fail must also included and the analyst must know all experienced failure-mode root cause and be able to segregate them.

In produce conforming product, the other way is using Statistical Process Control (SPC). By using statistical tools, this method monitors the performance level of manufacturing process. It will be able to predict if any significant deviations occurred that can lead to defect product [5-7]. It is statistical method regularly used in monitoring and control of a process to ensure it operates at full capabilities. By using a process behaves predictably, it able to produce as much conforming product with the least waste. This method can be applied to any process using measureable output even though it is usually applied to controlling manufacturing lines. The main tools in SPC methods are control charts and cause and effect diagrams. This tool usually focuses on quality continuous improvement. In order to prevent problematic product passed on to the customer as well as reducing waste product, variation in the process that may affect the quality of the end product need to be detected and corrected. Implementation of SPC Techniques in Automotive Industry: A Case Study by [5] conclude, SPC analysis have the capabilities in improving the efficiency of the production process. The improved manufacturing process resulted in decreasing number of defective end product thus reduce rework cost and time. However, SPC can only detect abnormality. If the process is consistently in bad zone, SPC will consider the process in good order and will not trigger early warning.

Japanese using Quality Assurance (QA) Network method to achieve 100% defect free product in the manufacturing process. This method create a network that traps manufacturing defects effectively thus resulting in free variation in end product [8]. [8] in Reduction of Defect through Quality Assurance research stated, this method can be implement in a few condition, when there is introduction of new car model, newer version of existing cars or any change in any of the operations or processes. This QA Network main aim is to achieve 100% defect free vehicle and one of the most important part to achieve that is maintaining the documentations process of the network created. QA Network method uses 5 stages of defect detection system. These 5 stages was design to make sure zero defect outflows. Stage 1 is, In Process with the aid of Pokayokes, Templates, Karakuri, Magic Checks and QC gates. Stage 2 is successive process which also involves Templates, Karakuri, Interdependency, Visual Checks and QC gates. Stage 3 is Within Group involving Interdependency, Visual, Functional and QC gates and the 4th stage is Within Shop involving Interdependent, Fitting, Functional and QC gates. The last stages are Outflow. When the defect is not detected anywhere, it outflows to the customer that result in degradation of company brand.

Figure 1 show how the implementation of QA Network in four stages indicated by the colour coding in the figure. The colour coding of this method start with the green line indicates the process, the blue line indicates process group, yellow line indicates process shop and the red line indicate the customer. The first stage of this method is to evaluating the process management sheet to verify the entire defect that occurred. The second step is to evaluate the check sheet very detail. This step introduces four group of occurrence prevention that is Facilities (Tools), Materials (Parts), Method, Person group. The third step in this implementation is filling and keeping this created document for further reference. The last stages will be to take counter measure to improve the condition of products.

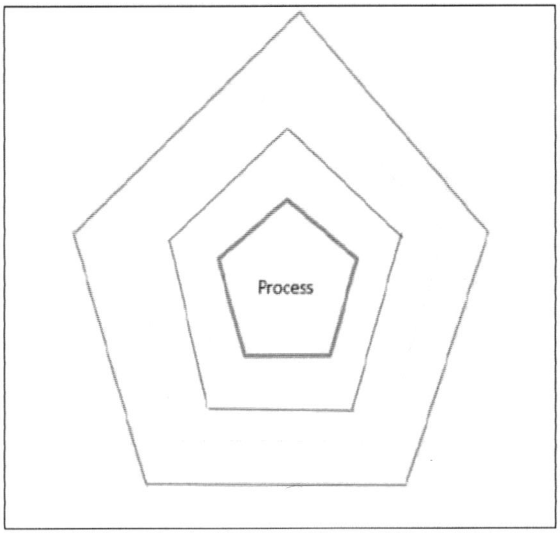

Legend:
Green Line – The Process
Blue Line - Process Group
Yellow Line – Process Shop
Red Line – The Customer

Figure 1. QA Network Web Diagram [8]

The QA Network when implemented with maximum accuracy capable in finding the origin of the specific kind of defects within a region in a very short time. However it needs to be updated when certain changes are made. This network is hard to maintain when some information is missed because manipulating the information is worse than no information. The difficulty in the implementation of QA Network is as good as its effectiveness as shown in figure 2. The blue line is before QA applied and red line after QA applied.

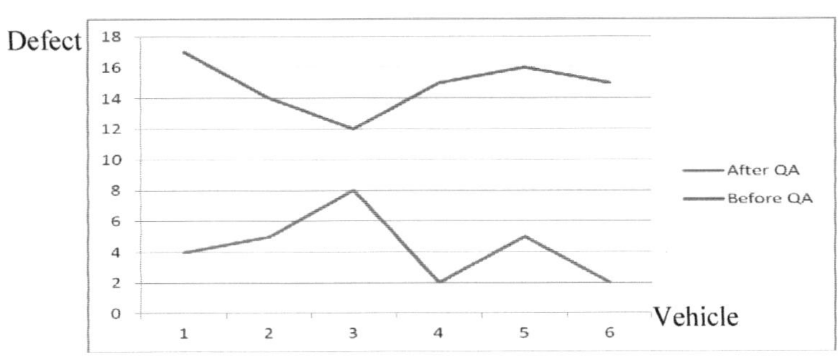

Legend:
Green Line – Before QA
Blue Line – After QA

Figure 2. Defects per Vehicle (DPV) before and after QA application [8]

[9] discussing about the application of Six Sigma using Taguchi approach to minimize defect In Application of Six Sigma in Automobile Industry. This research using automobile casting process at a foundry in North India as a subject to analyzes various significant process parameters. The objectives of [9] research are to develop a novel approach in creating Six Sigma projects and identify the critical parameters needed. Taguchi method used in testing the optimized parameters in an industrial case study and a trade off made to finalize the recommended process parameters used in manufacture automobile parts. [9] concluded, data have been collected and analyzed shows that their

standard deviation, represented by the Greek symbol σ value laid in between 3.0-3.16 generally. Therefore to increase the σ value of processes, most influencing parameters need to be identified. The effective range of the parameters identified to be studied for its effects on the σ performance of the processes. Further analysis to identify the levels of influencing parameters need to be conducted to increase the σ value of the processes and thereby result in the lower rejection of the components or parts. However this method is too extreme which can slow down the production leaving customer waiting for the product delivery.

Reduction of Rework in Bearing End Plate Using Six Sigma Methodology: A Case Study by [10] discussing the implementation of Six Sigma in reducing manufacturing defect of bearing end plate which occurred after machining process. Defect on the bearing end plate detected when there is variation in thickness and diameter of the bearing end plate thus a need of rework. The subject company use casting which is the raw material form supplier. When the machining process started, defects sign starting to show up. Because of the casting are supplied by the supplier, company just complain to the supplier for a new one. However, the reduction of the casting defects is not company responsibilities, so the focus of this project is to reduce the rework due to bearing end plate problem.

The first step of implementation of Six Sigma method is definition phase. Through this phase, project team selected, project charter created, project scope and project goal identify. Follow with problem statement definition, business case development and lastly the preparation of project schedule to be carried out. The second phase is measure phase. This phase is the project base lining including the identification of measurements factors which is vital in improving Critical to Quality (CTQ) characteristics, assessment of measurement system if required and quantification of the current status of the CTQ characteristics [10]. The third phase is analysis phase. In this phase, focus will be to identify the probable or potential causes of the problem. From this list, the root cause will be shortlisted using Gemba investigation. Gemba is a technique to verify whether a potential cause is a root cause or not through the physical observation of the process. The aim of this phase in the Six Sigma methodology is purely to come out with a list of root causes of the problem. The fourth phase is the improving phase. This phase is to identify the best resolution to overcome the root causes or minimizing it. The last phase is the control phase. In this phase, the process performance is evaluated after the implementation of the solution and compared with that at the start of the project. The achieved improvements are verified. The steps are taken to ensure that the improvements will be sustained in the process. However, Gemba investigation need a lot of time consuming to observe the whole production process thus limiting the application of this method to a very short production process or line.

A research conducted by [11] investigate suitable effective tools based on Define-Measure-Analyze-Improve-Control (DMAIC) using car audio product manufacturing as a model. For the time being, each industry has different versions implementation of DMAIC for every phase. Practitioners are using different tools for each phase of DMAIC and there are no standard in defining the verification phase. The implementation of DMAIC in case study and production reject rate in industry had been study to the detail. Research also included detailed adaptation process which being verify by publish literature.

This research focus is final assembly of the production of XXX product where the reject rate is analyzed. This company production shop floor consists of three different sections which is part preparation, surface mounting process and final assembly. The process of this product is sequential process, in batch, multi-patch and asynchronous involving Front of Line (FOL), End of Line (EOL) and Final Inspection process. This research conducted by monitoring reject rate for twelve months in the Final Inspection. By analyzing the data collected within that period, the solution to reduce the parts defect found.

DMAIC method used in solving this quality issue by divided into five phases. The first phase start by defining the status of quality issue process and the expected outcome after the completion of the improvement activity is derived. Define phase consists of the four tools which are Voice of Customer (VOC) or Voice of Process (VOP), Cross functional team forming, 5W1H (Why, When,

Who, Where, What and How) and scatter plot. This phase begins with solving the customer issues. VOC means the feedback from customer about the quality of the product, while, VOP means the process feedback about the previous process which is normally inspected by quality control (QC). This phase ends by applying the scatter plot to visualize the current situation of the reject rate. The plotting of the line chart will show the trend of the reject rate based on the observation time. From the scatter plot, setting the goals or objective of the improvement activity will then be initiated.

Second phase will be the measuring the defect of the product through application of the selected quality tools to determine and visualize the current quality issue. This phase consists of applying the Pareto chart, process mapping and process capability. Several types of defects captured and identified during observing the data and focusing on the problem solution. The Pareto chart is used to define and visualize the main contributing elements that contribute to the reject rate. Type of defect, defect quantity and cumulative percentages of defect quantity are usually used in plotting the Pareto diagram. Through the charting mechanism, it shows the weightage on how much effort need to be ratified which resulting in better use of limited resources by application of the 80/20 rules. Next, process mapping is applied as a tool by using basic concept of the process flow chart to mapping the potential processes contributing to the partial failure. Once the potential processes area has been detected, the process capability study will be conducted to calculate the process capability (Cp) and process capability index (Cpk). The purpose of this analysis is to identify the stability and accuracy of the process. Based on the quality standard, the value of Cp and Cpk more than 1.0 representing the process is stable.

Analyze phase consists of application of the quality tools to determine the possible causes of the defects and analyze all the possible causes to determine the main root causes. This phase consists of brainstorming by using Ishikawa diagram and a selection of the possible root causes through multi-voting to identify the possible root-causes of the potential causes as identified in Measure Phase. The brainstorming session using Ishikawa diagram involves cross functional team members to ensure all possible root causes will be recorded based on the four factors which are man, method, material and machine. The Ishikawa diagram can also allow the possible causes or problem to be identified structurally and can also determine the relationships between potential causes of problems.

Since the possible root-causes have been brainstormed and the possible root causes were recorded, the multi-voting analysis has been adopted to select the root causes based on mutual agreement among team members. From the all possible root causes from the Ishikawa diagram for stuck pins, three potential root causes have been selected based on the highest number of ratings according voting from the team members and the listed potential root causes according to the ranking one to three is a processes contributing into the defects.

Improve phase is the stage of implementation as a counter measure to solve the quality issues. The potential countermeasures will be identified through the observation and discussion, the hypothesis test conducted to ensure the counter measure really solves the quality issue. Once the process of improvement was validated by hypothesis test, the improvement activity will be implemented as per project planning.

The last phase is Control phase where this phase is about controlling the process and ensure the defects will not occur in future. This control phase completes an improvement activity ensures day to day processes stay in the meeting quality specification and become standard. This phase includes establishing a scatter plot, monitoring the process capability to ensure always meeting the specification and provide the proper training into workers to ensure the problem will not appear again in the future and highlighted the lesson learned in production floor to ensure workers not make the mistake. Scatter plot is the technique to monitor any abnormality of the data based on trend of the graph. Instability of the data can be recognized and process capability will be applied to monitor the consistency of the assembly processes. At the end of the improvements, visual of the lesson learned of defects will publish in production work station to alert the workers and ensure the defects will not happen again in future.The implementation of DMAIC with consist of 5 stages and must be supported by various quality tools may render the method becoming too complex and bound to overlook.

IOP Conf. Series: Materials Science and Engineering **530** (2019) 012034 doi:10.1088/1757-899X/530/1/012034

[12] in their paper title Defects Reduction in Manufacturing of Automobile Piston Ring Using Six Sigma, propose a method to reduce defects. It involves identify on the factors causing rejection of piston rings in the Automobile Piston Ring Manufacturing industry using a Six Sigma Methodology for problem solving and control. In the define phase, the aim is to identify the problems in the manufacturing line of the Molybdenum coated Piston rings which comprises of around 20 stations that each ring passes through to take form of the final product. By using Cost of Poor Quality (COPQ) analysis, each machine data in the production line is analyzed to find out the contribution of each machine in the rejection. Analyze also been made in term of process capabilities and all factors lead to defect product are listed. The measure phase helps in measuring the existing system and establish valid and reliable metrics to help monitor progress towards the goals defined in the previous step. Measurement System Analysis of the Manufacturing Process is carried out to compare the Standard Operating Procedures (SOP's) with the actual procedures being followed, the reason for deviation if any, sampling details, measuring instruments and their Calibration frequencies. In the analyze phase, Failure Mode and Effect Analysis (FMEA) is carried out for both the processes to find out the vital causes among the many causes listed in the cause and effect analysis. The vital causes for plasma spraying are listed as per its descending order of Risk Priority Number (RPN). Based on the finding in the analyze phase, improvements are carried out in this improve phase. This phase use all finding by quality tools in the previous phase to sort out solutions for the problems found out through the Analyse phase. But the most widely used tool is the Design of Experiments (DOE). It is a test or a series of tests which are intentionally performed on input variables in a system to see the effect in response variable. In the last phase, this control phase used to develop and implement process control plan to ensure sustenance of the improved process. This step to make sure that process stays in control after the solutions have been implemented. Control phase will detecting the out of control state and determines associated special causes so that actions can be taken to correct the problem before non conformances are produced. This method also suffer the same consequence as [9] which require too many quality tools that will render the process to becoming too complex .

In Improving Quality of Light Commercial Vehicle using PDCA Approach by [13], author study about the bits defect on the electrodeposited body is investigated. This research suggests a systematic quality improvement plan and optimization are performed. The application of the PDCA cycle improved 65% of bits and reduced 34% of sanding man hour. Quality improvement methods focus on reducing, defects or errors by eliminating variation in both manufacturing and service operations; hence, quality tools are often used to investigate problems and develop appropriate solutions to address these problems. The electrodeposition (ED) in the light commercial vehicle is a special coating method where the ED paint that is dispersed in water is electrically deposited on the substrate to form a uniform and water-insoluble. In order to coordinate continuous improvement effort, the PDCA cycle has been utilized. Figure 3 below show the propose implementation of PDCA methodology.

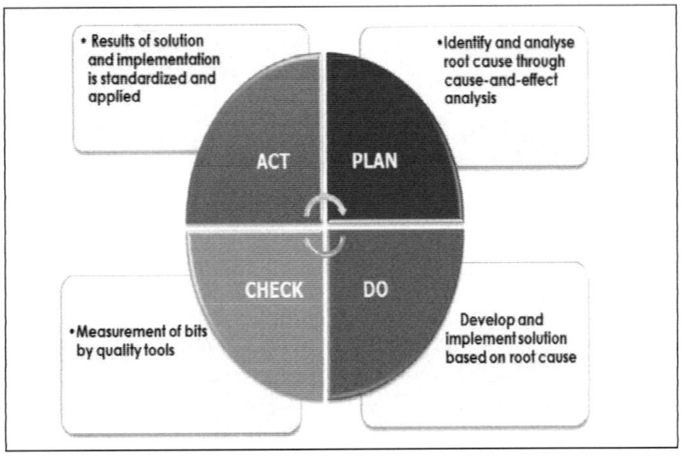

Figure 3. PDCA Methodology [13]

Result show that the main defect of the electrodeposited body was generation of bits. The systematic application of PDCA-cycle improved the generation of the bits. The average number of bits reduced was by 65%. This improvement shows the successful of this method in improving quality and reducing bits defect. However, this method may slow down process and do not allow variable that routinely crop up in a project.

3. Conclusion

A total of 5 basic quality systems have been reviewed ranging from Weibull Distribution Model, SPC, Six Sigma, Quality Assurance and PDCA. Researcher also using combination of quality tools in hybrid form to achieve optimum outcome in order to reduce defect of product manufacturing. Every method of quality management and continuous improvement has their own strength and weaknesses as detailed on table 1.

Table 1. Quality System Comparison

Type of Quality Tool	Advantages	Disadvantages
Weibull Distribution Model	Able to pin point root cause of defect. Scientific problem solving technique	Too complex and isoterix The accuracy of Weibull analysis depends on the quality of the data.
Type of Quality Tool	Advantages	Disadvantages
Statistical Process Control (SPC)	Able to predict if any significant deviations occurred that can lead to defect or rejected product Scientific problem solving technique	SPC can only detect abnormality. If the process is consistently in bad zone, SPC will consider the process in good order and will not trigger early warning.
Quality Assurance (QA)	Achieve 100% defect free product in the manufacturing process	Did not discuss the combination with other quality tools.

	Scientific problem solving technique	Hard to maintain when some information is missed.
Six Sigma	Able to reduce defect in each cases of case study Able to cooperate with various quality tools Scientific problem solving technique	This method is too extreme which can slow down the production leaving customer waiting for the product delivery
PDCA Cycle	A well know method for improvement of quality Scientific problem solving technique	This method may slow down process and do not allow variable that routinely crop up in a project

References

[1] M Raut and D D S Verma 2017 To Improve Quality and Reduce Rejection Level through Quality Control *Int. J. Recent Innov. Trends Comput. Commun.* Vol 5 No July pp 764–768

[2] S H Bambharolya and H R Thakkar 2015 Reducing Rejection Rate in Small Scale Machining Unit Using 7 Quality Control Tools - A Review *Int. J. Eng. Dev. Res.* Vol 3 No 4 pp 582–586

[3] N Hanifa, M Zaidin, M N Diah, P H Yee and S Sorooshian 2014 Quality Management in Industry 4 . 0 Era *J. Manag. Sci.* Vol 4 No 3 pp 82–91

[4] N I Idris, R Ahmad and N Muhammad November 2014 Product Reject Analysis using Weibull Distribution Model *8th MUCET 2014*

[5] D R Prajapati 2012 Implementation of SPC Techniques in Automotive Industry: A Case Study *Int. J. Emerg. Technol. Adv. Eng.* Vol 2 No 3

[6] R Godina, J C O Matias and S G Azevedo 2016 Quality Improvement With Statistical Process Control in the Automotive Industry *Int. J. Ind. Eng. Manag.* Vol 7 No 1 pp 1–8

[7] E Harpreet, S Oberoi, M Parmar, H Kaur and R Mehra 2016 SPC (Statistical Process Control): A Quality Control Technique for Confirmation to Ability of process *Int. Res. J. Eng. Technol.* Vol 3 No 6 pp 666–672

[8] M Noufil and M Ismail 2017 Reduction Of Defects Through Quality Assurance *Int. J. Mech. Prod. Eng.* Vol 5 No 7 pp 1–6

[9] S Kumar 2016 Application of Six Sigma in Automobile Industry *Int. J. Explor. Eng. Technol.* Vol 2 No 1 pp 39–44

[10] B John and A Areshankar 2018 Reduction of Rework in Bearing End Plate Using Six Sigma Methodology : A Case Study *J. Appl. Res. Ind. Eng.* Vol 5 No 1 pp 10–26

[11] M N C Ani, I Abdul Azid and S Kamarudin 2016 Solving quality issues in automotive component manufacturing environment by utilizing six sigma DMAIC approach and quality tools *Proc. Int. Conf. Ind. Eng. Oper. Manag.* Vol 8–10 March pp 1986–1997

[12] S Suresh, A L Moe and A B Abu 2015 Defects Reduction in Manufacturing of Automobile Piston Ring Using Six Sigma *J. Ind. Intell. Inf.*, Vol 3 No 1 pp 32–38

[13] A R Nabiilah, Z Hamedon and M T Faiz 2016 Improving Quality Of Light Commercial Vehicle *Manag. J. Adv. Manuf. Technol.*, no. Special Issue iDECON pp 525–534

General Process Change Qualification framework for Wafer Fabrication

D S Nair and J F Chin*

School of Mechanical Engineering, Engineering Campus, Universiti Sains Malaysia, 14300 Nibong Tebal, Penang, Malaysia

Corresponding author *: chinjengfeng@usm.my

Abstract. Wafer fabrication is the most value-added processes of semiconductor and electronics industries. The involved processes are state-of-the-art, most expensive, most complicated and highly sensitive. Any process changes require qualification dictated by different industry standards and also internal production performance specifications. The way a semiconductor industry performs qualification is often little known outside the industry and in contemporary literature. The paper offers insights into qualification process for wafer fabrication. Literature is reviewed on qualification in terms of reasons, stages and considerations. A general qualification framework currently employed in a foundry is shared as a valuable example.

Keywords : *Qualification, Semiconductor manufacturing, Framework*

1. Introduction

Wafer fabrication is the most expensive stage semiconductor manufacturing phase [1] and also the most complex manufacturing processes in existence [2]. It is the first basic stage of semiconductor manufacturing, preceding Sort (or Probe), Packaging (or Assembly) and finally Test [3-5]. Wafer fabrication builds integrated circuits (IC) such as electrical, photonic, or MEMS (Microelectromechanical Systems) circuits [6] in layers on silicon wafers normally 200mm or 300mm in diameter. Only some of these fabs, a common term for the factories that manufactures these products are considered to be pure play foundries. Foundries are wafer fabrication companies that produce custom made wafers for the different customers based on their product requirements, applications and specifications. They are the so called 'contract manufacturers' of semiconductor wafers.

Typically, steps ranging anywhere between 300 and 500 are involved [7], using high-tech machines, various gasses and chemicals. Wafer fabrication processes are done in (or a combination of various requirements) vacuum chambers or highly clean enclosed environment with tight control of particulate, ions, minerals, moisture, pressure, temperature, lighting and other forms of molecular level contaminants. Six main common modules of a fab are Diffusion, Implantation, Thin Films Deposition (Dielectric or Metal deposition processes), Chemical Mechanical Planarization (CMP), Photolithography and Etch [8]. In some process flows, wafers also undergo Epitaxial Growth and wafer backside grinding, also refereed as Thinning process. In each module, similar machines or tools sets (term used in fabs for machines or equipment) and processes are grouped together called tool group.

In wafer fabrication, tools and processes are closely monitored and maintained to ensure consistency of the output of every step. Any changes to any portion of process flows or tools cannot be arbitrarily done as it may affect final yield, electrical parameters and reliability of the devices. In technology development stage of a particular device, experimental changes in the process parameters and suitable tools are studied to optimize yields that suit the customers' electrical output parameters and reliability requirements. The running of confirmation pilots lots and engineering lots verify the stability of the entire flow and yield as per required. Next, the processes and the tools sets are fixed and monitored to be within agreed specifications and control limits.

The way a semiconductor industry performs qualification is often little known outside the industry and in contemporary literature. The paper offers insights into qualification process for wafer fabrication. The next two sections present literature review on qualification in terms of reasons, stages and considerations. In section 5, a general qualification framework currently employed in a foundry is shared as a valuable example. The framework is discussed in section 6 before a conclusion is given.

2. Qualification

Formally, qualification is a set of procedures of assurance for a specific system, premises or equipment to demonstrate the ability to achieve and fulfill specified requirements [14]. In addition, it provides documented evidence that a specific equipment, facility or system is fit or ready for the intended use [15]. If the manufacturer's specification offers a bigger range compared to the intended use or if a new process is developed to run beyond the current intended use, the equipment has to be qualified for the new range [16].

Graham et al. [15] identifies four stages of equipment qualification: design qualification, installation qualification, operational qualification and performance qualification. Design qualification (DQ) defines the functional and operational quality parameters requirements of the equipment and systems. The objective is to provide documented evidence that quality has been built into the design of the application. As a minimum this should include a listing of the manufacturer's instrument specification. The manufacturer also needs to demonstrate compliance with an appropriate quality system, during development and manufacture, and that the software source code is lodged with a secure third party. Installation qualification (IQ) intends to provide guarantee that the intended equipment is received as designed and specified. At this stage, the equipment is unpacked and checked against the order, ensuring that any pre-delivery qualification checks were made and recorded. Operational qualification (OQ) provides confirmation that the equipment functions as specified and operates correctly within the committed operation and functional limits. The equipment has to be calibrated before commencing this stage of qualification. Then appropriate qualification tests have to be done based on the intended use.

Performance qualification (PQ) is carried out to ensure the equipment consistently perform as required. Repeatability and reproducibility checking procedure including some tests applied during OQ is commonly involved. The checking may be carried out before use or included in a regular program, or combination of the two, as appropriate. Performance checks must be performed after maintenance and must be scheduled. The results must be fully documented.

Qualification begins with first identifying the reasons that warrant for changes. Once that is agreed upon, experiments and Designs of Experiment (DOE) are carried out using non-production wafers called Test Wafers (TW). Various measurements and cross-sections of structures that are of interest are done in the fab and Failure Analysis (FA) lab. Once the required criteria are met, engineering wafers (which are also non-production wafers) may then be processed through to confirm the electrical parameters. Finally test and control runs on actual production wafers are done to statistically verify that the changes do not alter the output and yield requirement of the current process. In each stage of the qualification, the results of DOEs and experiments have to be presented to a board of decision makers who review the data and allow or disallow the changes or request for more data to make decision. It is clear that senior representatives from each area of expertise and interest have to be present in reviewing the proposed qualifications and giving their relevant approval for the qualification proposal to allow the intended changes to take place. The review board members, at minimum, should consist of representatives from the Quality, Manufacturing, Technology, Yield and Process Engineering departments as each of them are important stakeholders to ensure the changes made are truly beneficial to the company without

IOP Conf. Series: Materials Science and Engineering **530** (2019) 012035 doi:10.1088/1757-899X/530/1/012035

jeopardizing the customers' interest. The collection of these representatives, make up the Change Review Board (CRB) or can be also called by various different names in different companies.

3. Reasons of change

Policy makers and top management alter and steer the direction of the original heading of the company to new placements to best satisfy future market demand. It is vital for operation management and is a strategic investment for the future (of a company). The cause for these changes can be termed as *Strategic Forecasting* and changes towards the new heading may require qualifications.

Further, continuous improvements are always needed for an organization to be competitive and successful. It is nearly impossible to maintain permanent condition over a long period of time as market conditions and technology evolves and forces changes to take place [9]. Improvements should be as a continuous process instead of being done when a crisis occurs.

Second, Capacity Improvement causes can bring about changes by the direct addition of new tools or process modules of an existing tool or cluster tool [10] and will increase the capacity. The new addition must statistically perform comparable, if not better than the existing unit. Third, when improvements are done on the existing process or tools to improve productivity using any known methods of lean manufacturing, changes occur. It is basically done by eliminating waste [11]. One of the key methods of gauging the improved productivity of a fab is by analyzing the fab's Cycle Time [12] which can be used to indicate shorter development time, faster time to market the products, lower WIP and thus better customer satisfaction. Productivity Improvement causes such changes to occur by means of improvement of the productivity

Fourth, the prospects to improve yields are constantly present at all times [13] in fabs due to the nature of relatively short life of devices, processes and equipment. Thus Quality Improvement changes are done to improve the quality and/or robustness of the current process or tools.

External forces also cause a huge need for changes. Cost Reduction is one of the most common and popular causes for changes too. These activities are done to improve the cost competitiveness and profitability of the current process or tools. Finding alternative source of supply to replace Original Equipment Manufacturer's (OEM's) parts and consumables have a huge impact to operational cost and profitability. However it is also one of the most risky changes and has to be dealt with caution. Common actions include proper statistical analysis and controlled release to production upon approval to mitigate risks. This lead yet to another cause of change called Risk Reduction. Since wafer fabrication is the highest contributor to the overall cost of semiconductor manufacturing, huge risks are at stake due to the high investment costs. Examples of these changes include reducing or eliminating risk to humans, profitability and loss of production time, tools or product. Changes caused by fulfillment of Environmental, Health and Safety (EHS) are done to improve environmental friendly manufacturing practices and methods of waste management of the current process or tools.

There are also other causes that contribute for changes to occur such as Customers' Request. Customers make certain changes for reasons known to them or their customers. These changes include alteration to certain parameters to suit their product requirements. Normally it involves recipe or process changes. The next drive for changes is caused by suppliers. Supply Issues causes changes triggered due to the unavailability or discontinuation of current supplies of process/production/spares or other materials which requires an alternative. There is no intend to bring about changes and it does not bring about any benefit to neither the fab nor the customers. However, due to unavailability or discontinuity of direct or indirect materials or parts, alternative source has to be used. Finally there are changes brought about by Governmental Policies. In order to abide to Governmental or other formal Global Policies, Rules and Regulations, forced changes have to be made in order for the company to continue supplying their customers who may be affected by such policy changes.

4. Factors affecting qualification

4.1. Industrial Standards

IOP Conf. Series: Materials Science and Engineering **530** (2019) 012035 doi:10.1088/1757-899X/530/1/012035

A qualification has to fulfill the requirements of accreditations, imposed by customers and the respective industry. To date, a number of accreditations are generally applied in wafer fabrication. These include MS ISO 9001:2008 [17], MS ISO 14001, OHSAS18001 [18], Sony Green Certificate, IECQ QC 080000 HSPM [19], SEMI standards [20] and JADEC [21].

MS ISO 9001:2008 [17] is a set of standards set by International Organization for Standardization which defines and specifies the requirements for an overall Quality Management System (QMS) of an organization. The focus is on effectively implementing QMS to meet customer requirements, it encompasses five major areas: QMS, Management responsibility, resource management, product realization, and measurement, analysis and improvement.

MS ISO 14001 and IECQ QC8000 HSPM [19] more focus on environment and hazardous material management. MS ISO 14001 governs and details the requirements for an overall Environment Management System (EMS) of an organization. The items stated here are specific to minimize the impact of the organizations' activities, products and services to the environment and strive for continuous improvement of an organization for its environmental performance. IECQ QC 080000 HSPM [20] lists the requirements for an electrical or electronic components manufacturer or supplier to exhibit to the international market place that the organization has developed, documented, and implemented processes for managing the production, selection and use of electronic components, assemblies, processes and related materials in accordance with customer, local, national and international HSF requirements. Sony Green has the similar intent of the previous two mentioned. Sony Green Certificate is a qualifications arrangement which is specific to Sony Corporation. In 2002, Sony established a set of "Management regulations for environment-related substances to be controlled which are included in parts and materials." Environment-related substances which are deemed harmful were immediately banned or reduced, as well as their applications. Since 2003, the corporation only purchases parts or materials for its products exclusively from suppliers that are certified as their Green Partners. To ensure that suppliers observe these regulations, the "Green Partner Environmental Quality Approval Program" was introduced. This qualification is a requirement for the suppliers and Green Partner approval needs to be renewed every two years.

OHSAS 18001:2007 is an international Occupation Health and Safety Assessment Series (OHSAS) for occupational health and safety (OH&S) management systems. It aims to help an organization to control OH&S risks for its employees and to improve their performance. The main aim of implementing the OHSAS standard is reducing and preventing accidents (including accidents related loss of live, equipment and time). It does not include product and service safety requirement nonetheless.

Apart from the general industrial standards, fabs are also governed by Semiconductor Equipment and Materials International (SEMI) standards [20]. There are many SEMI standards but the two most commonly used by fabs are the SEMIE10 and the SEMIE79. The former is the standards for specification for definition and measurement of equipment reliability, availability, and maintainability (RAM) and utilization, whereas the latter are standards for specification for definition and measurement of equipment productivity used in Overall Equipment Effectiveness (OEE) calculations for fab equipment.

The Joint Electron Device Engineering Council (JEDEC), was originally establishment in 1924 and after undergoing a few stages of revamps, was names as JEDEC in1958 [21]. As of current time, JEDEC emerges as one of the most dominant technical documentation reference in semiconductor industry.

4.2. Production Stability

The second aspect is to provide assurance to the production stability. The major indices commonly used to gauge the success of the manufacturing processes in the fab are done by the observant of other more general output parameters which are defined in the SEMI standards. Here are some of the indices:-

1. Moves - The number of wafers that successfully passes through a series of process steps clustered together as a stage. A fab which has the capacity to produce 24 thousand wafers a month may typically have to make 145,000 moves per day.

IOP Conf. Series: Materials Science and Engineering **530** (2019) 012035 doi:10.1088/1757-899X/530/1/012035

2. Utilization of tools – These manufacturing indices are derived based on the SEMI E10 and the SEMI E79. These standards among others show how many percent of the time the tool is being used for production activity with respect to the available of the tools.
3. Availability of tools – Also derived based on the SEMI standards for semiconductor manufacturing indicates how many percent of the time a tool is available for production regardless on whether or not there is material to process on that tool. This index is also commonly referred as the Uptime.
4. Turn Ratio – This index is common to reflect the number of moves generated compared to the ending of hand wip. It generally reveals how fast the production materials are moving through the various stages within the fab.
5. DPML – Days per mask layer. A normal industry standard for a 200mm fab running Aluminum metal layers can range anywhere from 1.6 to 2.0 DPML. It literally means from one mask layer to the immediate next mask layer it will take about 1.6 to 2.0 days. So for a 32 mask layer device, it will typically take about 64 days from start to complete if the declared mask layer is 2.0 DPML.

4.3. Process Stability
A process needs to be proven of its stability before release to production. To demonstrate process stability, all the response parameters of interest must display constant means and variances over a significant period. In-line product monitoring (ILM) systematically samples and scans production lots for defect levels based on individual rules for sampling as set and agreed by the customers' request, Technology department's and Quality department's respective representatives. The ILM is also done on tools that have recently undergone any form of maintenance that were scheduled or unscheduled, before being fully released back to production. Periodically performing qualification (qual) runs ensures the tools and their respective processing modules are kept into check for performance deterioration and stability monitoring. Quals are done using TW and the measurements are also monitored through sophisticated SPC systems as used for production wafers. It is also worth mentioning at this juncture that particulate contaminations and many physical forms of defects (such as scratches) are also monitored by performing particle quals using TW.

4.4. Customer Specifications
The control of quality, defect and yield supersedes all other priorities in any typical wafer fab. The reason being is, there is no point in producing defective wafers, which may fall below the acceptance yield or electrical test criteria of the customers. There are very strict acceptance criteria set by customers to ensure the chips in wafers are considered 100% good before they can be sawn into separate chips and be used in the semiconductor packaging process. Customers can easily reject a batch of wafers based on samples taken from the batch that fails to meet their criteria. These wafers will then have to be scrapped and thus a great loss of revenue to the company. So, sacrificing the speed of manufacturing to do inline monitoring and measurement as a measure of process stability monitoring and stopping the manufacturing line if certain tools fail their periodic qual runs or if the ILM fails the control limits or if the SPC charts for measurement parameters violates the Western Electric (WE) rules, is somewhat an acceptable practice and common in this industry.

5. A general framework for qualification in a foundry fab
Before the process of creating a generalized framework commence, a few factors need to be clearly defined and studied in order to make the step by step procedure optimal for the intended qualification. The first step is the establishment of a qualification review committee. As aforementioned the CRB members are formed and periodic congregations of such meetings are done to review the required changes. A facilitator from the quality department will arrange and document the proceeding of the meeting and follow up with the action items generated during the course of the change request.

IOP Conf. Series: Materials Science and Engineering **530** (2019) 012035 doi:10.1088/1757-899X/530/1/012035

The second step classifies the qualification based on the type of change required and further split it into the stages based on current organizational structure, urgency of the changes and risk mitigation. A set of guideline with standardized method of classifying the description of changes has to be devised. A typical example involving four classes based on the extent of changes required is given in Table 1.

Table 1: Class of changes

Class	Class 1	Class 2	Class 3	Class 4
Description	Major change	Minor change	Operational change	Unreleased technology change

Some very minimal Installation Qualifications (IQ) are being performed when new tools are brought into the fab during commissioning. However these works, mostly related to calibration, are done by equipment vendors to verify their tools' factory settings. These types of qualification works are normally go-no-go type or calibration whereby the tools must comply with the required calibration data before being handed over for further production qualifications. The majority of the qualifications that are performed on operational wafer fabs are the Operational Qualification (OQ) or the Performance Qualification (PQ). Other qualification, such as Design Qualification (DQ) is actually done by equipment manufacturers.

The third step of the framework relates to project-management and development of ownership. By looking into the overall qualification as a project that has clear sets of problem statement, goals, timeline and ownership, the entire qualification can be executed into few major stages as shown in the model framework of process change qualification in Figure 1 - first, the Pre-Change Qualification (PCQ) stage, where, non-destructive (to production wafers) tests and preliminary data are obtained to convince stake holders and CRB members to give approval to test the changes onto the production line using actual saleable wafers; Second, the Change Qualification (CQ) process being done on the production line in the fab using actual production wafers to obtain life data to make comparison of the required output parameters between wafers that have undergone the changes and wafers that are processed in standard conditions. Here, after reviewing the outcome of the production wafers' output parameters, the CRB members shall perform approval to the manner of which the changes are to be released to the fab based on the proposal of the change requestor. Some qualifications can be directly implemented some will have some form of amendments before being given approval for the third stage of Full Release (FR).

In these individual stages a cluster of activities have to be carried out in order to proceed to the next stage. At the PCQ stage, the requester defines the problem statement and conducts the tests and proposes the changes based on the results of the tests to the CRB member by preparing the PCQ presentation material, the CRB then makes decision to proceed as requested, proceed with the change request but the proposal need to be amended or the last option, reject the change request. They also review and in some case, when one or some of the CRB members deem fit, challenge the class of changes suggested by the requester based on the reasoning given and the individual fab's internal criteria requirements. If the CRB members approve the PCQ, then the evaluation with production lots are carried out at the CQ stage.

At the CQ stage, the requester who proposes the changes presents the results of the changes done on production units called split lots to the CRB members. Again at this stage, the CRB members can decide to either proceed with the proposal, proceed with the change with further changes to the proposal or they can also opt to reject the request. However, since at the PCQ stage, the CRB members already agreed to proceed to the CQ stage, rejecting the change request in this juncture is extremely rare. If the Class decision was not made in the PCQ session, then the CRB members can also decide what class of changes it is now and further decide if there is a need to update the customers of this change. If it is a class 1 or 2 change request, then customer approval is required before the changes can be actually implemented or tested on their wafers.

IOP Conf. Series: Materials Science and Engineering **530** (2019) 012035 doi:10.1088/1757-899X/530/1/012035

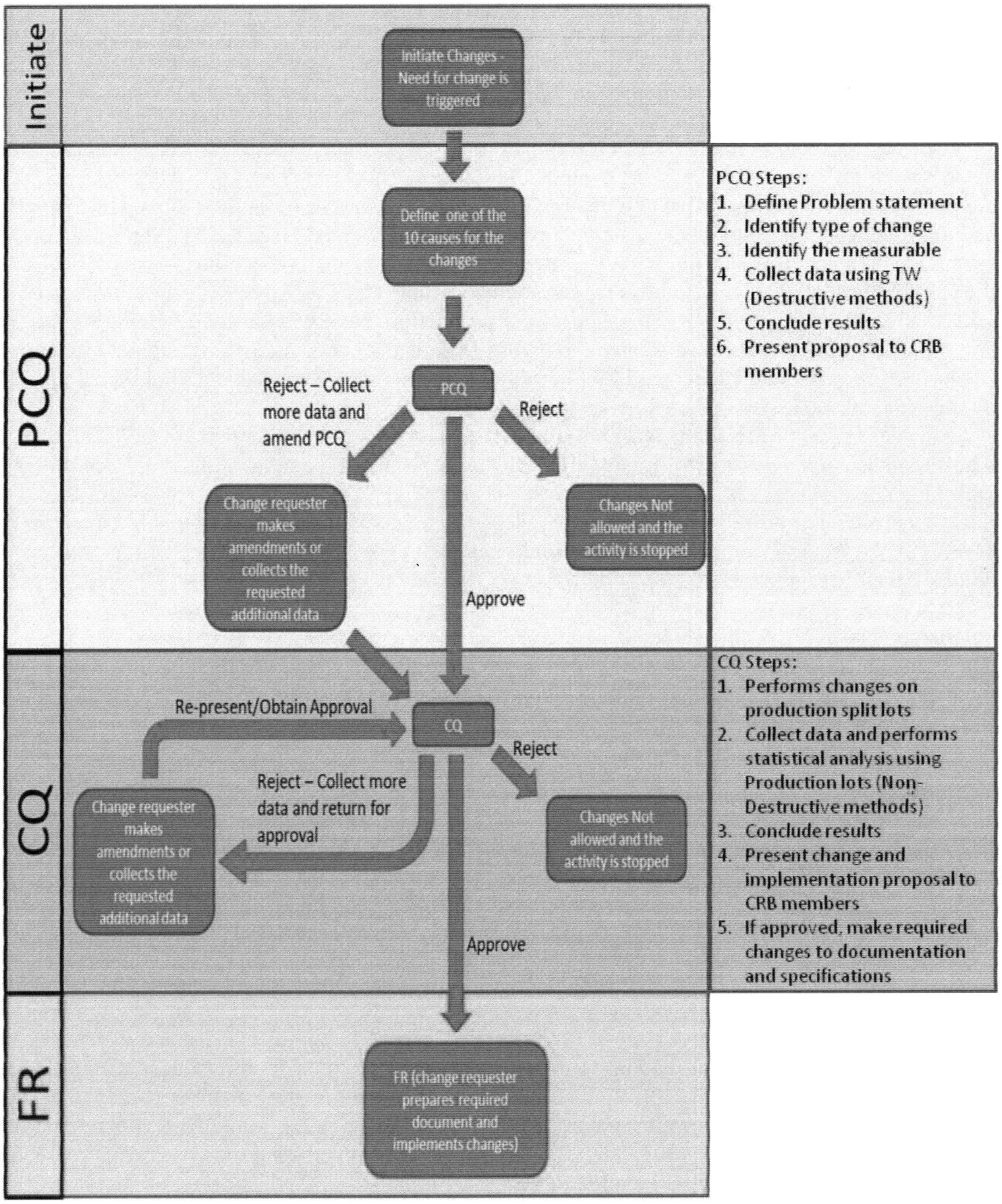

Figure 1: Model of a process qualification framework

The change requestor also proposes the manner in which the change will be implemented to the fab once the CQ is approved. In some the cases, CR may need some amendments before the FR is granted. The CQ data is usually reviewed for changes that may impact product quality and alignment with the relevant Technology or Yield Engineering department's person in charge is done and mentioned in the CQ presentation. Finally when FR is obtained, it also involves the amendments of all documentations and specification prior to execution of the qualified changes in the fab.

IOP Conf. Series: Materials Science and Engineering **530** (2019) 012035 doi:10.1088/1757-899X/530/1/012035

6. A case study of a process qualification in a foundry fab

The case study (Table 2) is for a change request for qualifying a new additional tool for Contact Etch Process. It is a repeat tool qualification, meaning that there already exists a similar tool of the same make and model that is qualified and running in the fab in stable mode. Thus this new tool is for a Capacity Improvement causes.

The case study shows the step by step action done by the change requester in order to release the new repeat too to the line. As can be seen, at the PCR stage, a lot of data from TW and engineering wafers are obtained and presented to the CRB members to convince the board that the tools is indeed reliable and qualified to proceed to the next step, CQ. Qualification runs are repeated to obtain as many as 10 separate data points at different intervals and timing to monitor the repeatability and reproducibility of the required results. The requester has to have enough data to statistically provide these evidences to the CRB. This is because, at the CQ stage, customer production lots are being used, thus if the tool is not able to perform as intended, it may cause deviations or defects on the production split lots and may cause scraps and yield loss. This will cause loss of money, man hours and delay in the tool being qualified to increase the fab capacity which is crucially required.

In the CQ stage, the actions are very straight forward, the requester runs the splits, collect the required monitoring data and statistically proof that the new tools performs the same as the POR tool by ensuring that there is not significant difference between the two sets of data. The results of electrical and yield parameters are also verified by the technology department's person in charge of Contact Etch process.

Finally the requester proposes the condition of release to the CRB and once the condition is accepted and met, the FR is granted.

7. Discussion on the framework

The framework clearly provides a strong visible guideline for change requesters. As the need to achieve manufacturing excellence and continuous improvements to move forward is ever demanding, it is important that such guidelines are available for change requester to ensure that they can perform their tasks in a predetermined step by step sequential procedure. It is self-explanatory and reduces time wasting to gather the methods from others who know the procedure due to past experiences.

Although passed all relevant product testing evaluations, changes frequently may not be feasible to be implemented throughout the production floor as FR. The reason being, different customers' devices have different aspect of sensitivity towards changes although they are built upon the same technology platform of fabrication. Thus, by cleverly adding a CR stage, where controlled production releases based on specific devices and customers that are often required to minimize risk both for the company and to the customers, can be done safely in a win-win manner.

Since the needs of data to be presented to the CRB can vary significantly depending on a variety of reasons and factors that have to be considered, it is difficult to actually lay out a specific requirement guideline for each type and category of qualifications. These requirements may be diverse based on different customers, devices, technology and applications. As such, the requirements are very subjective and are often discussed in the CRB forum to get consensus from attending members before the CQ split lots are run. Thus these prerequisites are not mentioned in qualification framework and are maintained as a subjective matter that are finally determined and agreed upon during the proposal is presented by the requestor. However a general guide as to what are normally required based on past experiences and feedback from the customers and device owners help the requesters to provide near accurate proposals for their change qualifications

An opportunity to further enhance the framework is identified by defining each step of the PCQ and the CQ in detailed sequential steps instead of being stated in a summarized manner. We also see that if each step can be additionally broken down, anyone following the guideline would know exactly what needs to be done at each stage within the step. The framework can be further be detailed out in a working step by step flowchart with decision making steps inserted to help the requester to make decisive decision at each juncture of the flow. In real practice, certain situations warrant conditional approval which requires minor modification or adjustment on the initial proposal, thus the usage of decision steps

IOP Conf. Series: Materials Science and Engineering **530** (2019) 012035 doi:10.1088/1757-899X/530/1/012035

as options should be explicit into the flowchart to further improve choice making with multiple plausible alternatives as accurate guidance for the requester.

In view of evolving nature of the qualification framework, a good documentation practice is vital such as adding clearer labels or identification tag numbers for each individual steps and further sub dividing them into groups of similar tasks. It also offers quick identification when search is required to be done in reference to a particular step of stage of qualification.

8. Conclusion

Qualification frameworks in fabs, till today, still remain a trade secret of each company. Different fabs in the world have their own methods of performing change qualifications but in general they all have the same considerations factors and requirements for similar processes, tools and application. Qualification methodologies are also some of the closely guarded knowhow of wafer fabrication industry alongside technical knowhow. The above qualification framework is the results of years of continuous improvements to enhance change qualification methods in a more efficient, manufacturing friendly and cost effective manner. It has worked out very well for the company. However, as steps are taken to move forward towards being competitive and positioning the company successfully ahead of other competitors, qualification framework remains a living document, open for multitudes of enhancement to add to our current best known methods of wafer fabrication change qualifications.

IOP Conf. Series: Materials Science and Engineering **530** (2019) 012035 doi:10.1088/1757-899X/530/1/012035

Table 2: Summary of the running of the case study

Stage	*Step*	*Action description*	*Day*
Initiate	1.0	A new additional tool for Contact Etch Process is brought in and has to be qualified to run production. It is a repeat tool qualification, meaning that there already exists a similar tool of the same make and model that is qualified	Day 1
PCR	2.1	Define the problem statement and what warrens the qualification, in this case, it is to support manufacturing capacity	Day 5
	2.2	Define the changes from the current condition to the proposed condition	Day 5
	2.3	Classify Changes to be undertaken based on DQ or OQ	Day 5
	2.4	Alignment check statement with other stakeholders (such as IE or Technology) and impact of changes to manufacturing indices (such as cycle time, capacity increase)	Day 10
	2.5	Define the methods and qualification matrix that was used to evaluate the new tool in comparison with the required specification and performance of an already existing tool. Qualification parameters are oxide etch rate, nitride etch rate, particle qual data, engineering wafers' contact diameter measurement also called Critical Dimension (CD) and engineering wafers' contact holes cross section view to measure and observe the profile of the hole	Day 10
	2.6	Present the evaluation and DOE results (preliminary results) of the TW and Engineering wafers' compared to the specification and current process of release (POR) tool to the CRB	Day 45
	2.7	Propose for some production split lots and request permission to proceed to the CQ stage where the declared the output parameters that will be checked on the split lots based on the relevant qualification matrix for the Contact Etch process	Day 45
CQ	3.0	Performs changes on production split lots the collect data and perform statistical analysis (using Non-Destructive methods)	Day 52
	3.1	Conclude results and prepare proposal including implementation strategies and timeline	Day 89
	3.2	Present the summary of the PCR and what was agreed upon and consented earlier (to refresh memory of the CRB members)	Day 92
	3.3	Present the split lots data of the parameters and statistically compare the data to the specification and current process of release (POR) tool to the CRB	Day 92
	3.4	Requester proposes the conditions for FR	Day 92
FR	4.0	FR Granted once the all condition meets the agreed criteria of the CQ proposal and requester updates all required documentations for the change to be implemented in the fab	Day 96

Acknowledgement
This work was supported by Research University Grant (RUI), Universiti Sains Malaysia [grant number 8014069].

References

[1] Chen J H, Fu L C, Lin M H and Huang A C 2001 Petri-net and GA-based approach to modeling, scheduling and performance evaluation for wafer fabrication *IEEE Transactions On Robotics And Automation* **17**(5) pp 619-636
[2] Mönch L, Fowler J W and Mason S J 2013 Production planning and control for semiconductor wafer fabrication facilities *Operation Research/Computer Science Interfaces* series 52 DOI 10.1007/978-1-4614-4472-5-1 (Springer; New York)
[3] Uzsoy R, Lee C Y and Martin-Vega L A 1992 A review of production planning and scheduling models in the semiconductor industry Part 1: system characteristics, performance evaluation and production planning *IIE Transactions on Scheduling and Logistics* **24**(4) pp 47-59
[4] Lin C H, Hwang S L and Wang E M Y 2007 A Reappraisal On Advanced Planning And Scheduling Systems *Industrial Management & Data Systems* **107**(8) Emerald Group Publishing Limited 0263-5577 DOI 10.1108/02635570710822822 pp 1212-1226
[5] Muhammad N A, Chin J F, Kamarrudin S and Chik M A 2015 Fundamental simulation studies of CONWIP in front-end wafer fabrication *Journal of Industrial and Production Engineering* **32**(4) pp 232-246
[6] Yao J J 2000 RF MEMS From A Device Perspective *Journal of Micromechanical Microengineering* **10** pp 9-28
[7] Intl. Optessa USA Inc., Semiconductor Manufacturing: Using Modern Scheduling Systems to Deliver to Customer Demand 2002 Available: http://www.optessa.com/assets/img/061102_Optessa_WhitePaper_Semiconductor.pdf
[8] Ibrahim K and Chik M A 2010 Variability due to tool configurations that impacts overall capacity in wafer fabrication facility *The 14th Asia Pacific Regional Meeting of International Foundation for Production Research* (Melaka)
[9] Oakland S J and Dotchin J A 1992 Theories and concepts in total quality management *Total Quality Management* **3**(2) pp 133-146
[10] Pierce N L and Drevna J M 1992 Development of generic simulation models to evaluate wafer fabrication cluster tools *Proceeding of the 1992 Winter Simulation Conference* pp 874-878
[11] Womack J P, Jones D T and Roos D 1990 The Machine That Changed the World *Productivity Press*
[12] Robinson J K 2002 Understanding and improving wafer fab cycle times *Semiconductor FabTech* www.SemiconductorFabTech.com *ICG Publishing* pp 1-6
[13] Leachman R C and Hodges D A 1996 Benchmarking semiconductor manufacturing *IEEE Transactions On Semiconductor Manufacturing* **9**(2) pp 158-169
[14] ISO 9000:2005 2009 Quality Management Systems – fundamentals and vocabulary
[15] Graham P, Broad N, Hailey P, Hardy A, Holland S, Hughes S, Lee D, Prebble K, Salton N and Warren P 2006 Guidelines for the development and validation of near infrared spectroscopic methods in the pharmaceutical industry *Handbook of Vibrational Spectroscopy, John Wiley & Sons Ltd* pp 1-21
[16] Burgess C and Jones D G 1998 Equipment qualification for demonstrating the fitness for purpose of analytical instrumentation *Analyst Issue* 9 Royal Society of Chemistry pp 1879-1886
[17] MS ISO 9001:2008 2009 Quality Management Systems (First Revision) (ISO9001:2008, IDT) *Department of Standards Malaysia, Malaysian Standard* pp 1-14
[18] OHSAS 18001:2007 2007 Occupation Health and Safety Management Systems – Requirements *The OHSAS Project Group*
[19] IECQ 08000 2012 IE commission quality assessment system for electronic components (IECQ System) & hazardous substance process management system requirement (HSPM) *IECQ*

Publications Edition 3 2012-05
[20] SEMI E10-0814 2014 Specification for Definition and Measurement of Equipment Reliability *Availability, and Maintainability (RAM) and Utilization* Semiconductor Equipment and Materials International
[21] Foundry Process Qualification Guidelines (Wafer Fabrication Manufacturing Sites) *JEDEC Solid State Technology Association* JEP001A 2014

IOP Conf. Series: Materials Science and Engineering **530** (2019) 012036 doi:10.1088/1757-899X/530/1/012036

Workload Analysis and Improvement of the Nurses Duty in the Hospital

H Soewardi and S R Kusuma*

Industrial Engineering Department, Faculty of Industrial Technology, Islamic University of Indonesia, Yogyakarta, Indonesia

Corresponding Author *: sasmitarahma15@gmail.com

Abstract. Nurse is someone who has a task as a medical asisstant in the hospital. They work for eight hours per day in serving patients and helping out the doctors such as, wound cares, infusions, bathing patients, injecting medical substances, giving medications, moving patients, changing patient clothes, etc, creating a lot of work for them. According to survey of the Indonesian National Nurses Association there are 50.9% of nurses experiencing work stress, dizziness, and fatigue because of high workloads. Therefore, it is necessary to be improved. Objective of this study is to analyse the mental workload of nurses and develop few of recommendations to reduce the mental workload. National Aeronautical and Space Administration Task Load Index (NASA-TLX) is used to evaluate the mental workload based on six variables: mental demand, physical demand, temporal demand, own performance, effort, and frustation level. While the developed improvement uses the participatory ergonomic method which involve nurse, doctor, ergonomist, and designer as stakeholder by implementing a group discussion to determine criteria of enhancement. Statistical analysis is applied to verify the hypothesis. Result of this study shows that the score of NASA-TLX is 62.46, which means the task is overload. As for the result, participatory ergonomic is verified at 5% of significant levels that requires some improvements, including but not limited to administrative activities in the form of making action invoices, physical activity that causes spinal cord injuries was lifting the patients and bending while injecting the patient, and work shift system where morning and afternoon shift longer than night shift.

Keywords : *Nurse, NASA-TLX, Participatory Ergonomic*

1. Introduction

Hospital is a public health service in the form of inpatient, outpatient and emergency. These services are carried out by medical personnels such as doctors, midwives, and nurses. They work in shift schedule for eight hours. The doctor is responsible for diagnosing the disease, giving medicine, performing surgery, etc. This task is assisted by nurses who have a task as a medical asisstant in the hospital [1] that includes wounds care, infusion, injecting, giving medication, changing patient clothes, and transfer of patient to another location. The Indonesian National Nurses' Assosiation in 2006 found that 50.9% of nurses get stress, dizziness, fatigue, and are unable to rest. It because of the high workload and time consuming. The workload causes psychological distress and feeling of entrapment to escape from an aversive situation, which both are related to fatigue[2-3].

IOP Conf. Series: Materials Science and Engineering **530** (2019) 012036 doi:10.1088/1757-899X/530/1/012036

Based on the preliminary study it is identified that 47.8% of nurses have complaints related to working hours, rest hours, work shifts, income, number of workers, and reward and punishment. This is encouraged by [4] who found 60.3% of nurse experiencing the high workload in doing the task. Study of [5] shows that nurses experiencing work stress directly related to work is caused by low income, heavy workload, night shift and the lack of interaction. The excessive workload can increase the occurrence of failure of collaboration between nurses and doctors, poor communication between nurses with patients, discharge nurses and nurse job dissatisfaction [6]. In addition, the impact of nursing workload can cause decreasing quality of care and safety patient, as well as level of anxiety, stress, burnout, and attrition of nurses[7-8]. This condition is significant to be improved in order to avoid the excessive workload.

The objective is to investigate the nurse workload and to recommend some improvement of causes the negative effect on nurse subjectively. The workload will be identified based on mental demand, physical demand, temporal demand, own performance, effort, and frustation level. As for the recommendation is resulted from focus group discussion among the stakeholders. They are nurse, doctor, ergonomist, and designer.

2. Methods

2.1. NASA-TLX Questionnaire

National Aeronautical and Space Administration Task Load Index (NASA-TLX) is a method developed based on subjective measurement needs consisting of nine factor scales, namely task difficulty, time pressure, type of activity, physical effort, mental effort, performance, frustration, stress and fatigue. The nine existing factors are simplified into six indicator factors, namely mental demand, physical demand, temporal demand, performance, effort, and frustation [9]. In this study, NASA-TLX questionnaires are distributed to 94 nurses from three hospital in Yogyakarta for assessing their mental workload.
The stages in measuring NASA-TLX are :

1. Weighting
   Respondents were asked to selecting one of two indicators considered as the dominant cause of workload mental at work .The number of the calculation of calculated from each indicators are most influential .The number of these calculations will be weights to every indicator mental burden .The following are table comparison indicators NASA-TLX :

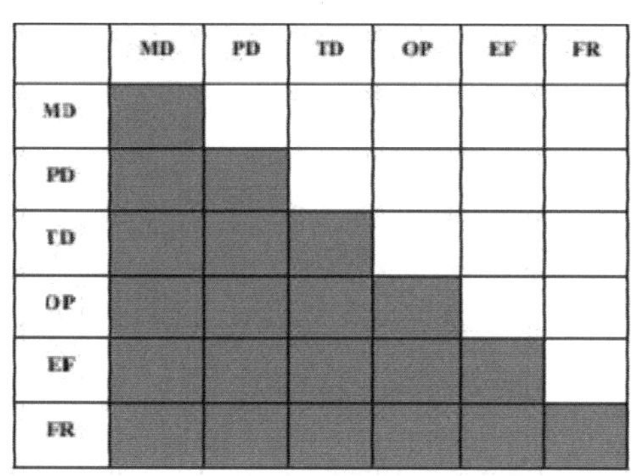

Figure 1. Weighting NASA – TLX Indicators
Source : [9]

2. Rating
   Respondents are asked to rate the six indicators of mental load. Subjective rating given is dependent on the mental burden felt by the respondents. The following rating indicator of NASA-TLX :

IOP Conf. Series: Materials Science and Engineering **530** (2019) 012036 doi:10.1088/1757-899X/530/1/012036

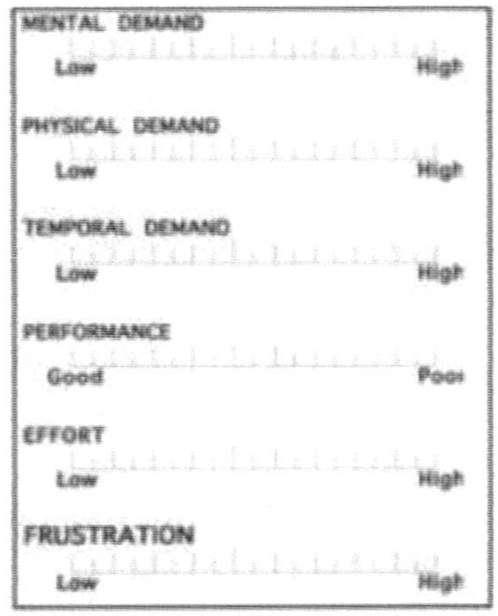

Figure 2. Rating Scale NASA – TLX Indicators
Source : [9]

3. Calculating the value of the product
   Represents the multiplication result of the indicator weight with the rating of each indicator :

$$\text{Product} = \text{rating x weighting factor}$$

4. Calculating Weighted Workload (WWL)
   WWL is obtained by summing the six products :

$$\text{WWL} = \sum \text{product}$$

5. Calculate the average WWL
   WWL is obtained by dividing the total number of weights :

$$\text{Score} = \frac{\sum product}{15}$$

6. Interpretation of score
   Classification of mental workload scores is explained by [9] as follows:

Table 1. Classification of Mental Workload

Workload Group	Score
Low	0 – 9
Moderate	10 – 29
Kinda High	30 – 49
High	50 – 79
Very High	80 – 100

In addition to research [10] the classification of mental workload is classified into:

IOP Conf. Series: Materials Science and Engineering **530** (2019) 012036 doi:10.1088/1757-899X/530/1/012036

Table 2. Classification of Mental Workload

Workload Group	Score
Underload	< 40
Optimal Load	40 – 60
Overload	≥ 60

2.2. Participatory Ergonomic

Participatory ergonomic is a process whereby employees participate together with management to strive to plan, implement, and evaluate processes with other resources to produce appropriate work design concepts so that systematically reduce workplace accidents [11-12],. In this study, participatory ergonomic is used to develop solution recommendations in reducing the mental workload experienced by nurses. Participatory ergonomics involve several stakeholders, namely nurses, doctors, ergonomist, and designer. There are three steps in applying participatory ergonomic according to [13] :

1. Participant selections
2. Conduct design and development based on the result of input provided by participants
3. Implementation of results for further evaluation by participants.

3. Result and Discussion

3.1 Result of NASA – TLX Questionnaire

The WWL average of 94 nurses is 62.46 where according to table 2 it is included in the overload category. The highest load category is at the medium level, namely effort with a value of 18.1704, then mental demand (MD) with a value of 10,737 and a physical demand (PD) indicator with a value of 10,133. The temporal demand (TD) indicator is 9.76, the frustation indicator (FR) 6.83 and the own performance (OP) indicator 6.82 which fall into the category of low workload.

Nurses have a high workload because they have a large responsibility in providing nursing services to patients [14]. In this study it is known that nurses need greater effort to handle patients because the number of nursing staff is not comparable with the large number of patients who enter. The tasks carried out require nurses to think, concentrate, and count, including giving drug doses, administering intravenous fluids and administering patient data. In administering doses of drugs and administering intravenous fluids, a thought process and proper calculation are needed from the nurse. In the patient administration process, the nurse carries out an action invoice that is the value of the services provided to the patient to be paid by the patient to the hospital. This activity becomes a problem for nurses because of the double job function.

Double functions occur because invoicing is an administrative task that can be done by admin workers without having to have nurse competencies. In research [15] even stated that indicators of mental needs become dominant factors in nurses because nurses are required to work quickly so that all patients can be served. Besides that, affecting the high mental needs are the number of patients and the unpredictable severity of the patient, the knowledge and skills possessed are not able to compensate for the difficulty of work, the pressure and demands to save the patient both the moral demands, the demands of the hospital leadership and the demands of the family. patients, are always faced with the right decision making and high responsibility in carrying out nursing care.

Nurse activities tend to be many and varied and involve physical involvement including encouraging beds, patient wheelchairs, transferring patients, and lifting patients on stretchers. Physical activity carried out by nurses has been greatly assisted by the existence of adequate assistive devices and facilitates the appointment, transfer, and encouragement. For example in the activity of pushing, the bed of a patient who has been equipped with a wheel so that there is no need for excessive energy to encourage, in the use of wheelchairs it is now easier for patients and nurses with the automatic buttons inside. While in the process of appointment and transfer such as the use of a stretcher, the scoop stretcher tool helps nurses because of its human adjusting principle so that it is only necessary to insert a stretcher to the bottom of the patient's body without having to lift the patient's body first. Even so, there are still complaints of spinal cord injury experienced by nurses in the appointment process.

The amount of work that exceeds the capacity causes the physical condition of the nurse to be easily tired and easily tensed [16]. This is because nursing services require technical skills and sufficient knowledge. Nurse activities can be categorized as monotonous work because of its routine nature such as sitting too long which causes the muscles to become stiff, standing and bending too long when doing nursing activities can cause pain in the waist. The impact experienced by nurses is spinal cord injury due to wrong lifting, this is often the case for emergency nurses. The increased risk of spinal cord injury occurs as a result of repetitive activities [17].

The shift system requires the nurse to complete the work on the shift to complete. This results in shift changes cannot be done in a timely manner. In other cases, nurses can "order" their shift schedules to the head of the room. But this causes a shortage of nurses in certain shifts. The existence of this "ordering" system also causes fatigue for nurses when the shift schedule is close so that the rest time is reduced. This is evidenced by the existence of complaints where nurses complain about working hours, rest hours, and work shifts where there is a correlation with the lack of available nursing staff.

3.2 Result of Participatory Ergonomic

Participatory ergonomic with nurses finds several problems in nurse activities including administrative problems in the form of invoicing action, problems due to physical work such as removing patients in moving patients resulting in spinal injury in nurses, as well as problems in the work shift system where there are deficiencies number of nurses in certain shifts.

Problems in administrative activities actually do not require the competence of a nurse, which is the making of an action invoice. Action invoices are used to enter the value of services used by the patient for the actions given by the nurse. So in this case it is necessary to identify the tasks performed by the nurse, whether the task is the main duty of the nurse or not. In this problem, a job redesign is needed for nurses. So that it can be identified as the main tasks that must be done by nurses and tasks that can be done by other people without having to have competence as a nurse. Job redesign is a process to get as much information as possible about the facts that occur in order to complete the tasks that are in a position [18]. Job redesign is a procedure that must be passed to determine the responsibilities and characteristics of people working in a position [19].

Physical activities carried out by nurses include transferring patients to bed to be taken to the nursing room, sewing wound activities, infusion, and measuring urine. In the activity of moving the patient to the bed to be taken to the ward there is a process of lifting, removing, and pushing a load. While in the activity of sewing the wound, infusion, and measuring urine there are jobs with a bent posture. All of these activities have the potential to cause nurses to experience injury. In overcoming the odd position when the posture is bent, it can be technically controlled. Technical control is carried out to reduce the impact of spinal injuries that can be experienced by nurses. The opportunity for spinal cord injury in nurses is 43.1 - 87% [20]. A bent position can not be avoided by nurses when providing nursing services with the patient lying in bed. Bending is a position of deflecting the spine in a frontal direction that can overload the intervertebral disc, and increase ligament contraction and spinal support muscles [21]. So if the bending work is done for a long time and with frequent frequency, the ligaments and supporting muscles of the spine can weaken and put pressure on the intervertebral disc.

Technical control is carried out by providing adjustable facilities and infrastructure such as chairs and beds that can be adjusted in height. When performing treatment in a patient's condition, the height of the nurse influences the magnitude of the arch's back angle [21]. The greater the arch angle of the back, the muscle and ligament contraction will increase so there can be an emphasis on the intervertebral disc. Analysis in the study [21] states that the curved angle of more than or equal to 600 has the potential

11 times more often to cause spinal cord injury. Further explained, the dominant work equipment causing spinal injury is the use of beds and gurney. By providing equipment that can be set the height of the nurse will be free to adjust the height of the body when conducting checks. Besides providing training on patient lifting techniques also controls the risk of injury experienced by nurses. Nurses must be trained so that minimal patient lifting and moving work are carried out by two nurses who are competent in patient transfer techniques, untrained nurses are proven to be significant risk factors for spinal cord injury [22]. The right recommendation is to remove or transfer the patient with a shrug method that is by using one hand to do the lifting and the other hand rests on the bed or base. Lifting with a shrug method has been used internationally for many years because this method produces lower intra-abdominal pressure compared with two-handed removal at the same time [17].

Determination of the work shift schedule is carried out by the head of the room with a system of "ordering" for nurses who want to determine the schedule on the desired day. The "ordering" system is common because it is a form of right given to nurses to be able to determine their own schedule so that in working nurses can be happier. But it cannot be denied that this system will result in a shortage of nurses on certain shifts. For this reason, it is necessary to do shift shift management by making a work shift system. The nurse's work shift system is as follows:

1. Shift rotation follows the sun
   The purpose of shift rotation following the sun is setting the shift according to the rotation experienced by the sun that is morning - evening - night. This is related to the human nervous system which has repulsion when sudden changes occur, so that with the shift of the sun's rotation there is no significant change in the nervous system [17].
2. The shift system uses continental patterns (2 - 2 - 3) or metropolitan patterns (2 - 2 - 2)
   A good work shift system according to [23] is continental and metropolitan patterns. Both patterns have differences in weekend cycles obtained, in continental patterns Saturday and Sunday weekends will be found in the fourth week. While in the metropolitan pattern the new weekend will be found in the eighth week, so that the continental pattern will be preferred by workers.

	Monday	Morning		Monday	Night
	Tuesday	Morning		Tuesday	Night
	Wednesday	Afternoon		Wednesday	-
Week I	Thursday	Afternoon	Week III	Thursday	-
	Friday	Night		Friday	Morning
	Saturday	Night		Saturday	Morning
	Sunday	Night		Sunday	Morning
	Monday	-		Monday	Afternoon
	Tuesday	-		Tuesday	Afternoon
	Wednesday	Morning		Wednesday	Night
Week II	Thursday	Morning	Week IV	Thursday	Night
	Friday	Afternoon		Friday	-
	Saturday	Afternoon		Saturday	-
	Sunday	Afternoon		Sunday	-

(a)

	Monday	Morning		Monday	Night
	Tuesday	Morning		Tuesday	Night
	Wednesday	Afternoon		Wednesday	-
Week I	Thursday	Afternoon	Week V	Thursday	-
	Friday	Night		Friday	Morning
	Saturday	Night		Saturday	Morning
	Sunday	-		Sunday	Afternoon
	Monday	-		Monday	Afternoon
	Tuesday	Morning		Tuesday	Night
	Wednesday	Morning		Wednesday	Night
Week II	Thursday	Afternoon	Week VI	Thursday	-
	Friday	Afternoon		Friday	-
	Saturday	Night		Saturday	Morning
	Sunday	Night		Sunday	Morning
	Monday	-		Monday	Afternoon
	Tuesday	-		Tuesday	Afternoon
	Wednesday	Morning		Wednesday	Night
Week III	Thursday	Morning	Week VII	Thursday	Night
	Friday	Afternoon		Friday	-
	Saturday	Afternoon		Saturday	-
	Sunday	Night		Sunday	Morning
	Monday	Night		Monday	Morning
	Tuesday	-		Tuesday	Afternoon
	Wednesday	-		Wednesday	Afternoon
Week IV	Thursday	Morning	Week VIII	Thursday	Night
	Friday	Morning		Friday	Night
	Saturday	Afternoon		Saturday	-
	Sunday	Afternoon		Sunday	-

(b)

Figure 3. (a) Continental Pattern (b) Metropolitan Pattern
Source : [17]

3. Work shift planning includes weekends with 2 consecutive holidays.
4. Break time between shifts is at least 11 hours.
5. Night work 2 days in a row must immediately be followed by a break of at least 24 hours.
   In designing work shift rotations, [17] explains two things that must be considered, namely the lack of rest or sleep should be kept as small as possible to reduce fatigue, and provide as much time as possible for family life and social contact. Therefore, giving of 2 holidays in a row can provide opportunities for personal life and social contact for nurses. Reducing fatigue can be done

by paying attention to giving a minimum of 11 hours of inter-shift time and resting at least 24 hours after night work for two consecutive days. In addition, the reduction in the number and quality of sleep of night shift workers will lead to reduced work performance [17].

6. Scheduling with a decentralized system.
   In carrying out the work shift scheduling system, it should be continued in a decentralized manner, which is given responsibility to each head of each room. Decentralized scheduling will make it easier for nurses to control the work environment and increase nurses' autonomy in scheduling and flexibility [24].

The three solution recommendations obtained from each expert are verified by all stakeholders as validation of whether the three recommendations are appropriate and applicable.

Table 3. Validity Test Results

Item-Total *Statistics*

	Scale Mean if Item Deleted	Scale Variance if Item Deleted	Corrected Item-Total Correlation	Cronbach's Alpha if Item Deleted
Recommendation_1	8.90	1.358	.591	.775
Recommendation_2	8.70	1.379	.665	.702
Recommendation_3	8.80	1.221	.672	.690

Validity test results are seen from the corrected item-total correlation which shows the value of 0.591 for recommendation 1, for recommendation 2 is 0.665, and for recommendation 3 shows the value of 0.672. All r counts $\geq$ r table is 0.4438, so the recommendation for the given solution is valid. In the reliability test, the value of cronbach's alpha is obtained at 0.797 where the consistency is included in the classification. Overall from the results of validity and reliability, the recommendations for the solutions provided can be applied to reduce the mental workload of nurses.

4. Conclusion

Based on the results of the processing and discussion carried out, it can be concluded as follows :

1. The mental workload of nurses experiences overload with an average value of weighted workload of 62.46. The highest indicator with medium workload is an effort of 18.17; mental demand of 10.737; physical demand of 10.133; and with a low workload namely 9.76 temporal demand; frustation level 6.83 and own performance 6.82.
2. Solution recommendations given for problems in administrative activities, spinal cord injury due to physical activity, and work shifts are as follows:
   a. Job redesign,
   b. Technical control
   c. Work shift system.
3. The recommended solution is valid with r count $\geq$ r table and reliable with a value of 0.797 so that recommendation solutions can be applied to reduce the nurse's mental workload.

Acknowledgement
This study was supported by Directorate of Student Development Islamic University of Indonesia, Yogyakarta, and Work System Design and Ergonomics Laboratory, Industrial Engineering Department, Faculty of Industrial Technology, Islamic University of Indonesia, Yogyakarta.

IOP Conf. Series: Materials Science and Engineering **530** (2019) 012036 doi:10.1088/1757-899X/530/1/012036

Reference

[1] R Maharja 2015 Analisis Tingkat Kelelahan Kerja Berdasarkan Beban Kerja Fisik Perawat di Instalasi Rawat Inap RSU Haji Surabaya *The Indonesian Journal of Occupational Safety and Health* pp 93-102

[2] K Sheppard 2015 Compassion Fatigue Among Registered Nurses : Connecting Theory and Research *Applied Nursing Research* pp 57-59

[3] A W Griffiths, A M Wood and S Tai 2018 The Prospective role of defeat and entrapment in caregiver burden and depression amongst formal caregivers *Personality and Individual Differences* pp 24-31

[4] F W Astuti, Ekawati and I Wahyuni 2017 Hubungan Antara Faktor Individu, Beban Kerja, dan Shift Kerja dengan Kelelahan Kerja Pada Perawat di RSJD Dr. Amino Gondohutomo Semarang Jurnal Kesehatan Masyarakat pp 163-172

[5] W Xinjuan, L Jiaqian, L Ge, L Ying, C Jing and J Zhaoxia 2018 The effect of emotional labor and competency on job satisfaction in nurses of China : a nationwide cross-sectional survey *International Journal of Nursing Sciences* p doi: 10.1016/j.ijnss.2018.08.001

[6] B L Marquis and J H Carol 2006 Leadership Roles and Management Functions in Nursing : Theory and Application (Philadelphia: Lippincott Williams & Wilkins)

[7] L H Aiken, D M Sloane, L Bruyneel, d H K Van, P Griffiths and R Busse 2014 Nurse Staffing and Education and Hospital Mortality in Nine European Countries : A Retrospective Observational Study *Lancet* pp 1824-1830

[8] C Delgado, D Upton, K Ranse, T Furness and K Foster 2017 Nurses' Resilience and the Emotional Labour of Nursing Work : An Integrative Review of Empirical Literature *International Journal of Nursing Studies* pp 71-88

[9] S G Hart and L E Staveland 1981 Development of NASA-TLX (Task Load Index Results of Empirical and Theoretical Research *In Human Mental Workload* pp 139-183

[10] M Syafei and W Katon 2011 Analisis Beban Kerja Pegawai Secara Subjektif dengan Menggunakan Metode NASA-TLX (Studi Kasus pada Bagian Proses Manufaktur di PT.Agronesia Divisi Industri Plastik-Bandung) *Proceeding 11th National Conference of Indonesian Ergonomics Society 2011*

[11] W Hendrick and B M Kleiner 2001 An Overview of Macroergonomics (USA: Lawrence Erlbaum Associates)

[12] P Sukapto 2007 Peran Participatory Ergonomics dalam Transfer Teknologi dan Implikasinya Terhadap Kecelakaan Kerja (Bandung: Universitas Katolik Parahyangan)

[13] J A M De 2004 A Three-Phased Model of Participatory Ergonomics Processes To Improve Work in The Construction Industry *Industrial Health Journal Delfin University of Technology* pp 383-387

[14] Y D Werdani 2016 Pengaruh Beban Kerja Mental Perawat Terhadap Tingkat Kepuasan Pasien di Ruang Rawat Inap Rumah Sakit Swasta di Surabaya *Jurnal Ners Lentera* vol. 4 pp 97-105

[15] S Susanti, A Pawennari, I N Afiah, M Dahlan and N Rauf 2017 Analisis Pengukuran Beban Kerja Mental Perawat Unit Gawat Darurat dengan Metode NASA-Task Load Index *Prosiding SNTI dan SATELIT* pp B324-328

[16] V V Runtu, L Pondang and R Hamel 2018 Hubungan Beban Kerja Fisik dengan Stres Kerja Perawat Diruang Instalasi Rawat Inap Rumah Sakit Umum GMIM Pancaran Kasih Manado *e-Journal Keperawatan* pp 1-7 2018

[17] E Nurmianto 2004 Ergonomi Konsep Dasar dan Aplikasinya (Surabaya: Prima Printing)

[18] W B Werther and D Keith 1996 Human Resources and Personal Management (USA: McGraw-Hill)

[19] G Dessler 2003 Manajemen Sumber Daya Manusia (Jakarta: PT. Indeks)

[20] A Schoenfisch, H Lipscomb, L Pompeii, D Myers and Dement J M 2013 Musculoskeletal Injuries Among Hospital Patient Care Staff Before and After Implementation of Patient Lift and Transfer Equipment *Scan J Work Environment Health* pp 27-36

[21] L M Kurniawidjaja, E Purnomo, N Maretti and I Pujiriani 2014 Pengendalian Risiko Ergonomi Kasus Low Back Pain pada Perawat di Rumah Sakit *MKB* pp 225-233

[22] T Wong, N Teo and M Kyaw 2010 Prevalence and Risk Factors Associated with Low Back Pain Among Health Care Providers in a District Hospital *Malaysian Orthopaedic Journal* pp 23-28

[23] E Grandjean 1986 Night Work and Shift Work in Fitting The Task To The Man : An Ergonomic Approach (London: Taylor)

[24] P Wright and S Mahar 2013 Centralized Nurse Scheduling to Simultaneously Improve Schedule Cost and Nurse Satisfaction *Omega* pp 1042-1052

IOP Conf. Series: Materials Science and Engineering **530** (2019) 012037 doi:10.1088/1757-899X/530/1/012037

Hybrid Methods of MOST and 5S for Reducing Time Processing and Waste Motion in Milk SMEs Industry: A Case Study

A D Sari[1], R Gumilar[1*], N Setiawan[1,2], M R Salleh[2], M R Suryoputro[1] and N Zhafira[1]

[1]Faculty of Industrial Technology Department of Industrial Engineering Islamic University of Indonesia

[2]Faculty of Manufacturing Engineering Universiti Teknikal Malaysia Melaka

Corresponding author *: retnogumilar90@gmail.com

Abstract. Motion is one type of waste and it is a great challenge for the Small and Medium Enterprises (SMEs) to enhance the efficiency and productivity of all processes. The integration between lean and ergonomics in waste reduction has been identified as a tool to boost up the performance of SMEs. The aim of this study is to reduce the processing time by eliminating the motion using the hybrid of MOST and 5S practices at CV. Sahabat Ternak. The data has been collected to determine and analyse the inefficient movements of operators at their work stations by comparing before and after MOST-5S implementation. The results had shown a significant reduction of processing standard time in each four division as follows: Receiving Div. is 5.7%, Processing Div. is 1.3%, Milling Div. is 5.8%, and Packaging Div. is 19.5%.

***Keywords**: 5S, MOST, Motion, Lean, Ergonomics*

1. Introduction

CV. Sahabat Ternak is one of the food and beverage manufacturing industry in Yogyakarta, Indonesia which transforms the Etawa goats milk into powder milk as the instant products. This industry has a great opportunity to provide milk products for sufficient community nutrition. Based on data from Yogyakarta Government, the current population of Etawa goats is approximately 5, 994. Each of female goat is able to produce 0.85 litres of milk per day [1]. The current amount is classified as ideal for daily needs [1]. The rapid development of livestock Etawa goats is one promising business potential if it can be processed and managed effectively and efficiently.

In global environment, Small and Medium Enterprises (SMEs) has become a key driver for economic growth through innovation process and commonly using manual or semi-automatic technology. However, many of them lack of the capability for developing continuous improvement process [2] . In this case study, the primary goal of company is to increase profit by reducing cost, increasing process efficiency and improving the productivity using the traditional process. It could be achieved by eliminating the waste in the production process with the integration of lean and ergonomics [3] and those wastes are including over processing, excessive motion, and transportation.

Motion is the most severe type of waste, because the extra movement will greatly affect the working time [4]. Several studies related to motion had been carried out, including [4] the motion analysis using work posture analysis with Ovako Working Assessment System (OWAS). The analysis of excessive motion using Rapid Entire Body Assessment (REBA) method also had been done by [5]. Then research conducted by [6] had shown that redesigning work stations, as well as the application of ergonomic aspects, especially Method Time Measurement (MTM) can reduce the number of lead times affected by work motion. According to [7] the indirect time measurement method with Maynard Operation Sequence Technique (MOST) can be done five times faster than MTM. Subsequent research was conducted by [8] which compared the time measurement method between Stopwatch method as well as MOST method. Measurement using MOST method had given a better result (i.e. 15% saving time) compared to Stopwatch method. Furthermore, research conducted by [9] suggested that identification of waste and making improvements with 5S can make the work more organized. The purpose of 5S is to reduce non value added activities, such as searching for something and reducing the distance to shorten work time [10].

This paper is presenting a case study about the difference before and after implementation of 5S in the production system at milk SMEs in Yogyakarta, Indonesia. The scope of this research will mainly focus on identifying the inefficient motion in the production process of milk powder.

2. Literature Review

Increasing business competition and high demands from consumers require companies to be able to manage the production process more efficiently and effectively [11]. Lean ergonomic is used to reduce activities and systems that are not in accordance with the principles of ergonomics [5]. Ergonomics is a key parameter of the assembly process, as well as parameters of lean manufacturing such as takt time, cycle time, and work in progress [12]. The linkage of lean production system with ergonomics is in human effort, worker autonomy, the risk of Work-related Musculoskeletal Disorder (WMSD), and participation or involvement of workers [13]. Employees assume that the ergonomic role of human factor and working conditions in lean manufacturing is very important for improvement in work stations, and its application can be more beneficial to the relationship between employees and the company [14].

Classification of waste is classified as transportation, over processing, waiting, motion due to movement, and waste of motion due to posture [5]. To be able to increase productivity and eliminate waste of motion, the proposed improvements including changing work positions, changing work layouts and eliminating work movements are necessary [5]. Motion can be eliminated by identifying and minimizing non value added activities (NVA), unnecessary movements, and identifying bottlenecks in the activity of taking and placing parts in Computer Numerical Control (CNC) machines using MOST [15]. MOST can determine activities that are not value added for some elements of work [16], and are able to minimize production costs [17]. This shows that MOST can reduce work movements in order to increase process productivity [18]. MOST can also be used for the implementation of standard time and setting the work station layout [19]. Positive changing in behavioural and environmental workplace can be provided through 5S practices. While, Radar chart is a tool in organizing tools and equipment, easy to find something based on function and place, better in control wastage, supporting for continuously improvement and have better mind set and awareness for important healthy and safety on working [9]. The combination of MOST and 5S aims to measure the standard time needed by the operator to work and save the processing time by applying the 5S method.

3. Methodology

In the process of data collection, sample selection was conducted in each related division. In the milk receiving division, there is only 1 operator involved. In the processing division, there are 9 operators, which then selected one operator using a rating factor as a sample. In the milling and packaging division there is only 1 operator for both divisions. Operators who are used as research samples are the operators who have an average ability with a minimum of 1 year work experience. The object of research in this problem is the production process in SMEs processing milk powder. This study will identify the activities that cause ineffective motions happened in the milk production process using MOST and 5S. Data collection was done as follows:

1. Activity Identification
   Observe the overall activities that take place in production activities.
2. Motion Classification
   Identify all movements performed by the operator using an ergonomics approach.
3. Movement Identification
   Find out which ineffective movements are classified as non-value-added activities and calculate the processing time by MOST method.
4. Proposed improvements
   Propose the improvements. Re-layout the work station and apply 5S at the working area.

4. Result and Analysis

The unnecessary motion is particularly related to ergonomics. Excessive motion consists of bending, twisting, lifting, reaching, and walking. These should be dealt with as soon as they are recognized [20]. Accordingly [21], ineffective and inefficient movements can be arising in activities such as finding and selecting to object, putting on right position, etc. Based on the observations, the following are the analysis of ineffective motion elements that has been carried out by the operator during performing their tasks.

Table 1. Ineffective motion elements performed by the operator.

Division	Ineffective Motion
Milk Receiving	Searching for the filters that match the size of the milk tank. Selecting the marker which was used to write the name of the milk depositor in the plastic packaging.
Milk processing	Searching for equipment when starts to cook.
Milling	There is no ineffective movement by the operator. This is because the operator had worked in accordance with the economic principles of the movement such as working with both hands, both hands are started and ended simultaneously, and there is no process of finding and choosing.
Packaging	Repeating the inspection procedure when the operator wants to pack milk into plastic.

After identifying the ineffective movement carried out by the operator, then the standard time has been calculated. From the standard time obtained, the analysis had been done to identify the potential movements to be minimized. The efforts to improve work methods are carried out by applying 5S to each operator's work area. The proposed changes that have been made are shown in Table 2:

Table 2. The proposed 5S implementation at the working areas

No	5S Method	Treatment
1	Seiri (Sort)	The first step is to do a large cleaning that serves to remove dust and dirt on the equipment. The second stage discards equipment that is unnecessary. The third stage is to separate the defective equipment and placed in temporary storage. Set aside the damaged milk filter so that it is not mixed with a good milk filter. Therefore, the operator does not necessary to select and inspect the filter to be used.

No	5S Method	Treatment
2	Seiton (Set-in Order)	The first stage is grouping the equipment, so that only the necessary equipment will be available in the work area. The second stage prepares the equipment storage. The third stage is to provide a boundary mark on the storage area that serves as a limit of goods. The last stage is labelling the storage place and equipment and it will facilitate the operator in storing and returning equipment.
3	Seiso (Shine)	The first stage is to carry out the micro level cleaning activities at the entire work area environment. The second stage is the cleansing at the individual level, cleaning carried out in the specific work area of the operators. The last stage is micro phase cleaning where the cleaning is done on equipment used by each operator such as milk mixers, knives, pans and machines used.
4	Seiketsu (Standardize)	Form a standard operating procedure e.g. sequence task, job description, person scheduling task per day for processing operators. Receipt, milling and packaging operators must mop and sweep the workspace before returning home. The awareness of employees to maintain Seiri, Seiton and Seiso is constantly being formed because it has begun to get used to new methods and ways of working.
5	Seitsuke (Sustain)	Counselling to operators about the importance of cleanliness and neatness of equipment and workplaces.

5S activities such as cleaning, sorting items, giving labels and restrictions had been adopted and implemented.

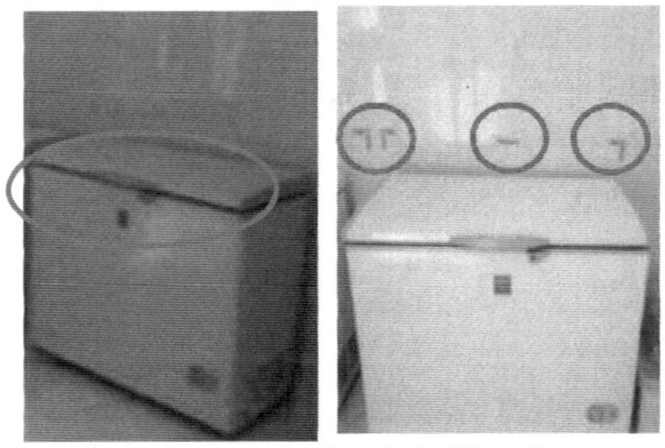

Figure 1. Before and after labelling freezer in 5S

Figure 2. Before and after arrange the content of freezer in 5S

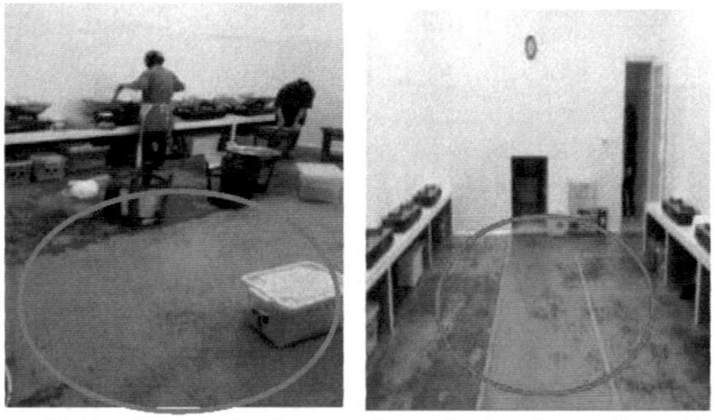

Figure 3. Before and after make a limitation of work area in 5S

Figure 1-3 shows the improvement ideas that had been adopted at CV. Sahabat Ternak Industry. As shown in Figure 1, the position among the freezers are more fixed after separated with restriction sign, while Figure 2, shows better organizing on the freezer contents, and Figure 3, shows the equipment are easier to identify and closer to reach. Several improvement conditions can lead to some positive impacts such as easier to calculate and identify the liquid milk packs based on date or suppliers so that it could be easy for tracing the quality of material related some defects, speed up the retrieval process and simplify the work flow with yellow line guidance.

After improving the work method with 5S, the new standard time measurement was done. Standard time is the time required by an operator who has an average level of ability to complete a job normally in the best study system by considering allowance [21]. Standard time measurement with MOST is carried out in four stages. The initial stage was done before implementing 5S (before). After implementing 5S, data collection was done with MOST 16 days after 5S implementation (after 1). Subsequent data collection was carried out 40 days after the 5S implementation (after 2). Data is taken 2 months after the implementation of 5S (after 3). Standard time measurement with MOST is done by the following formula.

$$Standard\ Time = Normal\ time\ X \frac{100}{100-\%\ Allowance} \quad (1)$$

Based on the results of the standard time calculation with MOST performed on the work process, the results of standard time between before and after 5S are shown as in Table 3.

Table 3. The difference standard time

	Before	After 1	After 2	After 3
Milk receiving	0,87 minutes/liter	0,83 minutes/liter	0,8 minutes/liter	0,82 minutes/liter
Milk processing	23,56 minutes/liter	23,51 minutes/liter	23,49 minutes/liter	23,25 minutes/liter
Milling	4,78 minutes/kg	4,63 minutes/kg	4,5 minutes/kg	4,5 minutes/kg
Packaging	2,82 minutes/pcs	2,54 minutes/pcs	2,34 minutes/pcs	2,27 minutes/pcs

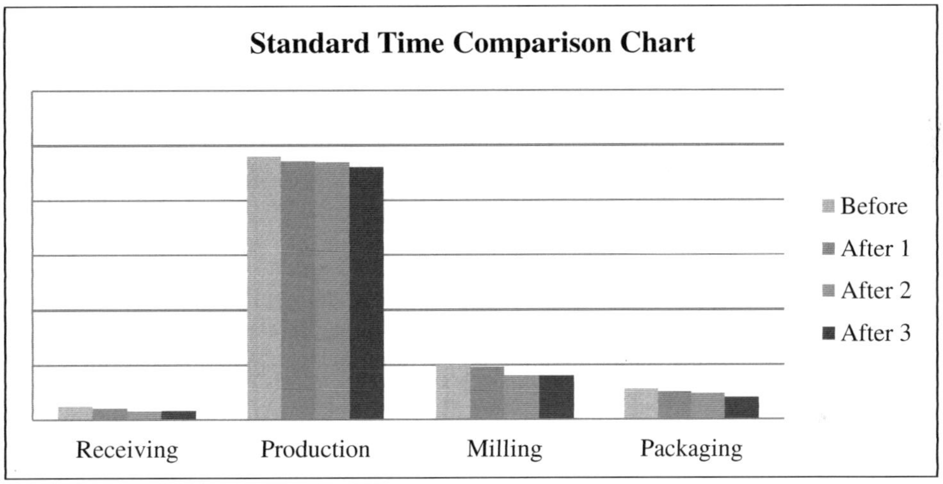

Figure 4. Standard time comparison chart

The results of the standard time calculation using MOST indicate that the standard time value has decreased from after 5S application. The decrease in standard time was caused by the successful application of 5S to reduce ineffective motions and speed up the standard time of a job. Although the decline in standard time experienced is not significant but can affect the amount of output per day produced.

In this stage, statistical processing uses Friedman's non-parametric test, it is because the comparison test consists of more than 2 groups and the data used. Friedman test aims to find out whether there are differences in effects between treatments. The treatment here is the application of 5S in different time periods. An average comparison is made on the data group so that it can be seen whether there is a difference between the standard time before, after 1, after 2 and after 3.

Hypothesis:

H0: There is no significant difference between the standard time before and after the 5S application
H1: There is a significant difference between the standard time before and after the 5S application
Significance level: $\alpha = 5\%$ or 0.05
If the p-value is <0.05 then H0 is rejected and H1 is accepted. Following are the results of Friedman's testing using SPSS software on four different treatments.

IOP Conf. Series: Materials Science and Engineering **530** (2019) 012037 doi:10.1088/1757-899X/530/1/012037

Table 4. Standard time difference test

Test Statistics[a]

N	4
Chi-Square	11.154
df	3
Asymp. Sig.	.011

a. Friedman Test

Friedman test results show a significance value of p-value is 0.011. The p-value 0.011 is smaller than 0.05, so the conclusion is H0 rejected and H1 is accepted. It means that there are significant differences in the comparison of the measurement results at the standard time, in the conditions before and after the 5S implementation.

5. Conclusion and Future Research

The ineffective motion identification had been carried out at milk receiving, processing, milling, and packaging division. The percentage of decreasing in standard time using the hybrid of MOST and 5S practices are 5.7% at the receiving division, 1.3% at the processing division, 5.8% at the milling division and 19.5% at the packaging division. Based on statistical calculation using Friedman's testing, the result (p-value = 0.11<0.05) and it shows that there are significant differences between standard time and standard output before and after the application of 5S.

For further research, it is recommended to carry out data collection in a gradual and relatively long period of time, so that the changes that can be seen significantly. Subsequent research is expected not only to measure the standard time produced, but also to calculate the value of productivity resulting from the changes that occur.

Acknowledgments

This study was supported by Directorate of Student Development Universitas Islam Indonesia, Yogyakarta; Industrial Engineering Department, Faculty of Industrial Technology, Universitas Islam Indonesia; Faculty of Manufacturing Engineering, Universiti Teknikal Malaysia Melaka; Government of Industrial, Trading and Cooperative of Sleman Yogyakarta District; and CV. Sahabat Ternak (Etawa Goat Milk SMEs).

Reference

[1] D Agus 2017 "Radar Jogja," [Online]. Available: https://www.radarjogja.co.id/2017/03/30/susu-kambing-pe-potensi-baru-di-sleman/. [Accessed 11 September 2017].

[2] I L Nunes 2015 Integration of Ergonomics and Lean Six Sigma A Model Proposal *Procedia Manuf* Vol 3 No Ahfe pp 890-897

[3] J P E Souza and J M Alves 2017 Lean Integrated Management System : A Model For Sustainability Improvement *International Converence on Production Research (ICPR)* No Icpr pp 678-648

[4] G Mulyati, Suharno and M Muharom 2015 An Implementation of Lean-ergonomic Approach to Reduce Ergonomic Parameter Waste in the Manufacture of Crackers *ICoA Conference Proceedings ISSN 2413-0877* Vol 3 pp 21-24

[5] Sumiyanto and N. Rizani 2017 Analisis Ergowaste pada Proses Produksi Yoke dengan Pendekatan Lean Ergonomics Di PT.X *Prosiding SNTI dan SATELIT Malang: Jurusan Teknik Industri* pp B272-277

[6] P Kuhlang, T Edtmayr and W Sihn 2011 Methodical approach to increase productivity and reduce lead time in assembly and production-logistic processes *CIRP Journal of Manufacturing Science and Technology* p 24–32

[7] B Niebel and A Freivalds 2009 Niebel's Methods, Standards, and Work Design 12th ed New York: McGraw-Hill

[8] J Senthil and G Haripriya 2016 Time Analysis With MOST Technique *Int. J. Chem. Sci 14(S2) ISSN 0972-768X* pp 519-526

[9] L Huda 2016 Lean Ergonomic with 5S Concept : A Case Study in Small Scale Industry *4th SEANES International conference on human factors and ergonomics in South-East Asia*

[10] A Sigh and I Ahuja 2015 Review of 5S methodology and its contributions towards manufacturing performance *towards manufacturing performance* Vol 5 No 4

[11] I N Pujawan 2009 Ekonomi Teknik, Edisi Kedua Jilid Pertama ed., Surabaya: Guna Widya

[12] L Botti, C Mora and A Regattieri 2017 Integrating ergonomics and lean manufacturing principles in a hybrid assembly line *Computers & Industrial Engineering 111* p 481–491

[13] P Arezes, J Carvalho and A Alves 2014 Workplace ergonomics in lean production environments: A literature review p 57–70

[14] Z Santos, L Vieira and G Balbinotti 2015 Lean Manufacturing and ergonomic working conditions in the Automotive Industry *6th International Conference on Applied Human Factors and Ergonomics (AHFE)* p 5947 – 5954

[15] S Gunjar and A Pandey 2016 Application of Maynard Operation Sequence Technique (MOST)-A case study *International Journal of Innovations in Engineering and Technology (IJIET) ISSN: 2319-1058* Vol 6 No 3

[16] Belokar R M, Y Dhull, S Nain and S Nain 2012 Optimization of Time by Elimination of Unproductive Activities Through MOST *International Journal of Innovative Technology and Exploring Engineering (IJITEE) ISSN: 2278-3075* Vol 1 No 1

[17] P K Gupta and S S Chandrawat 2012 To Improve Work Force Productivity in a Medium Size Manufacturing Enterprise by MOST Technique *IOSR Journal of Engineering (IOSRJEN) e-ISSN: 2250-3021, p-ISSN: 2278-8719* Vol 2 No 10 pp 8-15

[18] Deshpande V A 2007 M.O.S.T-The Most Advanced Work Measurement Technique *Journal of Engineering & Technology S. P. University* Vol 20 pp 109-113

[19] T Yadav 2013 Measurement Time Method for Engine Assembly Line with Help of Maynard Operating Sequencing Technique (MOST) *International Journal of Innovations in Engineering and Technology (IJIET) ISSN: 2319-1058* Vol 2 No 2

[20] J Walder and C Kerk 2007 Integrated Lean Thinking & Ergonomics : Utilizing Material Handling Assist Device Solutions for a Productive Workplace, South Dakota School of Mines

[21] R Barnes 1980 Motion and Time Study: Design and Measurement of Work, New York: John Wiley and Sons

IOP Conf. Series: Materials Science and Engineering **530** (2019) 012038 doi:10.1088/1757-899X/530/1/012038

A Collaborative Scheme with Vendor in Kaizen

A D Peterson and J F Chin*

School of Mechanical Engineering, Engineering Campus, Universiti Sains Malaysia, 14300 Nibong Tebal, Pulau Pinang, Malaysia.

Corresponding author *: chinjengfeng@usm.my

Abstract. The paper investigates a novel collaborative Kaizen scheme that involves vendor as the volunteered countermeasure provider in exchange for a fair sharing in gain. The skeleton of the scheme is made up of Plan-Do-Check-Act (PDCA). In Plan, the required improvement is first assessed by the Kaizen team to gauge the need to bring in vendor. The proposal must be scrutinized and undergoes negotiations for acceptable terms to both parties, which ultimately helps to build a long-term partnership. The scheme is presented as a case study in an Original Equipment Manufacturer (OEM) where energy consumption and maintenance cost for workstation lighting system at the production were reduced through the implementation of countermeasure proposed by vendor. The implementation cost was fully borne by the vendor and in return, the vendor would be paid in amount equivalent to the saving on the energy bill for the next nine months.

1. Introduction

The success of Japanese automotive industry has for the past decades attracted various studies on their manufacturing operation and management. One of the attributing factors is lean manufacturing. The concept was made known and popularized by Womack et al. [1] in their book titled "The machine that changed the world". The essence is to streamline process and eliminate waste "Muda" [2-4]. Muda generally describes activities not adding value from the customer's perspective. According to Ohno[5], there are seven classes of Muda: overprocessing, overproduction, defect, motion, work-in-process (WIP), idling and transport. The elimination of Muda calls for continuous improvement (CI) or Kaizen [6]. It is a team-based, goal-oriented and systematically deployed activity, coordinated at the enterprise level. A common structure of Kaizen is based on PDCA, which breaks down the Kaizen into four stages, as the name implies: PLAN – detain planning, DO – implementation of works, CHECK – measure the changes and ACT – refine the improvement [7-8].

Evident by the prevalence of more sophisticated supply chains, business activity outsourcings and technology development partnerships, the manufacturing sector has grown increasingly collaborative where independent parties with complementary strengths, work together to more effectively pursue on shared interests. Yasin et al. [9] estimated that as much as 75% of production cost for purchasing goods and services originated from external sources. Such phenomenon affects all levels of manufacturing organizations, and therefore calls for a new paradigm of management. Turiera and Cros [10] discussed successful partnerships between major organizations, for example, Coca-Cola & Heinz who developed 100% recyclable "Plant Bottle" which reduces carbon footprint by 19% in manufacturing. Microsoft & Toyota developed energy management system which connect hybrid Toyota vehicle and an electric system together to maximize the efficiency of charging. Many researches show engagement with external parties through lean practices enhances its success. Chong et al. [11] presented a case study

IOP Conf. Series: Materials Science and Engineering **530** (2019) 012038 doi:10.1088/1757-899X/530/1/012038

where good practices were transferred to suppliers, in hope to improve their product quality and process capacity. Southard et al. [12] implemented Poka-yoke to track pharmaceutical products and achieved a better coordination with supplier in replenishment while reducing operational cost. Adopting Lean during vendor collaboration can reduces movement of raw materials, WIP storage, rework time etc. This could be achieved by using Lean tools such as seven wastes, 5S, JIT (Just-in-time), Poka-yoke etc [13-17].

The research investigates a special form of collaborative Kaizen scheme, where vendor acts as the volunteered countermeasure provider in a Kaizen in exchange for a fair sharing in gain from the project. This form of collaboration is interesting in view that conventional Kaizen is often deployed using internal capacity and resources of an organization. In many occasions, this would limit the options or countermeasures in the Kaizen, especially if new or unfamiliar technology is involved. On the other hand, vendor, owning and with better knowledge on the technology would be keen to promote their countermeasure and secure the business. The major concern by the organization however would be on the risk accompanying to the amount of investment, the success rate and the effectiveness of the countermeasure. The scheme presents an alternative paradigm whereby vendor is persuaded to significantly absorb the countermeasure cost (hence the risk) and in return being offered a share in the financial benefits. This could be considered a win-win scenario.

The arrangement of the paper follows as: Section 2 introduces the collaboration Kaizen scheme with vendor. Section 3 validates the scheme in an OEM electronic measuring system manufacturer. Section 4 discusses the collaboration and the implementation and Section 5 provides a conclusion.

2. Collaborative Kaizen scheme with vendor

The collaborative Kaizen scheme (as shown in Figure 1) adopts PDCA as the methodological skeleton. It begins with the Plan stage which defines the problem. A Kaizen team is formed to grasp the current condition and assess in-house problem-solving capability before deciding to engage a vendor. To start the vendor engagement, suitable vendors will be formally invited to submit their proposals. Next, the Kaizen team reviews these proposals, makes selection and negotiates on price, lead time, risk and reward sharing etc. The outcome of the review would have to be presented for management approval so that contract could be issued.

Do stage is where the countermeasure is implemented by vendor, and progress to be monitored. The Kaizen team would provide the necessary supervision. Check stage compares the initial and improved stages to gauge the degree of improvement achieved. Act stage refines and standardizes the improvement. Upon the project's closure, the overall engagement is reviewed with achievement and shortcoming documented. The knowledge obtained would be shared within the organizations.

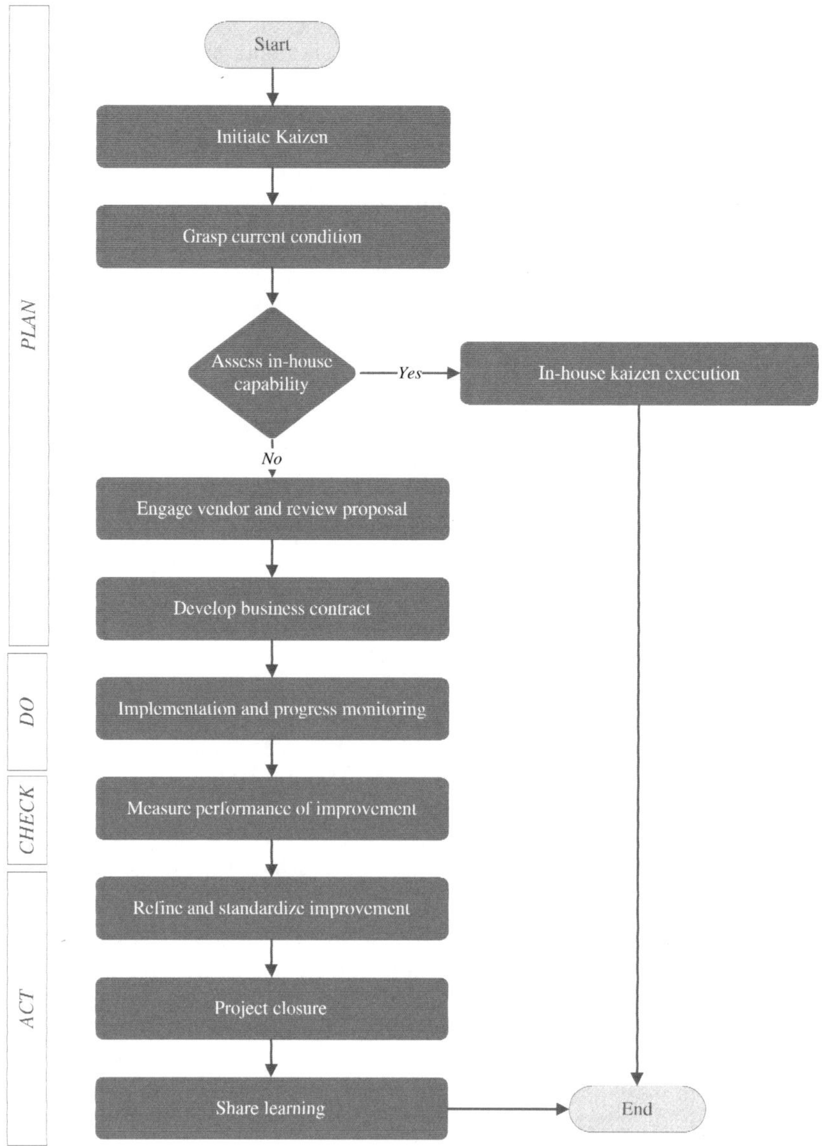

Figure 1. Collaborative Kaizen scheme with vendor

3. Collaborative Kaizen scheme with vendor

Company A is an OEM which develops electronic measuring systems and is currently based in Penang, Malaysia. Its products focus on technology fields such as Cellular (5G), Aerospace & Defense. The company adopts Lean practices and tools for daily troubleshooting and continuous improvement. Company A has also an established vendor data system which categorizes vendors according to the types of engagement support given to production floor.

3.1. Plan stage

There were 128 workstations in the production lines catered for processes such as assembly, debugging, testing, repair, inspection, button up etc. A Kaizen team was formed to investigate ways to reduce expenditures on workstation lighting. A workstation lighting comprises two lighting holders, two 36-watts fluorescent tubes, two electronic ballasts, switches and wirings. The Kaizen team consisted of one champion, one mentor, one team leader, one co-leader, and two representatives from production lines.

IOP Conf. Series: Materials Science and Engineering **530** (2019) 012038 doi:10.1088/1757-899X/530/1/012038

A SMART goal was determined that is to reduce the expenditure associated to the lighting system of the 128 workstations by 30% within the duration of two weeks.

To grasp the current condition, fishbone analysis was performed through the brainstorming amongst the Kaizen team members, as illustrated in Figure 2. Multiple causes were identified and using CNX analysis the findings were further classified into Constant (requires standard operating procedures), Noise (uncontrollable) and X (variables considered as key processes). X variable will be investigated for improvements [18]. Two main causes to focus on were high energy consumption and maintenance cost.

The electricity and maintenance costs from 2014 – 2016 are as shown in Table 1. Fluorescent tubes were no longer considered energy efficient and the maintenance cost was high due to multiple parts and replacements. Around 30% of workstations required servicing or repair yearly due to wear and tear. The works must be carried out by external vendor to comply with the company safety policy. The average replacement cost for a set of lighting was RM 90 per trip, which includes material (36-watt fluorescent tubes and ballast), wiring and workmanship.

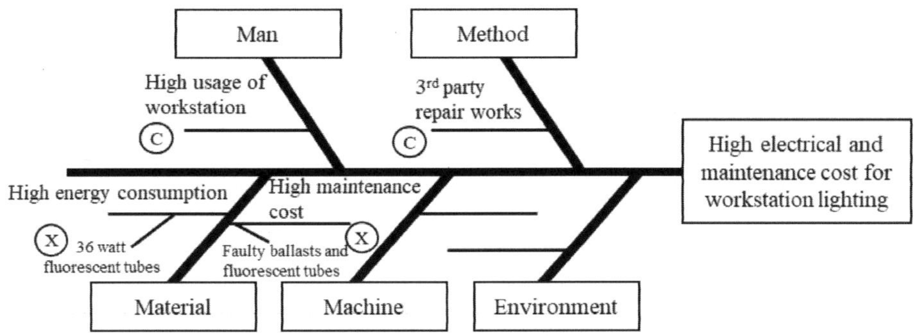

Figure 2. Fishbone analysis

Table 1. Energy bill and maintenance cost from 2014 to 2016

No.	Year	Electric bill (RM)	Maintenance cost (RM)
1.	2014	9,411.36	3,310
2.	2015	9,485.67	3,380
3.	2016	9,501.87	3,510

Potential countermeasures were identified by the Kaizen team, however they were deemed unsatisfactory. The Kaizen team initiated vendor engagement on the ground that mass installation of lighting needed expert knowledge unavailable to the company. Several potential vendors were formerly liaised. A draft contract and reviewing criteria were established and distributed to them. They were allowed an escorted Gemba walk to the production and data collection. Their proposals would be submitted within two days. After reviewing the proposals, Vendor X was selected. The countermeasure proposed by Vendor X was summarized in Table 2.

Table 2. Summary of the countermeasure

Countermeasures & Principle	• To convert conventional fluorescent lighting tubes into light-emitting diode (LED) tubes.
Improved/innovated ideas	• Less energy required changing from 36-Watt to 16-Watt. • Do not require electronic ballast.
Installation method & Sample	• The current lighting setting (the fluorescent tubes, electronic ballast, tube holder and wiring) would be removed and replaced with new LED tubes, wirings and tube holders.

IOP Conf. Series: Materials Science and Engineering **530** (2019) 012038 doi:10.1088/1757-899X/530/1/012038

An independent evaluation on the installation cost would be approximately RM5,000. The replacement cost for LED tubes would reduce approximately to RM 60/trip due to a lower material cost. In addition, LED is significantly more reliable and with a longer life than fluorescent tubes. It estimated that the failures of the lighting would reduce by 10%. Negotiation with Vendor X was done for better pricing, lead time and risk sharing. The final price of LED tubes is RM 25/2pcs, with one-year warranty. The vendor absorbed the whole installation cost in exchange for payment equivalent to the resultant energy cost saving over a fixed period. The saving would be calculated using the bill calculator by Tenaga Nasional Berhad (TNB) official website. The resultant saving is RM 836.93/month. For the next nine months, this amount would be paid to vendor on monthly basis upon confirmation of such saving by the finance department. The details of the business contract also include the current organization policy such as the vendor must have a valid business registration number, technical certification and payment term etc. The final decision, business contract details and implementation plan were discussed, complied for management and finance approval.

3.2. Do stage

Installation was performed in one day during the weekend when the production was not running. The workstations were cleared and sensitive equipment were covered to prevent dust and damage from the installation. The Kaizen team was stationed in the facility to monitor the implementation. Thirty minutes was required to install LED on a workstation as shown in Table 3. Lux meter was used to measure the brightness after the several installations. The LED tubes were slightly brighter than the fluorescent tubes but within the acceptable lighting criteria of 420 +/- 5% Lux. Production resumed as normal the next day without any disruption reported.

Table 3. Steps for workstation lighting installation

Step 1 The protective sleeve was removed, exposing the faulty fluorescent tube.  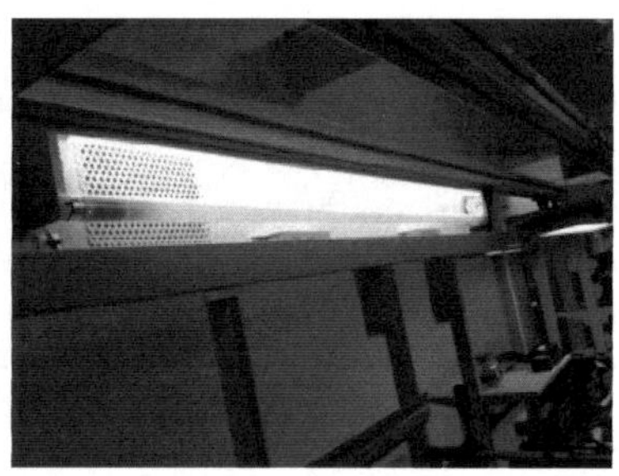
Step 2 The fluorescent tubes, electronic ballasts, tube holders and wiring were removed. 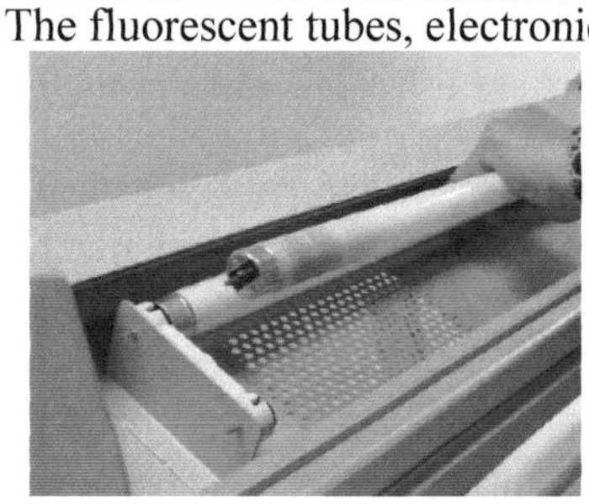 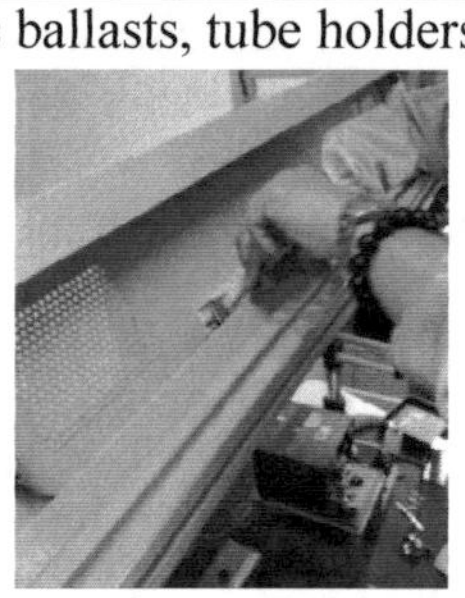 

Step 3 New tube holders were drilled into the lighting frame and LED tubes were inserted.

Step 4 New connectors for wiring and LED tubes were installed.

Step 5 The new lighting system was tested and a protective sleeve was installed.

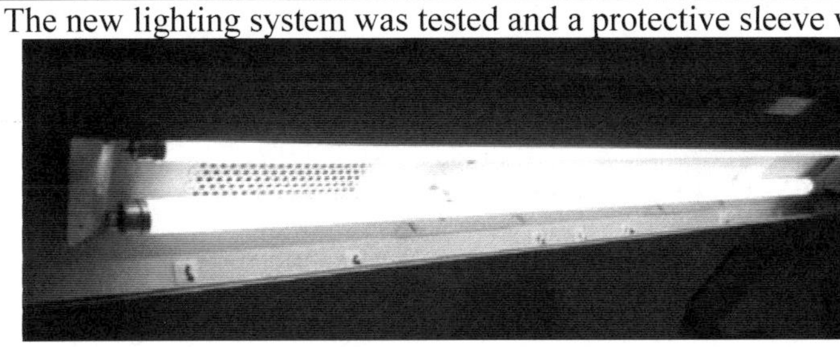

Observation:
The difference after the modification cannot be notice after the protective sleeve was place back.

3.3. Check stage

A comparison between initial and improved stage for the implemented countermeasure is shown in Table 4. The workstation lighting has been monitored for the next two months after the installation. Evidently, the implementation improved the efficiency and operational cost of the workstation. This can only be achieved with the support and expertise of Vendor X.

Table 4. Comparison between initial and improved stages

Initial stage	Improved stage
Cost concern: *High energy consumption and maintenance cost due to faulty ballasts and fluorescent tubes.*	Cost concern: *Achieved more than targeted reduction of 30% savings by using LED tubes without ballasts.*
Efficiency of workstation lighting: *Failure report shows approximately 30% of workstations required yearly repair works.*	Efficiency of workstation lighting: *Reduce yearly failure of workstations lighting by 10% due to longer life span of LED tubes.*
Financial validation: - *Finance department validated yearly expenses.*	Financial validation: - *Finance department validated cost reduction.*

Other complaints: - *Messy wiring of workstation lighting.*	Other complaints: - *Messy wiring of workstation lighting.*
	Post implementation monitoring: - *Workstation lighting conversion did not disrupt operations and no failures were detected within two months monitoring period.*

3.4. Act stage

The countermeasure improved the overall efficiency of the workstations. The changes did not disrupt operations schedule and no failures were detected. Refinement for workstation lighting cables further improved the tidiness of the cables using both cable ties and cable tie mounts for better 5S control. This improvement was shared with other departments for standardize of workstation lighting system and reductions of operation cost. RM 7,532.37 was released to vendor over the course of 9 months, as stipulated by the contract. The detail of Vendor X is updated into vendor engagement system. Kaizen team's experience was shared within the organization as shown in Table 5.

Table 5. Knowledge sharing

Knowledge Sharing
Objective: To reduce expenditure associated to the lighting system of the 128 workstations by 30% within the duration of two weeks.
Root causes: High energy consumption and maintenance costs.
Project timeline: 2 weeks
Duration of implementation: 1 day
Vendor's countermeasure
Vendor: X
Countermeasure: Convert workstation lighting system from 36-watt fluorescents tubes to 16-watt LED tubes.
Cost bearing: RM 4,605
Gain received: RM 2,927.37
Kaizen Team Financial Gain
Project review: 100% success and no disruption to production.
Finance gain: Electricity saving Year 1 – RM 2,510.79 Year 2 – RM 10,043.16 Maintenance cost saving Year 1 – RM 3,510 Year 2 – RM 1,950
Technical Knowledge
Conversion from fluorescent light to LED tube reduced electricity usage, maintenance costs.

4. Discussion

In this section, several aspects are discussed. First, the scheme offers detail planning, execution, control and review when vendor is involved. The purpose of the engagement is in line with the objective, namely to reduce both energy and maintenance costs. The scheme helps the Kaizen team to obtain external expertise to solve the problem. Decision makings at the planning stage were based on consensus, facilitated by several lean tools such as fishbone analysis. By outsourcing the problem-solving, the vendor became the execution party while the Kaizen team drives the engagement progress and monitoring. Flexibility was given to vendors in brainstorming for countermeasures and risk sharing criteria. Countermeasure proposed by vendor to fit all workstations with 16-watt LED tubes reduced the energy consumption significantly while the maintaining the lux value within acceptable range. The implementation cost was fully borne by Vendor X in return for the energy savings for nine months. A rough estimate is the vendor gained additional 63.6% from the deal.

Several key considerations were not adequately addressed in this case study but worthwhile for further investigation. First is a mechanism to decide on conditions suitable to approach vendor. This also needs to be in line to the organization common direction. Second is finding the best way to choose vendors perhaps based on their proposal, experience (track record), post-project service level and also other strategic reasons. The decision must be reached through consensus and fact-based analysis. Third involves the development of a sound assessment method to decide the attribution of cost and gain from the improvement between the organization and the vendor. Fourth, the contract must be made explicitly with the approval from management and stakeholders. Fifth is the coordination of the Kaizen especially during the implementation and project closure to avoid any complication. Lastly is the adoption of lean manufacturing concepts such as the learning and vendor development. Knowledge sharing enhances the knowledge competence within the organization [19].

5. Conclusion

This paper presents a preliminary scheme to deploy Kaizen with the aid of vendor. PDCA is a well-defined methodology in promoting continuous improvement. The engagement has shown result in a win-win agreement. The countermeasures provided by Vendor X reduced both energy and maintenance costs. A few lessons were learned during the engagement. First, vendor was willing to absorb implementation cost to secure the business and greater reward. Second, the collaborative scheme gave Kaizen team an option to outsource tasks beyond their expertise to external party.

Acknowledgement

This work was supported by Research University Grant (RUI), Universiti Sains Malaysia [grant number 8014069].

References

[1] Womack J and Jones D 1996 Lean Thinking: Banish Waste and Create Wealth in Your Corporation *New York NY*: Simon & Schuster
[2] Womack J, Jones D and Ross D 1990 The Machine that Changed the World *New York NY*: Rawson Associates
[3] Hines P, Holweg M and Rich N 2004 Learning to evolve *International Journal of Operations & Production Management* 24(10) pp 994-1011
[4] Che Ani M N and Chin J F 2016 Self-reinforcing mechanisms for cellularisation: a longitudinal case study *International Journal of Production Research* 54(3) 696-711
[5] Ohno T 1988 Toyota Production System: beyond large-scale production *productivity press*
[6] Tan P Y, Chin J F, Prakash J and Loh W P 2012 Reducing Electronic Component Losses in Lean Electronics Assembly with Six Sigma Approach *International Journal of Lean Six Sigma* Vol 3 Iss: 3 pp 206 – 230
[7] Singh J and Singh H 2012 Continuous improvement approach: state of art review and future implications *International Journal of Lean Six Sigma* 3(2) pp 88-111
[8] Chong M Y, Chin J F and Loh W P 2013 Lean incipience spiral model for small and medium enterprises *International Journal of Industrial Engineering* 20
[9] Yasin M, Small M and Wafa M 1997 An empirical investigation of JIT effectiveness: an organizational perspective *Omega* 25(4) pp 461-471
[10] Turiera T and Cros S 2013 CO Business: 50 examples of business callaboration *2nd ed.*
[11] Chong M Y, Chin J F and Hamzah H S 2012 Transfer of total productive maintenance to supply chain *Total Quality Management & Business Excellence* 23 467-488
[12] Southard P, Chandra C and Kumar S 2012 RFID in healthcare: a Six Sigma DMAIC and

simulation case study *International Journal of Health Care Quality Assurance* 25(4) pp 291-321

[13] Rheaume J 2003 High-Mix, Low-Volume Lean Manufacturing Implementation and Lot Size Optimization at an Aerospace OEM *Massachusetts Institute of Technology*

[14] Shah R and Ward P 2007 Defining and developing measures of lean production *Journal of Operations Management* 25(4) pp 785-805

[15] Kumar S, Wolfe A and Wolfe K 2008 Using Six Sigma DMAIC to improve credit initiation process in a financial services operation *International Journal of Productivity and Performance Management* 57(8) pp 659-676

[16] Eroglu C and Hofer C 2011 Lean, leaner, too lean? The inventory-performance link revisited *Journal of Operations Management* 29(4) pp 356-369

[17] Gnanaguru R, Puvaneswari K, Mallick J, Jegadheesan C, Sivakumar V and Devadasan S 2011 Toyota's A3 reports for improving 6-S activities: an aeronautical industry case study *International Journal of Services and Operations Management* 10(2) pp 239-254

[18] Martin G, Barnett K, Burgess C, Curry P, Ermer J, Gratzl G, Hammond J, Herrmann J, Kovacs E, LeBlond D, LoBrutto R, McCasland-Keller A, McGregor P, Nethercote P, Templeton A, Thomas D and Weitzel J 2013 Lifecycle Management of Analytical Procedures: Method Development, Procedure Performance Qualification, and Procedure Performance Verification [online] Uspnf.com. Available at: http://www.uspnf.com/sites/default/files/usp_pdf/EN/USPNF/revisions/lifecycle_pdf.pdf [Accessed 7 Oct. 2018]

[19] Cabrera E and Cabrera A 2005 Fostering knowledge sharing through people management practices *The International Journal of Human Resource Management* 16(5) pp 720-735

Application of 5S and Visual Management to Improve Shipment Preparation of Finished Goods

L K Yik and J F Chin*

School of Mechanical Engineering, Engineering Campus, Universiti Sains Malaysia, 14300 Nibong Tebal, Pulau Pinang, Malaysia.

Corresponding author *: chinjengfeng@usm.my

Abstract. The purpose of this paper is to present a lean case study to resolve issues related to shipment preparation in the finished goods store of a manufacturing industry. The process was first investigated through cause and effect diagram. Several root causes detected, including unorganized finished goods inventory, the absence of first-in-first-out (FIFO) rule in process flow, lack of visibility and inadequate inventory data recording and keeping. The two lean manufacturing tools, notably 5S and visual management were deployed in the finished goods store as the countermeasures. This leads to a significant improvement where the shipment preparation was seen reduced by 50%. The offshoot benefits include increased awareness and motivation for lean manufacturing by the involved staff.

Keywords : *Lean manufacturing, 5S, Visual system, Shipment, Due date*

1.0 Introduction

Inventory is defined as the stock of any resource used in manufacturing. Inventory levels can be viewed as indicators of process capability and production efficiency [1]. Inventory ill-managed however would bring forth storage issue, poor response to customer demand, hiding production problems etc [2]. A crude categorization of inventory would be raw materials, work-in-process (WIP) and finished goods. WIP represents products that yet to be completed in the production. The primary functions of WIP are to cushion against production disruption and to respond to fluctuations in customer demand. Another more special form of inventory is safety stock. Safety stock (also called buffer stock) is a term used to describe inventory in keeping to prevent stock out and back order situations. It protects against deviations in delivery date, meeting customer requirements, delivery quantity and other types of inventories [3].

The paper aims to present a framework to improve the shipment preparation of finished goods by deploying lean manufacturing principles, techniques and tools. The second aim would be to demonstrate the deployment of framework using existing resources (people, technology etc.) in a manufacturing organization without incurring additional cost to achieve the improvement.

A single case study approach has been adopted in this research. It has been proposed in the literature that case studies can be applied to the area of theory development as well as problem solving [4]. In general, case studies are often preferred when researchers have little control over the event and when the focus is on a contemporary phenomenon in some real-life context [5]. Case study approach is selected after careful consideration of several issues. First, one key aim of the study is to empirically identify inventory related difficulties. Manufacturing take place in a complex environment. Hence, it is critical to capture the experiences of the relevant people and the context of their actions to better understand inventory management practices and related difficulties. Second, as the research deals with the difficulties and challenges manufacturer are currently facing, this research deal with a contemporary event [5-6]. This research study begins with intensive review of relevant literature and the past research study. This is followed by the selection of company to carry out the implementation. Then, the existing finished goods storing process was studied. The related problems were identified and the corresponding countermeasures were formulated. This was followed by implementation.

IOP Conf. Series: Materials Science and Engineering **530** (2019) 012039 doi:10.1088/1757-899X/530/1/012039

The paper organization is as follows. In Section 2, the relevant literature on lean manufacturing, tools specifically 5S and visual management were reviewed. In Section 3, the problem and the solution of inventory management of the case study company would be discussed in details. The result and discussion presented in Section 4. The research study conclusion is the last section of this paper.

2.0 Literature review

2.1 Lean manufacturing

Lean manufacturing, one of the popular management philosophies, has been adopted by manufacturing and servicing organizations since the last two decades. It can be distinguished in two perspectives:- philosophical perspective or a practical perspective. The philosophical perspective relates to the guiding principles or overarching goals. Lean manufacturing has been defined as an integrated manufacturing system intended to maximize capacity, utilization and minimize buffer inventories through the minimization of system variability. The tenet is to do more with less, in terms of human effort, equipment, time, and space [7-8]. On the other hand, practical perspective involves implementing a set of shop floor tools and techniques aiming to reduce waste within the plant and supply chain [9-10].

One of the signature concepts of lean manufacturing is the identification and elimination of wastes throughout a product's entire value stream [11]. A waste is any activity in the manufacturing processes that does not create value to the customer. Seven wastes are identified in the lean manufacturing [12]. First, defect implies parts that cannot be sold "as is" or that must be reworked. Second, overproduction means making number of parts exceeding the quantity that can be sold. Third, waiting is a stage whereby parts are idling between processes. Forth, transportation indicates moving parts or materials to various storage locations, from process to process, etc. Fifth, inventories are parts not sold representing committing money and storage space. Sixth, motion indicates moving parts more than the minimum needed to complete and ship them. Finally, extra processes are doing more "work" to a part than is required.

In reducing these wastes, lean manufacturing involves deployment of lean tools. Common lean tools include value stream mapping, kanban, kaizen, heijunka, poka-yoke, etc. According to Che Ani and Chin [13], lean tools in application provide a complementary effect as they are based on a same underlying philosophy. This ensures the sustainability of the practice. Two of the fundamental lean tools related to this research is 5S and visual management. Originated from Japan, 5S comes from five Japanese words- Seiri (Sort), Seiton (Set in order), Seiso (Shine), Seiketsu (Standardize) and Shitsuke (Sustain) [14]. The main idea is to organize workplace to reduce time wasted by looking for things and to maintain process stability. Several literatures [15-17] reported that implementing 5S brings forth multiple benefits to the organization such as improved usage of working area, work environment improvement, prevention of tools losing, reduction in pollution, employee discipline, employee awareness and moral, improvement in the internal communication and internal human relation. In transfering the knowledge of autonomous maintenance to vendor, Chong et al. [18] showed how 5S is developed by the case study companies. Tan et al. [19] devised a 5S system including a checklist for electronics assembly plant to reduce the material losses at the front-end process. Chong et al. [20] implemented 5S in a packaging company through their lean incipience framework. Vipulkumar and Hemant [21] applied 5S in a ceramic industry and increases the process efficiency of the company. In another application of 5S, Agrahari et al. [8] achieved additional 30% of storing place and reduce unproductive time with 10%.

Table1. Terms and feature of 5S [22]

Japanese Term	Equivalent "S" term (5S)	Equivalent "C" term (5C)	Features
Seiri	Sort	Clear	• Search useful and useless items. • Separate both items & Classify them. • Critical items should be kept for use nearby.
Seiton	Systematize or Set in order	Configure	• Arrange useful items properly. • Discard useless items. • Give color according to their use.
Seiso	Sweep or Shine	Clean and Check	• Clean and polish all working palces. • Remove trash or dirt properly.
Seiketsu	Standardize	Confirmity	• Maintain cleanliness. • Tested/Inspected randomly. • Convert into high standard.
Shitsuke	Self-Discipline	Custom and Practice	• Motivate employee towards work. • Generate self discipline.

Another common lean tool is visual management. Visual management is defined as a set of techniques to create visual communication, visual control and ultimately a visual workplace. Visual management also defines performance improvement in any type of organization using strong graphic visualization techniques [23]. Tezel et al. [24] identified nine functions of visual management i.e transparency, discipline, continuous improvement, job facilitation, on-the-job training, creating shared ownership, management by facts, simplification, and unification. In this sense, visual management facilitates employee autonomy, serves to eliminate waste, fosters continuous improvement, allows for quick response & recovery, leads to information sharing, exposes abnormalities and maintains gains. For an example, a visual board installed at the workplace displays performance of production, attendance, quality concern, etc. Goh et al. [25] developed a visual board to regulate the flow and WIP between processes between manufacturing cells. This reveals key production statuses in a glance [26].

3.0 Case study

A manufacturer of ferrite was selected in this research. The case study company is a make-to-order (MTO) manufacturer of ferrite core product. The company produces four types of products, namely product R, product C, product T and product D. The monthly order volume is 100 tons to 120 tons. The percentage of the each product type is 65%, 20%, 10% and 5% respectively. Product R was selected because it significantly impacts performance of the company.

3.1 Improvement framework

The existing shipment preparation process was analyzed and described as below. Informal interviews, personal observations, and archival documents (finished goods despatched note, stock card etc) were the sources of data. PDCA cycle was used to systematically improve shipment preparation process. The four underlying phases are P(Plan), D(Do), C(Check) and A(Act). In "Plan" phase, the improvement framework starts with forming a multidisciplinary team. It was leaded by a project manager, with team members including three storekeepers, one store clerk, one selection and packing supervisor, and one internship student. Next, the improvement goal was defined. The existing process and material flow

IOP Conf. Series: Materials Science and Engineering **530** (2019) 012039 doi:10.1088/1757-899X/530/1/012039

were analysed by using tools such as process mapping, value stream mapping etc. The root cause is determined through applying cause and effect diagram. Suitable countermeasures to resolve the root cause are proposed and the details of the improvement steps are formulated. In the "Do" phase, the proposed countermeasure is implemented by using various strategies and tactics in order to achieve the project objectives. The project team members would be trained through on-job-training (OJT) approach. The OJT is the practical training and the training was carried at the real life environment rather than theory explanation in the classroom. In the "Check" phase, the performance of the implementation is evaluated in order to seek the further refinement. The "Act" phase is the area where the improvement cycle would be repeated in case the goal is not attained. The project would be closed if the goal is attained.

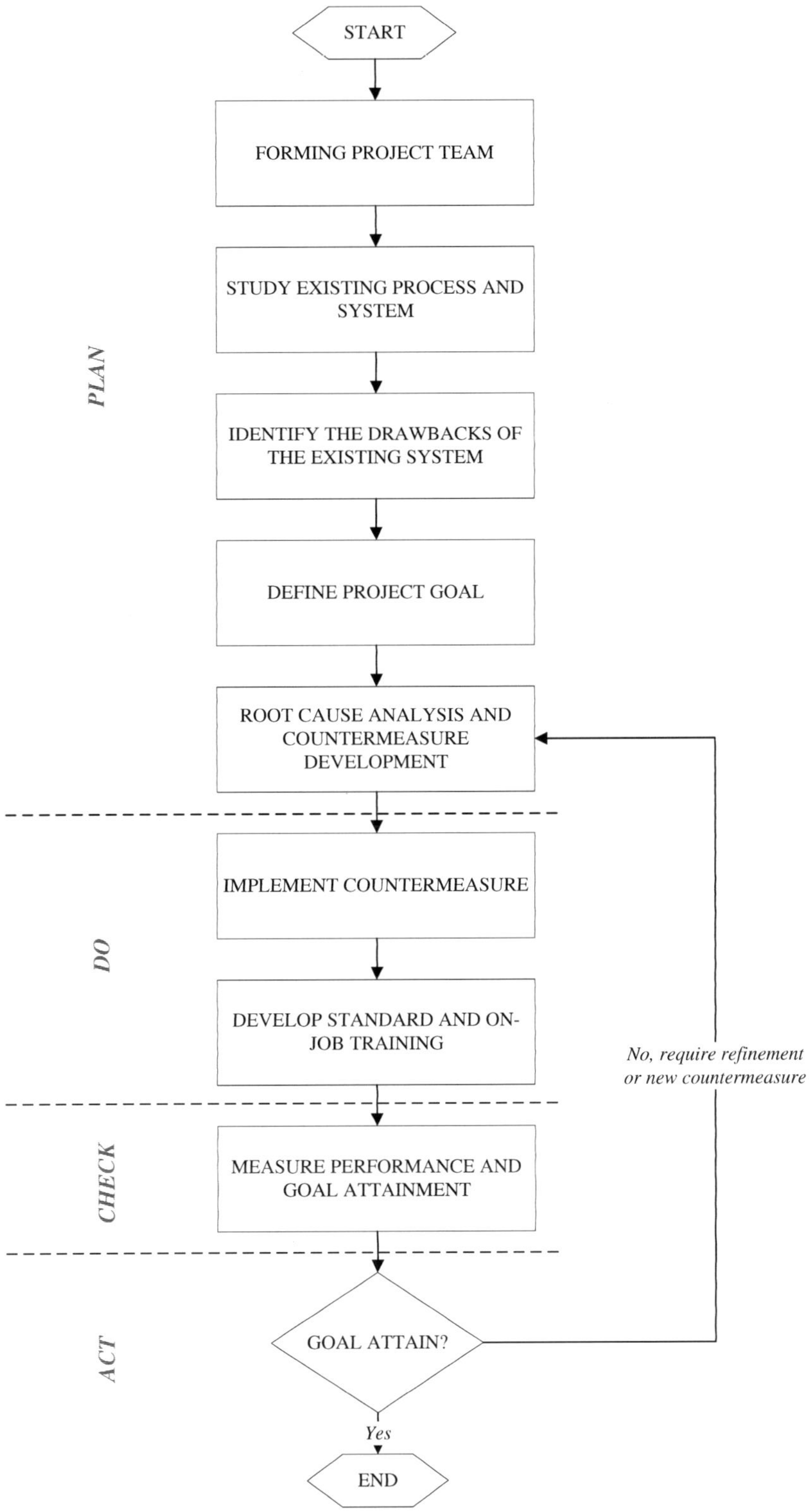

Figure 1. Shipment preparation process improvement framework

IOP Conf. Series: Materials Science and Engineering **530** (2019) 012039 doi:10.1088/1757-899X/530/1/012039

3.2 Plan stage

The shipment preparation process started with the storekeeper receiving shipment advice from the sale department. The storekeeper checked the stock card and finished goods despatch note file to verify the product quantity in the store against demand stated in the shipment advice. The storekeeper then brought the finished goods despatch note to the storage area to retrieve the product lots for shipment. The product lots would then be placed and arranged onto plastic pallets. Then, the storekeeper recorded the products lot number and quantity (kpcs) into the shipment traceability form. After recording, the product lots would be transferred to a wooden crate. The lot was then placed at a temporary storage area to wait for delivery. The completed shipment traceability form was kept in the associated folder. The storekeeper spent 6.5 days to complete the shipment based on completion rate three wooden crates per day. Twenty wooden crates were needed for each Friday weekly shipment. Shipment postponement to the following Monday happened incidental to insufficient pack to store quantity. This affects three forth of the shipments.

The existing shipment preparation process was broken down into sub activities of value-added (VA) and non-value added (NVA). Non-value added activities could be eliminated or replaced with more efficient methods. The existing shipment preparation process of a wooden crate (288 cartons) consists of twelve activities contributing to 3.58 hours. The details of these activities are described in Figure 2.

Cause and effect diagram was developed. Several causes related to method were unorganized inventory, lack of visibility, and inadequate inventory data recording and keeping. A cause related to man was failed to follow first-in-fist-out (FIFO). The product lots received from upstream (selection and packing) was stored at the area without clear location and indication. This entailed the storekeeper to spend on average 40 min to search and retrieve product during shipment preparation. The third cause was lack of visibility. The inventory details were kept in the finished goods despatch note file and stock file. The storekeeper compared the stock file record to determine the product quantity available with the finished goods despatch note. After that, the details were filled into shipment traceability form. The forth cause was inadequate inventory data recording and keeping. The store clerk spent significant amount of time to perform double entry works. She keyed in the product lots quantity (kpcs) into the Excel spreadsheet and the stock file separately. The data required for shipment preparation process effectively which was required data like lot number of the carton box. The storekeeper needed to obtain the lot number from finished goods despatch note.

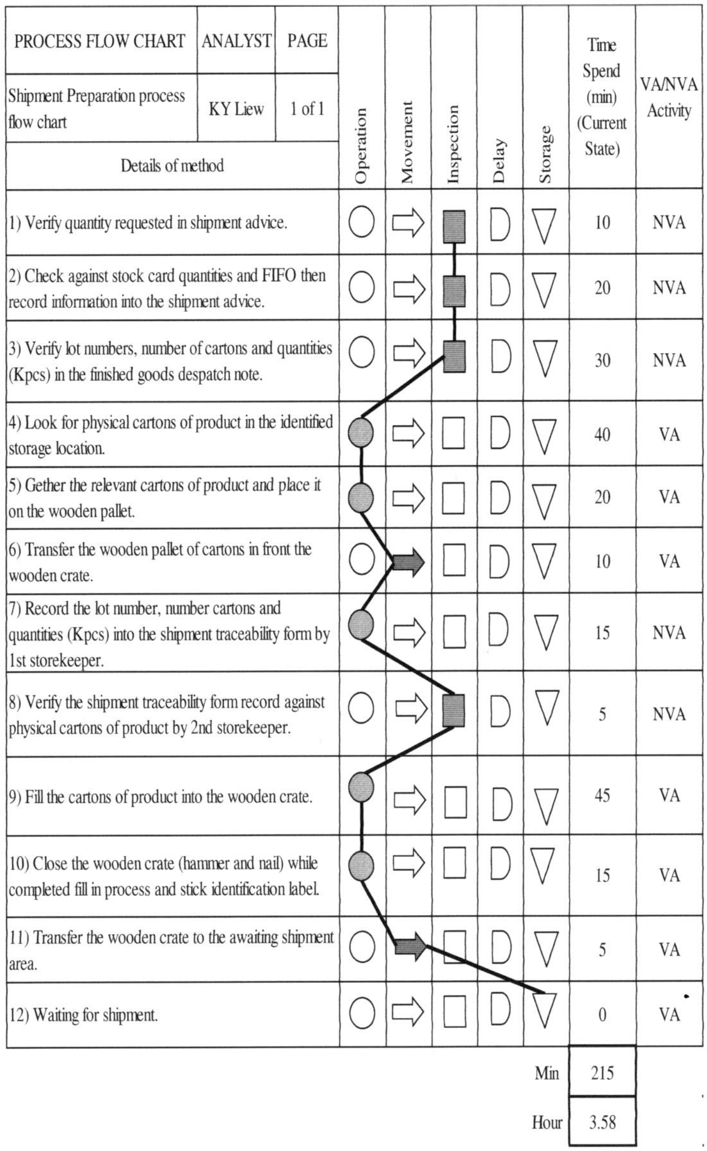

Figure 2. Current shipment preparation process flow chart

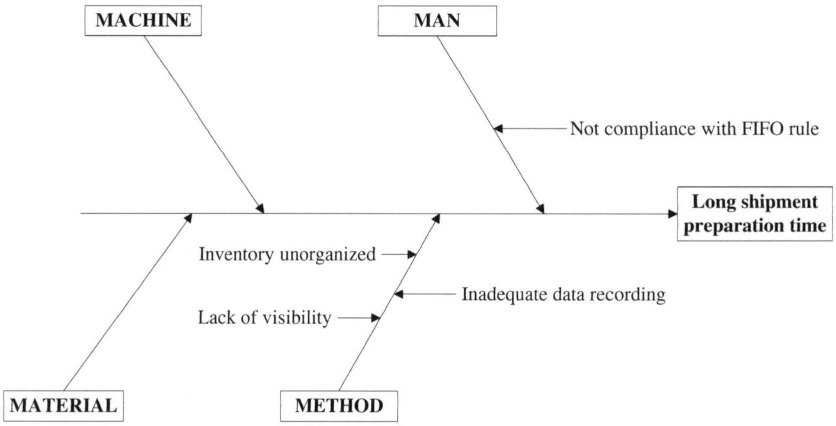

Figure 3. Cause and effect diagram of long shipment preparation time

Several countermeasures are developed in corresponding to the root causes defined previously. As aforementioned, the key concepts to drive the development are 5S and visual management. They are shown in Table 2.

Table 2. Countermeasure development

Root Cause	Description	5S Tools	Visual Management Tools	Countermeasure
Inventory Unorganized	The fresh product lots and the old product lots were placed together at the store area without status indicator.	Seiri Seiton Seiso	Yellow outline	• Redesign the finished goods store layout. • Re-layout the products location in the finished goods store, controlling the left over product from the previous years.
Incompliance to FIFO rule	No clear indication of the product lots received from the upstream (selection and packing).	Seiketsu Shitsuke	Red sticker	• Indicating the product lots with red sticker written STORE DATE after completed packing process.
Lack of visibility	Product lots placed at the storage area without clear information of each carton product.	Seiketsu Shitsuke	Stock list	• Prepare a store list for each location to indicate the details like lot number and number of carton.
Inadequate data recording	Use inefficient method to data collection and storage cause	Seiri Seiton Seiso	Excel spreadsheet	• Convert the hardcopy of finished goods despatch note into

	redundant work and can provide information effectively.			softcopy of Excel spreadsheet. The softcopy contents all information required for shipment preparation which include product code, lot number, quantity, net weight, production date and location.

3.3 Do stage

3.3.1 Implementing countermeasure for lack of visibility. The existing finished goods store was re-laid out, as shown in Figure 4. The storage column was designed according to the size of the pallet (4ft (L) x 1.5ft (W)) and ten pallets of product lots can be placed in each storage column. The column was identified by a location indicator with alphabet in ascending order from A to Z and associate with customer name. A stock list displayer (made of clear folder and metal plate) was installed in front of the storage column to keep the stock list. The information in the stock list includes customer name, product code, lot number, number of carton and quantity in kilo-piece (kpcs). The storekeeper could refer to the stock list at ease to find the location of the product lots to be retrieved.

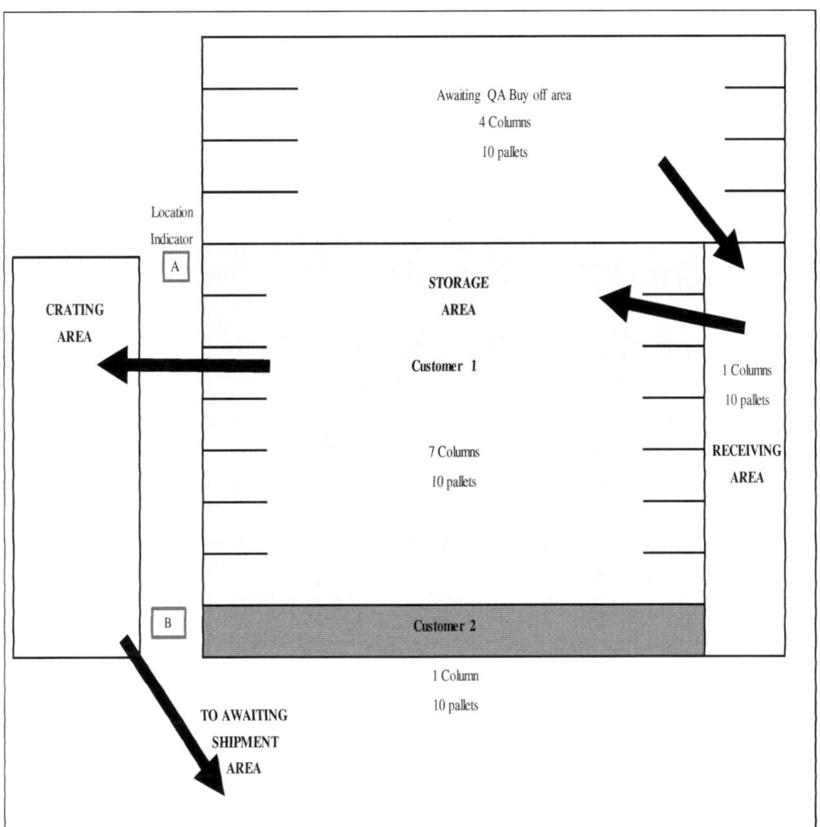

Figure 4. Improved layout of finished goods store.

3.3.2 Implementing countermeasure for unorganized inventory unorganized. As aforementioned, inventory was now to be arranged on predetermined columns. Furthermore, each pallet of products would carry an individual stock list holding the information about the product lots and the assigned storage column. Clear directive was given to the storekeeper to abide to the new layout and inventory storing arrangement.

3.3.3 Implementing countermeasure for incompliance to FIFO rule. The product lot was labelled with a red sticker written with store date after completed packing process to facilitate storekeeper to comply with FIFO.

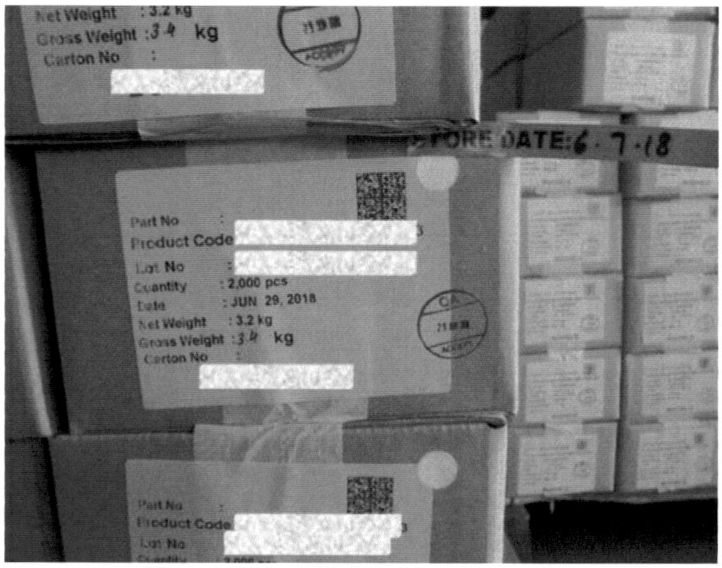

Figure 5. Store Date sticker on a carton

3.3.4 Implementing countermeasure for inadequate inventory data recording. Excel spreadsheet (Figure 6) replaced the manual recording of finished goods in hardcopy despatch note. The spreadsheet contains the information of products, lot numbers, quantities, net weights, production date, and location. The spreadsheet search function aids storekeeper to identify product lot location.

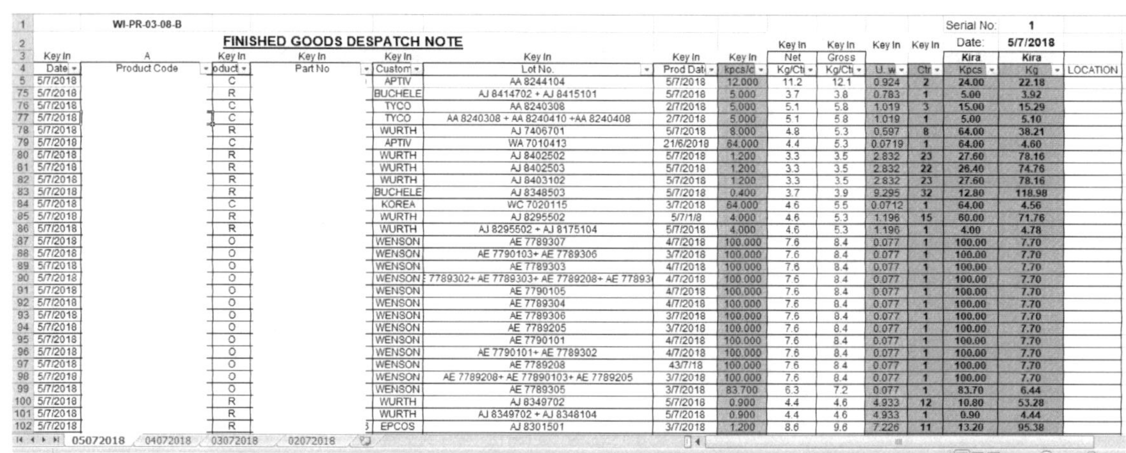

WI-PR-03-08-B   Serial No: 1

FINISHED GOODS DESPATCH NOTE   Date: 5/7/2018

	Key In Date	A Product Code	Key In Product	Key In Part No	Key In Customer	Key In Lot No.	Key In Prod Date	Key In kpcs/ctn	Key In Net Kg/Ctn	Key In Gross Kg/Ctn	Key In U. w	Key In Ctn	Kira Kpcs	Kira Kg	LOCATION
5	5/7/2018		C		APTIV	AA 8244104	5/7/2018	12.000	11.2	12.1	0.924	2	24.00	22.18	
75	5/7/2018		R		BUCHELE	AJ 8414702 + AJ 8415101	5/7/2018	5.000	3.7	3.8	0.783	1	5.00	3.92	
76	5/7/2018		C		TYCO	AA 8240308	2/7/2018	5.000	5.1	5.8	1.019	3	15.00	15.29	
77	5/7/2018		C		TYCO	AA 8240308 + AA 8240410 +AA 8240408	2/7/2018	5.000	5.1	5.8	1.019	1	5.00	5.10	
78	5/7/2018		R		WURTH	AJ 7406701	5/7/2018	8.000	4.8	5.3	0.597	8	64.00	38.21	
79	5/7/2018		C		APTIV	WA 7010413	21/6/2018	64.000	4.4	5.3	0.0719	1	64.00	4.60	
80	5/7/2018		R		WURTH	AJ 8402502	5/7/2018	1.200	3.3	3.5	2.832	23	27.60	78.16	
81	5/7/2018		R		WURTH	AJ 8402503	5/7/2018	1.200	3.3	3.5	2.832	22	26.40	74.76	
82	5/7/2018		R		WURTH	AJ 8403102	5/7/2018	1.200	3.3	3.5	2.832	23	27.60	78.16	
83	5/7/2018		R		BUCHELE	AJ 8348503	5/7/2018	0.400	3.7	3.9	9.295	32	12.80	118.98	
84	5/7/2018		C		KOREA	WC 7020115	3/7/2018	64.000	4.6	5.5	0.0712	1	64.00	4.56	
85	5/7/2018		R		WURTH	AJ 8295502	5/7/1/8	4.000	4.6	5.3	1.196	15	60.00	71.76	
86	5/7/2018		R		WURTH	AJ 8295502 + AJ 8175104	5/7/2018	4.000	4.6	5.3	1.196	1	4.00	4.78	
87	5/7/2018		O		WENSON	AE 7789307	4/7/2018	100.000	7.6	8.4	0.077	1	100.00	7.70	
88	5/7/2018		O		WENSON	AE 7790103+ AE 7789306	3/7/2018	100.000	7.6	8.4	0.077	1	100.00	7.70	
89	5/7/2018		O		WENSON	AE 7789303	4/7/2018	100.000	7.6	8.4	0.077	1	100.00	7.70	
90	5/7/2018		O		WENSON	7789302+ AE 7789303+ AE 7789208+ AE 77893	4/7/2018	100.000	7.6	8.4	0.077	1	100.00	7.70	
91	5/7/2018		O		WENSON	AE 7790105	4/7/2018	100.000	7.6	8.4	0.077	1	100.00	7.70	
92	5/7/2018		O		WENSON	AE 7789304	4/7/2018	100.000	7.6	8.4	0.077	1	100.00	7.70	
93	5/7/2018		O		WENSON	AE 7789306	3/7/2018	100.000	7.6	8.4	0.077	1	100.00	7.70	
94	5/7/2018		O		WENSON	AE 7789205	3/7/2018	100.000	7.6	8.4	0.077	1	100.00	7.70	
95	5/7/2018		O		WENSON	AE 7790101	4/7/2018	100.000	7.6	8.4	0.077	1	100.00	7.70	
96	5/7/2018		O		WENSON	AE 7790101+ AE 7789302	4/7/2018	100.000	7.6	8.4	0.077	1	100.00	7.70	
97	5/7/2018		O		WENSON	AE 7789208	43/7/18	100.000	7.6	8.4	0.077	1	100.00	7.70	
98	5/7/2018		O		WENSON	AE 7789208+ AE 77890103+ AE 7789205	3/7/2018	100.000	7.6	8.4	0.077	1	100.00	7.70	
99	5/7/2018		O		WENSON	AE 7789305	3/7/2018	83.700	6.3	7.2	0.077	1	83.70	6.44	
100	5/7/2018		R		WURTH	AJ 8349702	5/7/2018	0.900	4.4	4.6	4.933	12	10.80	53.28	
101	5/7/2018		R		WURTH	AJ 8349702 + AJ 8348104	5/7/2018	0.900	4.4	4.6	4.933	1	0.90	4.44	
102	5/7/2018		R		EPCOS	AJ 8301501	3/7/2018	1.200	8.6	9.6	7.226	11	13.20	95.38	

05072018 04072018 03072018 02072018

Figure 6. Excel Spreadsheet of finished goods despatch note

IOP Conf. Series: Materials Science and Engineering **530** (2019) 012039 doi:10.1088/1757-899X/530/1/012039

3.4 Check phase

3.4.1 On job training. On-job training (OJT) was introduced to the finished goods store personnel. It was done when the project manager accompanied the store personnel to carry out the improvement activities. Any concern raised during implementation and training would be resolved immediately if possible. This motivated store personnel and built trust, and hence provided the impetus for future improvement.

3.5 Act phase

After implementing the countermeasures, the time spent for shipment preparation reduced to 1.83 hours per wooden crate. Also, activities were reduced to eight from the previous twelve. Data entry process was shrunk to 2 hours. No further modification is required, as the system has been running smoothly as designed over the next two months.

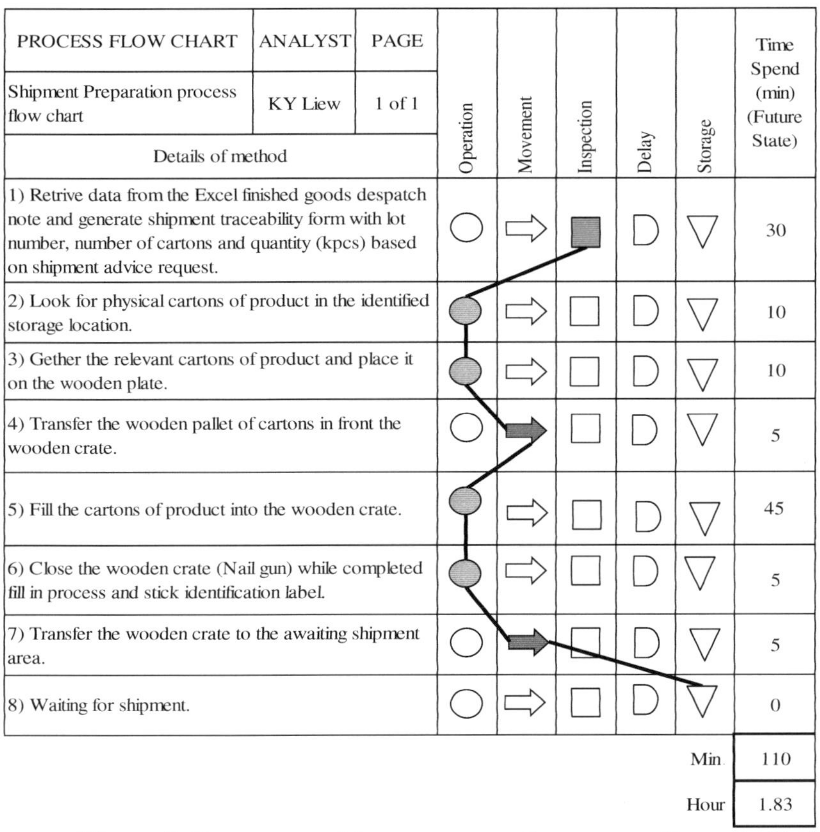

Figure 7. Future state shipment preparation process flow chart

4.0 Conclusion and future work

Through the project, the shipment preparation process was significantly improved by using the developed framework. This provided sufficient time for storekeeper to carry out stock-taking daily. The workload of the store clerk was also streamlined by the removal of the work redundancy in recording. PDCA ensures systematic running of the project. Two primary tools deployed in the framework are 5S and visual management. Evidently, 5S allows system inefficiencies to be categorized and addressed separately. As the shipment preparation process is manual, the installation of visual management provides instant vital information in task execution. This also stabilises the process when the variability was reduced. The framework also promotes human touch and team spirit as problem solving was demonstrated as a shared effort. The future improvement of task is using barcode system to manage the flow of the product lots in the finished goods store.

Acknowledgement
This work was supported by Research University Grant (RUI), Universiti Sains Malaysia [grant number 8014069].

References

[1] Lieberman M B and Demeester L 1999 Inventory reduction and Productivity growth: Linkages in the Japanese automotive industry *Management Science* Vol 45 No 4
[2] Ziukov S 2015 A literature review on models of inventory management under uncertainty *ISSN 2029-8234* Verslo Sistemos ir Ekonomika Business System and Economics Vol 5 (1)
[3] Radasanu A C 2016 Inventory management, service level and safety stock *Journal of Public Administration Finance and Law*
[4] Eisenhardt K M 1989 Building theories from case study research *Academy of Management 710 Review* 14 (4) 532–550
[5] Yin R K 1994 Case study research: design and methods *2nd ed* Thousand Oaks CA: Sage Publications
[6] Cavaye A L M 1996 Case study research: a multi-faceted research approach for IS *Information Systems Journal* 6 (3) 227–242
[7] Womack J and Jones D 1996 Lean Thinking: Banish Waste and Create Wealth in your Corporation, Simon $ Schuster *New York NY*
[8] Agrahari R S, Dangle P A and Chandratre K V 2015 Implementation of 5S methodology in the small scale industry: A case study *International Journal of Scientific & Technology Research* 4(4) pp 180-187
[9] Shah R and Ward P T 2007 "Defining and developing measures of lean production" *Journal of Operations Management* Vol 25 No 4 pp 785-805
[10] Vienazindiene M and Ciarniene R 2013 Lean Manufacturing Implementation And Progress Measurement ISSN 2029-9338 *Economic and Management:2013* 18 (2)
[11] Rathje M S, Boyle T A and Deflorin P 2009 Lean take two reflections from the second attempt at Lean implementation *Business Horizons* 52: 79-88
[12] Hines P and Rich N 1997 The Seven Value Stream Mapping Tools *International Journal of Operations & Production Management* 17 46-64
[13] Che Ani M N and Chin J F 2016 Self-reinforcing mechanisms for cellularisation: a longitudinal case study *International Journal of Production Research* 54(3) 696-711
[14] Bayo-Moriones A et al.2010 5S use in manufacturing plants: contextual factors and impact on operating performance *International Journal of Quality & Reliability Management* Vol 27 Iss:2 pp 217-230
[15] Ghodrati A and Zulkifli N 2013 The impact of 5S implementation on industrial organizations' performance ISSN 2319-801X *International Journal of Business and Management Invention* Vol 2 Issue 3 pp 43-49
[16] Purohit S R and Shantha V 2015 Implementation of 5S methodology in a manufacturing industry ISSN 2229-5518 *International Journal of Scientific & Engineering Research* Vol 6 Issue 8
[17] Kobarne A R 2016 Implementation of '5S' technique in a manufacturing organization: A case study ISSN 2278-8808 *Scholarly Research Journal For Interdisciplinary Studies* Vol 3/23 Mar-April 2016
[18] Chong M Y, Chin J F and Hamzah H S 2012 Transfer of total productive maintenance to supply chain *Total Quality Management & Business Excellence* 23 467-488
[19] Tan P Y, Chin J F, Prakash J and Loh W P 2012 Reducing Electronic Component Losses in Lean Electronics Assembly with Six Sigma Approach *International Journal of Lean Six Sigma*

IOP Conf. Series: Materials Science and Engineering **530** (2019) 012039 doi:10.1088/1757-899X/530/1/012039

Vol 3 Iss: 3 pp 206 – 230

[20] Chong M Y, Chin J F, Loh W P 2013 Lean incipience spiral model for small and medium enterprises *International Journal of Industrial Engineering* 20

[21] Vipulkumar C P and Hemant T 2014 A case study: 5s implementation in ceramic manufacturing company ISSN 2277-5056 Bonfring *International Journal of Industrial Engineering and Management Science* Vol 4 No 3 pp 132-138

[22] Sidhu B S, Kumar V and Bajaj A 2015 The "5S" Strategy by using PDCA Cycle for Continuous Improvement of the Manufacturing Processes in Agriculture Industry *International Journal of Research in Industrial Engineering* Vol 2 pp 10-23

[23] Liff, Stewart and Pamela A P 2004 Seeing is believing: how the new art of visual management can boost performance throughout your organization *AMACOM Div American Mgmt Assn*

[24] Tezel B A, Koskela L J and Tzortzopoulos P 2009 The functions of visual management *International Research Symposium* Salford UK

[25] Goh W J, Chin J F, Kamaruddin S, Muhammad N A 2014 A Visual Board to Facilitate Production Flow Regulation at Sequential Processes *International Journal of Industrial Engineering and Management* 1-9

[26] Brien M O 2014 6S and visual management *LBSPartners, Nexus innovation centre* Tierney building *University of Limerick Limerick Ireland*

IOP Conf. Series: Materials Science and Engineering **530** (2019) 012040 doi:10.1088/1757-899X/530/1/012040

Deployment of an FMEA-Integrated Framework to Improve Operational Performance in Semiconductor Manufacturing: A Case Study

C F Liew[1], J Prakash[1], S Kamaruddin[2*] and K S Ong[1]

[1]Faculty of Engineering and Green Technology, Universiti Tunku Abdul Rahman, Jalan Universiti Bandar Barat, 31900 Kampar, Perak, Malaysia

[2]Mechanical Engineering Department, Universiti Teknologi Petronas, Bandar Seri Iskandar 32610, Perak, Malaysia

Corresponding author *: shahrul.k@utp.edu.my

Abstract. The failure mode effect analysis (FMEA) aims to strengthen the operational performance of a process of a product. However, current FMEA practice does not quantify the achievement in operational performances. Risk priority number (RPN), the only success indicator of FMEA, is independent of operational performance monitoring. Thus, the capability of FMEA to strengthen operational performance may be ineffectual unless it demonstrates a quantifiable improvement in operational performance. This study offered guidelines for manufacturing industry to quantify the operational performance within the FMEA methodology by operational performance indicator (OPI). In tandem with RPN, the OPI in FMEA ascertains the priority set by RPN and defines more refined priority than RPN. Design for manufacturing and assembly (DFMA) and Poka-yoke provide systematic guidelines for developing corrective actions for eliminating or reducing the occurrence of failure while strengthening current control to prevent the delivery of any non-conformance to customers. Application of the new FMEA-based integrated framework in a semiconductor manufacturer in Malaysia demonstrated that 69.2% improved the overall equipment efficiency (OEE).

1. Introduction

Failure mode and effect analysis (FMEA) aims to improve operational performance (OP) of a product or process [1]. However, risk priority number (RPN), a product of severity (S), occurrence (O), and detection (D), is the only indicator of FMEA, has no relationship with operational performance (Puente et al., 2002). RPN can be improved by reducing S, O, or D due to three risk factors are weighed equally. For instance, RPN can be reduced by reducing D, which is a solution for containing failure [2]. However, the OP, such as defect rate or scrap cost, may not decrease. Thus, the capability of current FMEA practice to improve OP cannot be validated unless it demonstrates a quantifiable operational improvement [3].

Operational performance indicator (OPI), which is an accurate indicator to quantify the OP of a product or process, can be used to determine the success of FMEA in tandem with RPN [4]. The priority of goal of corrective action in FMEA is to eliminate the occurrence of failure and enhance current control [2] [5]. Both OPI and RPN can be improved concurrently by optimising the product design [2], which, in turn, is correlated with the occurrence of failure [3]. In reality, design modification is not always feasible due to manufacturing constraints, cost concerns, or rigid customer requirements. Thus, an alternative approach is to develop a more effective error-proofing system for preventing delivery of non-conformance to customers [2] [6]. Design for manufacturing and assembly (DFMA) and Poka-yoke provide systematic guidelines for optimising design and developing a more robust error-proofing system.

DFMA is the amalgamation of design for manufacturing (DFM) and design for assembly (DFA) [7] to increase manufacturability and reduce assembly difficulty [7-8]. DFM streamlines and standardises the design and manufacturing activities of a product concerning the integration of function, form, or fabrication [9-10]. DFA simplifies the product and enhances assembly quality by minimising part count and integrating parts based on function [9]. Poka-yoke is an error-proofing method for preventing the occurrence of failure or facilitating error detection [6]. Poka-yoke has two approaches [10], which are: (1) preventive-based, a control method for stopping the process whenever the error that creates the failure is encountered; and (2) detection-based, a warning method for preventing non-conformance from flowing to the downstream process.

This paper presents a framework for improving FMEA by integrating the OPI, DFMA, Poka-yoke with FMEA. This paper is organised as follows. Section 2 presents the methodology of the proposed integrated framework. Section 3 presents the case study, which is used to validate the new integrated framework. Section 4 discusses the implementation results. Section 5 concludes the paper.

2. Methodology

A new integrated framework, namely FODP framework is constructed by integrating the methodology of **F**MEA, **O**PI, **D**FMA, and **P**oka-yoke. As shown in Figure 1, the proposed FODP framework consists of 10 steps. Details of each step are discussed below.

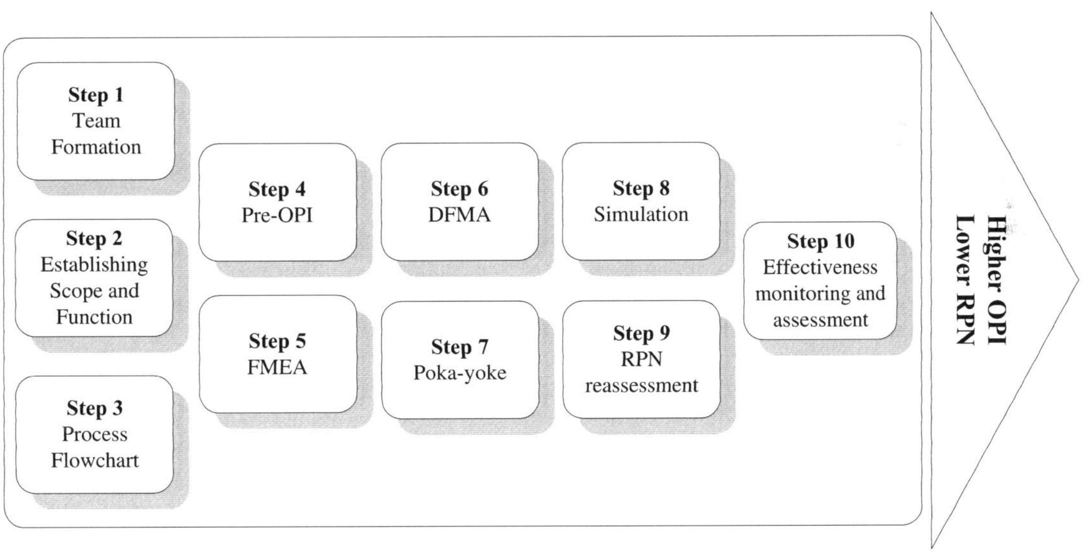

Figure 1. Illustration of FODP framework

- Step 1: Team formation – form a cross-functional team to work towards a common goal.
- Step 2: Scope and function establishment – determine the scope to establish the boundaries of the initiative. Consequently, understand how the product or process functions within requirements.
- Step 3: Process flowchart – provide a logical and visual depiction for better visualisation and comprehension of the entire assembly or manufacturing processes to be analysed.

IOP Conf. Series: Materials Science and Engineering **530** (2019) 012040 doi:10.1088/1757-899X/530/1/012040

- Step 4: Pre-OPI – measure the OPI that relates to quality, cost, and delivery time for all processes.
- Step 5: FMEA - analyse the failure of all processes from start to finish. Different from conventional FMEA, FODP framework ascertains the priority that set by RPN by OPI. In other words, failure mode with higher RPN has a lower priority if the OPI is not critical.
- Step 6: DFMA – when a change in product design is feasible, practitioners optimise the product design by DFMA guidelines for improving the RPN and OPI by mitigating or minimising the occurrence of a failure.
- Step 7: Poka-yoke – when the product design change is impractical or unable to eliminate the occurrence of a failure, prevention-based or detection-based Poka-yoke should enhance current control.
- Step 8: Simulation – evaluate the effectiveness of corrective actions by assessing the preliminary OPI when testing the physical working prototype. If simulation result fails to meet the expectation, practitioners are required to return to Steps 6 and 7 to develop more effective corrective action.
- Step 9: RPN reassessment – re-evaluate the RPN when a positive result is revealed in Step 8. In case the new RPN is not reduced, practitioners are required to return to Steps 6 and 7 to develop more effective corrections.
- Step 10: Effectiveness monitoring and assessment – monitor the effectiveness of corrective actions after they are implemented by measuring the same OPI as Step 2 for the critical process. Consequently, comparing the pre- and post-OPI to confirm the effectiveness of the corrective actions. If a high OPI is achieved, the success of the initiative is confirmed.

3. Case Study

The FODP framework was validated in company S, a semiconductor manufacturer located in Ipoh, Malaysia. Product B, an integrated circuit (IC), was selected to demonstrate the FODP framework. Product B encountered low overall equipment efficiency (OEE), which is the product of yield, equipment availability, and equipment performance [11]. The application of FODP framework is intended to increase the OEE.

A multi-discipline team was formed. Consequently, process flow chart (Figure 2) was constructed for illustrating the process and its function in the testing department. Table 1 showed that for before improvement, the electrical testing process had the lowest yield, availability, and OEE. Further equipment downtime analysis before improvement (Table 2) showed that the significant downtime was relevant to the test contactor, which constituted the test probe and elastomer, as shown in Figure 3(a). FMEA analysis results were summarised in Table 3. The process of electrical testing, vision and tape and reel, and visual inspection had the same RPN. However, the electrical testing process had the highest priority for improvement due to it had the highest criticality in pre-OPI. The test probe connects the IC and printed circuit board. The elastomer disseminates the spring force to test probe to prevent the damage of IC and reducing the wear of test probe, which, in turn, leads to a longer lifespan. However, after specific insertions, the elasticity of elastomer started deteriorating, which eventually lead to the lower spring force. Test probe started deforming (Figure 3b) with lower spring force. Thus, the underperformed test contactor resulted in: (1) higher yield loss due to weak contact established between IC and the printed circuit board; and (2) higher equipment downtime for frequent test contactor maintenance.

IOP Conf. Series: Materials Science and Engineering **530** (2019) 012040 doi:10.1088/1757-899X/530/1/012040

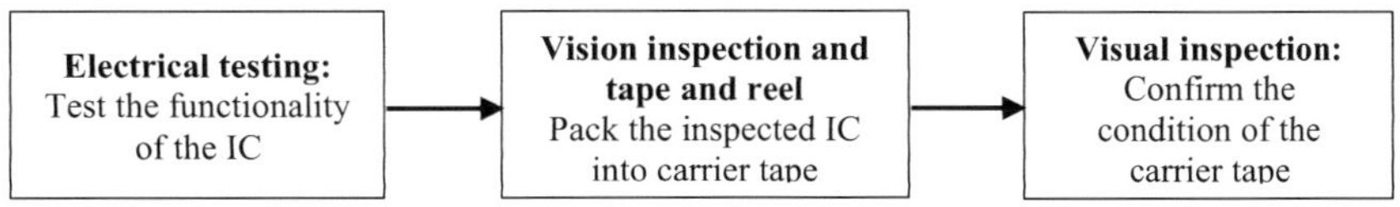

Figure 2. Process flowchart in the testing department

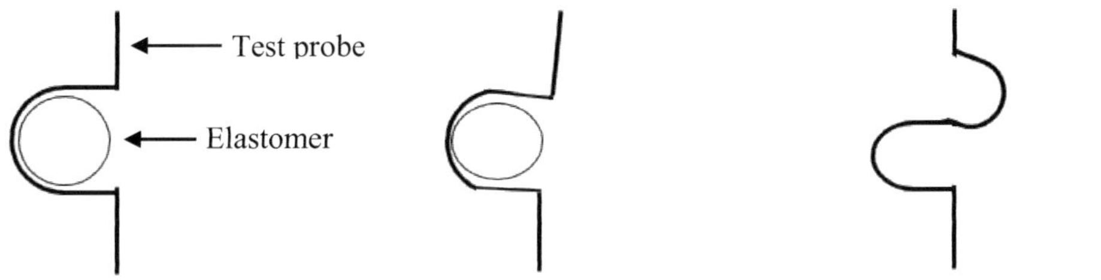

(a) Existing test contactor (b) Deformed test contactor (c) New test contactor

Figure 3. Test contactor

The current underperformance test contactor was resulted by the inappropriate profile design and coating material. Two modifications were subsequently proposed to enhance current test contactor, which is: (1) modified the test probe profile for higher spring force and longer lifespan, as shown in Figure 3(c). The need of elastomer could be eliminated by the enhancing the spring force on the new test probe; and (2) coated a layer of durable alloy on test probe for reducing the cleaning frequency by minimising the solder migration. The new test contactor prolonged the cleaning and lifespan. However, routine maintenance, such as cleaning and replacement were still required due to the test contactor is a consumable part. Current manual control of manually monitoring the yield and tracking the insertion was automated by developing a system. When the yield and insertions hit the predefined limit, the system automatically stops the equipment for the operator to conduct the routine maintenance.

Table 1 and 2 (after improvement) show that the test contactor significantly improved the OEE and equipment downtime with a new profile and coating material. As shown in Table 3, by the same corrective actions reduced the RPN from 280 to 24. By comparing the OEE before and after improvement, the OEE of an electrical testing process was improved from 45.7% to 82.7%.

4. Discussion

The FODP framework successfully quantified the OPI before and after the improvement. Furthermore, by comparing the OPI before and after improvement, the success of FMEA initiative was qualitatively quantified in term of operational performance. The FODP framework ascertained the electrical testing process as the most critical process due to it has the highest criticality in RPN and OPI. Besides improving the RPN and OPI, the modification of test probe profile also simplified the test contactor design by eliminating the need of elastomer. However, routine maintenance, such as cleaning and replacement were still required due to the test contactor is a consumable part. Thus, a new system was developed to automate the manual yield monitoring and lifespan tracking process.

5. Conclusion

By applying the OPI in tandem with RPN in FODP framework, the applicability of FMEA is improved in few aspects: (1) the success of FMEA is quantified in operational performance by comparing the OPI before and after improvement; (2) besides RPN, the priority that is determined for improvement is also

IOP Conf. Series: Materials Science and Engineering **530** (2019) 012040 doi:10.1088/1757-899X/530/1/012040

critical in OPI; (3) the corrective actions are planned to improve the RPN and OPI by mitigating or reducing the occurrence of failure; (4) the effectiveness of corrective actions in OPI is assured before and after the implementation. The orientation toward OPI improvement and the conformance of results arising from the FODP framework are demonstrated through the case study.

Table 1. Average yield, availability, performance, and OEE for two consecutive weeks.

OPI	Before improvement			Simulation			After improvement		
	ET	VTR	VI	ET	VTR	VI	ET	VTR	VI
Y (%)	70.8	98.7	99.4	92.5	98.3	98.4	94.0	98.4	98.9
A (%)	65.7	85.0	86.0	80.7	84.4	85.5	83.7	84.7	85.8
P (%)	98.3	98.6	98.8	98.3	98.4	98.3	98.3	98.4	98.2
OEE (%)	45.7	82.7	84.5	73.4	81.6	82.7	77.3	82.0	83.3

Y: yield; A: availability; p: performance; OEE: overall equipment efficiency; ET: electrical testing; VTR: vision inspection and tape and reel; VI: visual inspection

Table 2. Average equipment downtime for two consecutive weeks (hours).

Downtime	Before improvement	Simulation	After improvement
Test contactor cleaning	25.8	2.8	2.5
Test contactor replacement	8.6	0.8	0.5
Handler jamming	1.3	1.2	1.2
Cover and carrier tape changeover	0.9	1.5	1.3
Total downtime	36.5	6.3	5.4

IOP Conf. Series: Materials Science and Engineering **530** (2019) 012040 doi:10.1088/1757-899X/530/1/012040

Table 3. Integrated FMEA analysis table

Process	Pre-OPI (OEE)	Potential failure mode	Potential effect(s)	Potential cause(s)	Current process control	Existing rating				Corrective action(s)		Post-OPI	New rating			
						S	O	D	RPN	DFMA	Poka-yoke		S	O	D	RPN
ET	45.7%	Low OEE in electrical testing process	Failing customer's delivery plan	Underperformed test contactor with improper profile design and coating material results in high yield loss and equipment downtime for regular test contactor maintenance	1) Hourly yield monitoring and insertions tracking 2) Test contactor cleaning after every 5,000 insertions 3) Replace new test contactor after every 30,000 insertions	8	7	5	280	Change the test contactor profile and coating material	Develop a system to automate the manual yield monitoring and insertion tracking process	77.3%	8	3	1	24
VTR	82.7%	Low OEE in vision and tape and reel process	Failing customer's delivery plan	High equipment downtime is resulted by: 1) Reteach the reference in vision system for detecting the IC marking defect 2) Incorrect carrier and cover tape are used 3) The cover tape is not sealed properly on the carrier tape	1) Provide relevant training to technicians, who do the reference teaching 2) Specify the part number of the carrier and cover tape on traveller for reference 3) Visual inspection to check the condition of the sealing	8	5	7	280			82.3%				

IOP Conf. Series: Materials Science and Engineering **530** (2019) 012040 doi:10.1088/1757-899X/530/1/012040

Process	Pre-OPI (OEE)	Potential failure mode	Potential effect(s)	Potential cause(s)	Current process control	Existing rating				Corrective action(s)		Post-OPI	New rating			
						S	O	D	RPN	DFMA	Poka-yoke		S	O	D	RPN
VI	84.5%	Low OEE in the visual inspection process	Failing customer's delivery plan	High equipment downtime is resulted in 1) Reteach the reference in vision system for detecting the IC marking and sealing defect 2) Not able to roll the carrier tape due to motor faulty	1) Provide relevant training to technicians, who do the reference teaching 2) Provide relevant training to technicians, who maintain the roller 3) Preventive maintenance to check the roller module	8	5	7	280			83.6%				

Acknowledgements
The authors acknowledge the YUTP-FRG grant (0153AA-E36) provided by Yayasan UTP for funding the study that resulted in publishing this article.

References

[1] Liu H C, Liu L and Liu N 2013 Risk evaluation approaches in failure mode and effects analysis: A literature review *Expert systems with applications 40*(2) pp 828–838
[2] Puente J, Pino R, Priore P and de la Fuente D 2002 A decision support system for applying failure mode and effects analysis *International Journal of Quality & Reliability Management 9*(2) pp 137–150
[3] C F Liew, Joshua Prakash and K S Ong 2018 Deployment of operational performance indicators in failure mode effect and analysis (FMEA) application: A literature review and classification *International Journal of Productivity and Quality Management* (forthcoming)
[4] Intra G, Alteri A, Corti L, Rabellotti E, Papaleo E, Restelli L, Biondo S, Garancini M P, Candiani M and Viganò P 2016 Application of failure mode and effect analysis in an assisted reproduction technology laboratory *Reproductive BioMedicine Online 32*(2) pp 132–139
[5] Maria Laura Chiozza and Clemente Ponzetti 2009 FMEA: A model for reducing medical errors *Clinica Chimica Acta 404*(1) pp 75–78
[6] Puvanasvaran A P, Jamibollah N and Norazlin N 2014 Integration of POKA YOKE into process failure mode and effect analysis: a case study *American Journal of Applied Sciences 11*(8) pp 1332
[7] Boothroyd G, Dewhurst P and Knight W A 2010 Product Design for Manual Assembly *Product Design for Manufacture and Assembly* pp 73-131
[8] Robert Bogue 2012 Design for manufacture and assembly: background, capabilities and applications *Assembly Automation 32*(2) pp 112–118
[9] Lu Q and Wood L 2006 The refinement of design for manufacture: inclusion of process design *International Journal of Operations & Production Management 26*(10) pp 1123-1145
[10] M Dudek-Burlikowska and D Szewieczek 2009 The Poka-Yoke method as an improving quality tool of operations in the process *Journal of Achievements in Materials and Manufacturing Engineering 36*(1) pp 95-102
[11] Chong K E, Ng K C and Goh G G G 2015 Improving Overall Equipment Effectiveness (OEE) Through Integration of Maintenance Failure Mode and Effect Analysis (Maintenance – FMEA) in a Semiconductor Manufacturer: A Case Study *Industrial Engineering and Engineering Management (IEEM) 2015 IEEE International Conference* pp 1427–1431

IOP Conf. Series: Materials Science and Engineering **530** (2019) 012041 doi:10.1088/1757-899X/530/1/012041

The Application of DMAIC to Improve Production: Case Study for Single-Sided Flexible Printed Circuit Board

N Mohamad[1], S Ahmad[1], H A Samat[2], C K Seng[2] and F M Lazi[1]

[1]Faculty of Technology Management and Technopreneurship, Universiti Teknikal Malaysia Melaka, Melaka

[2]School of Mechanical Engineering, University Sains Malaysia, Penang

Corresponding author *: norizahm@utem.edu.my

Abstract. In this study the DMAIC (Define, Measure, Analyse, Improve and Control) technique is applied at a company producing Flexible Printed Circuit Board (FPCB) in Malaysia. It is an initiative for continuous improvement by the company. The aim of the study is to reduce the number of open defects during the production of the single sided FPCB which has substantial impact on the profit margin of the company. The study examines one of the production line which has the highest defect rate. The study has utilised the Six Sigma tools such as brainstorming, process mapping, fish bone diagram, Pareto chart and control charts. From the analysis, it is found that the major contributor to the open defect problems are contamination of the separator during production. The causes of the contamination were identified and action plans were implemented. These actions have resulted in an overall reduction of defect rate and cost saving for the company. For the duration of the study, the overall defect rate for the particular line was reduced from 0.6% to 0.37% with a total saving of RM 1423.60. The case study showed that DMAIC process is an effective approach that can solve what seems to be a major problems using simple solutions. If conducted properly and efficiently, it has the potential to give good returns to the company.

1. Introduction and Background

Manufacturing company all around the world are facing the turbulent economic condition which include global competition, customer demand for high quality product and reduced in lead time. To remain competitive companies are continuously looking for ways to reduce production cost, enhance productivity and improve product quality. In response to these needs, companies need to have improvement strategies that utilise the available resources efficiently and effectively [1]. Many leading companies are leveraging on lean tools and techniques to enhance productivity and quality performance. One of the widely used performance strategy improvement is the utilisation of Six Sigma. Six Sigma efforts are directed towards the minimisation of waste and non-value added activities (NVA) and maximising customer satisfaction. Six Sigma drives on substantial improvement to the bottom line by designing and monitoring everyday business activities [2]. The DMAIC Six Sigma has the potential to realise cost savings and improved quality on the shop floor [3]. The DMAIC methodology follows the phases of: define, measure, analyse, improve and control. The Six Sigma DMAIC process is an improvement system for existing processes falling below specification and looking for incremental improvement. Although, the DMAIC stage model is a structured problem solving device, with the original task of variation reduction, the method are now popular for more general tasks, such as quality improvement, efficiency improvement, cost reduction, and other pursuits in operations management, and beyond manufacturing in services, healthcare, and other types of operations [4].

The Japanese word 'Muda' meaning waste is referred to the human activities that exploit the resources in the company but does not create any profit or value. This concept of 'Muda' in the

IOP Conf. Series: Materials Science and Engineering **530** (2019) 012041 doi:10.1088/1757-899X/530/1/012041

manufacturing industry was presented by Taiichi Ohno, a Toyota executive [5]. In particular, he defined seven classes of waste that typically affect a manufacturing process:

Table 1. Seven Waste of Lean Manufacturing

Types of waste	Explanations
Over processing	Adding more value to a product than the customer actually require such as painting area that will never be seen or be exposed to corrosion.
Waiting	It is the act of doing nothing or working slowly whilst waiting for a previous step in the process. How many times you seen operators stood waiting for a previous operation, a delivery products to arrive or just slowly working for lengthen the time since run out of material.
Transportation	It is the movement of products from one location to another. Eg: This could be from the machining shop to the welding shop, or from the production facility in china to the assembly line in America. This transportation add no value to the product.
Inventory	Is the raw materials, work in progress (WIP) and finished goods stock is held, we often hold far more than is required to produce goods and services when the customer wants them using Just In Time (JIT).
Motion	Waste being a process step that is not value adding, moving is not necessarily working. The main causes of the waste of motion are with regards to cell layout, placing product at floor level on pallets, poorly arranged space, tools that are disorganized, lack of space and organization for component parts and so on.
Defect	Defect are when products or service deviate from what customer requires or the specification. Defect can be cause by many kind of problems, some should be avoidable with a little thought when designing the product, non-standard operation, different in the way that processes are undertaken by different operator on different shift.
Overproduction	Overproduction is making products in too great a quantity or before it is actually needed leading to excessive inventory.

In this project, DMAIC (Define- Measure- Analysis- Identify- Control) tool is utilise at the production floor. DMAIC is a very structured approach with a very detail analysis process of implementing the improvement. DMAIC also provides a business road map for solutions. This helps the business to solve the problems from the start of the manufacturing process till finish while producing bottom line results. Besides, DMAIC also support an analytical approach, allowing the business to use the collected data. This help the business ensure the accurate baseline. This paper aims to investigate the non-value added activity and waste of the single sided flexible Printed Circuit Board process using DMAIC. The aim is to reduce open defect from 0.5% to 0.4% by end of December 2017 and to propose a standardise documentation for the control of the open defect for the single-sided flexible circuit board.

IOP Conf. Series: Materials Science and Engineering **530** (2019) 012041 doi:10.1088/1757-899X/530/1/012041

2. DMAIC methodology

DMAIC is a closed-loop process, which aim to eliminate those non-value added step and focus on new measurements and applies technology for continuous improvement [6]. Figure 1 shows the DMAIC methodology.

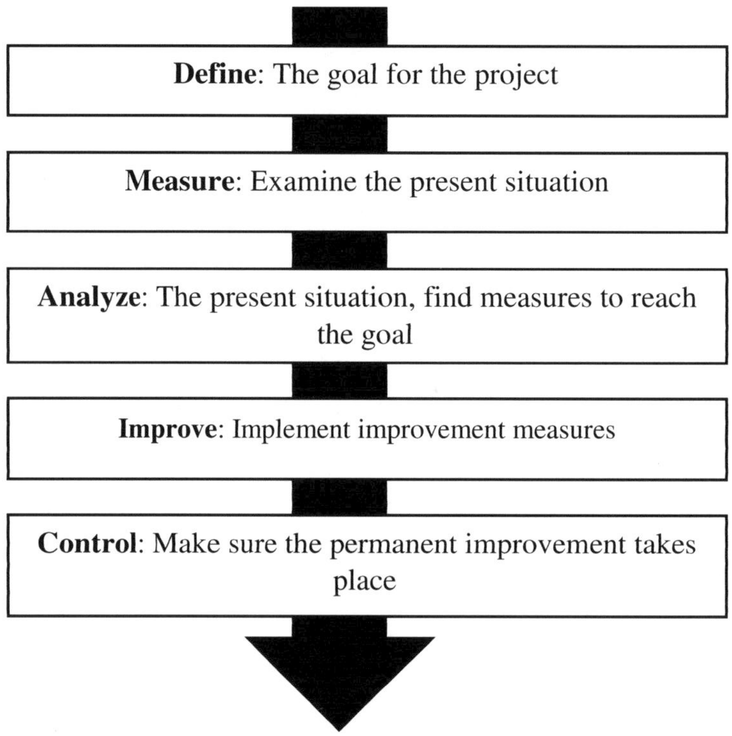

Figure 1. DMAIC methodology

DMAIC is regulated under five phases [7-8] as:

1. Defining the goal and its requirements: Identifying the problem which needs a solution and ends with understanding the issues.
2. Measuring the current process: Understand all the processes in organization, customers' expectations, suppliers' specifications and identification of the possible places where a problem may occur
3. Analysing the results of measurements, determining the causes of process imperfections and possible solutions for them: confirm the analysis some samples should be performed and potential problems have to be proven to be real problems.
4. Improving the process, implementing the changes, which eliminates the imperfections: to take necessary information to create and develop an action plan in order to improve the functioning of the organization, financial aspects and customer relationship issue.
5. Controlling of the improved process, monitoring the results in a continuous way: controls the future state of the process in order to minimize deviation from the objectives and ensure that the correction is implemented before it would have bad influence on the result in the process.

IOP Conf. Series: Materials Science and Engineering **530** (2019) 012041 doi:10.1088/1757-899X/530/1/012041

3. DMAIC Implementation at Case Study Company

The DMAIC methodology is performed via the Roadmap of DMAIC as shown in Figure 2.

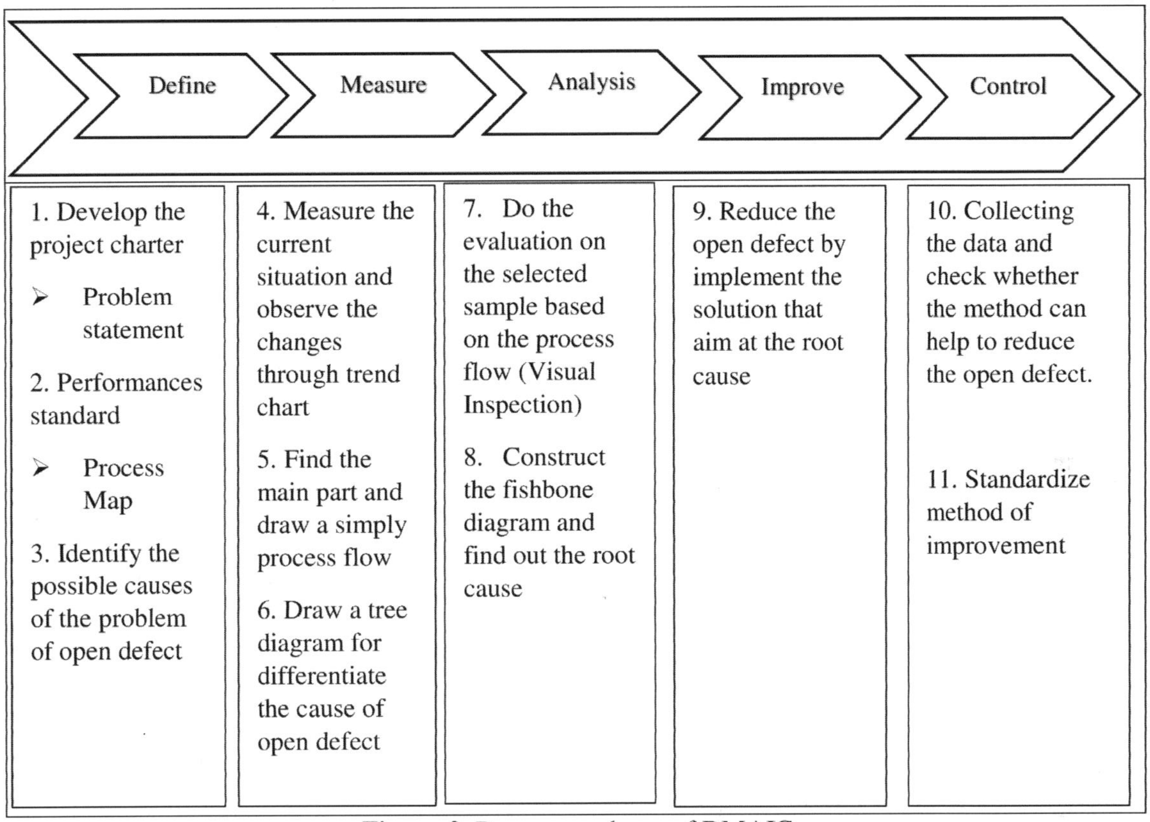

Figure 2. Process roadmap of DMAIC

Define Phase

When implementing the Six sigma DMAIC approach, the most important thing is to define the scope of the problem. The production line that has the highest number of defect was chosen for process improvement. Firstly, a project charter was developed. The project is led by the project manager and the team member consists of line leader and six team members. The smart goal of the project was to reduce open defect from 0.57 % to 0.4 % by end of December 31st 2017 with start date of July 1st during this stage, to further understand the process, Gemba walk was performed. Figure 3 shows the process flow of the production of the single sided FPCB.

IOP Conf. Series: Materials Science and Engineering **530** (2019) 012041 doi:10.1088/1757-899X/530/1/012041

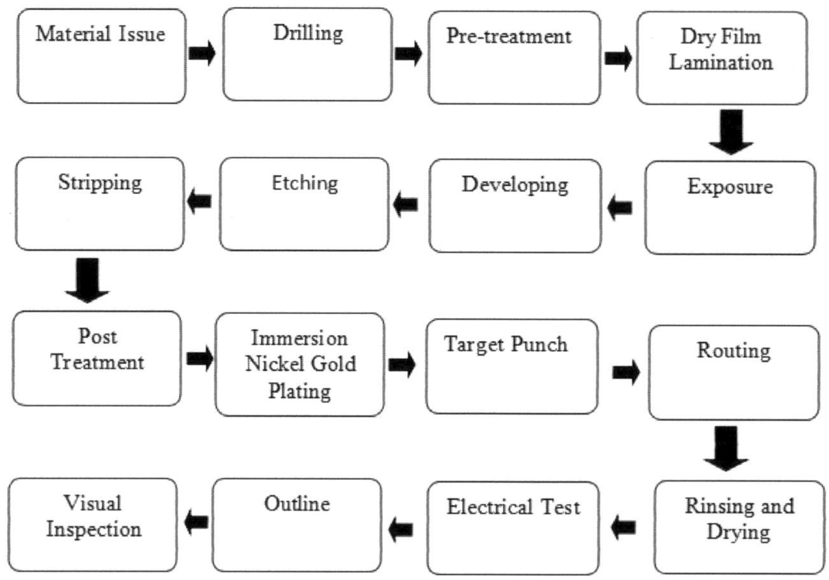

Figure 3. Process flow of the manufacturing process of single sided

Measure Phase

In this phase, the performances of the manufacturing process of single-sided FPCBs for the past 8 months is analysed using the Visual Inspection Report. The overall trend chart was plotted and shown in Figure 4.

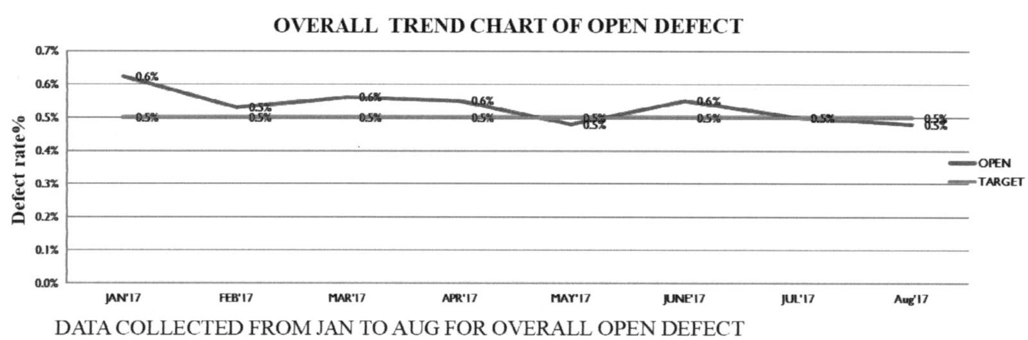

Figure 4. Graph of Overall Open defect of Single-sided FPCBs (Day &Night Shift) for January 2017- August 2017

From the graph, the rate for target of the defect is set at 0.5% but as we can see from the graph the average defect rate of the open defect is around 0.62% from January 2017 till August 2017. The production for Single-sided FPCBs for 1 day is 23 lots, 1 lot have 80 panels, and 1 panel have 512 pieces. So the production volume for Single sided FPCBs for every month is round 28262400 pieces. It is interesting to note that, since the production volume is large so the difference of 0.1% can help company save cost and improve the quality of the FPCB.

Based on the calculation of the yield report for open defect of single-sided FPCBs from January 2017 until August 2017, a Pareto chart was plotted as shown in Figure 5. Part 13236 is chosen as the main focus on reducing the rate of open defect. Then, the trend for the performance of the open defect rate for part 1326 was plotted as shown in Figure 6.

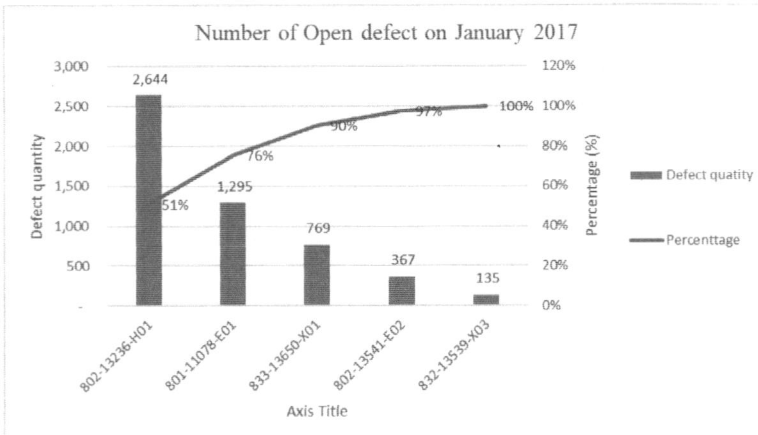

Figure 5. Pareto chart of Open defect for 5 types of single-sided FPCs for January 2017

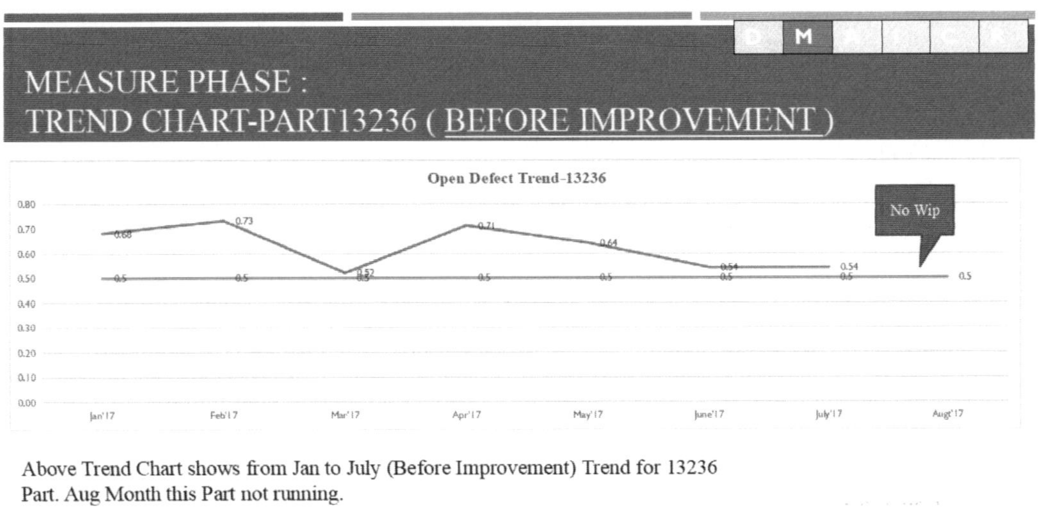

Figure 6. Trend chart of Open defect for parts 13236 for January 2017

The trend chart of Figure 6, shows the performances of defect rate for part 13236 from January 2017 to August 2017. The straight line is the target (indicator) of open defect that is set by the company. Whilst, the fluctuating line indicates the actual situation of defect rate from January 2017 until August 2017. The fluctuation of the data, shows that the process is not stable. Hence, these process can be further improved.

In the next step, process mapping for part 13236 was performed and presented in Figure 7. Process mapping is one of the most effective tools for the improvement of production and business performances. Mapping the process can help the team to identify each step in the process, help in selecting what to measure, and where (and how) to focus on actual process. Besides, process mapping also helps to visualise the process. In this project from the 17 activities, we will focus on the first 7 activities as shown in Figure 6. The In Process Quality Check (IQPC) activity is able to detect any error of abnormality during manufacturing process. The number of defect at every process from activity 1 to 6 were recorded and results from July 2017 to Oct 2017 are as shown in Figure 6. It is observed that the Open defect of the circuit for part 13236 have 2 main causes; problem of contamination (0.57%) and pin hole (0.53%). Since, contamination problem has a higher percentage of defect, thus this project would focus on problem of contamination.

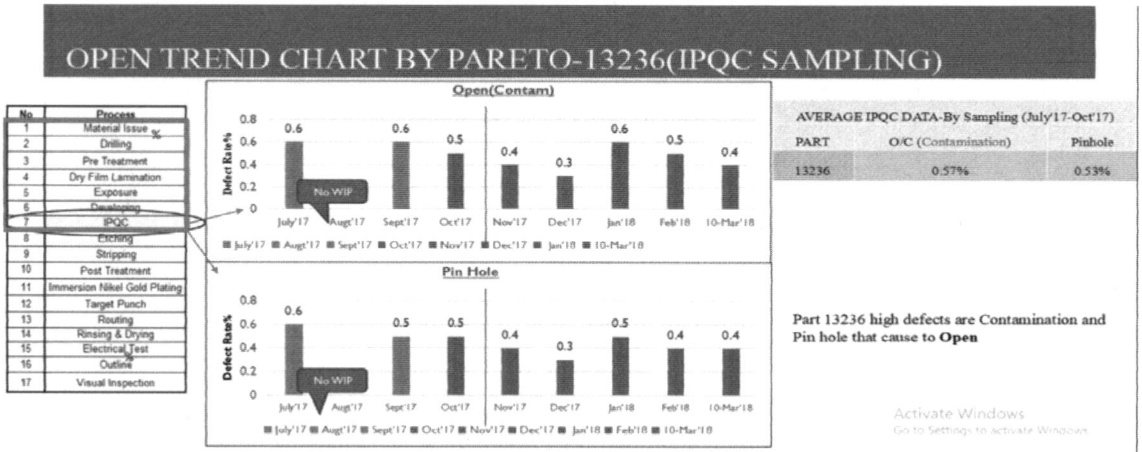

Figure 7. The process mapping and the IQPC results

Analyse Phase

The evaluation is conducted on 20 panels (10240 pieces) of the single-sided FPCBs. The process includes Post treatment, Dry film lamination, exposure, develop and IPQC. From this evaluation we can identify the processes that are contaminated. The process uses visual checking and IPQC (In-Process-Quality-Control). Table 2 shows the results of the evaluation.

Table 2. Evaluation record for checking open defect of selected 20 panel (sample)

Process		Activity	Test Result	Defect %	Date	Time	Remark
		Visual checking before Post Treatment	16 Pnls found contamination out of 20 panels	80%	9/1	8pm	
Post Treatment		-Visual checking after post Treatment -Arrange panels with Separator	No contamination	0%	9/1	9pm	
		-Remove Separator and Visual check	10 panels found contamination out of 20 panels	50%	9/1	9.30pm	
	Panel Cleaning	After panel cleaning, visual checking	0 contamination	0%	9/1	11pm	
Dry Film Lamination		After Dry Film lamination, visual checking	0 contamination, found dent mark	0%			Panels arranged in Tray/bin. Found bin not clean –has cured dry film chips
		Before exposure panels taken out from Bin, visual checking	11 Panels found dry film chips stick on Panel surface	55%	9/1	11.30pm	
Exposure		Visual checking after Exposure			10/1	12.30am	
Develop		After Develop IPQC checking					
IPQC			FOUND OPEN: 17pcs	0.16%	10/1	8am	17 PCS OUT OF 10240 PCS OPEN

From Table 2, it is seen that after drilling process, the visual checking results before the post treatment process, established that 16 panels (80%) out of 20 panels were contaminated. After the post treatment process, the visual checking results found 0% contamination and the panels were arranged with the separator. After visual checking, it was found that 10 (50%) panels were contaminated. The panels were cleaned and visual checking found that no panels were contaminated. The separators were identified as one of the causes of contamination. The 20 panels were arranged back in the Bin for the next process. Before the exposure process, the panels were taken out from Bin and visual checking was conducted. Dry film chips stick was found sticking on the surface of 11 panels. All the panels underwent the exposure process and the developing process and IQPC checking was conducted. It is found that 17 out of 10240 pieces have Open defect problem. From this evaluation, the contamination that occurs at the separator is the main issue that needs to be addressed.

Brainstorming session was conducted to demine the main causes of the contamination at the separator. There were many possible causes of contamination at the separator and they are described in a fishbone diagram as shown in Figure 8.

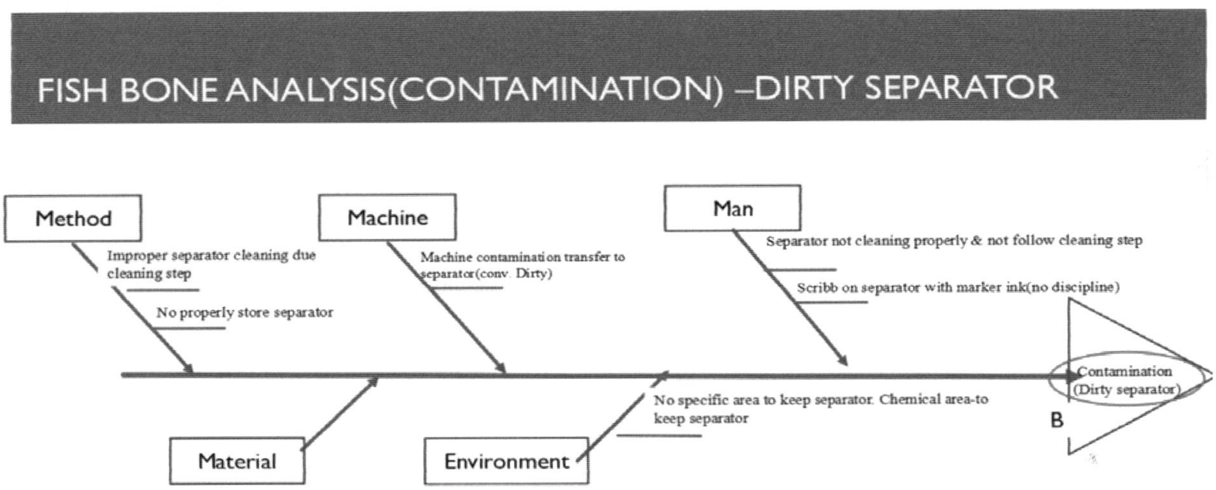

Figure 8. The fishbone diagram for causes of existing contamination at the separator

Improved Phase

The next step is to determine the action that needs to be taken to reduce the open defect due to contamination of the separator. All the group members especially those who have experience in the manufacturing process in FPCBs identified possible improvement actions, prioritize them, tested the improvements, and finalize the improvement action plan. Table 3 shows the action plan that were implemented by the group. Based on the identified root causes, suitable improvements were proposed. A structured improving effort can lead to innovative solution that can improve the baseline measure. Figure 10 shows the improvement with a substantial decrease in defects with highest in February 2017 to the lowest in February 2018. The overall open defect has decreased from 17781 units (0.6%) in January 2017 to 4766 unit (0.37%) in January 2018. Whilst part 13236 which has the highest defect also decreases from 2644 units (0.68%) to 132 units (0.26%). Besides, the setting target was reduced from 0.5 % to 0.4 %. The accumulated cost saving for the company from September 2017 till March 2018 is RM 1423.60. This is as shown in Figure 10.

IOP Conf. Series: Materials Science and Engineering **530** (2019) 012041 doi:10.1088/1757-899X/530/1/012041

Table 3. Action plan for implementing the solutions

IMPROVEMENT :
ACTIONS PLAN TAKEN

Defect	Category	Cause	Short term Action	Control (Standardisation)	Date Start	PIC	Long term Action	PIC	Date
Contamination	Method	A Improper cutting method	To change cutting knife every 3 lots	Recording in First Piece Checklist	4-Oct'17	Kalai / Yaakob/ Alim	To fix auto cutting machine	Yap	2nd week Mar'18
	Material	B Dirty Separator	Implement separator cleaning by daily	Recording in separator cleaning Checklist	12-Oct'17	Production (Mariappan/ Sanjeev/Sim)	To fix separator cleaning machine	Tew	TBA
		C Panel Contamination	Go through panel cleaning machine	Recording in Panel Cleaning Checklist	6-Sept'17	Kalai / Yaacob/ Alim	NA	-	
		D Dirty Tray/Bin Due to dry film chips	Imaging PM – Tray/Bins cleaning by weekly	Recording in Tray Cleaning Checklist	14-Oct'17	Kalai / Yaacob/ Alim	N/A	-	
	Man	E Exposure glass not clean every 10 cycle	To fix cleaning glass every 10 cycles	To record in Expose glass cleaning checklist	22-Jan'18	Kalai/Yaacob/ Alim	To modify machine- add in buzzer to trigger every 10 cycles of exposure.	Yap	End of Mar'18

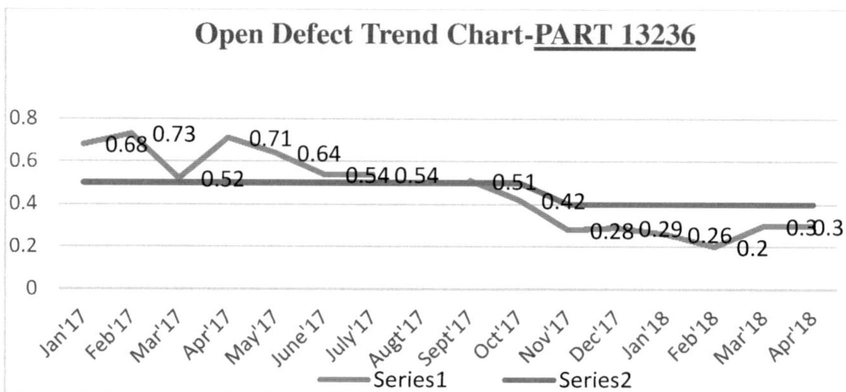

Figure 9. Open trend chart of part 13236

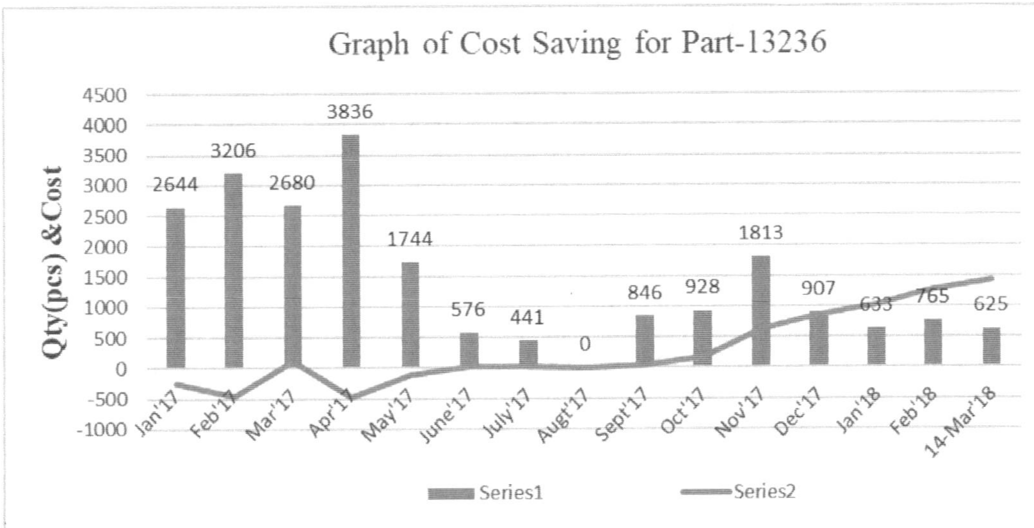

Figure 10. Graph of the cost saving from January 2017 until March 2018

Control Phase

In the phase, we need to ensure that the full-scale implementation of the improvement action plans are carried out. Controls need to be set up controls to monitor the system so that gains are sustained. Specifically, in this phase, a new work instruction was created for separator sheet cleaning procedures. Work instruction are develop to guide operator in 4 keys areas that are training, reference, problem solving and continuous improvement. A check list needs to be prepared to record the operators who are in-charge on the process of checking of the condition of separator. This can help supervisors, identify operators that need to be given more trainings.

In this phase, one point lesson (OPL) or Attention Note was created. OPL is a simple operational tool used to educate operators in a company to help to improve the quality of the product and service quality. The main aim of OPL is to achieve the similar standard of work among operators and shift. Detailed working procedures would be provided so that it can be a guideline for the operators. This is necessary as some improper production processes would be eliminated and some new procedures or steps are introduced. Operators that are more familiar with the old process steps may need some times to fully follow the new changes. A control plan was also provided to monitor the ongoing performance using statistical tools. This is to monitor the stability of the new process.

4. Conclusion

DMAIC technique is a strategic tool to improve the business performance by improving the manufacturing or operation process through effective application of both statistical and non-statistical tools/techniques. The study has demonstrated that the widely promoted DMAIC technique is able to improve the performances of case study company and achieve customer satisfaction. As overall, the defect rate has decreased whilst it has also deliver cost savings to the company. The DMAIC is an excellent continuous improvement technique that will succeed if supported by top management. However, companies need to train staff to conduct DMAIC as projects chosen for implementing DMAIC need to justify the investments, time and effort put into the project. If conducted properly and efficiently, it has the potential to give good returns to the company.

IOP Conf. Series: Materials Science and Engineering **530** (2019) 012041 doi:10.1088/1757-899X/530/1/012041

References

[1] Gupta N 2013 An application of DMAIC Methodology for Increasing the Yarn Quality in Textile Industry *IOSR Journal of Mechanical and Civil Engineering* 6 (1) 50-65

[2] Tushar N Densai and Shrivastava R L 2008 Six Sigma – A New Direction to Quality and Productivity Management *Proceedings of the World Congress on Engineering & Computer Science* Oct 22-24 San Francisco

[3] Kumar S and Sosnoki M 2009 Using DMAIC Six Sigma to systematically improve shop floor production quality and costs *International Journal of Productivity and Performance Management* 58 (3) 254-273

[4] de Mast J and Lokkerbol J 2012 An analysis of the Six Sigma DMAIC method from the perspective of problem solving *International Journal of Production Economic* 139(2) 604-614

[5] T Ohno 1988 Toyota production system: Beyond large-scale production *Productivity Press* New York

[6] Anup A and P N Shende 2011 Minimization of Rework In Belt Industry Using DMA *International Journal of Applied Research in Mechanical Engineering* 1 (1) 54

[7] Antony J and Banuelas R 2001 Six Sigma: A Business Strategy for Manufacturing Organizations *Manufacturing Engineering* 8 (3) 119-121

[8] Wang H 2008 A Review of Six Sigma Approach: Methodology, Implementation and Future Research *Zhejiang Normal University* Jinhua City China 1-4

Common Mistakes in Running PDCA: A Survey on University Student PDCA Projects

Z K Wani*, J F Chin and N A Muhammad

School of Mechanical Engineering, Universiti Sains Malaysia, Penang, Malaysia

Corresponding author *: zarirah.karrim@yahoo.com

Abstract. The main purpose of this paper is to present common mistakes in running Plan-Do-Check-Act (PDCA) project by using case study. New practitioners with petite knowledge are having a high tendency to perform simple mistakes in conducting PDCA project. The study focused on the projects conducted by Universiti Sains Malaysia (USM) manufacturing students as approaches. The mistakes are collected during the project presentation and finalized by the Lean Six Sigma (LLS) experts. A questionnaire survey was sent to LSS experts to rate the high impact mistakes according to their experiences and knowledge. In total, there are 12 mistakes listed coverall to all PDCA phases. Five mistakes are highlighted as high impact mistakes according to 80-20 rules of Pareto. The findings emphasize some ideas that might help new practitioner to avoid such mistakes. However, the case study presented only focus on the mistakes in project conducted by university students not a practitioner in industries.

Keywords: *PDCA, Lean Six Sigma, Problem solving, High impact mistakes*

1. Introduction

Lean Six Sigma (LSS) is a widely-accepted methodology for continuous improvement [1-3]. Normally, a mature LSS organization would deploy a structured training scheme to impart relevant knowledge to its staff. An example would be a four-level certification system, starting from Yellow Belt (YB) and gradually proceeds to Green Belt, Black Belt and Master Black Belt [4]. Each level would have the staff acquires different LSS methods and tools which increase in complexity. It is also a norm for the staff to go through a practical session, e.g. leading and direct involving an improvement project. For example, in YB level, participant familiarizes with PDCA. PDCA is one of the fundamental and methodological tools introduced in the structured training scheme. It divides an improvement project into stages of deployment and defines steps to facilitate the project running. According to Sokovic [5] and Eirin et al. [6], the success of PDCA heavily relies on the support of management and the knowledge about how to use the method, when to apply, efficient performance and activities which can help to focus on the topic.

Similar to PDCA, Kaizen is also known as a continuous improvement which is a process of changing thing towards better. A common structure of Kaizen is based on four different stage of PDCA namely: PLAN – detain planning, DO – implementation of works, CHECK – measure the changes and ACT – refine the improvement [7-8]. In relating to this research, previous researchers has discussed about the common mistakes in executing Kaizen such as lacking a charter, unable to identify critical success factors, having too large scope, not function team members, less supports from top management and etc. Some of the mistakes listed are related to this research output where the mistakes also made by participant while conducting their PDCA project. According to Emiliani [9], the lack of involvement from the executives in performing Kaizen lead to the failure of a project. The issue is then highlighted as one of the challenge in executing kaizen [10]. By not seeing waste, unevenness, and unreasonableness, first-hand, they are not going to believe it hence, not giving supports to the project executed.

The next section briefly reviews PDCA cycle. Section three outlines the case study and methodology. Section four lists down the pitfalls made by participants in undertakings of PDCA followed by the discussion. Lastly, the conclusion will be made in section five.

2. PDCA Cycle

PDCA cycle is a method used to coordinate a continuous improvement project. The evolution of PDCA is summarized in Figure 1. It was originated from Walter Shewhart (Shewhart Cycle) and later popularized by Edward Deming (Deming Wheel) [2]. Deming wheel constantly iterates four phases namely design, production, sales and research, with quality and service [11]. Deming Wheel was gradually replaced by PDCA cycle due to similarity [12] and preference. PDCA cycle is not only prevalent in manufacturing industries but also in other fields such as medical [13], agriculture [14] and service [15].

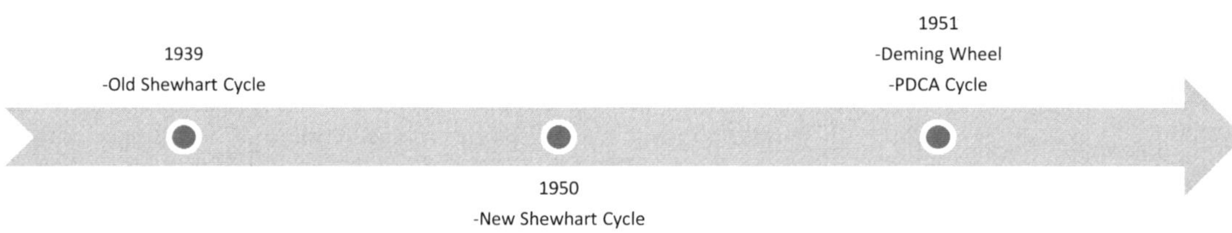

Figure 1. Evolution of PDCA

PDCA cycle helps practitioner to carry out improvement systematically. It leads the improvement towards the direction of preventing error recurrence by establishing standards and continuing standardization [11]. In "Plan" phase, the goal statement are clearly stated to highlight the main purpose of the project. By addressing the grasp of current condition, the practitioner able to obtain the understanding of the present working method and environment. Setting the target condition for a project describes a desire future state of which the project would be resolved adequately. In determining the root cause of the problem, practitioner might apply several LSS tools such as cause and effect diagram, 5-WHY analysis, and Value Stream Mapping (VSM) to conduct the root cause analysis. The suitable countermeasures are selected via relevant LSS tools such as prioritization matrix, FMEA, and Poka-Yoke. "Do" phase covers the process implementation of selected countermeasures. Naturally, it takes a comparatively longer time, partially due to the pilot run or test before full implementation. "Check" phase involves the monitoring and reflection of the improvement to detect any possible refinement. The results of before and after implementation are compared in an appropriate visual or graphical analysis. To sustain the project improvement, "Act" phase plays the important role among all the phases. Project standardization and monitoring after implement the countermeasure to keep the assurance to the workers in following the implementation made. Upon the project closure, practitioners are encouraged to perform Yokoten which is to share the experiences and knowledge gained throughout the project to other related department. PDCA cycle repeats until the goal is achieved. It is not unusual that PDCA go through several iterations before the goal is achieved [6].

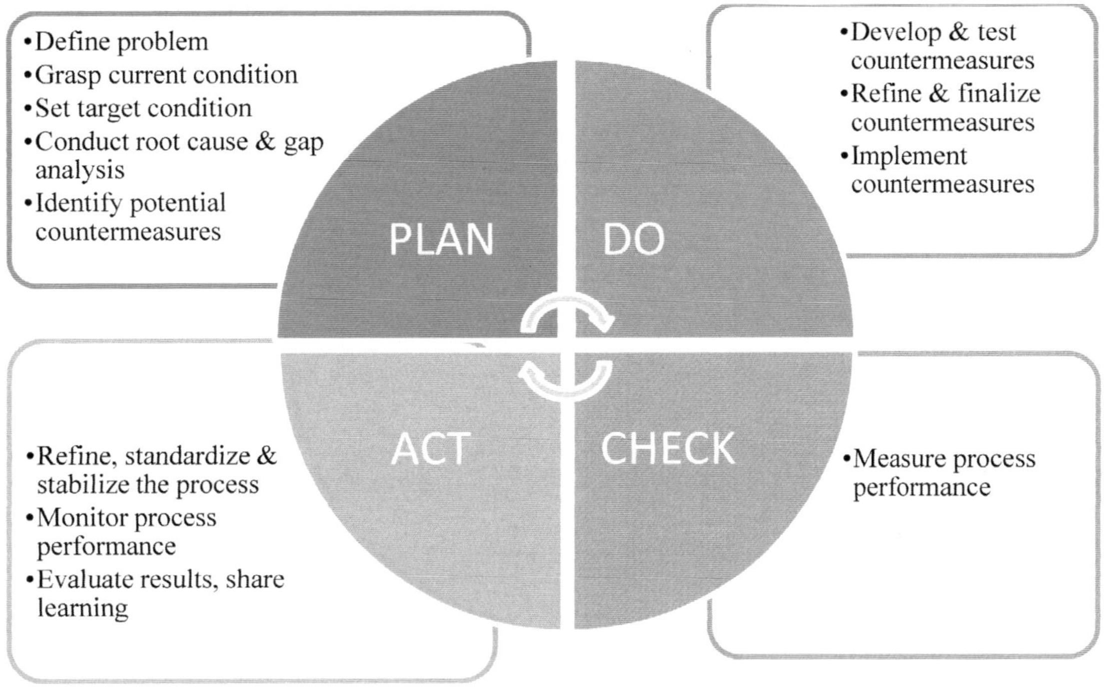

Figure 2. PDCA Cycle [5-6]

3. Case study and methodology

This section outlines the case study and research approach. PDCA projects of practitioner from fifth batch of YB training program were studied. They have completed the training prior to the project. These projects related to the improvement conducted in industries. Twelve PDCA projects each runs with a group of two participants and takes place at different industries including packaging industries, food industries, paper industries, groceries store and etc. One mentor from academic field are assigned to assist three groups. The completed project duration is around 10-12 weeks.

The project's outcomes were presented to LSS experts from various levels of LSS and the mistakes are identified through the project presentation. A questionnaire developed to rate the mistakes which can cause high impact to the successful of a project. The questionnaire was distributed and filled by ten LSS practitioners in industries and academic field. An interview session was conducted to rank the mistakes, obtain justifications and suggestions to avoid the high impact mistakes.

4. Results and discussion

Twelve mistakes (Table 1) were gathered during the project presentation. Eight mistakes are identified in Plan phase (66.67%), one mistake each in Do and Check phases (8.33%), and two mistakes in Act phase (16.67%). Plan phase is most decision intensive where the path of the project is determined.

4.1. List of mistakes from PDCA project conducted by USM students

Table 1. List of mistakes

Phase	No	Mistakes
PLAN	M1	Did not provide a specific, measurable, attainable, relevant and time bound (SMART) goal.
	M2	Poor study on current condition.

IOP Conf. Series: Materials Science and Engineering **530** (2019) 012042 doi:10.1088/1757-899X/530/1/012042

	M3	Mismatching target conditions and goal.
	M4	Project obstacle is not properly defined and made explicit.
	M5	Erroneous data collection.
	M6	Improper use of LSS tools
	M7	Insufficient refinement in analysis
	M8	Weak relationship between root cause and countermeasure.
DO	M9	Wrong application of tools during implementation phase
CHECK	M10	Results before and after the implementation are not properly compared.
ACT	M11	No evidence on process standardization
	M12	Learning experience is not shared (no Yokoten)

4.2. High impact mistakes

From the survey result, M6 has been rated as the highest impact mistake. According to the Pareto chart shown in Figure 3, five mistakes occupy 80% of the votes.

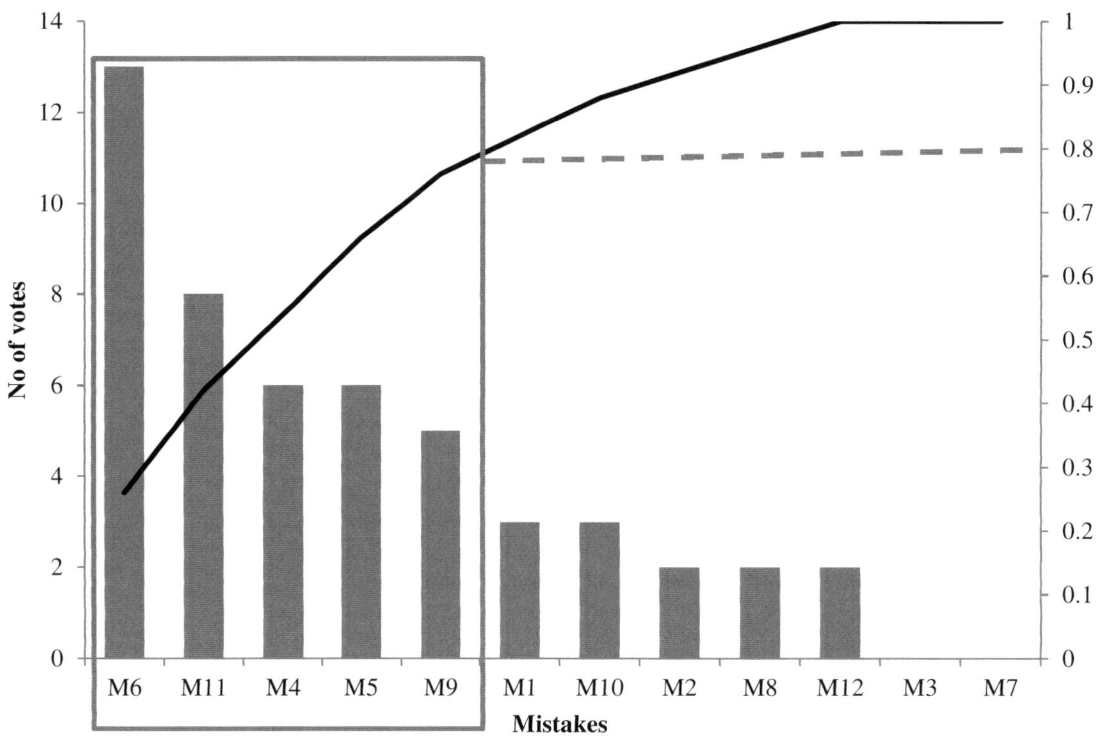

Figure 3. 80/20 rule of Pareto chart voted by respondent

The first high impact mistake highlight is improper use of LSS tools in "Plan" phase. Improper use of LSS tools means the tools is applied in inappropriate way and not used thoroughly. For example, the practitioner applied cause and effect diagram to identify the root cause but the critical root cause is not specified. Failing to do so potentially leads to undertaking of wrong countermeasure and hence poor attainment of goal. This might be because of the lack of understanding on the use of LSS tools among the practitioners. Nordin et.al [14] states that the lack of understanding the LSS tools is a main barrier for a successful improvement project. Improper use of LSS tools also occurred when multiple similar tools applied especially in identifying root cause. One of the group identify the root cause on the high replacement order in packaging factory using both 5-WHY analysis as well as cause and effect diagram. In 5-WHY analysis, the root cause founded are corrugated cutting operation, paper debris not cleaned regularly and inspection work after finishing not carried out all the time. However, a different set of the root causes were listed in cause and effect diagram. This leads to confusion; hence additional step of screening is required.

The second high impact mistake is no evidence on process standardization after the implementation in "Act" phase. Practitioner run the countermeasures in project improvement without notify the workers regarding the improvement made. Practitioners are required to establish a standard work such as Work instruction (WI), Standard Operating Procedure (SOP), Leader Standard Work (LSW) and Visual Management in their project for standardization purpose. However, the documentation work unable to be prepared due to no evidence in process standardization while conducting the project. Fundamentally, process standardization boosts workers quality and productivity while reducing ambiguity in executing task. In the findings, 25% of the groups did not provide evidence on process standardization; SOP and LSW. In one of the PDCA projects, creating a standard way to load carton is listed as the countermeasure but no SOP provided. Apart from that, another one group added steps in manual sealing process but no SOP was established to acknowledge the changes. Consequently, the worker would not be able to capture the altered steps and tend to turn to the original practices.

The third mistake is practitioner unable to properly define and make explicit to the project difficulties such as obstacles and problems faced in "Plan" phase. Practitioner only specified the difficulties in general such as the environment is not good without unambiguously state details condition of the environment faced. Able to anticipate, define and acknowledge the obstacles help to produce a better improvement plan. By making them explicit to the stakeholders, certain obstacles could be discussed, addressed and resolved at the early stage, so that they are not becoming hindrances during the project deployment. To illustrate the point in other view, a process improvement project in a contract manufacturer plant needs customer approval. In addition, products from the pilot run needs verification by the customer. With such aspect not acknowledging as an obstacle, the progress of the project suffered two weeks delay.

The fourth mistake is the erroneous of data collection in "Plan" phase. The mistake occurs when the data collected contradict with the goal statement. Practitioner may not clearly understand the project goal thus lead to collect wrong data. The information from the workers are not sensible to be use as data collected. It is essential to collect the actual and correct data by our own because data collection aids in decision making process, changes or pursuing a particular course or action [17]. In one of the projects, the data presented is not accurate when the total time consumed for the task of loading carton in lorry not include the entire task such as loading and unloading carton from trolley. Wrong data collection able to cause the project deviated from its direction and the goal unable to accomplish.

The fifth mistake is wrong application of tools during implementation in "Do" phase. As mentioned earlier, various LSS tools applicable in implementing the countermeasures such as visual management, experimentation, and four-steps training. The availability of many compatible tools promoted in different literature worsens the situation, especially for novice practitioner [18]. In this case, to acquire a decent knowledge transfer, the practitioner required to choose the suitable LSS tools to run the implementation. According to James and Pavnaskar [19, 20], misunderstanding of the tools uses can causes the misapplication of LSS tools. For instance, one of the countermeasures listed in a project is four steps training for newly joined worker to familiarize them with the process. However, only one to two times training conducted thus, it shows that the practitioner not applied the correct application of four steps training. As a result of unable to apply the LSS tools correctly, the purpose of the improvement will not be achieved.

IOP Conf. Series: Materials Science and Engineering **530** (2019) 012042 doi:10.1088/1757-899X/530/1/012042

Suggestion to avoid high impact mistakes

In conjunction with the high impact mistakes, the author and team has come out with the suggestions to avoid them. The mistakes can be prevented by having a frequent Gemba walk, attending more specific training especially on interrelationship of tools, close mentorship especially for novice practitioner, goal alignment upon completion of the key steps, checklist of good practices or mistake avoidance and progress meeting. The suggestions are explained in details below and may be effective for several mistakes.

Frequent Gemba walk is suggested for practitioner to gain direct exposure to the situation. In LSS principle, Gemba walk is refer to the visit at real place where improvement will be conducted. Practitioner will gained valuable insight into the value stream in the particular process based on their own observation rather than relying on others. It take place on a regularly schedule basis, usually at one week intervals [21]. A structured plan is required before starts Gemba walk in order to collect correct data. The planning should include few aspects such as what, how, when and where to collect the data. According to John [22], the data collection consist of two different purpose and stage which are, collecting the data before and after improvement for comparison purpose. Practitioner also encouraged to documenting the observations made during Gemba walk by recording and jot down the task carried out by workers. If the camera is not allowed in the workplace, sketching on a piece of paper would be helpful. Other than planning and documentation, mix up schedule for Gemba walk also helps the practitioner to get a complete sense of what happens in the particular process with the different workers nature at different time. Eventually, M5 can be avoided with the frequent Gemba walk with a truthful data gathered.

Practitioners are suggested to attend specific training for LSS tools before conducting a PDCA project. Mistakes involving the application of LSS tool (M6 and M9) can be avoided through the provision of suitable and topical training especially on interrelationship of LSS tools. By attending the training, practitioners will have an idea in deciding the appropriate tools to be use in their project. Improper use and wrong application of the LSS tools shows the lack of knowledge among the practitioners. Knowledge and information about LSS tools uses is necessary due too high number of tools available. Langlay [23] said, educational actions is highly emphasize in order to get knowledge about the method so that practitioner are able to perform continuous improvement initiative based on PDCA. This is supported by Yan and Makinde [24] where the management must prioritize training opportunities to all in assuring the level of understanding on LSS tools usage. There are several type of LSS tools training available including online and offline training. However, the training attended should consist the details on lists, uses, and how to apply the LSS tools in a project.

The third suggestion to avoid mistakes in PDCA project is to have a close mentorship especially for novice practitioner. Mentorship in the context of this research is a person who is an expert in LSS will guide the new practitioner throughout their PDCA project. Close mentorship means the practitioner must having a regular face-to-face meeting in a sustained period of time with the mentor. The encouragement and guidance from a mentor in planning, running experiment and making decision for the duration of conducting project helps new practitioner. It is require for them to have a close mentorship moreover further intensive mentoring would be preferable to assist them in project deployment. In addition to the tool options, mentoring also helps to make decision to choose the key obstacles as well as in outlining the data needed in the project. All mistakes listed in high impact mistakes can be avoid by having a close mentorship except for M11. However, the compatibility between mentor and practitioner is vital for a successful mentorship [25].

The fourth suggestion is to make a goal alignment upon completion of the key steps. Goal alignment is the process of aligning the work with the appropriate determination after went through few process. The aforementioned goal is to keep the practitioner to work towards the main goal and guide for collecting data. It is significant to specify the goal statement prior data collecting. This is supported by Grace and John in their research work [26]. In order to align the goal statement, practitioners are suggested to present the obstacles and difficulties faced in the project to the team members. When the people understand what the constraint are, they can helps to define it clearly and suggest on how they align with the goals of their project. By having discussion about this issues, M4 and M5 can be avoided. Without the goal alignment, it is impossible for the practitioner to gather correct data and properly define the obstacle.

Creating a checklist of a good practices or mistake avoidance is the fifth suggestion to avoid mistake in PDCA project especially M11. Good practice checklist is a list to decrease failure by compensating the limitations of human memory and attention. Checklist is necessary in project improvement due to several countermeasures made in the meantime. Providing checklist to the workers able to sustain the improvement made. The main intention is to avoid the negligence in providing the evidence for process standardization. Besides, with the checklist for all improvement conducted, there is no chances for some information to be left behind. A standard will be formally establish in "Check" phase to keep track on the standardization made so that it can help to notify the new workers join the company about the changes. Hence, the practitioners are suggested to provide checklist so that it can be translated into a documented work such as SOP, WI and LSW before the project enclosed.

In avoiding perform high impact mistakes in PDCA project, the progress meeting among the team are necessary. The progress meeting is to update and discuss with the team members on the project development. The effective oversight into an improvement process able to help the goal is well accomplish. With regular progress meeting, team members will be more perceive on the improvement progress and partake for giving ideas in problem solving. All cross-functional members in team able to give ideas for any matter in their area of expertise. The team members should settle into an effective routine after the agendas are produced ahead of time and assignments are conducted between meetings [26]. All mistakes listed in high impact mistakes can be avoided with the aid of progress meeting because of practitioners will become more cautious on the steps of improvement. The progress meeting involving all stakeholders also required to address the obstacles and problems of the project. This method is helpful for the practitioner to explicit the obstacle faced in the improvement project to the top management. Difficulties encountered can be eliminate with the management acknowledgement. Instruction and support from top management are urgently needed especially the difficulties is involving the commitment from the employees, customers and constraint in the financial aspect.

5. Concluding remark

This study shows that "Plan" phase play an important role in PDCA project. Mistakes occurred in in planning the project is more than half of the total mistakes. To avoid all high impact mistakes, a regular progress meeting is advices. However, the lack of knowledge in LSS tools will gave huge impact in an improvement project.

The extensions of this research include the study into the synergy of LSS tools in PDCA project. Each PDCA phase has several suitable LSS tools to be use. By recognizing the LSS tools to each phase, it would be easier for the practitioner to apply them when conducting a project. The PDCA project conducted easily with a deeper explanation on how LSS tools can be used in each phases. The practitioner of YB program acquires their first time experience to conduct project in industries. With the bits of the help of this paper, there is no doubt if the mistakes come from them can be eliminated in the future.

References

[1] Salah S, Rahim A and Carretero J A 2010 The integration of Six Sigma and lean management *International Journal of Lean Six Sigma 1*(3) 249-274

[2] Antony J 2004 Some pros and cons of six sigma: an academic perspective *The TQM magazine 16*(4) 303-306

[3] Salah S, Carretero J A and Rahim A 2010 The integration of quality management and continuous improvement methodologies with management systems *International Journal of Productivity and Quality Management* 6(3) 269-288

[4] De Koning H, Verver J P, van den Heuvel J, Bisgaard S and Does R J 2006 Lean six sigma in healthcare. Journal for Healthcare Quality 28(2) 4-11
[5] Sokovic M, Pavletic D and Pipan K K 2010 Quality improvement methodologies–PDCA cycle, RADAR matrix, DMAIC and DFSS *Journal of achievements in materials and manufacturing engineering* 43(1) 476-483
[6] Lodgaard E, Gamme I and Aasland K E September 2012 Success factors for PDCA as continuous improvement method in product development *In IFIP International Conference on Advances in Production Management Systems* pp 645-652 *Springer* Berlin Heidelberg
[7] Singh J and Singh H 2012 Continuous improvement approach: state of art review and future implications *International Journal of Lean Six Sigma* 3(2) pp 88-111
[8] Chong M Y, Chin J F and Loh W P 2013 Lean incipience spiral model for small and medium enterprises *International Journal of Industrial Engineering* 20
[9] Emiliani B 2008 Attention: We Can Stop Making the Same Mistake Now
[10] Emiliani B, Stec D J, Grasso L and Stodder J. 2007 Better thinking, better results: case study and analysis of an enterprise-wide lean transformation *Center for Lean Business Management*
[11] Moen R and Norman C 2006 Evolution of the PDCA cycle
[12] Imai M 1886 Kaizen: The Key to Japan's Competitive Success *New York: Random House* pp 6
[13] Li J S, Zhang X G, Chu J, Suzuki M and Araki K 2012 Design and development of EMR supporting medical process management *Journal of medical systems 36*(3) 1193-1203
[14] Sidhu B S, Kumar V and Bajaj A 2013 The" 5S" Strategy by Using PDCA Cycle for Continuous Improvement of the Manufacturing Processes in Agriculture Industry *International Journal of Research in Industrial Engineering 2*(3) 10-23
[15] Nakashima H, Fujii H and Suwa M 2014 Designing methodology for innovative service systems *In Serviceology for Services* pp 287-295 *Springer Tokyo*
[16] N Nordin, M B Deros and D A Wahab 2010 "A Survey on Lean Manufacturing Implementation in Malaysian Automotive Industry" *International Journal of Innovation, Management and Technology* Vol 1 pp 374 -381
[17] Wazed M A and Shamsuddin A 2009 Theory Driven Real Time Empirical Investigation on Joint Implementation of PDCA and 5S for Performance Improvement in Plastic Moulding Industry *Australian Journal of Basic and Applied Sciences 3*(4) 3825-3835
[18] Pavletić D, Soković M and Paliska G January 2008 Practical application of quality tools *In 2nd International Quality Conference* Kragujevac
[19] James T 2006 "Wholeness as well as leanness [lean production]" *Manufacturing Engineer* Vol 85 No 5 pp 14-17
[20] Pavnaskar S J, Gershenson J K and Jambekar A B 2003 "Classification scheme for lean manufacturing tools" *International Journal of Production Research* Vol 41 No 1 pp 3075-3090
[21] Mann D 2014 Creating a lean culture: tools to sustain lean conversions *Productivity Press*
[22] Ovretveit J 1999 A team quality improvement sequence for complex problems *BMJ Quality & Safety* 8(4) 239-246
[23] Langley G J 2009 The improvement Guide. A practical Approach to Enhancing Organizational Performance *Jossey-Bass San Francisco*
[24] Yan B and Makinde O D 2011 Impact of Continuous Improvement on New Product Development within SMEs in the Western Cape, South Africa *African Journal of Business Management* 5 2200-2229
[25] Keyser D J, Lakoski J M, Lara-Cinisomo S, Schultz D J, Williams V L and Zellers D F 2008 Advancing institutional efforts to support research mentorship: a conceptual framework and self-assessment tool *Acad. Med.* 83(3) 217-25
[26] Gorenflo G and Moran J W 2010 The ABCs of PDCA *Public Health Foundation* http://www. phf. org/resourcestools/Pages/The_ABCs_of_ PDCA. aspx.

IOP Conf. Series: Materials Science and Engineering **530** (2019) 012043 doi:10.1088/1757-899X/530/1/012043

Heuristic Based Scheduling for Toothpaste Filling Problem

S Promjun* **and P Jarumaneeroj**

Industrial Engineering Department, Chulalongkorn University, Bangkok,10330, Thailand

Corresponding author *: Suwatthiphong.P@student.chula.ac.th

Abstract. This research focuses on the development of a heuristic method for the Toothpaste Filling Problem (TFP) whose complication lies on a vast variety of products (formulae and sizes), together with specific restrictions on filling machines and storage tanks. More specifically, filling machines are non-identical, each of which has different speeds and capacities for filling toothpaste of various sizes. As unloading requires storage tanks, each with much larger capacity when compared to one particular production order, a tank will be then matched with several production orders and be repositioned among filling machines to avoid unnecessary changeovers. We solve the TFP by means of a two-phase method, where the initial solution is constructed by an Earliest Due Date (EDD) dispatching rule. Once completed, the initial solution is then iteratively improved by a series of improvement operators, mimicking the concept of Variable Neighbourhood Search (VNS), until no improvement could be found. We test our heuristic on a set of sample instances acquired from one of the biggest consumer products manufacturers in Thailand. Impressively, when compared to the current practice, our proposed heuristic could help reduce makespan by 32% on average, or equivalently an annual saving of $0.6 million, mainly from the reduction on overtime payment and changeovers.

1. Introduction

Currently, the competition among consumer products businesses is quite intense, in terms of price, quality, and marketing strategy. As such, the companies need to introduce new products much more frequent in order to satisfy the needs of diversified customer segments. While these wide ranges of products may help strengthen the companies' business, they however pose a serious problem on the manufacturing side as more products imply more complicated production planning – and so higher production cost from additional changeovers.

Typically, a changeover is required when we change the production sequence from one to another order. And, the times required by changeovers may differ as they depend on the similarity – such as formulae and container sizes – between two consecutive orders. Production could have become worse with inefficient planning, as it may lead to plans with longer makespans, and so delivery delay. In order to avoid such circumstances, many need to rely on ad hoc schedules of additional work shifts, which are mere short-term solutions to this problem.

To help solve this problem in longer term, many dispatching rules are then devised [1]. But, these rules, unfortunately, work well under certain assumptions – and they might be even inapplicable due to the existence of some practical restrictions like compatibility and storage tank constraints found in the underlying Toothpaste Filling Problem (TFP) investigated in this paper.

Toothpaste is produced from a continuous line of two processes, that is, mixing and packing as shown in Figure 1. All ingredients will be first mixed and loaded into storage tanks of various sizes in mixing department. Once completed, these storage tanks will be, in turn, unloaded by filling machines, where the resulting products will be subsequently packed and stored in the company's warehouse at the end of production line.

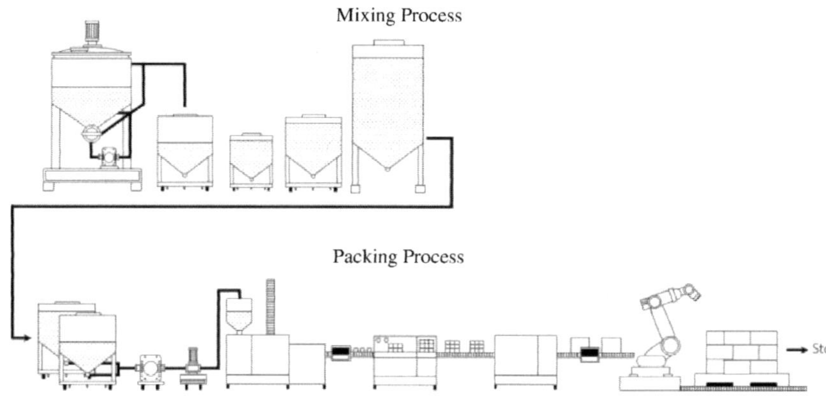

Figure 1. Toothpaste manufacturing process.

Based on our case study company's production profile, toothpaste filling process is quite rich in the sense that there are about 30 different toothpaste formulae of various sizes to be produced each month – accounting for an approximately 80 SKUs in total. Since (i) a changeover may involve a change of toothpaste formulae, or a change of container sizes, or both, depending on the sequence of orders assigned to each machine, and (ii) filling machines are of different speeds, each with different production capacities, deciding which order to be placed in which machine is clearly not an easy task – as the finishing time of an order may vary and inevitably affects the finishing times of the rest. Moreover, as storage tank's capacity is considerably large when compared to the amount required by one particular order, the tank will be then matched with several orders; but, only be unloaded by one machine at a time. Consequently, the tanks will be typically repositioned among filling machines to avoid unnecessary changeovers. This problem is made more complicated when there are emergency orders and ingredient shortages that prevent some orders to be fulfilled.

Due to the aforesaid complication, an efficiently thorough production plan is hard to be made resulting in high amount of overtime payment and number of overdue jobs. Based on previous statistics, we found that approximately 23 orders were overdue monthly; and, the company had to pay around $0.43 million annually for additional work shifts.

In order to help reduce these figures in longer term, a two-phase heuristic is herein proposed, where we construct an initial solution based on an easy-to-implement rule; and, once completed, such a solution is then iteratively improved by a series of improvement operators, namely Swap (SWAP) and Moving Exchange (MOVEX), until no improvement could be found. The implementation of our proposed heuristic is quite similar to that of the Variable Neighborhood Search (VNS); but, with no solution shaking as we repeatedly execute such a heuristic every time production information has been updated.

2. Literature Review

Scheduling is a common process found in most manufacturing; but, its detail may vary depending on characteristics of the underlying production processes. For example, the machines may be either identical or non-identical, each of which may be compatible with all or a subset of products. The jobs may also vary and require different operational setups as production sequences have been altered. In the literature, these problems are known as sequence-dependent scheduling problems with parallel machines [1] [2-3], which are proven to be NP-hard problems [4-5]. Thence, solving these instances of practical sizes to optimality is seemingly implausible.

In order to address this issue, many efficient heuristics are then devised and applied – though the finding of optimal solution is not guaranteed. Reference [6] transformed a variant of sequence-dependent scheduling problems with single machine, called the Float Glass Manufacturing Problem (FGMP), into the Travel Salesman Problem (TSP), and solved the resulting problem by means of the Variable Arc Exchange Heuristic [7]. As an extension to the FGMP, where multiple machines were brought into consideration, [8] investigated and transformed a so-called Tablets Film Coating Problem (TFCP) into the Vehicle Routing Problem with Time Window (VRPTW). They found that small TFCP instances, with up to 25 production orders, could be solved to optimality; but, unfortunately, not for practical TFCP instances comprising of at least 50 production orders. Accordingly, a two-phase heuristic was then devised. In their setting, the initial solutions were first constructed by easy-to-implement dispatching heuristics; and, once completed, such solutions were then iteratively improved by a series of local search operators, mimicking the concept of the Variable Neighborhood Search (VNS), until no improvement could be found.

Similar to [8], [9] solved practical crew transportation problems arisen in most oil and gas exploration companies by means of the VNS; but with different local search operators. More formally, the authors constructed an initial solution based on Solomon's Push Forward Insertion Heuristic (PFIH), and iteratively improved such a solution by λ-interchange and cyclic transfer algorithms until one of the stopping criteria had been met. The VNS was also employed in the study of [10] who investigated the integrated production-inventory problem, known as the Production Routing Problem.

3. Problem Description

The Toothpaste Filling Problem (TFP) is a variant of sequence-dependent scheduling problems with non-identical parallel machines, as it requires different setup times for assignments of the same production sequence on different filling machines. And, in this case, two types of setups are involved, one for the changes of container sizes and another for those of toothpaste formulae. As changing container size requires more operational steps when compared to those of the formulae, the setup time is thus higher – as reported in Table 1.

Table 1. Setup times for changes of container sizes and toothpaste formulae on different filling machines.

Machine Names	Setup Times (Min.)		
	Container Sizes	Formulae	Both Container Sizes and Formulae
A	40	15	55
B	60	15	75
C	60	15	75
D	90	60	150
E	45	10	55

While planners may be able to reduce setup times by minimizing total number of setups, this strategy, however, may lead to plans with longer makespans – and delivery delay – as machines need to wait for other orders with the same formula and container size. In order to avoid such a circumstance, storage tanks are instead repositioned among these machines. However, as storage tanks could only be unloaded by one machine at a time – but, each may be used for filling toothpaste of various sizes (orders) – production planning is thus complicated and made more difficult for planners to find the optimal filling sequence on a daily basis. This situation might have become worse when there are emergency orders and other disruptions, including ingredient shortages, preventing orders to be fulfilled.

More importantly, as we may transform the TFP into the VRPTW, where a tour is defined by a sequence of orders assigned to each filling machine, solving practical TFP instances to optimality in a timely manner is less likely.

4. Methodology

Our proposed heuristic is a two-phase one, where the initial solution is constructed based on a simple and easy-to-implement dispatching rule, i.e. EDD; and, once done, it will be further improved by a series of improvement operators, including Swap (SWAP) and Moving Exchange (MOVEX), mimicking the concept of the Variable Neighborhood Search (VNS), until no improvement could be found. While the structure of such a heuristic is quite similar to those of [9] and [6], its implementation is far more complicated due to a large variety of toothpaste formulae and container sizes, together with storage tank and compatibility constraints.

More formally, while storage tanks can only be unloaded by one machine at a time, each can be used for filling toothpaste of various sizes. Additionally, not all products could be filled by all machines due to the existence of compatibility between container sizes and filling machines. Consequently, every time an order is assigned, we need to check whether such an action is possible. If not, the next move in queue will be called; and repeat until all orders are completely assigned.

4.1. Preprocessing Steps

At the beginning of planning process, related input data will be acquired from the database, including Semi-Finished Product Codes (S/FC), Finished Product Codes (F/C), semi-finished product quantities, finished product quantities, delivery schedules, number of available storage tanks, and the amount of packing material. We also need to collect the times required by all changeovers (Table 1), as well as speeds and capacities of all filling machines as reported in Table 2 below.

Table 2. Speeds and product compatibility on different filling machines.

Machine Names	Speed (Pcs./Min.)	Container Sizes)Gram(													
A	80	-	-	-	-	-	80	90	100	-	140	150	160	180	-
B	80	-	-	25	40	50	-	90	100	130	140	150	160	180	200
C	160	15	20	25	40	-	80	-	-	-	-	-	-	-	-
D	70	15	-	25	40	-	80	90	-	-	-	-	160	-	-
E	160	-	-	-	-	-	80	90	-	-	140	150	160	-	200

4.2. Initial Solution Construction

Step 1 Check the current list of orders whether there are backorders from the previous planning cycles, maybe from ingredient shortages. If ones exist, they will be first scheduled based on the same rules as incoming production orders.

Step 2 Schedule the incoming orders based on their due dates, i.e. place those with the earliest due dates in the production queue first (EDD). In this step, we need to check whether the amount of semi-finished products is sufficient for all production orders. If they are not enough, those associated orders will be removed from the production queue. Similarly, as completed orders will be subsequently packed downstream, orders with insufficient packing material will be also removed from the production queue until the next planning cycle begins.

IOP Conf. Series: Materials Science and Engineering **530** (2019) 012043 doi:10.1088/1757-899X/530/1/012043

Step 3 Check filling machines' status, and so the ready times of all machines. If the machine is currently idle, the ready time will be set to the current planning cycle's starting time; else, it will be set to the current order's finishing time.

Step 4 Allocate orders in the production queue to each machine based on similarity between the current order and the next one in queue. Since changing container sizes requires more setup times, an order with the same container size and formula will be first allocated, followed by an order with the same container size but different formula, an order with different container size but same formula, and an order with different container size and formula, respectively.

Step 5 Consequently allocate orders to machines based on the above rules until the production queue is empty.

Step 6 Calculate makespan, lateness, and number of tardy jobs from such an allocation.

4.3. Improving the Solutions with Improvement Operators

Once an initial solution is constructed, it will be iteratively improved by two improvement operators, i.e. SWAP and MOVEX, until no improvement could be found, as described by the following pseudo code.

```
Pseudo Code of the Proposed Heuristic
1: Input data: S/PC, F/C, Quantities of Semi-finished products and finished products, Delivery schedules,
               Number of available storage tanks, Amount of packing material, Production orders.
2: Construct an initial solution using EDD and a set of predefined rules
3. z = true
4: while z = true
5: Call SWAP
6: Select a pair of orders from different machines
7: Check compatibility between orders and filling machines
8:        if compatibility condition is violated, discard such a pair and select the new one (Back to Line 6)
9:        else
10:       Check the availability of storage tanks and other components
11:            if an order is not ready at the time of swapping, discard such a pair and select the new one
               (Back to Line 6)
12:            else
13:            Check the availability of storage tanks for the whole production sequences
14:                    if swapping is possible for the whole production sequences on both machines,
                       swap the positions of such a pair and compute makespan, lateness, and number
                       of tardy jobs.
15:                    else
16:                    Discard such a pair and select the new one (Back to Line 6)
17:                    end if
18:            end if
19:       end if
20: if no better solution is found within r iterations
21: Call MOVEX (the implementation of MOVEX is quite similar to that of SWAP)
22:       if no better solution is found within r iterations, z = false
23:       else
24:       Return a better tour as new solution and back to Line 6
25:       end if
26: end if
27: end while
28: return Best-found solution
```

Figure 2. Pseudo code of the proposed heuristic.

IOP Conf. Series: Materials Science and Engineering **530** (2019) 012043 doi:10.1088/1757-899X/530/1/012043

4.3.1. SWAP Sub-routine. SWAP is an easy, but powerful, improvement operator, where we swap a pair of orders between two different machines and check whether a better solution is found. While this could be easily done in typical VRP-like problems, care should be however taken in this problem as not all pairs of orders could be swapped due to compatibility and storage tank constraints.

Step 1 Select a pair of orders from different machines to be swapped.

Step 2 Check whether these orders are compatible with the selected filling machines. If not, discard such a pair and proceed to the next one.

Step 3 Check the availability of both storage tanks and the amount of semi-finished products. If at least one of them is not available at the times of swapping, discard such a pair and proceed to the next one.

Step 4 Check whether these orders are currently unloaded by other machines at the times of swapping. If yes, discard such a pair and proceed to the next one.

Step 5 Swap the positions of the selected order pair and calculate the resulting makespan, lateness, and number of tardy jobs.

Step 6 If the new solution is better than the original solution, store this new solution as the current one and repeat from Step 1 until no improvement could be found.

4.3.2. MOVEX Sub-routine. If SWAP could not find a better solution, Moving Exchange (MOVEX) will be subsequently called. The concept of MOVEX is quite simple as we try to reduce makespan by moving an order from a machine with the longest processing time, i.e. the one that defines makespan, to others with less processing times. As usual, we need to check whether such an order could be moved; and if a better solution is found, SWAP will be called and repeat until MOVEX terminates.

5. Heuristic Results

In order to evaluate the performance of our proposed heuristic, we have computed and compared the makespans from such a heuristic with those of current planning and EDD-based planning over a 30-day period as reported in Table 3. All information was acquired from one of the biggest consumer products manufacturers in Thailand; and the experiments were conducted on a computer with Intel® Core™ i5-6200U CPU 2.30 GHz. Ram 8.00 GB. using Python as a programming language. Impressively, our proposed heuristic could reduce production makespan and lateness by 32% and 53% on average; and, it takes only few seconds to perform the whole computation.

6. Conclusion and Future Works

This paper focuses on the development of a heuristic for toothpaste filling process so that makespan, lateness, and number of tardy jobs, could be reduced. The devised heuristic is a two-phase one, where the initial solution is first constructed based on EDD and a set of allocation rules; and, once completed, it will be iteratively improved by a series of improvement operators, namely SWAP and MOVEX, until no improvement could be found. Based on our experiments, the proposed heuristic could potentially reduce production makespan by 32% on average, or equivalently a saving of $0.6 million per year — mainly from the reduction of overtime payment and changeovers. Additionally, lateness could be reduced by 53%.

When compared to the current planning, or other simple dispatching rules, our proposed heuristic is much more efficient and it takes just few seconds to complete the whole computation. Accordingly, we can execute this heuristic every time the system has been updated, either by the arrival of emergency order or other disruptions. This heuristic could also be adapted to the planning of other consumer products having the same production process, like shampoo or liquid soap — as these products are typically produced by the same manufacture; but, maybe, at different sites. As such, substantial saving could be realized.

Table 3. A comparison of makespans from different methodologies.

Sample No.	Makespan (Min.)				% Improvement
	Current Plan	EDD	Heuristic	Time (Sec.)	
1	2,904	2,447	1,979	6.90	32%
2	2,891	2,081	1,591	1.71	45%
3	1,858	1,261	999	2.80	46%
4	3,182	3,316	2,128	4.85	33%
5	2,780	2,044	1,634	6.52	41%
6	1,789	2,010	1,238	5.19	31%
7	1,976	1,854	1,832	2.73	7%
8	1,594	2,431	1,570	4.81	2%
9	1,962	1,826	1,365	3.26	30%
10	1,645	1,867	1,482	3.40	10%
11	1,771	2,981	1,319	4.21	26%
12	2,030	2,497	1,427	4.90	30%
13	1,915	2,905	1,279	3.60	33%
14	2,126	1,814	1,467	5.78	31%
15	2,206	2,935	2,030	7.26	8%
16	2,465	3,213	1,349	4.95	45%
17	2,760	1,329	1,255	4.30	55%
18	1,625	1,821	1,297	6.54	20%
19	3,082	1,868	1,381	8.79	55%
20	2,115	1,263	1,080	3.92	49%
21	2,565	2,541	1,873	5.20	27%
22	1,833	1,576	1,315	3.10	28%
23	1,665	2,071	1,123	7.35	33%
24	1,705	1,733	865	3.10	49%
25	4,337	3,466	2,371	4.54	45%
26	1,680	1,502	1,261	4.17	25%
27	1,772	2,357	1,504	4.13	15%
28	1,730	1,227	1,050	2.82	39%
29	1,165	2,036	1,042	3.92	11%
30	2,470	2,059	1,319	3.84	47%

References

[1] Pinedo M 2008 Scheduling: Theory, Algorithms, and Systems. *Springer Publishing Company* NY

[2] Bilge U, Furkan K, Kurtulan M, Pekgun,P 2004 A Tabu Search Algorithm for Parallel Machine Total Tardiness Problem *Computers & Operations Research* pp 397–414

[3] Zadeh B S, Moghaddam R T, Moghadam A T, Rastgar I 2017 Solving a bi-objective unrelated parallel batch processing machines scheduling problem A comparison study *Computers and Operations Research. Conference on Industrial Engineering and Applications* pp 71–90

[4] Karabulut K and Tasgetiren F 2014 A variable iterated greedy algorithm for the traveling salesman problem with time windows *Information Sciences* pp 279,383-395

IOP Conf. Series: Materials Science and Engineering **530** (2019) 012043 doi:10.1088/1757-899X/530/1/012043

[5] Zhang D, Cai S, Ye F, Si Y W, Nguyen T T 2017 A hybrid algorithm for a vehicle routing problem with realistic constraints *Science Direct Information Sciences* pp 167–182
[6] Kitisook K and Jarumaneeroj P 2017 A Variable Arc Exchange Heuristic for Float Glass Manufacturing Problems *4th International Conference on Industrial Engineering and Applications*
[7] Helsgaun K 2000 An Effective Implementation of The Lin-Kernighan Traveling Salesman Heuristic *European Journal of Operational Research* pp 106–130
[8] Pontrakul C and Jarumaneeroj P 2018 Heuristic Based Scheduling for Tablets Film Coating Process *ICACS '18 Proceedings of the 2018 2nd International Conference on Algorithms, Computing and Systems* pp 238–242
[9] Jarumaneeroj P and Kunaporn S 2017 A Variable Neighborhood Search Approach for Crew Transportation Problems *4th International Conference on Industrial Engineering and Applications*
[10] Qiu Y, Wang L, Xu X, Fang X, Pardalos P M 2018 A variable neighborhood search Heuristic algorithm for production routing problems *Applied Soft Computing* pp 311–318

IOP Conf. Series: Materials Science and Engineering **530** (2019) 012044 doi:10.1088/1757-899X/530/1/012044

Performance Evaluation of Continuous and Discrete Particle Swarm Optimization in Job-Shop Scheduling Problems

N I Anuar[1*], M H F M Fauadi[2] and A Saptari[3]

[1] Faculty of Engineering & Technology, Multimedia University, Jalan Ayer Keroh Lama, 75450 Ayer Keroh, Melaka, Malaysia

[2] Faculty of Manufacturing Engineering, Universiti Teknikal Malaysia Melaka, Hang Tuah Jaya, 76100 Durian Tunggal, Melaka, Malaysia

[3] Department of Industrial Engineering, President University, Jl. Ki Hajar Dewantara, Kota Jababeka, Cikarang Baru, Bekasi 17550 - Indonesia

Corresponding author *: nurulizah.anuar@mmu.edu.my

Abstract. The Particle Swarm Optimization (PSO) is an optimization method that was modeled based on the social behavior of organisms, such as bird flocks or swarms of bees. It was initially applied for cases defined over continuous spaces, but it can also be modified to solve problems in discrete spaces. Such problems include scheduling problems, where the Job-shop Scheduling Problem (JSP) is among the hardest combinatorial optimization problems. Although the JSP is a discrete problem, the continuous version of PSO has been able to handle the problem through a suitable mapping. Subsequently, its modified model, namely the discrete PSO, has also been proposed to solve it. In this paper, the performance of continuous and discrete PSO in solving JSP are evaluated and compared. The benchmark tests used are FT06 and FT10 problems available in the OR-library, where the goal is to minimize the maximum completion time of all jobs, i.e. the makespan. The experimental results show that the discrete PSO outperforms the continuous PSO for both benchmark problems.

1. Introduction

The Job-shop Scheduling Problem (JSP) is described as consisting of a set of jobs and a set of machines, where each job consists of a set of operations. Each operation of a job needs to be processed on a certain machine following a specified order, where this machine order is unique for each job. Each operation also has a processing time, specific on each machine. The order of machines that a job must visit or the specific order of operations is the precedence constraints for that job. These precedence constraints impose some difficulty to the JSP [1].

The Particle Swarm Optimization (PSO) is a population-based heuristics that is used to optimize the solution of many engineering problems. It is one of the evolutionary computation search algorithms, developed by Kennedy and Eberhart in 1995, which refers to a family of algorithms used to find optimal or near optimal solutions to numerical or qualitative problems. The original PSO algorithm was discovered through simulations of social behaviour; it is related to the bird flocking, fish schooling and swarm theory. PSO simulates this behaviour, learns from the scenario and uses it to solve the optimization problems [2].

PSO was first introduced to solve continuous optimization cases, and it has been extended to solve problems in discrete spaces as well. This is done by implementing a particle representation procedure that can encode the particle's position in the continuous PSO with the solution of discrete problems [3].

For scheduling problems like JSP, the particle representation procedure specifies the mapping between the position of the particle in PSO and the scheduling solution in JSP. Each particle in PSO represents its position in the continuous search space, but not the actual potential solution of the discrete problems. Thus, the mapping is established so that each particle in PSO can represent a schedule in JSP.

On the other hand, in the discrete PSO, each particle directly represents the candidate solution. For scheduling problems like JSP, this means each particle represent the sequence of operations, instead of its position in the search space. Thus, it eliminates the need to encode the particles to the potential scheduling solutions as compared to the continuous PSO.

In this paper, the continuous and discrete PSO are employed separately to find the values of the minimum makespan for FT06 and FT10 instances. Subsequently, the performances of these two versions of PSO are evaluated and compared to distinguish which version produces a better performance. The rest of this paper is organized in the following manner: first, the standard PSO is described along with a brief review of the discrete PSO. This is followed by a description of the continuous and discrete PSO implemented in this research. Lastly, the experimental results and conclusions are presented.

2. Particle Swarm Optimization

In a PSO algorithm, each agent is called a 'particle' and each particle corresponds to a position in the continuous, multi-dimensional search space, where it flies around with a velocity. It can be described as having an initial of randomized swarm of particles (a population of agents) and then searches for optima by updating each generation. The candidate solutions travel through the problem space by following the current optimum particles. All of the particles have fitness values which are evaluated by the fitness function to be optimized. All of these particles also have velocities which direct the flying of the particles [2].

This velocity is constantly updated by the particle's own experience and the experience of the particle's neighbours or the experience of the whole swarm. Thus, in every iteration, each particle is updated by following two 'best' values. The first 'best' value is the best solution (fitness) each particle has achieved so far. This value is called pBest. Another 'best' value is the best value obtained so far by any neighbour of the particle in the population. This value is called nBest. When a particle takes all the population as its neighbours, the best location is a global best. This value is called gBest.

After finding the two best values, the particle updates its velocity and position using the following equations:

$$v_{id} = w^* v_{id} + c_1 r_1 (p_{id} - x_{id}) + c_2 r_2 (p_{gd} - x_{id}) \tag{1}$$

$$x_{id} = x_{id} + v_{id} \tag{2}$$

for $i = 1, 2, 3, \ldots, N_s$ and $d = 1, 2, 3, \ldots, D$
N_s = the total number of particles in the swarm
D = the problem dimension, i.e. the number of parameters of the function being optimized
w = inertia weight, to control the impact of the previous history of velocity on the current velocity
c_1 and c_2 = positive constants, called acceleration coefficients or learning factors
r_1 and r_2 = independently generated random numbers in the range [0, 1]
$x_{id} = [x_{i1}, x_{i2}, \ldots, x_{iD}]$, represents the position of the ith particle
$v_{id} = [v_{i1}, v_{i2}, \ldots, v_{iD}]$, represents the velocity of the ith particle
$p_{id} = [p_{i1}, p_{i2}, \ldots, p_{iD}]$, represents the best position of the ith particle
g represents the index of the best particle among all particles in the population
$p_{gd} = [p_{g1}, p_{g2}, \ldots, p_{gD}]$ is the best position found among all particles in the population

The two equations above are the equations describing the flying trajectory of a population of particles. Equation (1) is applied to compute the new velocity of each particle based on its previous velocity and the distances of its current position from its own best experience and the neighbourhood's best experience. The particle will then fly towards a new position as per equation (2). Each particle's performance is assessed corresponding to a pre-defined fitness function, which is associated to the problem being solved.

PSO has been presented as an optimization method in the continuous domain. However, many optimization cases are established in the discrete domain. Typical examples in the discrete domain include problems that involve sequencing or permutation, such as in JSP. Consequently, the continuous PSO has been modified to operate on discrete problems; there are already several variants of discrete PSO developed to achieve this purpose. Based on the literature, discrete PSO techniques can be categorized into either binary numbers or positive integers [4].

In 1997, Kennedy and Eberhart extends their work of the continuous PSO to a discrete binary version of PSO for solving De Jong's suite of test functions [5]. In this binary version, the velocity equation remains the same as equation (1) except for p_{id} and x_{id} are integers in {0, 1}. v_{id} is constrained to the interval [0.0, 1.0] using the sigmoid function and it then represents the probability of bit x_{id} taking the value of 1. Thus, the position update for each particle does not make use of equation (2). Instead, a random number is generated within the interval of [0.0, 1.0] and if its value is less than the value of the transformed v_{id} (using the sigmoid function), x_{id} is updated to bit 1. Otherwise, it is updated to bit 0.

For the second category of discrete PSO approaches, Hu, Eberhart and Shi in 2003 proposed a PSO for permutation optimization in solving n-queens problems [6]. Each particle represents the permutation of numbers of D dimensions, i.e. it is made up of positive integers. The velocity equation also remains the same as equation (1), like in the binary version, but it now represents the possibility that the particle changes, i.e. a larger velocity will cause the particle to more likely change to a new permutation sequence. The velocity is normalized to the range of 0 to 1 by dividing it by the maximum range of the particle. According to Wang, Lin and Yang [7], a random number is set between 0 and 1. If the normalized velocity is less than or equal to the random number, a position will be swapped, i.e. the position will be set to the value of the same position in the global best position by swapping the operations. Otherwise, the position is maintained. Thus, the position update for each particle also does not make use of equation (2), similar to the binary version of PSO discussed previously.

Zhang, Li, Li and Huang proposed in 2005 a method to solve the resource-constrained project scheduling problem, where they considered two forms of solution representations, i.e. priority-based representation and permutation-based representation [8]. The first form operates on the continuous domain, where each particle represent the priority values of activities and the values can be ranged from between 0 and 1. Each particle is also randomly initialized with the velocity of continuous values. The update of velocity and position are similar to equation (1) and (2) as well. On the other hand, the second form of solution representations operates on the discrete domain, where each particle represents the permutation or sequence of activities and its value must be in the form of positive integers. Each particle is also randomly initialized with the velocity of continuous values, but later rounded up to integers and limited to the number of dimensions or activities in a project. The velocity equation also remains the same as equation (1), but it now corresponds to the distance or gap between the current permutation or sequence and its own local or global best permutation found so far. It represents the probability that the permutation will be changed, i.e. a larger gap will cause the particle to more likely change to a new permutation sequence.

Lian, Jiao and Gu in 2006 proposed a similar particle swarm optimization algorithm (SPSOA) to solve JSP [9]. In their research, each particle represents a sequence of working procedures of jobs. Then, a coding scheme is used to make every particle to have a valid scheduling. Instead of using equation (1) and (2), various new mutation and crossover operators are introduced in order to perform the velocity update and the position update afterwards.

Zhang, Wang and Ji provided a comprehensive survey on PSO and its applications which also includes the field of binary optimization and discrete optimization in [10]. In this research, the second category of discrete PSO which consists of positive integers will be used, instead of binary numbers, since it is more suitable for the permutation optimization problems like JSP.

3. Continuous and Discrete PSO

For the two versions of PSO, the main differences lie in the initialization and the update of both position and velocity of the particles. The equation for velocity is the same for both versions of PSO. However, in discrete PSO, the velocity does not represent the change in position; it represents the probability that the positions are swapped to produce a new permutation sequence.

For the continuous PSO, the position and velocity are initialized in a continuous search space. Then, the velocity and position are updated according to equation (1) and (2), respectively. For the discrete PSO, the position is initialized as a sequence of operations of jobs, while the velocity is initialized in a continuous search space. Then, the velocity is updated according to equation (1), while the position is updated by swapping the operations. The details of the position update are as follows:

First, the velocity is normalized to the range of 0 to 1. In this paper, the sigmoid function is used to compress the velocity value to be between 0.0 and 1.0. Second, a random number between 0 and 1 is generated such that, if the normalized velocity is greater than or equal to the random number, the position will be set to the value of the same position in the global best position by swapping the operations.

For instance, given a JSP of 3 jobs being processed on 2 machines, the total operations will amount to 6, which will be the number of dimensions of the particle. For the continuous PSO, the position and velocity are initialized randomly in continuous values from x_{min} to x_{max} and $-v_{min}$ to v_{max}, respectively as shown in table 1:

Table 1. Example of position and velocity initialization for continuous PSO.

Dimension	Initial Position	Initial Velocity
1	3.4299	−0.2464
2	0.3782	−2.0930
3	5.8050	3.7116
4	4.6567	1.7110
5	2.4170	4.8089
6	1.6693	−5.1933

For the discrete PSO, the position is initialized randomly in discrete values as a sequence of operations of jobs from 1 to the number of operations. The velocity is initialized randomly in continuous values from $-v_{min}$ to v_{max} as shown in table 2:

Table 2. Example of position and velocity initialization for discrete PSO.

Dimension	Initial Position	Initial Velocity
1	4	0.9142
2	3	−4.9549
3	1	−3.2301
4	5	5.9768
5	6	−1.4690
6	2	2.6561

For the continuous PSO, the velocity and position are updated according to equation (1) and (2), respectively. For the discrete PSO, after the velocity has been updated according to equation (1), it will be normalized by using the sigmoid function to compress the velocity value to be between 0.0 and 1.0, as shown in figure 1. A sigmoid function is given by the following equation:

$$s(v_{id}) = \frac{1}{1 + \exp(-v_{id})} \tag{3}$$

Dimension	1	2	3	4	5	6
Current Velocity	−4.4162	−1.6928	5.7734	−0.3526	3.1054	−2.6846

Normalized Velocity	0.0119	0.1554	0.9969	0.4128	0.9571	0.0639

Figure 1. Example of velocity normalization for discrete PSO.

For the position, in this paper, it will not be updated according to equation (2) since the meaning behind the update in position now refers to the swapping of the operations in each particle. This is done by comparing the position with the global best position and swapping the operation to match the operation in the global best position.

A random number is generated between 0 and 1 and each element in the normalized velocity is compared against the random number. If the normalized velocity is greater than the random number, that element in the current position will be swapped with another element so that it matches the element in the global best position. For example, in figure 2, assume that the normalized velocity is greater than the random number in Dimension 4, thus the value 5 in the current position will be swapped. Since the global best position in Dimension 4 contains the value 6, thus the value 5 will be swapped with the value 6 in Dimension 5.

Dimension	1	2	3	4	5	6
Global best position	5	1	2	6	3	4
Current Position	4	3	1	5	6	2

Global best position	5	1	2	6	3	4
Next position	4	3	1	6	5	2

Figure 2. Example of position update for discrete PSO.

IOP Conf. Series: Materials Science and Engineering **530** (2019) 012044 doi:10.1088/1757-899X/530/1/012044

4. Experimental Results

In the experiments, the continuous and discrete PSO for JSP is programmed using MATLAB on an Intel Core notebook with 4GB of RAM running Windows 10. The performance of both versions of PSO are evaluated using two benchmark problems which are accessible from the OR-Library website [11]. The first problem, called FT06, consists of 6 machines and 6 jobs with a total of 36 operations. While the second problem, called FT10, consists of 10 machines and 10 jobs with a total of 100 operations. During the experimental process, the parameters used are given as follows: The size of population is set to 45 and 50 for FT06 and FT10 problems, respectively. The inertia weight is linearly decreased from 0.9 at the start of the run and 0.4 at the end of the run. Both acceleration constants are set to 2.0. The maximum of iterative generations is set to 3000 for FT06 and 6000 for FT10, where each instance is performed randomly 20 times.

The computational results for continuous and discrete PSO are presented in table 3 to be compared. The quality of solution is evaluated in terms of the minimum makespan, the percentage of deviation from the optimal solution and the computational time in seconds. In table 3, best solution indicates the best minimum makespan obtained and mean deviation indicates the average percent relative increase in minimum makespan with respect to the optimal solution after 20 runs. The mean computational time denotes average computational time in seconds which is obtained after 20 runs.

The optimal solution to be compared for FT06 problem is 55, which is obtained from the paper by Klemmt, Horn, Weigert, and Wolter [12] who used exact methods to solve the problem. The best solution attained by both versions is 55 as well; therefore, both versions are able to achieve the optimal solution. However, in terms of mean deviation, the discrete PSO performs better with a mean deviation of 0%; in other words, the best solution is equal to the optimal solution in all 20 runs. For the mean computational time, the discrete PSO performs faster with 20 seconds.

The optimal solution to be compared for FT10 problem is 930, which is obtained from the paper by Carlier and Pinson [13] who used exact methods to solve the problem. The best solution attained by the discrete PSO is 965 which outperforms the continuous PSO; hence, the discrete PSO manages to attain a near-optimal solution with 3.76% deviation from the optimal solution. In terms of mean deviation, the discrete PSO performs better with a mean deviation of 9.62%. For the mean computational time, the discrete PSO performs faster with 536 seconds.

Table 3. Computational results for continuous and discrete PSO in solving FT06 and FT10 problems.

Problem instance	Optimal solution	Continuous PSO			Discrete PSO		
		Best solution	Mean deviation (%)	Mean computational time (sec)	Best solution	Mean deviation (%)	Mean computational time (sec)
FT06	55	55	1.36	57	55	0	20
FT10	930	982	12.47	724	965	9.62	536

The significant differences of these experimental results could be due to the fact that the continuous PSO needs to implement the particle representation procedure to establish the mapping between the position of the particle in the continuous search space of PSO and the scheduling solution in the discrete space of JSP. Thus, this extra step leads to additional processing effort and computational time. On another hand, for the discrete PSO, the particle's position directly represents its scheduling solution in JSP. This approach is more straightforward since it eliminates the need to encode each particle to a schedule. Therefore, in contrast to the continuous PSO, it spends less effort and time to find the best solution.

5. Conclusion

In this paper, two versions of PSO in the case of solving JSP are studied. In the continuous PSO, the initialization and the update of both position and velocity of the particles are carried out using the standard PSO algorithm. For the discrete PSO proposed in this paper, the position is initialized as a sequence of operations of jobs, while the velocity is initialized in a continuous search space. Then, the velocity is updated according to the velocity equation of the standard PSO algorithm, while the position is updated by swapping the operations.

The performance of continuous and discrete PSO are tested on FT06 and FT10 instances, which is taken from the OR-library. The performance measure considered in this study is the makespan minimization with the use of MATLAB software. Based on the experimental results, it is found that the discrete PSO produces better performance in solving both problem instances, as compared to the continuous PSO.

During the course of evaluation and comparison of these two versions of PSO, the authors are able to obtain some insight into different PSO versions in terms of how it can considerably have some bearing on the performance of PSO in solving JSP.

The versions are shown to yield substantially contrasting degree of performances for similar benchmark problems, where the discrete PSO is able to outperform the continuous PSO with a significant improvement in attaining the best solutions of the two problem instances as well as the mean computational times. Thus, the types of problem space should be contemplated as one of the main factors when working with PSO in solving JSP.

Future efforts will include further investigation of the mechanism of each PSO version in order to thoroughly examine what makes one version better than the other. Since only two problem instances are experimented on, more problem instances in the OR library will be applied to assess the continuous and discrete PSO in order to verify whether the performance is subject to the problems being solved.

References

[1] Błażewicz J, Domschke W and Pesch E 1996 The job shop scheduling problem: Conventional and new solution techniques *Eur. J. Oper. Res.* 93 pp 1–33

[2] Kennedy J and Eberhart R 1995 Particle swarm optimization *Neural Networks, 1995. Proceedings., IEEE International Conference on* vol 4 pp 1942–8

[3] Liao C J, Chao-Tang Tseng and Luarn P 2007 A discrete version of particle swarm optimization for flowshop scheduling problems *Comput. Oper. Res.* 34 pp 3099–111

[4] Anghinolfi D and Paolucci M 2009 A new discrete particle swarm optimization approach for the single-machine total weighted tardiness scheduling problem with sequence-dependent setup times *Eur. J. Oper. Res.* 193 pp 73–85

[5] Kennedy J and Eberhart R C 1997 A discrete binary version of the particle swarm algorithm *1997 IEEE International Conference on Systems, Man, and Cybernetics. Computational Cybernetics and Simulation* vol 5 pp 4104–8

[6] Hu X, Eberhart R C and Shi Y 2003 Swarm intelligence for permutation optimization: A case study of n-queens problem *2003 IEEE Swarm Intelligence Symposium, SIS 2003 - Proceedings* pp 243–6

[7] Wang Y R, Lin H L and Yang L 2012 Swarm refinement PSO for solving n-queens problem *Proceedings - 3rd International Conference on Innovations in Bio-Inspired Computing and Applications, IBICA 2012* pp 29–33

[8] Zhang H, Li X, Li H and Huang F 2005 Particle swarm optimization-based schemes for resource-constrained project scheduling *Autom. Constr.* 14 pp 393–404

[9] Lian Z, Jiao B and Gu X 2006 A similar particle swarm optimization algorithm for job-shop scheduling to minimize makespan *Appl. Math. Comput.* 183 pp 1008–17

[10] Zhang Y, Wang S and Ji G 2015 A Comprehensive Survey on Particle Swarm Optimization Algorithm and Its Applications *Math. Probl. Eng.* 2015 pp 1–38

[11] Beasley J E 1990 OR-Library

[12] Klemmt A, Horn S, Weigert G and Wolter K-J 2009 Simulation-based optimization vs. mathematical programming: A hybrid approach for optimizing scheduling problems *Robot.*

Comput. Integr. Manuf. 25 pp 917–25
[13] Carlier J and Pinson E 1989 An algorithm for solving the job-shop problem *Manage. Sci.* 35 pp 164–76

IOP Conf. Series: Materials Science and Engineering **530** (2019) 012045 doi:10.1088/1757-899X/530/1/012045

Profit Optimization Through the Application of Theory of Constraints (TOC): A Case Study in Printing Company

C Saleh[1*], T Immawan[1], A B Hassan[2] and M N Zakka[1]

[1]Department of Industrial Engineering, Faculty of Industrial Technology, Islamic University of Indonesia, Yogyakarta, Indonesia

[2]Electrical, Electronic and Automation Section, Universiti Kuala Lumpur Malaysian Spanish Institute (UniKL – MSI), Kulim High Technology Park, 09000, Kulim, Kedah, Malaysia

Corresponding author *: chairul@uii.ac.id

Abstract. Globalization has made industrial competition to become intensified and more stringent. As technology advances, traditional management is deemed insufficient to meet the needs of many companies in order to achieve their goals. The theory of constraints is considered to be a pertinent method to overcome the faced challenges. The theory is aimed to effectively manage constraints that prevent companies from achieving their goals. It is focused on enhancing system capacity constraints and minimizing inventory and operating costs to increase the profit by managing company's resources efficiently. The purpose of this study is to apply the theory on a case study in a printing company to manage obstacles in an effective way by defining constraints that prevent the company from achieving its targets. Thus, it can increase the company's profitability. Results show that there are capacity constraints on glue and laminating machine causing the company to solely produce 72.06% of the total demand. It is suggested that if the company adds one more laminating glue machine, an increasing profit as much as 22.33 % can be attained.

1. Introduction

Limitation of resources to produce products in order to meet market demands is considered to be a constraint that may prevent many companies from achieving their targets. The theory of constraints (TOC) is considered to be a pertinent approach to help companies to improve their performances and efficiency [1].

TOC is very suitable to handle production capacity constraints by identifying bottlenecks and increasing capacity on the identified bottleneck system. Thus, it may benefit companies to increase their profits [2].

According to the theory, company's performance can be measured through three parameters, namely throughput, inventory, and operating costs [3]. Throughput is the rate of product being produced per unit of time. Throughput rate is inversely proportional to the cost associated with raw materials and supplies. When the throughput increases, the cost associated with raw materials and supplies will decrease. This could happen due to the throughput is obtained from the reduction of selling price and materials and supplies cost [2]. Inventory cost is the total of production and labor costs [2]. Whereas operating cost is cost expended to undertake production activities apart from material costs [2].

IOP Conf. Series: Materials Science and Engineering **530** (2019) 012045 doi:10.1088/1757-899X/530/1/012045

2. Literature Review

TOC was first introduced by Eliyahu Goldratt [4] and has become one of the most important factors in business management ever since. The TOC philosophy emerged in the 1980s which was an evolution of the previous version in the field of production/operations, called Optimized Production Technology (OPT). The OPT was originally created as a scheduling software program (with a secret algorithm). TOC can be used to improve the performance of complex systems [5]. TOC works to influence the performance from two different perspectives, namely the management system and continuous quality improvement [6].

Performance measurement in TOC is divided into 2 categories, i.e. financial measurement and activity measurement. Financial measurement consists of net income, ROI and cash flow. Activity measurement consists of throughput/purchase, stock/raw material costs, and operating costs.

According to the theory, there is at least one obstacle in each company that prevents them from achieving its objectives. The TOC basically claims that the limitation of company's capacity is considered as a constraint in the production process. Therefore, it is necessary to define and eliminate boundaries to increase the capacity, for example by eliminating congestions during the production activities.

The following is several main assumptions of Theory of Constraints (TOC):

- The main objective of a company is to make profit: If the company is assumed as a chain, the strongest chain is the weakest child chain. From this point of view, the weakest child chain need be found and strengthened.
- Direct labor cost is considered as a part of operating cost: according to the theory, all costs are associated with the operational cost unless the direct materials and inventory costs. Profitability is measured by calculating the selling price, number of sales and direct materials cost and inventory costs. In addition, direct labor costs and manufacturing overhead costs are defined as fixed costs.
- There are external and internal factors that prevent the company from making a profit. It is easy to identify capacity constraints because it generally occurs in the production process.
- Planning the production flow is a difficult task since constraints/bottlenecks must be defined and changes in products, and product mixes must be considered.

The Theory of Constraint (TOC) has 3 focus parameters namely throughput (outcome), stock (raw material costs), and operating costs [3]. The following is an explanation of these parameters.

- Throughput (Outcome) is found from the results of a reduction between the selling price of the product and the cost of raw materials. Contribution margins are found by reducing all variable costs (direct raw materials and inventory costs, direct labor costs, variable parts of overhead costs) from the selling price. Contribution of margin and throughput are different although seems similar. Louderback & Patterson (1996) says that throughput is found by reducing the cost of direct raw materials and inventory costs from the selling price. Because only variable costs contain direct raw material costs and inventory costs in the TOC (Okutmus et al., 2015). Okutmus et al., (2015) said that output or throughput is a measure found by reducing raw material costs and inventory costs from the selling price of the product (Tiryakigil, 2011)
- Stock or Inventory is defined as money invested for producing a product, which is the total of production cost and labor cost [2].
- Operating cost is defined as the cost expended regarding the running of business's core operations. Operation costs includes all production costs except the cost associated with raw materials [2].

IOP Conf. Series: Materials Science and Engineering **530** (2019) 012045 doi:10.1088/1757-899X/530/1/012045

The application of TOC consists of 5 steps as know 5FS:

- Identifying the constraints
- Exploiting the constraints effectively
- Subordinating every related decision to the constraints
- Elevating the constraints
- Repeat the process

3. Application of TOC

3.1. Company's Information

This research took place in a local printing company, in Jakarta, Indonesia. The company applies 7 hours of work on Monday to Friday, and 5 hours of work for Saturday. The company recently has a bottleneck problem within its production system. It has 5 work centers, i.e. cutting, printing, ponding, varnishing, and laminating. All work centers have the same time capacity every month. During May 2018 there were 576,000 seconds time capacity. Additionally, the total demands on May 2018 were 136,000 unit, and there were 27 types of product had different demand quantities.

The company produces a product, which consists of two parts, namely the top and bottom. The upper part passes through all work stations except the vertical engine. Varnishing machines are used to varnish the bottom part. After the upper part is finished on the glue and laminating machine, the product will return to the ponding machine for finishing the ponding process. Then it will be cleaned and packed with the bottom. While the bottom does not pass through the process in the printing machine and glue & laminating machine.

3.2. TOC Running

3.2.1. Constraint Identification. The first step of the TOC is to identify constraints. Identification is done by comparing the theoretical capacity and the actual capacity. Theoretical capacity is the available capacity of active working time. While the actual capacity is the capacity of active work time that is actually needed. If actual capacity exceeds the theoretical capacity, then constraints will occur. The following table shows the comparison between theoretical capacity and actual capacity of each machine.

Table 1. Constraints Identification

No	Code	Work Centre (second) (demand x production time)				
		Cutting	Printing	Pond	Vernish	Laminating
1	A1	750	5,000	20,000		
2	A2	750	5,000	20,000		60,000
3	A3	750	5,000	20,000		60,000
4	A4	750	5,000	20,000		60,000
5	B1	1,125	7,500	30,000		90,000
6	B2	450	3,000	12,000		36,000
7	B3	450	3,000	12,000		36,000
8	B4	450	3,000	12,000		36,000
9	B5	750	5,000	20,000		60,000
10	B6	750	5,000	20,000		60,000
11	B7	750	5,000	20,000		60,000
12	B8	750	5,000	20,000		60,000
13	B9	750	5,000	20,000		60,000
14	B10	450	3,000	12,000		36,000
15	B11	450	3,000	12,000		36,000

No	**Code**	**Work Centre (second) (demand x production time)**				
		Cutting	**Printing**	**Pond**	**Vernish**	**Laminating**
16	B12	225	1,500	6,000	9,000	18,000
17	C1	2,400	16,000	64,000		192,000
18	D1	1,500	10,000	40,000	60,000	
19	E1	1,500	10,000	40,000	60,000	
20	F1	450	3,000	12,000	18,000	36,000
21	G1	450	3,000	12,000		36,000
22	H1	1,500	10,000	40,000	60,000	
23	J1	450	3,000	12,000		
24	K1	450	3,000	12,000	18,000	
25	L1	450	3,000	12,000		
26	M1	450	3,000	12,000		
27	N1	450	3,000	12,000	18,000	
Actual Capacity (second)		20,400	136,000	544,00 0	243,000	1,032,000
Theoretical Capacity (second)		576,000	576,000	576,00 0	576,000	576,000
Differences		555,600	440,000	32,000	333,000	**(456,000)**
Capacity Usage Rate		4 %	24 %	94 %	42 %	179 %

As an example the product A2 works in a cutting station with a capacity of 200 sheets/second or 0.15 seconds/sheet. The request of A2 product is 5,000 units. So that at the cutting work station, the actual capacity is 0.15 x 5,000 = 750 seconds. At A2 print work station requires actual capacity of 1 x 5,000 = 5,000 seconds. The ponding work station requires actual capacity of 4 x 5,000 = 20,000 seconds. Whereas glue & laminating work stations require actual capacity of 12 x 5,000 = 60,000. The calculation is carried out on all products so that the actual capacity required by each work station will be obtained. After that, a comparison is made between the actual capacity and the theoretical capacity. For example, the actual working capacity of glue and laminating stations is 1,032,000 seconds. This exceeds the existing theoretical capacity of 576,000 seconds. So there is a difference of 456,000 seconds. In other words, the glue & laminating machine is jammed. This is called a constraint (bottleneck). The source of these constraints is called capacity constraints.

3.2.2. Exploiting the Constraints Effectively. At this stage, the throughput rate is calculated by subtracting the raw materials and supplies costs with the selling price of product [2]. Table 2 below shows the throughput rate calculation for each product.

Table 2. Throughput Rate/Product

No	Code	Selling Price (Unit)	Raw Material Cost (Unit)	Throughput Rate (Rp/Unit)
1	A1	1,200	700	500
2	A2	1,450	640	810
3	A3	1,450	550	900
4	A4	1,800	900	900
5	B1	1,450	550	900
6	B2	2,400	800	1,600
7	B3	2,850	1,000	1,850
8	B4	2,850	1,000	1,850

No	Code	Selling Price (Unit)	Raw Material Cost (Unit)	Throughput Rate (Rp/Unit)
9	B5	1,450	640	810
10	B6	1,450	640	810
11	B7	2,400	800	1,600
12	B8	2,400	800	1,600
13	B9	3,000	1,250	1,750
14	B10	2,400	800	1,600
15	B11	2,400	800	1,600
16	B12	8,000	1,900	6,100
17	C1	1,150	600	550
18	D1	1,250	600	650
19	E1	1,250	600	650
20	F1	2,250	1,125	1,125
21	G1	2,850	800	2,050
22	H1	1,800	1,100	700
23	J1	1,350	750	600
24	K1	2,250	1,150	1,100
25	L1	1,700	800	900
26	M1	1,700	800	900
27	N1	2,250	1,000	1,250

Then calculate the constraint rate per product based on the time limit on the work stations which exhibit the bottleneck (glue & laminating). Table 3 shows the constraint rate of the product based on the time constraint.

Table 3. Constraint Rate/Product Based on Time Constraint

No	Code	Constraints Rate (Unit)	Time Constraints (second)	CR/CT (Rp/sec)
2	A2	810	12	68
3	A3	900	12	75
4	A4	900	12	75
5	B1	900	12	75
6	B2	1,600	12	133
7	B3	1,850	12	154
8	B4	1,850	12	154
9	B5	810	12	68
10	B6	810	12	68
11	B7	1,600	12	133
12	B8	1,600	12	133
13	B9	1,750	12	146
14	B10	1,600	12	133
15	B11	1,600	12	133
16	B12	6,100	12	508

No	Code	Constraints Rate (Unit)	Time Constraints (second)	CR/CT (Rp/sec)
17	C1	550	12	46
20	F1	1,125	12	94
21	G1	2,050	12	171

Based on the table above, product B12 have the highest constraint rate. So that the product is prioritized to produce. The next priority sequence is sorted by the value of the constraint rate from the highest to the lowest.

Based on table 3 it can be seen that from 18 types of the bottleneck product, only 12 types can be produced. And A4 products can be produced as many as 500 units of the total 5,000 unit requests. This happens due to the insufficient production time capacity which causes the company fail to meet all customer demands [2].

The next step is calculating the profit based constraints. Table 4 below shows the company's attainable profit. Total selling is obtained by multiplying the selling price with demand. Total raw material cost is obtained by multiplying the raw material cost with demand. Total running expenses are obtained by multiplying running expenses with demand.

Table 4. Calculate profit based constraints

No	Code	Total Selling	Total Raw Material Cost	Running Expenses		Net Profit
				Direct Labor Cost	Operation Cost	
3	A3	7,250,000	2,750,000	725,000	1,725,000	2,050,000
4	A4	900,000	450,000	67,500	152,500	230,000
6	B2	7,200,000	2,400,000	532,500	1,386,000	2,881,500
7	B3	8,550,000	3,000,000	682,500	2,194,500	2,671,500
8	B4	8,550,000	3,000,000	682,500	2,194,500	2,671,500
11	B7	12,000,000	4,000,000	890,000	2,310,000	4,800,000
12	B8	12,000,000	4,000,000	890,000	2,310,000	4,800,000
13	B9	15,000,000	6,250,000	990,000	2,710,000	5,050,000
14	B10	7,200,000	2,400,000	534,000	1,386,000	2,880,000
15	B11	7,200,000	2,400,000	534,000	1,386,000	2,880,000
16	B12	12,000,000	2,850,000	633,750	1,749,500	6,765,750
20	F1	6,750,000	3,375,000	441,000	1,087,500	1,846,500
21	G1	8,550,000	2,400,000	570,000	1,500,000	4,080,000
Total Income		113,150,000	39,275,000	8,172,750	22,091,500	43,610,750

Net income in May 2018 based on constraints is IDR 43,610,750. Net income is obtained by subtracting the Total Sales with the Total Cost of Raw Materials and Operation Costs.

3.2.3. Subordinate Every Related Decision to The Constraints. This research will provide three possible decision alternatives, namely adding one more of both glue & laminating machines, implementing an outsourcing for the remaining products that cannot be done to other companies, and holding overtime hours. In addition, special shifts can also be made. However, it is difficult to do so due to the difficulty of finding labors to work for the additional shift and it would expend very large operational costs.

IOP Conf. Series: Materials Science and Engineering **530** (2019) 012045 doi:10.1088/1757-899X/530/1/012045

3.2.4. Elevating The Constraints. The first alternative decision that the company can make is to buy an addition machine for gluing & laminating. A Chinese brand machine costs approximately Rp 150,000,000. So the theoretical capacity will be twice as large as before. But the consequence of buying a machine is the increasing of operation cost. This is due to three things: the addition of operators, the burden of electricity, and the depreciation of the new machines. Depreciation is a reduction in the value of fixed assets within certain period of time based on their economic age. The depreciation cost of the machines based on the straight-line method is IDR 367,000 per month (150 million - 40 million / 25/12). 40 million is the remaining machine value after 25 years later.

The second alternative decision is implementing an outsourcing to other companies. This alternative is feasible to implement since it does not require a high investment, unlike selecting the first alternative which requires a considerably high investment.

The third alternative is to hold overtime. This alternative tends to be the most feasible solution because it has the smallest risk.

3.2.5. Back to Step 1. The first step is to re-identify constraints. Table 5 shows the overview of the constraints reidentification.

Table 5. Constraints Reidentification

No	Code	Work Centre (second)				
		Cutting	Printing	Ponding	Varnishing	Laminating
1	A1	750	5000	20000		
2	A2	750	5000	20000		60000
3	A3	750	5000	20000		60000
4	A4	750	5000	20000		60000
5	B1	1125	7500	30000		90000
6	B2	450	3000	12000		36000
7	B3	450	3000	12000		36000
8	B4	450	3000	12000		36000
9	B5	750	5000	20000		60000
10	B6	750	5000	20000		60000
11	B7	750	5000	20000		60000
12	B8	750	5000	20000		60000
13	B9	750	5000	20000		60000
14	B10	450	3000	12000		36000
15	B11	450	3000	12000		36000
16	B12	225	1500	6000	9000	18000
17	C1	2400	16000	64000		192000
18	D1	1500	10000	40000	60000	
19	E1	1500	10000	40000	60000	
20	F1	450	3000	12000	18000	36000
21	G1	450	3000	12000		36000
22	H1	1500	10000	40000	60000	
23	J1	450	3000	12000		
24	K1	450	3000	12000	18000	

No	Code	Work Centre (second)				
		Cutting	Printing	Ponding	Varnishing	Laminating
25	L1	450	3000	12000		
26	M1	450	3000	12000		
27	N1	450	3000	12000	18000	
Actual Capacity (second)		20,400	136,000	544,000	243,000	1,032,000
Theoritical Capacity (second)		576,000	576,000	576,000	576,000	1,152,000
Difference		555,600	440,000	32,000	333,000	120,000
Usage		4 %	24 %	94 %	42 %	90 %

After reidentifying the repetitive constraints, no bottleneck found within work station. So that the proposed alternative succeeded to overcome the bottleneck problem.

The next stage is to calculate the income that will be obtained by purchasing an additionl machine for both gluing & laminating. After purchasing the machines, the net profit that can be achieved is Rp. 56,151,000. With a note of being exposed to the depreciation expenses of Rp. 367,000. (150 million - 40 million / 25/12). This gain increases from the constraint gain. The gained profit may increase by 22.33% from Rp. 43,610,750 to Rp. 56,151,000.

After implementing the outsourcing, the company can make profit as much as Rp. 55,668,000. In other words, outsourcing alternatives can be used to overcome the bottleneck. The obtained benefits are lower than the first alternative. The gained profit may increase by 21.66% from Rp 43,610,750 to Rp 55,668,000.

Implementing the overtime can also reduce the bottlenecks. Profits may also increase as much as 21.45% from Rp 43,610,750 to Rp 55,518,000. This alternative has a lower advantage compared to the two previous alternatives. But this alternative is the most affordable than the two previous alternatives.

4. Discussion

Based on the data processing and analysis, each machine has a capacity of 576,000 seconds. This capacity is insufficient to meet customer demands, which requires 1,032,000 seconds for the total production time. Bottleneck as a capacity constraint is found in the glue & laminating sections which causes the company to solely meet 72.06% of the total demand. In addition, this also makes the company to only produce 12 out of 18 total types of bottleneck products whereas there are requests for 27 types of products, 18 of which are bottlenecks and 9 are non-bottlenecks. Based on the TOC analysis there are 12 bottleneck products which are among the production priorities. The decision is made through a throughput rate calculation to determine production priority.

There are 3 alternatives that companies can do to handle the bottleneck problem: adding one more unit for both glue & laminating machines, implementing outsourcing, and preventing overtime hours. Each alternative has advantages and disadvantages. The first alternative tends be difficult to realize since it would spend a considerably high investment. However, this alternative would increase the production process and may reduce the bottleneck instantly. The second alternative is considered to be more feasible to implement, even though it is fairly difficult, to find the outsourcing companies. The implementation of outsourcing may benefit the company to save the supply of glue and plastic. The last alternative is to hold overtime which tends to be the easiest scenario to implement. However, this alternative may pay larger costs.

After identifying the constraints, the calculation of obtained net profit associated with the bottleneck is carried out. During the bottleneck condition, the company makes profit of Rp. 43,610,750 from the total sales of 12 types of products and 500 units of A4 products. While the remaining 6 types of product will be done with the 3 alternatives above.

5. Conclusion

Production capacity is onc of the identified constraints. Based on the data analysis results, glue & laminating machines are sources of the production process constraints. This causes the company to solely produce 72.06% of the total demands or 98,000 units out of 136,000 units in total.

Efforts to optimize the bottleneck in glue & laminating machines are based on throughput rate per second. It is suggested to arrange some product mixes to enhance the profitability. The best product mixing arrangements are suggested as follow: 1,500 units of Maulins Black Forest, 3,000 units of Han's 26, 3,000 units of Maulins Lapis Eko Surabaya, 3,000 units of Maulins Bolu Eko, 5,000 units of Maulins Sipon, 3,000 units of Maulins Lapis Surabaya Slope, 5,000 units of Maulins Roll Steam, 5,000 units of Maulins Surabaya Steamed Bandung, 3,000 units of Maulins Cake Tape, 3,000 units of Maulins Roll Cake, 3,000 units of Dian Chatering, 5,000 units of Aldi Brownies, and 500 items of Aldi Bolu Hijau.

There are 3 alternatives that can be done to optimize the bottleneck system, i.e. adding one more unit of glue & laminating machines, outsourcing, and holding overtime hours.

Based on the calculation results, an increasing profit could be achieved. Adding one more unit of glue & laminating machines may increase the profit by Rp. 56,151,000 or 22.33%. Without depreciation the company may increase the profit as much as Rp. 56,018,000. While by implementing the outsourcing, Rp. 55,668,000 or 21.66% rising profit is achievebale. By holding overtime, the company may achieve an increasing of Rp. 55,518,000 of 21.45% of the profit. From the description above the maximum profit enhancement that can be generated is as much as 22.15% or Rp. 56,018,000.

6. Future Research

For the future work, performing a simulation study, and comparing the simulation result with the TOC results is highly recommended. It is expected to provide a broad insight of method to handle bottleneck problems.

References

[1] V Šukalová and P Ceniga 2015 Application of the Theory of Constraints Instrument in the Enterprise Distribution System *Procedia Econ. Financ.* Vol 23 pp 134–139

[2] E Okutmuş, A Kahveci and J Kartašova, Aug 2015 Using theory of constraints for reaching optimal product mix: An application in the furniture sector *Intellect. Econ.* Vol 9 No 2 pp 138–149

[3] Z T Şimşit, N S Günay and Ö Vayvay Sept 2014 Theory of Constraints: A Literature Review *Procedia - Soc. Behav. Sci.* Vol 150 pp 930–936

[4] E M Goldratt 1984 *The Goal: A Process of Ongoing Improvement.* Massachusetts: North River

[5] M Naor, E S Bernardes and A Coman Jan 2013 Theory of constraints: is it a theory and a good one? *Int. J. Prod. Res.* Vol 51 No 2 pp 542–554

[6] M C Gupta, G K Sahi and H Chahal July 2013 Improving market orientation: the theory of constraints-based framework *J. Strateg. Mark.* Vol 21 No 4 pp 305–322

IOP Conf. Series: Materials Science and Engineering **530** (2019) 012046 doi:10.1088/1757-899X/530/1/012046

Integrating Activity-Based Costing and Computer Simulation to Investigate Thailand-Malaysia Cross Border: A Case Study of Sadao Customs House

P Suttishe, W Kongkaew, L Treeranurat and N Sirivongpaisal*

Business Logistics and Supply Chain Management Research Team, Department of Industrial Engineering, Faculty of Engineering, Prince of Songkla University, Hat Yai, Songkhla, 90112, Thailand

Corresponding author *: nikorn.s@psu.ac.th

Abstract. This research aims to study the cross border service between Thailand-Malaysia at Sadao Customs House, Songkhla, Thailand. Initial problem was stated as severe traffic jam at border and lots of trucks waiting for custom processes. With the purpose of investigating for solution, vital information such as cost of service was needed. Nevertheless; under the current costing system, it cannot lead to the improvement direction as it has not revealed the efficiency of resources consumption which are used to drive service activities. Also, it cannot indicate wastes or unproductive activities operating under the service. Hence this research proposed activity-based costing (ABC) and computer simulation as an integrated tool to investigate the cross border service in order to acquire service cost information which will be used for improvement stage. Results of study can be expressed in terms of cross border service unit cost and activities analysis.

1. Introduction

Thailand is bordered with four neighbouring countries which are Myanmar, Laos, Cambodia, and Malaysia. Thailand's trade with neighbours during the past three years, 2014-2016, has grown steadily with average 0.78%, annually. In 2016, the trade value was worth 1,013,389.20 million THB (125,123.16 million MYR), increased 0.71% from year 2015 (Note 1 THB is 0.12347 MYR on 17 Aug 2018). Then in 2017, it was totally valued 1,076,389.37 million THB (132,901.80 million MYR), which were export value 649,926.84 million THB (80,246.47 million MYR), and import value 426,462.53 million THB (52,655.33 million MYR). Among Thailand's neighbours, Malaysia is a key market with the highest value, accounted for 52.48% of total border trade.

Thailand and Malaysia border trade in 2017 had entire value 564,657.27 million THB (69,718.23 million MYR), increased by 12.61% to 2016, consisting of export value of 312,551.93 million THB (38,590.79 million MYR), and import value of 252,105.34 million THB (31,127.45 million MYR). Main exported products include rubber, rubber products, computer equipment and components, automobile parts and accessories, and particle boards, etc. Also, imported items are computer equipment and components, computer magnetic disk, industrial machinery, etc. Border trade of Thailand and Malaysia can be made through 9 permanent cross border customs houses which are: (1) Sadao Customs House, Sadao District, Songkhla Province; (2) Padang Besar Customs House, Sadao District, Songkhla Province; (3) Su-ngai Kolok Customs House, Su-ngai Kolok District, Narathiwat Province; (4) Tak Bai Customs House, Tak Bai District, Narathiwat Province; (5) Ban Pra Kob

Customs House, Nathawee District, Songkhla Province; (6) Wangprachan Customs House, Kuan Don District, Satun Province; (7) Satun Customs House, Muang District, Satun Province; (8) Betong Customs House, Betong District, Yala Province; and (9) Buketa Customs House, Wang District, Narathiwat Province. Each location of customs house can be illustrated in Figure 1.

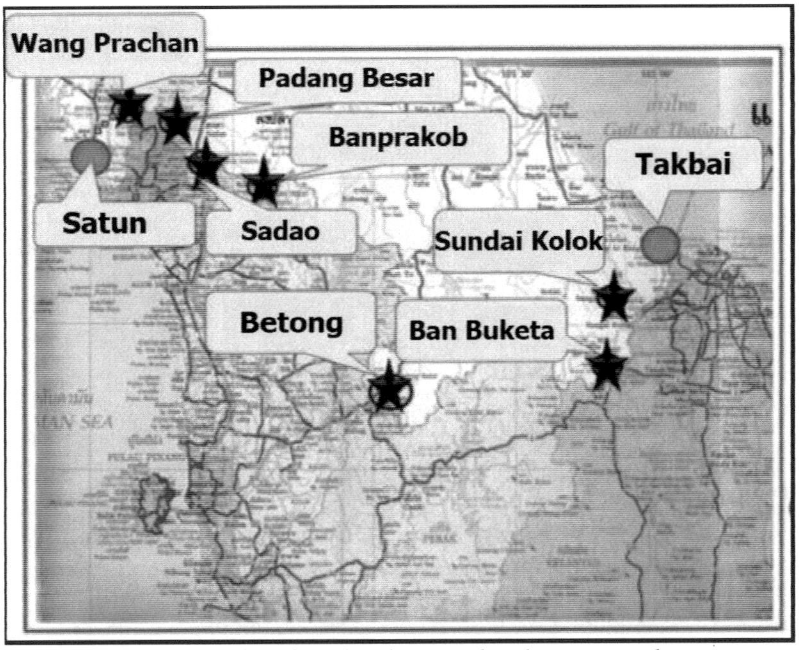

Figure 1. Thailand-Malaysia cross border customs house

With the highest trade value, Sadao Customs House had a trade volume more than 300,000 million THB during 2014-2017. Especially in 2017, its trade value was 348,902.79 million THB and still has the potential to increase unceasingly. Statistics of trade value at Sadao Customs House are illustrated in Table 1.

Table 1. Thailand-Malaysia Border Trade Value at Sadao Customs House.

	Export (Million)		Import (Million)		Total (Million)	
Year	THB	MYR[a]	THB	MYR[a]	THB	MYR[a]
2014	156,064.95	19,269.34	192,711.94	23,794.14	348,776.89	43,063.48
2015	143,474.24	17,714.76	181,459.78	22,404.84	324,934.02	40,119.60
2016	147,495.00	18,211.21	185,497.60	22,903.39	332,992.60	41,114.60
2017	165,682.32	20,456.80	183,220.47	22,622.23	348,902.79	43,079.03

[a] 1 THB is 0.12347 MYR on 17 Aug 2018.

Key factor such highway infrastructure connecting central Thailand to ports in Malaysia is contributing to the highest trade value of Sadao Customs House compared to others. Consequently, the number of trucks crossing border at Sadao Customs House is increasing every year, resulting in congestion at border. As well as, customs clearance procedure at the border may cause delay and lead to more congestion.

The field survey was conducted during January 2018 and it was realized that the average number of trucks in queue was 308 trucks per day in regular time but it could be up to 370 trucks per day in festival period or long weekend, resulting in not only traffic jam at border but also missed shipments

IOP Conf. Series: Materials Science and Engineering **530** (2019) 012046 doi:10.1088/1757-899X/530/1/012046

to port. This delay problem has a negative impact on customs operations, rising in increased workflows, unnecessary overhead costs, which can affect to the efficiency of sector based on cost effectiveness. So information about costs is compulsory to investigate cost effectiveness. Currently, the costing system used in public sector is established on resources basis categorized into personnel costs and non-personnel costs. Personnel costs are such salaries, wages, and remunerations. Non-personnel costs include machinery and equipment cost, material cost, and overhead cost. It does not reflect to activity running in that sector. And there are not much details linking to processes and operations. Therefore this research targets to analyze cost of cross border service at Sadao Customs House based on activity or so called "Activity-Based Costing (ABC)" integrating with computer simulation. The key mechanism of ABC is cost allocation to activities. Besides, simulation model has beneficial role in this study by providing the reflection of cross border activities which are subjected to change under the uncertain working environment. Study objective is to realize the actual cost of cross border service of customs house by ABC and to identify value of improvement opportunities in Sadao Customs House.

This paper is organized into 5 sections, which are (1) Introduction, (2) Literature review, (3) Research methodology, (4) Results and discussion, and (5) Conclusions and future research.

2. Literature review

In this section, some related theories and research works were reviewed and presented. Beginning with ABC model, and computer simulation technique is reviewed after that. At last, former researches are wrapped up in order to give some ideas for all readers.

2.1. Activity-based costing

Activity-based costing (ABC) is the concept of cost management system, which is aimed for administrators focusing on managing activities and associated costs; therefore it aims directly to activities. Activities cause expenses and cost, and product is stemmed from those activities. Activity is the process of transforming an organization's resources into a product and it drives cost incurred in organization. ABC system is an important and indispensable for cost and service analysis. It can illustrate cost of service and identify factors affecting to the cost and nature of activities. Besides, it can be used as a tool for decision making to improve operations or activities to be more efficiency thru waste identification or waste reduction. Three basic principles of ABC [1] can be stated as: (1) activities consume resources; (2) activities have a reason; and (3) customers, products, or channels cause different levels of activity. Hence ABC mainly spots on the activities of process and endeavour to identify the cost of each activity driven by resources in order to calculate product cost.

Difference between traditional costing system and ABC system can be stated in two key points. First, ABC allocates costs based on activity cost pool basis rather than cost pool basis. Second, activity or resource driver, which is used as a basis for calculating product cost, is structured differently from the traditional costing system. Each activity must be analysed what is the resource driver or the root cause that drives the cost of activity. This makes ABC more proper for organization having variety of products. Albeit product diversity is not the existing condition for organization; somehow, ABC still offers cost allocation principle which is in line with the proportion of activities that they actually consume contrasting to traditional costing system which ignores the principle that resources are actually consumed by activities.

A number of terms commonly used in ABC which is referred from [1] are presented in Table 2.

Table 2. Definition of ABC Terms.

Term	Definition
Resource	A resource used by the business in carrying out its business
Resource driver	A measurable quantity used to allocate resources to activities
Process	A series of connected activities required to achieve an outcome

Activity	A series of related tasks carried out repeatedly
Task	An element of work, a series of which make up an activity
Activity driver	An event or factor that causes an activity to be performed
Cost object	The entities that are to be costed

2.2. Computer simulation technique

Simulation is the method used to simulate real situation or system behaviour on a computer. With the ability of computer software, it can imitate the flow of activities by collecting and analysing data, building computer model, using it as a tool for investigating existing system and for improving future one without interfering the real system. Steps in conducting computer simulation study [2] can be summarized in the following steps: (1) define objective and plan the study, (2) collect and analyse system data, (3) build the model, (4) validate the model, (5) conduct experiments, and (6) present the results.

The great advantage of simulation is the power to deal with dynamic situation of system such as rapid change and high uncertainty. It makes system improvement more suitably. Nevertheless; there are still some drawbacks for computer simulation such (1) training and time exhausting for model building, (2) underlying comprehension of the system that decision maker must have, (3) necessity to have validated model, otherwise answers are hardly representative of system.

2.3. Related researches

In this section, earlier ABC research studies [3-5], focusing on methodology and applicability, are presented. Bokor and Markovits-Somogyi [3] studied logistics service providers who require reliable and true information about service costs. Distortions of costing information shall be minimized and the drivers of costs shall be accurately explored. Methodology of ABC and industry specific adaptations may be used to improve the accuracy of logistics service providers' costing procedures. Additionally, it can deliver information on cost efficiency as well. This paper aims to analyse the applicability of ABC for the case of logistics service providers.

While Ping and Yu [4] explored scientific and technological levels of Chinese society which had a substantial increase and domestic logistics industry had also achieved a rapid development. However, compared with the logistics enterprises of other countries, Chinese logistics enterprises costing system was still lagging behind and cost management awareness was relatively weak. This study discussed the practical application of the ABC system in the logistics business enterprise based on the analysis of cost management status of logistics from the microscopic point of view.

Prior that Baykasoglu and Kaplanoglu [5] investigated modern costing approaches including ABC for in-land transportation companies. One of the main difficulties of in-land transportation companies is to determine and evaluate true cost of their operations and services. If used and implemented properly, ABC can be very helpful for transportation companies to determine cost of their operations with higher accuracy. In this paper, an application of ABC to a land transportation company that is located in Turkey was presented in detail. In order to improve the effectiveness of the ABC, an integrated approach that combines ABC with business process modeling and analytical hierarchy approach was proposed. It was figured out that the proposed approach was quite effective in costing services of the land transportation company compared to the existing traditional costing system which was in use.

Additionally, simulation integrated to ABC works were reviewed in a number of works such as [6] developed a simulation model to measure costs in an emergency department setting for patients presenting with possible cervical-spine injury who needed radiological imaging. Simulation was introduced as a realistic means to perform ABC analysis, because traditional ABC methods had difficulty coping with process variation in healthcare. Simulation was found to be an accurate and viable means to conduct an ABC analysis. Research output provided more complete information than could be achieved through other conventional analysis, which extra influenced management to negotiate contractual reimbursements better. Also, Spedding and Sun [7] illustrated how discrete event simulation may be used to evaluate the ABC of a manufacturing system. A visual interactive simulation software WITNESS was used to model a semi-automated printed circuit board (PCB) assembly line. The PCB assembly line case study demonstrated how ABC can be applied to a manufacturing system using simulation modelling techniques. Lastly, Takakuwa [8] studied a framework to design simulation models in order to perform ABC for flexible manufacturing systems. In this study, a simulation model for a FMS was constructed. Then a procedure for cost accounting was developed for obtaining the unit cost of the products through simulation experiment. It was shown that precise cost accounting can be performed compared to actual manufacturing activities, if kinds of designated products and their production quantities were specified.

3. Research methodology

In this section, the research methodology is presented. Actually, it can be derived from two parts which are ABC and simulation integrated to methodology having details as follows:

- Step 1: identify cross border system including with processes, activities, resources costs.
- Step 2: analyse cost of specific activities which are classified into six activities (A1-A6) for cross border system.
- Step 3: allocate resources into all activities in proportion to time spent.
- Step 4: analyse cost of activities.
- Step 5: collect, analyse data and develop computer simulation model by ProModel, also verify and validate the model.
- Step 6: develop activity cost drivers and allocate cost of activities to total cost of cross border service by running simulation model.

It can be illustrated in Figure 2.

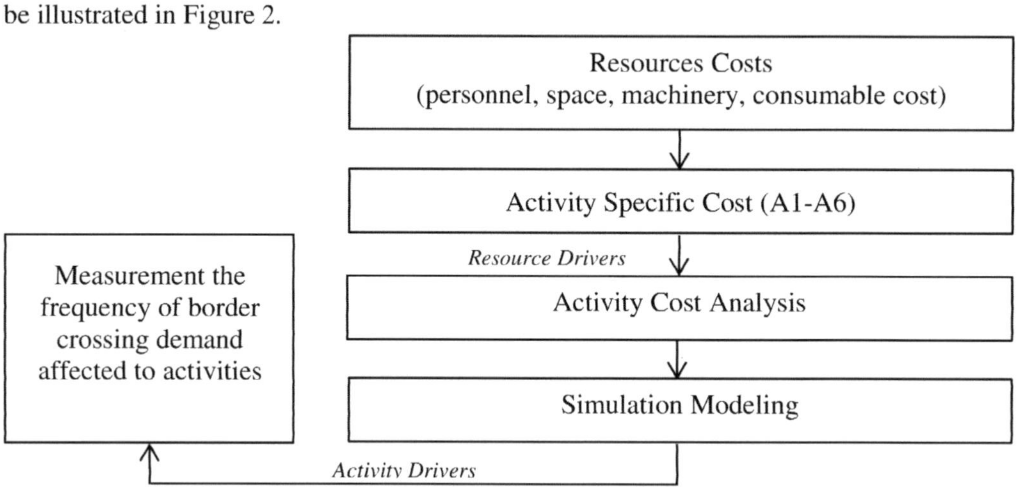

Figure 2. Research methodology (adapted from [1])

4. Results and discussion

According to research methodology illustrated from the last section, results and analysis will be discussed in the following. This study is mainly focused on all activities of Thailand-Malaysia cross border, especially import and export. Data were collected and analysed from primary and secondary sources.

4.1. Activities analysis

Cross border service at the Sadao Customs House can be divided into four main processes, which can be described as follows. When a truck arrives at the Customs House, the driver submits documents for custom inspection. After documents are approved, X-Ray inspection is the next process. And then, the following stage is release inspection process for export. At last, the truck will drive to border for final release inspection process for crossing border. Entire processes of cross border, drawn in dash rectangle, can be illustrated in Figure 3.

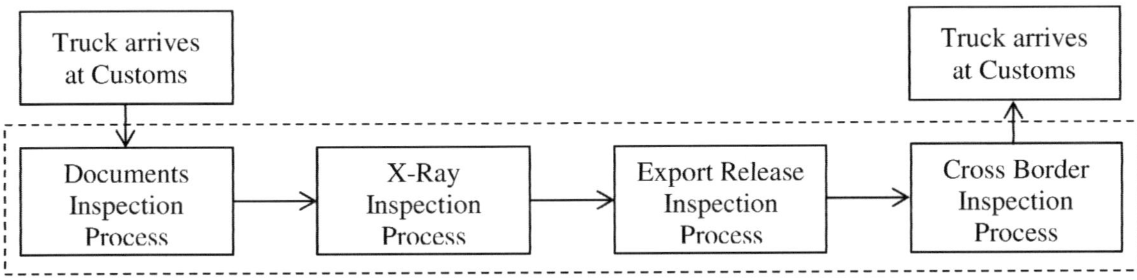

Figure 3. Processes of cross border service

From the four main cross border processes previously illustrated, they can be shown that the activities and the tasks embracing to these processes in Table 3.

Table 3. List of activities and tasks in cross border system

Activity Code	Activity	Task Code	Task
A1	Receive bill of lading	A11	Transact data to e-Export system
		A12	Authenticate information with shipping company by e-mail
A2	Inspect bill of lading	A21	Assign code number to bill of lading
		A22	Check the validity
		A23	File document to the system
A3	Validate the documents for X-Ray inspection	A31	Prepare documents for cargo inspection and identify type of cargo
		A32	Prepare documents for X-Ray inspection
		A33	Arrange the order for X-Ray inspection
A4	Inspect cargo by X-Ray	A41	Inspect cargo
		A42	Scan bill of lading
		A43	Arrange documents and submit X-Ray inspection results to e-Export system
A5	Prepare customs export's documents	A51	Verify X-Ray inspection results
		A52	Prepare export's documents

Table 3. (continued) List of activities in cross border process

Activity Code	Activity	Task Code	Task
A6	Validate documents for releasing export	A61	Check the authenticity of export's documents and bill of lading
		A62	Check truck's license plate and match with filed documents

4.2. Resources cost allocation

Resources cost data during January through December 2017 of Sadao Customs House was sourced for calculation the activity specific cost, as shown in Table 3. These resource costs are personnel cost, space cost, machinery and equipment cost, and consumable cost. In this study, the operating time of each activity and number of usages were used as resource driver. Allocation of resources cost framework can be illustrated in Figure 4.

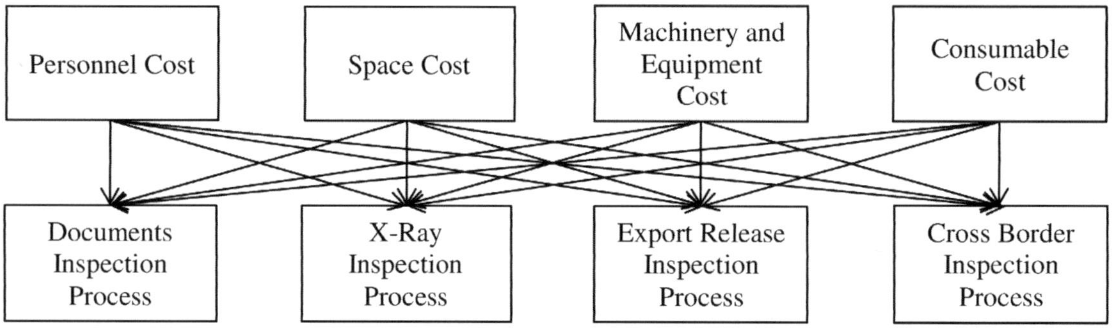

Figure 4. Resources cost allocation framework

Result of resources allocation to specific activities (tasks) and uniting them into the activity cost can be demonstrated in the following Figure 5.

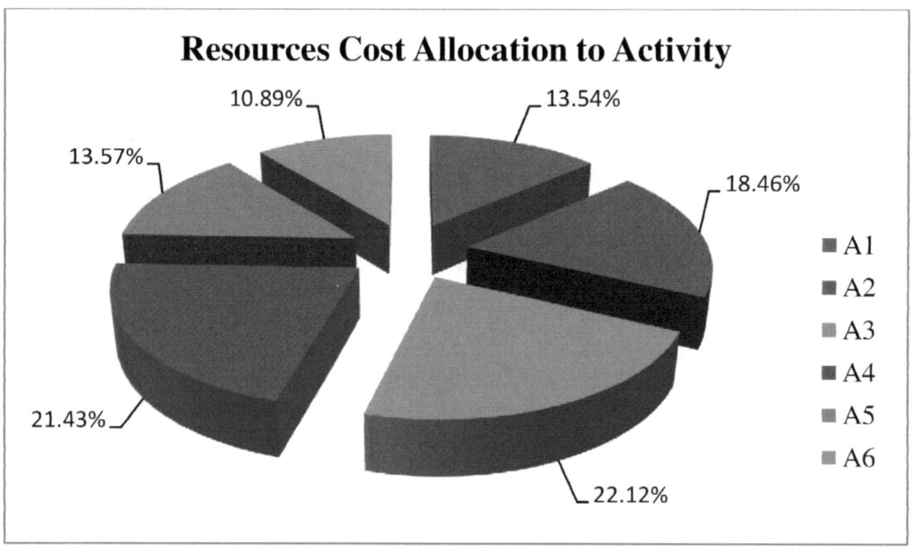

Figure 5. Resources cost allocation

4.3. Simulation modeling of cross border system

Cross border simulation model was handled by ProModel® 2016 software package and sequentially developed along Figure 3 above. Data collection and data analysis such truck arrival time, data inspection time, data inputting time, X-Ray inspection time, inspection time of releasing export, etc. were carried out. Data analysis can be presented the results in Table 4.

IOP Conf. Series: Materials Science and Engineering **530** (2019) 012046 doi:10.1088/1757-899X/530/1/012046

Table 4. Distribution and parameter estimation for cross border process

Process	Distribution	Parameter	Unit
Truck arrival time (05:00-23:00)	Weibull	W(0,1.26,2.82) [a]	minutes/truck
Data inspection and data inputting time	Weibull	W(3,1.75, 0.197) [a]	minutes/truck
X-Ray inspection time	Fixed	5	minutes/truck
Export release inspection time	Weibull	W(4,7.65,0.496) [a]	minutes/truck
Cross border inspection time	Weibull	W(2,2.44,0.191) [a]	minutes/truck

[a] Weibull (min, alpha, beta)

Subsequently, cross border model was built, verified, and validated. In order to ensure that the simulation model can be the representative of the real system, it can be shown that the *P-value* of hypothesis testing is 0.437 which indicates no rejection on hypothesis. Illustration of cross border simulation model can be presented in Figure 6.

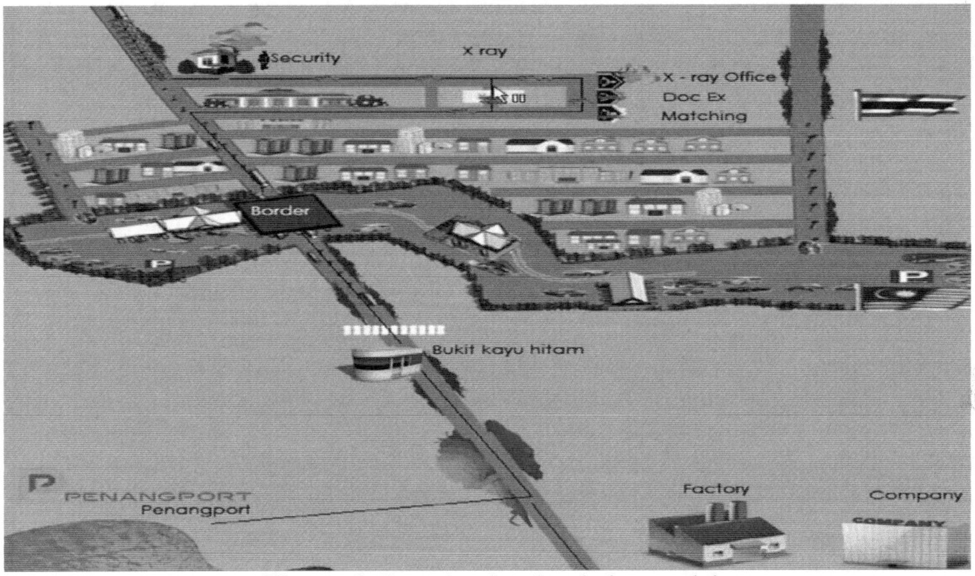

Figure 6. Cross border simulation model

4.4. Cost object of cross border service

As described in Figure 3, this study was integrated with simulation modeling to measure the frequency of border crossing demand affected by activities, which simulation results will serve as activity drivers for ABC. Activity cost was analyzed by tracking resources cost to each task and combining all together to activity cost. And simulation model was built and run for 15 replications to collect number of trucks that cross border through Sadao Customs House. The simulation outputs can be exhibited in Table 5.

Table 5. Simulation outputs and cost object

Activity Code	Activity Cost (Baht/Year)	Activity Driver (Number of Trucks)			Activity's Unit Cost [a] (Baht/Truck)
		$\bar{x}$	$S.D.$	95% CI	
A1	4,290,188.58	316,989	249	(316,852 ; 317,127)	(13.540 ; 13.528)
A2	5,849,288.35	474,173	44	(474,149 ; 474,197)	(12.336 ; 12.335)
A3	7,010,145.84	109,705	6	(109,702 ; 109,708)	(63.902 ; 63.898)
A4	6,791,095.42	109,705	6	(109,702 ; 109,708)	(61.905 ; 61.902)
A5	4,301,518.70	315,623	9	(315,618 ; 315,628)	(13.629 ; 13.628)
A6	3,450,073.44	314,341	6	(314,337 ; 314,344)	(10.976 ; 10.975)
Total					(176.288 ; 176.267)

[a] equal "Activity cost is divided by number of trucks"

From Table 5, each activity cost resulting from resource allocation is presented and can be realized that the highest and second highest activity cost are from X-Ray inspection process. The key item which contributes to these costs is machine depreciation. Moreover, number of trucks which is obtained from running simulation is used as activity driver to calculate unit cost and eventually compute the total cost of cross border service.

Also, it can be recognized that high variation in number of trucks is only arisen at A1 activity, which is the first operation of Customs House where truck entering. This phenomenon is compliant to the field survey which was conducted before the study, which there is truck jam at the border. Trucks could be queued up, approximately to 300 trucks per day and it could be more severe during festival period or long weekend. Regularly, office hours of Sadao Customs House are between 05:00-23:00. The main reason of border jam is truck arriving border continually and concurrently since every single truck expects to obtain the first queue. However, some of them may have trouble with information authentication task between shipping company and Customs House, which can cause delay and queue up.

Next, each activity's unit cost is also calculated in Table 5. All are alike between lower and upper unit costs, except A1's unit cost is slightly different. Lastly, cost of cross border service at Sadao Customs House is 176.28 Baht/Truck.

5. Conclusions and future research

This study has main objective to investigate the cost of cross border service at Sadao Customs House, Songkhla, Thailand by integrating ABC and simulation. Major advantage of integration two techniques is from the outstanding aspect of each technique. ABC can accurately reflect the product or service cost based on resources consumed by activities running into that product or service. Simulation model can imitate the working environments that surround the activities which operate for product or service. Future research is to find the solution to improve cross border service focusing on operations or activities and making them to be more efficiency thru waste identification.

Acknowledgments

This project is financially supported by STEM Workforce Project of National Science and Technology Development Agency, Thailand and Graduate Scholarship Program of Faculty of Engineering, Prince of Songkla University. Also authors are thankful to officers and staffs of Sadao Customs House who facilitated and supported to us until this research can be completely fulfilled.

References

[1] Barrett R 2004 Introduction to activity-based cost management *http://www.algsoftware.com*
[2] Harrell C, Ghosh B K and Bowden R O Jr. 2012 *Simulation using ProModel* (New York: McGraw-Hill)
[3] Bokor Z and Markovits-Somogyi R 2015 Applying activity-based costing at logistics service providers *Periodica Polytechnica Transportation Engineering* 43(2) 98-105
[4] Ping L C and Yu S 2013 A study on the application of activity based costing in Chinese logistics enterprises *Journal of Applied Sciences* 13(16) 3189-3195
[5] Baykasoglu A and Kaplanoglu V 2008 Application of activity-based costing to a land transportation company: a case study *International Journal of Production Economics* 116 308–324
[6] Glick N D, Blackmore C C and Zelman W N 2000 Extending simulation modeling to activity-based costing for clinical procedures *Journal of Medical Systems* 24(2) 77-89
[7] Spedding T A and Sun G Q 1999 Application of discrete event simulation to the activity based costing of manufacturing systems *International Journal of Production Economics* 58(3) 289-301
[8] Takakuwa S 1997 The use of simulation in activity based costing for flexible manufacturing system *Proceeding of the 1997 Winter Simulation Conference* p 793 – 800

IOP Conf. Series: Materials Science and Engineering **530** (2019) 012047 doi:10.1088/1757-899X/530/1/012047

Maintenance for Energy efficiency: A Review

N Firdaus[1,2,*], H A Samat[2] and N Mohamad[2]

[1]Department of Mechanical Engineering, Sekolah Tinggi Teknik PLN, Indonesia

[2]School of Mechanical Engineering, Universiti Sains Malaysia, Malaysia

Corresponding author *: nofirman@sttpln.ac.id

Abstract. Energy efficiency program is used to reduce energy cost that leads to a reduction in production or overhead cost. Energy saving can be obtained by application of energy-efficient technologies, operational improvement, and effective maintenance. However, maintenance and energy efficiency is usually researched separately. The purpose of this paper is to investigate the role of maintenance in energy saving and available maintenance approaches for energy consumption reduction. Various literatures and publication on the research areas were reviewed and summarized to show the importance on maintenance and approached commonly used for energy efficiency.

1. Introduction

Energy is one the vital factors in our world, not only for the industrial sector but also to commercial and residential sectors. Our world has been heavily depending on energy for a long time, especially fossil fuel. As the fossil fuel production and reserves deplete, energy consumption increases. Consequently, fuel price direction is always going up in the long term. On the other hand, renewable energy is still trying hard to fill the hole left by fossil fuel. Therefore, energy management has shown up to be a partial solution for an increase in energy price and a decrease in fossil fuel production. The objective of energy management is to manage the energy usage effectively and efficiently. Many sectors such as industry, commercial, and residential have been planning and then implementing energy efficiency and energy management program with one goal, to reduce energy cost.

The primary benefit of application of energy-efficient technologies (EET) is a reduction in energy cost [1-2]. Generally, energy efficiency is usually related to the environment. To support this, Pons pointed out that energy efficiency had a strong relationship with environmental performance [2]. But these are not the only benefits. The application of EET also contributes to an increase in productivity that may lead to lower the production cost [1]. Other research found out that energy intensity was proportional to productivity that 1% rise in productivity may lead to more than 1% increase in energy efficiency [3]. Moreover, a study conducted in China's iron and steel industries also revealed that improvement in energy efficiency also improved productivity [4]. But the benefits do not end here. EET also generates non-energy benefits that may outweigh the energy-saving benefits [5]. The non–energy benefits may include the following: reduced production costs, reduced cost of environmental compliance, reduced waste-disposal costs, better product quality, higher utilization factor, higher reliability and better worker safety [5]. The bottom line is that energy-efficient technologies generate not only energy-benefits but also non-energy benefits.

In general, there are two ways to save energy: application of energy efficient technology, and implementation of energy management practices. Energy efficiency research mainly focuses on efficient

IOP Conf. Series: Materials Science and Engineering **530** (2019) 012047 doi:10.1088/1757-899X/530/1/012047

technology application and diffusion [6]. But, this is not the only way to save energy. Energy management practices could contribute to energy saving either. Energy management practices include housekeeping, maintenance, energy monitoring, energy training, energy audit, and etc [6-7]. In addition, several researchers also stated the importance of housekeeping and maintenance in energy savings [6] [8-11]. To make an energy efficiency program much more cost-effective, one needs to combine energy efficient technology application with energy management practices [6]. An effective maintenance program will play a very important role in energy management practices.

One of energy management practices highlighted by Backlund is maintenance and monitoring [6]. Without performing maintenance and monitoring, the potential energy savings generated by energy efficient technologies will not be achieved [6]. Moreover, by implementing proper maintenance and energy monitoring continuously, one can detect inefficiencies and malfunctioning equipment [6], and then any necessary actions could be taken to restore the equipment energy performance to an acceptable level by a user. Even, Simple maintenance actions can have a dramatic effect in maintaining energy performance of an equipment [12]. In addition, energy management practices such as maintenance and energy monitoring are less capital intensive and rather requires knowledge, attention, and awareness [6]. In contrast, energy efficient technologies investment is relatively more capital intensive. Therefore, maintenance could be used as one of the cost-effective strategies in energy efficiency program.

Although many people believe that maintenance is important for sustainable of energy efficiency project, maintenance, and energy management are two topics which are often researched independently. Therefore, there is a need to combine these two fields to gain more energy efficiency potential in its application.

2. Maintenance and Energy

Knowles has tried to investigate the relationship between maintenance and energy in a refrigeration system by surveying the literature for evidence of a link because of lack of real-world data [12]. He found out that maintenance had a major role to make sure refrigeration system operates at maximum efficiency. He also stated that detailed energy monitoring was required to improve maintenance procedures. Monitoring the condition of individual main components are also needed to obtain optimal performance [12]. He also addressed that a new method of maintenance was required [12]. This approach should apply a predictive maintenance that uses inexpensive technologies such as infrared thermography, vibration sensor, and oil analyzer [12]. The justification of monitoring technology application is based on several criteria such as investment cost, and the simplicity of use [12]. Finally, he concluded that good maintenance practice could improve energy efficiency, increase the reliability of refrigeration system and reduce the leakage of greenhouse gasses [12].

Another researcher, Lewis, tried to demonstrate the interdependencies of energy and maintenance management based on three case studies in facilities in North America [13]. Based on the case studies, there was an interdependent link between energy and maintenance management, but this interdependent link was not widely applied in practice because reactive maintenance and energy management were commonly applied [13]. If the link were widely applied, both the energy and maintenance program could be more productive. Lastly, he also concluded that tools were required to assist facility managers in planning and implementing energy and maintenance management programs [13].

In addition to Knowles and Lewis's works, Al-Ghanim developed a quantitative model in the form of multiple linear regression that related energy consumption and maintenance factor (Failure rate) and production factor (Production rate) [14] in an energy-intensive plant. This model was used and tested in an aluminum profile manufacturing plant in Middle-East. He proved that there was a significant relationship between these three factors. A proper preventive maintenance program and other factors such spare parts availability, and skill of maintenance personnel would result in lower failure. When the number of failure decrease, energy consumption per unit production would decrease because of less energy consumed for fewer production startup. This method could generate significant energy savings at low cost [14]. The same result also obtained by Nofirman, who performed a case study analysis in chiller system [15]. The condenser cleaning scheduled was optimized based on an increase in energy consumption and condenser cleaning cost. By finding an optimal interval for condenser cleaning, there would be large energy savings potential [15]. The potential savings were obtained at low cost [15].

3. Maintenance for Energy Efficiency

Energy efficiency program is important to reduce energy consumption and cost. Energy efficiency improvement can be achieved by application of energy efficiency technologies [2], operational or production improvement [16], and proper housekeeping and maintenance practices [11] (Figure 1). Application of energy efficiency technology can range from application of energy-efficient motor, application of variable speed drives (VSD), application of waste heat recovery technlogies and etc. Energy consumption reduction also can be achieved by operational or production improvement [16] in many ways, such operational or production improvement, economic scheduling of equipment, economic scheduling of chiller plants and etc. Our focus is on reducing energy consumption through proper maintenance strategies (Figure 2). Maintenance strategies for energy efficiency can be classified into housekeeping, preventive maintenance, predictive maintenance (PdM), and condition-based maintenance (CBM), and advanced condition monitoring (Fault detection and diagnosis and etc) (Figure 2). Potential reduction of energy consumption through maintenance strategis is discussed in the next several sections.

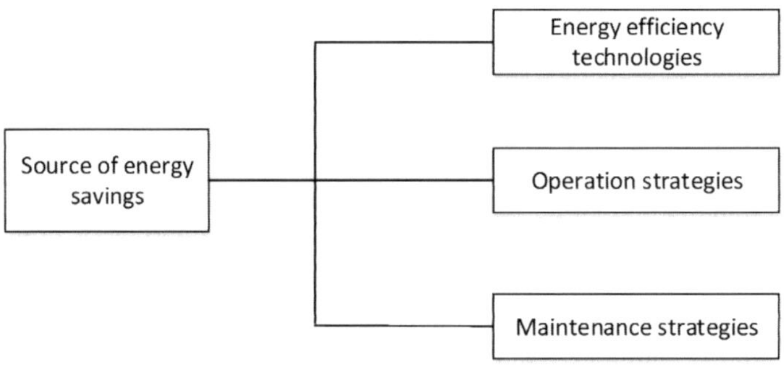

Figure 1. Sources of energy saving opportunities

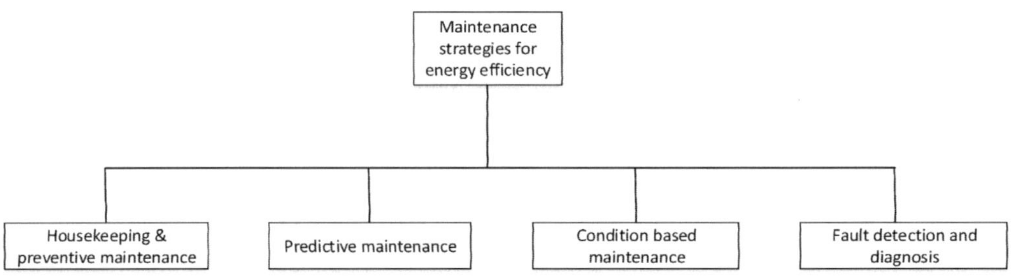

Figure 2. Maintenance strategies for energy efficiency

IOP Conf. Series: Materials Science and Engineering **530** (2019) 012047 doi:10.1088/1757-899X/530/1/012047

3.1. Housekeeping and Maintenance
Housekeeping is usually concerned with aesthetic, organizational aspects, and safety hazards [17]. However, there is no single accepted definition for housekeeping. The example of housekeeping for safety consideration is such as identification of spills, leaks, damaged equipment, faulty and etc. A steam leak does not have an impact on safety only but also on energy consumption. Preventing or repairing steam leak would minimize the negative impact both on safety and energy consumption. Nowadays, the effective energy management program incorporates housekeeping into the program. In energy efficiency perspective, housekeeping are things related to energy loss identification by walking through or performing a visual inspection on a facility [18]. In practice, implementation of housekeeping cannot be separated from maintenance.

Housekeeping and maintenance are recognized as energy saving practices that can be implemented at no or low cost [11]. Identification of (potential) energy loss is usually part of an energy audit. The basic level of energy audit is walk-through energy audit. During the walk-through audit, an auditor always performs a visual inspection to identify (potential) wasted energy in a facility [18]. Some energy audits finds energy savings potential through housekeeping and maintenance in areas of steam distribution system, compressed air distribution system, electric motor and etc [10-11] [43] .

3.1.1. Housekeeping and Maintenance in Industrial Facilities A walk-through energy audit on steam distribution found improper insulation, poor insulation, worn-out insulation, and no insulation on some part of steam distribution [10] [19]. The estimate of energy loss was equal to 250 MWh annually [10]. A large part of steam distribution was not properly installed which lead to steam leakage and energy loss [19]. Repairing steam leak has payback period within 4 months [11]. Corroded pipes and valves were also discovered because of improper maintenance [19]. In addition, inspection of steam traps was also potential for energy savings. During an energy audit, most of the steam traps are in bad condition, some of them did not function properly, some of them were damaged, and few of them did not function at all [10-11] [19]. Checking and maintaining steam traps regularly has payback period within 5 months [11]. Based on the result, it seems that a good housekeeping and maintenance to prevent energy loss in steam distribution system was ignored.

Lighting energy audit is performed to evaluate the lighting system in a facility while production running. During an energy audit, it was found that windows and roof-light were difficult to access and rarely cleaned, and the workers were not satisfied with the lighting level [10]. To solve this issue, regularly scheduled maintenance was proposed. Providing adequate lighting level can increase productivity. Lumen level of lighting system degrades over time by 30-50% [18]. Additional lumen degradation may be caused by accumulated dirt on the lighting equipment. It is believed that there is a relationship between lighting level and productivity. The worker might prefer higher lighting level than the minimum required by standard [20]. One study showed that a proper lighting level might increase worker productivity up to 7.7% [20]. Another study demonstrated that changed the light level from 30 fc to 50 fc might increase the productivity by 8% [21]. In contrast, reducing lighting level from 100 fc to 50 fc may decrease productivity up to 28% [18]. Therefore, a good maintenance program to maintain the lighting level is necessary

Electric motor energy consumption is responsible for 30-80% of total industrial consumption around the world [22]. Life cycle cost (LCC) of electric motors indicate that 92% of the cost spent on energy consumption, 5% on capital, and the rest 3% on maintenance [23]. To lower motor LCC, one needs to reduce its energy consumption. One energy audit found out that the motors were 25 years old, rewound several times, and fortunately well-maintained [10]. However, when rewinding the motor, the cost of rewinding and reduction in motor efficiency should be taken into consideration. First, the cost of motor rewinding is between 17.5% (for large motor) and 88% (for small motor) of the initial cost of high efficient- motor [22]. For a small motor, replacing the motor with a high-efficient motor may be the best option. In contrast, rewinding larges motor may be more cost-effective. Second, the motor rewinding may reduce its efficiency in the range of 1-2% [22]. Rewound the motor several times may further reduce motor efficiency.

In addition to the application of energy-efficiency technology to motors, there are other opportunities to save energy by implementing a good housekeeping and maintenance program. Firstly, turning off the motor when it is not needed. Zohir recommended to switch-off the 20 motors when they were not needed in a factory in Egypt [11]. It could save energy by 84,000 MWh annually without additional investment cost [11]. Secondly, further savings could be achieved by performing proper cleaning on the motor [22]. The failure modes of motor failure consist of overloading (25%), contamination and dirt (43%), single phasing (10%), broken insulation (12%), deterioration (5%), and other (5%) [24]. Any dirt covers the motor surface and parts may lead to insulation damaged, contamination of lubricant (a damaged to bearing), and higher motor temperature by blocking the motor cooling process. Thirdly, to perform proper lubrication on the motor. Fourthly, to make sure proper installation and alignment of the motor [22]. All of these factors, if they are not done properly, may cause motor failure or at least a performance deterioration. Proper maintenance program may prevent or reduce the consequences of motor failure.

Compressed air system (CAS) is widely used in industrial or manufacturing process because it is easy to use and has low investment cost. But it is also one of the most expensive utilities in an industrial facility. The largest part of CAS life cycle cost goes to energy cost (78%) and then followed by capital and maintenance cost (16% and 6%) [25]. The CAS consist of the supply-side and demand-side system. It is believed that 50-70% of energy saving potential may come from the demand side and the rest saving potential may come from the supply-side [26]. The greatest energy loss of compressed air system comes from air leaks. Air leaks in compressed air system have no use and waste 20-50% of a compressor output [25] [27-28].

Air leaks also contribute to operating losses, reduced reliability, lowered production, and increased maintenance requirement and cost [25]. The best way to detect leaks is to use an ultrasonic leak detector. By knowing the size and the location of the leaks, immediate action to seal the leaks can be performed to mitigate further energy losses. Preventing and repairing the leaks are very cost-effective. Several studies has revealed that it had a payback period of less than one year [27]. Consequently, reduced air leaks can increase production and decrease operating and maintenance cost [27]. However, total prevention of air leaks is impractical, but 15% leakage rate is acceptable [28]. In addition to reducing leaks, housekeeping and maintenance can have a positive impact on energy efficiency on compressed air system by minimizing or eliminating inappropriate use of compressed air, switching-off air-using equipment when not in used, changing air filter regularly, and keeping the compressor and intercooling surface clean [27] [29].

The operation and maintenance (O&M) personnel also play a very important role in energy management practice. Incapability of O&M personnel in maintaining and operating the equipment could prevent or at least lower the projected energy savings. Shirley has demonstrated the importance of energy management practices related to O&M [30]. One study showed that 80% of energy savings could be caused by energy management practices of O&M personnel [30], what's more, many energy audits focus on technical and economic analysis and usually does not consider the operation and maintenance factors. These factors included evaluation of past maintenance practices, maintenance cost, preventive maintenance techniques and tools, staff skills and attitudes [31]. Therefore, for a successful energy efficiency programs, operation and maintenance factors should be addressed accordingly.

3.2. Condition Based Maintenance (CBM) For Energy Efficiency

Less than a decade ago, the term of factories 4.0, also known as smart factory, has been introduced. The concept of a smart factory is the future of manufacturing. It also indicates that the era of 4th industrial revolution has just begun. The factories 4.0 concept combines the automation and data exchange in manufacturing technologies. It includes the application of internet of things (IoT), cloud computing, cyber-physical system (CPSs), big data analysis (BDA), and information and communication technology (ICT) [32]. These key technologies enable intelligent manufacturing. The emerge of the new manufacturing technology also affects the energy management and maintenance strategy and practice. It leads to an advanced application of condition monitoring technology. It may enables maintenance strategy to shift from traditional preventive and predictive maintenance to condition-based maintenance [33-34].

The application of IOT in energy management might enable manufacturing to measure and record energy consumption at near real-time at the component, sub-system, equipment, production line, and plant level [35-37]. By monitoring energy consumption in near real time and with the supporting tool for visualization of energy consumption pattern and profile, one might be able to notice any abnormal behavior in increased energy consumption. The increased energy consumption profile may indicate fault or malfunction of the equipment [38-39]. With the help of data and decision analysis for maintenance, immediate and proper maintenance actions can be performed to tackle the problem [35&38]. Monitoring energy consumption also enables maintenance personnel to detect the source of failure (At what machine and its location) and diagnose the type of failure [36]. To enable maintenance personnel to detect and diagnose faults correctly, the measurement of other parameters may be required. The parameters need to be measured divided into two categories but not limited to equipment and work environment parameter [34]. The equipment parameter consists of speed, displacement, position, vibration, flow, temperature, voltage, power and etc. On the other hand the. The work environmental parameter comprises of noise, temperature, humidity, electromagnetic radiation and etc [34].

3.2.1. Application of CBM in Industrial Manufacturing

Endo conducted field experiments of energy monitoring for maintenance of two press machine, a molding machine and four compressors in a discrete manufacturing plant [40-41]. The monitoring collected two kinds of data for two press machines, and a molding machine. Firstly, a behavior data (Equipment behavior data) in the form of the waveform of energy. Secondly, a target data in the form of energy [40]. In addition, it was important to define the operational condition of equipment: Normal energy and wasted energy. The normal energy represented the energy consumption during the production of a normal product (Non-defected product) on the designed tact time. The wasted energy is classified into stopped condition (Anomaly stop, and tooling change), performance decrement condition (Idling and tact delay), and defective condition [40].

With the help of visualization tool, the operator could decide the proper action to restore the performance when the abnormal behavior of energy consumption happens for each operational condition type. In this case, they did two improvement action to reduce energy consumption. First improvement action was to reduce the tact delay by maintaining the components of the molding machine and a press machine and replacing a mold in the molding machine. Second improvement action was to reduce energy consumption by turning off the press machine when it was not needed and replacing a mold in a molding machine. The energy saving obtained from the improvement was around 10% [40].

The same procedure was applied to compressor system. Endo defined compressed air flow and waveform of energy as behavior data and energy as target data. The wasted energy occurred during stop condition. To improve the energy consumption, several improvement actions were performed based on the visualization of behavior and energy data pattern, and operational condition of equipment. First, lowering the set pressure of the compressor system. Second, improving pipes around the dicing machine. The result of the improvement was 7% energy savings in term of kWh consumption and 20% energy savings in term of specific energy consumption (kWh/m3) [41]. The improvement result for machine tools and compressors was not only in term of reduced energy consumption but also an increase in productivity [40-41].

Motor one of the largest energy-consuming equipment in industries [22]. Most of the energy efficiency programs for the motor is a replacement with more efficient more [22][42]. Second is by using variable speed drive whenever possible [42-43]. However, the motor efficiency deteriorates overtime and wastes energy. To overcome this issue, Singh proposed a methodology for motor efficiency motoring and maintenance optimization with cost-benefit analysis [44]. Simulation was performed for three type of efficiency drop curves: Linear, exponential, and quadratic curves [44]. On the other hand, traditional condition monitoring usually defines the critical limit, when a parameter reaches the limit, it is time to perform maintenance. Unfortunately, it is not always economically to

IOP Conf. Series: Materials Science and Engineering **530** (2019) 012047 doi:10.1088/1757-899X/530/1/012047

perform maintenance since it does not consider the wasted energy during the performance degradation since most motor faults lead to efficiency degradation. Consequently, the need for condition monitoring such as vibration, oil, current, and thermography analysis can be minimized. In addition, by performing correct maintenance based on efficiency monitoring, the wasted energy can be minimized [44]. Therefore, performing efficiency monitoring based maintenance can reduce wasted energy and maintenance cost.

3.2.2. Maintenance Policies with Energy Efficiency Consideration

Maintenance policy definition is "... set of rules describing the triggering mechanism for the different maintenance action. It is a question of what triggers maintenance action ?..." [45]. Nowadays, maintenance policy in literature is more related to maintenance optimization model. Maintenance policies are classified into corrective maintenance, preventive maintenance, predictive maintenance, autonomous maintenance, and design out maintenance [46]. Unfortunately, there are only few maintenance policies with energy saving consideration available in the literature [47-48]. Combination of condition monitoring and maintenance policies create a condition-based maintenance (CBM). Some of available energy-oriented maintenance policies is discussed in the next few paragraphs.

First, Hoang introduced energy efficiency indicator (EEI) and mathematical formulation for calculating the proposed EEI at component and system level [49]. He also proposed a concept of remaining efficiency-efficient lifetime (REEL). REEL provided the information regarding efficient lifetime before the efficiency degrades below an allowable level at component or system level. REEL may be a useful tool to assist maintenance decision. The proposed concept was tested on a fan blower system. Hoang also tested the proposed EEI based-CBM model on TELMA platform software to predict the EEI and REEL at both component and system level. The results showed that the proposed model could predict the EEI and REEL when the motor experienced efficiency degradation on their two main components; accumulator and motor drive [50]. Then He compared the proposed EEI based-CBM model with conventional CBM Model (Based on deterioration level). The result demonstrated that EEI based-CBM could surpass conventional CBM in term of cost and implementation [51].

Second, Xu proposed a periodic maintenance model that considered productivity and energy efficiency. In a special case, this model could lead to maximum average productivity and average efficiency. But it is likely that in most cases, there must be a trade-off between average productivity and efficiency [52]. Furthermore, Yildrim proposed a combined model for production and maintenance problem with minimal repair and energy consumption consideration to minimize the total cost [53]. In addition, a study by Yan showed that there was a relationship between energy consumption and reliability of a machine tool. Yan also proposed a methodology to assess the impact of maintenance on energy consumption. The result demonstrated that as the reliability decreases, energy consumption increases dramatically [54].

Next, Zou proposed an opportunity window model to optimize energy saving by turning off a certain machine and performing preventive maintenance at the same time without sacrificing the production output [47]. The trade-off between energy saving and production output may occur. Finally, Xia developed an energy-oriented opportunistic maintenance [48]. The concept is different from traditional opportunistic maintenance that only considers maintenance cost and equipment downtime [55]. The proposed methodology used energy saving window (ESW) policy to reduce energy consumption, downtime, and simplify scheduling for a batch production line in series [48].

3.3. Faults detection and diagnosis, and thermoeconomic diagnosis

Fault detection and diagnosis (FDD) and thermoeconomic diagnosis are one of advanced condition monitoring methods for an energy system. An energy system usually consists of several subsystems or components that work in series. One fault or malfunction on a component may affect the performance of the next component and so on. Therefore, fault or malfunction diagnosis for this type of system is relatively difficult to detect what component causes faults since a fault in one component may affect many other component performance and the entire system. Several chiller FDDs and thermoeconomic diagnosis are briefly explained in the next two paragraphs.

Chiller faults detection and diagnosis (FDD) is used to detect a fault on the system and then diagnose the cause of the fault. Therefore, necessary maintenance actions can be performed to restore the performance. Chiller FDD is one type of condition monitoring (CM) techniques. In general, the chiller FDD is classified into quantitative-model based, qualitative-model based, and process history based [56]. There are eight common faults occur on the chiller system [57]: reduced condenser water flow, reduced evaporator water flow, refrigerant leak, refrigerant overcharge, excess oil, condenser fouling, non-condensables in the refrigerant, and defective expansion valve. Each of the common faults contributes to reduced chiller performance, in other words, it increases chiller energy consumption. The examples of chiller FDD method are ruled-based diagnostician [58], simple thermodynamic model [59], thermoeconomic monitoring [60], and etc. However, the integration of the chiller FDD into current maintenance program has never been available in the literature yet.

A thermoeconomic diagnosis was first theoretically developed by Valero [61]. Thermoeconomic analysis can be applied to the design, operational, and maintenance stage of an energy systems [62]. It is as important as energy analysis. Thermoeconomic diagnosis can detect a malfunction of an equipment and then quantity the wasted energy because of the malfunction, and therefore proper maintenance action can be taken on the right equipment and with the right maintenance solution [62-65]. By considering maintenance cost and wasted energy cost, an optimal maintenance result can be obtained. In addition, the effect of maintenance on an equipment of an energy system can be validated by thermoeconomic diagnosis [62]. Some case studies applied thermoeconomic diagnosis on refrigeration system [66], steam power plant [62], combined cycle power plant [67], gas turbine system [68], desalination plant [69] and etc. However, it seems that the application of thermoeconomic diagnosis in practices is not widely used in practice.

4. Maintenance concept: RCM, TPM, and ECM

A maintenance concept is defined as the set of various maintenance strategies (Corrective, predictive, CBM, etc) and the general structure in which these strategies are selected and implemented [29&70]. The maintenance concept forms the framework from which maintenance strategies are developed [70]. The well-known maintenance concepts are reliability centered maintenance (RCM) and total productive maintenance (TPM) (Figure 3). RCM was born in airlines industries, on the other hand, TPM was born in manufacturing industries. The RCM is an asset-centered methodology, on the other hand, TPM is a people-centered methodology [71]. However there could be a relationship between TPM and RCM [72]: RCM could enhance TPM implementation, and vice versa. These two approaches are widely used in industrial and manufacturing plants around the world. However, there is another maintenance concept developed for energy efficiency called energy centered maintenance [ECM]. These maintenance concepts will be briefly reviewed in the next several paragraphs.

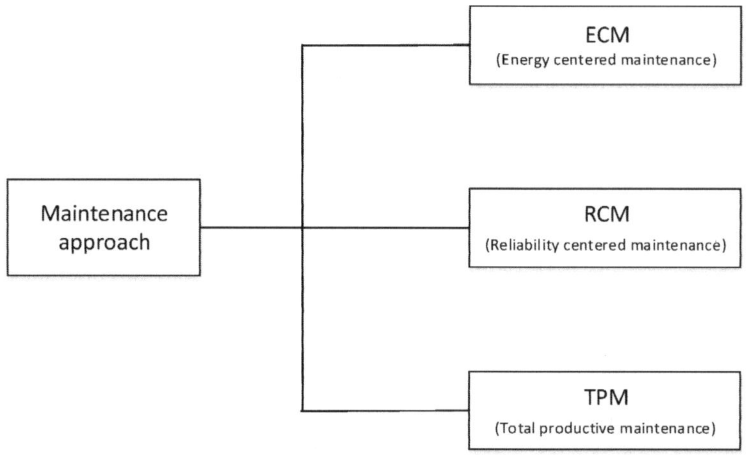

Figure 3. Potential maintenance approaches for energy efficiency consideration

First is reliability-centered maintenance (RCM). RCM is a process used to determine the maintenance requirements of any physical assets in its operating context [73]. The primary objective of RCM is to maintain system function, not on the system hardware [74]. The RCM was born for the first time in the airline's industry. RCM can be applied to the component, sub-system, and system level. Since then, the RCM has been applied to many different industries such as marine industry [75], US Navy ships [76], steel industry [77], electric distribution systems [78], automotive industry [79], refining industry [80], real estate industry [81], process industry [82], airline industries [83], power plant [84], and etc. Moreover, RCM can also be applied for specific consideration such as safety [85] and reliability [86]. The RCM could also be adapted and modified for specific needs such as maintenance design for a paint-spraying robot [87]. Unfortunately, it seems that specific application of RCM for energy efficiency consideration has never been done.

Next is total productive maintenance (TPM). TPM according to Nakajima [88] is ..." *productive maintenance carried out by all employees through small group activities....*". TPM also has a goal to maximize equipment effectiveness. To improve equipment effectiveness, TPM starts by identifying the six losses with respect to equipment [88]. The overall equipment effectivenes (OEE) is used a key performance indicator in TPM implementation. However, in recent development and implementation of TPM, there are 16 losses types that fall into four categories: losses that impede overall equipment efficiency, equipment loading time, worker efficiency, and efficient use of resources [89]. The last category includes yield loss, consumable loss, and energy loss. Implementation of TPM in a manufacturing company in Japan could reduce energy consumption by 30% [88]. A case study of TPM in a steel manufacturing plant showed that TPM implementation could reduce the energy cost between 8% and 27% [90].

A greater savings might be achieved by more focusing on the maintenance with energy consideration. Unfortunately, there are only few researches focusing on the application of TPM with energy consideration. Al-Homoud introduced the concept of total productive energy management (TPEM) [91]. TPEM approach combined the principle of total quality management (TQM) and total productive maintenance (TPM) to reduce energy consumption. The objective of TPEM is to motivate all members involves in the operation and maintenance of a facility to take part in energy conservation activities. He also emphasized that energy savings through operation and maintenance practices could be successfully achieved and more feasible to implement than those achieved through major retrofits [92].

To improve TPM process, Barletta proposed an Energy Overall Equipment Effectivenes (Energy OEE) as key performance indicator (KPI) for a discrete manufacturing firms [93]. Energy OEE was used to assess the impacts of energy consumption losses instead of time losses. The application of Energy OEE was tested in a virtual manufacturing facility modeled by discrete event simulation (DES) on a CNC machine [92]. The result demonstrated that the use of Energy OEE along with DES might improve both energy efficiency (Reduced energy consumption) and productivity (Reduced defective parts) [92]. The introduction of Energy OEE on TPM process could improve energy efficiency factor in manufacturing industries.

The last one is energy centered maintenance (ECM). Howell introduced ECM concept to increase energy efficiency of a facility through maintenance [93]. ECM is a maintenance concept focusing on energy consumption reduction. Energy consumption excesses or energy waste are used as the primary criterion in determining the specific maintenance or repair needs [93]. ECM consists of six steps: First step is equipment identification that energy related. Second is data collection and equipment baseline prediction. Third is development of ECM inspection plan. Fourth is equipment performance monitoring and then comparing with the baseline. Any deviation with baseline needs root cause analysis for improvement. Next is the selection of maintenance action to restore equipment efficiency. The last step is the improvement of maintenance plans. A case study of ECM application in a facility claimed that ECM could help reducing energy consumption by as much as 30% [93]. It seems that ECM can be a prospective maintenance concept focusing on energy consumption reduction.

5. Discussion

The heart of maintenance for energy efficiency is predictive maintenance. The first strategy of maintenance for energy efficiency is housekeeping. Housekeeping identifies a potential energy loss by performing a visual inspection of an industrial facility. In several circumstances, the use of a predictive technology such as ultrasonic leak detector, light meter and other may be required during the housekeeping activity. From an economic point of view, housekeeping is usually cost-effective for energy management program since it can be done at no or low cost. In addition, more saving can be obtained from non-energy benefits such as an increase in productivity.

The second maintenance strategy is condition-based maintenance (CBM). CBM techniques are performed by monitoring energy in real time with the help of IoT technologies. It may need many sensors to monitor many key parameters of in-service equipment. The key performance indicator of a fault is the increase in energy consumption. Any abnormal profile of energy consumption indicates that a fault on an equipment has occurred. Since many equipment faults lead to performance degradation, the need for other predictive maintenance tasks such as vibration analysis, oil analysis, and thermography analysis can be minimized. Therefore, CBM application may generate two main benefits; lower energy and maintenance cost. Since the CBM may require many measurement instruments and supporting system, the investment cost tends to be higher. Unfortunately, the cost-effective of CBM application is rarely discussed in the literature. In addition, there may be a need for a methodology to assess the economics of CBM application in practice.

The third strategy is advanced condition monitoring for an energy system such as FDD and thermoeconomic diagnosis. The investment cost for the application of advanced condition monitoring might be low since it usually only needs existing measurement instruments. But it may need a human resource with high knowledge and skills to develop the system and interpret the result of condition monitoring into maintenance action. The human factor may cause the application of these methods is not widely applied in practice. In addition, the Integration of FDD and thermoeconomic diagnosis into current maintenance program have never been available in the literature yet.

The application of these maintenance strategies can be optimized by applying maintenance concepts such as RCM and TPM. The use of maintenance concept may help to select the appropriate maintenance strategy and to implement the proactive task effectively and efficiently. Unfortunately, the application of RCM and TPM with energy efficiency consideration is rarely available in the literature. In addition, a novel maintenance framework may be needed to integrate available maintenance concepts and strategies to achieve more energy-saving potential cost-effectively.

6. Conclusion

Proper maintenance is essential to reduce energy losses. Energy saving through maintenance can be obtained by housekeeping and simple maintenance, condition based maintenance and advanced condition monitoring. Proper housekeeping and simple maintenance can reduce energy losses at low cost and relatively cost-effective (Shorter payback period). Application of condition-based maintenance also can generate energy savings. But the economic analysis of CBM implementation for energy efficiency may be required to justify the cost-effective of CBM application.

The trend of maintenance strategy for energy efficiency is condition-based maintenance. First is by combining energy monitoring with IoT technologies. Second is by introducing an energy-related performance indicator for condition-based maintenance. Next is the use of advanced condition monitoring method such as fault detection and diagnosis (FDD), and thermoeconomic diagnosis. However, maintenance for energy efficiency cannot be separated from operational improvement. During an inspection or condition monitoring, energy can be saved by both proper maintenance and operational improvement.

There are some maintenance concepts available in the literature, the well-known concepts are RCM and TPM. Both RCM and TPM have been widely used in many application. Unfortunately, RCM and TPM application are not fully explored and utilized for energy reduction consideration. In addition, a new maintenance framework may be required to integrate available maintenance concepts and strategies to achieve more energy-saving potential.

References

[1] Worrell E, Laitner J A, Ruth M and Finman H 2003 Productivity benefits of industrial energy efficiency measures *Energy* 28 p 1081–98
[2] Pons M, Bikfalvi A, Llach J and Palcic I 2013 Exploring the impact of energy efficiency technologies on manufacturing firm performance *J. Clean. Prod.* 28 p 289-96
[3] Boyd G A and Pang J X 2000 Estimating the linkage between energy efficiency and productivity *Energy Policy* 28 p 289-96
[4] Zhang J and Wang G 2008 Energy saving technologies and productive efficiency in the Chinese iron and steel sector *Energy* 33 p 525–37
[5] Pye M and McKane A 2000 Making a stronger case for industrial energy efficiency by quantifying non-energy benefits *Resources, Conservation and Recycling* 28 p 171–183
[6] Backlund S, Thollander P, Palm J and Ottosson M 2012 Extending the energy efficiency gap *Energy Policy* 51 p 392–96
[7] Jovanović B and Filipović J 2016 ISO 50001 standard-based energy management maturity model - Proposal and validation in industry *J. Clean. Prod.* 112 Part 4 p 2744-55
[8] Abdelaziz E A, Saidur R and Mekhilef S 2011 A review on energy saving strategies in industrial sector *Renew. Sustain. Energy Rev.* 15 p 150–68
[9] Kannan R and Boie W 2003 Energy management practices in SME - Case study of a bakery in Germany *Energy Convers. Manag.* 44 p 945–59
[10] Gordić D, Babić M, Jovičić N, Šušteršič V, Končalović D and Jelić D 2010 Development of energy management system - Case study of Serbian car manufacturer *Energy Convers. Manag.* 51 p 2783–90
[11] Zohir A E 2010 Energy efficiency improvement by housekeeping measures. Thermal Issues in Emerging Technologies, ThETA 3, Cairo, Egypt, Dec 19-22 nd
[12] Knowles M and Baglee D 2012 The role of maintenance in energy saving in commercial refrigeration *J. Qual. Maint. Eng.* 18 No. 3 p. 282-94
[13] Lewis A, Elmualim A and Riley D 2011 Linking energy and maintenance management for sustainability through three American case studies *Facilities.* 29 Issue: 5/6, p.243-254
[14] Al-Ghanim A 2003 A statistical approach linking energy management to maintenance and production factors *J. Qual. Maint. Eng.* 9 Issue: 1 p.25-37

[15] Firdaus N, Prasetyo B T and Luciana T 2016 Chiller: Performance deterioration and maintenance *Energy Eng. J. Assoc. Energy Eng.* 113
[16] Plambeck E L 2012 Reducing greenhouse gas emissions through operations and supply chain management *Energy Econ.* 34 p S64–S74
[17] Dufort V M and Infante-Rivard C 1998 Housekeeping and safety: An epidemiological review *Saf. Sci.* 28 p 127–38
[18] Capehart B L, Turner W C and Kennedy W J 2006 *Guide to energy management*
[19] Ozturk H K 2005 Energy usage and cost in textile industry: A case study for Turkey *Energy.* 30 p 2424–46
[20] Juslën H 2007 *Lighting , Productivity and Preferred Illuminances - Field Studies in the Industrial Environment*
[21] van Bommel W and van den Beld G 2004 Lighting for work: a review of visual and biological effects *Light. Res. Technol.* 36(4) p 255 – 69
[22] Hasanuzzaman M, Rahim N A, Saidur R and Kazi S N 2011 Energy savings and emissions reductions for rewinding and replacement of industrial motor *Energy* 36 p 233-240
[23] Krings A, Boglietti A, Cavagnino A and Sprague S 2017 Soft Magnetic Material Status and Trends in Electric Machines *IEEE Trans. Ind. Electron.*
[24] McCoy GA, Litman T and Douglass JG 1003 *Energy-efficient electric motor selection handbook*
[25] Saidur R, Rahim N A and Hasanuzzaman M 2010 A review on compressed-air energy use and energy savings *Renew. Sustain. Energy Rev.* 14 p 1135–53
[26] Nehler T, Parra R and Thollander P 2018 Implementation of energy efficiency measures in compressed air systems: barriers, drivers and non-energy benefits *Energy Effic.* 11 p 1281 – 1302
[27] Saidur R and Mekhilef S 2010 Energy use, energy savings and emission analysis in the Malaysian rubber producing industries *Appl. Energy* 87 p 2746–58
[28] Dindorf R 2012 Estimating potential energy savings in compressed air systems *Procedia Engineering* 39 p 204 – 11
[29] Alsyouf I 2007 The role of maintenance in improving companies' productivity and profitability *Int. J. Prod. Econ.* 105 p 70–78
[30] Hansen S J 1999 Investment Grade Energy Audits *Cogeneration and Competitive Power Journal* 14 pp 63-70
[31] Hansen S J and Brown J W. *2003* Investment Grade Energy Audit CRC Press
[32] Zhong R Y, Xu X, Klotz E and Newman S T 2017 Intelligent Manufacturing in the Context of Industry 4.0: A Review *Engineering* 3 p 616–630
[33] Shrouf F and Miragliotta G 2015 Energy management based on Internet of Things: Practices and framework for adoption in production management *J. Clean. Prod.* 100 p 1-12
[34] Xiaoli X, Yunbo Z and Guoxin W 2011 Design of intelligent internet of things for equipment maintenance *Proceedings - 4th International Conference on Intelligent Computation Technology and Automation, ICICTA 2011*
[35] Shrouf F, Ordieres J and Miragliotta G 2014 Smart factories in Industry 4.0: A review of the concept and of energy management approached in production based on the Internet of Things paradigm *IEEE International Conference on Industrial Engineering and Engineering Management*
[36] Brizzi P, Conzon D, Khaleel H, Tomasi R, Pastrone C, Spirito A M, Knechtel M, Pramudianto F and Cultrona P 2013 Bringing the Internet of Things along the manufacturing line: A case study in controlling industrial robot and monitoring energy consumption remotely *IEEE International Conference on Emerging Technologies and Factory Automation, ETFA*
[37] Vijayaraghavan A and Dornfeld D 2010 Automated energy monitoring of machine tools *CIRP Ann. - Manuf. Technol.*
[38] Karnouskos S, Colombo A W, Lastra J L M and Popescu C 2009 Towards the energy efficient future factory IEEE International Conference on Industrial Informatics (INDIN)

[39] Vikhorev K, Greenough R and Brown N 2013 An advanced energy management framework to promote energy awareness *J. Clean. Prod.*
[40] Endo M, Tsuruta K, Saitoh Y, Nakajima H and Hata Y 2011 Toward an optimal QCDE in manufacturing by health monitoring of equipment energy consumption *Conference Proceedings - IEEE International Conference on Systems, Man and Cybernetics*
[41] Endo M, Nakajima H and Hata Y 2012 Simplified Factory Energy Management System based on operational condition estimation by sensor data *IEEE International Conference on Automation Science and Engineering*
[42] Saidur R 2010 A review on electrical motors energy use and energy savings *Renew. Sustain. Energy Rev.* 14 p 877–98
[43] de Almeida A T, Fonseca P and Bertoldi P 2003 Energy-efficient motor systems in the industrial and in the services sectors in the European Union: Characterisation, potentials, barriers and policies *Energy* 28 p 673–90
[44] Singh G, Anil Kumar T C and Naikan V N A 2019 Efficiency monitoring as a strategy for cost effective maintenance of induction motors for minimizing carbon emission and energy consumption *Reliab. Eng. Syst. Saf.* 170
[45] Pintelon L, Gelders L and Puyvelde F V 2000 *Maintenance Management* 2nd Edition (Acco)
[46] Ding S H and Kamaruddin S 2014 Maintenance policy optimization—literature review and directions *Int. J. Adv. Manuf. Technol.* 76 p 1263 – 83
[47] Zou J, Arinez J, Chang Q and Lei Y 2016 Opportunity Window for Energy Saving and Maintenance in Stochastic Production Systems *J. Manuf. Sci. Eng.* 138
[48] Du S, Xia T, Xi L, Pan E and Xiao L 2018 Energy-Oriented Maintenance Decision-Making for Sustainable Manufacturing Based on Energy Saving Window *J. Manuf. Sci. Eng.* 140
[49] Hoang A, DO P, Iung B, Levrate E and VOISIN A 2014 Prognostics and Energy Efficiency: Survey and Investigations *Annual conference of the prognostics and health management society*
[50] Hoang A, Do P and Iung B 2015 Prognostics on energy efficiency performance for maintenance decision-making: Application to industrial platform TELMA *Proceedings of 2015 Prognostics and System Health Management Conference, PHM 2015*
[51] Hoang A, Do P and Iung B 2016 Investigation on the use of energy efficiency for condition-based maintenance decision-making *IFAC-PapersOnLine* 49:28 p 073–078
[52] Xu W and Cao L 2014 Energy efficiency analysis of machine tools with periodic maintenance *Int. J. Prod. Res.* 52:18 p 5273 – 85
[53] Yildirim M B and Nezami F G 2014 Integrated maintenance and production planning with energy consumption and minimal repair *Int. J. Adv. Manuf. Technol.* 74 p 1419 – 30
[54] Yan J and Hua D 2010 Energy consumption modeling for machine tools after preventive maintenance *IEEM2010 - IEEE International Conference on Industrial Engineering and Engineering Management*
[55] Ab-Samat H and Kamaruddin S 2014 Opportunistic maintenance (OM) as a new advancement in maintenance approaches: A review *J. Qual. Maint. Eng.* 20 No. 2 pp. 98-121
[56] Katipamula S and Brambley M R 2005 Review article: Methods for fault detection, diagnostics, and prognostics for building systems—A review, part I *HVAC R Res.* 11:1 p 3-25
[57] Comstock M C, Braun J E, Groll E A and Danks R 2002 A survey of common faults for chillers *ASHRAE Transactions.* 108 p 819
[58] Comstock M C, Braun J E and Groll E A 2002 The sensitivity of chiller performance to common faults *ASHRAE Transactions.* 7:3 p 263-79
[59] Saththasivam J and Ng K C 2008 Predictive and diagnostic methods for centrifugal chillers *ASHRAE Transactions.* 114 p 282-87
[60] Piacentino A and Talamo M 2013 Innovative thermoeconomic diagnosis of multiple faults in air conditioning units: Methodological improvements and increased reliability of results *Int. J. Refrig.* 36:8 p 2343-65

[61] Valero A, Serra L and Lozano M A 1993 *Structural Theory of Thermoeconomics ASME Winter Annual Meeting Symposium on thermodynamics and the design, analysis and improvement of energy systems.*
[62] Ray T K, Datta A, Gupta A and Ganguly R 2010 Exergy-based performance analysis for proper O&M decisions in a steam power plant *Energy Convers. Manag.* 51:6 p 1333–44
[63] Usón S, Valero A and Correas L 2010 Energy efficiency assessment and improvement in energy intensive systems through thermoeconomic diagnosis of the operation *Appl. Energy* 87 p 1989–95
[64] Verda V, Serra L and Valero A 2005 Thermoeconomic Diagnosis: Zooming Strategy Applied to Highly Complex Energy Systems. Part 1: Detection and Localization of Anomalies *J. Energy Resour. Technol.* 127 p 42-49
[65] Verda V 2006 Accuracy level in thermoeconomic diagnosis of energy systems *Energy* 31 p 3248–60
[66] D'Accadia M D and De Rossi F 1998 Thermoeconomic optimization of a refrigeration plant *Int. J. Refrig.* 39:12 p 1223–32
[67] Zaleta-Aguilar A, Olivares-Arriaga A, Cano-Andrade S and Rodriguez-Alejandro D A 2016 β-characterization by irreversibility analysis: A thermoeconomic diagnosis method *Energy* 111 p 850-58
[68] Orozco D J R, Venturini O J , Palacio J C E and Del Olmo O A 2016 A new methodology of thermodynamic diagnosis, using the thermoeconomic method together with an artificial neural network (ANN): A case study of an externally fired gas turbine (EFGT) *Energy* 123 p 20-35
[69] Uche J, Serra L and Valero A 2006 Exergy Costs and Inefficiency Diagnosis of a Dual-Purpose Power and Desalination Plant *J. Energy Resour. Technol.* 128:3 p 186-93
[70] Waeyenbergh G and Pintelon L 2002 A framework for maintenance concept development *Int. J. Prod. Econ.* 77 p 299–313
[71] Jardine A K S and Tsang A H C 2013 *Maintenance, replacement, and reliability: Theory and applications, second edition*
[72] Ben-Daya M 2000 You may need RCM to enhance TPM implementation *J. Qual. Maint. Eng.*
[73] Moubray J 1997 *Reliability centered maintenance*
[74] Rausand M 1998 Reliability centered maintenance *Reliab. Eng. Syst. Saf.* 60:2, pp 121-132
[75] Conachey R M and Montgomery R L 2002 Application of reliability-centered maintenance techniques to the marine industry *ABS Technical Papers*
[76] Jacobs K S 1998 Reducing maintenance workload through Reliability-Centered Maintenance (RCM) *Nav. Eng. J.* 110:4 pp 89-97
[77] Deshpande V S and Modak J P 2002 Application of RCM to a medium scale industry *Reliab. Eng. Syst. Saf.* 77 pp 31-34
[78] Goodfellow J W 2000 Applying reliability centered maintenance (RCM) to overhead electric utility distribution systems, *Power Engineering Society Summer Meeting IEEE*
[79] Ramli R and Arffin M N 2012 Reliability centered maintenance in schedule improvement of automotive assembly industry *Am. J. Appl. Sci.* 9:8 p 1232-36
[80] Deepak P and Jagathy Raj 2013 A New Model For Reliability Centered Maintenance In Petroleum Refineries *Int. J. Sci. Technol. Res.* 2:5 p 54-64
[81] El-Haram M A and Horner M W 2002 Practical application of RCM to local authority housing: A pilot study *J. Qual. Maint. Eng.* 8 p 135-43
[82] Afefy I H 2010 Reliability-Centered Maintenance Methodology and Application: A Case Study *Engineering* 2 p 863-73
[83] Matteson T D 1985 Airline experience with reliability-centered maintenance. *Nuclear Engineering and Design* 89 p 385-90
[84] Bhangu N S, Singh R and Pahuja G L 2011 *Reliability centred maintenance in a thermal power plant: a case study. Int. J. Productivity and Quality Management.* 7 p 209-28
[85] Deshpande V S and Modak J P 2002 Application of RCM for safety considerations in a steel plant

Reliab. Eng. Syst. Saf. 78 p 325–34
[86] Deshpande V S and Modak J P 2003 Maintenance strategy for tilting table of rolling mill based on reliability considerations *Reliab. Eng. Syst. Saf.* 80 p 1–18
[87] Pintelon L, Nagarur N and Van Puyvelde F 2002 Case study: RCM – yes, no or maybe? *J. Qual. Maint. Eng.* 5 p.182-92
[88] Nakajima S 1998 *Introduction to TPM: Total Productive Maintenance* Productivity Press
[89] Ahuja I P S and Khamba J S 2008 Total productive maintenance: Literature review and directions *Int. J. Qual. Reliab. Manag.* 25 p 709-56
[90] Ahuja I P S and Khamba J S 2007 An evaluation of TPM implementation initiatives in an Indian manufacturing enterprise *J. Qual. Maint. Eng.* 13 p. 338-52
[91] Al-Homoud M S 2000 Total productive energy management *Energy Engineering,* 97 p 21-38
[92] Barletta I, Andersson J, Johansson B, May G and Taisch M 2015 Assessing a proposal for an energy-based Overall Equipment Effectiveness indicator through Discrete Event Simulation *Proceedings - Winter Simulation Conference*
[93] Howell M T, Alshakhshir F 2017 *Energy Centered Maintenance: A Green Maintenance System* Fairmont Press, Incorporated, CRC Press, Taylor & Francis Group

IOP Conf. Series: Materials Science and Engineering **530** (2019) 012048 doi:10.1088/1757-899X/530/1/012048

A Review on Joint Optimization of Maintenance with Production Planning and Spare Part Inventory Management

J Q Hwang and H A Samat*

School of Mechanical Engineering, Universiti Sains Malaysia, 14300 Nibong Tebal, Pulau Pinang, Malaysia

Corresponding author *: hasnida@usm.my

Abstract. In maintenance system, the increasing complexity of the manufacturing system and the joint optimization of maintenance is catching the attention of researches. Therefore, the aim of the paper is to review the literatures on joint optimization of maintenance with other interrelated functions in the manufacturing system. The classification is made based on the type of decision variables, which are production planning (production scheduling, lot sizing and buffer) and spare part inventory management. The relationships and dependencies are explain in the discussion to show the significance of the integrated model. Most studies considered two functions in their integrated model, which is either maintenance policy with production planning or maintenance policy with spare part management. It can be observe that research on optimization of more than two functions is still new and yet to be explore. Moreover, most researches focused at the decision-making phase without extensive discussion in various functions that might involve. The objective function which is the cost is also presented in the paper. The objective function is categorized into two criteria: maintenance related cost and also inventory and spare part related cost.

1. Introduction

Nowadays, the role of maintenance seems to be more vital than ever due to the increase of automation and mechanization in the industry. Companies are also focusing on being highly flexible but at the same time optimizing their speed, performance and quality. As the output of the maintenance system is in service form, therefore this function should be integrated with all other departments in the organization to ensure a world-class performance. For large scale enterprise, maintenance cost may consume up to 40% of the total operating budget [1-3]. The main function of maintenance is to preserve the ability of the asset to continue performing with minimal failure [1]. Therefore, lack of maintenance to the asset may cause disaster and losses in the plant.

A well-engineered maintenance management is one of the keys for the organization to achieve their competitive strength in the global market. Well-performed maintenance helps to achieve lowest total life-cycle cost of an asset and hence leads to the reduction of long-term cost and hazard level [1]. Maintenance minimizes the possibility of downtime which creates losses due to the cost of idle production, penalty of late deliveries, cost of overtime to complete the order and lost sales [4]. The new paradigm of business that is focusing on continuous improvement on their process, the advancement in Internet of Things (IoT), rapidly changing technology in the 21st century and the transition into Industry 4.0 force the maintenance to improve so that all the functions are on the same pace. The maintenance planning should not be conducted independently as old practice. Integration with other functions is advisable for risk and cost minimization as well as performance optimization [1][5].

There are factors that are closely related to maintenance engineering which include scheduling, buffer, inventory, routing, spare part and lastly lot sizing. All these factors can be categorized into two main functions which are production and spare part inventory. Production, spare part inventory and maintenance are interrelated functions although in the past they are treated independently. All three functions contribute significantly to the total operating budget, profit and losses of the organization. Production produces output as demanded and therefore contributes to the manufacturing cost and profit of the organization. Another equally important function which is spare part inventory relates to the inventory storage cost and shortage cost. Subsequently, the significance of maintenance to the financials of the organization was proven by the study by Chan et al [6]. According to the author, maintenance cost is about 15% to 30% of total manufacturing cost [6]. The optimization of only one aspect does not mean the lowest cost and best performance for the whole system [5][7][8]. Unilateral planning may end up higher cost and resources to sustain over-assumed system [9].

Collaboration between departments is increasingly essential. As consequence, joint optimization of more than one functions has witnessed a considerable growth in these few years. In the review paper by Ruschel et al. [8] on the topic of industrial maintenance decision making, literature relate to joint optimization has a considerable percentage among 16 applications area. There are a few papers that included quality control in their integrated model. The examples are the studies done by Bouslah et al.[10-12], Guo et al. [13] and Fakher et al. [14]. All of the mentioned authors joint optimized quality control together with production planning and maintenance. Their purpose is to create a feedback system that enables the optimization of maintenance level and period with the result of quality control. This function is not included in the paper as the quality control is not the decision variable in the research. The quality control is integrated to determine the optimal planning of maintenance and production.

Considering the potential of joint optimization of maintenance in the future, a review was done on the joint optimization of maintenance with production planning or spare part inventory management. A general overview of the past researches on this topic, a brief illustration on the maintenance strategies and the objective function will be reviewed thoroughly in the paper. This analysis aims to provide an in-depth understanding of the importance of integrating optimization of maintenance with production planning including the production scheduling, lot sizing and the buffer or spare part inventory management. This review focuses on past researches that give the outcome of joint optimized solution on more than one function whether by two stages or simultaneously.

The scope is limited to papers that include costs as the objective function which include inventory cost and maintenance cost. Although some researches might use downtime, environmental impact, service level or safety as their objective function, total cost has the highest popularity among the researchers. Certainly, we cannot doubt that cost minimization is the common goal for all the organizations. They are striving to produce good products with the least resources and waste to gain more competitive advantage than their competitors. Hence, it is crucial to propose the cost model that is able to screen the real cash flow in the industry. In the paper, a comprehensive review of cost elements in the cost model is provided as a guideline for replication. The paper is structured as following: type of maintenance strategy is presented in Section 2; joint optimization of maintenance with production planning is illustrated in Section 3, a discussion of joint optimization of maintenance with spare part inventory is in Section 4; Section 5 is the discussion of objective function which is the total cost and finally the conclusion of the review is in Section 6.

2. Maintenance Strategy

The maintenance strategies can be categorized into three types which include corrective/ breakdown maintenance (CM), preventive maintenance (PM) and conditional-based maintenance (CBM). The three strategies have their benefits and limitations. Although they have a different concept, they

IOP Conf. Series: Materials Science and Engineering **530** (2019) 012048 doi:10.1088/1757-899X/530/1/012048

involve similar procedure which is the inspection of the status of machines and replacement or repair of the failed part [5].

Corrective Maintenance (CM) is the maintenance task conducted when the asset degrades until it fails unexpectedly. CM is an unscheduled maintenance task and thus requires highest priority. The replacement or repair work should be done as soon as possible as the machine failure may interfere the production planning For this reason, CM is known as firefighting approach [7][15][16]. CM tasks may consist of fail-repair, salvage, rebuild, overhaul and servicing [7]. CM is an important factor in the performance of maintenance as CM consumed more effort from the organization [7]. This strategy is suitable to be used for only some cases. For example, when the hazard rate is constant, the failure does not bring serious effect or losses and the breakdown equipment has low priority among other machines. The limitation of CM is that it has a high possibility to halt the whole production especially when the shortage of maintenance resources including spare parts and technician occur [5].

Preventive Maintenance (PM) is then introduced to minimize the drawback of stochastic breakdown. PM is the planned maintenance task in which the task is conducted on a scheduled basis and interval. The planning of PM may be based on the recommendations of equipment manufacturers and the failure distribution data of the machines [17-18]. The tasks that will be conducted during PM are inspection servicing, calibration, testing, alignment, adjustment and installation [7]. PM brings benefits by increasing the system availability, balancing the workload, reducing the overtime, increasing the revenue, ensuring stable product quality and improving safety. However, the implementation of PM may not be cost effective if the planning is not optimal. PM has disadvantages as well. The spare part consumption is higher and thus increase in cost. The equipment may be exposed to possible damage between the interval of PM. Although PM may reduce the possibility of machine failure, there is still the possibility of machine breakdown unexpectedly. Therefore, the role of CM cannot be neglected as well during stochastic machine breakdown [7].

For traditional concept for PM, the machine is assumed to be as good as new after the maintenance. However, in the real case, the state of the machine is between as good as new and as bad as old after PM. This concept is called imperfect maintenance. It assumes that the failure rate decreases after each PM. This concept had been used by many authors so that the proposed model will be more realistic and relevant to the industry [10][13][19-20].

Conditional-based maintenance (CBM) is another more advanced maintenance strategy had grown its popularity among the industry nowadays. Many authors used CBM in their studies [21-24]. For CBM, the maintenance task is conducted based on the comparison of the condition of the assets to the threshold. Therefore, diagnostic software, sensors or cameras are required to monitor routinely the condition of the machine and then compare with the pre-set threshold. The variables that are collected include the machines' vibration, temperature, oil analysis, pressure and so on [17]. CBM is suitable to manage critical wear [25]. This maintenance strategy brings maintenance closer to the production. However, its drawbacks are that high skill and high technology equipment are needed. CBM can be planned but cannot be scheduled [1]. For the current trend, CBM has the highest interest from the researchers. 40% of the literature reviewed by Ruschel et al. [7] are on CBM. In this case, the need for real-time monitoring technologies is increasingly vital. Whereas for preventive maintenance, it has 23% of all the reviewed literature which is the second highest from the maintenance strategies [8].

3. Joint Optimization of Maintenance with Production Planning

Production and maintenance planning are among the most common and significant problems faced in the manufacturing industry [26]. In the past, production and maintenance planning were treated separately and independently in the production line. Most of the past studies on scheduling assumed the machine never breakdown [5], [27]. There are plenty of researchers that studied the maintenance planning method. Review done by Ernnie et al. had provided comprehensive information on preventive maintenance planning and the methods used in the industry [18]. Ruschel et al. [8] also presented a review paper about industrial maintenance decision-making, the proposed models and the application of methods and tools.

However, in reality, the maintenance activities may interfere with the production output and performance. The delay of maintenance may cause the stoppage of the line and lead to decrease in availability and scheduling disorder. Meanwhile, the duration of the maintenance task will affect the system availability. Most of the time, maintenance department and production department have different objective and therefore cause conflicts. For example, the production manager does not willing to stop the production for planned maintenance whereas the maintenance manager would prefer to increase the frequency of maintenance to reduce the breakdown of machines. The machine is unavailable during the implementation of scheduled PM task or unscheduled CM task [5][26][28-29]. Due to the interdependence between them, the joint optimization of production and maintenance has become a vital issue. Recently, there is growing interest in optimizing maintenance policy together with production planning. The scope of production planning is limited in the integrated model that include production scheduling, lot sizing and buffer.

3.1. Maintenance with Production Scheduling

There are a few researchers try to integrate the planning and optimization of maintenance and production scheduling. In 1996, Asharyeri et al.[30] had developed a mixed-integer linear programming model to optimize the preventive maintenance and scheduling of the production job in order to minimize the production cost and maintenance cost. Cassady and Kutanoglu [31], Wang and Liu [32], Benmansour et al. [27] and Liu et al. [28] presented integrated maintenance and production schedule for a single machine system. The objective of the study of Cassady and Kutanoglu [31] is to optimize the Preventive Maintenance action and job sequence with the hope to minimize the total weighted expected completion time. Wang and Liu [32] had proposed an integrated optimization model for production scheduling and preventive maintenance with its time to failure subject to a Weibull probability distribution. Benmansour et al. [27] aimed to determine the optimal period to perform preventive maintenance and the sequence of the job to minimize the cost. This is the first study that considers tardiness cost problem in their optimization study. In the model proposed by Liu et al. [28], dummy age (system existing age) of the machine is introduced to show the effect of the maintenance action. Dummy age decreases when implementing a maintenance task. This integrated model beside solving job scheduling and maintenance planning problem, resource planning had been determined simultaneously as well [28].

Joint optimization of maintenance and production scheduling for a multi-machine system is more complex than a single machine system. Optimal maintenance policy for a single machine may not be optimal for the multi-machine system. More constraints must be considered in the model. The study of Xiao et al. [26] had optimized the production scheduling and preventive maintenance in a series system to minimize total cost including production cost, preventive maintenance cost, minimal repair cost for unexpected failures and tardiness cost. The decision variables to be determined are the optimal group preventive maintenance interval for machines and the assignment of the PM task which is the starting and ending time of the maintenance activities. However, the assumption causes the imperfectness of the model. The author assumed the machine is as-good-as new after the maintenance [26]. On the other hand, Portioli-Staudacher and Tantardini [29] assumed that the whole production line will stop during maintenance. Hence, their proposed mathematical model is able to model the multi-machine system as a single machine with a single production rate. Their mathematical model joint planned the production and preventive maintenance activities that can be used for rescheduling purpose. Their objectives are to minimize production cost, maintenance cost, inventory carrying cost, workforce cost and rescheduling cost [29].

3.2. Maintenance with Lot Sizing

Traditionally, maintenance and lot sizing are planned separately as they are related to different cost elements. The optimization of the maintenance schedule is to reduce the cost of maintenance, downtime and failure whereas the optimization of production lot size is closely related to the cost of inventory, setup and lost sales. In fact, the two parameters are closely related too. The failure may

IOP Conf. Series: Materials Science and Engineering **530** (2019) 012048 doi:10.1088/1757-899X/530/1/012048

occur in the middle of production of a lot and so the integrated model of maintenance and lot sizing is necessary as well. The failure of the machine causes stoppage which then leads to lost sales. Furthermore, to minimize the interruption to the production, it is advisable that the maintenance task is conducted during the idle time of the machine or after the production of a complete lot [22].

There are papers found in the literature that jointly optimized lot sizing and maintenance planning. Bergeron et al. [33] had conducted a study to joint optimize lot-sizing and preventive maintenance periodicity of each machine in a serial production line to minimize the sum of production and maintenance cost. The machines are separated by the buffers and the optimal buffer size will be considered in the model as well. The optimization method used is the combination of Tabu Search and Scatter Search algorithms [33]. The study of Nourelfath et al. [20] integrated production, maintenance and quality in their model in order to minimize the total cost, but at the same time complete the demand of all products. They considered a multi-period multi-product capacitated lot sizing problem rather than the EPQ models which only apply to single inventory item [20].

In 2015, Jafari and Makis [23] joint optimized economic manufacturing quantity (EMQ) and preventive maintenance for a production facility. Their objective is to minimize the long-run expected average cost per unit time which include inspection cost, lost production cost, inventory holding cost, preventive maintenance cost, corrective maintenance cost and setup cost. The optimal decision variables to be determined are production lot size and preventive maintenance level [23]. Beheshti-Fakher et al. [34] also joint optimized maintenance plan, lot sizing and quality decision for a single machine producing multiproduct. Their study determined the optimal PM level to be performed at the beginning of each period and the optimal lot size to be processed in each period [34]. In 2016, Bouslah et al. [11] integrated production, preventive maintenance and quality control for a stochastic production system while meeting a predefined restriction on the average outgoing quality limit (AOQL). The study determined the optimal production lot size, inventory threshold, the sampling plan parameters and the overhaul threshold [11].

In the study conducted by Peng and van Houtum [22], the production lot size was optimized by taking the CBM activities into account to minimize the average long-run cost rate. The proposed model used the renewal theory. Control limit policy was used to determine the time to conduct the predictive maintenance and they assumed that maintenance tasks are conducted during machine idle state The purpose is to reduce the interruption to the production plan [22]. Guo et al. [21] also conducted a study on optimization of production lot size and CBM for a multi-component production system to minimize the total cost including preventive maintenance cost, corrective maintenance cost, setup cost, inspection cost inventory holding cost and the shortage cost. The objective of the study is to optimize production lot size and preventive maintenance threshold by using the coupling model and genetic algorithm [21].

In another paper published by Guo et al. [13], the model proposed integrated the production lot sizing, quality control and maintenance planning. The maintenance strategy used for this production line is condition-based maintenance. The decision variables to be optimized in the study are lot size, inventory threshold, preventive maintenance and overhaul thresholds. Inspection is carried out after every production run. Maintenance is implemented if detected degradation level exceeds the threshold. The paper explained the influence of parameters on system performance. The summary of the influence of parameters to the decision variables is summarized as Table 1. The parameters included in the test are production setup cost, PM cost, CM cost, overhaul cost, inventory holding cost, rejection cost of defective item and shortage cost. The author studied the effect of the increment of the parameters on the optimal lot size, inventory threshold, overhaul threshold and PM threshold [13].

IOP Conf. Series: Materials Science and Engineering **530** (2019) 012048 doi:10.1088/1757-899X/530/1/012048

Table 1. Relationship of the parameters on decision variables

	Optimal lot size	Optimal inventory threshold	Optimal overhaul threshold	Optimal PM threshold
Production setup cost increase	Increase	Increase	Decrease	Decrease
PM cost increase	Decrease	Increase	Decrease	Increase
CM cost increase	Decrease	Decrease	Decrease	Decrease
Overhaul cost increase	Decrease	-	Increase	Decrease
Inventory holding cost increase	Decrease	Decrease	Decrease	Increase
Rejection cost of a defective item increase	Decrease	Decrease	Decrease	Increase slightly
Shortage Cost increase	Decrease	Increase	Decrease	Decrease

3.3. Maintenance with Buffer

The significant of the buffer cannot be ignored also in maintenance planning. The effect of stoppage due to failure and maintenance can be alleviated with buffers built between machines. Therefore, the buffer level and buffer inventory holding cost are closely related to the maintenance frequency. To study the relationship of maintenance and buffer, D.Meller [35] and S.Kim [36] conducted a study on the impact of preventive maintenance on system cost and buffer size. Fitouhi et al. [37] also developed a methodology to analyze the tradeoff between preventive maintenance and buffer to the system performance.

Some researches had been done on the joint optimization of maintenance and buffer. Ribeiro et al. [38] jointly optimized the maintenance of a capacity-constrained resourced, its feed machine and inlet buffer size with a mixed integer linear programming model. The study of Bergeron et al. [33] as mentioned in section 3.2 also considered optimal buffer sizing in the proposed model. Buffer sizing was optimized to reduce the blockage and starvation of machines on the production line [33]. Gan et al. [35] focused on the joint optimization for maintenance, buffer and spare parts for a production system with genetic algorithm. The buffer inventory affects the inventory holding cost. In order to determine the significance of optimizing four variables related to maintenance, buffer and spare parts simultaneously, comparisons were made among four-variable optimization, three-variable optimization and two-variable optimization. The optimization results are different. When only two of them are optimized, the decision on the third variable may be improper, therefore leads to undesired optimization result [35].

4. Joint Optimization of Maintenance with Spare Part Inventory Management

Joint optimization of maintenance and spare part or inventory management is also an interesting topic for research. The performance of maintenance has a high dependency on the availability of the spare parts. The shortage of spare parts leads to delay of maintenance work, thus prolong the downtime of the line. In contrast, excessive spare parts may end up with high holding cost. Hence, an optimum level of spare part is necessary to ensure the smoothness of the maintenance task and minimal inventory cost [5][39].

For past studies in old days, the planning and optimization of spare parts and maintenance activities were done separately also. For spare part inventory management, it relies on past spare part demands rather than the maintenance planning [24][39][40]. The researchers assumed 100% spare parts' availability during the maintenance task. To fulfill this assumption, the spare parts should be highly standardized for easy procurement, or low cost so that large quantities can be stored to avoid stock-out [41]. However, this is not applicable in the real case as many of the parts are highly customized. Also, the ordering of spare parts from suppliers often requires certain lead time [42]. The oversimplified assumptions will definitely require high inventory holding cost to store extra spare parts.

Therefore, the optimization of maintenance activity should be done simultaneously with spare part inventory management for timely supply of resources. For this topic, Adrian et al. [9] had done a

IOP Conf. Series: Materials Science and Engineering **530** (2019) 012048 doi:10.1088/1757-899X/530/1/012048

review paper on joint optimization of inventory and maintenance that include costs. A framework was proposed in this paper to classify all characteristics which are important [9]. The two inventory review policies include continuous review policy and periodically review policy. For continuous review policy, the inventory level is checked continuously and the order is made when certain condition occurs. The two most common policies are (s, S) and (s, Q) policy. Whereas when the inventory is reviewed periodically, the order is made based on the forecast demand at the beginning of each cycle. For example, (R, S) policy. From the review paper, the first policy is more common as 15 out of 23 papers applied continuous review policy in their studies. Please refer to the review paper for more information on inventory review policy [9].

The study on this topic is developing rapidly. The recent studies include the work presented by Zahedi-Hosseini [43] which is also the first study on this topic that involves a parallel system. The planned maintenance inspection interval, T, review period R, and the order-up-to level S of the periodic review inventory replenishment policy were optimized. The researchers utilized the ProModel Discrete Event Simulation package and the optimization tool SimRunner to determine the optimum policy [43]. Olde Keizer [20], Jiang [41] and Gan [32] joint optimized the maintenance and inventory control of spare parts for a multi-component system because in real case the systems contain multiple components. Optimal maintenance and spare part inventory management for a single component is not always optimal for a multi-component system [24][35][44]. Gan's work is the first optimization work that considers maintenance, spare part inventory and buffer. Four control variables include the preventive maintenance age, the spare part arrival time, buffer size, and the initiating time of buffer accumulation was optimized to obtain minimal long-term expected cost [32].

Panagiotiduo [45] on the other hand derived optimal inspection interval and provided sufficient spare parts to support both corrective and preventive maintenance task in order to get the minimal expected total maintenance and inventory cost per time unit. Mjirda [46], Kader [47] and Wang [48] conducted joint optimization for preventive maintenance and spare parts inventory problem. Mjirda [46] determined an optimal maintenance schedule with taking account the inventory management of spare parts whereas Kader [47] determined the optimal maintenance plan and order quantity of spare parts simultaneously to minimize the global cost and carbon footprint. Wang [48] on the other hand optimized the order interval, PM interval and order quantity.

In 2011, Wang [39] had also conducted a study to optimize three decision variables which are ordering quantity and interval and also inspection interval using the Delay-Time concept. Same to Zahedi-Hosseini [43] and Wang [48] who also adopt Delay-time concept in their studies. Meanwhile, Olde [24] and Zhang [49] conducted studies that consider condition-based maintenance. Olde saved significantly from optimizing both maintenance decisions and the timing of ordering spare parts [24].

5. Objective Function

For most past studies done, the most common objective of joint optimizing for maintenance and other function is the minimization of cost [50]. All the decision made related to the line from production planning to maintenance policy and spare part management will affect the total operating budget. Hence, there is a need to seek for the balance between these functions for a minimal operational budget. Gan et al. [35] explained about the relationships between maintenance, spare part and buffer and their effect to the system cost which is the sum of inventory holding cost, spare part ordering cost, maintenance cost and shortage cost. The figure provided in the paper can be a guideline to illustrate the relationship of the functions to the total cost. The relationship of cost elements with the maintenance planning was also explained in the paper by Bouslah [12]. In Levitt's book [51], the author mentioned the cost area that can be decreased with good maintenance practices are. The mentioned cost element in his book can also be included as the cost model so that the model will be more realistic [51].

The cost elements chosen are dependent on the parameters involved in the joint optimization with maintenance. In this paper, the cost factors are categorized into two groups which are maintenance related cost, and inventory related cost. These two categories are the most commonly included in the proposed cost model. The maintenance-related cost includes all the cost element for the input to the maintenance task and the losses due to the failure in repairing the breakdown asset immediately whereas the inventory related cost is the cost of stocking and purchasing the spare part for maintenance and the production materials. Some author also includes the production related costs in their cost model. However, most of them do not describe in detail the cost elements involved in this category. A variable rate of production cost was added into the model without giving further explanation. Hence, this category is not included in the discussion.

5.1. Maintenance related cost

Most of the constructed cost model will consider a fixed value for corrective maintenance or preventive maintenance and cost of sampling or inspection. Some may include also the cost of labour, cost of downtime and material cost. Further information related to the cost of spare part will be further illustrated in the next section. The two hidden costs that should be considered when comparing the maintenance policies are the cost of downtime and cost of emergency repair. The cost of downtime is usually the highest core cost area. Its calculations are difficult as they are not definable cost and their reason is sometimes confusing. Meanwhile, the emergency action to restore the breakdown unit leads to disruption to the production, hence reducing the output of the production. These losses should also be included in the model [1].

The model is not able to represent the total cost of the maintenance system without considering the penalty cost. In the cost model by Xiao et al. [26] , Benmansour et al. [27] , Liu et al. [28], Pandey el al. [52] , Dellagi et al. [53] and Fakher et al. [14], they included penalty cost associated with schedule delay or some known as tardiness cost and backorder cost. Penalty cost is incurred only when the order is not completed on time. This is because the decision on maintenance policy may affect the production scheduling[14][26-28][52-53]. Pandey et al.' [52] model also included the cost of rejection which is caused by the performance degradation of the machine. The poor quality of the products produced by such machine will end up with rejected item. This cost element varies depending on the type of production system and it may be criteria for maintenance planning. If the cost of rejection is too high, the cost of corrective maintenance is higher than the cost of preventive maintenance. Therefore, the period of preventive maintenance should be reduced [52].

The other rare cost elements that are included in the cost model are as follows. Restoration cost which is a function of the detection delay. Separation cost is the cost to separate conforming and nonconforming unit [14]. In Hennequin and Ramirez Restrepo 's [54] cost model, they included hardship cost which is related to losses due to machine ergonomics and accident in the line.

5.2. Inventory related cost

There are two types of inventory that are considered in the cost model which are the production inventories and maintenance spare part inventory. The production inventory consists of raw material, subcomponents and the finished product that have not been shipped. Whereas for the maintenance spare part inventory, it relates to the spare part needed for repairing and the scheduling of maintenance task. Both serve the different departments and thus should be separated in inventory management. To calculate the cost of inventory, it is very significant to include the factor of investment cost, expenses of warehousing, taxes, insurance, labour cost, cost of deterioration, shrinkage, obsolescence and return. The inventory cost is usually unique for each operation [51].

The cost of maintenance spare part inventory is included in most of the proposed cost model especially in the joint optimization model related to spare part management. The maintenance material is considered a significant cost element in total maintenance cost. It's percentage is between 40% to 50% in the usual US industry [7]. Beside the cost of the spare part itself, the spare part holding cost is the most common inventory related cost included in the cost model. This cost element appeared in the researches done by Gan et al. [35], Wang [39], Jiang et al. [44], Panagiotidou [45], Mjirda et al. [46], Wang [48] and lastly Zhang and Zeng [49]. In most of the paper, they set a fixed rate for holding cost per unit time. The assumption made is all the spare part will be using the same pre-determined holding cost rate [35][39][44-46][48-49].

Some researchers considered the ordering cost for spare part in their cost model as well. This cost element can be found in proposed cost model by Gan et al. [35], Wang [39], Jiang et al. [44], Wang [48] and also Zhang and Zeng [49]. Subsequently, the insufficient spare part affects the performance of the maintenance which leads to losses. This cost element attracts the attention of researchers recently. The spare part shortage is incurred when the consumption rate is more than replenishment of spare part inventory, causing the spare part shortage and thus delay the maintenance task. This cost element was included in the studies by Gan et al. [35], Jiang et al. [44], Panagiotidou [45] and Mjirda et al. [46].

In the studies that joint optimized maintenance with production planning including scheduling, lot sizing and buffer, the holding cost for inventory is added into the cost model. The examples are the studies done by Ben-Daya and Noman [19], Bouslah et al. [10], Cheng et al. [21], Peng and Van Houtum [22], Jafari and Makis[23][55], Gan et al. [35] , Suliman and Jawad [56] . In the cost model proposed by Gan et al. [35] also, shortage cost was included. This cost element is incurred when the precedence machine breakdown and the buffer is exhausted. Restoration rate is slower than the consumption rate, end up the possibility of stoppage of the whole line [35]. For illustration, this cost element was also discovered in researches by Jafari and Makis [55] and also Suliman and Jawad [56].

Some researches feel that adding in backlog cost is significant. Bouslah et al are among them as backlog cost can be found in their three papers [10-12]. Backlog cost occurs when the customers need to wait until the complete batch of demand is completed. To minimize the total cost, the organizations need to find a balance point between backlog cost and inventory holding cost [57]. The paper by Bouslah et al. in 2013 had given a clear picture on the effect of backlog cost to the safety stock level and optimum lot size [10]. Lastly, the cost of disposing surplus buffer inventory is very rare as this cost element can be found only in the study by Gan et al [35].

5.3. Discussion

From the discussion from section 5.2 and 5.3, it's clear that there is no standard on the cost element that should be included in the cost model for the joint optimization of maintenance. Besides dependent on the parameters for joint optimization, the decision is also on the hand of stakeholders. However, a brief guideline on the common cost element to be included is provided in Table 2 for reference.

Table 2. Parameters for joint optimization and common cost elements in maintenance

Parameters for Joint optimization with maintenance	Common Cost Element included in the cost model	Other Cost Element included in the cost model (optional and dependent on the stakeholders)
Production scheduling or job sequencing	Maintenance Cost (PM and CM), Inspection Cost, Setup Cost, Tardiness Cost or Penalty Cost	Labour Cost, Cost of Lost Sales, Backorder Cost, inventory holding cost
Lot Sizing, Buffer and Spare Part Inventory	Maintenance Cost (PM and CM), Inspection Cost, Setup Cost, Downtime Cost, Spare Part Cost, Inventory Holding Cost, Spare part Ordering Cost, Spare part Shortage cost,	Labour Cost, Cost of Rejection, Backlog Cost, Inventory Holding Cost

Most of the literatures discussed about preventive maintenance policies. For PM, the concept of imperfect maintenance is gaining the interest from the researchers. Conditional based maintenance is also comparatively popular in the past research. The maintenance-related decision variables that are usually found in the past studies are scheduling of maintenance task, maintenance period, maintenance level and maintenance threshold. Similarly, the most common cost elements that are included in the cost model are the PM or CBM cost, spare part cost, labour cost and setup cost (appear occasionally).

The other cost elements are included depending on the objective and the scope of the study. However, it is encouraged that the losses and penalty due to faulty planning are added into the cost model so that the model is as convincing as possible. For example, shortage cost of the spare part, cost of downtime, penalty cost for incomplete or late order and cost of lost sales. A few researches had added the penalty cost element into their model. Cost of rejection, separation cost, hardship cost, surplus cost and backlog cost are comparatively less frequent in the past papers.

The report of the state of art research in joint optimization for maintenance provides direction for future research. The review paper acts as a guideline to help the industries in improving their integrated model. Close collaboration between departments brings down risk in decision making and improves communication within the organization. The joint optimization concept increases the flexibility of the organization by saving the time in their planning, besides result promising. This effort is definitely stepping stone for them to achieve world-class OEE. Although there are attempts done that consider more than one functions out of three discussed functions, more effort is needed. From the review conducted, it was found that researches on joint optimization of multi-unit system is limited. Most of the paper which study multi-unit system treated the whole system as single-unit system due to the assumption made.

As reviewed by Van Horenbeek [8], most publications on joint optimization of maintenance are on the tactical level of decision making [9]. The researches on the implementation of the selected strategies are infrequent, thus raising the doubt to the feasibility of the decision made. Future research on this area is urged to validate the chosen solution. Therefore, researches must make effort to bridge experimentation model or project with real implementation to validate the success of decision made and the proposed methodology. Lastly, for the objective function, there is still lack of clear guideline on the cost element that should be included in the cost model, end up we have lack of cost model that consider the penalty cost and losses due to faulty planning. These hidden costs are advisable to be included in the model although the maintenance does not directly relate to the penalty cost of losses sales and shortage cost.

6. Conclusion

This paper presented the significance of joint optimization of maintenance policy with production planning (production scheduling, lot sizing and buffer) and spare part management. Their relationship and effect to the total cost were explained based on review of various publications and literatures in this research area. Many publications had explored the integrated model of maintenance with production planning and maintenance with spare part. However, there are limited research that joint optimized more than two functions in the model. Therefore, this area shows high potential due to the need for more realistic model to replicate complex manufacturing system. The review shows that maintenance, production and spare part are three interdependent functions. Neglecting any one of them may end up extra operating budget to sustain the system with oversimplified assumption.

References

[1] Levitt J 2009 *The handbook of maintenance management* (Industrial Press Inc)
[2] Niebel B W 1994 *Engineering maintenance management* (CRC Press)
[3] Dunn S 1998 *Maintenance Conf Proc, Queensland, Australia*
[4] Anderson D and Analytics O E 2003 Reducing the Cost of Preventive Maintenance *Maintenance J.* 15
[5] Ben-Daya M, Duffuaa S O and Raouf A 2012 *Maintenance, modeling and optimization* (Springer Science & Business Media)
[6] Chan F T S, Lau H C W, Ip W L, Chan H K and Kong S 2005 Implementation of total productive maintenance: A case study *Int. J. Prod. Econ.* 95 71–94
[7] Dhillon B S 2002 *Engineering maintenance: a modern approach* (CRC Press)

[8] Ruschel E, AlvesE, Santos P, De Freitas E and Loures R 2017 Industrial maintenance decision-making : A systematic literature review *J. Manuf. Syst.* 45 180-194
[9] Van Horenbeek A, Bure J, Cattrysse D, Pintelon L and Vannsteenwegen P 2013 Joint maintenance and inventory optimization systems: A review *Int. J. Prod. Econ* 143 499-508
[10] Bouslah B, Gharbi A and Pellerin R 2013 Joint optimal lot sizing and production control policy in an unreliable and imperfect manufacturing system *Int. J. Prod. Econ* 144 143–156
[11] Bouslah B, Gharbi A and Pellerin R 2009 Integrated production, sampling quality control and maintenance of deteriorating production systems with AOQL constraint *Omega (United Kingdom)* 61 110–126
[12] Bouslah B, Gharbi A and Pellerin R 2018 Joint production, quality and maintenance control of a two-machine line subject to operation-dependent and quality-dependent failures *Int. J. Prod. Econ.* 195 210–226
[13] G. Q. Cheng, B. H. Zhou and L. Li 2018 Integrated production, quality control and condition-based maintenance for imperfect production systems *Reliab. Eng. Syst. Saf.* 175 251–264
[14] Fakher H B, Nourelfath M and Gendreau M 2018 Integrating production, maintenance and quality: A multi-period multi-product profit-maximization model *Reliab. Eng. Syst. Saf.* 170 191–201
[15] Dhillon B S 2006 *Maintainability, maintenance, and reliability for engineers* (CRC press)
[16] Stephens M P 2010 *Productivity and reliability-based maintenance management* (Purdue University Press)
[17] Ben-Daya M, Ait-Kadi D, Duffuaa S O, J. Knezevic and A. Raouf 2009 *Handbook of maintenance management and engineering* (Springer) 7
[18] Ernnie Illyani Basri S K, Izatul Hamimi Abdul Razak and Hasnida Ab-Samat 2017 Preventive maintenance (PM) planning : a review *J. Qual. Maint. Eng* 23 114–143
[19] Ben-Daya M and Noman S A 2006 Lot sizing, preventive maintenance, and warranty decisions for imperfect production systems *J. Qual. Maint. Eng* 12 68–80
[20] Nourelfath M, Nahas N and Ben-Daya M 2016 Integrated preventive maintenance and production decision for imperfect processes *Reliab. Eng. Syst. Saf.* 148 21–31
[21] Cheng G Q, Zhou B H and Li L 2017 Joint optimization of lot sizing and condition-based maintenance for multi-component production systems *Comput. Ind. Eng* 110 538–549
[22] Peng H and Van Houtum G J 2016 Joint optimization of condition-based maintenance and production lot-sizing *Eur. J. Oper. Res.* 253 94–107
[23] Jafari L and Makis V 2015 Joint optimal lot sizing and preventive maintenance policy for a production facility subject to condition monitoring *Int. J. Prod. Econ.* 169 156–168
[24] Olde Keizer M C A, Teunter R H and Veldman J 2017 Joint condition-based maintenance and inventory optimization for systems with multiple components *Eur. J. Oper. Res.* 257 209–222
[25] Levitt J 2003 *Complete guide to preventive and predictive maintenance* (Industrial Press Inc)
[26] Xiao L, Song S, Chen X and Coit D W 2016 Joint optimization of production scheduling and machine group preventive maintenance *Reliab. Eng. Syst. Saf.* 146 68–78
[27] Benmansour R, Allaoui H, Artiba A, Iassinovski S and Pellerin R 2011 Simulation-based approach to joint production and preventive maintenance scheduling on a failure-prone machine *J. Qual. Maint. Eng* 17 254–267
[28] Liu Q, Dong M and Chen F F 2018 Single-machine-based joint optimization of predictive maintenance planning and production scheduling *Robot. Comput. Integr. Manuf.* 51 238–247
[29] Portioli-Staudacher A and Tantardini M 2012 Integrated maintenance and production planning: A model to include rescheduling cost *J. Qual. Maint. Eng.* 18 42–59
[30] Ashayeri J, Teelen A and Selenj W 1996 A production and maintenance planning model for the process industry *Int. J. Prod. Res.* 34 3311–3326
[31] Cassady C R and Kutanoglu E 2005 Integrating preventive maintenance planning and production scheduling for a single machine *IEEE Trans. Reliab* 54 304–309

[32] Wang S and Liu M 2013 A branch and bound algorithm for single-machine production scheduling integrated with preventive maintenance planning *Int. J. Prod. Res.* 51 847–868
[33] Bergeron D, Jamali M A and Ait-Kadi D 2009 Coupling simulation and optimization for an integrated production and maintenance planning in manufacturing systems *IFAC Proceedings* 42
[34] Beheshti-Fakher H,Nourelfath M and Gendreau M 2016 Joint planning of production and maintenance in a single machine deteriorating system *IFAC-PapersOnLine* 49 745–750
[35] Gan S, Zhang Z, Zhou Y and Shi J 2015 Joint optimization of maintenance, buffer, and spare parts for a production system *Appl. Math. Model* 39 6032–6042
[36] Meller R D and Kim D S 1996 The impact of preventive maintenance on system cost and buffer size *Eur. J. Oper. Res.* 95 577–591
[37] Fitouhi M C, Nourelfath M and Gershwin S B 2017 Performance evaluation of a two-machine line with a finite buffer and condition-based maintenance *Reliab. Eng. Syst. Saf.* 166 61–72
[38] Ribeiro M A, Silveira J L and Qassim R Y 2007 Joint optimisation of maintenance and buffer size in a manufacturing system *Eur. J. Oper. Res.* 176 405–413
[39] Wang W 2011 A joint spare part and maintenance inspection optimisation model using the Delay-Time concept *Reliab. Eng. Syst. Saf.* 96 1535–1541
[40] Marseguerra M, Zio E and Podofillini L 2005 Multiobjective spare part allocation by means of genetic algorithms and Monte Carlo simulation *Reliab. Eng. Syst. Saf.* 87 325–335
[41] Barlow R E and Proschan F 1965 *Mathematical Theory of Reliability* (New York: John Wiley and Sons)
[42] Brezavscek A and Hudoklin A 2003 Joint optimization of block-replacement and periodic-review spare-provisioning policy *IEEE Trans. Reliab.* 52 112–117
[43] Zahedi-Hosseini F, Scarf P and Syntetos A 2018 Joint maintenance-inventory optimisation of parallel production systems *J. Manuf. Syst.* 48 73–86
[44] Jiang Y, Chen M and Zhou D 2015 Joint optimization of preventive maintenance and inventory policies for multi-unit systems subject to deteriorating spare part inventory *J. Manuf. Syst.* 35 191–205
[45] Panagiotidou S 2014 Joint optimization of spare parts ordering and maintenance policies for multiple identical items subject to silent failures *Eur. J. Oper. Res.* 235 300–314
[46] Mjirda A, Benmansour R, Allaoui H and Goncalves G 2016 On the joint optimization of the periodic preventive maintenance and the spare parts inventory problem, *IFAC-PapersOnLine* 49 881–885
[47] Kader B, Sofiene D, Nidhal R and Walid E 2015 Ecological and joint optimization of preventive maintenance and spare parts inventories for an optimal production plan *IFAC-PapersOnLine* 28 2139–2144
[48] Wang W 2012 A stochastic model for joint spare parts inventory and planned maintenance optimisation *Eur. J. Oper. Res.* 216 127–139
[49] Zhang X and Zeng J 2017 Joint optimization of condition-based opportunistic maintenance and spare parts provisioning policy in multiunit systems *Eur. J. Oper. Res.* 262 479–498
[50] Van Horenbeek A, Pintelon L and Muchiri P 2010 Maintenance optimization models and criteria," *Int. J. Syst. Assur. Eng. Manag.* 1 189–200
[51] Levitt J 2004 *Managing factory maintenance* (Industrial Press Inc).
[52] Pandey D, Kulkarni M S and Vrat P 2011 A methodology for joint optimization for maintenance planning, process quality and production scheduling *Comput. Ind. Eng* 61 1098–1106
[53] Dellagi S, Chelbi A and Trabelsi W 2017 Joint integrated production-maintenance policy with production plan smoothing through production rate control *J. Manuf. Syst.* 42 262–270
[54] Hennequin S and Ramirez Restrepo L M 2016 Fuzzy model of a joint maintenance and production control under sustainability constraints *IFAC-PapersOnLine* 49 1216–1221
[55] Jafari L and Makis V 2012 Optimal lot-sizing and maintenance policy for a partially observable

production system *Comput. Ind. Eng.* 93 88–98
[56] Suliman S M A and Jawad S 2012 Optimization of preventive maintenance schedule and production lot size *Int. J. Prod. Econ.*137 19–28
[57] Choi T M 2014 *Handbook of EOQ inventory problems* (AMC) 10

IOP Conf. Series: Materials Science and Engineering **530** (2019) 012049 doi:10.1088/1757-899X/530/1/012049

Knowledge Base and Rule Base Safety Management Systems Development Strategies and Applications for Petrochemical Industries: A Brief Outline

M M Asad[1*], R Hassan[1], B Sherwani[2] and F Sherwani[3]

[1]Faculty of Technical and Vocational Education, Universiti Tun Hussein Onn Malaysia, Batu Pahat, Malaysia.

[2]Faculty of Electronics Engineering, NED University of Engineering and Technology, Karachi, Pakistan.

[3]Faculty of Electrical and Electronics Engineering, Universiti Tun Hussein Onn Malaysia, Batu Pahat, Malaysia.

Corresponding author*: mujtabaasad11@gmail.com

Abstract. This review paper discussed about the development tactics and applications of knowledge based and rule-based decision support systems for health and safety in oil and gas sector. It recommends the users on the extent of health and safety concerns for risk and hazard investigation and prevention in which the system developers and researchers mostly face issues during selection and adaptation of development approaches. A complete and systematic review and classification of articles from last seven years has been conducted and briefly analysed, to investigate various methodologies employed in previously developed rule based (RB) and knowledge based (KB) decision support systems. All major bibliographic databases covering a broad series of health and safety fields were searched in various oil and gas production sector related studies. Eighteen articles met the KB and RB study. Twelve of these met the methodological prospective of required investigation. In this review paper, researchers briefly explain about the methodological aspects and industrial applications of rule and knowledge-based decision support systems for reducing work place risks and hazards in safety and health domain. Conclusions that can be extracted from this analytical review paper of the methodological variation and applications of rule based and knowledge-based decision support system is that both inferences are interrelated to each other. RB and KB inferences techniques and methods are used in both interpretations for developing decision support systems in the context of oil and gas extraction and production sector. But safety and health consider KB decision systems as more effective for taking suitable decisions due to real time information and experts' opinions which facilities safety professionals to take appropriate decision then RB reasoning decision support systems.

1. Introduction

Decision support systems are computer-based systems providing an appropriate decision to the users in a particular domain [1]. Development of framework of decision support systems consists of three major phases; Intelligence, choice, implementation [2]. Also, decision support systems are classified in four components such as: input, user expertise, output and decision. Decision support systems (DSS) provide prevailing and flexible approach for extracting solutions to a variety of problems and issues in multiple industrial and non-industrial settings [3].

Especially in health and safety domain of manufacturing and production industries, decision support systems play an important and efficient role for reducing risks and hazards at workplace [4].Fire, fall, slip and trip, hazardous chemical and other such incidents are the most common cause of major injuries for workers and for the prevention of these incidents and hazards decision support and expert systems are widely used in different oil extraction industries for taking decisions for hazard prevention in real time. According to the available literature on the development of industrial based decision support systems specifically for oil and gas drilling safety is quite limited for knowledge base and rule base decision support systems. Therefore, this review paper has been carried out to briefly discuss and highlight the design and development methodologies and industrial applications of existing industrial safety and health decision support system with knowledge base and rule base inferences.

2. DSS Based on Rule Base Interpretation

The Rule based decision support system can be defined as, a particular system which collects required information and knowledge from the experts of that discipline in a constructive approach, and interprets that knowledge in the form of rules, such as IF and THEN statements [5]. These rules are also used to perform operations on data implication in order to reach suitable conclusions or recommendations. Normally, a rule-based decision support system consists of four main elements such as: detail record and list of rules, interpretational engine, working memory and user interface [6].

2.1 Detail Record

First main component of rule-based support systems is the record and list of detail rules which are specific and related with the particular domain of required field [7]. Example of a rule-based system is shown in Figure 1.

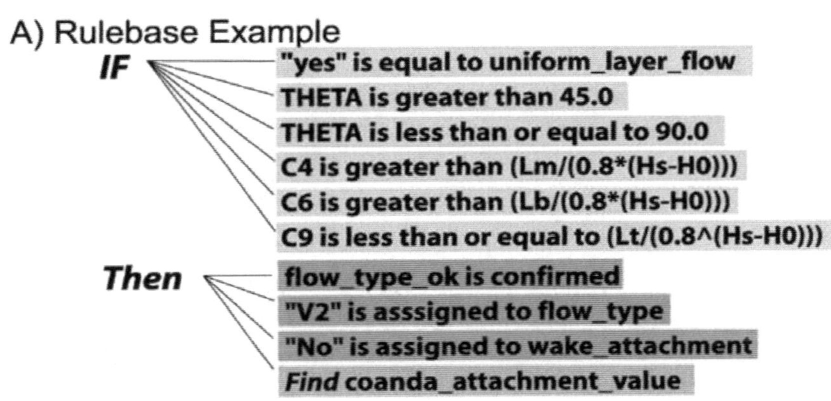

Figure 1. If-Then Rule Base Example

2.2 Interpretational Engine

The second component of rule base decision support system is the interpretational engine which takes on user requirement and enquiry regarding particular information with the help of defined rules [8]. Execution programs are run by performing the following match-resolve-act cycle.

a) Match: In this initial stage, all production will coordinate in opposition to the operational memory [8]. As a result, a conflict set is obtained. An instantiation of a formation is a requested rundown of operational memory components that fulfills the left-hand side of the production.
b) Conflict-Resolution: In this stage, one of the formation instantiations in the conflict set is selected for execution. On the off chance that no measures are fulfilled, the mediator stops.

IOP Conf. Series: Materials Science and Engineering **530** (2019) 012049 doi:10.1088/1757-899X/530/1/012049

c) Act: In this third stage, the activities of the formation chosen in the conflict determination stage are executed [9]. These activities may change the body of operational memory. Towards the end of this stage, execution comes back to the initial stage.

2.3 Temporary Working Memory/User Interface

These two components of RB decision support system are one of the main elements because without the working or operational memory and user system interface, system is not able to connect with the different user's domain in all over the world for sending and receiving input and output signals [10]. These interpretations are basically a computer program that provides method for analysis about information in the rule base, and for formulating respective conclusions. Figure 2 shows the architecture of a rule base system.

Figure 2. Architecture of Rule Base System

3. Implication of Rule Base Decision Support System in Health and Safety

Mainly, rule-based (RB) decision support systems for health and safety in oil and gas sector are: knowledge verification system, weather and environmental forecast strategy, knowledge acquisition, knowledge demonstration, bio separation, resource deployment, biochemical, Permit Control & Monitoring, working at height to confined spaces, training system [11]. Rule base reasoning systems are quite common in safety and health domain of small industrial organization which deals with limited data and specified it with several rules and conditions [12]. According to the detailed review it has been identified that the petrochemical industries prefer rule base reasoning systems for dealing with safety in standardized operations [13].

Alternatively, some experts have not emphasized on rule base systems due to limited knowledge and information to overcome life threatening hazards and risks [14]. But on the other hand, system developers consider rule base systems as easy to maintain decision support systems compared to knowledge-based systems [15]. According to the previous researchers, confined space advisor 1.1 and PCM systems in the domain of confined space and working at height of occupational and safety at oil and gas industries are considered as most effective decision support systems. Decision support systems based on rule-based approach are depicted from Table 1 [17].

Table 1. Approaches and Applications of Exciting RB Decision Support Systems

System	Safety Domain	Approach
Confined Spaces Advisor 1.1	Confined spaces that have hazardous conditions	Rule Base
DUST-EXPERT	Explosive dusts	Rule Base
Preventing falls from slips and trips	Preventing falls from slips and trips	Rule Base
PCMS	The system covers a range of safety domains from working at height to confined spaces, and excavation to electrical work.	Rule Base

IOP Conf. Series: Materials Science and Engineering **530** (2019) 012049 doi:10.1088/1757-899X/530/1/012049

Furthermore, as per comprehensive literature review its has been recommended that that rule-based systems are more appropriate if them merge with knowledge based systems and there is a sheer need of such systems which can also used for vestibule safety training activities at oil and gas industries to train workforce for safe operations [15-16].

4. Decision Support System Based on Knowledge Base Interpretation

Knowledge base decision support systems are artificial intelligence-based applications that use a database of knowledge. KB decision support systems are also called systematic form of database systems. These systems contain structured and unstructured information and knowledge [12]. Knowledge base support system has two major constructive parts such as knowledge base and inference engine. Knowledge base represent about the facts and particular knowledge in the form of data base and inference engine represent the logic and rules to the knowledge and information stored in database [13-18]. The term KB systems include all the organizational information and technology applications that may be helpful to manage the knowledge assets of an organization. Figure 3 explains the conceptual diagram of KB system.

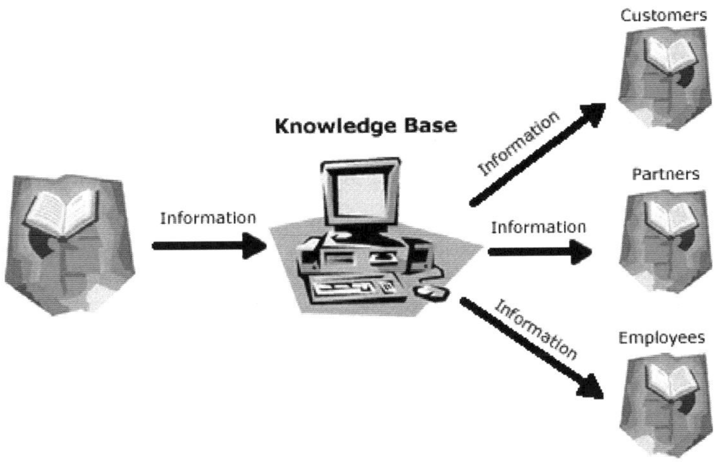

Figure 3. Conceptual Diagram of Knowledge Base system

Generally, the main applications of knowledge-based decision support systems for health and safety for reducing work place risk and hazards in oil and gas sector are: Fall and slip advisor system , hazard and risk analysis, workplace safety decision support system, knowledge management system, knowledge representation system, buildings environment evaluation system, chemical hazard incident management, fire risk assessment system, climate forecasting, safety strategic management, environmental protection, wastewater treatment in industries and incidents controlling system [12]. Based on detailed review it has been analyzed that most of the safety and health management systems are knowledge based and take decision according to the recommendation and knowledge provided by field experts [12].

Some safety and health experts suggested that those systems which contain real time accident investigations feedback from field experts are more effective than other systems. But also, regular maintenance and upgrading in the knowledge base of decision support systems are required on weekly or monthly bases which is considering to be bit costly and time taken. Most of the existing knowledge base decision support systems are used for accident prevention for fire and chemical awareness and construction safety at oil and gas industries worldwide as shown in Table 2 [17]. Such as Fire safety Advisor 1.0 a and E-Tool developed by occupational safety and health administration is working on knowledge base principles and reviewed and updated based on latest experts' knowledge on monthly bases [15-18].

IOP Conf. Series: Materials Science and Engineering **530** (2019) 012049 doi:10.1088/1757-899X/530/1/012049

Table 2. Approaches and Applications of Exciting KB Decision Support Systems

System	Safety Domain	Approach
SPONCOM	Fire prevention	Knowledge Base
Fire Safety Advisor 1.0a	Fire Safety	Knowledge Base
Hazard Awareness Advisor 1.0	Hazard Awareness	Knowledge Base
CSHM	Construction & Oil Gas	Knowledge Base
Hearing Protector Device Compendium	Noise	Knowledge Base
E TOO: Oil and gas well drilling and servicing.	Oil and gas industry work	Knowledge Base

5. Discussion

Approaches and applications of rule base and knowledge base decision support systems inferences for health and safety in oil & gas sector are technologically most important area of investigation from reducing and controlling work place risk and hazards. Quite a few conceptual methods are reported in this paper. Throughout this briefed review researcher analyze that knowledge and rule base interpretations are commonly used in many decision support system and expert systems due to the simplicity and convenience for solving the health and safety. For KB both knowledge and rules or logic are used for the development of decision support system [15-16].

This review it shows that in KB systems developers must have to relate the knowledge with rule base interpretation. In addition, Knowledge based systems also process data and rules to output information and make decisions. On the other side, Rule-based systems process data and output information, but they also process rules and make decisions [19]. The decision support systems in health and safety for reducing the risks provide a safe and comfy atmosphere to the workers by adopting these methodological approaches in development of system.

6. Conclusion

This paper discussed about the development approaches and applications of rule based and knowledge-based decision support systems inferences for health and safety in oil & gas industries in order to investigate which methodology in rule base and knowledge base decision support systems have been effective for safety and health related management systems development for petrochemical industries. Major bibliographic databases covering a wide series of safety and health discipline were explored in different oil and gas production sector related studies. For KB and RB 18 articles met the study's relevance criteria. Conclusion that can be extracted from this analytical review paper of the methodological variation and applications of Rule base (RB) and knowledge base (KB) decision support systems is that both inferences are used at petrochemical industries in different domains of occupational safety and health.

This brief literature review also highlighted that the knowledge-based decision support systems are considered as more effective for accident prevention then rule based decision support systems due to its dynamic nature and real time hazard controlling recommendations. Moreover, it is also recommended by the safety experts to develop such decision systems for accident prevention which are based on KB and RB inferences for vestibule training activities of oil and gas employees who are supposed to expose with hazardous conditions.

Acknowledgement
This research work and manuscript publication has been sponsored by Universiti Tun Hussein Onn Malaysia (UTHM) under the TIER 1(H209) grant.

References

[1] Chou S Y and Chang Y H 2008 A decision support system for supplier selection based on a strategy-aligned fuzzy SMART approach *Expert systems with applications* 34(4) 2241-2253
[2] Filippoupolitis A and Gelenbe E 2009 A distributed decision support system for building evacuation *In Human System Interactions, 2009 HSI'09* 2nd Conference on (pp. 323-330) Ieee
[3] Beliën J, Demeulemeester E and Cardoen B 2009 A decision support system for cyclic master surgery scheduling with multiple objectives *Journal of scheduling* 12(2) 147
[4] Ramirez-Rosado I J, Fernandez-Jimenez L A, Monteiro C, Sousa J and Bessa R 2009 Comparison of two new short-term wind-power forecasting systems *Renewable Energy* 34(7) 1848-1854
[5] Wikström P, Edenius L, Elfving B, Eriksson L O, Lämås T, Sonesson J, and Klintebäck F 2011 The Heureka forestry decision support system: an overview
[6] Qi H and Altinakar M S A 2012 GIS-based decision support system for integrated flood management under uncertainty with two dimensional numerical simulations *Environmental Modelling & Software* 26(6) 817-21
[7] Gutierrez J 2014 Am I safe? A Personal Safety Multicriteria Spatial Decision Support System *(Doctoral dissertation, The Claremont Graduate University)*
[8] Samimi E, Shahosseini M A, Abasaltian A and Shafaghi S 2015 Identifying and prioritizing critical success factors (CSFs) in retaining and developing knowledge workers in oil and gas project-based companies *Indian Journal of Science and Technology* 8(11)
[9] Li Y, Ryan P B, Wei Y and Friedman C 2015 A method to combine signals from spontaneous reporting systems and observational healthcare data to detect adverse drug reactions *Drug safety* 38(10) 895-908
[10] Asad M M, Hassan R B, Sherwani F, Ibrahim N H and Soomro Q M 2018 Level of Satisfaction for Occupational Safety and Health Training Activities: A Broad-Spectrum Industrial Survey *In Journal of Physics: Conference Series (Vol 1049 No 1 p 012021)* IOP Publishing
[11] Asad M M, Asad M M and Hassan R 2015 Development of KBES with hazard controlling factors and measures for contracting health and safety risk in oil and gas drilling process: a conceptual action plan
[12] Asad M M 2014 A Systematic Review: Development Techniques and Utilization of Expert Systems Inferences for Health and Safety Environment in Oil & Gas and Petroleum Industries *In Malaysia University Conference Engineering Technology*
[13] Clark D M 2013 Recovering Errors in System Safety Analyses Through Quality Checks, Part 2 *Journal of System Safety* pp 12-19
[14] Podtschaske B, Fuchs D and Friesdorf W 2013 Integrated therapy safety management system *British journal of clinical pharmacology* 76(S1) 5-13
[15] Asad MM 2014 A Systematic Review: Development Techniques and Utilization of Expert Systems Inferences for Health and Safety Environment in Oil & Gas and Petroleum Industries *In Malaysia University Conference Engineering Technology 2014 Sep 11*
[16] Asad M M, Hassan R B, Ibrahim N H, Sherwani F and Soomro Q M 2018 Indication of Decision Making through Accident Prevention Resources among Drilling Crew at Oil and Gas Industries: A Quantitative Survey *In Journal of Physics: Conference Series (Vol 1049 No 1 p 012022)* IOP Publishing
[17] Asad M M, Hassan R B, Sherwani F and Siming I A 2014 Preliminary Survey on MATLAB Learning among Power Electronics Students in Technical Education: A Case Study *World Academy of Science, Engineering and Technology* International Journal of Social,

Behavioral, Educational, Economic, Business and Industrial Engineering 8(11) 3655-3659
[18] Asad E M M, Hassan R B and Sherwani E F 2014 Instructional models for enhancing the performance of students and workforce during educational training *Academy Arena* 6(3) 27-31
[19] Asad M M, Hassan R B and Sherwani F 2014 An Analytical Comparison between Open Loop, PID and Fuzzy Logic Based DC-DC Boost Convertor *World Academy of Science, Engineering and Technology* International Journal of Electrical, Computer, Energetic, Electronic and Communication Engineering 8(11) 1780-1785

IOP Conf. Series: Materials Science and Engineering **530** (2019) 012050 doi:10.1088/1757-899X/530/1/012050

Conceptual Analysis and Survey of Total Productive Maintenance (TPM) and Reliability Centered Maintenance (RCM) Relationship

N A A Azid[1] , S N A Shamsudin[1], M S Yusoff[2] and H A Samat[1*]

[1]School of Mechanical Engineering, Engineering Campus Universiti Sains Malaysia, 14300 Nibong Tebal, Pulau Pinang, Malaysia.

[2]Manufacturing Section, Universiti Kuala Lumpur Malaysian Spanish Institute, Kulim Hi-Tech Park, 09000 Kulim, Kedah, Malaysia.

Corresponding author: hasnida@usm.my

Abstract. Maintenance is an important issue that needs to be addressed during the design and manufacturing or building of a system. Both Total Productive Maintenance (TPM) and Reliability Centered Maintenance (RCM) have its own concepts and principles as a maintenance strategy to give impact to the performance and reliability of the system. Therefore, in this research, the TPM will be studied with RCM to show their relationship. Based on the conceptual analysis, the author will analyze publications from journals, books, and articles to find all related information, such as genealogy, definition, concepts, and principle, and implementation of both strategy. The conceptual method used is Näsi's four elements of conceptual analysis and a questionnaire survey method which has been answered electronically to study the current maintenance strategies in the real industries and to relate it to TPM and RCM. The result of the study shows that TPM had started in 1970 while RCM started in the 1960s, but the foundation of TPM and RCM was started in 1950s with Breakdown Maintenance (BM). TPM is basically about elimination of faults through day-to-day activities involving the entire force work while the concept of RCM is to improve equipment reliability. The objective of TPM is to achieve zero breakdowns, zero defects, and zero accidents while TPM is to preserve the functions. Although both are using different tools but both are linked together through the Lean Tool and can increase product quality, equipment reliability, increase safety, and increase profit. From the survey result, most of the respondent comes from the companies that implementing traditional maintenance (PM and CM) rather than TPM and RCM. TPM mostly can be implemented in big plant industries while RCM can be applied in small or medium size plant.

1. Introduction

Along with rapid changes of technology and competition among industries, organizations employ different strategies and policies to increase productivity and decrease costs. Nowadays, companies have to meet competition from all over the world. Maintenance is one of a policy which is used to cut costs, increase productivity and to continue with the global competition. A lot of maintenance strategies have been developed over the years. Maintenance plays an important role in the production system because it is a largest controllable cost and status quo represents a challenge for leading managements to re-evaluate their maintenance strategies and decision making work attempts to understand different maintenance policies making for better maintenance of assets. So, the efficiency of maintenance systems can affect the profit of a plant. Therefore, selecting appropriate maintenance policies are vital for each manufacturing company. An organization can save huge amount of time, money, and the useful resources in dealing with reliability, maintainability, and performances.

A company can use an innovative approach which is a Total Productive Maintenance (TPM) to maintenance and optimizes equipment efficiency, eliminates faults and promotes autonomous maintenance between operators, through day-to-day activities involving the entire force work [1]. According to Nakajima (1988), as cited by Lycke [2], TPM is an approach that involves employees from the maintenance department and the production department to equipment management. An autonomous maintenance by operators and small group activities in every department and at every level. All improvements should be standardized and maintained because TPM is continuous improvements focusing on the equipment and the production.

Maintenance concepts such as Reliability Centered Maintenance (RCM) have been successfully applied in the process industry to reduce unnecessary preventive maintenance actions and come up with a systematic and efficient maintenance plan. RCM is a most systematic and efficient process to address an overall programmatic approach to the optimization of plant and equipment maintenance. Maintenance needs of the process plants equipment can be better addressed with RCM approach. Failure Mode and Effect Analysis (FMEA) help in identifying all possible causes of failure with a specific reference to the component of systems and sub-systems [3-4]. RCM is a technique for prioritizing the maintenance strategy in a systematic and logical manner. The primary objective of RCM is to preserve functions [5].

Worldwide competition is dependent, among other things, on deregulation, technical development, increased global trading, internationalization, and new principles for organization and management. This paper focus on a conceptual analysis of TPM and RCM, the similarities and differences between both strategies. The relationship between TPM and RCM will help organizations to simplify the process of implementing TPM and RCM. The analysis of these findings will form the basis for recommendations and guidance for organizations, which intend to implement both TPM and RCM. So, the success implementation of both maintenance can be achieved.

Näsi's four elements of concept analysis were employed to investigate the relationship between TPM and RCM. A questionnaire survey technique has been developed in the present study to gather informations and establish the relationship of TPM and RCM. The structured questionaire was sent to manufacturing companies around Pulau Pinang to find the correlation and level of implementation of TPM and RCM.

2. Methodology

This research uses conceptual analysis method of Näsi's four elements of concept analysis published by Anita Nuopponen [6] to study the conceptual analysis of TPM and RCM and to investigate the relationships of TPM and RCM. The published journals from several of authors that is related to this topic was studied from different publication years. Subsequent, survey method has been studied to choose the right survey technique to study the relationship of TPM and RCM with respect to the real application in industry. A structural questionnaire was sent to manufacturing companies around Pulau Pinang to gather information on the topic. Then, all the data will be analysed before come out with a conclusion.

2.1. *Conceptual analysis method*

In Näsi's four elements of concept analysis, there are four elements in the model as depict in Figure 1. The first element is creating a knowledge foundation, the second one is external analysis, followed by internal analysis and the last element is forming conclusions.

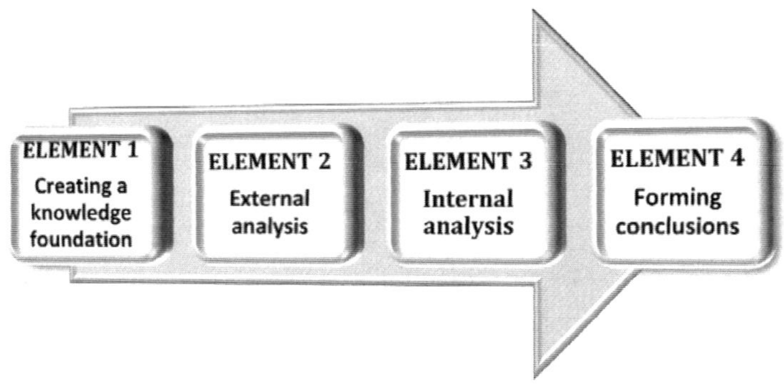

Figure 1. Näsi's four elements of concept analysis

2.1.1. Element 1: Creating a knowledge foundation. Firstly, all the information on relevant research and its results about TPM and RCM is gathered. The common information about genealogy, definition, objective and purpose of implementation of TPM and RCM in the literature were gathered. The literature had been limited to only from journals, books, and articles from the different author. In this methodology, it is divided into two parts which is TPM and RCM. Figure 2 shows the element 1 in Näsi's four elements of concept analysis.

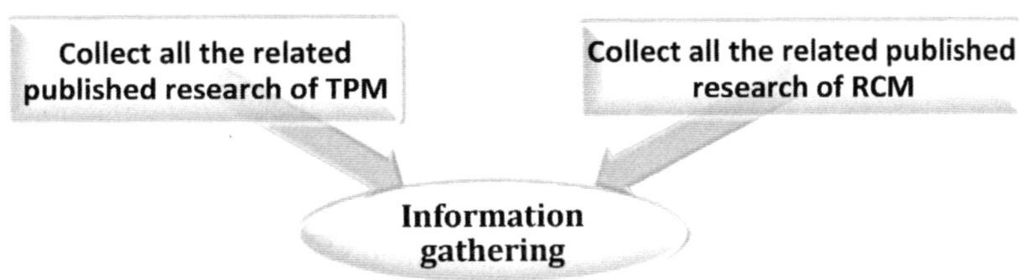

Figure 2. Element 1 (creating a knowledge foundation)

2.1.2. Element 2: External analysis. Element 2 involves distinguishing and delimiting the concepts to be studied from their superordinate concepts and other related concepts, mean all the concept about TPM and RCM from the literature has been distinguished. In this element, it is focused on separating and understands the concept of TPM and RCM. The authors conducted reviews on previous research about TPM and RCM. All the information from Element 1 such as definition, history, objective, implementation and all principle that already mentioned in the paper is distinguished and the characteristic reviewed. Then the data was analysed from each perspective. This element also applied for both of TPM and RCM.

2.1.3. Element 3: Internal analysis. In Näsi's internal analysis Element 3, the concepts between both TPM and RCM are analysed in different views and the concepts are broken down into their parts. This element reanalysing and combining all research that discussed both TPM and RCM to differentiate between the terms discussed. The author discussed the link between TPM and RCM. Element 2 focussed only on the view of TPM and the view of RCM individually from different kinds of literature but in element 3, the focused is more on the relationship between TPM and RCM as visualised in Figure 3, the input from Element 2 will helps in reanalysing steps in Element 3 to give an input of Element 4.

IOP Conf. Series: Materials Science and Engineering **530** (2019) 012050 doi:10.1088/1757-899X/530/1/012050

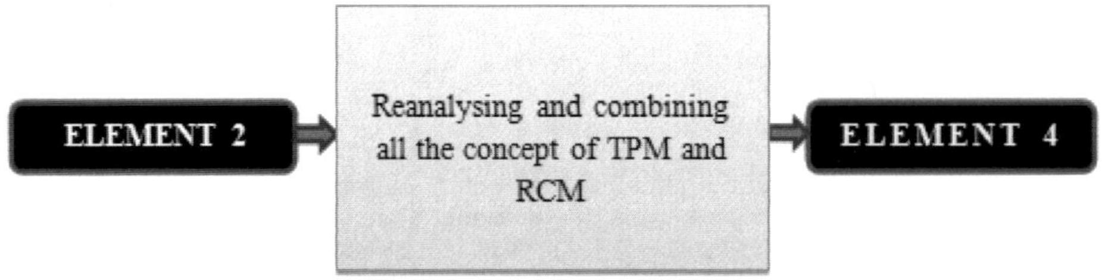

Figure 3. Element 3 (internal analysis)

2.1.4. Element 4: Forming conclusions. Finally, forming conclusions about the conceptual analysis of TPM and RCM. In this step, the solutions to concept problems are offered. Solutions could be either modifying or accepting the old concepts, or even forming new concepts. Then, the similarity and the differentiation between TPM and RCM are documented. As shown in Figure 4 an input from Element 3 further clarified before proceed with the documentation.

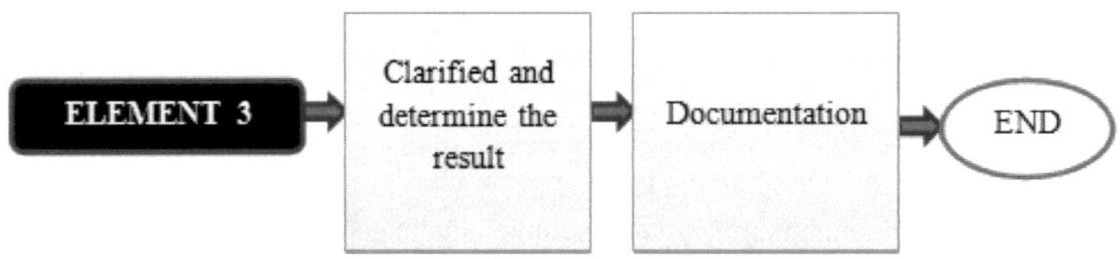

Figure 4. Element 4 (forming conclusion)

2.2. *Survey method*
Survey method had been employed to collect information in an organised and methodical manner about the implementation of TPM and RCM in industry located in Pulau Pinang. This survey had been broken down into several phases. The first is the planning phase, which is followed by the second phase of the design and development, followed by implementation phase. Finally, the entire survey process is reviewed and evaluated. The phases of the survey are as depicted in Figure 5 [7].

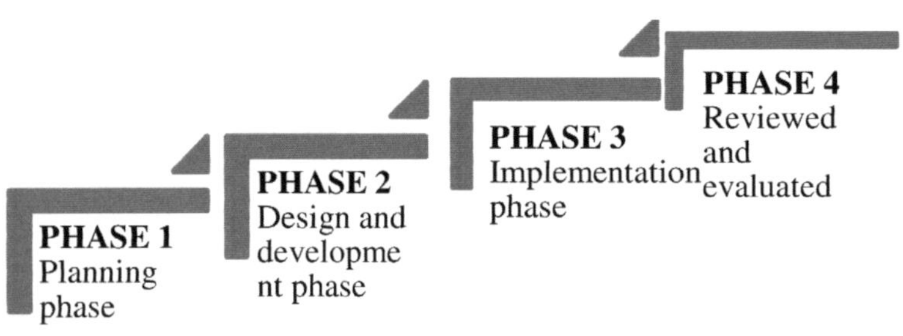

Figure 5. The phase of the survey

2.2.1. Steps of survey method. A survey is a much more complex procedure than simply asking questions and compiling answers to produce statistics. To yield accurate information, numerous steps must be carried out following precise procedures and method. These steps include formulating the survey objectives, determining the survey frame, determining the sample design, designing the questionnaire, performing data collection, data analysis, and documentation. The steps are as follows.

(i) Formulation of the Statement of Objectives

Statement of Objectives is one of the most important tasks during conducting a survey. This establishes not only the survey's broad information needs, but the specific topics to be addressed, the

IOP Conf. Series: Materials Science and Engineering **530** (2019) 012050 doi:10.1088/1757-899X/530/1/012050

operational definitions to be used, and the analysis plan. In this paper, the main objective for this survey is to study the current maintenance strategies in the real industries and to relate it with the TPM and RCM.

(ii) Selection of a Survey Frame

The survey frame provides the means of contacting and identifying the units of the survey population. It is important to decide who will be included in the survey population. The survey frame should include some or all of the following items such as identification data, contact data and classification data. Identification data are the items on the frame that uniquely identify each sampling unit, such as name, exact address, and a unique identification number. Contact data are the items required to locate the sampling units during collection, for such as mailing address or telephone number. Classification data are useful for selection of sample and possibly for estimation.

(iii) Determination of the Sample Design

This survey can be split into two type which is census surveys and sample surveys. In census survey, data are collected for all units in the population while, in a sample survey, data are collected for only a fraction (typically a very small fraction) of units of the population. Two types of sampling exist which is probability sampling and non-probability sampling. Non-probability sampling provides an easy, inexpensive and fast way of selecting units from the population but uses a subjective method of selection. The data analyst must assume that the sample is representative of the population in order to make inferences about the population from a non-probability sample. This is often a risky assumption given the subjective method of selection. Probability sampling is more costly, more complex, and takes longer time than nonprobability sampling.

(iv) Questionnaire Design

A questionnaire or form can be defined as a group or sequence of questions designed to obtain information on a subject from a targeted respondent. Questionnaires can either be in computerised or paper format. The goal is to obtain information in such a way that survey respondents understand the questions and can provide the correct answers easily in a form that is suitable for subsequent processing and analysis of the data. The questionnaire design process starts with the formulation of survey objectives and information requirements and continues with consult with data users and respondents. Next, review previous questionnaires followed draft the questionnaire before review and revise the questionnaire. Then the questionnaire must be tested and revised and lastly, the questionnaire must be finalized.

(v) Data Collection Methods

Data collection is the process of gathering the required information for each selected unit in the survey. The method of data collection should be chosen in order to achieve a high participation rate and to collect data as complete and accurate as possible while minimising the burden to the respondent and operational constraints. The basic methods of data collection can be divided into two which is Self-Enumeration and Interviewer-assisted (Personal Interviews or Telephone Interviews).

With self-enumeration, the respondent completes the questionnaire without the assistance of an interviewer. There are a variety of ways that the questionnaire can be delivered to and returned by the respondent: by post or facsimile, electronically (including the Internet) or by an enumerator. Self-enumeration methods require a very well-structured questionnaire, simple and easy to follow questionnaire with clear instructions for the respondent. A telephone number detail of respondent must be included to obtain help completing the questionnaire. There is two type of self- Enumeration which is when computer-based it is called Computer-Assisted Self Interviewing (CASI) and when paper-based, this method is called Paper and Pencil Interviewing (PAPI).

The interviewer-assisted consist of two ways. The first one is personal Interviews. These personal interviews involve interviewer and respondent. The interview is conducted in person, usually at the respondent's residence or work place, although it can be conducted in a public place. The interviewer will assist the respondent to complete the questionnaire. Secondly, an interviewer assists the respondent to complete the questionnaire over the telephone and this is called as Telephone Interviews.

This paper focussed on self-enumeration which required the respondent completes the questionnaire via form created and distributed in internet. This is why simple and easy to understand questionnaire with clear instructions is important to ensure respondents did not experience any confusions and difficulty in understanding the survey questions regarding to the RCM and TPM implementation level in their company.

(vi) Data Analysis

Data analysis must relate the survey results to the questions and issues stated by the statement of objectives. This is one of the most crucial steps in a survey since the quality of the analysis can substantially affect the usefulness of the whole survey. In this step, the data collection will be summarized and interpreted their meaning in a way that provides clear answers to questions that initiated the survey related to TPM and RCM view. This data analysis may be restricted to the survey data alone or it may compare the survey's estimates with results obtained from other data or survey sources.

(vii) Documentation

Documentation is a final step in survey method. This step provides a record of the survey of TPM and RCM and should encompass every survey step and every survey phase. It may record different aspects of the survey and be aimed at different groups, such as management, engineer, technical staff, and operator. For example, a report on data quality provides users a context for the informed use of the data. A survey report that includes not only what decisions were made, but also why they were made provides management and technical staff with useful information for future development and implementation of similar surveys. During implementation, documentation of procedures for staff helps to ensure effective implementation of TPM and RCM in the real industry.

3. Results

3.1. Conceptual analysis result

In conceptual analysis method, all related information from the selected journals, articles and book were reviewed for both TPM and RCM with respect to the Näsi's four elements of concept analysis. Table 1 shows the concept and the relationship of TPM and RCM.

Table 1. Conclusion table of all elements

Characteristics	Total Productive Maintenance (TPM)	Reliability Centered Maintenance (RCM)
History	i. TPM concept was first introduced in the year of 1970 ii. Developed based on the concepts and methodologies of Productive Maintenance. iii. The concept first introduced by M/s Nippon Denso Co. Ltd, a supplier of M/s Toyota Motor Company, Japan.	i. Founded in the 1960s ii. Initially oriented towards maintaining airplanes and used by aircraft manufacturers, the government and airlines
Theory	i. Promotes autonomous maintenance ii. Eliminates faults through day-to-day activities involving the entire force work iii. Optimizes equipment efficiency	i. Identifying effective and efficient preventive maintenance ii. Improving equipment reliability
Objective	i. Zero defects ii. Zero breakdowns iii. Zero accidents	i. Preserve the functions
Process View	i. Decrease process delay ii. Improve process flow	i. The system reliability linked with maintenance process of system

Methodology	i. Understanding the eight pillar of TPM, Six losses,	i. Analyzing failures of equipment ii. Assessing the consequences of each failure iii. Choosing the best maintenance action to ensure that the desired overall level of system performance is met.
Tools	i. Autonomous Maintenance ii. 5S iii. OEE iv. JIT v. LMC vi. SMED	i. FMEA ii. FMECA iii. PHA iv. FTA v. OMF vi. HAZOP Analysis
Type of sector	i. Maintenance ii. Service iii. Manufacturing (Big plant industries) iv. Process	i. Maintenance ii. Manufacturing iii. Service
Benefits	i. Reduction in delays and downtime losses ii. Reduction in speed losses iii. Reduction in total manpower iv. TPM ensures better and more energy saving v. Greater cost effectiveness	i. Improve system reliability ii. Greater safety and environmental integrity iii. Improved operating performance iv. Greater cost effectiveness, v. The greater motivation of individuals

From all the definitions, to make it simple to understand, TPM best described as an innovative approach to company maintenance that promotes autonomous maintenance between operators, eliminates faults through day-to-day activities involving the entire force work and optimizes equipment efficiency in order to maximise overall equipment effectiveness, OEE. Autonomous Maintenance (AM) is a program to reduce technicians' workload through operators' involvement. Thus, the involvement of operators in maintenance tasks can reduce the workload of technicians [8].

Lean manufacturing is a system that is focused on the processes which minimising variations and facilitate waste reduction thereby facilitating cost reduction in plants. There are various tools and techniques to implement lean principles to an industry which is 5S, Value Stream Mapping (VSM), Total Productive Maintenance (TPM), Failure Mode and Effect Analysis (FMEA), Total Quality Management (TQM), Quality Function Deployment (QFD), Kaizen and Kanban [9]. TPM is one of the tools used in Lean manufacturing.

The principle of RCM can be described from the definition of RCM. The theory of RCM described by Gajanand Gupta and Rajesh P. Mishra in the year of 2016 [3]. They defined RCM as the function of the operating system to recognize the consequences of the failure by the system function and failure analysis. Various tools used in RCM include Physical Hazard Analysis (PHA), Failure mode and effect analysis (FMEA), Failure mode effect and criticality analysis (FMECA), Optimizing Maintenance Function (OMF) , Fault Tree Analysis (FTA), and Hazard and Operability (HAZOP) Analysis [10].

3.2. Survey results

The authors conducted a questionnaire survey with the title "Conceptual Analysis of Total Productive Maintenance (TPM) And Reliability Centered Maintenance (RCM)". The main objective of this survey is to study the current maintenance strategies in the real industry mainly focussed in Pulau Pinang. This survey involved all people from the top managements to the operator by using self-enumeration data collection method. The survey has been answered through electronically (including the Internet) using the application of Google Forms. The target of respondent for this survey is 21 respondents. All the result of the survey shown in Figure 6 and Figure 7:

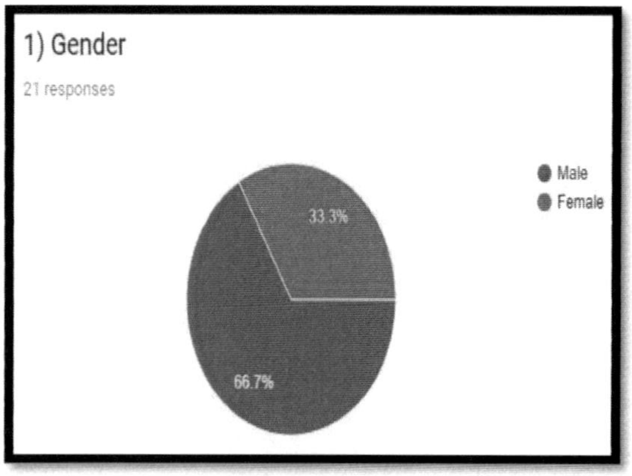

Figure 6. Gender type.

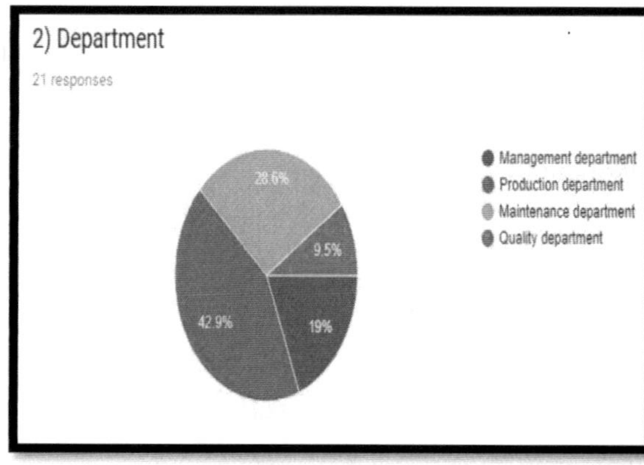

Figure 7. Type of department.

The survey is not biased to one gender only but included both male and female. The percentage of male respondents is 66.7% and is higher than the female that just only 33.3% shown in Figure 6. An organization consists of several functional departments works in a team to sustain and develop a successful plant. Figure 7 shows four main type of department that most contribute to the company. The highest respondent percentage which is 42.9% is from production department and the lowest one is 9.5% from quality department.

This survey focused on two level of the business type which is Multi – National Corporation (MNC) and Small and Medium size Enterprises (SME). MNC is corporate giants, typically houses of multiple departments and consist a large pool of staff who generally keeps to their own departments, while an SME is relatively small in size and has employees frequently working across divisions to make end goals meet. From result in Figure 8, most of the respondents come from MNC with percentage of 57.1%.

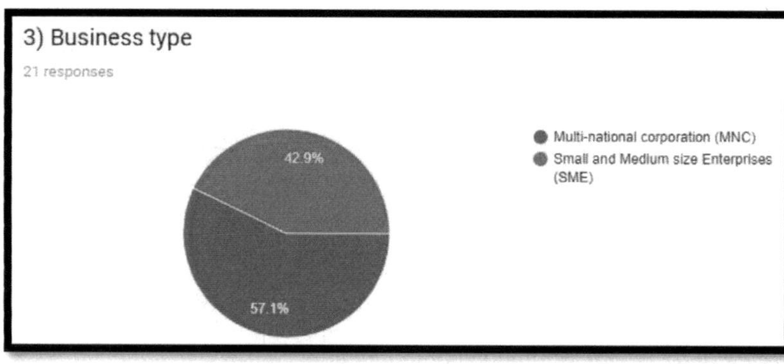

Figure 8. Type of business involved

Maintenance strategies one of the important factors that contribute to increase the production and decrease the production cost. Based on the Figure 9, most of the respondent comes from the companies that implementing Preventive Maintenance and Corrective Maintenance rather than RCM.

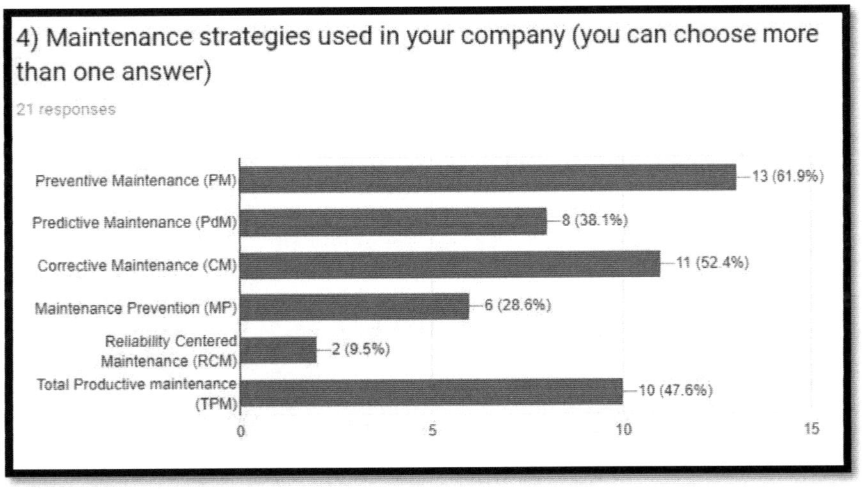

Figure 9. Type of maintenance strategies used in an organization.

In overall employee involvement plays important role in developing a strong maintenance system. Figure 10 shows most of the industries involved all the employees in their maintenance activity with resulted in 71.4%. Safety is the first factor need to be considered in choosing a job place. Based on Figure 11, most of the respondent comes from the organization with safe in work environment.

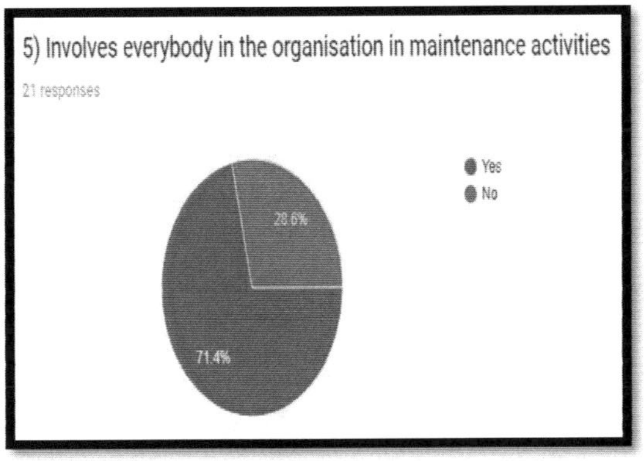

Figure 10. Employees involvement

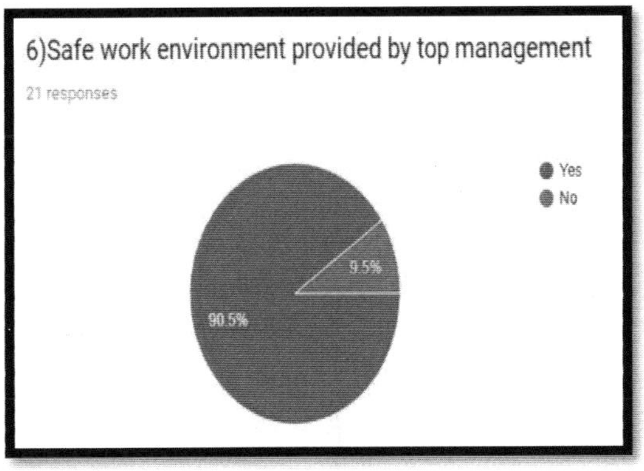

Figure 11. Safety of work environment

Management commitment to education and training of employees is a key factor in developing an excellent organization system. Based on the Figure 12, 95.2% of the respondent get education and training in their organization.

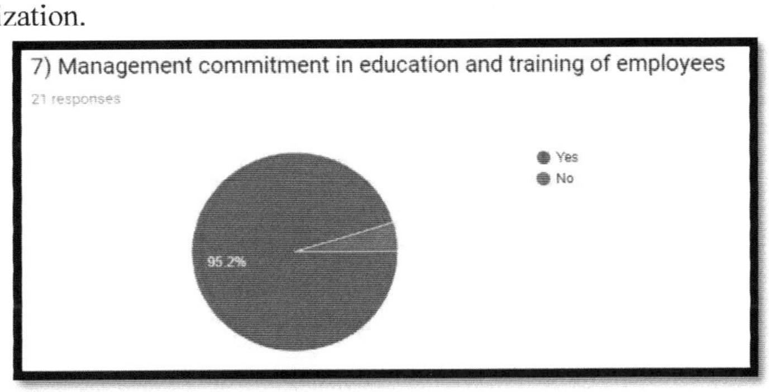

Figure 12. Management commitment to education and training of employees.

The tool that commonly used in TPM is 5S, OEE, SMED, JIT, and LMC While for RCM is FMEA, PHA, FTA, and HAZOP. All of this tool contribute to the excellent implementation of of TPM and RCM. Based on the Figure 13, the highest tool usage in the most respondent organization is 5S with the percentage of 85.7 %.

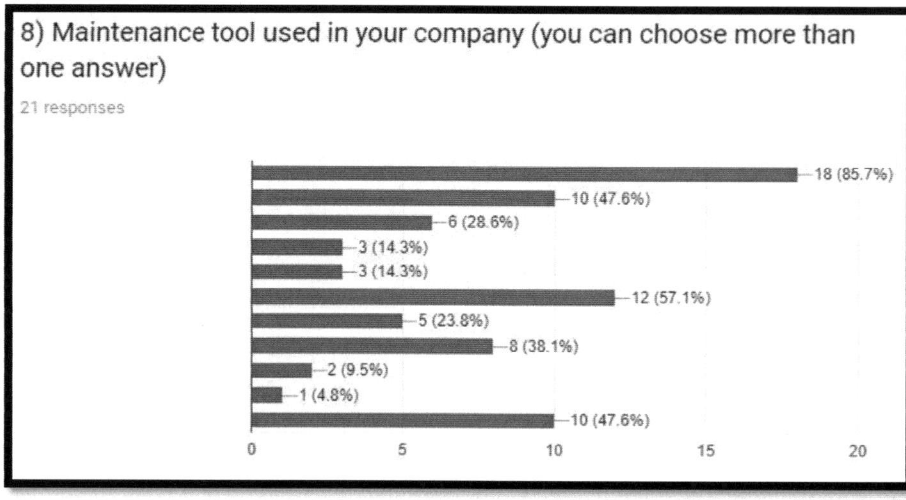

Figure 13. Type of maintenance tools used in an organization

Determination of the effectiveness of maintenance strategies in a certain place can be divided into three stages which are good, fair and bad. Figure 14 shows the percentage of all the stages from all of 21 respondent. The result showed the effectiveness maintenance strategies most of the respondents in a fair stage. This mean, the organization still working towards the good stages.

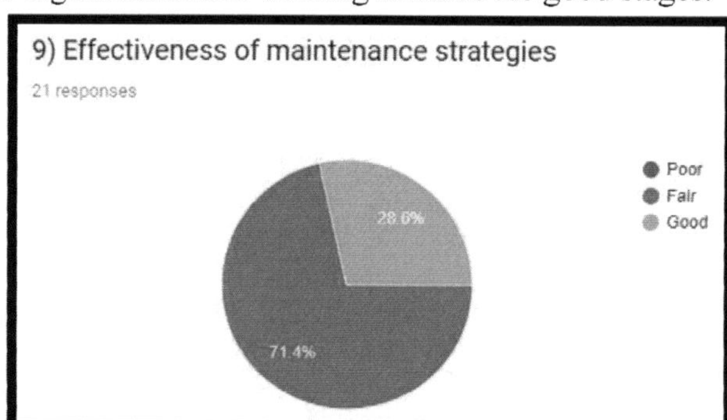

Figure 14. Effectiveness ranking of maintenance strategies.

4. CONCLUSIONS

Näsi's four elements of concept analysis had been chosen because it is a step by step procedure and very simple to understand compare than other conceptual analysis methods. It has four elements in the model which are Element One: Creating a knowledge foundation, Element Two: External analysis, Element Three: Internal analysis and Element Four: Forming conclusions. For this paper, in element one, found several kinds of literature that related to TPM and RCM. In Element 2, authors breakdown the information in the literature into several parts which is genealogy, definition, principles, and concepts, and implementation of TPM and RCM. Element 3 linked the formation of the Element 2, while Element 4 described and concluded the relation of the information into more simple to understand. Genealogy of TPM is basically, was first introduced in the year of 1970, as a unique

IOP Conf. Series: Materials Science and Engineering **530** (2019) 012050 doi:10.1088/1757-899X/530/1/012050

Japanese philosophy, which has been developed based on the concepts and methodologies of Productive Maintenance and its concept first introduced by M/s Nippon Denso Co. Ltd, while RCM was founded in the 1960s and initially oriented towards maintaining airplanes and used by aircraft manufacturers, the government and airlines.

There are many definitions of TPM and RCM had been described from various literature. TPM is an innovative approach to company maintenance that promotes autonomous maintenance between operators, eliminates faults through day-to-day activities involving the entire force work and optimizes equipment efficiency in order to maximise overall equipment effectiveness, OEE while RCM is a structured approach to blends failure and risk analysis with preventative and condition based maintenance, improving equipment reliability and ensures that only the maintenance that matters is carried out. TPM and RCM can be linked together because the tool used in RCM which is FMEA included together with TPM as a tool of Lean Manufacturing. Both have the same concept which wants to increase product quality, equipment reliability, increase safety, and reduce operation cost through waste elimination. Mostly TPM is described to be implemented in big plant industries while RCM can be applied in small or medium size plant. In nutshell, the result of the conceptual analysis is illustrated in Figure 15.

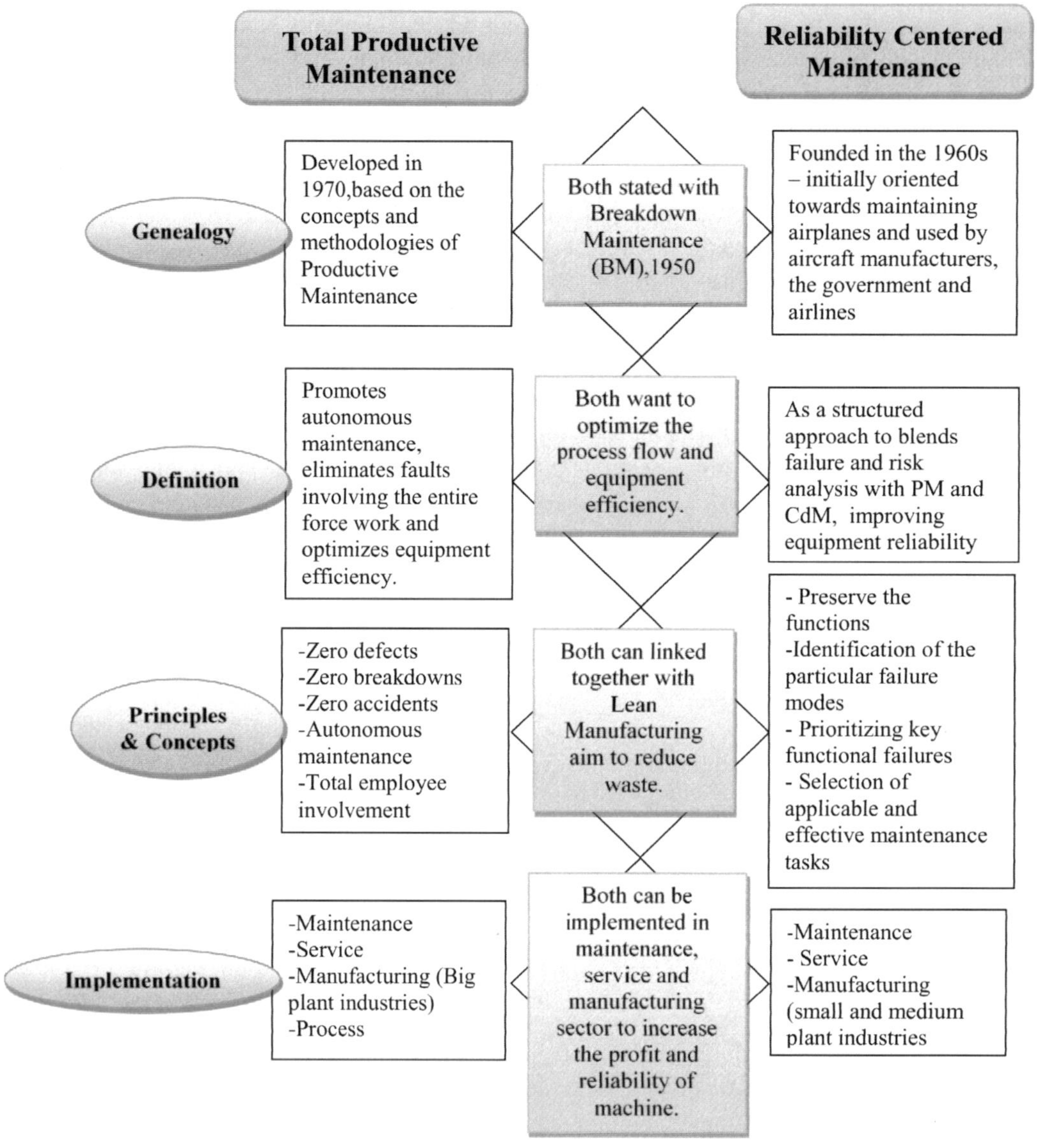

Figure 15: The relationship of TPM and RCM

Questionnaire survey method is done and this survey involved all people from the top managements to the operator by using self-enumeration data collection method. In this method, the respondent completes the questionnaire without the assistance of an interviewer. The survey has been answered electronically (including the Internet). So, the survey is done using the application of Google Forms. Based on the survey result, most of the respondents come from Multi- National Corporation (MNC) with percentage of 57.1%. Question four asked about the Type of maintenance strategies used in an organization. From the result, most of the respondent comes from the companies that high implementing Preventive Maintenance and Corrective Maintenance rather than TPM and RCM. So, it is clear that most of the companies still used the traditional maintenance. Most of the respondent ranked their organization effectiveness maintenance strategies in a fair stage. As an addition, in future the analysis in this study can include continuous improvement together with RCM and TPM.

Acknowledgement
The authors would like to thanks Universiti Sains Malaysia for Short Term Grant (304.PMEKANIK.6315112) resulting of the success of this project.

References

[1] Pinto H, Pimentel C and Cunha M 2016 Implications of total productive maintenance in psychological sense of ownership *Procedia-Social and Behavioral Sciences* **217** pp 1076-1082
[2] Lycke L 2000 Implementing total productive maintenance: driving forces and obstacles (Doctoral dissertation, Luleå tekniska universitet)
[3] Gupta G and Mishra R P 2016 A SWOT analysis of reliability centered maintenance framework *Journal of Quality in Maintenance Engineering* **22(2)** pp 130-145
[4] Backlund F 2003 Managing the introduction of reliability-centred maintenance, RCM: RCM as a method of working within hydropower organisations (Doctoral dissertation, Luleå tekniska universitet)
[5] Chopra A, Sachdeva A and Bhardwaj A 2014 Importance of Training and Data management Issues in implementing Reliability Centered Maintenance (RCM)
[6] Nuopponen A 2011 Methods of concept analysis-tools for systematic concept analysis (part 3 of 3) *LSP Journal-Language for special purposes, professional communication, knowledge management and cognition* 2(1)
[7] Yin R K and Heald K A 1975 Using the case survey method to analyze policy studies *Administrative science quarterly* pp 371-381
[8] Ab-Samat H, Khoe P G, Liau W S, Tan H F, Yap W Y and Kamaruddin S 2012 Implementation of autonomous maintenance in semiconductor industry: a case study *In Advanced Materials Research* (Vol 591 pp 708-711) Trans Tech Publications
[9] Vinodh S and Joy D 2012 Structural equation modeling of sustainable manufacturing practices *Clean Technologies and Environmental Policy 14*(1) pp 79-84
[10] Ahuja I P S and Khamba J S 2008 Total productive maintenance: literature review and directions *International Journal of Quality & Reliability Management 25*(7) pp 709-756

IOP Conf. Series: Materials Science and Engineering **530** (2019) 012051 doi:10.1088/1757-899X/530/1/012051

Paper Machine Breakdown Reduction by FMEA and Preventive Maintenance Improvement: A Case Study

S Wannawiset[1*] and S Tangjitsitcharoen[2]

[1]Graduate Student, Department of Industrial Engineering, Faculty of Engineering, Chulalongkorn University, Bangkok, Thailand

[2]Associate professor, Department of Industrial Engineering, Faculty of Engineering, Chulalongkorn University, Bangkok, Thailand

Corresponding author *: wannawiset007@gmail.com

Abstract. FMEA is a widely selected tool for the application of machinery breakdown prevention. This research adopted FMEA technique to reduce and prevent the breakdown of the important equipment of a large commercial paper machine that has capacity of 200,000 tonnes per annum located in the Northeast area of Thailand. This research accomplished the improvement of preventive maintenance in other machines for rapid implementation and improved machine inspections. After implementation, the downtime of paper machine belonging to maintenance department responsibility reduced significantly 5 hours per month by average, machine availability, OEE, and MTBF increased 0.46%, 0.51%, and 10 hours respectively. The machine downtime has been on the downward trend.

1. Introduction

In the present, pulp and paper industry has high competition [1-2]. To make the company profitable, stay competitive, survive in the global marketplace, companies need to improve and find the way to develop to excellence continuously to promote the success and sustainable growth of the company in the long run. Many companies around the world including studied pulp and paper mill experienced machine breakdown problems [3-4]. The problems lead to losses of production opportunity, repairing costs and product quality costs resulting in the rising production cost.

PM1 paper machine is the paper machine with 200,000 TPA (Tonnes Per Annum) of capacity located on one of the largest pulp and paper mills in Thailand. It started producing commercial paper in 2008. In the past three years, PM1 suffered from many breakdown problems which impacted production volume, production costs and made maintenance costs higher. So this paper focused on breakdown reduction of PM1 paper machine.

Failure Mode and Effect Analysis (FMEA) is an effective and accepted tool in the application to prevent or reduce the breakdown problems of many machines. It considers potential failure modes, causes of failure mode and effects and represents the risk in the RPN (Risk Priority Number) form which can be determined by the product of Severity (S), Occurrence (O) and Detection (D). Once the risk assessment is done, the next step is to find corrective or preventive actions to cope with causes of failure or failure mode. After that, the RPN will be re-calculated.

FMEA was found to be successfully used in both aerospace and automobile industries for enhancing manufacturing reliability of product [5]. It was widely accepted and implemented later in many businesses in the application of maintenance field. FMEA was applied to reduce failure of boiler in sugar mill [6] and improve the reliability of wind turbine system [7]. With the application of the suitable FMEA software package to 2 MW wind turbine, the reliability of generator and gearbox increased. H R Feili et all applied FMEA to identify failure risk of Geothermal Power Plants (GPPs) [8]. K Cicek et all applied FMEA in concepts to identify possible risks of failure of vessel oil system [9]. Moreover, FMEA was used to improve the reliability of thin-film transistor liquid crystal display products [10]. E Rungsa et all used FMEA as a tool in the application of developing computerized preventive maintenance management system for CNC machine. After improvement and implementation, the availability, quality rate, and performance rate increased significantly, resulting in the increase of OEE [11]. S Tangjitsitcharoen et all also successfully applied FMEA techniques to new elevator for preventive maintenance improvement. The countermeasures to failure causes and failure mode resulted in the reduction of a number of failures and the incretion of MTBF [12].

The purpose of this study is mainly to apply FMEA technique along with the improvement of preventive maintenance and machine inspection to reduce PM1 paper machine breakdown loss time belonging to maintenance department responsibility.

2. Problem identification

This paper summarized the breakdown of the PM1 paper machine from January 2017 to December 2017. The Pareto diagram was based on the equipment type to determine the type of machine to be improved and focused on.

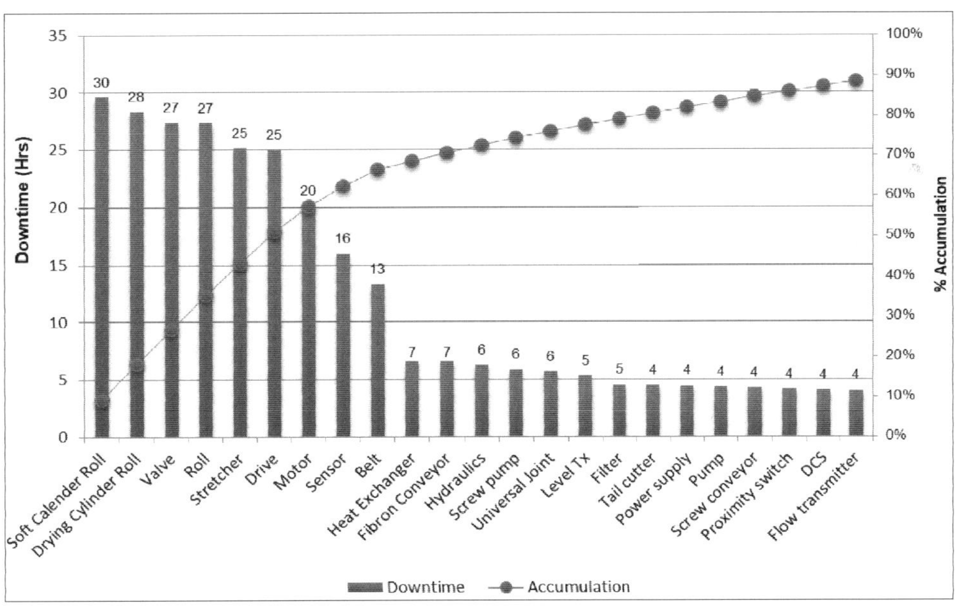

Figure 1. Pareto diagram of PM1 paper machine breakdown loss time

It can be seen from the picture 1 that 80% of breakdown loss time occurred with 17 types of equipment. According to the 80/20 rule of pareto diagram, all the 17 types were selected to be improved for breakdown loss time reduction. The Soft Calender roll and Drying cylinder roll are found to be very chronically problematic thus are important and the solution is more difficult than others, so FMEA is a way to find actions to prevent breakdowns and proper maintenance plans. Most of the problems found on other machines can be easily solved by creating new preventive maintenance plans.

3. Methodology

This research applied FMEA to Soft Calender roll and Drying Cylinder roll and improved preventive maintenance for other equipment with the following steps:

1. Review production processes and machines.
2. Brainstorm for the possible failure modes of machine and potential effects.
3. Identify possible cause of failure and assign severity, occurrence, and detection criteria.
4. Evaluate risk priority number (RPN).
5. Identify corrective or preventive actions to cope with failure causes or failure modes.
6. Assess risk priority number (RPN) after actions
7. Revise preventive maintenance programs for other equipment

4. Failure mode and effect analysis of Soft Calender roll and Drying Cylinder roll

4.1. Criteria for assessing RPN

The criterion for assessing RPN was determined by expert team at the studied mill and the agreement with the production department.

4.1.1. Criteria for severity assessment. This criterion is mostly based on the paper machine downtimes occurring in this paper mill. The severity criterion is shown in the table 1.

Table 1. Severity criteria for FMEA

Effect	Severity Criteria	Rank
Hazardous without warning	Very high severity ranking - affects operator, plant or maintenance personnel and safety and/or effects non-compliance with government regulations without warning.	10
Hazardous with warning	High severity ranking - affects operator, plant or maintenance personnel and safety and/or effects non-compliance with government regulations with warning.	9
Very high	Downtime more than 8 hours.	8
High	Downtime between 4 hours and 8 hours.	7
Moderate	Downtime between 1 hour and 4 hours.	6
Low	Downtime between 30 minutes and 1 hour.	5
Very low	Downtime up to 30 minutes	4
Minor effect	Process parameters variability exceeds upper/lower control limits; adjustments or process controls need to be taken. No downtime and no defective parts.	3
Very minor	Process parameters variability within upper/lower control limits; adjustments or process controls need to be taken. No downtime and no defective parts.	2
None	Process parameters variability within upper/lower control limits; adjustments or process controls not needed or can be taken between shifts or during normal maintenance. No defective parts.	1

4.1.2. Criteria for occurrence assessment. It is the criteria for the assessment to find chance that it may occur. It used the data from the past to present to apply in such assessment.

For example, the problem of sensor malfunction causes the machine to order the paper to be cut, which usually occurs 2 times a month by average so in this case the MTBF of sensor is 2 weeks. In another case, the failure of Drying Cylinder roll's bearing occurred every 8 years by average, so the MTBF of this bearing was about 8 years or 416 weeks.

The criterion for occurrence assessment is shown in the table 2.

Table 2. Occurrence criteria for FMEA

Probability of Failure Occurrence	Possible Failure Rates Criteria	Rank
Very high: Failure is almost inevitable	MTBF less than 2 weeks	10
	MTBF between 2 and 4 weeks	9
High: Repeated failures	MTBF between 5 and 12 weeks	8
	MTBF between 13 and 24 weeks	7
Moderate: Occasional failures	MTBF between 25 and 52 weeks	6
	MTBF between 53 and 104 weeks	5
	MTBF between 105 and 208 weeks	4
Low: Relatively few failures	MTBF between 209 and 312 weeks	3
	MTBF between 313 and 520 weeks	2
Remote: Failure unlikely	MTBF greater than 520 weeks	1

4.1.3. Criteria for detection assessment. In the preparation of the detection criteria, we considered the ability to detect the failure of the machine, equipment or parts before the machine malfunction or breakdown. In case where the failure of the machine is difficult to be detected, the score will be high, such as the visually examination of the A53 carbon steel pipe that transports water may be unable to identify whether the pipe is damaged, or whether the pipe thickness still in an acceptable value or not so the score is high unlike the use of Ultrasonic Thickness Measurement or X-ray to inspect the pipe or sealing weld. In case if the failure detection of equipment or parts of the machine are relatively simple, the score is less. For example, a detection of bearing failure on a rotating machine with 3,000 cycles per minute of speed by using a vibration analyser that can analyse the spectrum when combines with severity level, it can identify whether the bearing is damaged or not so the score in this case is quite low.

Table 3. Detection criteria for FMEA

Detection	Detection Criteria	Rank
Extremely Unlikely	Machinery controls or inspections cannot detect a potential cause and subsequent failure, or there is no machinery inspections	9
Low Likelihood	Low chance that machinery inspections will early detect a potential cause and subsequent failure mode before the machinery functional failure occurs.	7
Medium Likelihood	Medium chance that machinery inspections will early detect a potential cause and subsequent failure mode before the machinery functional failure occurs. Machinery controls will prevent imminent failure.	5
High	High chance that machinery inspections will early detect a potential cause and subsequent failure mode before the machinery functional failure occurs. Machinery controls will prevent an imminent failure and isolate the cause.	3
Almost Certain	Machinery controls or inspections almost certain to detect a potential cause and subsequent failure mode before the machinery functional failure occurs.	1

4.2. FMEA result

As the calendar section and dryer section comprised of many machines/equipment, this paper represented only the analysis of soft calendar roll which was shown in the table 4.

Table 4. FMEA result of Soft Calender roll

Possible Failure Mode(s)	Potential Failure Effect(s)	S 1	Possible Failure Cause(s)	O 1	Current Failure Prevention(s)/ Detection Method(s)		D 1	RPN₁	Recommended Action(s)		S 2	O 2	D 2	RPN₂
High Vibration	PM1 may shutdown	7	Defective or damaged roll bearing	3	- Vibration analysis - Oil analysis	1M 3M	3	63	- Vibration analysis/Oil analysis - Check plain bearing conditions - Replace plain bearing - Check roller bearing conditions	1M/3M 3M 4Y 4Y	7	2	3	42
		6	Bearing housing hold-down bolt looseness	2	- Vibration analysis - Check bolts/Tighten all bolts	1M 1M/1Y	3	36	- Vibration analysis/Check loose bolts - Tighten all bolts	1M/1M 1Y	6	2	3	36
Bearing Failure	PM1 must shutdown	7	Roll misalignment	3	- Vibration analysis	1M	5	105	- Vibration analysis/Roll alignment	1M/3Y	7	2	5	70
		7	Dirty oil lubrication	2	- Oil analysis	3 M	3	42	- Oil analysis	3M	7	2	3	42
			Roll misalignment	3	- Vibration analysis	1M	5	105	- Vibration analysis/Roll alignment	1M/3Y	7	2	5	70
			Deterioration	3	- Replace bearing as CBM		3	63	- Replace plain bearing	4 Y	7	2	3	42
Abnormal roll surface wear	PM1 may shutdown	7	Roll misalignment	3	- Vibration analysis	1M	5	105	- Vibration analysis/Roll alignment	1M/3Y	7	2	5	70
Poor Paper Quality	PM1 may shutdown	7	Roll misalignment	3	- Vibration analysis	1M	5	105	- Vibration analysis/Roll alignment	1M/3Y	7	2	5	70
			Defective or damaged bearing	3	- The same as in high vibration case		3	63	- The same as in high vibration case		7	2	3	42
			Roll face damaged	4	- Roll grinding/Check roll face	3M	3	84	- Roll grinding/Check roll face - Roll recover	3M 4Y	7	4	3	84
		6	Roll unable to maintain required internal pressure	5	- Check oil seal conditions - Check DPR - Check directional valve	3M 3M 2M	1	30	- The same as current failure prevention - DPR overhaul - Valve overhaul	 1Y 2Y	6	4	1	24
		5	Spurious oil pressure	8	- Online oil pressure monitoring		1	40	- Calibrate and service pressure transmitter - Control valve overhaul	1Y 2Y	5	5	1	25
		6	Defective DPR	5	- Check DPR	3M	3	90	- Check DPR/DPR overhaul	3M/1Y	6	4	3	72
		7	Worn/damaged longitudinal seals or end seals	2	- Check seal conditions	3M	3	42	- Check seal conditions/Replace seals	3M/3Y	7	2	3	42
		6	Hydraulic cylinder leakage	5	- Check hyd. cylinder conditions	2W	3	90	- Check conditions/Replace hyd. cylinder	1D/2Y	6	4	3	72
Fail to cycle	PM1 must shutdown	7	Helper drive belt torn	5	- Visual inspection/Replace helper drive belt	2W/1Y	3	105	- Visual Inspection - Replace helper drive belt	1D 6 M	7	3	3	63
		7	Universal joint shaft sheared	3	- Regrease	45D	7	147	- Regrease/Check U-joint conditions - Replace U-joint	45D/6M 5Y	7	2	5	70
		6	Motor burnt	4	- Check motor temperature - Check motor current - Vibration analysis - Clean motor externally	1M 1M 1M 1Y	5	120	- Check motor current/Check temperature - Vibration analysis/Check voltage balance - Clean motor/Leak to ground test - Open check terminal box and clean - Check, clean and tighten all electrical connections - Motor overhaul and electrical testing	1M/1M 1M/3M 3M/1Y 1Y 1Y 3Y	6	3	5	90

IOP Conf. Series: Materials Science and Engineering **530** (2019) 012051 doi:10.1088/1757-899X/530/1/012051

After analysing possible failure mode, potential effect and possible failure cause, risk assessment was performed in the form of RPN. This research mainly reduced the risk of failure mode caused by failure cause with RPN greater than 100. This figure was agreed by mill expert team. After evaluating the RPN, the action was determined. After implementing actions, the RPN was subsequently assessed and found to be reduced.

5. Improvement of preventive maintenance programs for general equipment and inspections

Because PM1 is a large paper machine and consists of a large number of smaller machines, the FMEA is done only in the most important and problematic areas, namely the Calender section and the Dryer section. Others can be easily improved by creating and revising the PM plans. Equipment that needed preventive maintenance improvement was mainly equipment that produced the breakdown loss time according to the Pareto diagram in Figure 1 and other equipment to prevent breakdown.

The equipment/machines that were improved in preventive maintenance tasks and plans were shown in the table 5.

Table 5. Selected equipment for preventive maintenance improvement

No.	Equipment for PM Improvement	No.	Equipment for PM Improvement
1	Control valve	11	Screw pump
2	Rolls	12	Universal joint
3	Stretcher	13	Level transmitter
4	Variable speed drive	14	Filter
5	Electric motor	15	Tail cutter
6	Sensor	16	Vacuum pump
7	Belt	17	Gear reducer
8	Heat exchanger	18	Fan/Blower
9	Fibron conveyor	19	Multi-stage pump
10	Hydraulics	20	Switchgear

In addition to improving the preventive maintenance tasks and plans, improvement of the inspection of the machine was also done. In the past, machine inspection was done every 2 weeks for each type of machine with unclear inspection order and inspection route. As described, there was one inspection sheet for each type of the machine, but the machines are scattered throughout the mill thus making inspection route very far. After the improvement, the machinery inspection route was divided into two routes and many types of machines were examined orderly in the same inspection sheet to minimize the distance of the inspection. Moreover, the new inspection is risk-based inspection. Frequency of the inspection was revised from the traditional inspection performed every two weeks to daily inspection, weekly, bi-weekly, and monthly inspection as machine ranking and risk of failure. More than 25,000 inspection items per month were increased and more basic inspection tools were deployed thus making machine inspection more effective and better.

6. Implementation of actions from FMEA and preventive maintenance improvement

In May 2018, planned shutdown of the paper machine was performed to improve and maintain according to the plans. There were replacement of new plain bearings and Differential Pressure Regulator (DPR) of Soft Calender roll, change of Soft Calendar roll and pressure transmitter. Moreover there was roll alignment checking for soft calendar roll during this shutdown. The maintenance of Drying Cylinder roll and the maintenance according to the improved preventive maintenance plans were performed as well. Furthermore, there were corrections of machine abnormalities found on the inspection of the machine and from the notice of the production department.

IOP Conf. Series: Materials Science and Engineering **530** (2019) 012051 doi:10.1088/1757-899X/530/1/012051

In addition, a new inspection pattern was introduced in early May 2018 with the introduction of a new inspection sheet to inspect machines daily, weekly, bi-weekly and monthly. Moreover, the old inspection sheets and plans for some types of machines that were not revised in inspection tasks were increased of the frequency from two weeks to one week per month.

Figure 2. 200,000 TPA PM1 paper machine

Figure 3. Soft Calender roll changing during planned shutdown of May 2018.

Figure 4. Calendar roll alignment checking

Figure 5. Dryer steam joint services

7. Result

After implementation of actions from FMEA, starting preventive maintenance according to the revised maintenance plan, and new inspections, it appears that the breakdown loss time of the paper machine between before improvement and after improvement, the breakdown loss time decreased significantly as shown in the table 6. The availability, OEE, and MTBF of PM1 paper machine in the three months following the improvement are also increased as shown in the table 7 below.

Table 6. Breakdown loss time comparison

Before Improvement (105 Days)		After Improvement (107 Days)	
Period	**Breakdowns (Hrs.)**	**Period**	**Breakdowns (Hrs.)**
16- 28 Feb. 2018	60	1-30 Jun. 2018	52
1- 31 Mar. 2018	21	1-31 Jul. 2018	35
1-30 Apr. 2018	37	1-31 Aug. 2018	13
1- 31 May 2018	57	1-15 Sep. 2018	11

Table 7. Availability, OEE and MTBF comparison

Before Improvement				After Improvement			
Month	**A (%)**	**OEE (%)**	**MTBF (Hrs.)**	**Month**	**A (%)**	**OEE (%)**	**MTBF (Hrs.)**
Mar'18	91.17%	83.04%	32	Jun'18	86.33%	76.82%	28
Apr'18	87.72%	79.48%	14	Jul'18	88.74%	81.34%	30
May'18	87.89%	78.21%	27	Aug'18	92.91%	83.95%	35
Three Months	88.95%	80.26%	21	Three Months	89.41%	80.77%	31

Table 8. Percentage of difference of OEE and MTBF

	OEE	**MTBF**
Before Improvement	80.26%	21 Hrs.
After Improvement	80.77%	31 Hrs.
Difference	0.51%	10 Hrs.
Percent of Difference	0.64%	47.6%

8. Conclusion

This research accomplished the application of FMEA technique to the PM1 paper machine in dryer and calendar section, the creation and revision of preventive maintenance for both mechanical and electrical equipment and the improvement of machine inspections. After implementation, the PM1 breakdown loss time that belongs to maintenance department responsibility was reduced by 5 hours per month from 38 hours on average during the three months prior to improvement from March to May 2018, down to 33 hours on average over the period of 3 months after improvement. The availability increased 0.46% from 88.95% to 89.41%, the OEE increased 0.51% from 80.26 % to 80.77% and the MTBF increased 10 hours from 21 to 31 hours.

References

[1] Toppinen A, Patari S, Tuppura A and Jantunen A 2017 The European pulp and paper industry in transition to a bio-economy : A Delphi study *Futures* Vol 88 pp1-14

[2] Patari S, Tuppura A, Toppinen A and Korhonen J 2016 Global sustainability megaforces in shaping the future of the European pulp and paper industry towards a bioeconomy *Forest Policy and Economics* Vol 66 pp 38-46

[3] Chiu Y S P, Chen K K and Ting C K 2012 Replenishment run time problem with machine breakdown and failure in rework *Expert Systems with Applications* 39(4) pp 1291-1297

[4] Chakraborty T, Giri B C and Chaudhuri K S 2008 Production lot sizing with process deterioration and machine breakdown *European Journal of Operational Research* 185(2) pp 606-618

[5] Prasad S 1991 Improving Manufacturing Reliability in IC Package Assembly Using the FMEA Technique *IEEE Transactions on Components, Hybrids, and Manufacturing Technology* 14(3) Pp 452-456

[6] Mariajayapratash A and Senthilvelan T 2013 Failure detection and optimization of sugar mill boiler using FMEA and Taguchi method *Engineering Failure Analysis* Vol 30 pp 17-26

[7] Hoseynabadi A H and Oraee H 2010 Failure Modes and Effects Analysis (FMEA) for wind turbines *International Journal of Electrical Power & Energy Systems* 32(7) pp 817-824

[8] Feili H R, Akar N, Lotfizadeh H, Bairampour M and Nasiri S 2013 Risk analysis of geothermal power plants using Failure Modes and Effects Analysis (FMEA) technique *Energy Conversion and Management* Vol 72 pp 69–76

[9] Cicek K, Turan H H, Topcu Y I and Searslan M N 2010 "Risk-based preventive maintenance planning using Failure Mode and Effect Analysis (FMEA) for marine engine systems" *Second International Conference on Engineering System Management and Applications* Sharjah United Arab Emirates pp 1-6

[10] Su C T, Lin H C, Teng P W and Yang T 2014 Improving the reliability of electronic paper display using FMEA and Taguchi methods : A case study *Microelectronic Reliability* 54(6-7) pp 1369-1377

[11] Rungsa E and Tangjitsitcharoen S 2014 Development of Computerized Preventive Maintenance Management System with Failure Mode and Effect Analysis for CNC Machine *Applied Mechanics and Materials* Vol 627 pp 365-371

[12] Tangjitsitcharoen S, Phummanee K and Lohasiriwat H 2016 "Improvement of preventive maintenance plan of new elevator product" paper presented at The International Conference on Flexible Automation and Intelligent Manufacturing *Korea Institute of Science and Technology Seoul*

IOP Conf. Series: Materials Science and Engineering **530** (2019) 012052 doi:10.1088/1757-899X/530/1/012052

Identification of Potential Ergonomic Risk Factors and Mitigating Measures for Malaysian Oil and Gas Drilling Industries: A Conceptual Research Proposition

M M Asad[1*], R Hassan[1], K Latif[2] and F Sherwani[3]

[1]Faculty of Technical and Vocational Education, Universiti Tun Hussein Onn Malaysia, Batu Pahat, Malaysia.

[2]Faculty of Engineering, DRB-HICOM University of Malaysia, Pekan 26607, Malaysia

[3]Faculty of Electrical and Electronics Engineering, Universiti Tun Hussein Onn Malaysia, Batu Pahat, Malaysia.

Corresponding author*: mujtabaasad11@gmail.com

Abstract. This research article briefly highlights the proposed research framework for identification of potential ergonomic risk Factors and mitigating measures for Malaysian oil and gas drilling industries. Because ergonomic hazard is considered as one of the major causes of lifedraining injuries and life time disabilities among oil and gas drilling crew. In this proposed research study, mix method research design will be adopted with explanatory research design. Whereas, industrial survey and semi structure interviews will be conducted to from oil and gas drilling crew (150) and safety and health experts (09) for identifying the potential ergonomic hazards with their effective mitigating measures. Whereas, descriptive statistics, What-If analysis and thematic analysis approach will be used in this proposed research study. The expected outcome of this proposed research framework is to provide the most potential ergonomic risk factors associated with oil and gas drilling operations at onshore and offshore domain with their most suitable hazard controlling measures at Malaysian industries. Similarly, this proposed study will provide a comprehensive mitigating approaches and strategies to safety and health officials to design effective safety trainings and instructional designs based on ergonomic safety professionals' opinions.

1. Introduction

In the Malaysian economy oil and gas industries play a vital role [1]. Therefore, in Malaysian GDP exploration and production of oil and gas operations have significant contributions [2]. oil and gas exploration and operation are one of tough job, because it has involved a lot of physical and ergonomic activities at onshore and offshore drilling domains [4]. Due to the nature of this job there are several physical and ergonomic hazards are involved during well drilling, dripping and cementing operations which can cause critical injuries such as falling hazards, tripping and ergonomic hazards among older oil and gas workers [3]. According to the previous studies, it has been identified that the ergonomic hazards are badly affect the oil and gas drilling workers as compare to others because they need to perform performed they activities for large span of time [5]. Similarly, the findings in various studies

IOP Conf. Series: Materials Science and Engineering **530** (2019) 012052 doi:10.1088/1757-899X/530/1/012052

have revealed that the long-term ergonomic activities practicing have been associated with high prevalence of musculoskeletal disorder (MSD) [6].

Occasionally the oil and gas workers are confronted with the risk of muscle and spinal cord injuries and accidents. Also, majority of worker are facing problems of low back and spinal strain [1-3]. Because they expose with awkward postures most of their working times [2]. This adds further complexity on the problem of workers' health. Similarly, the recovery time of the ergonomic injuries effected by oil and gas drilling crew are longer and painful [4]. Therefore, this proposed study tends to focus on the identification of those potential ergonomic risk factors which cause major health problems to aging workers at Malaysian oil and gas industries along with their preventive controlling measures. In this study exploratory research design will be adopted and the gathered data will be identified based on qualitative and quantitative findings.

2. Problem Statement

Onshore and offshore oil and gas drilling operations which are associated with several safety, ergonomic and environmental hazards and therefore they are considered as most challenging profession worldwide [2]. There are high injuries and accidents rates among oil and gas drilling crew has been reported for more than a decade, due to the lack of effective health and safety awareness of safe oil and gas drilling activities [5]. In the years 2007–2012, the occupational fatality rate of the oil and gas drilling industry was 2.5 times higher than the construction industry and 7 times higher than general industry [6]. Furthermore, there are many underlying risk factors such involved in high rate of oil and gas drilling fatalities, critical accidents and life-threatening injuries [1-2]. Whereas, due to the rapid change in environment and advancement in drilling technology drilling crew and safety professionals have to face new challenges every day regarding health and safety [5-7]. Oil and gas drilling workforce have to deal with life-threatening conditions and injuries while performing their jobs [8]. In which ergonomic hazards are most prominent and common due to the large number of injuries cases and disabilities recorded.

Moreover, most of the oil and gas drilling crew reported the ergonomic injuries such as back stain, damage of back bone, muscle injury and sever fractures during rig assembling, tripping and cementing operations at both drilling domains [6-9]. There are some underlying causes which has been reported by previously are due to wrong posture while lifting objects, lack of safety and health practices and insufficient engineering and personal protective requirements. According to the report published by Department of Occupational health and Safety Malaysia in 2014, the number of musculoskeletal diseases cases has been increased in oil and gas and manufacturing industries workforce from last four years due to ergonomic industries and accidents at sites as shown in Figure 1 [3].

Therefore, there is an urgent need to identify the most effective preventive measures to address the potential ergonomic risk factors associated with oil and gas drilling operations for reducing the work place risk at onshore and offshore oil and gas industries in a resourceful manner for the suitable decision making in hazardous conditions [6]. Since the improved and suitable occupational health and safety surveillance measures can help to reduce and eliminate the on-job injury and illness during well drilling operations [5-7].

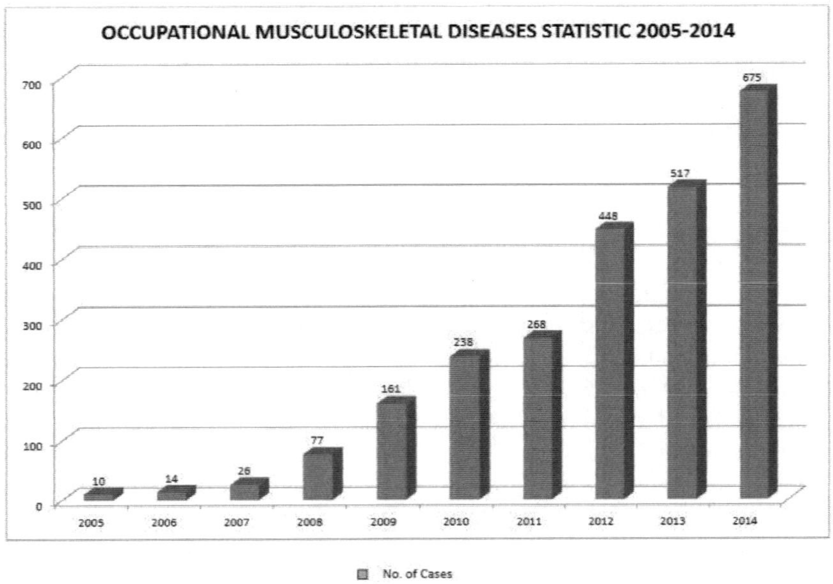

Figure 1. Occupational Musculoskeletal Diseases Statistics 2005-2014 (DOSH 2014)

3. Research Objectives
This study will aim to achieve following research objectives:
1. To identify the potential ergonomic risk factors associated with Malaysian oil and gas drilling industries.

2. To recognize the effective mitigation measures for potential ergonomic risk factors associated with Malaysian oil and gas drilling industries.

4. Research Questions
This proposed study will focus to answer listed research questions.
1. What are the potential ergonomic risk factors associated with Malaysian oil and gas drilling industries?

2. What are the mitigation measures for potential ergonomic risk factors associated with Malaysian oil and gas drilling industries?

5. Significance of Research
The significance of this proposed study is:
1. This proposed research will facilitate health and safety officials and Malaysian oil and gas drilling industries to provide maximum ergonomics protection to workforce.
2. This study will provide detailed experts recommendations for controlling potential ergonomic risks according to the hierarchy of controls for Malaysian oil and gas drilling industries.
3. This study will provide awareness to oil and gas workers for protecting themselves against ergonomic hazards for controlling spinal cord related injuries.

6. Methodology
The population of this proposed study will be drilling crew and health and safety experts from Malaysian (Petronas) oil and gas industries. Whereas, explanatory mix method approach will be adopted for this study, to achieve the aforementioned research objectives, for quantitative study 150 respondents (oil and

IOP Conf. Series: Materials Science and Engineering **530** (2019) 012052 doi:10.1088/1757-899X/530/1/012052

gas drilling crew) will be selected for answering the survey research questionnaire for identifying the potential ergonomic risk factors among drilling crew of Malaysian oil and gas industries. Similarly, for qualitative study, from Malaysian palm oil industrial sector, nine 09 Health and safety expert will participate in this proposed study for semi structure interview.

6.1 Data Analysis

For analyzing the quantitative data, descriptive statistic methods such as mean and standard deviation and independent T-Test will be utilize through SPSS 22. Whereas, for the identification of potential ergonomic risk factors researcher will select What-if Analysis approach for extracting meaningful and detailed information through in-depth semi structured interviews from health and safety experts. What-If Analysis is an effective method for determining the hazard controls by asking questions on the basis of potential hazards for making judgments regarding the acceptability of those risks and determining a recommended action for those risks judged by field experts. Because of the nature of this potential ergonomic risk identification technique researcher also required to identify the associated risk factors related with ergonomics.

Further, In the reference of identifying potential ergonomic risks and their mitigations thematic analysis approach will be adopted for interpreting the findings of this proposed study. Main themes will be developed on the bases on the characteristics of hazard and their appropriate controls. Similarly, sub themes for each activity and their controls will be sorted out on the bases of identified ergonomic risks and their recommended and suggested mitigating measures by experts from Malaysian onshore and offshore oil and gas industries as shown in Figure 2.

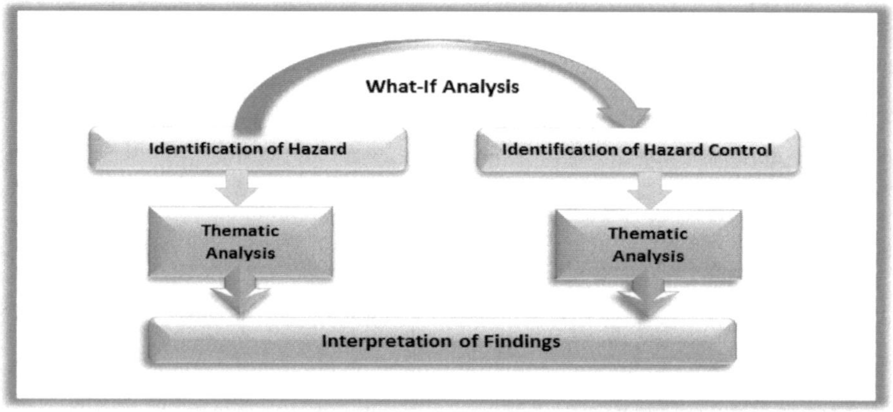

Figure 2. Proposed Data Analysis Framework

6.2 Mixed Method Research Design

Mixed methods research is a research design with philosophical assumptions as well as methods of inquiry (Creswell, 2003). It involves philosophical assumptions that guide the direction of the collection, analysis of data and the mixture of qualitative and quantitative approaches in many phases in the research process. As a method, it focuses on collecting, analysing, and mixing both quantitative and qualitative data in a single study or series of studies [2]. Its central premise is that the use of quantitative and qualitative approaches in combination provides a better understanding of research problems than either approach alone [7].

6.3 Explanatory Research Design of Proposed Study

In this proposed study, the sequential explanatory research design will be used of different methods rather than the integration of data analysis. The sequential (Explanatory, Exploratory, and Embedded) method uses both quantitative and qualitative data, which are implemented and connected at different phases [5]. In sequential explanatory research design, quantitative follows the qualitative findings. It means that quantitative research is conducted firstly and based on the attained findings, qualitative research is conducted for the in-depth understanding and validity of quantitative research findings [6]. This design is relatively effective to implement and describes findings well where mixed methods are employed [7]. Furthermore, a sequential explanatory mixed-methods research design is chosen because multiple methods and techniques can be used to provide the most complete understanding of the research problems [1-2]. The block diagram of explanatory research design as shown in Figure 3.

Figure 3. Flow of Explanatory Research Design

6.4 Approach for Semi- Structure Interview

The detailed information obtained from the semi-structured interviews helps to confirm and improve the data findings from the survey questionnaire. However, in this proposed study, semi-structured interviews from health and safety experts will be used to further investigate and contrast the information gained in the survey questionnaire analysis for the identification of potential ergonomic risk factors and their most suitable hazard controls for accident prevention at onshore and offshore drilling sites at Malaysian oil and gas industries. To obtain more in-depth information and data that could not be gathered from the survey instrument has been collected using semi- structured interviews [6]. Whereas, the respondents of these interviews were total of nine (09) health and safety practitioners from Malaysian oil and gas industry based on their expertise in ergonomics and working experiences. While, semi-structured questions will be asked during the interview because it helps to define the areas to be explored, but also allows the interviewer or interviewee to diverge in order to pursue an idea or response in more detail on specific issue and problem [6]. Moreover, the semi structured interviews will be conducted with the help of What- If analysis approach. This approach is considered as a most suitable quantitative approach for the identification of industrial hazards and risks in sufficient way [7].

Therefore, What- If analysis approach will be used for the identification of potential ergonomic hazards associated with drilling operation along with their effective hazard controls. The interviews will be conducted in offices and meeting rooms of targeted industries. While, the average amount of time for a one interview was from 45 to 90 minutes respectively [5]. Prior to the commencement of each interview, the researcher will elaborate the purpose and objectives of the study in order to set the mood of participants. Likewise, even though the consent forms and interview information sheets were also being provided two weeks in advance and signed prior to the interview. Correspondingly, in this study all participants will be notified that the discussions would be audio recorded to enable the researcher to improve the accuracy of the transcripts. All interviews will be conducted in the English language. The interviews will be transcribed manually by using Microsoft excel and coded to identify the main and sub themes.

7. Expected Research Outcome

This proposed study will identify the potential ergonomic risk factors associated with Malaysian onshore and offshore oil and gas industries along with their mitigations and controlling measures. This study will be beneficial for oil and gas industries for overcoming the ergonomic hazards and issues and among drilling crew in their respective industries. This proposed study will provide a detailed strategic information and recommendation to safety and health officials to develop appropriate training programs according to the expert's recommendations. Similarly, this study will provide awareness to Malaysian oil and gas workforce for taking appropriate measures to prevent ergonomic hazards and risk which may cause them to sever back pains, trains and spinal cord problems by taking appropriate safety and health preventive measures.

Acknowledgement
This research work and manuscript publication has been sponsored by Universiti Tun Hussein Onn Malaysia (UTHM) under the TIER 1(H209) research grant.

References

[1] Hassan R B, Asad M M, Soomro Q M and Sherwani F 2017 Severity of the Casing and Cementing Operation with Associated Potential Hazards in the Drilling Process in the On and Offshore Oil and Gas Industry: A Cross-Sectional Investigation into Safety Management *Pertanika Journal of Social Sciences & Humanities*
[2] Asad M M, Asad M M and Hassan R 2015 Development of KBES with hazard controlling factors and measures for contracting health and safety risk in oil and gas drilling process: a conceptual action plan
[3] DOSH 2014 Annual Safety and Health Report
Retrieved from: http://www.dosh.gov.my/index.php/en/archive-statistics/2014
[4] Nolan D P 2014 Handbook of fire and explosion protection engineering principles: for oil, gas, chemical and related facilities *William Andrew*
[5] Asad M M, Hassan R B, Ibrahim N H, Sherwani F and Soomro Q M July 2018 Indication of Decision Making through Accident Prevention Resources among Drilling Crew at Oil and Gas Industries: A Quantitative Survey *In Journal of Physics: Conference Series (Vol 1049 No 1 p 012022)* IOP Publishing
[6] Asad M M, Hassan R B, Sherwani F, Ibrahim N H and Soomro Q M July 2018 Level of Satisfaction for Occupational Safety and Health Training Activities: A Broad-Spectrum Industrial Survey *In Journal of Physics: Conference Series (Vol 1049 No 1 p 012021)* IOP Publishing
[7] Asad E M M, Hassan R B and Sherwani E F 2014 Instructional models for enhancing the performance of students and workforce during educational training *Academy Arena* 6(3) pp 27-31
[8] Clarke S and Cooper C 2004 Managing the risk of workplace stress: Health and safety hazards *Routledge*
[9] Salvendy G and Karwowski W 2010 Advances in human factors, ergonomics, and safety in manufacturing and service industries *CRC Press*

IOP Conf. Series: Materials Science and Engineering **530** (2019) 012053 doi:10.1088/1757-899X/530/1/012053

Development of the Ergonomic Wrist Posture Range for Indonesian in Typing Activity Using Electromyograph

H Soewardi and N Zhafira*

Industrial Engineering Department, Faculty of Industrial Technology, Universitas Islam Indonesia, Yogyakarta, Indonesia

Corresponding Author *: nadiazhafiranaufal@gmail.com

Abstract. Carpal Tunnel Syndrome (CTS) is one of the most common neuropathies which occurs due to repetitive activity of hand using non-neutral wrist posture whether in flexion, extension, or ulnar and radial. Even though this posture may be inevitable, however it should be confined in maximum range. Thus, this is very significant to develop the allowable of maximum deviation of wrist. Objective of this study is to determine the maximum range of wrist postures in such four motions with evaluating the muscle contraction at wrist. A case study applied is to type a manuscript using computer. Experimental study was conducted by using electromyograph to identify Maximum Voluntary Contraction (MVC) on *Flexor Digitorum Superficialis* (FDS) and *Abductor Pollicis Brevis* (APB). Ten healthy university students (5 females, 5 males) were participated with age between 20 to 23 years old. The statistical analysis was conducted to test the hypothesis. Result of this study shows that the suggested maximum wrist posture range is not greater than 30° extension, 20° flexion, 20° ulnar, and 20° radial, although the neutral posture of 0° was highly suggested. This guideline is proven to be valid for both male and female.

Keywords: *Wrist posture, Typing, Electromyography, Flexor Digitorum Superficialis, Abductor Pollicis Brevis*

1. Introduction

The use of personal computer (PC) is more and more widespread in recent decades for completing the task in a company [1]. In 2017, 13.7% of Indonesia population use personal computer and 22.52% use laptop and it increases gradually [2]. The main industrial activity done by using the PC is typing and adjustment the movement of cursor during the work day. These activities that is carried out frequently and repetitively will induce the risk of Musculoskeletal Disorders especially in upper extremity, including hand, wrist, and arm [3]. The disorder is also called the Carpal Tunnel Syndrome (CTS) which occurs as a result of the compressed median nerve attaching on the carpal tunnel at wrist [4-5]. These symptoms can disturb the ability to perform a task and potentially resulting in disability of hand [6-7] identified that prevalence of CTS is around 10.5% for frequent computer users in general. In Indonesia, [8] reported 97.3% of computer operators with non-ergonomic hand positions experiencing CTS.

Several studies have summarized that the well-known risks of work-related CTS is caused by forceful and repetitive movement [9], prolonged use of the hands and wrists [10], exposure to vibration [11] as well as non-neutral wrist posture [12]. For the last cause whether in flexion, extension, or ulnar and radial, it is always taken place excessively that is over a threshold of posture. This condition results in fatigue on muscle. Thus, this is crucial to know the maximum threshold of the wrist posture safely.

IOP Conf. Series: Materials Science and Engineering **530** (2019) 012053 doi:10.1088/1757-899X/530/1/012053

Electromyography (EMG) is one of electrodiagnosis techniques to record, analyze and display the muscle contraction [13-14]. This technique is valid to diagnose CTS by identifying the muscle contraction at wrist [15] such that it is reasonable to use this EMG. This study aims to define wrist posture maximum range in flexion, extension, and radial as well as ulnar by evaluating the muscle contraction at wrist.

2. Methods

Ten university students, consisting of five males and five females, were enrolled in this study. All participants are experienced in working on computer or portable computer for minimum 6 months [16], minimum fifteen hours as the median hours for computer use [17], and had no evidence of Carpal Tunnel Syndrome and musculoskeletal disorders conditions on upper extremity based on physical examination and medical history [14].

A within-subject design experiment was applied to all subjects by performing fifteen typing task, with five minutes duration of each task. The prior pilot study shows the significant difference in muscle activity for five minutes typing compared with one minute and three minutes typing. The participants were allowed to practice the typing for five minutes to become acclimated before the experiment began. The keyboard angle on each task were adjusted in order to achieve a desired wrist angle, with 10° of angle increment. The determined angles for extension were 40°, 30°, 20° and 10°, the determined angles for flexion were 40°, 30°, 20° and 10°, the determined angles for radial deviation were 30°, 20° and 10°, and the determined angles for ulnar deviation were 30°, 20° and 10°, and 0° for the natural posture as shown in figure 1. Goniometer is used to measure the desired angle. Flexion and extension measurement was conducted by centralizing the goniometer fulcrum on the lateral aspect of the wrist and aligning the goniometer distal arm towards the fifth metacarpal lateral plane. Ulnar and radial measurement was conducted by centralizing the goniometer fulcrum on the dorsal aspect of the wrist while the wrist in pronation position and align the goniometer distal arm towards the third metacarpal dorsal plane [18]. At the end of the experiment, the participants performed maximum static contraction against manual resistance for five seconds in kneeling position for obtaining Maximum Voluntary Contraction (MVC) value [19] [20].During performing typing task, participants were not allowed to rest the hands or wrists on a surface in order to prevent palm contact pressure [12].

Figure 1. Experimental Design

The experiment apparatus consisted of two personal computers for diplay EMG signal data recorded by LabQuest 2 (Vernier Software and Technology, USA) and to display the typing task, EMG Sensor (Vernier Software and Technology, USA), and pre-gelled surface electrode as shown in figure 2. The standard workstation were given with 500mm – 1000mm monitor distance range [21] and the upright position chair which gives horizontal thigh and vertical lower leg posture [22]. The software used to display the typing task was GS Typing Tutor version 2.96 (GrassSoftware).

Figure 2. Experimental Environment

The information regarding the experiment procedure were explained to the study participant and Phalen Test performed in order to diagnose the carpal tunnel syndrome indication [23].The surface electrodes were attached afterwards in parallel to the muscle fibers with 2 cm distance between one electrode to another into the sterilized skin area, and one reference electrode on nearby but unaffected area [19]. *Flexor Digitorum Superficialis* (FDS) and *Abductor Pollicis Brevis* (APB) were two muscles used as the research object which innervated by median nerve, the only nerve passes through carpal tunnel [24] [25]. The location of attached surface electrodes on both muscles is illustrated in Figure 3. Electromyograph signal was sampled at 500Hz with high-pass and low-pass amplifier bandpass filter are set to 10Hz - 500Hz with 100ms Root Mean Square to smooth the signal and Maximum Voluntary Contraction applied to normalized the data [19].

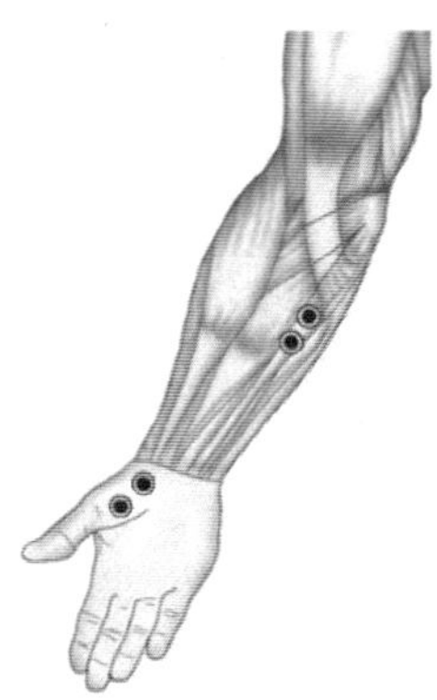

Figure 3. FDS and APB Electrode Placement

3. Results and Discussion

The results of the experiment show that the muscle contraction increases along with the angle increment in flexion, extensions, ulnar and radial posture, and the lowest muscle contraction is in 0° position. A rank-based nonparametric test was performed to determine whether there was significant difference on one posture treatment with another. The probability value (p value) on Kruskal-Wallis test shows both the posture differences on *Flexor Digitorum Superficialis* (FDS) and *Abductor Pollicis Brevis* (APB) muscles were significant (p: 0.00). Figure 4 and 5 illustrate the average data of the normalized EMG signal shown with the percentage of Maximum Voluntary Contraction (%MVC) of these muscles. The %MVC of FDS muscle on 0° is 6.623%, on 10° extension is 13.181%, on 20° extension is 16.224%, on 30° extension is 24.267%, on 40° extension is 27.758%, on 10° flexion is 12.849%, on 20° flexion is 19.676%, on 30° flexion is 25.298%, on 40° flexion is 30.828%, on 10° ulnar is 12.255%, on 20° ulnar is 16.896%, on 30° ulnar is 20.707%, on 10° radial is 11.582%, on 20° radial is 15.308%, and on 30° radial is 23.465%.

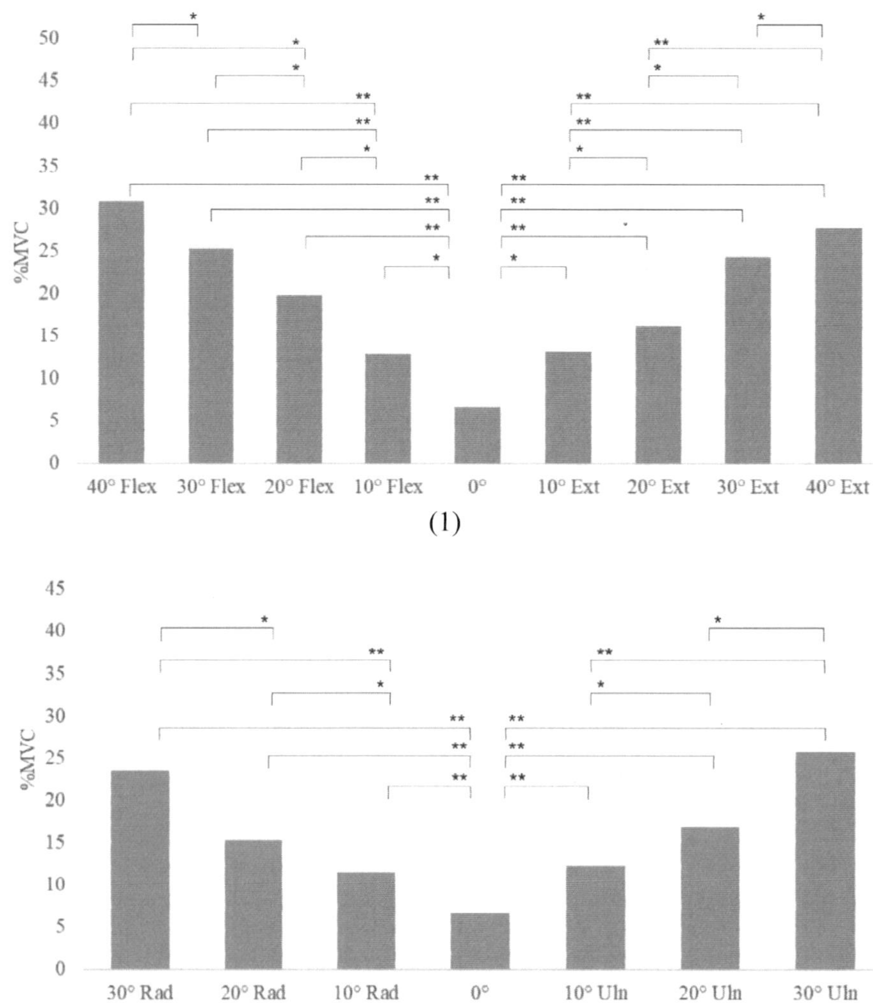

*$p > 0.05$ (not significant), **$p < 0.05$ (significant)

Figure 4. %MVC of *Flexor Digitorum Superficialis* Muscle at different wrist postures; (1) Flexion and Extension, and (2) Ulnar and Radial

Mann Whitney U Test was conducted to compare differences between two postures whether in flexion, extension, ulnar, or radial. Figure 4 above shows the results of Mann Whitney U Test on *Flexor Digitorum Superficialis* muscle. There is a significant difference between wrist posture of 0° and 20° flexion, 0° and 30° flexion, 0° and 40° flexion, 10° flexion and 30° flexion, 10° flexion and 40° flexion, 20° flexion and 40° flexion, as well as 0° and 20° extension, 0° and 30° extension, 0° and 40° extension,

10° extension and 30° extension, 10° extension and 40° extension, 20° extension and 40° extension. On the other hand, there is significant difference between wrist posture of 0° and 10° radial, 0° and 20° radial, 0° and 30° radial, 10° radial and 30° radial, as well as 0° and 10° ulnar, 0° and 20° ulnar, also 0° and 30° ulnar, 10° ulnar and 30° ulnar.

Whereas the %MVC of APB muscle on 0° is 11.423%, on 10° extension is 14.612%, on 20° extension is 14.869%, on 30° extension is 21.725%, on 40° extension is 27.573%, on 10° flexion is 16.176%, on 20° flexion is 20.231%, on 30° flexion is 24.371%, on 40° flexion is 29.843%, on 10° ulnar is 17.002%, on 20° ulnar is 20.358%, on 30° ulnar is 23.197%, on 10° radial is 17.792%, on 20° radial is 23.280%, and on 30° radial is 25.793% as shown in figure 5.

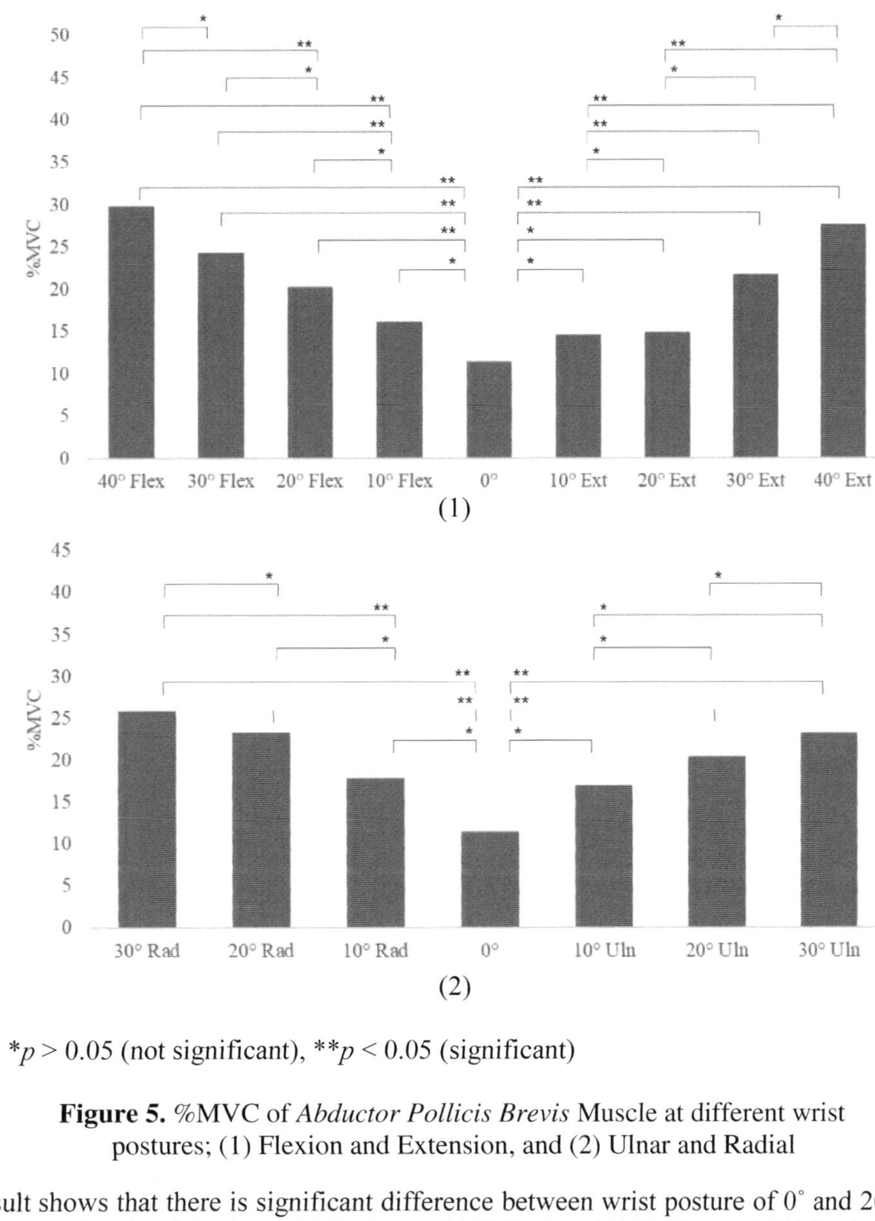

*$p > 0.05$ (not significant), **$p < 0.05$ (significant)

Figure 5. %MVC of *Abductor Pollicis Brevis* Muscle at different wrist postures; (1) Flexion and Extension, and (2) Ulnar and Radial

The result shows that there is significant difference between wrist posture of 0° and 20° flexion, 0° and 30° flexion, 0° and 40° flexion, 10° flexion and 30° flexion, 10° flexion and 40° flexion, 20° flexion and 40° flexion, as well as 0° and 30° extension, 0° and 40° extension, 10° extension and 30° extension, 10° extension and 40° extension, 20° extension and 40° extension. On the other hand, there is significant difference between wrist posture of 0° and 20° radial, 0° and 30° radial, 10° radial and 30° radial, as well as 0° and 20° ulnar, also 0° and 30° ulnar on *Abductor Pollicis Brevis* muscle.

The FDS muscle contraction on extension motion shows the significant difference at 20° angle increment (0° - 20°) while in APB at 30° angle increment (0° - 30°). The muscle contraction on 20° FDS extension is 16.224 %MVC, while in APB muscle of 30° extension is 21.725 %MVC. Since the %MVC value is still below 50-60 %MVC, it has not shown a major disturbance as muscle failure and blood circulation disruption [26] [13].Based on the effort scale developed by [27], the %MVC value is still categorized as light to moderate effort. Thus, the extension wrist postures no greater than 30° is acceptable. On flexion motion, both FDS and APB muscle contraction show the significant difference on 20° of angle increment in flexion. The radial and ulnar posture on FDS muscle shows the significant difference on 10° angle increment for both radial and ulnar (0° - 10°), and 20° angle increment on radial and ulnar for APB muscle (0° - 20°). An adequate radial and ulnar maximum range is 20° since the %MVC value still in acceptable value. However, the sustained and prolonged muscle contraction will disclose the sign of muscle fatigue shown by the reduced conduction velocity of muscle fibres due to lack of energy source, and begins at 11%MVC [28]. Thus, the neutral posture is highly suggested to avoid muscle fatigue.

Some studies show that gender differences influence physical measurement, including body mass and height, blood pressure [29], and anthropometric size [30]. Based on the data obtained, the wrist muscle contraction on male tends to be greater than female had. However, the p value of non-parametric independent difference test shows no significant differences at between (FDS $p = 0.221$, APB $p = 0.254$). Therefore, the threshold is possibly applied for both males and females.

4. Conclusion

Based on the experimental results, it is concluded that as the angle increment rises, there is an increase in wrist muscle contraction which leads to muscle fatigue. The suggested ergonomics wrist angle for typing is close to 0° as the neutral posture with considering the maximum range to avoid excessive muscle fatigue. The maximum range which needs to be avoided is no higher than 30° extension, 20° flexion, 20° ulnar, and 20° radial especially in prolonged use of computer.

Acknowledgement

This study was supported by Directorate of Student Development Universitas Islam Indonesia, Yogyakarta, and Work System Design and Ergonomics Laboratory (DSK&E), Industrial Engineering Department, Faculty of Industrial Technology, Universitas Islam Indonesia, Yogyakarta.

References

[1] Conte C, Ranavolo A, Serrao M, Silvetti A, Orengo G, Mari S and Draicchio F 2014 Kinematic and Electromyographic Differences between Mouse and Touchpad Use on Laptop Computers *International Journal of Industrial Ergonomics* 44 413-20

[2] KEMKOMINFO 2017 *Survey Penggunaan TIK 2017 serta Implikasinya Terhadap Aspek Sosial Budaya Masyarakat* (Jakarta: Pusat Penelitian dan Pengembangan Aplikasi Informatika dan Informasi dan Komunikasi Publik)

[3] Qin J, Chen H and Dennerlein J 2013 Wrist Posture Affects Hand and Forearm Muscle Stress During Tapping. *Appl. Ergon.* 44 1-8

[4] Latinovic R, Gulliford M and Hughes R 2006 Incidence of Common Compressive Neuropathies in Primary Care *J Neurol Neurosurg Psychiatry* 77 263-265

[5] Salawati L and Syahrul 2014 Carpal Tunel Syndrome *Jurnal Kedokteran Syiah Kuala* 14 29-37

[6] Turner J, Franklin G, Fulton-Kehoe D, Sheppard L, Wickizer T and Wu R 2007 Early Predictors of Chronic Work Disability Associated with Carpal Tunnel Syndrome: A Longitudinal Workers' Compensation Cohort Study *Ind. Med.* 50 489-500

IOP Conf. Series: Materials Science and Engineering **530** (2019) 012053 doi:10.1088/1757-899X/530/1/012053

[7] Stevens J, Witt J, Smith B and Weaver A 2001 The Frequency of Carpal Tunnel Syndrome in Computer Users at a Medical Facility *Neurology* 56 1568-70

[8] Suherman B, Maywati S and Fatturahman Y 2012 *Beberapa Faktor Kerja Yang Berhubungan Dengan Kejadian Carpal Tunnel Syndrome (CTS) Pada Petugas Rental Komputer Di Kelurahan Kahuripan Kota Tasikmalaya* (Tasikmalaya: Universitas Siliwangi)

[9] Septiawati D, Hasyim H and Najmah 2013 Faktor Risiko Ergonomi Saat Mengetik dan Hubungannya dengan Carpal Tunnel Syndrome *Jurnal Ilmu Kesehatan Masyarakat* 4 237-44

[10] You D, Smith H and Rempel D 2014 Meta-Analysis: Association Between Wrist Posture And Carpal Tunnel Syndrome Among Workers *Saf Health Work* 5 27-31

[11] Sharma P 2016 Study of CTS Caused by Vibration Exposure on The Workers Engaged in Manufacturing Sector *International Journal of Advances in Science Engineering and Technology* 4 90-6

[12] Rempel D, Keir P and Bach J 2008 Effect of Wrist Posture on Carpal Tunnel Pressure while Typing *J. Orthop Res.* 26 1269-73

[13] Criswell E 2011 *Surface Electromyography, Second Edition* (Canada: Jones and Bartlett)

[14] Roman-Liu D and Bartuzi P 2013 The Influence of Wrist Posture on The Time and Frequency EMG Signal Measures of Forearm Muscles *Gait & Posture* 37 340-4

[15] Wang L 2013 Electrodiagnosis of Carpal Tunnel Syndrome *Physical Medicine Rehabilitation* 24 67-77

[16] Kumar S, Sambyal S and Vij J 2013 Analysis of Disability and Pain in Computer Professionals with Neck Pain *IOSR JNHS* 3 6-9

[17] Cook C, Burgess-Limerick R and Papalia S 2004 The Effect of Wrist Rests and Forearm Support during Keyboard and Mouse Use *International Journal of Industrial Ergonomics* 33 463-72

[18] Nsorkin C and White D 2009 *Measurement of Joint Motion : A Guide to Goniometry* vol 4 (Philadelphia: F.A. Davis)

[19] Konrad P 2005 *The ABC of EMG* (Noraxon Inc.: USA)

[20] Jepsen J, Laursen L, Larsen A and Hagert C 2009 Manual Strength Testing in 14 Upper Limb Muscles, A Study of Inter-rater Reliability *Acta Orthopaedica Scandinavica* 4 442-8

[21] HFES 2001 *Human Factors Engineering of Computer Workstation 100* (Santa Monica: Human Factors and Ergonomics Society)

[22] Kroemer K, Kroemer H and Kroemer-Elbert K 2001 *Ergonomics: How to design for ease and efficiency* Vol 2 (New Jersey: Prentice Hall)

[23] Brüske J, Bednarski M, Grzelec H and Zyluk A 2002 The Usefulness of The Phalen Test and The Hoffmann-Tinel Sign in The Diagnosis of Carpal Tunnel Syndrome *Acta Orthopaedica Belgica* 68 141-145

[24] Butler T, Kilbreath S, Gorman R and Gandevia S 2005 Selective Recruitment of Single Motor Units in Human Flexor Digitorum Superficialis Muscle During Flexion of Individual Fingers. *J Physiol* 567 301-309

[25] Soewardi H, Anugraheni A and Shabrina N 2005 Analysis of Electromyography on Computer Interaction Devices to The Risk of Carpal Tunnel Syndrome *JCP* 10 347-353

[26] Grandjean E 1986 *Fitting The Task To The Man : An Ergonomic Approach* (London: Taylor and Francis)

[27] Bernard T 2012 *Static Work* Version 3.1 (Tampa: College of Public Health University of Florida) pp 1-5

[28] Stalling L M, Fetcher P and Vorro J 1976 Premature Occlusal Contact Influence on Mandibular Kinesiology (Baltimore, MD: University Park Press)

[29] Sakurai M, Miura K, Takamura T, Ota T and Ishizaki M 2006 Gender Differences in Association between Anthropometric Indices of Obesity and Blood Pressure in Japanese. *Hypertens Res* 29 75-80

[30] Fragala M, Clark M, Walsh S, Kleppinger A, and Judge J 2012 Gender Differences in Anthropometric Predictors of Physical Performance in Older Adults *Gender Medicine* 9 445-56

Food Container with Expiration Date Reminder (OPTIMA) as the Solution to Minimize Leftover Food

O R Pertiwi*, N Zhafira, A M Ma'arij and A D Sari

Industrial Engineering Department, Industrial Technology Faculty, Universitas Islam Indonesia, Yogyakarta

Corresponding author *: oktavira.revi@gmail.com

Abstract. As time goes by, food waste has become one of the global issues that needs to be addressed due to its adverse impact on the economy, environment, and social. Preliminary study shows 77% of respondents leaves the food until it expires. Moreover, a study shows that most of the reason consumers discard the food is to avoid the food which almost and has expired. OPTIMA is a food container with expiration reminder product design which aims to avoid expired food disposal. This innovation can be integrated with mobile phone application so that consumers can control the expiration period of stored food in order to be optimized before the expiration date. The expiration date will be displayed on the LED display and indicator lights using the battery power source. It was designed with PDCA process (Plan, Do, Check, Action). Based on usability test through the prototype, 60% of respondents only had less than or equal to one error to use the product for the first time, 80% of respondents stated that the product can be used easily, 80% of respondents were satisfied with the product. It is expected to be used as one of the medias of actualization and development of technology to find alternative and innovative solutions to the problem of food waste stacking and expected as the innovation for household technology.

Keywords: *Food waste, expiration date, food container*

1. Introduction

Food waste is a significant problem in the food supply chain [1]. Over time, food waste is one of the global issues that needs to be addressed because it has a negative impact on the economic, environmental and social aspects. In developed countries, household is one of the largest sources of food disposal [2]. A study shows that 31% of food is wasted; with a percentage of 21% by consumers and 10% by food producers. Figure 1 below shows the percentage of food waste produced by consumers.

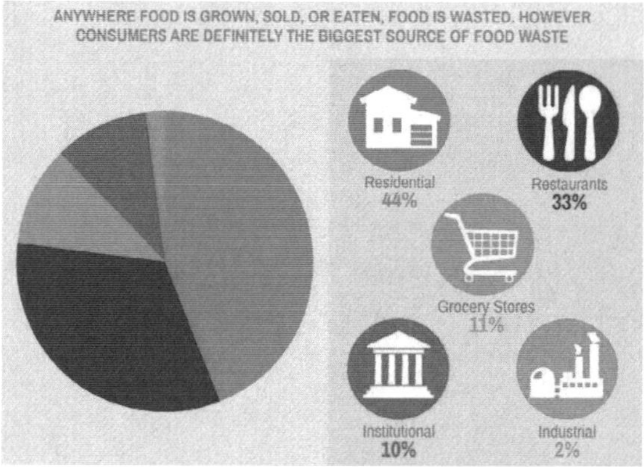

Figure 1. Percentage of Food Waste

Since most of food wastes come from households, studies are conducted to observe the consumer behavior. The results show that the reason consumers discard their food is to avoid the expiration date and lack of food storage management skills which becomes concerning [2]. Food containers made from plastic become one of important things of household lately. The function of plastic as a food and beverage container gets the biggest attention due to its versatile function [3].In this recent decade, customers are willing to have products which have more value added [4]. The value added can be in form of either non-wasteful process, multifunctional, or technology presence.

This paper presents the product innovation of food container with expiration date reminder which can be integrated with applications on mobile phones so that consumers can control the food expiration period that is stored and can be optimized before the expiration period.

2. Methods

Kaizen philosophy was used in designing the product, which includes the stage of plan, do, check, and action (PDCA). A quality management applied in these stages will define the structural problem solving in response to continuous customer feedback [5]. It has been extensively used in many aspects; business process, manufacturing, and product development. There are four stage in PDCA. Plan is the first stage, which is about to define purpose, goals, and objectives based on data collection earlier [6]. At plan phase, 75 questionnaires were distributed to housewives, students, employees and teaching staffs to perceive the voice of customers which then translates into customer needs, so that the attributes that need to be in the product design will be known.

Do is the second stage which is to identify needs, change, and implementation. In this research, Do stage is to make the prototype of the product used in the manufacture of prototypes include iron plate, RTC, LED lights, and the application based on the output of the questionnaire for the size and product material needed by consumers.

Check is the third stage which is used for monitor, evaluate, and analyze the change which compares previous and new data [6]. One of the methods to evaluate product is usability testing. Usability is a related but complementary, concept that defines the quality of the interaction between people and systems. If ergonomics and human-centered design are the means to create products that fit people, usability is how we measure the fit [7]. Usability is a critical dimension of product quality that affects product success [8]. It is defined in terms of efficiency, effectiveness, user satisfaction, and whether specific goals can be achieved in a specified context of use [9]. Usability testing is done by asking the respondents to operate the prototype directly with certain tasks. The task is started with turn the product on, put the food container in the particular place, input the data needed to the OPTIMA application in the phone, read the information from the LCD screen, end with interview to obtain the specified target.

Act is the last stage of PDCA. It is to adjust strategies for improvement and refine the product. Based on the stages of planning, prototyping, and evaluation, the inspection and further research are done to improve the tools to reduce errors and failures in the product so the product can be operated according to its function. Then, the granting of Patent Rights is carried out in an effort to maintain authenticity and avoid acts of plagiarism.

3. Results and Discussion

3.1. Questionnaire Results

The questionnaire involved 75 respondents in which 56% of respondents were housewives. The results of the questionnaire are shown in the table below:

Table 1. Questionnaire Results

No.	Question	Answer	Percentage (%)
1.	Intensity of Leaving Food Through the Expiry Period	Ever	77
		Never	23
2.	Type of Food Stored	Dry Food	37
		Cake, cookies	30
		Fresh Food	25
		Cooked Ingredients	5
		Herbs	3
3.	Desired Food Container Material	Plastics	60
		Glass	27
		Acrylic	13

The results of the questionnaire can then be summarized to make the following voice of customers:

1. As many as 58 of the 75 respondents or 77% of respondents stated that they had left food until they passed the expiration period.
2. The type of food most frequently stored by respondents in a food container is dry food, which is as much as 37% of respondents.
3. The food container material needed by respondents is made from plastic as much as 60% of the respondents.

3.2. Product Design

Based on the results of the voice of the customer, a food container design is made as shown below:

IOP Conf. Series: Materials Science and Engineering **530** (2019) 012054 doi:10.1088/1757-899X/530/1/012054

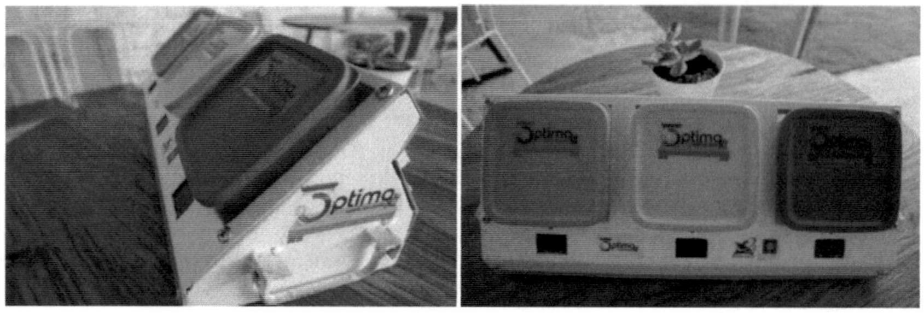

Figure 2. OPTIMA Product

The specifications of OPTIMA are:

1. Handle on the side of the products; serve to facilitate the transfer of products.
2. Medium-sized container, as food containers that can replace as desired.
3. LCD screen to indicate the time and type of food
4. Indicator light, which lights up according to the expiration date. Green light means there are 3 days or more left before the expiration date, yellow light means less than 3 days before the expiration date, red light means the expiration date already exceeded.
5. The power button on the back of the product.
6. Integrated with applications on smartphones using the Bluetooth feature shown in figure 3.

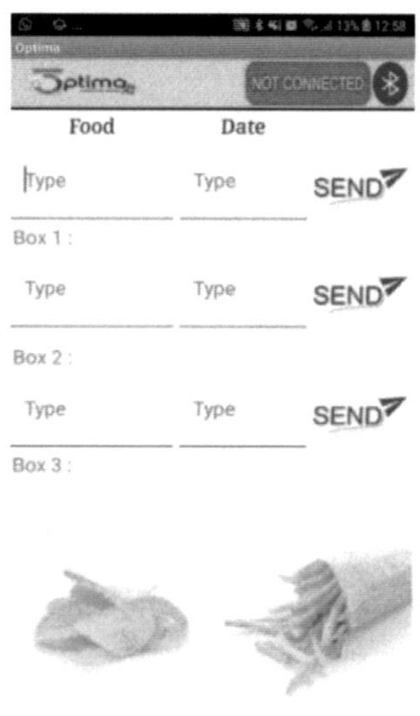

Figure 3. Application Interface

3.3. *Work Mechanism*

The working mechanism of the OPTIMA product is explained in the scheme below:

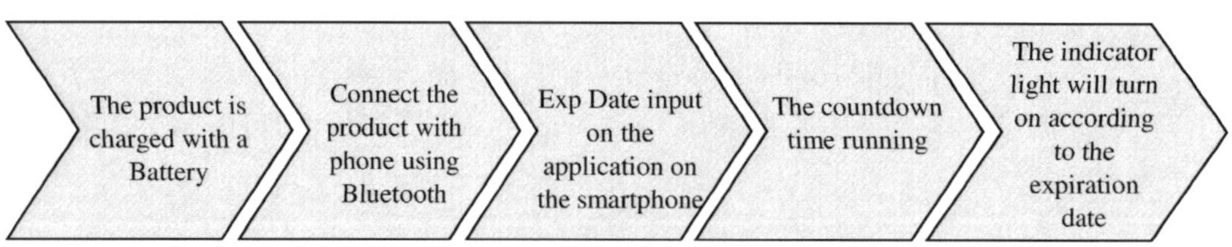

Figure 4. OPTIMA Work Mechanism

This product uses 2 dry batteries as its resources. As it has been filled with a dry battery, the product can be turned on by clicking on the power button on the back of the product. Then the user enters the data in the form of expiration date and type of food on the OPTIMA application that has been integrated in the smartphone. After inputting, the countdown time will run and the indicator light will light up according to the expiration date. The existence of these features is expected to be able to minimize food expiration by paying attention to the indicator lights on the product.

3.4. Usability Testing Result
Usability testing is done by giving the task to respondents in using OPTIMA products. The task given is starting from when turning on the product to reading the information on the LCD screen. The attributes examined were Learnability, Satisfaction and Effectiveness with the following results:

Table 2. Usability Testing Result

No	Learnability	Satisfaction	Effectiveness
1	5	5	2
2	4	5	0
3	4	4	2
4	4	5	1
5	4	5	1

Based on Table 2, it was found that as many as 60% of respondents claims that this product is effective to use with an error rate of less than 1 time, 80% of respondents stated that the product could be used easily, and 80% of respondents stated that respondents were satisfied with OPTIMA products.

4. Conclusion
OPTIMA is a product design in the form of metal plate containing 3 pieces of food container that can be replaced as needed which designed by PDCA aspects. There are 3 LCD screens to display the food name and expiration date that has been inputted through the application on the smartphone with the help of the Bluetooth feature. OPTIMA testing has been carried out with usability testing which resulted that this product is usable to be used.

However, the improvement of this product is needed along with the development of technology. Thus, the suggestions given for the further research is to maximize information technology so that the application connected with the product can be better by adding pop up reminder features on smartphones and improving the interface to be more user-friendly. Ergonomics aspect also should be considered by improve the handle, product material, and brightness level of the indicator light.

Acknowledgement
This study was supported by Directorate of Student Development Universitas Islam Indonesia, Yogyakarta, and Ministry of Research, Technology and Higher Education of the Republic of Indonesia.

References

[1] Amani P and Gadde L 2015 Shelf Life Extension and Food Waste Reduction *Proceedings in System Dynamics and Innovation in Food Networks 2015* Igls Austria

[2] Aschemann-Witzel J, Hooge I, Amani P, Bech-Larsen T and Oostindjer M 2015 Consumer-Related Food Waste: Causes and Potential for Action *Sustainability* **7** 6457-77

[3] Nurullaili and Wijayanto A 2013 Analisis Faktor-Faktor yang Memengaruhi Loyalitas Konsumen Tupperware *Jurnal Administasi Bisnis* 2 89-97

[4] Gidey E, Jilcha K, Beshah B and Kitaw D 2014 The Plan-Do-Check-Act Cycle of Value Addition *Ind Eng Manage* 4 1-5

[5] Ralph K and Allen T 1984 *Organizational Issues in The Introduction of New Technologies* (USA: Cambridge Mass)

[6] Knight J E 2012 Applying The PDCA Cycle to The Complex Task of Teaching and Assessing Public Relations Writing *Int Journal of Higher Education* 1 67-83

[7] van Kujik J, van Driel L and van Ejik D 2015 Usability in Product Development Practice; An Exploratory Case Study Comparing Four Markets *Appl Ergon* 47 308-23

[8] Mack Z and Sharples S 2009 The Importance of Usability in Product Choice: A Mobile Phone Case Study. *Ergonomics* 52 1514-28

[9] NSAI 2018 *Ergonomics of human-system interaction - Part 11: Usability: Definitions and concepts.* (Brussels: CEN-CENELEC Management Centre)

EASILY (Easy to Use Kitchen Utilities) Designing Cutting Boards Sets for Elderly and Arthritis Sufferers with Quality Function Deployment

A M F Zubaidi*, M K Sumitro and A S Husna

Industrial Engineering Department, Faculty of Industrial Technology, Universitas Islam Indonesia, Yogyakarta, Indonesia

Corresponding author *: ariefzubaidi23@gmail.com

Abstract. Based on the Indonesian Population Census in 2015, the number of elderly people aged over 60 years in Indonesia reached 2.1 million people. Elderly women do ADL (Activities Daily Living) is higher than the population of elderly men. One ADL that is still often done by the elderly woman is cooking while the process of manually using a cutting board. The purpose of this research is to developed cutting board set named EASILY (Easy-to-Use Kitchen Utilities) which is a set of cutting tools consisting of ergonomic knife and cutting board that comfortable to use and also it does not need much power from operator to do the cutting activities. The steps taken with the method of Quality Function Deployment, to process and interpreting consumer need for product specifications. Usability test is applied with satisfaction attribute performed to determine the level of user satisfaction through interview when subject using EASILY product. The result of this reserach show that EASILY have 4 components there are Cutting Board, Knife Holder, Food Holder, and Trash Tray. The subject feel comfortable while using EASILY after doing the usability testing. Recommendation for further reserach the product development can be done using another product design method, and the usability testing should using more respondent.

Keywords: *QFD, kitchen utilities, elderly, usability*

1. INTRODUCTION

According to the Indonesian population census conducted in 2015, the number of elderly people over the age of 60 who are in Indonesia reached 2.1 million people spread throughout Indonesia with the highest value for the number of elderly people in the Special Province of Yogyakarta [1]. This amount is included in the high category, considering that the percentage of elderly people in the Indonesian population is 10%. The number of elderly people in Indonesia itself is dominated by the female gender which has a comparison with the number of elderly women in the female sex that is equal to 9: 8 [2]

From the number of elderly people in Indonesia, there are several health problems suffered by the elderly, from census data, the list of diseases that are suffered by the elderly is as follows:

Table 1. Number of Elderly Patients [2]

No.	Health Problems	Prevalence		
		55 – 64 Years Old	**65 – 74 Years Old**	**75+ Years Old**
1.	Hypertension	45.9	57.6	63.8
2.	Arthritis	45	51.9	54.8
3.	Stroke	33	46.1	67
4.	PPOK	5.6	8.6	9.4
5	DM	5.5	4.8	3.5
6.	Cancer	3.2	3.9	5
7.	Coronary Heart Dusease	2.8	3.6	3.2
8.	Kidney Stones	1.3	1.2	1.1
9.	Heart Failure	0.7	0.9	1.1
10.	Kidney Filure	0.5	0.5	0.6

When viewed from table 1.1 above, information can be obtained that arthritis is the number 2 disease most commonly suffered by elderly people in Indonesia, Gout Arthritis Disease (GA) according to the American Collage of Rheumatology is a disease and potential disability due to arthritis that has long been known, its symptoms usually consist of severe episodic from one joint inflammation pain. Gout is a form of chronic inflammation of the arthritis, swelling and the most common pain in the big joints of the big toe, but can affect other joints and can become more severe. Gout is a term used by a group of metabolic disorders characterized by increased levels of uric acid. [3]

Gout arthritis is one of the rheumatic diseases which ranks third after arthrosis and rheumatoid arthritis, rheumatic patients in Indonesia are estimated to be almost 80% of the population aged 40 years or older [4]. The World Health Organization (WHO) estimates that around 335 million people in the world suffer from rheumatic diseases. This amount corresponds to an increase in the number of elderly people. Musculoskeletal problems are the most common chronic problem occurs in the elderly, with around 49% of elderly experiencing some form of arthritis [5].

Most of the elderly in Indonesia, especially for elderly women, are still doing work that is closely related to the use of hand power, especially household work and cooking. This can increase the risk of the elderly being affected by arthritis, coupled with the strength of the elderly who have been reduced so that the elderly need tools that can help to do work, especially domestic work. Especially in this study will provide innovation 1 set of knives and ergonomic cutting boards and also can reduce the energy needed without reducing the results of cuts from the work of the elderly.

2. METHODS

2.1 Quality Function Deployment

QFD was first developed in Japan by Mitshubishi's Kobe Shipyard in 1972, which was later adopted by Toyota. Quality function Development (QFD) is a method for product planning and development allows the team developer to set clearly customer desires and needs, and also to evaluate each product proposal or service capability systematically in relation to the impact on achieving these needs. According to Gaspersz [6] The essence of QFD is a large matrix that connects what the customer needs (What) and how a product will be designed and produced to meet the customer's wishes (How). The most commonly used analogy to describe the structure of QFD is a matrix in the form of a house. The term often used is House of Quality (HOQ) [7].

The first method of data collection for the design of EASILY product was a questionnaire by distributing 30 respondents. Respondents are elderly people aged above 65 years old which are mainly female. There are 3 stages of questionnares distributions. The design of this product begins with the

IOP Conf. Series: Materials Science and Engineering **530** (2019) 012055 doi:10.1088/1757-899X/530/1/012055

distribution of questionnaires to get Voice of Customer, namely identification of customer needs and then proceed with the creation of Technical Response. The next step is to disseminate the second questionnaire regarding the scale of interest, and the results of the questionnaire are tested for reliability and product design. Next, in designing this product is the dissemination of a third questionnaire on the comparison of these products with existing products (competing products) and filling in the House of Quality (HOQ).

2.2 Handtool Design

In designing this product, an ergonomic guidlines for hand-tool design is considered. In designing hand tools, the level of strength and durability of a grip (grip) needs to be considered [8] . While grip endurance is the length of time a person needs to maintain a certain level of pressure.

Ergonomic guidlines for hand-tool design [9]:

a) Avoid rigid designs by giving contours to the tool in accordance with the shape of the finger. (using 50 percentile)
b) Avoid the use of tools that demand unusual movements or cannot be used effectively with normal posture and low pressure.
c) Avoid tools with sharp edges.
d) The surface of the tool must be safe, non-conductive and smooth.
e) Make sure the size of the tool used is right (not too small) because the device is not suitable for size requires more power in its use.
f) The equipment used should have the capability to be used with either the right hand or the left hand.
g) Tools should be effective for both men and women.
h) Tools should have special objectives / tools only used for certain activities.
i) Consider the grip angle and minimize excessive wrist rotation.
j) Ensure security in the use of tools.
k) Avoid using tools that have excessive vibration.

2.3 Usability

Usability is a way to find out the ease of a product used by the intended user. Satisfaction is an indicator of user satisfaction when using a product. This indicator is obtained by directly offering EASILY products for use by targeted users, in this case the elderly and people suffering from arthritis.

The usability testing is conducted after the design is prototyped. There are 5 respondents of elderly female users of daily kitchen tools. The usability factor considered in this product development is the satisfaction aspect. Usability satisfaction data retrieval is in addition to satisfaction points, users can also provide input, to the product as additional points in the hope that they can develop EASILY as a product that is truly needed and desired by the target, namely the elderly and arthritis sufferers.

3. RESULT AND DISCUSSIONS

The first method of data collection for the design of EASILY products was a questionnaire by distributing 30 respondents. The results of the data will be processed by the QFD method in identifying consumer desires. The following is the result of processing questionnaire data with the QFD method. In this figure 2 below is the result of House of Quality, to decide action that should be taken.

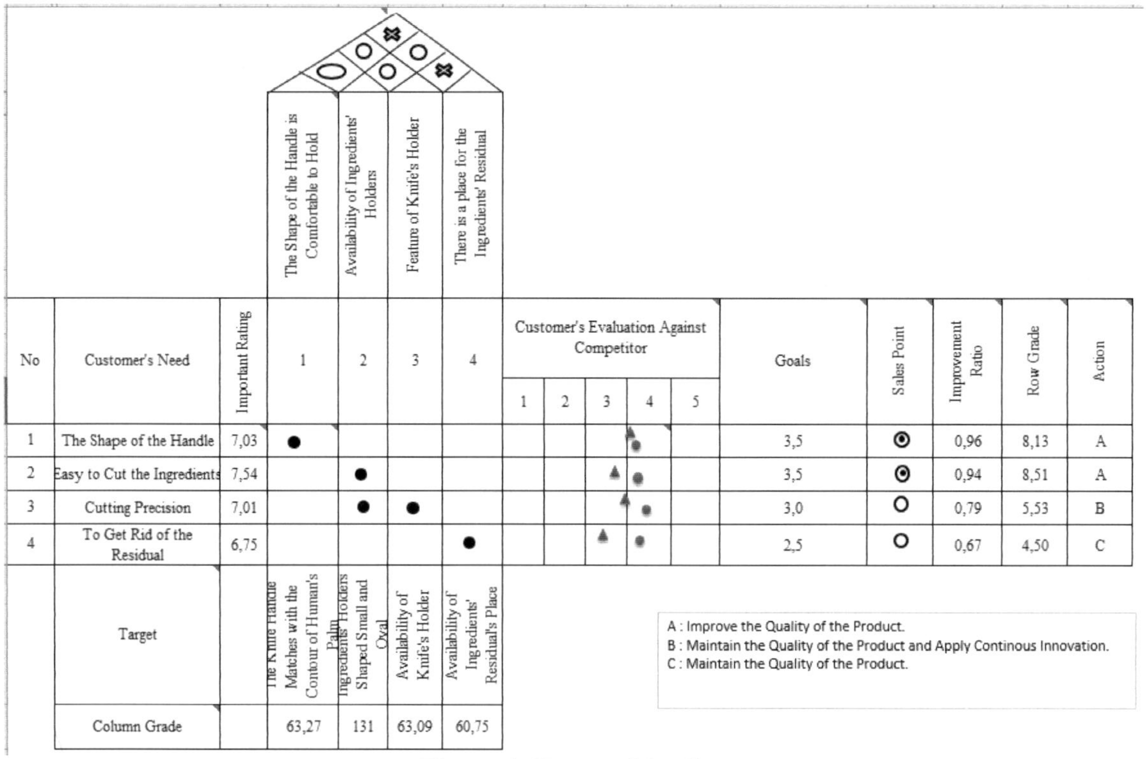

Figure 1. House of Quality

From the results of QFD analysis, it can be seen that there are 4 aspects that become consumer needs, namely the shape of the handle, the ease of cutting material, the level of precision in cutting and the ease of removing residue. Based on the values obtained in the row weights, these values are used for selecting actions for the relevant criteria. There are three types of actions that can be carried out for the related criteria namely action A (Improving product quality), B (Maintaining product quality and continuously innovating products), and C (Maintaining product quality). After identifying the actions that must be carried out based on the house quality, the data obtained is that the action that must be taken is the screen on the cutting board is oval and thin, then the handle of the knife that will be made adjusts to the contours of the human hand, then has a blade support with the aim that the user do not have to hold the knife with great power, then the last is the presence of a residue from the cutting. This residual place is temporary and if it is full it must be disposed of manually by the user.

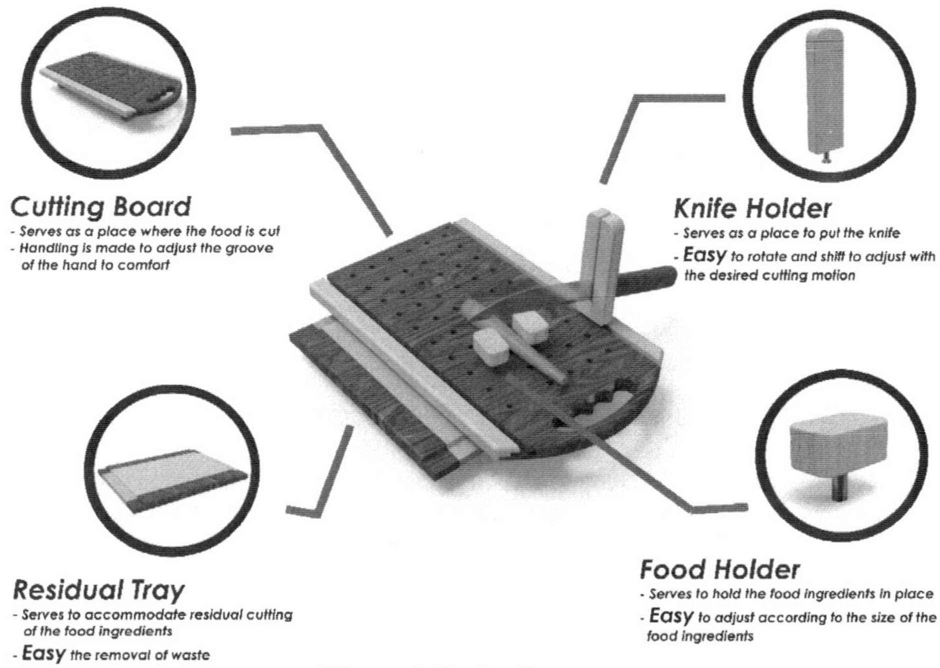

Figure 2. Design Prototype

After making the identification process for the specifications of the cutting board. The next step is the process of usability testing after the researcher has carried out the prototyping process. Attributes that will be used for usability testing are satisfaction. In usability testing, interviews will be conducted with users, namely the elderly and arthritis sufferers. Interviews are conducted by asking several questions related to user satisfaction when using cutting boards. The reusability testing process was carried out in 5 elderly people.

The results of the usability test showed that the five users felt high satisfaction during the process of product testing, especially with the residual tray feature, making it easier for users to store trash temporarily when cutting processes. However, the user provides input so that the blade support section is made more comfortable by replacing the material, and can be rotated 360° so that it can be used to cut various types of food ingredients.

4. CONCLUSION

Easily (easy-to-use kitchen utilities) is a set of kitchen equipment consisting of 3 main components. The three components are knives, cutting boards and knife supports. The three components use an anthropometric approach so that it matches the user's body dimensions and can directly increase comfort when used. The knife support is useful to facilitate users who are elderly people and arthritis sufferers who have limitations in moving the joints of the hand. This buffer functions grip and guide so that the user does not need a very tight grip when using a knife and can be more precise in cutting. Cutting boards in this set of kitchen utensils contain movable bulkheads or position. The insulation on this cutting board serves to make it easier for the user so that they do not need to hold tightly or even do not need to hold the material to be cut.

IOP Conf. Series: Materials Science and Engineering **530** (2019) 012055 doi:10.1088/1757-899X/530/1/012055

REFERENCES

[1] B Yogyakarta 2016 Jumlah Lansia di Indonesia *Badan Statistik Indonesia Yogyakarta*

[2] K K R Indonesia 2016 Jumlah Penderita Penyakit Lansia *Kementrian Kesehatan Republik Indonesia Jakarta*

[3] D Anggraeni 2016 pengaruh budaya keselamatan pasien terhadap sikap melaporkan insiden pada perawat di instalasi rawat inaprumah sakit tk. Ii dr. Soepraoen. *Jurnal aplikasi manajemen.* Vol 14 No 42

[4] I Junaidi, Rematik Dan Asam Ura 2013 Jakarta: PT. Bhuana Ilmu Populer

[5] S Aisyah 2017 Manajemen Nyeri Pada Lansia Dengan Pendekatan Non Farmakologi *Jurnal Keperawatan Muhammadiyah* Vol 2 No 1 pp 178 - 182

[6] V Gaspersz 2002 Pedoman Implementasi Program Sis Sigma Terintegrasi dengan ISO 90001:2000 MBNQA dan HACCP *Jakarta: PT. Gramedia Pustaka Utama*

[7] A Zohir, C I Parwati and J Susetyo 2017 INTEGRASI METODE SERVQUAL, QFD, DAN KANO UNTUK MENINGKATKAN KUALITAS PELAYANAN PADA PERGURUAN TINGGI IAIM NU KOTA METRO *Jurnal REKAVASI* Vol 5 No 1 pp 10 - 17

[8] A Bazar 1978 Grip strength of cerebral palsied. Human Factors 20 pp 741-744

[9] A Mital 1991 Hand Tools: Injuries, Illnesses, Design, and Usage *Advances in Human Factors/Ergonomics* Vol 15 pp 219 - 256

Ergonomics Redesign of the Effective Chopping Tool for Harvesting the Eucalyptus Plant

H Soewardi[1*] and A S Husna[2]

Industrial Engineering Department, Faculty of Industrial Technology, Universitas Islam Indonesia, Yogyakarta, Indonesia

Corresponding author *: hartomo@uii.ac.id

Abstract. Eucalyptus plant is a plant producing an essential oil. It is beneficial oil for fabricating the cosmetic and aroma therapy. In consequence, a quality of the oil is very substantial to ensure the quality of product. One of factor affecting it is the degree of purity of the oil. In the interest of producing the high purity, a good harvesting technique should be considered particularly with a design of the chopping tool. Preliminary study was found that the quality and number of the oil are still not able to reach the target by using the existing tool. Thus this is significant to improve the tool which is more effective and safer. The purpose of this study is to redesign the ergonomic and innovative chopping tool. TRIZ used to determine the inventive principles for producing the specific solutions or design on basis a contradiction between improving and worsening features. Fuzzy linguistic concept is implemented to specify the design attribute that is vague. Result of this study finds four specifications of a purposed design. They are sharp with specifications thickness of the tool at the tip is 0.2 cm and the base is 0.37 cm, then strong with specifications the gilding is done for 2.63 minutes with 1216ºC temperature, and for durable is the use of "*Per*" Steel as the material, and for comfortability is the size of the handle that corresponds to the dimensions of the human hand namely the diameter of the handle is 3.25 cm, the length of the handle is 12 cm, the width of each contour of the fingers is 2 cm using silicon rubber material. Signed-rank test results in that the proposed design is valid specifically it is more comfortable, more powerful, sharper, and more durable by using the existing design.

Keywords*: Manual harvesting tool, Eucalyptus, TRIZ, Fuzzy Linguistic*

1. INTRODUCTION

The eucalyptus plant (Melalauca leucadendron Linn.) is one of the plants that produces essential oils that are widely used and needed by the industry in Indonesia for manufacturing cosmetics and aromatherapy. This oil has a large opportunity to be developed as the high economic value of product by cultivating eucalyptus the plants [1]. According to [2], production capacity of eucalyptus leaf in Java is as much as 53.760 tons per year for producing 300 tons of the essential oil with a total land area of 24.255,56 ha [3]. There is company in East Java that has eucalyptus plantation land that uses "*rimbas*" method to harvest eucalyptus leaf and eucalyptus oil processing plant, having land area as big as 8.121 ha. In this area, harvesting process of eucalyptus leaves uses the "*rimbas*" method there are time and cost efficient. By this method eucalyptus plants grow until the stem reaches the age of 9 months, then the stems are cut to as high as 110 cm and new branches are harvested by the workers. This company currently has several problems related to the amount of eucalyptus leaf production and the purity degree of eucalyptus oil, which have not reached the target. Factors that are considered the most influential is chopping tools factor, because of this background, preliminary study is conducted using Nordic body maps questionnaire, collected data in Figure 1 as follows

IOP Conf. Series: Materials Science and Engineering **530** (2019) 012056 doi:10.1088/1757-899X/530/1/012056

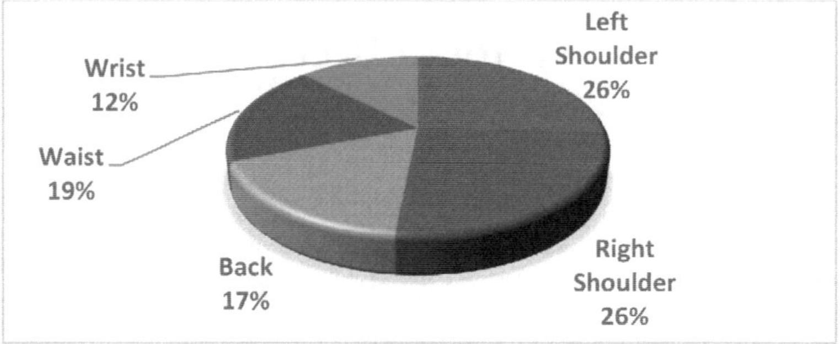

Figure 1. Nordic Body Maps Result

In Figure 1, it is the result from 20 of the 30 workers who complained related to the use of existing chopping tools, with very sick levels in some parts of the body. One of the causes of high complaints of pain in the body parts of workers is caused by harvesting equipment that is currently used is not in accordance with the contour of the woody plant and is not comfortable to use. As many as 25 out of a total of 30 workers, want a change in chopping tool that is more comfortable to use and in accordance with the contours of eucalyptus plants.

With a background in the amount of production and yield that have not met the target with the main causes of workers' complaints about the tools used and workers who want chopping tools that are in accordance with the contour of eucalyptus plants, there is a need for innovations related to manual harvesting devices that can replace the sickle , to improve the quality of eucalyptus oil by using the TRIZ method. TRIZ is discovered by G.Altshuller from Russia through the study of the principle sources used to obtain ideas derived from patents contained in the TRIZ method [4]. In this study, the TRIZ Method is used as it provides a solution from the root cause of the problem objectively and responds to consumer desires from repairing manual chopping tools.

2. METHOD

2.1 Survey

A survey by interviewing the field supervisor, and field workers finds out the root of problems related to the use of existing harvesting equipment using the function analysis, and cause and effect chain analysis of TRIZ.

Further survey will be done by distributing consumer attribute questionnaires to the proposed manual chopping tools, that delivered to 30 workers in the field, and then the second questionnaire will contains weighting for each consumer attributes. After that the result will tested for validity and reliability.

2.2 Theory of Inventive Problem Solving (TRIZ)

TRIZ tools are employed as a knowledge-based and function-centered support for generating innovative product ideas [5]. In this study, TRIZ will be used in the process of finding root causes and resolving the contradictions that appear using the 40 inventive principles and 76 standard principles. Three TRIZ models general problems are physical contradiction, engineering contradiction, and the susceptibility of the field model. Selected inventive principles are chosen that can be most applied to solve problems, and determine the actions.

IOP Conf. Series: Materials Science and Engineering **530** (2019) 012056 doi:10.1088/1757-899X/530/1/012056

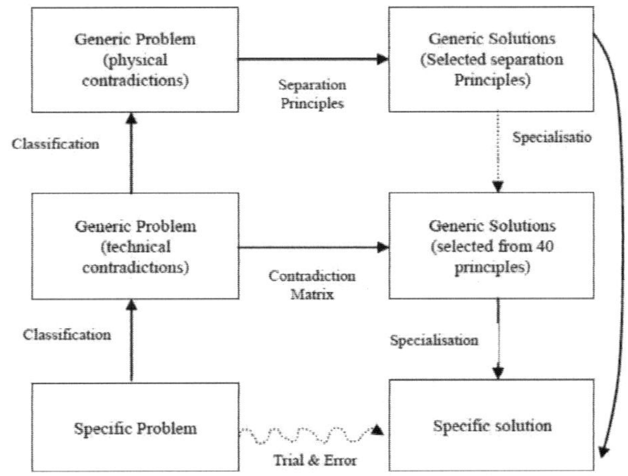

Figure 2. TRIZ Problem Solving Steps[6]

2.3 Fuzzy Linguistic

Fuzzy set theory is a method used to overcome uncertainty. Fuzzy logic provides a tool that is able to capture blurred information, generally explained in natural language, and convert it to a numeric format in the membership function . Linguistic variables are variables whose values are words or sentences in a natural or artificial language. In other words, they are variables with lingual expression as their values [7]. Fuzzy linguistic mamdani will be used to determine the specifications of attributes that are still vague, ie sharp and strong. By using the steps as follows [8]:

1. Define variables that are vague or not specific and linguistic terms based on experts.
2. Build membership functions for each variable.
3. Make a rule base that will be used.
4. Convert input data into fuzzy using the fuzzification process into Matlab software.
5. Evaluate rules on membership functions.
6. Combine the results and each membership function, and get a specific value for each attribute that is still vague.

3. RESULT AND DISCUSSION

3.1 Collecting Consumer Attribute

After distributing questionnaires to the attributes of consumer desires on 30 workers, six attributes were tested for validity which was conducted to determine the level of accuracy of the instruments used in the study [9]. And reliability tests were used to see the stability and constancy of measuring instruments used in a study [10], the most consistent value is shown by the reliability value $\alpha \geq 0.8$ [11]. Obtained the valid and reliable attributes in table 1 as follows:

Table 1. Final Consumer Attribute

No.	Attribute
1.	Strong
2.	Sharp
3.	Comfortable
4.	Durable

3.2 Determine Product Specification Using TRIZ

Specific problems obtained from expert interviews are supervisor and workers in which the workers do not comfortable when using existing chopping tool. TRIZ general problem steps will use function analysis and cause and effect chain analysis tools to identify the potential root cause from the spesific problem. The two tools can be seen in Figure 3. And Figure 4. There will be a potential root cause that will be combined between potential root cause and consumer attribute. The next step is to enter into the TRIZ problem modeling seen in table 2. which contains the distribution of potential root causes based on the TRIZ problem model, before the completion of the TRIZ general solution.

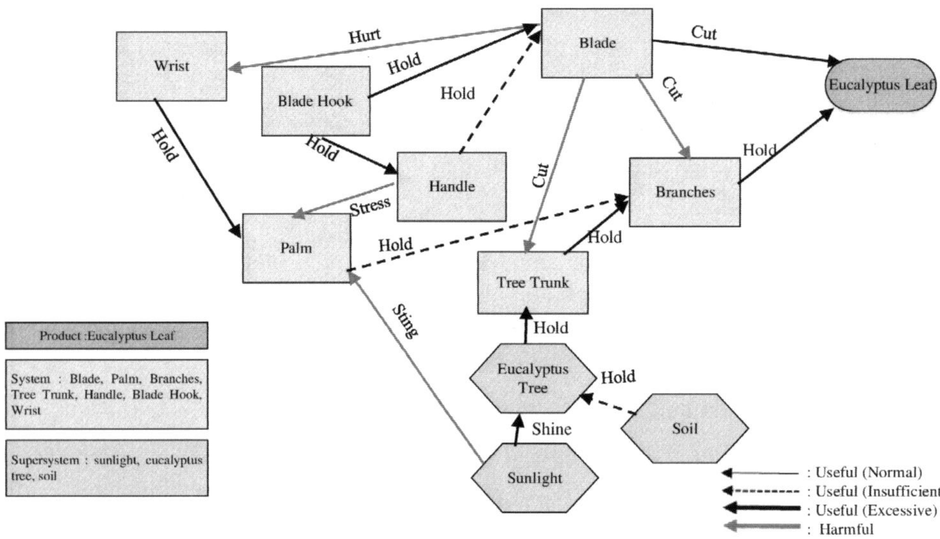

Figure 3. Function Anlaysis

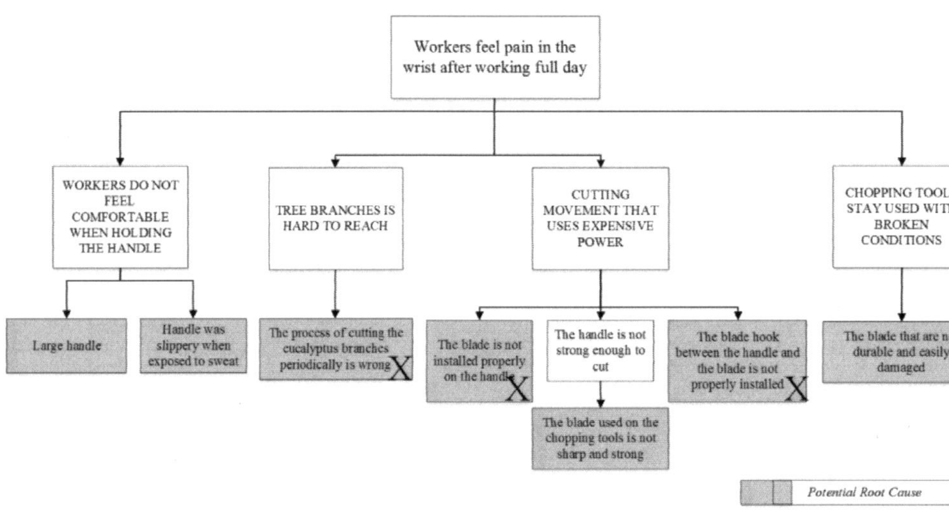

Figure 4. Cause and Effect Chain Analysis

Table 2. TRIZ General Problem

No.	Attribute	Root of the Problem	Identification	Model of Problem
1.	Strong & sharp	The blade used on the chopping tools is not sharp and strong	Inventive problem	Engineering Contradiction
2.	Comfortable	Handle was slippery when exposed to sweat	Inventive problem	Substance Field Model

No.	Attribute	Root of the Problem	Identification	Model of Problem
3.	Durable	Large Handle size The blade that are not durable and easily damaged	Inventive problem	Engineering Contradiction Physical Contradiction

After knowing several options from TRIZ specific solutions given in table 40 inventive principles, the next step is to determine the inventive principles that can be applied to solve problems according to model of problem, which is physical contradiction, engineering contradiction, and substance field model. Thus, the following table 3 the action taken in this study will be presented for each potential root cause and consumer attribute.

Table 3. TRIZ Results

Attribute	Inventive Princples	Action
Sharp & Strong	Principle 10 Prior Action	change the gilding technique and thickness of the sickle eye so that it has a better level of sharpness and strength than before and the contour of the sickle eye that adapts to the branches of eucalyptus plants.
Comfortable (1) Slippery Handle	Principle 1.2.2 Increase Su – Field Effectiveness	Add silicon rubber to the sickle handle
Comfortable (2) Handle Size	Principle 35 Parameter Changes	By changing the physical state of the size of the sickle handle according to hand anthropometry, the use of the dimension of the width of the index finger (WIF), maximum grip (DGMAX), and the width of the palm to the thumb (WPT) so that it matches the size of the hand grip of the worker.
Durable	Principle 33 Homogeinity	Change the metal components of the blade with better quality, namely "*per*" steel

3.3 Determine Fuzzy Attributes

When determining the fuzzy attribute, an interview was conducted with an expert, namely a blacksmith related to the attributes of consumer desires that still have detailed specifications for the prototype making process [12]. The membership function process and rule making are based on the results of interviews with 2 steel experts. The study fuzzy mamdani will be used with fuzzy triangular number, linguistic variables are variables in which the value is in the form of words or sentences in nature or artificial. developing basic concepts and fuzzy integral theorems Fuzzy numbers are valuable in fuzzy sets. There are two fuzzy attributes that are strong and sharp with sharp and strong specifications as shown in Table 4, as follows:

Table 4. Fuzzy Attribute Identification

No.	Attribute	Spesific Attribute	Action	Description
1.	Strong	Blade Thickness	Thick	Fuzzy
2.	Sharp	Gilding technique	Fair	Fuzzy
3.	Comfortable	Handle size and component	Size according to hand anthropometry and using silicon	Non-fuzzy
4.	Durable	Blade componnet	Using "per" steel as blade material	Non-fuzzy

Strong attributes are determined by blade thickness, based on interviews with experts, the thickness of the blade will be divided into two parts, which are at the base and tip of the blade. After the membership function process, and the assessment rule by using Matlab R2013a software generate the optimum value for blade thickness is as follows:

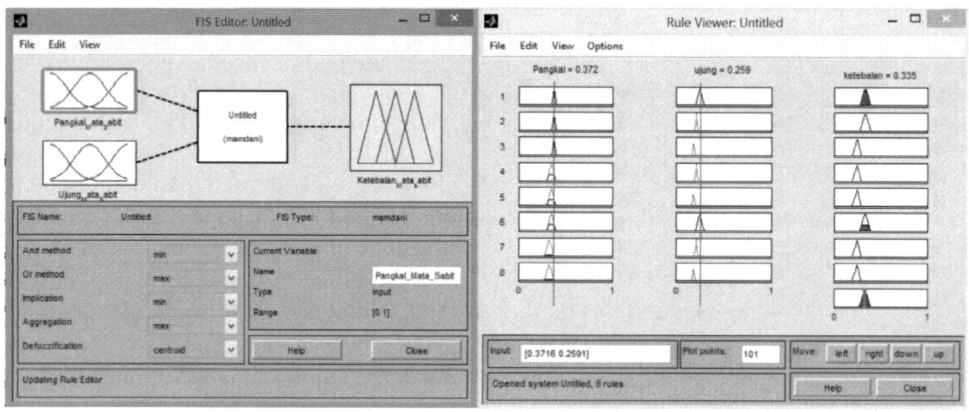

Figure 5. Stronng Attribute Membership Function

Table 5. Optimum Value for Strong Attiribute

Element	Optimum Value (cm)	Selected Design (cm)
Blade's base	0,372	0,37
Blade's tip	0,259	0,2

And for sharp attributes, it is influenced by the blades gilding technique, namely the time and temperature of sickle eye plating. With the optimum value as follows:

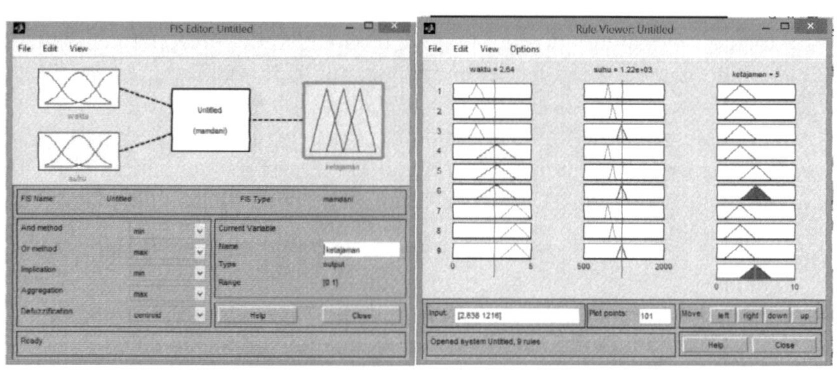

Figure 6. Sharp Attribute Membership Function

IOP Conf. Series: Materials Science and Engineering **530** (2019) 012056 doi:10.1088/1757-899X/530/1/012056

Table 6. Optimum Value for Sharp Attiribute

Element	Optimum Value	Selected Design
Time	2,63	2,63 minute
Temperature	1216	1216 °C

3.4 Comfortable Attribute Specification

In the comfortable attributes, the executed action is to make a handle in accordance with the anthropometry of the human hand and use materials that can reduce the strength of the hand grip of the hand when the sweaty hands. The handle size will be modified using the three dimensions of the body and its application, as follows

Table 7. Handle Dimension

Dimension / Percentile	Application	Size	Allowance	Design Size
The width of the index finger	Finger Contour	2,00 cm	-	2 cm
The width of the palm until the thumb	Handle length	11,13 cm	1 cm	12,13 cm
Maximum Grip Diameter	Handle Diameter	2,95 cm	0,3 cm	3,25 cm

In the handle length and maximum handle diameter, the workers will use the dimensions of the body width of the index finger and the width of the palm to the thumb [13]. While the use of allowances according to [14] large allowance for the length of the andle is 1-1.25 cm and the handle diameter is 0.1-1 cm. The prototyping process will be carried out using solidwork software to make prototypes in 3D and a real prototype will be made and will be used for the validity and signed-rank testing process.

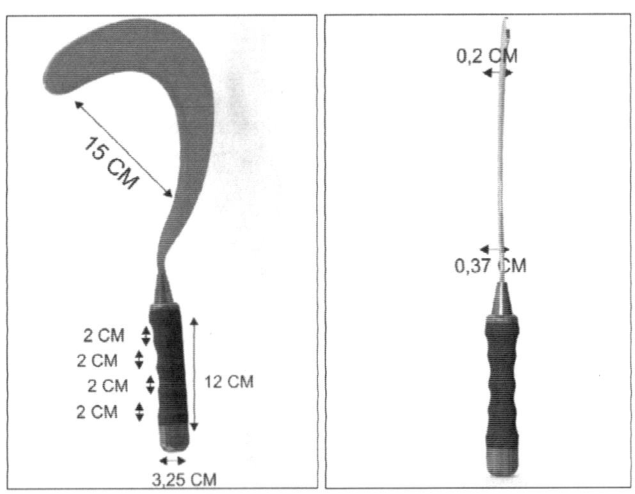

Figure 7. 3D Mock-Up

3.5 Validating The Product

Validation test will be carried out using a marginal homogeinity test which is a test of two related samples to determine whether there are differences in responses between two interconnected groups of data [15] and Wilcoxon signed-rank will be conducted to find out whether the design is in accordance with the wishes of consumers and there are differences that are felt by workers when using old tools and proposals. This testing process will be carried out on 22 people harvesting the eucalyptus leaves results

IOP Conf. Series: Materials Science and Engineering **530** (2019) 012056 doi:10.1088/1757-899X/530/1/012056

showing that the proposed chopping tool has greatly fulfilled the consumer's desire with indigo sig. 2 tailed in 4 attribute have value of $\geq$ 0,05 and in the signed-rank test it is known that there is a difference felt by workers when using the proposed chopping tool as evidenced by the sig.2 tailed value $\leq$ 0.05.

4. CONCLUSION

Based on the results of the processing and discussion carried out, it can be concluded as follows:

The study using TRIZ method to identify the potential root cause of the main problem of workers who feel uncomfortable when using the exsting chopping tool. After determining the potential root cause, the completion of potential root cause is conducted using 40 inventive principles and the principles that is applied most to solve problems from the four attributes are strong, sharp, comfortable, and durable. In the comfortable attribute, a handle will be made that matches the dimensions of the human body, for the contour of the finger on the handle 2 cm wide, the handle length 1 cm, and the diameter of handle 3.25 cm. Lastly, the durable attribute will be indicated by a blade material that is "*per*" steel.

In strong and sharp attributes that still do not have specifications for the proposed chopping tool, it is identified using fuzzy linguistic, with the results on the strong attribute indicated by the blade thickness at the tip of 0.2 cm and at the base of 0.37 cm. As well as the sharp attributes indicated by the gilding technique which is 2.63 minutes with a temperature of 1216ºC. Validation and signed-rank tests performed indicate that the proposal design has fulfilled the consumer's desires and there is a difference felt by consumers when using existing tools and proposals.

The recommendation for this study is the possibility of using different methods to further review the statistical aspects. In addition, automatic harvesting or machine-powered tools can also be developed so that worker fatigue can be minimized

Acknowledgement

This study was supported by Directorate of Student Development, also Industrial Engineering Department, Faculty of Industrial Technology, Universitas Islam Indonesia.

REFERENCES

[1] Mulyadi 2005 *Studi pengelolaan kayu putih Melaleuca leucadendron Linn. Berbasis ekosistem di BDH Karangmojo*, Gunung Kidul, Yogyakarta (Yogyakarta: UGM Press)

[2] Utomo and Mardi P 2012 *Mode Produksi Daun Pada Hutan Tanaman Kayu Putih (Melaleuca cajuputi)* (Bogor: Institut Pertanian Bogor)

[3] Wijayani D 2015 *Analisis Penentuan Harga Pokok Minyak Kayu Putih di Pabrik Minyak Kayu Putih Sukun* (Universitas Gadjah Mada, Yogyakarta)

[4] Azammi A M N, Sapuan S M, Ishak M R and Sultan M T H 2018 Conceptual design of automobile engine rubber mounting composite using TRIZ-Morphological chart-analytic network process technique *Defence Technology* I pp 1-10

[5] Chou J R 2014 An ideation method for generating new product ideas using TRIZ, concept mapping, and fuzzy linguistic evaluation techniques *Advanced Engineering nformatics* 28 pp 441-54

[6] Stratton R and Mann D 2001 The Theory of Inventive Problem Solving (TRIZ) and Sstematic Innovation-a Missing Link in Engineering Education *Research Gate* 1 pp 1-15

[7] Li W 2017 A Hybrid Approach Based on fuzzy AHP and 2-tuple fuzzy liguistic method for evaluation in-fligh service quality *Journal of air transport management* 60 pp 49-64

[8] Maulidyawati S B 2018 *Desain Boneka Edukasi Sebagai Media Belajar Lagu Daerah indonesia Menggunakan Pendekatan Affective Design* (Universitas Islam Indonesia, Yogyakarta)

[9] Putra Z F S, Sholeh M and Widyastuti N 2014 Analisis Kulaitas Layanan Website BTKP-DIY Menggunakan Metode Wequal 4.0 *Jarkom* 1 pp 174-85

[10] Widjaja F N and Sandjaja S S 2009 Uji Validitas dan Reliabilitas Index of Teaching Stress (ITS) *NOETIC Psychology* 36 pp 10135-47

[11] Namagachi M and Lokman A M 2010 *Innovations of Kansei Engineering* (Ohio: CRC Press)

[12] Wu H Y, Tzeng G H and Chen Y H 2009 A fuzzy MCDM Approach for evaluating bangking performance based o balanced scorecard *xpert systems with applications* Vol 36 pp 10135-47

[13] Kumar R, Kumar A, Rani V and Kumar N 2015 Anthropometry and agriculture tools : design aspect ; a review *International Ergonomic Association Cnference* pp 1-11

[14] Strasser H 2007 *Assesment of the ergonomic quality of hand-tools computer input devices* (Amsterdam: IOP Press)

[15] Yamin S and Kurniawan H 2009 *SPSS Complete* (Jakarta: Salemba Infotek)

IOP Conf. Series: Materials Science and Engineering **530** (2019) 012057 doi:10.1088/1757-899X/530/1/012057

Significance of Gender in Badminton Lunge Classification

J J J Lee, W P Loh* and Z W Ho

School of Mechanical Engineering, Engineering Campus, Universiti Sains Malaysia, 14300 Nibong Tebal, Penang, Malaysia.

Corresponding author*: meloh@usm.my

Abstract. Most badminton lunge studies considered only the male players. No study has investigated the effect of gender on the badminton lunge movement. Past researches presented the statistical analysis approach on badminton lunge motion study. However, no study has classified lunge motion specifically by gender segregation. This paper investigates the gender effects in distinguishing the five lunge directions; center-forward (CF), left-forward (LF), right-forward (RF), left-lateral (L) and right-lateral (R) lunges using classification approach. The case study involved video captures of five-male and six-female university-level players lunging in the badminton 21-point singles. A total of 23 attributes of 10894 instances were extracted to classify lunges into CF, LF, RF, L and R lunges using 21 classification algorithms of the WEKA tool. Lunge direction classifications were performed on unsegregated and segregated genders datasets. Lunge classification on segregated gender datasets showed improvement over the unsegregated gender on 17 out of 21 classification algorithms. The notable improvements were on Trees-Hoeffding Tree (+8.67% on male, +3.08% on female) and Functions-Multilayer Perceptron algorithms (+4.35% on male, +9.37% on female). The study shows that gender segregation improves the lunge classification accuracies.

Keywords: *attribute, badminton, classification, gender, lunge direction*

1. Introduction

The badminton lunge is a frequently performed footwork skill accounting for 15% of the entire movement in the badminton singles [1]. Good lunging skills enable efficient shot execution with good body posture balancing within minimum time and effort.

Cronin et al. [2] investigated the power determinants for a good lunge performance. Lunge attributes were reviewed in the design of conditional training programs for improving lunge skills [3]. Lunge motion can be categorised based on the movement direction, namely the center-forward (CF), left-forward (LF), right-forward (RF), left-lateral (L), right-lateral (R), left-backward (LB) and right-backward (RB) lunges. The single lunge [1] [4-8] and multi-directional lunges [9-12] had also been considered in past studies.

Ahlawat et al. [13] and Fernandez-Fernandez et al. [14] reported the difference in badminton sports performances by male and female players. However, case studies involving both the male and female players are limited. Majority case studies merely considered the male players [1][4-9][12]. The effect of gender factor on badminton lunge performances had not been investigated, especially for the lunge motion classification. Thus, this study contributes the novel idea on examining the gender attributes in distinguishing the CF, LF, RF, L and R lunges using the data classification approach.

IOP Conf. Series: Materials Science and Engineering **530** (2019) 012057 doi:10.1088/1757-899X/530/1/012057

The experimental case study involved five-male and six-female players in the 21-point badminton singles. The spontaneously performed lunge movement attributes values were extracted from video-image-numeric transformation using the Kinovea V0.7.10 software. The data were classified into five lunge directional classes using 21 classification algorithms aided by the Waikato Environment for Knowledge Analysis (WEKA) tool. The lunge classifications were performed on both genders (unsegregated gender) followed by the segregated gender dataset. The difference in classification performance (measured in percentage accuracy) was evaluated.

2. Methodology

Data mining analysis was performed on an experimental case study concerning the badminton lunge motion in 21-point singles. Lunge classification was performed on (i) unsegregated genders and (ii) segregated gender. The entire study framework is shown in figure 1.

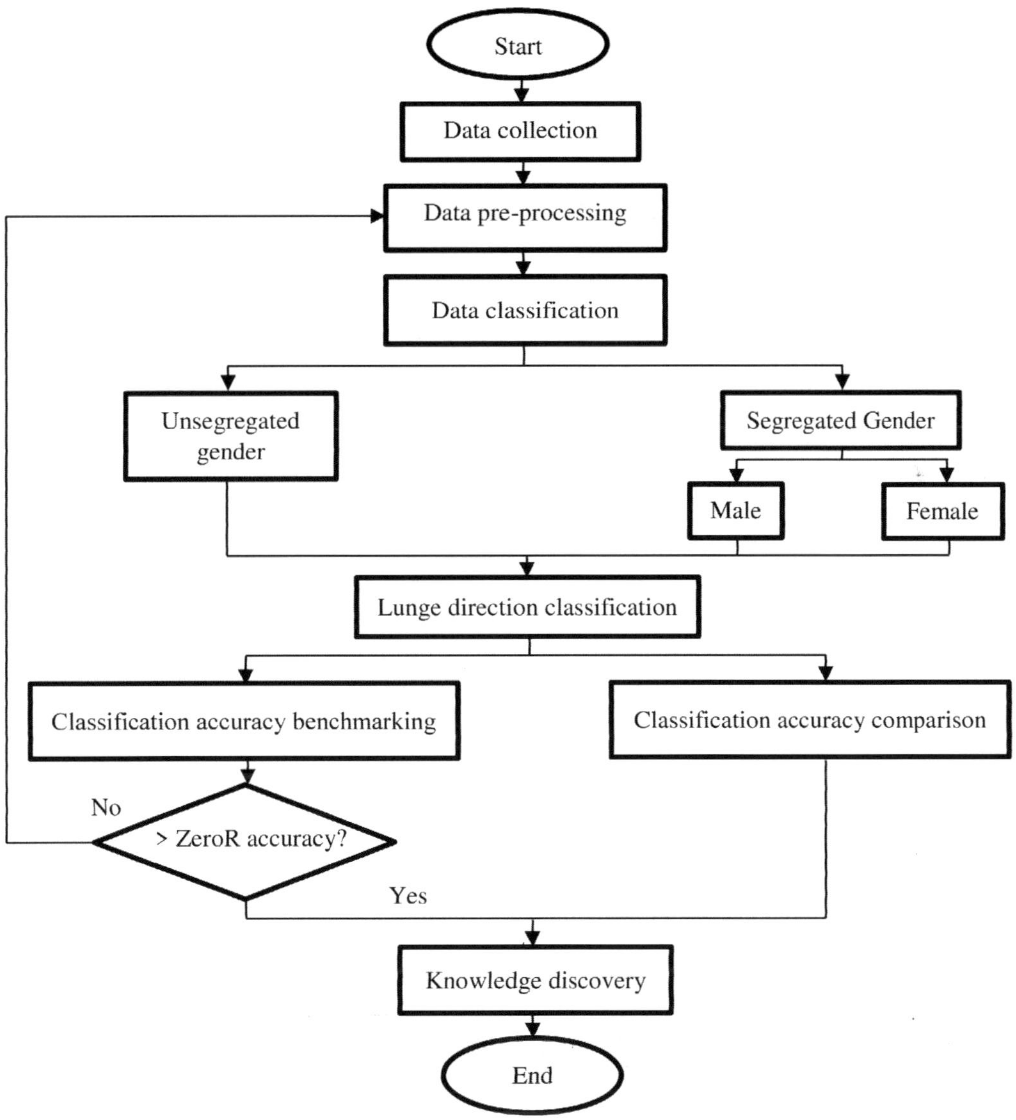

Figure 1. Flowchart of the study.

2.1. Participants
Eleven right-handed badminton players (5M, 6F, 22.3±1.1 years old; 166.6±7.5 cm; 56.1±6.7 kg) of Universiti Sains Malaysia (USM) were recruited for this study. The players have at least five years' experience in badminton tournaments and are free from lower extremity injuries. The players received a detailed explanation of the experimental procedure and provided their consents to participate in the experiment.

2.2. Experiment
The experiments were conducted at the indoor badminton court of Azman Hashim USM Sports Arena, USM. Prior permission was obtained from the Sports and Recreation Center, USM. A five-camera recording system was set up as shown in figure 2. The players were required to play the standard 21-point badminton singles spontaneously without specific restrictions. The natural spontaneous lunge motions were performed at five identified directions: 0° (center-forward, CF), 0° < lunge angle ≤ 45° (left-forward, LF or right-forward, RF) and 45° < lunge degree ≤ 90° (left-lateral, L or right-lateral, R) based on the players' foot orientations (lunge angle) (figure 2).

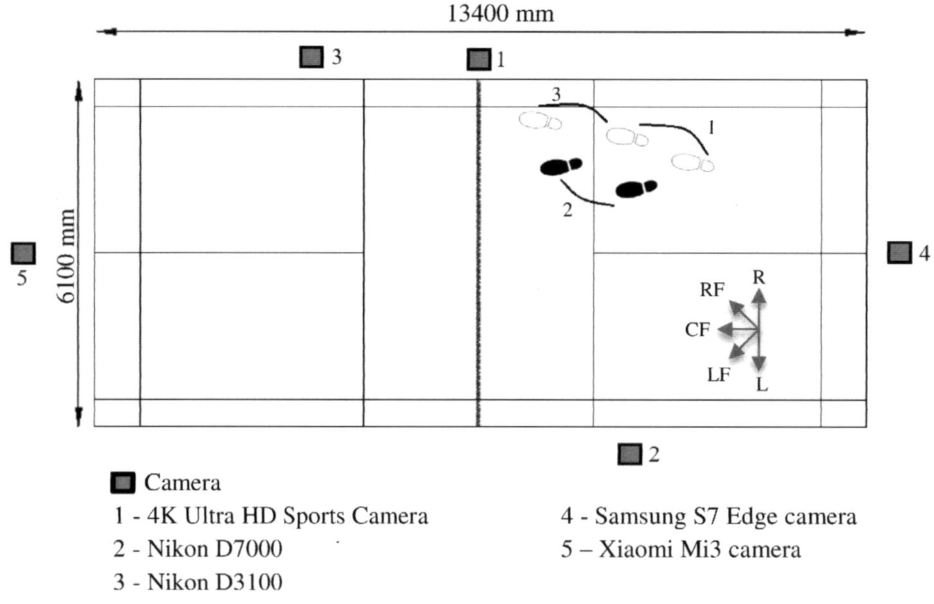

Figure 2. Experimental setup, lunge direction and footsteps (an example of three-step lunge).

2.3. Data pre-processing
Two-level data transformations: video-image and image-numeric were performed using the Kinovea V0.7.10 software: "Track Path" function at default 24 frames per second (fps) setting. From the images, 23 attributes were extracted as shown in table 1. The numeric data were presented in comma-separated values (CSV) file format readable by the Waikato Environment for Knowledge Analysis (WEKA) tool for data classification.

Qualitative data inspection was performed by investigating the potential outliers and extreme values following equations (1) and (2). The detected outliers and extreme values will be discarded from the datasets.

IOP Conf. Series: Materials Science and Engineering **530** (2019) 012057 doi:10.1088/1757-899X/530/1/012057

$$Q_1 - 6 * IQR \leq Outlier < Q_1 - 3 * IQR \quad \text{Or}$$
$$Q_3 + 3 * IQR < Outlier \leq Q_3 + 6 * IQR \quad (1)$$
$$Extreme\ value < Q_1 - 6 * IQR \quad \text{Or}$$
$$Extreme\ value > Q_3 + 6 * IQR \quad (2)$$

where
$Q_1 = first\ quartile$
$Q_3 = third\ quartile$
$IQR = interquartile\ range, Q_3 - Q_1$

The demographic (age and gender) and anthropometric (height, weight, true leg length and apparent leg length) factors, along with the lower extremity kinematics were considered (table 1). The lunge directions (CF, LF, RF, L and R) were assigned as the class attributes for classification analysis aided by the WEKA tool.

Table 1. Study attribute description.

No.	Attribute: Description	Unit
1	A: Age	years
2	G: Gender	-
3	H: Height	cm
4	W: Weight	kg
5	TLL: True leg length, the distance between the pelvis and the bottom of the heel [15]	cm
6	ALL: Apparent leg length, the distance from the umbilicus to the medial malleolus [15]	cm
7	LT: Lunge time, the time with respect to the beginning of the lunge movement	s
8	GP: Game phase, the time expressed in percentage with respect to the overall duration of the match	%
9	xL: x-coordinate of left ankle	m
10	yL: y-coordinate of left ankle	m
11	dxL: Change in x-coordinate of left ankle with respect to previous frame	m
12	dyL: Change in y-coordinate of left ankle with respect to previous frame	m
13	dL: Distance by left ankle with respect to previous frame (equation (3))	m
14	sL: Speed of left ankle (equation (4))	ms^{-1}
15	xR: x-coordinate of right ankle	m
16	yR: y-coordinate of right ankle	m
17	dxR: Change in x-coordinate of right ankle with respect to previous frame	m
18	dyR: Change in y-coordinate of right ankle with respect to previous frame	m
19	dR: Distance by right ankle with respect to previous frame (equation (3))	m
20	sR: Speed of right ankle (equation (4))	ms^{-1}
21	NOS: Number of steps performed from start until completion of the lunge movement (Figure 2)	-
22	Sc: Score point of player at that point of time	-
23	LD: Lunge direction	-

$$Distance, D_i = \sqrt{(dx_i)^2 + (dy_i)^2} \tag{3}$$

where
$dx_i = x_i - x_{i-1}$
$dy_i = y_i - y_{i-1}$

$$Speed, s_i = \frac{D_i}{FD_i} \tag{4}$$

where
D_i =lunge distance

$$Frame\ duration, FD_i = \frac{1}{Frame\ per\ second\ (fps)}$$

2.4. Data classification
Data classification was performed at three levels using five classifiers' algorithms: Bayes (Bayes Net, Naïve Bayes, Naïve Bayes Multinomial Text, Naïve Bayes Updateable), Functions (Logistic, Multilayer Perceptron, Simple Logistic), Lazy (IBk, Kstar, LWL), Rules (Decision Table, JRip, Part, ZeroR as baseline classifier) and Trees (Decision Stump, Hoeffding Tree, J48, LMT, Random Forest, REP Tree) on ten-fold cross-validation mode in WEKA. At the first level, lunge direction classification was performed on the study attributes and instances considering both genders (unsegregated gender).

At the second level analysis, the effect of gender on lunge direction classification was investigated with the gender being defined as the class attribute. The demographic (age) and anthropometric (height, weight, true leg length and apparent leg length) measurable attributes were excluded for gender classification to avoid biases due to their possible correlational relationship with gender. This was to investigate the interaction between lunge kinematics and gender classification, without the influence of demographic and anthropometric factors.

At the third level analysis, lunge direction classification was segregated by gender and executed separately for male and female players. The lunge classification accuracies were compared between the segregated and the unsegregated gender dataset.

At all levels, the classification accuracies resulted from the five classifiers algorithms were also compared with the ZeroR (baseline classifier) for benchmarking purpose. The ZeroR algorithm is commonly used as the standard baseline classifier for cross-validation classifications [16]–[19]. The classification accuracy was deemed reliable if the classification accuracies are above the ZeroR accuracy.

3. Results and discussion
There were 23 attributes of 10894 instances extracted from the video-image-numeric transformation at data pre-processing stage for the data classification analysis. Following outliers and extreme values inspections, no outliers or extreme values were detected from the datasets.

For the first level lunge classification, Bayes, Functions, Lazy, Rules and Trees algorithms showed accuracies of at least equal to or above the baseline classifier, ZeroR (36.2585%). The best-performed algorithms for each classifier were: Bayes-Bayes Net (83.4863%), Functions-Multilayer Perceptron (81.2282%), Lazy-IBk (95.8693%), Rules-Decision Table (99.1004%) and Trees-Random Forest (98.0632%) (figure 3). The top range of accuracies were 81.2282 - 99.1004% indicating that the lunge motions were well segregated into five directions (CF, LF, RF, L and R) with an average minimal 8.4505% error.

There were only 18 attributes of 10894 instances considered for the gender classification. The gender classification accuracies were equal or higher than ZeroR (58.6929%) (figure 3). The best-performed algorithms for each classifier were Bayes-Bayes Net (87.3692%), Functions-Multilayer Perceptron (96.5118%), Lazy-IBk (98.4303%), Rules-Part (98.6231%) and Trees-Random Forest (99.7246%) (figure 5). These results observe that the lunge kinematics attributes efficiently segregated the gender, suggesting an important role of the gender factor contribution to lunge classification.

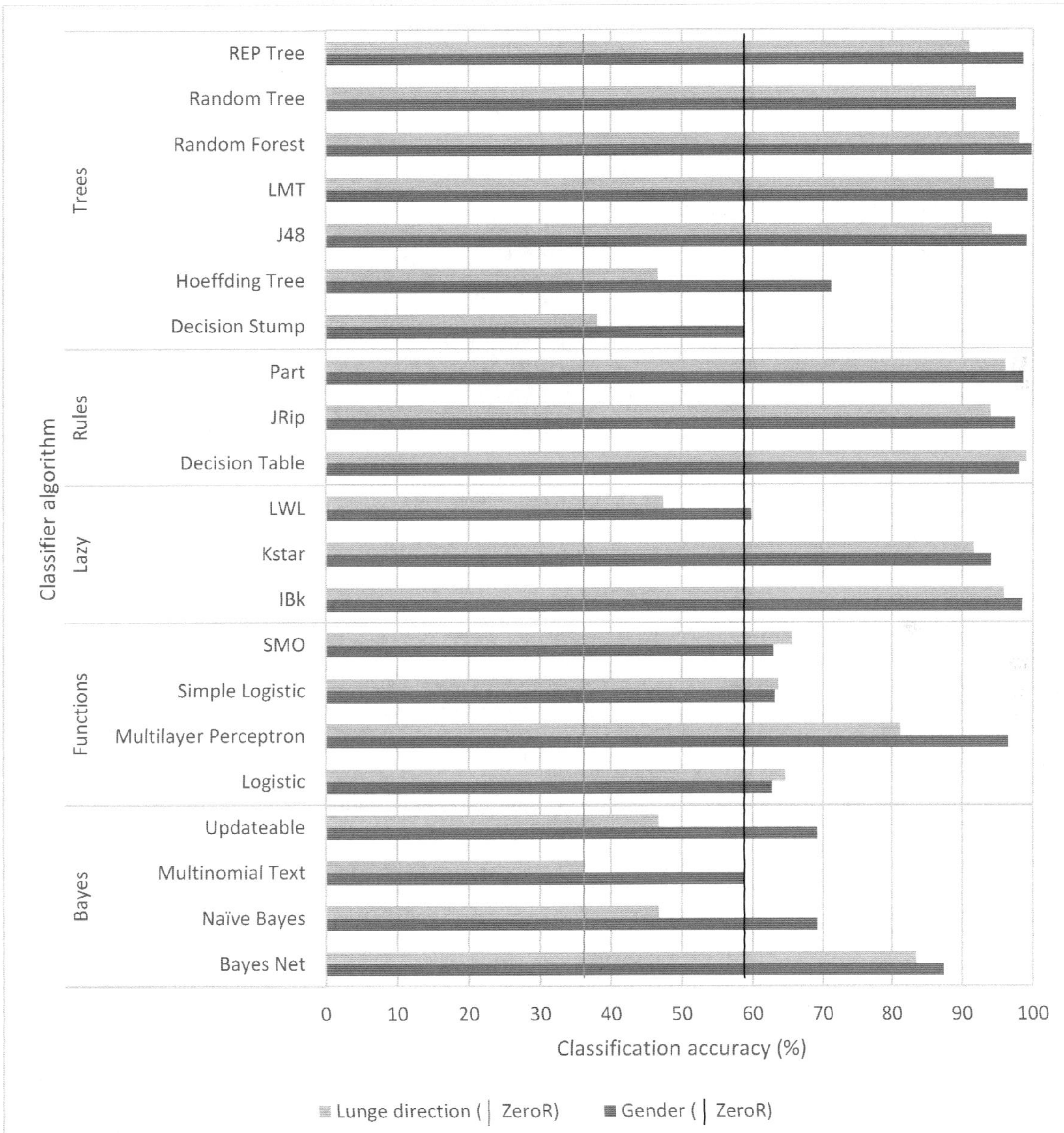

Figure 3. Classification accuracy by lunge direction and gender on Bayes, Functions, Lazy, Rules and Trees classifiers.

On subsequent segregation by gender, a total of 22 attributes of 6394 instances and 4500 instances were considered for the male and female respectively. There were improvements in lunge classification

IOP Conf. Series: Materials Science and Engineering **530** (2019) 012057 doi:10.1088/1757-899X/530/1/012057

accuracies in 17 out of 21 classifier algorithms by segregated gender (figure 4). The notable accuracy improvements were on Trees-Hoeffding Tree (+8.67% on male, +3.08% on female) and Functions-Multilayer Perceptron (+4.35% on male, +9.37% on female). On the other hand, Lazy-Ibk and Kstar showed minimal decrease of 0.4361 and 0.8347% respectively for the male segregated dataset. Lazy-LWL and Bayes-Multinomial Text whereas, showed a decrease of 0.0361 and 2.2585% respectively for the female segregated dataset. Aside for Lazy-Ibk, Kstar, LWL and Bayes-Multinomial Text, the findings suggested that executing lunge classification with gender segregation yields a better classification performance for the majority algorithms.

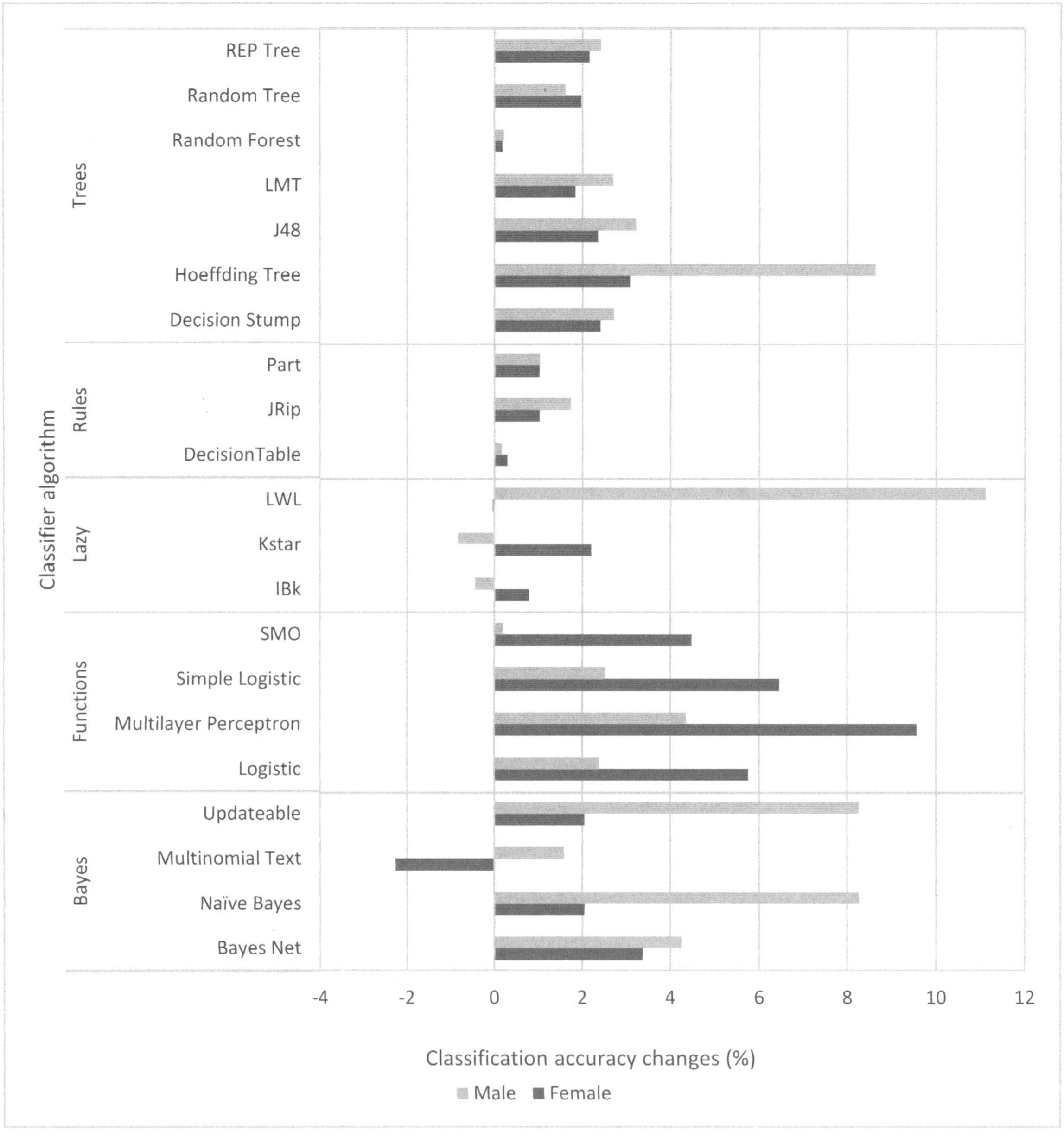

Figure 4. Lunge direction classification accuracy difference between the segregated and unsegregated gender dataset.

The confusion matrix for Functions-Multiplayer Perceptron algorithm, which showed notable accuracy improvement in segregated gender datasets was presented in figure 5. The i-th row, j-th column element of confusion matrix show the percentage instances of i-th class (mis)classified as j-th class. The shaded diagonal elements presented the correctly classified percentage instances. The difference in percentage instances for the male and female segregated dataset respectively was calculated from equation (5).

$$\Delta\%_G = \%_G - \%_U \tag{5}$$

where
G = male or female
$\%_G$ = percentage instances for gender segregated dataset
$\%_U$ = percentage instances for unsegregated dataset

For the unsegregated dataset, the highest error contribution of 4.36% was due to the misclassification of RF instances as CF. Classification with gender segregation has reduced this error by 3.52% in male and 3.23% in female segregated data as shown in figure 5.

Overall findings show that segregating gender enhances the classification accuracy. This suggested the male and female performed different lunge postures [13]. Therefore, inclusive of both genders in the datasets could hardly distinguish the badminton lunge direction.

CF	L	LF	RF	R	<--- classified as
32.37 (2.43) -0.77*	0.58 (-0.12) 0.02*	1.16 (1.17) -0.27*	1.75 (-1.60) -1.02*	0.40 (-0.29) -0.23*	CF
1.14 (0.32) -0.12*	**6.63 (8.53) 3.88***	0.65 (0.46) -0.21*	0.66 (-0.40) -0.64*	0.34 (-0.07) -0.01*	L
2.34 (1.35) -1.41*	0.48 (0.16) -0.14*	**14.96 (12.61) 3.55***	0.93 (-0.61) -0.68*	0.39 (-0.06) -0.12*	LF
4.36 (-3.52) -3.23*	0.37 (-0.09) -0.23*	0.96 (-0.07) -0.30*	**20.96 (-15.76) -3.56***	0.49 (-0.33) -0.38*	RF
0.50 (-0.14) -0.16*	0.37 (-0.34) -0.12*	0.39 (0.23) -0.01*	0.52 (-0.41) -0.35*	**6.32 (-3.45) 6.46***	R

#unsegregated dataset (#changes for male segregated dataset) # changes for female segregated dataset*

Figure 5. Confusion matrix showing percentage instances classified into lunge direction using the Functions-Multilayer Perceptron.

4. Conclusion

This paper distinguishes lunge motion into five lunge directional classes, namely the CF, LF, RF, L and R using 21 classification algorithms of the WEKA tool. Particularly, the effect of gender attribute on lunge direction classification was investigated. Classifications were performed on three levels: lunge direction classification considering unsegregated gender, gender classification, and lunge direction classification on the segregated gender datasets.

The classification results were reliable with accuracies obtained above the ZeroR baseline classifier in all 21 classification algorithms. There were improvements in lunge classification accuracy with segregated gender dataset in the majority 17 out of 21 algorithms. Therefore, lunge classification is ought to be performed separately on male and female segregated dataset for improved accuracy.

This study provides a guideline for conditional training programme design to improve the badminton lunging performance. Specifically, different conditional training has to be designed for the male and female players. The current study framework considered only the university-level badminton players, thus have limited potential for generalization. Future works could consider multiple case studies involving badminton players of different skills background.

IOP Conf. Series: Materials Science and Engineering **530** (2019) 012057 doi:10.1088/1757-899X/530/1/012057

References

[1] G Kuntze, N Mansfield, and W Sellers 2010 A biomechanical analysis of common lunge tasks in badminton *J. Sports Sci.* Vol 28 No 2 pp 183–191
[2] J Cronin, P J McNair and R N Marshall 2003 Lunge performance and its determinants *J. Sports Sci.* Vol 21 No 1 pp 49–57
[3] S J Maloney 2018 A Review of the Badminton Lunge and Specific Training Considerations *Strength Cond. J.* pp 1
[4] H -W Lin, K -S Huang, K -M Pan and C -L Tsai 2015 Biomechanical Analysis of Badminton Different Forward Steps *ISBS - Conf. Proc. Arch.* Vol 1 No 1 pp 1066–1069
[5] Q Mei, Y Gu, F Fu and J Fernandez 2017 A biomechanical investigation of right-forward lunging step among badminton players *J. Sports Sci.*, Vol 35 No 5 pp 457–462
[6] W K Lam, J Ryue, K K Lee, S K Park, J T M Cheung and J Ryu 2017 Does shoe heel design influence ground reaction forces and knee moments during maximum lunges in elite and intermediate badminton players? *PLoS One* Vol 12 No 3 pp 1–13
[7] W K Lam, R Ding and Y Qu 2017 Ground reaction forces and knee kinetics during single and repeated badminton lunges *J. Sports Sci.* Vol 35 No 6 pp 587–592
[8] L Fu, F Ren and J S Baker 2017 Comparison of Joint Loading in Badminton Lunging between Professional and Amateur Badminton Players *Appl. Bionics Biomech.* Vol 2017
[9] Y Hong, S J Wang, W K Lam and J T -M Cheung 2014 Kinetics of Badminton Lunges in Four Directions *J. Appl. Biomech.* Vol 30 No 1 pp 113–118
[10] M T Huang, H H Lee, C F Lin, Y J Tsai and J C Liao 2014 How does knee pain affect trunk and knee motion during badminton forehand lunges? *J. Sports Sci.* Vol 32 No 7 pp 690–700
[11] C F Lin, S H Hua, M T Huang, H H Lee and J C Liao 2015 Biomechanical analysis of knee and trunk in badminton players with and without knee pain during backhand diagonal lunges *J. Sports Sci.* Vol 33, No 14 pp 1429–1439
[12] X Hu, J X Li, Y Hong and L Wang 2015 Characteristics of plantar loads in maximum forward lunge tasks in badminton *PLoS One* Vol 10 No 9 pp 1–10
[13] U K R Ahlawat, N Yadav and D Shaw 2018 Kinematic differences among the players / repetitions and between the gender in regard to left (subordinate) leg lunges exercise for lower extremities with 15 RM load *Int. J. Acad. Res. Dev.* Vol 3 No 1 pp 167–173
[14] J Fernandez-Fernandez, J G De La Aleja Tellez, M Moya-Ramon, D Cabello-Manrique and A Mendez-Villanueva 2013 Gender differences in game responses during badminton match play *J. Strength Cond. Res.* Vol 27 No 9 pp 2396–2404
[15] S Sabharwal and A Kumar 2008 Methods for assessing leg length discrepancy *Clin. Orthop. Relat. Res.* Vol 466 No 12 pp 2910–2922
[16] R Sood, I Koprinska and V G Agelidis 2010 Electricity Load Forecasting Based on Autocorrelation Analysis *The 2010 International Joint Conference on Neural Networks (IJCNN)* pp 1–8
[17] N Lavesson, M Boldt, P Davidsson and A Jacobsson 2011 Learning to detect spyware using end user license agreements *Knowl. Inf. Syst.* Vol 26 No 2 pp 285–307
[18] K Palaniappan and S Mukherjee 2011 Predicting 'essential' genes across microbial genomes: a machine learning approach *10th International Conference on Machine Learning and Applications and Workshops* pp 189–194
[19] M Schuchhardt, B Scholbrock, U Pamuksuz, G Memik, P Dinda and R P Dick 2012 Understanding the Impact of Laptop Power Saving Options on User Satisfaction Using Physiological Sensors *Proceedings of the 2012 ACM/IEEE international symposium on Low power electronics and design* pp 291–296

IOP Conf. Series: Materials Science and Engineering **530** (2019) 012058 doi:10.1088/1757-899X/530/1/012058

Segregating Offline and Online Handwriting for Conditional Classification Analysis

L C Wong and W P Loh*

School of Mechanical Engineering, Engineering Campus, Universiti Sains Malaysia, 14300 Nibong Tebal, Penang, Malaysia.

Corresponding author: meloh@usm.my

Abstract. Offline and online handwriting patterns vary when exposed to external factors. Recent researches on both type handwriting analysis were focused on the image feature extraction, pattern recognition, and classification approach. However, no studies have considered segregating the conditional effects of handwriting patterns on two categories: offline vs. online and normal vs. vibration. Hence, the main goal of the study was to investigate the effects of classifying induced vibration on the offline and online handwriting patterns. There were 25 experimental handwriting samples retrieved under four pre-defined class: FLN, FLV, NLN, and NLV. Extracted data attributes consisted of seven handwriting size metrics and two demographic data. JRip algorithm and Decision Stump algorithm of WEKA tool were employed on tenfold cross-validation mode for handwriting classification and a further segregation by offline and online handwriting. JRip and Decision Stump merely resulted in 54% and 44% classification accuracy respectively. On data categorization between the offline and online handwriting, both algorithms achieved classification accuracies of 56% and 76% respectively. Findings showed that the misclassifications between offline and online were reduced to 5 and 7 instances for JRip and Decision Stump algorithms respectively, whereas the total misclassification between normal and vibration were reduced to 22 and 13 instances.

Keywords. *Classification, Handwriting, Offline, Online, Vibration*

1. Introduction

Handwriting is the output of the writing process using simple writing tools such as a pen and pencil. Each subject's handwriting gives an individualistic unique identity. The handwriting feature extraction and recognition have captured diverse attention from forensic science, computer science, psychology, or pattern recognition areas. There are two types of handwriting recognition: offline and online. The offline handwriting recognition involves the interpretation and conversion of handwritten characters from hard copy documents. Online handwriting recognition, whereas, involves digitized representation of a digital pen movement, where the data describes sequential information about a position, velocity, and acceleration of the handwriting. However, both handwriting data patterns are prone to external factors' impact such as the writing surface, pen or pencil used, and the writing force.

Recent researches on offline handwriting analysis were reported in Liwicki et al. [1], Kumar et al. [2], and Mukherjee and De [3]. Offline handwriting works in Surinta et al. [4] concerned extracting high dimensional feature vector with local gradient feature descriptors for handwritten character recognition. In addition, the authors compared the classification accuracy performance between K-Nearest Neighbour (K-NN) and Support Vector Machine (SVM) classifiers. Meanwhile, Wakahara et al. [5], Mandal et al. [6], and Verma et al. [7] focused on the online handwriting. In Samanta et al. [8], the authors approached online handwriting recognition through the application of von Mises and Gaussian distribution on circular and linear feature components respectively. The circular, linear and combined features were fed into a fully connected non-homogeneous hidden Markov model (HMM) for classification. Handwriting researches on both offline and online handwritings were considered in Liwicki and Bunke [9], Plamondon and Srihari [10], and Zhang et al. [11]. In Bunke [9], a multiple classifier systems (MCS), which was based on HMM and bidirectional long short-term memory (BLSTM) classifier for both offline and online handwritten text-line recognition. The output sequences of the MCS were combined with Recognizer Output Voting Error Reduction (ROVER) combination strategy. External factor's effect on handwriting analysis was discussed in Schomaker and Plamondon [12], and Chang et al. [13]. Schomaker and Plamondon [12] studied the relationship between axial pen force and pen-point kinematics. Meanwhile, the effects of hand soreness in handwriting patterns evaluations were also a concern [13]. The soreness was induced through prolonged writing and measured with a visual analogue scale. In general, researches mainly focused on the efficiency of feature extraction and classification, while the studies on the effects of an external factor in the handwriting patterns were relatively scarce.

Hence, this study adds to the body of knowledge by investigating the effects of induced vibration on the offline and online handwriting patterns. The goal of the study was to identify the classification accuracies of offline and online handwriting under the influence of vibration and to improve the classification system through data segregation. Quantitative attributes (letter width, spacing, size, inclined angle, and slant) were extracted to facilitate information analysis. Study data involve offline and online experimental handwritings of 25 participants under the normal and vibration exposure condition. The Rule-based classifier: JRip and Tree-based classifier: Decision Stump algorithms were used for handwriting conditions classification on tenfold cross-validation mode. The pre-defined conditions are offline normal (FLN), offline with vibration (FLV), online normal (NLN), and online with vibration (NLV). Data segregation into offline and online handwriting was performed based on the confusion matrices evaluation.

Section 2 describes the research methodology that includes data collection approach, pre-processing, classification and knowledge discovery. The analysis and findings are discussed in Section 3. Section 4 concludes the main findings from the paper.

2. Methodology

2.1. Experimental data collection

A total of 25 right and left-handed university students (15 males, 10 females, 22 ± 2 years old) were involved on a voluntary basis with signed informed consents. The subjects were required to write the phrase "Sphinx of black quartz, judge my vow" under two writing conditions: normal and vibration exposed, using the Pilot G2 05 gel ink rollerball pen on a sheet of white paper followed by the stylus (S Pen) on Samsung Tablet A.

The experiment was performed on a desk at 0.74 m height (elbow height) for the normal condition. Subsequently, the Mondial Slim Beauty Fitness Massager (100V – 240AC50/60HZ) was strapped on the desk to simulate the vibrational effects of the vibration exposed condition.

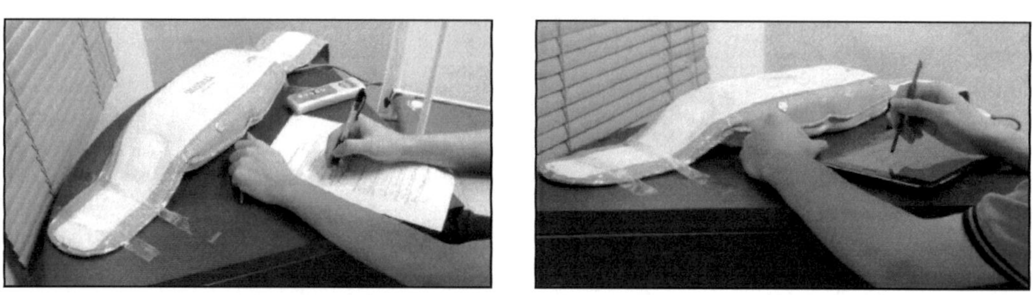

(a) (b)

Figure 1. The phrase "Sphinx of black quartz, judge my vow" written under vibrational effects in (a) offline and (b) online handwriting.

IOP Conf. Series: Materials Science and Engineering **530** (2019) 012058 doi:10.1088/1757-899X/530/1/012058

(a)

(b)

Figure 2. Sample image for the (a) normal (N) and (b) with vibrational effect (V) by the same subject.

2.2. Pre-processing

The recorded data were essentially made up of 25 samples of handwritings with 9 attributes and 100 instances through the image to numeric transformation (Table 1). Handwriting size metrics include average letter width (W), average spacing (S), size (SZ1-SZ3), inclined angle (IA), and slant (SL) was extracted as shown in figure 3.

Table 1. Descriptive summary of data attributes.

Attributes	Description	Range	
		Offline	Online
W	Average letter width (cm)	[0.16 – 0.32]	[0.26 – 0.54]
S	Average spacing width (cm)	[0.22 – 0.57]	[0.21 – 0.44]
SZ1	Size of "tall" letter (cm)	[0.30 – 0.78]	[0.53 – 1.20]
SZ2	Size of "short" letter (cm)	[0.20 – 0.55]	[0.28 – 0.75]
SZ3	Size of "tail" letter (cm)	[0.40 – 0.91]	[0.65 – 1.55]
IA	Inclined angle (°)	[-2.00 – 3.00]	[-1.50 – 1.25]
SL	Slanting of writing	{L, S, R}	{L, S, R}
H	Handedness	{L, R}	{L, R}
G	Gender of subject	{M, F}	{M, F}

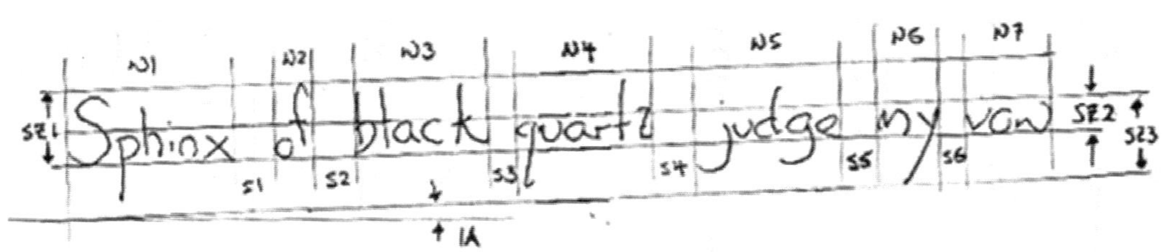

Figure 3. Handwriting size metrics (w_i, s_i, IA, SZ1-SZ3) features.

$$W = \frac{\sum_{i=1}^{n} w_i}{k}, \quad n = 7\ (total\ number\ of\ words),\ k = 29\ (total\ number\ of\ letters) \tag{1}$$

$$S = \frac{\sum_{i=1}^{N} s_i}{N}, N = 6\ (total\ number\ of\ spacings) \tag{2}$$

where: w_i= width word number i, and
s_i= spacing between word number i

IOP Conf. Series: Materials Science and Engineering **530** (2019) 012058 doi:10.1088/1757-899X/530/1/012058

The attributes were represented in interval scales, except for SL, handedness, and gender represented in the nominal scale: {left (L), right (R), and straight (S)}, {left (L), and right (R)} and {male (M), and female (F)}. The W, S, and SZ1-SZ3 were measured with Endo Keiki 6"/15cm Steel Ruler (in cm), whereas, the IA was measured with a protractor (in °). The computation of W and S attributes were based on Equations (1) and (2).

The outlier and extreme value screenings were based on equations (3) and (4). The mean substitution was used to treat the missing values, whereas instances with outliers or extreme values present were removed.

$$Q3 + 3 * IQR < Outliers \leq Q3 + 6 * IQR \quad \text{or} \tag{3}$$

$$\mathrm{Q1} - 6 * \mathrm{IQR} \leq Outliers < \mathrm{Q1} - 3 * \mathrm{IQR}$$

$$Extreme\,Values < Q_1 - 6 * IQR \quad \text{or}$$

$$Extreme\,Values > Q_3 + 6 * IQR \tag{4}$$

where: Q_1 = lower quartile (25% data)
Q_3 = upper quartile (75% data)
IQR = interquartile range

2.3. Data Classification
This process involved recognizing handwriting features extracted into four predefined data classes; FLN, FLV, NLN, and NLV using WEKA tool. JRip algorithm from rule-based classifier and Decision Stump algorithm from tree-based classifier were adopted to recognize the handwriting data under tenfold cross-validation. This training method allows determination of robustness of general models in predicting the classes of new handwriting data. ZeroR algorithm was adopted as the benchmarking algorithm to evaluate the classification performance.

2.4. Knowledge Discovery
At this level, the confusion matrices generated by the JRip and Decision Stump algorithms could be mined by inspecting the misclassification between offline and online, normal and vibration. Based on the confusion matrix analysis, appropriate data segregation can be performed on the handwriting data to improve the classification performance of both algorithms.

3. Results and Discussion
The data contained neither missing values nor outliers or extreme values in the pre-processing stage. Hence, the handwriting data was confirmed clean.

The accuracy of the all conditions handwriting data categorized into four pre-defined classes: FLN, FLV, NLN, and NLV using JRip and Decision Stump algorithms on tenfold cross-validation were 54% and 44% respectively. Both classification accuracies were found reliable (above 20% accuracy found with ZeroR as a benchmark) as shown in figure 4. The confusion matrices generated from both algorithms are shown in table 2.

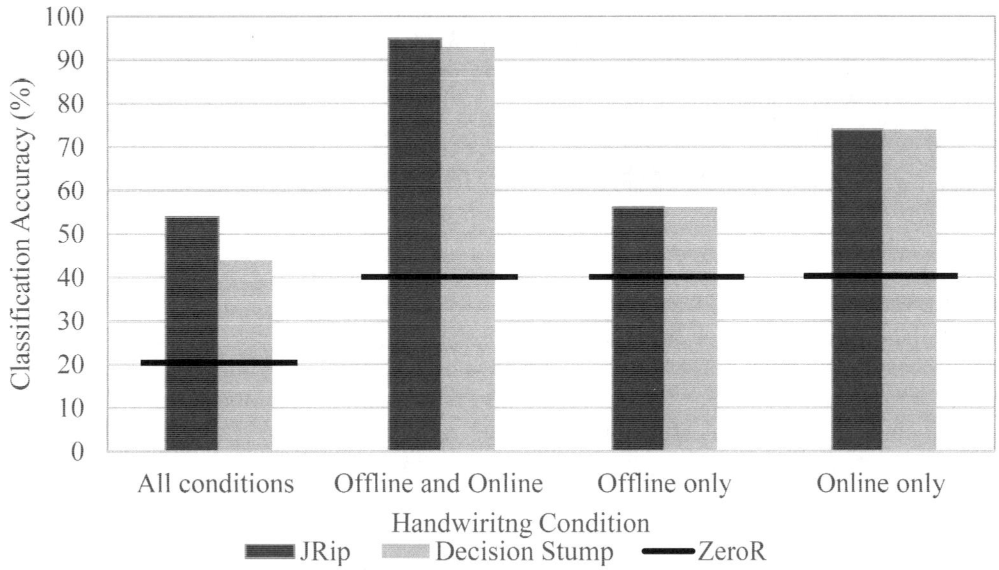

Figure 4. Classification accuracy on all conditions, offline and online, offline only, and online only handwriting with JRip and Decision Stump algorithm.

Table 2. Confusion matrix showing the number of instances classified into FLN, FLV, NLN and NLV on JRip / *Decision Stump* classification.

FLN	FLV	NLN	NLV	← Classified as
17 / 21	6* / 2*	1 / 1	1* / 1*	FLN
11* / 22*	10 / 1	2* / 1*	2 / 1	FLV
4 / 4	2* / 0	7 / 9	12* / 12*	NLN
1* / 1*	1 / 0	3* / 11*	20 / 13	NLV

correctly classified instances
#* misclassified between normal and vibration
misclassified between offline and online

Based on table 2, the number of misclassified instances were located outside the matrix diagonal. It can be observed that there were 14 instances misclassified between offline and online, whereas 38 instances were misclassified between normal and vibration condition, amounting to 46 misclassified instances for the JRip algorithm, where there were 6 instances misclassified between all four conditions. As for the Decision Stump algorithm, the misclassifications between offline and online, and between normal and vibration conditions were 9 and 50 instances respectively, amounting to 56 misclassified instances, with 3 instances misclassified between all four conditions.

Findings show that there were fewer instances misclassified between offline and online handwriting as compared to between normal and vibration conditions. This led to the further refining classification analysis by categorizing handwriting data into two pre-defined conditional classes: offline and online. The results obtained as depicted in figure 4 showed that the JRip and Decision Stump algorithms were able to achieve the percentage classification accuracy of 95% and 93% respectively.

The results showed that the JRip algorithm performed better than Decision Stump in classifying between offline and online handwriting data. Based on table 3, the number of misclassification instances between offline and online for both algorithm was reduced from 14 and 9 instances to 5 and 7 instances respectively on JRip and Decision Stump.

Table 3. Confusion matrix showing the number of instances classified into offline (FL) and online (NL) on JRip / *Decision Stump* classification.

FL	NL	← Classified as
49 / *46*	1 / *4*	*FL*
4 / *3*	46 / *47*	*NL*

correctly classified instances
misclassified between offline and online

Tenfold cross-validation classification was employed by both JRip and Decision Stump algorithms to categorize the offline and online handwriting data separately into two pre-defined classes: N and V. Classification accuracy of JRip and Decision Stump achieved were 56% for offline handwriting, and 74% for online handwriting. All accuracies were found reliable (> 40% on ZeroR as a benchmark). The hierarchy sequence of the split classifications beginning with the type of handwriting (offline vs. online), followed by the presence of vibration (normal vs. vibration) were summarized in figure 5.

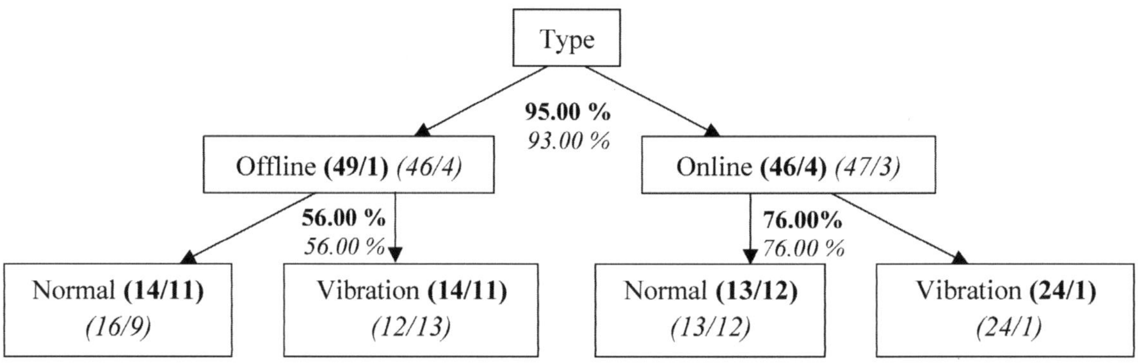

Legend: Offline (49/1) = Type of handwriting (# correctly classified instances / # incorrectly classified instances)

Figure 5. Hierarchy breakdown of conditional handwriting classification by percentage using **JRip** and *Decision Stump*.

Although the classification accuracies by JRip and Decision Stump algorithms were equivalent for the offline handwriting data, JRip algorithm had categorised more normal instances into vibration condition correctly as compared to the Decision Stump (figure 5). Meanwhile, more than half of vibration instances were categorised into normal instances by the Decision Stump algorithm (figure 5). For the online handwriting, both algorithms gave similar misclassification confusion between normal and vibration condition; 12 normal instances were misclassified into vibration, and only one vibration instances were misclassified into normal.

Through the further classification split of handwriting data into offline and online, JRip and Decision Stump algorithms were able to improve the classification accuracy by 2% to 30%. The number of misclassification between normal and vibration were reduced from 38 and 50 instances to 22 and 13 instances respectively on JRip and Decision Stump. This suggests that both algorithms provide better results for conditional classification of offline and online handwriting data.

Conclusion

This study considered experimental offline and online handwriting patterns analysis that were exposed to normal and vibration conditions. A total of nine attributes: seven handwriting (W, S, SZ1, SZ2, SZ3, IA, and SL), and two demographics (gender and handedness) data were extracted from the handwriting. The tenfold cross-validation was employed with JRip and Decision Stump algorithms used to classify the handwriting into four pre-defined conditional classes (FLN, FLV, NLN, and NLV). The classification accuracy of both algorithms though low but was proven reliable (above ZeroR benchmark). An examination of instances classified into the confusion matrix analysis from the JRip and Decision Stump algorithms gave hint that the offline and online were better distinguished compared to between normal and vibration conditions.

Hence, handwriting data were segregated into offline and online to exhibit improved handwriting classification into N and V classes. In particular, both types of handwriting under vibrational stress was tested and proven better for conditional classification. The approach of conditional classification reveals that through segregation of handwriting data between offline and online, the percentage classification accuracy was improved. Future works could be extended to include compatible public handwriting case study data to validate the findings. Other handwriting segregation such as between gender or handedness can be adapted as added factors into the handwriting conditions for analysis.

References

[1] M Liwicki, S Ebert and A Dengel 2014 Bridging the Gap between Handwriting Recognition and Knowledge Management *Pattern Recognit. Lett.* Vol 35 No 1 pp 204–213

[2] G Kumar, P K Bhatia and Indu 2013 Analytical Review of Preprocessing Techniques for Offline Handwritten Character Recognition Vol 3 No 3

[3] S Mukherjee and I De 2017 Feature Extraction from Handwritten Documents for Personality Analysis *Int. Conf. Comput. Electr. Commun. Eng. ICCECE 2016*

[4] O Surinta, M F Karaaba, L R B Schomaker and M A Wiering 2015 Recognition of Handwritten Characters using Local Gradient Feature Descriptors *Eng. Appl. Artif. Intell.* Vol 45 pp 405–414

[5] T Wakahara, H Murase and K. Odaka 1992 On-line handwriting recognition *Proc. IEEE* Vol 80 No 7 pp 1181–1194

[6] S Mandal, S R M Prasanna and S Sundaram 2018 GMM Posterior Features for Improving Online Handwriting Recognition *Expert Syst. Appl.* Vol 97 pp 421–433

[7] B Verma, J Lu, M Ghosh and R Ghosh 2004 A Feature Extraction Technique for Online Handwriting Recognition *Neural Networks* 2004 Proceedings *2004 IEEE Int. Jt. Conf.* Vol 2 pp 1337–1341

[8] O Samanta, U Bhattacharya and S K Parui 2014 Smoothing of HMM Parameters for Efficient Recognition of Online Handwriting *Pattern Recognit.* Vol 47 No 11 pp 3614–3629

[9] M Liwicki and H Bunke 2009 Combining Diverse On-line and Off-line Systems for Handwritten Text Line Recognition *Pattern Recognit.* Vol 42 No 12 pp 3254–3263

[10] R Plamondon and S N Srihari 2000 On-Line and Off-Line Handwriting Recognition : A Comprehensive Survey *IEEE Trans. Pattern Anal. Mach. Intell.* Vol 22 No 1 pp 63–84

[11] X Y Zhang, Y Bengio and C L Liu 2017 Online and Offline Handwritten Chinese Character Recognition: A Comprehensive Study and New Benchmark *Pattern Recognit.* Vol 61 pp 348–360

[12] L R B Schomaker and R Plamondon 1990 The Relation between Pen Force and Pen-point Kinematics in Handwriting *Biol. Cybern.* Vol 63 No 4 pp 277–289

[13] S H Chang, C L Chen and N Y Yu 2015 Biomechanical Analyses of Prolonged Handwriting in Subjects with and without Perceived Discomfort *Hum. Mov. Sci.* Vol 43 No 8 pp 1–8

Training Performance Measurement with Schema and Mental Rotation Test

S Kaewprasert* and A Jiamsanguanwong

Department of Industrial Engineering, Faculty of Engineering, Chulalongkorn University, Bangkok, 10330, Thailand

Corresponding author *: suwat.k@student.chula.ac.th

Abstract. Training was a crucial technique for skill development and basic foundation of learning in manufacturing while the traditional way of training selection was to trial-and-error those training methods to see its performance which take cost and time. The purpose of present study was to preliminary investigate the training performance measurement using cognitive schema and mental rotation test in camera assembly tasks with basic training and physical training condition for production staffs. 180 novice participants performed in this study. Results revealed that both cognitive schema and mental rotation test could significantly distinguish training performance between basic and physical training condition in the same direction. The implication of this study was on the measurement that could shorten the time and reduce cost in estimating the performance of each training method which could be used to support decision making. Discussion was provided.

Keywords: *Training, Assembly, Mental Rotation, Schema, Cognition, Skill acquisition.*

1. Introduction

In the manufacturing, training was a crucial technique for skill development and basic foundation of learning in organizations [1-2], since it could improve productivity, increase job performance, and reduce the cost of manufacturing process [3]. Besides, it also helped employee acquire new knowledge, skills, attitude and behaviors [4]. Hence, training becomes a mandatory process for newly hired employees/workers in various position in many organizations [5].

There were various methods of training available to use such as on-the-job-training, lecture-based training, and simulation training etc., while the traditional way of selection was to trial-and-error those training methods to see its performance that fit most with the context of working condition with acceptable training cost. This would take a period of time (weeks or months) to reveal the training performance in term of productivity [6], cost and time [7], and the quantity of defect and its effectiveness [3]. Thus, it would be worth to investigate the possible training performance measurement to overcome the traditional trial-and-error method to save cost and time for support manager decision making.

The schema script measurement was found to correlate with memory performance [8], and its accuracy could measure learning performance of people (e.g. Kraiger, et al. [9]). It is a framework developed in people's mind to solve a problem, organize knowledge, and support future instruction and learning [10]. The schema script usually represents people's understanding of the orderly step of any processes such as the working process [11]. Accordingly, schema test might be able to distinguish training performance since it relates to human learning.

IOP Conf. Series: Materials Science and Engineering **530** (2019) 012059 doi:10.1088/1757-899X/530/1/012059

The mental rotation ability was also found to be an essential ability that helps people integrate and understand cognition process from multiple viewpoints after training [12]. It is the ability to make the mental image of a given 2D or 3D object turning in space [13], and it could be improved by training [14]. This ability would be important in skill-based training such as in the manufacturing tasks. The mental rotation test (MRT) could be used as the paper-and-pencil based test which people must match 20 rotated figures to target figures [15]. Therefore, MRT might be able to distinguish skill-based training performance, especially in manufacturing.

Both schema test and MRT were used for measuring training performance in several areas. For example, Moreau & Clerc [16] investigated gender differences in sports training performance by measuring MRT. Stransky & Wilcox [12] revealed that students who received surgical training program had higher MRT scores which lead to better skill acquisition. Arthur [17] also used schema test to compare its similarity between trainees and expert. He found that it had a correlation with skill acquisition and could be used for prediction of skill retention and transfer. To the best of our knowledge, there was no research apply schema test and MRT in manufacturing context of training performance measurement.

Thus, in this study, we separated training conditions into basic training and physical training in camera assembly tasks. According to Bloom taxonomy [18], the basic training was focusing on remembering and understanding by lecture-based with the demonstration, while the physical training was recognized as a higher level of training than those basic ones by focusing on remembering, understanding, and applying. The physical training used VDO training together with real simulation on real work. Both training condition had 1 hours in each condition. According to literature reviews [1, 18, 19], the physical training condition would provide better skill acquisition than basic training condition.

The purpose of present study was to preliminary investigate the training performance measurement using schema and mental rotation test in camera assembly tasks with basic training and physical training condition for production staffs. The results of this study could support for the future study of using schema and MRT to estimate the effect size from each training condition to support decision making in training selection if schema test and MRT could clearly differentiate the work skill acquisition from the different level of training condition.

2. Methods

2.1. Participants

Participants performed in this experiment were 180 Thai associate degree students trainee (Male = 51, Female = 129) and with no previous experiences in assembly tasks. The age ranged between 17 and 22 years (M = 19.16, SD = 1.39).

2.2. Stimulus Materials

2.2.1. Basic training materials. The hard copies with color printed of work instructions were provided to each participant to read follow the process while listening to supervisors during explain and demonstrate each step of assembly tasks at the workstation. This work instruction had specified caution-point into work instruction.

2.2.2. Physical training materials. The VDO clip of the procedure was provided with a standard with high resolutions (1920x1080) and the average length of 66 seconds. All steps in VDO were demonstrated with captions and had no sound. Participants could stop and replay any time. Moreover, the real parts (5 sets) for assembly were provided for participants to practice follow the VDO instruction at the workstation.

IOP Conf. Series: Materials Science and Engineering **530** (2019) 012059 doi:10.1088/1757-899X/530/1/012059

2.3. Apparatus
The places for training were conducted at the digital camera assembly workstation. It was equipped with all tools used for assembly such as Tweezers, Bit-torque, Screw-driver, Electric-driver, Jig, Fixture etc. The computer systems were installed at all stations of use for the training with monitor screen 16-inch. Finally, closed-circuit television (CCTV) and stopwatch were used for recording time in this study.

2.4. Task
The camera assembly process was divided into 9 different tasks according to the standard procedure of assembly line to achieve its task's standard time (66 seconds) from parts to finish goods. All participants were randomly assigned to only one task, between-subject condition.

2.5. Measurement

2.5.1. Schema test. This test consisted of 6 assembly steps of their own task to ask participants to re-order these steps into the correct work order. The score was only counted if the order was correct and continue sequenced with the max score of 6.

2.5.2. Mental Rotation Test. Participants were given the 3 color-printed pictures in the different visual angle of work-in-process regarding their task to answer whether it was right or wrong in assembly task at their station within 5 minutes. The correct answer would be scored as 2 points for each picture. The scores would range from 0 to 6.

2.6. Procedure
Prior to beginning the experiment, participants were given a brief explanation of objective of the research and sign a consent form. See Figure 1. Participants were then randomly assigned one training condition (Basic training or Physical training) to acquired assembly skill. The experiments were designed to random assignment with both training conditions and tasks. In each training condition, participants were received the one-hour training. During the training period, all the process and equipment were observed and controlled by the certificated training skilled supervisor from Technical Training Center of the company case study. After training, participants were asked to complete schema test and MRT, respectively. Then, they were thanked and dismissed.

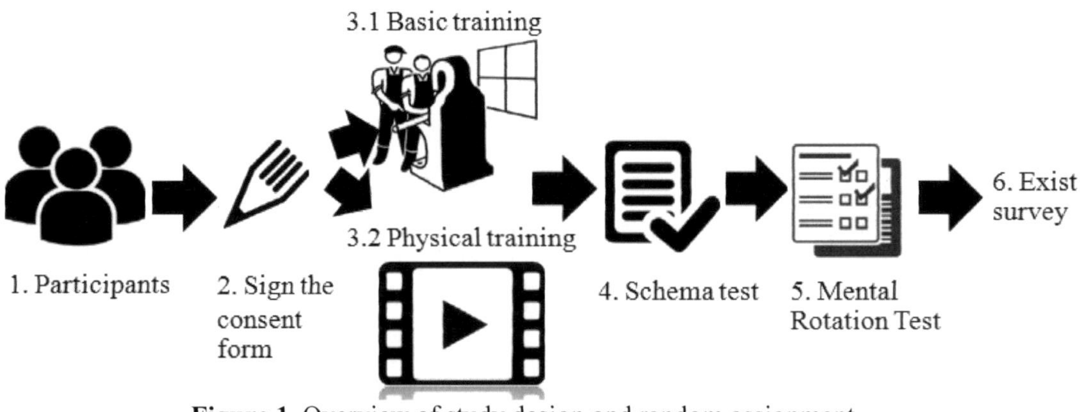

Figure 1. Overview of study design and random assignment

IOP Conf. Series: Materials Science and Engineering **530** (2019) 012059 doi:10.1088/1757-899X/530/1/012059

3. Results

The descriptive statistics of basic and physical training condition on schema and MRT scores were shown in Table 1. A multivariate analysis of variance (One-way MANOVA) was conducted to assess if there were differences between the two training condition (basic training and physical training) on schema and mental rotation test scores.

The results revealed the significant difference between the training condition (Wilk's $\Lambda = .79$, $F(2, 177) = 23.51$, $p = .00$, $\eta 2 = .21$). Examination of the coefficients for the linear combinations distinguishing training methods indicated that schema and mental rotation scores contributed most to distinguish the method training. The analysis of variance (ANOVA) indicated that participants with physical training condition got higher schema test score than those with basic training condition ($F(1, 178) = 35.175$, $p < .05$). In the same direction, participants with physical training condition also got higher MRT score than those with basic training condition ($F(1, 178) = 10.770$, $p < .05$). Games-Howell post hoc tests revealed that, of the schema test with the basic training did seem to be associated with differences in physical training ($p<.01$), also of the mental rotation test with the basic training did seem to be associated with differences in the physical training ($p<.01$) as shown in Figure 2.

Table 1. Means and standard deviations of basic and physical training condition on schema and mental rotation test

Training Conditions	N	Schema Test		Mental Rotation Test	
		M	SD	M	SD
Basic training	90	3.34	1.49	3.11	1.50
Physical training	90	4.07	1.47	4.40	1.41

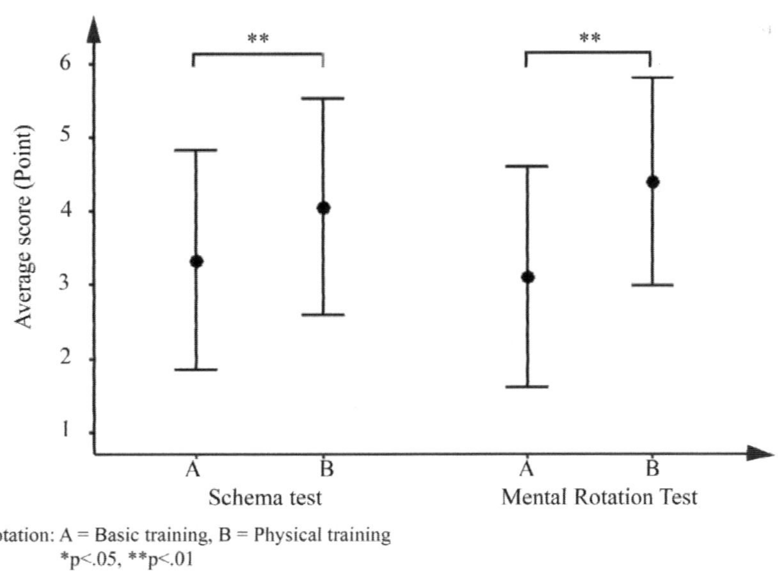

Figure 2. The interaction plot of mean and standard deviation of basic and physical training condition on schema and mental rotation test scores.

4. Discussions and Conclusions

The purpose of present study was to preliminary investigate the training performance measurement using cognitive schema and mental rotation test in camera assembly tasks with basic training and physical training condition. The experiments with 180 participants were conducted. The results test revealed that both MRT and schema test could differentiate the performance of two types of training in the same direction. People who received physical training condition got the significantly higher scores from both schema test and MRT.

The results of this study showed that schema test and MRT could be used to level performance of the training condition instead of conduct full trial-and-error as the traditional method in the manufacturing context. However, future study should investigate the effect size as well as the accuracy of using these two measurements and conduct the real experiment in order to calibrate these measurements and its validity. Since human cognition schema and MRT are both individual assessment level, while some type of manufacturing such as continuous assembly line may require measuring performance as a group-based performance. This would be worth investigating in the further study whether the individual performance measurement remains valid in those group-based contexts of manufacturing.

Finally, the schema and mental rotation test might be another method that could help the manager in making the decision in training condition of manufacturing context. This could shorten the time and cost comparing to conduct trial and error. While the future study may consider investigating its sensitivity of these measurements whether it applies to all types of training.

References

[1] L A Kidwell, D G Fisher, R L Braun and D L Swanson 2013 Developing learning objectives for accounting ethics using Bloom's taxonomy *Accounting Education* Vol 22 pp 44-65

[2] E M Dar-El 2013 *Human learning: From learning curves to learning organizations* Vol 29 Boston: Springer Science & Business Media

[3] D M Edralin 2011 Training and development practices of large Philippines companies *Asia pacific business review* Vol 17 pp 225-239

[4] I LGoldstein and J K Ford 2002 *Training in Organizations: Needs Assessment, Development, and Evaluation (4th ed.)* Belmont, CA, US: Wadsworth

[5] S Wagner 2009 Hands around the Job *Mechanical Engineering* Vol 131 p 28

[6] K Karpinska, K Henkens, J Schippers and M Wang 2015 Training opportunities for older workers in the Netherlands: A Vignette Study *Research in Social Stratification and Mobility* Vol 41 pp 105-114

[7] A Chatzimouratidis, I Theotokas and I N Lagoudis 2012 Decision support systems for human resource training and development *The International Journal of Human Resource Management* Vol 23 pp 662-693

[8] F De Brigard, T F Brady, L Ruzic and D L Schacter 2017 Tracking the emergence of memories: A category-learning paradigm to explore schema-driven recognition *Memory & cognition* Vol 45 pp 105-120

[9] K Kraiger, E Salas and J A Cannon-Bowers 1995 Measuring knowledge organization as a method for assessing learning during training *Human factors* Vol 37 pp 804-816

[10] S P Marshall 1995 *Schemas in problem solving* New York: Cambridge University Press. doi:10.1017/cbo9780511527890

[11] Chapter 7: Scripts and Schemata 2014 *Critical Photography Series,* pp. 129-141, 05//

[12] D Stransky, L M Wilcox and A Dubrowski 2010 Mental rotation: Cross-task training and generalization *Journal of Experimental Psychology: Applied* Vol 16 pp 349-360

[13] N Hoyek, C Collet, P Fargierand and A Guillot 2011 The use of the Vandenberg and Kuse Mental Rotation Test in children *Journal of Individual Differences* Vol 33 pp 62-67

[14] M Peters, B Laeng, K Latham, M Jackson, R Zaiyouna and C Richardson 1995 A redrawn

Vandenberg and Kuse mental rotations test-different versions and factors that affect performance *Brain and cognition* Vol 28 pp 39-58

[15] S G Vandenberg and A R Kuse 1978 Mental rotations, a group test of three-dimensional spatial visualization *Perceptual and motor skills* Vol 47 pp 599-604

[16] D Moreau, J Clerc, A Mansy-Dannay and A Guerrien 2012 Enhancing spatial ability through sport practice *Journal of Individual Differences* Vol 33 pp 83-88

[17] E A Day, W Arthur Jr and D Gettman 2001 Knowledge structures and the acquisition of a complex skill *Journal of applied psychology* Vol 86 p 1022

[18] L W Anderson, D R Krathwohl and B S Bloom 2001 *A taxonomy for learning, teaching, and assessing: A revision of Bloom's taxonomy of educational objectives*: Allyn & Bacon

[19] B S Bloom, M D Engelhart, E J Furst, W H Hill and D R Krathwohl 1956 Taxonomy of educational objectives, handbook I: The cognitive domain *ed: New York: David McKay Co Inc*

IOP Conf. Series: Materials Science and Engineering **530** (2019) 012060 doi:10.1088/1757-899X/530/1/012060

Running and Cycling Induced Fatigue on Wrapper vs. BLR Feature Selection for IBk Classification

S Tang[1], W P Loh[1*] and C F Lin[2]

[1] School of Mechanical Engineering, Engineering Campus, Universiti Sains Malaysia, 14300 Nibong Tebal, Penang, Malaysia.

[2] Department of Physical Therapy, National Cheng Kung University, Tainan City, Taiwan.

Corresponding author *: meloh@usm.my

Abstract. Running and cycling fatigue causes muscle pains, cramps and accidental injuries. Previous studies had considered the importance of tri-axial accelerometer to detect fatigue motion in stability, balance and postural deviation aspects. While tri-axial accelerometer is important, the capability to predict running and cycling fatigue from the biomechanical attributes were unclear. Therefore, the study aims to (i) compare the featured attributes selected from wrapper approach and Binary Logistic Regression (BLR) on running and cycling datasets and (ii) perform IBk classification accuracy comparison on the feature selection attributes. Public running, experimental running and cycling induced fatigue datasets were employed to test the analysis. The most significant attributes identified in the public running was RMS_ML, followed by Range_ML and the cycle frequency in experimental running and cycling respectively. On 10 folds cross-validation classification test using the IBk algorithm in WEKA, accuracies for experimental running and cycling datasets were 93.1% and 90.5% from wrapper method, 65.6% and 76.2% from BLR respectively. Wrapper method performs better than BLR in data overfitting phenomenon. Findings reveal that the mediolateral variation at body trunk motion plays a major impact to predict fatigue running but fatigue cycling shows cycling frequency as the main attribute in fatigue cycling prediction.

1. Introduction

Continuous endurance sports like running and cycling under fatigue state contributes to muscle soreness and cramp. Running and cycling comparisons had been investigated from the aspect of muscle fatigue on peripheral neuromuscular [1]. Cycling is generally better than running considering the low impact and less eccentric muscle motion [2]. Running-induced fatigue attributes included lengthy training duration on muscle damage [3], plantar pressure distribution on different foot structures [4], foot strike pattern for impact force reduction [5] or surface type (treadmill or overground) on acceleration impact [6]. On the other hand, the cycling-induced fatigue attributes as reported in Raymond et al. [7] were riding posture, saddle height, muscle fatigue or cadence.

Identifying significant attributes contributing to running and cycling fatigue were crucially important for injury risk prevention. Past studies have reported tri-axial accelerometers on fatigue detection [8–11], in which, the stability and movement balancing were focused. Knowing the importance of tri-axial accelerometer in fatigue detection, researchers attempted to investigate the posture deviation through

IOP Conf. Series: Materials Science and Engineering **530** (2019) 012060 doi:10.1088/1757-899X/530/1/012060

acceleration root mean square (RMS) ratio, step frequency, steps and stride regularity and sample entropy [11-12]. Running and cycling are the repetitive nature of activities that involve lower limb motions. However, the comparison between running and cycling using the significant attributes of accelerometry derived quantities and sacral trajectory measurements for fatigue determination was not reported. The accelerometry derived quantities might be redundant as each derived quantity is repeatedly measured in the vertical (VT), mediolateral (ML) and anteroposterior (AP) directions. Sacral trajectory data collection in the three directions is not only tedious but also time-consuming. Thus, feature selection approach to determine the important attributes that differentiate the fatigue levels in running and cycling is pivotal to reduce the data complexity for class prediction.

Therefore, the objectives of this study are mainly i) to compare the featured attributes selected from wrapper approach and Binary Logistic Regression (BLR) on running and cycling datasets and ii) perform IBk classification accuracy comparison on the feature selection attributes. Wrapper method of feature selection with IBk supported classifier was applied to select the attributes in Waikato Environment for Knowledge Analysis versions 3.8 (WEKA) tool. BLR statistical analysis on IBM Statistical Package for the Social Sciences Statistic versions 23 (SPSS) was applied to compare the results. The IBk classification algorithm was used to test the classification performances on selected attributes. Three datasets will be employed for the wrapper and BLR feature selection for the IBk classification analysis: public running, experimental running and experimental cycling.

2. Methodology

2.1. Dataset

Three datasets: public running, experimental running and experimental cycling were used to test the feature selection methods using the wrapper and BLR approaches.

Public running: Dataset were retrieved from Schütte et al. [8] as the benchmark data. Dataset involved 20 athletes (11 males, 9 females, 21.1±2.1 years old, 177±8cm, 66.1±6.2kg) whom participated in continuous fatigue run on a treadmill (until Borg RPE reaches minimum 17) based on individual time-trial speeds [8]. The time-trial speed of each participant was obtained through the recorded time on 3.2km run a day before. A single tri-axial accelerometer (X16-2 wireless accelerometer, 3 axis ±16g, 15-bit resolution, Gulf Coast Data Concepts, MS) was attached to the participants' waists by double-sided tape and elastic straps to measure the acceleration of trunk in three directions: VT, ML and AP during the treadmill run. Running motion was recorded on ten Vicon system cameras (Vicon®, Oxford, Metrics UK) for sacral trajectory measurements.

Experimental running: The study subjects involved were non-athletes, involving 14 males and 6 females (22.5±3.9 years old, 167.0±8.5cm, 62.9±18.6kg) whom were tasked to perform running on a treadmill (Cybex Model 700T). Participants have signed an informed consent at their voluntary basis prior to the experiment. Instruction was first given to run 6km/hr for 6 minutes on the treadmill for familiarization [13]. After some warming up, the treadmill speed was increased by 1km/h every 2 minutes till 70% of maximum heart rate was achieved (Borg RPE scale reaches 13) [14]. The heart rate was monitored on the treadmill throughout the experiment. The similar speed at Borg RPE scale-13 was maintained till 90% of maximum heart rate (Borg RPE scale reaches 17). Participants continued to run at Borg RPE scale-17 for another 2 minutes and the termination time was recorded. An iPhone6 installed with SensorLog versions 2.0 APP (to function like a tri-axial accelerometer) was attached to the back of the body trunk (with a waist bag) throughout the experiment to measure the trunk acceleration. Two cameras (Canon EOS 30D) for sacral trajectory measurements were stationed at the side and the back of the treadmill to capture the participants' motions.

Experimental cycling: There were 16 males and 4 females (26.1±4.3 years old, 168.7±8.9cm, 66.0±12.1kg) participated in the cycling induced fatigue experiment on an ergometer bike (Cateye Upright Bike Model EC-2300). The participants have signed their voluntary consent prior to the experiment. Similar experimental procedure to the experimental running was employed. The experiment

IOP Conf. Series: Materials Science and Engineering **530** (2019) 012060 doi:10.1088/1757-899X/530/1/012060

began with 6km/h warm up for 6 minutes [13] followed by 1km/h speed increment at every 2 minutes till 70% of maximum heart rate was achieved (Borg RPE scale reaches 13) [14]. The heart rate was monitored on the ergometer bike itself during the experiment. The similar speed at Borg RPE scale-13 was maintained till 90% of maximum heart rate (Borg RPE scale reaches 17). Participants continued the cycling motions at scale-17 for another 2 minutes and the termination time was recorded. An iPhone6 with SensorLog versions 2.0 APP was attached to the back of the body trunk throughout the experiment for acceleration data recording. Two cameras (Canon EOS 30D) were stationed at the side and the back of ergometer bike to record the cycling motions.

The recorded running and cycling data consist of 880 records each (22 attributes along with a class attribute x 40 instances). The 22 attributes describe the accelerometry-derived measures and sacral trajectory measures in three directions: VT, ML and AP. Acceleration root mean square (RMS), ratio RMS, step regularity, stride regularity, step frequency or cycle frequency and sample entropy were attributes of accelerometry measures while the sacral trajectory measures included the mean and range of displacement. There were two predefined class attributes: fatigue and pre-fatigue. Following Schütte et al. [8], the prior 10 seconds motions at Borg RPE scale-17 is identified as pre-fatigue, while the last 10 seconds motions prior to termination is defined as fatigue.

2.2. Feature selection

The data features, which provides the relevant information that enhances classification analysis were considered at this level.

The datasets were tested on the wrapper method of WEKA and BLR using SPSS tool to select the significant attributes. Wrapper method applied Wrapper Subset Evaluator with BestFirst searching method in forward direction. IBk (Instance-bases) learner was selected as the supported classifier which employed the nearest neighbour technique to form the attributes subsets. The importance of the generated subset was evaluated on merit point basis (ranged from 0 to 1) [15]. It was noted that the higher the merit the better will be the subset in distinguishing fatigue and pre-fatigue classes.

The BLR approach was later used to compare selected attributes with those featured attributes identified from the wrapper approach. The BLR approach was deemed suitable as the class predicted is categorical (fatigue vs. pre-fatigue). BLR applied the forward stepwise (Wald) method to search the attributes in the forward direction. Wald Chi-Squared forward search method is desirable to ensure the attributes selected in a model are significant. BLR presents the coefficient value (B), standard error (S.E.), Wald statistic (Wald), Sig. value and the odds ratio (Exp(B)) of each attribute selected. The statistical significance of attributes was determined by lower S.E. and Sig. value < 0.5. Meanwhile, Exp(B) examine the relationship of attributes with the fatigue and pre-fatigue classes following [16]:

$$\text{Exp(B)} = 1\text{: no relationship}$$
$$\text{Exp(B)} > 1\text{: positive relationship}$$
$$\text{Exp(B)} < 1\text{: negative relationship}$$

On feature selection approach, both wrapper and BLR approaches returned subsets of data features to be classified using 10 folds cross-validation on IBk algorithm. The significant attribute selections were evaluated by the percentage data classification accuracies in using the IBk algorithm. ZeroR classification results are referred as the baseline classifier to verify informative classification. The classification accuracy must be higher than ZeroR accuracy to ensure informative results obtained, else the results are deemed meaningless for interpretation.

3. Results and discussions

Findings from the study datasets experimented were segregated by the wrapper method, BLR approach and their comparisons at feature selection level. This is followed by classification accuracies to judge the feature selection approaches.

3.1. Feature selection

3.1.1. Wrapper method. The attributes selected of each dataset under evaluation of wrapper subset evaluator with Best First in forward searching method was tabulated in Table 1.

Table 1. Attributes selected from wrapper method

Dataset	Total number of subset		Merit of best subset found		Number of attributes selected		Attribute selected
	N	Percent difference (%)	Merit	Percent difference (%)	N	Percent difference (%)	
Public Running (benchmark)	184	-	0.786	-	5	-	*RMS_AP,* RMS_ML, ***SampEntropy_AP,*** StrideReg_AP, ***StrideReg_VT.***
Experimental Running	214	16.3	0.890	13.2	8	60	Disp_VT, **Range_AP,** Range_ML, ratio RMS_VT, ***SampEntropy_AP,*** **SampEntropy_VT**, StepReg_VT, ***StrideReg_VT.***
Experimental Cycling	216	17.4	0.876	11.5	6	20	CycleFreq, **Range_AP**, ratio *RMS_AP,* RMS_AP, RMS_VT, **SampEntropy_VT**.

bold- similar attribute selected for experimental running and cycling
italic- similar attribute selected for public running and experimental cycling
bold & italic- similar attribute selected for public and experimental running

Table 1 shows that the experimental datasets: running and cycling resulted in higher number of subsets evaluated (Running: 16.3%, Cycling: 17.4%) with the merit of the best subset (Running: 13.2%, Cycling: 11.5%) and number of attributes (Running: 60%, Cycling: 20%) selected as compared to public running data (benchmark). The attributes selected on feature selection approaches differ on datasets basis. The commonly selected attributes were *RMS_AP,* ***SampEntropy_AP*** and ***StrideReg_VT***. *RMS_AP* was similarly selected in public running and experimental cycling while ***SampEntropy_AP*** and ***StrideReg_VT*** were selected for public and experimental running. **Range_AP** and **SampEntropy_VT** were the attributes resulted in experimental running and cycling. However, there was no similar attribute selection from three different datasets.

3.1.2. Binary logistic regression (BLR). On BLR, the attributes were selected using the Wald forward searching method as shown in Table 2. B and Exp(B) indicate the attributes strength and relationship direction with the data class. S.E. is referred as the indicator of coefficient's precision.

Table 2. Attributes selected from BLR on three datasets

Dataset	B	S.E.	Wald	df	Sig.	Exp(B)	Attribute selected
Public Running (benchmark)	-6.481	2.882	5.056	1	0.025	0.002	RMS_ML
	-5.774	2.909	3.939	1	0.047	0.003	SampEntropy_AP
	5.381	2.060	6.823	1	0.009	217.329	Constant
Experimental Running	-8.314	4.348	3.657	1	0.056	0.000	StepReg_ML
	-0.913	0.443	4.256	1	0.039	0.401	Range_ML
	9.508	3.733	6.489	1	0.011	13472.313	Constant
Experimental Cycling	-0.060	0.029	4.274	1	0.039	0.942	CycleFreq
	7.888	3.799	4.311	1	0.038	2664.600	Constant

There were lesser attributes selected using the BLR approach. Only one to two attributes were identified being important in the three datasets. RMS_ML and SampEntropy_AP were selected in public running, StepReg_ML and Range_ML were seen in experimental running whereas only one single attribute, CycleFreq was selected in experimental cycling. All attributes selected were found statistically significant (*Sig.* < 0.05) except for StepReg_ML.

In public running (benchmark), RMS_ML and SampEntropy_AP were statistically significant (*Sig.* < 0.05). RMS_ML demonstrated major impact compared to SampEntropy_AP due to lower S.E. (2.882 < 2.909) and Sig. value (0.025 < 0.047) (Table 2). The RMS_ML showed B value = -6.481<1 indicating the negative relationship exists between RMS_ML with the pre-fatigue class. This is visualized in Figure 1(a), as RMS_ML increases, amount of data belonging to pre-fatigue decreases. The Exp(B)=0.002 in RMS_ML implied that when the RMS_ML increases by 1, probability to have pre-fatigue class reduces by 99.8%.

While in experimental running, only the Range_ML was statistically significant (*Sig.* < 0.05). Lower standard error (0.443 < 4.348) in Range_ML supports the results that the attribute is significant compared to StepReg_ML to distinguish between fatigue and pre-fatigue classes. The relationship of Range_ML and the pre-fatigue class is negative (B = -0.913, Exp(B) = 0.401). When Range_ML increases by 1cm, the probability to have pre-fatigue class decreases by 59.9%. This can be observed in Figure 1(b), the number of data accumulated at pre-fatigue class decreases as Range_ML increases.

Similar observation was shown in experimental cycling, the relationship between CycleFreq with the pre-fatigue class is negative (B = -0.060, Exp(B) = 0.942). CycleFreq is statistically significant (*Sig.* < 0.05). The probability to have pre-fatigue class reduces by 5.8% when CycleFreq increases in 1 cycle/min (Figure 1(c)).

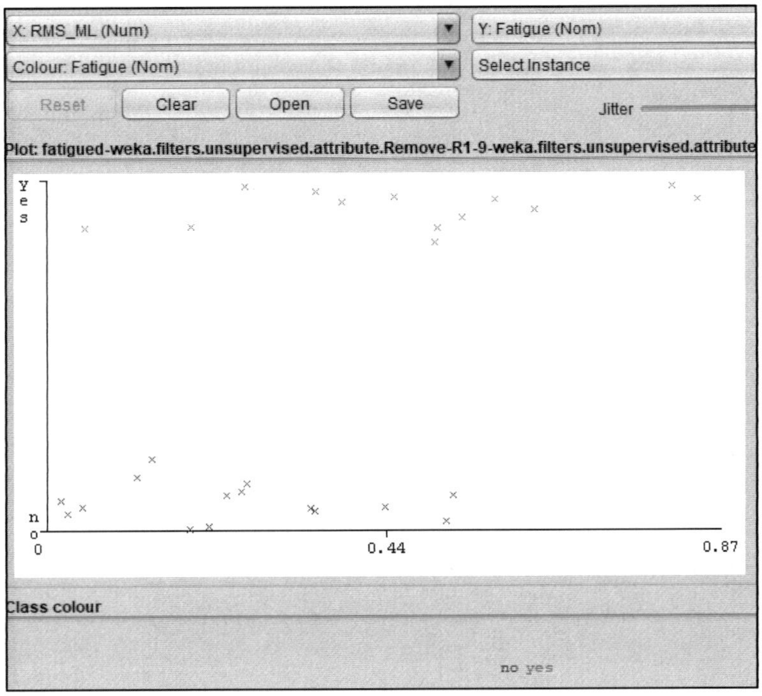
X: RMS_ML (Num)
Y: Fatigue (Nom)
Colour: Fatigue (Nom)
Select Instance
Reset
Clear
Open
Save
Jitter
Plot: fatigued-weka.filters.unsupervised.attribute.Remove-R1-9-weka.filters.unsupervised.attribute
yes
no
0
0.44
0.87
Class colour
no yes

(a)

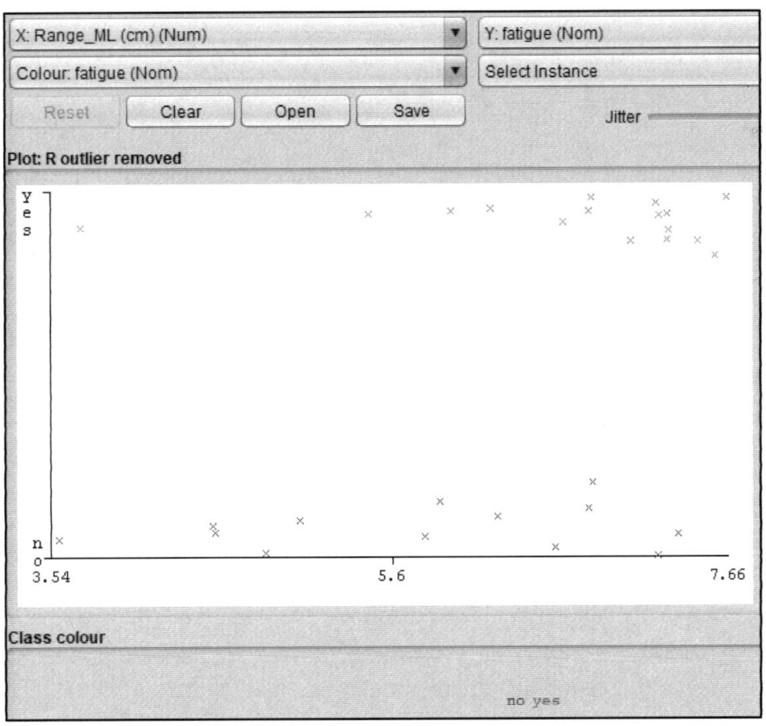
X: Range_ML (cm) (Num)
Y: fatigue (Nom)
Colour: fatigue (Nom)
Select Instance
Reset
Clear
Open
Save
Jitter
Plot: R outlier removed
yes
no
3.54
5.6
7.66
Class colour
no yes

(b)

IOP Conf. Series: Materials Science and Engineering **530** (2019) 012060 doi:10.1088/1757-899X/530/1/012060

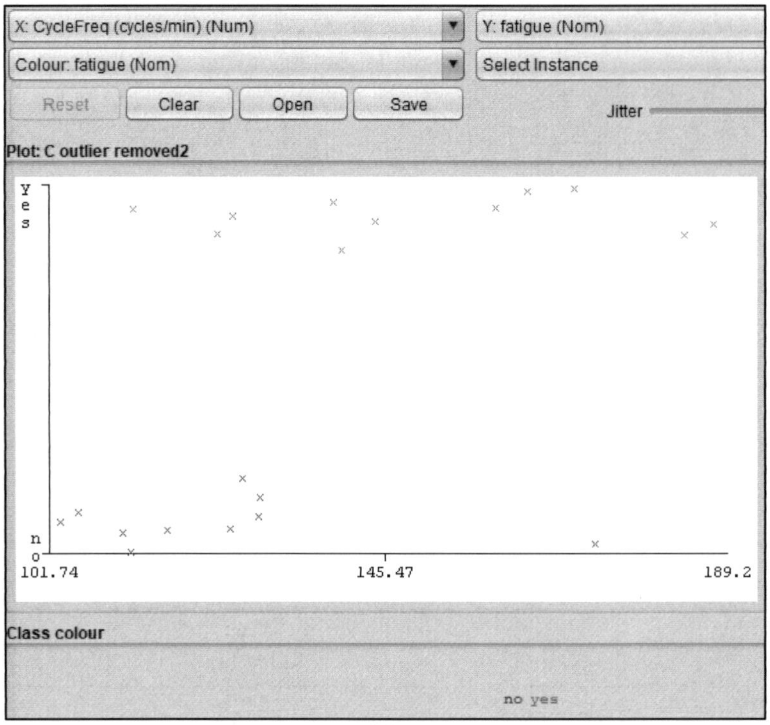

(c)

Figure 1. The data classes (yes = fatigue, no = pre-fatigue) on (a) RMS_ML (b) Range_ML (c) cycle frequency attributes

3.1.3. Wrapper method vs. BLR. Table 3 compiles the attributes selected using wrapper method and BLR. There were five to eight attributes selected for the three datasets using the wrapper method. Meanwhile, the BLR shows the presence of merely one or two attributes for each dataset that were identified being significantly sufficient (Sig < 0.5) to classify data into fatigue or pre-fatigue classes.

Table 3. Selected attributes using *wrapper method and **BLR for three datasets

Selected Attribute	Public Running (benchmark)	Running	Cycling
RMS_VT			*
RMS_ML	* **		
RMS_AP	*		*
Ratio RMS_VT		*	
Ratio RMS_ML			
Ratio RMS_AP			*
StepReg_VT	*	*	
StepReg_ML			
StepReg_AP	*		
StrideReg_VT		*	
StrideReg_ML			
StrideReg_AP			

StepFreq or CycleFreq			* **
SampEntropy_VT		*	*
SampEntropy_ML			
SampEntropy_AP	* **	*	
Disp_VT		*	
Disp_ML			
Disp_AP			
Range_VT			
Range_ML		* **	
Range_AP		*	*

Wrapper method and BLR apply the forward search method for attribute selection. The wrapper is aided by the IBk classifier that applies the nearest neighbour technique to select the most efficient subset containing the relevant attributes. BLR, whereas, selects the significant attributes based on Wald statistics. Findings in Table 3 shows eight attributes (36.36%): Ratio RMS_ML, StepReg_ML, StrideReg_ML, StrideReg_AP, SampEntropy_ML, Disp_ML, Disp_AP, and Range_VT were redundant as the attributes were not selected in both wrapper method and BLR on the three datasets. The public running benchmark data illustrates RMS_ML and SampEntropy_AP were the two similar attributes detected between wrapper method and BLR. Meanwhile, on experimental running and cycling only one similar attribute identified from both feature selection approaches: Range_ML and CycleFreq respectively.

3.2. Classification

The attributes selected from wrapper method and BLR were tested on data classification into two predefined data classes: pre-fatigue and fatigue. This is to investigate the effects of feature selection on the classification accuracy between wrapper method and BLR on the three datasets.

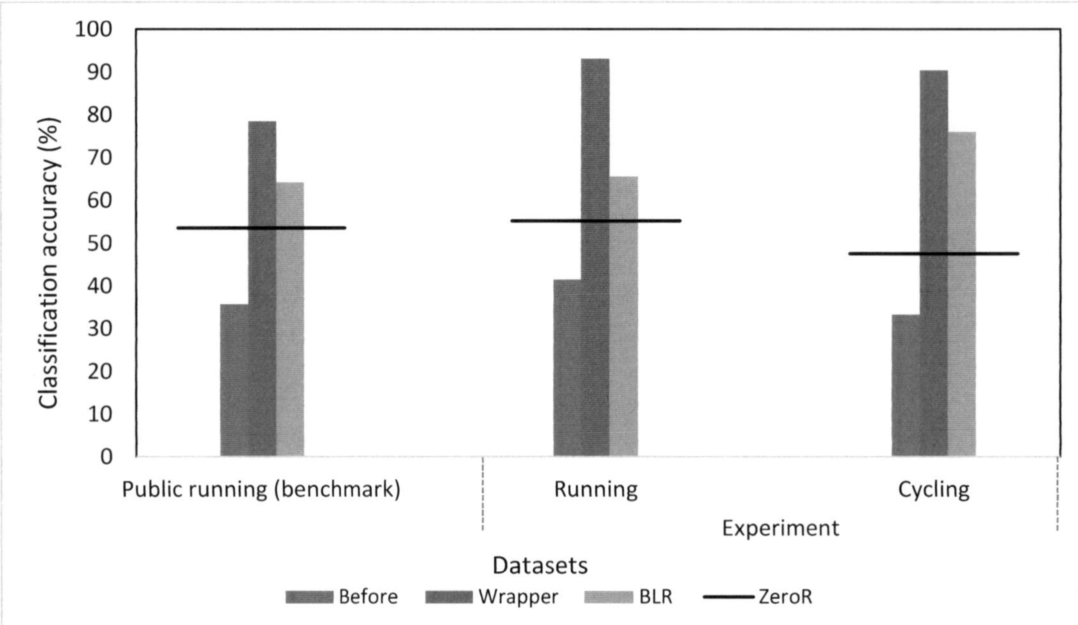

Figure 2. Percentage classification accuracy on three datasets before and after feature selection

Figure 2 shows the accuracy of classification on IBk algorithm of the three datasets before and after the feature selection approach using wrapper method and BLR. The baseline classifier (ZeroR) classifies the data into fatigue or pre-fatigue: 53.6% in public running, 55.2% in experimental running and 47.6% in experimental cycling. Before selecting the attributes, the classification accuracies resulted were lower than the baseline, ZeroR (< 47.6%) in the three datasets. This shows that the classification performances obtained are not reliable.

In public running, upon wrapper approach feature selection, the classification accuracy improves to 78.6%, while BLR approach improves to 64.3% classification accuracy (Figure 2). Contrary to the public running (benchmark), experimental running shows eight main attributes selected using wrapper method (Table 3) with 93.1% classification accuracy (Figure 2). However, BLR approach picked only a single Range_ML attribute (Table 3) resulting in 65.6% accuracy (Figure 2).

Experimental cycling, whereas, shows 90.5% classification accuracy on six attributes selected using wrapper method while the BLR selected only a single attribute; cycle frequency with 76.2% accurate classification (Table 3 & Figure 2). The classification accuracy of wrapper and BLR feature selection successfully improves for all datasets classification accuracy. Overall classification accuracy in experimental running and cycling were found higher than public running; showing 14.5% and 11.9% improvement on the wrapper method feature selection but merely 1.3% and 11.9% improvement in BLR approach respectively. The higher classification accuracy enhancement of wrapper indicates that attribute selection approach is critical in not only model complexity reduction but also address to data overfitting phenomenon. The BLR was found to have overfitted data causing the effect of information loss. As compared to the wrapper, BLR loses about 60-87.5% selected data. Thus, BLR is inappropriate for feature selection in the study datasets. However, its statistical significance test helps to identify which attribute having a better impact over the other on classification analysis.

The RMS_ML was the most significant attribute in public running data (Table 2). RMS_ML shows the variability of acceleration in mediolateral trunk motion. The results implied that the variation of acceleration dispersion in mediolateral trunk acceleration has major contribution to predict running fatigue. This finding was agreed in Schütte et al. [8] that the fatigue participants were identified from increased variability in horizontal plane trunk acceleration. In other words, when the value of RMS_ML increases, the participant has a higher tendency to experience the fatigue.

On the other hand, the feature selection approaches indicated Range_ML had a major contribution for experimental running in distinguishing the fatigue classes. However, feature selection on the public dataset confirms RMS_ML being significant for distinguishing fatigue classes. Although the identified attribute differs for two running datasets (public vs. experimental), the main emphasis was similar i.e. the mediolateral variation. Such finding indicates either variability of acceleration in ML or ML deviation of trunk motion displacement is desirable in fatigue run prediction. ML variation is crucially important in fatigue run detection. The variability of ML increases the fatigue tendency, therefore, causing the loss of stability to maintain the running posture.

The feature selection attributes in experimental cycling differ from public and experimental running. CycleFreq was the most significant attribute demonstrated (Table 2). Cycling speed plays a major impact in fatigue classification because fatigue condition causes high-speed motion difficult to sustain. Cycling in a seated position is more stable and balanced, thus ML variation was not identified in this dataset. VT and AP variations of trunk acceleration motion are critical only when body posture change from sitting to standing. Thus, the six attributes: Range_AP, ratio RMS_AP, RMS_AP, RMS_VT and SampEntropy_VT selected from wrapper including CycleFreq are important for cycling fatigue prediction.

4. Conclusion

This study considers the attributes selection comparison between wrapper method and BLR on IBk classification performance. As per the results obtained, the significant attributes selected from wrapper method are found better than the BLR resulting in 93.1% classification accuracy for the experimental running (an improvement of 51.7% from the original dataset) and 90.5% classification accuracy for cycling (an improvement of 57.2% from the original dataset). The most significant attributes resulted in the three employed datasets using wrapper approach verified on BLR showed different significant attributes selected; RMS_ML for public running, Range_ML for experimental running, and CycleFreq for experimental cycling. Conclusively, the fatigue condition in running can be predicted when the mediolateral variation at body trunk increases. Meanwhile, the cycling fatigue level can be determined from the cycling speed impact on lower limb muscles.

References

[1] Rampinini E, R Darragh D, Ferioli D, La Torre A, Alberti G and Bosio A 2014 Peripheral Neuromuscular Fatigue Induced by Repeated-sprint Exercise: Cycling vs Running *J. Sports Med. Phys. Fitness*

[2] Chan H C K, Ho W K Y and Yung P S H 2017 Sprint cycling training improves intermittent run performance *Asia-Pacific J. Sport Med. Arthrosc Rehabil Technol* [Internet] 11 p 6–11 Available from: https://doi.org/10.1016/j.asmart.11.001

[3] Quinn T J and Manley M J 2012 The impact of a long training run on muscle damage and running economy in runners training for a marathon *J. Exerc. Sci. Fit* 10 p 101–6

[4] Anbarian M and Esmaeili H 2016 Effects of running-induced fatigue on plantar pressure distribution in novice runners with different foot types *Gait Posture* 48 p 52–6

[5] Daoud A I, Geissler G J, Wang F, Saretsky J, Daoud Y A and Lieberman D E 2012 Foot strike and injury rates in endurance runners: A retrospective study. *Med Sci Sports Exerc.* 44(7) p 1325–34

[6] García-Pérez J A, Pérez-Soriano P, Llana Belloch S, Lucas-Cuevas A G, Sánchez-Zuriaga D, Lucas-Cuevas Á G and Sánchez-Zuriaga D 2014 Effects of treadmill running and fatigue on impact acceleration in distance running *Sport Biomech* [Internet] 13(3) p 259–66. Available from:
http://www.tandfonline.com/doi/abs/10.1080/14763141.2014.909527%5Cnhttp://www.ncb

IOP Conf. Series: Materials Science and Engineering **530** (2019) 012060 doi:10.1088/1757-899X/530/1/012060

i.nlm.nih.gov/pubmed/25325770
[7] So R C H, Ng J K F and Ng G Y F 2005 Muscle recruitment pattern in cycling: A review *Physical Therapy in Sport* 6 p 89–96
[8] Schütte K H, Maas E A, Exadaktylos V, Berckmans D, Venter R E and Vanwanseele B 2015 Wireless tri-axial trunk accelerometry detects deviations in dynamic center of mass motion due to running-induced fatigue *PLoS One* [Internet]. 10(10) p 1–12. Available from: http://dx.doi.org/10.1371/journal.pone.0141957
[9] Bris R Le, Billat V, Auvinet B, Chaleil D, Hamard L and Barrey E 2006 Effect of fatigue on stride pattern continuously measured by an accelerometric gait recorder in middle distance runners *J Sports Med Phys Fitness* 46(2)
[10] Evans S and Winter S 2018 Effects of fatigue on center-of-mass acceleration during a prolonged overground run *J Phys Ther Sport Med.* 2(1)
[11] Winter S C, Gordon S, Brice S M and Lindsay D 2018 Centre of Mass Acceleration-Derived Variables Detects Differences between Runners of Different Abilities and Fatigue-Related Changes during a Long Distance Over ground Run. *J Phys Fitness, Med Treat Sport* 4(2)
[12] Schütte K H, Seerden S, Venter R and Vanwanseele B 2018 Influence of outdoor running fatigue and medial tibial stress syndrome on accelerometer-based loading and stability. *Gait Posture* [Internet]. 59 p 222–8. Available from: http://dx.doi.org/10.1016/j.gaitpost.2017.10.021
[13] Lavcanska V, Taylor N F and Schache A G 2015 Familiarization to treadmill running in young unimpaired adults *Hum. Mov. Sci.* 24(4) p 544–57
[14] Borg G 1990 Psychophysical scaling with applications in physical work and the perception of exertion *Scand J Work Environ Heal* 16 p 55–8
[15] B Demisse G, Tadesse T and Bayissa Y 2017 Data Mining Attribute Selection Approach for Drought Modelling : A Case Study for Greater Horn of Africa *Int. J. Data Min. Knowl. Manag. Process* [Internet]. 7(4) p 01-16. Available from: http://airccoline.com/ijdkp/V7N4/7417ijdkp01.pdf
[16] Szumilas M 2010 Explaining odds ratios *J. Can Acad. Child Adolesc. Psychiatry* 19(3) p 227–9